Crashworthiness of Transportation Systems: Structural Impact and Occupant Protection

NATO ASI Series

Advanced Science Institutes Series

A Series presenting the results of activities sponsored by the NATO Science Committee, which aims at the dissemination of advanced scientific and technological knowledge, with a view to strengthening links between scientific communities.

The Series is published by an international board of publishers in conjunction with the NATO Scientific Affairs Division

A Life Sciences **B Physics**	Plenum Publishing Corporation London and New York
C Mathematical and Physical Sciences **D Behavioural and Social Sciences** **E Applied Sciences**	Kluwer Academic Publishers Dordrecht, Boston and London
F Computer and Systems Sciences **G Ecological Sciences** **H Cell Biology** **I Global Environmental Change**	Springer-Verlag Berlin, Heidelberg, New York, London, Paris and Tokyo

PARTNERSHIP SUB-SERIES

1. Disarmament Technologies	Kluwer Academic Publishers
2. Environment	Springer-Verlag / Kluwer Academic Publishers
3. High Technology	Kluwer Academic Publishers
4. Science and Technology Policy	Kluwer Academic Publishers
5. Computer Networking	Kluwer Academic Publishers

The Partnership Sub-Series incorporates activities undertaken in collaboration with NATO's Cooperation Partners, the countries of the CIS and Central and Eastern Europe, in Priority Areas of concern to those countries.

NATO-PCO-DATA BASE

The electronic index to the NATO ASI Series provides full bibliographical references (with keywords and/or abstracts) to more than 50000 contributions from international scientists published in all sections of the NATO ASI Series.
Access to the NATO-PCO-DATA BASE is possible in two ways:

– via online FILE 128 (NATO-PCO-DATA BASE) hosted by ESRIN, Via Galileo Galilei, I-00044 Frascati, Italy.

– via CD-ROM "NATO-PCO-DATA BASE" with user-friendly retrieval software in English, French and German (© WTV GmbH and DATAWARE Technologies Inc. 1989).

The CD-ROM can be ordered through any member of the Board of Publishers or through NATO-PCO, Overijse, Belgium.

Series E: Applied Sciences - Vol. 332

Crashworthiness of Transportation Systems: Structural Impact and Occupant Protection

edited by

Jorge A. C. Ambrósio

Manuel F. O. Seabra Pereira

and

Fernando Pina da Silva

Instituto de Engenharia Mecânica,
Instituto Superior Técnico,
Lisboa, Portugal

Springer-Science+Business Media, B.V.

Proceedings of the NATO Advanced Study Institute on
Crashworthiness of Transportation Systems: Structural Impact and Occupant Protection
Tróia, Portugal
July 7–19, 1996

A C.I.P. Catalogue record for this book is available from the Library of Congress

ISBN 978-94-010-6447-7 ISBN 978-94-011-5796-4 (eBook)
DOI 10.1007/978-94-011-5796-4

Printed on acid-free paper

NATO-ADVANCED STUDY INSTITUTE

Crashworthiness of Transportation Systems: Structural Impact and Occupant Protection

TRÓIA, PORTUGAL

JULY 7 - 19, 1996

MAIN SPONSOR:

NATO: North Atlantic Treaty Organization

SPONSORS

Associação do Comércio Automóvel de Portugal
European Research Office of the US Army
Ford Motor Company
Fundação Luso Americana para o Desenvolvimento
Indústria de Componentes Mecânicos
Instituto de Engenharia Mecânica pólo IST
Junta Nacional de Investigação Científica e Tecnológica
National Science Foundation
National Institute for Aviation Research
Prevenção Rodoviária Portuguesa
Office of Naval Research European Office
Ordem dos Engenheiros da Região Sul
Renault Portuguesa
Sociedade Gráfica da Paiã
TAP Air Portugal

DIRECTOR

Jorge A. C. Ambrósio, IDMEC-Instituto Superior Técnico, Portugal

CO-DIRECTOR

Wlodek Abramowicz, Polish Academy of Sciences, Poland

SCIENTIFIC COMMITTEE:

Norman Jones, University of Liverpool, United Kingdom
David Viano, General Motors, USA
Jac Wismans, T.N.O., The Netherlands

CONTENTS

PREFACE

During the last few years, major scientific progress has been achieved in fields related to the area of crashworthiness and injury biomechanics. In view of this progress and given the fast rate of technical developments in vehicle crash protection, a need to access the current state-of-art and results from different groups and schools of thought was felt with the objective of focusing trends in future research and contributing to new legislative developments in safety terms.

With these purposes, the Advanced Study Institute on *Crashworthiness of Transportation Systems: Structural Impact and Occupant Protection* took place in Tróia, Portugal, during the period July 7-19, 1996. Eighty five participants, from fifteen countries, representing academia, industry, government and research institutions attended this ASI, organized by the Instituto de Engenharia Mecânica, Instituto Superior Técnico. In addition to the lectures which were delivered by prominent researchers, the full program of the Institute also included contributed presentations by participants. These papers have been reviewed and a selection is being published in the International Journal of Impact Engineering and in the International Journal of Crashworthiness. Interchange of experiences between leading scientists, young scholars and specialists was greatly encouraged and discussions were promoted in order to generate new ideas and define directions for future research and developments.

Crashworthiness insures vehicle structural integrity and its ability to absorb crash energy with minimal diminution of survivable space. Restraint systems limit occupant motion mitigating injuries that may result from contact with vehicle interior during sudden acceleration conditions. Both structural crashworthiness and occupant protection technologies are multi-disciplinary and highly specialized, including complex technical fields spanning from the areas of mechanics to biological sciences. This book brings together, in a tutorial and review manner, a comprehensive summary of current work and is therefore suitable for a wide scope of interests, ranging from advanced students to researchers and developers concerned with advanced theoretical and design issues in crashworthiness of transportation systems. The applications will help manufacturers and design engineers to apprise the different approaches available today and their use and suitability as efficient design tools. This book is organized into eight parts, the first seven addressing the state-of-art techniques and methods in the principal areas of study of the Institute and the final part with the a review of the state-of-art and a summary of the directions for future research and developments proposed in the panel session at the end of the Institute.

The biomechanics of injuries has been extensively addressed in recent years by the engineering and medical communities. The tolerance of various components of humans has been investigated extensively and is now represented by a variety of parameters which exhibit different levels for specified degrees of injuries. Major efforts are still required in order to have results that can be directly applied in the design of vehicles, so as to mitigate trauma during secondary collisions. In Part I, the aspects related with the occupant kinematics and injury biomechanics are presented. The mechanisms for injury of the head, neck and spine, in particular, are analyzed in terms of the mechanical response of the tissues involved. Life threatening injuries or serious disabilities and their relation with the excessive brain loading during crash are presented and discussed here. Finally, the pathophysiological responses to trauma are considered against the forensic evidence of possible injury causation.

Occupant safety in transportation systems, and ultimately, the potential survival in the event of a vehicle crash is determined by two factors: structural crashworthiness and restraint systems in the passenger compartments. In Part II of this book the reader will find a review of the biomechanics of impact in road accidents and their relation with factors such as structural integrity, compatibility and retrain systems for different types of impacts. A variety of technological solutions for passive and active restrain systems with specific energy absorbing characteristics and different motion limiting mechanisms adaptable to the passenger and to the crash conditions are included. Compatibility issues raising from the heterogeneous nature of the masses and structures of vehicles involved in car-to-car or car-to-truck collisions or contact with vulnerable road users are also discussed in this part of the book. Special attention is paid not only to the existing research and testing procedures currently used but also to the design and test methodologies used by car manufacturers.

A wide variety of crash tests, including a large family of dummy tests, are being used by different safety programs. Some seem to be more relevant to real world accident scenarios than others.

Part III includes information concerned with the development and use of crash test dummies and their numerical modeling. The models of dummies based on multibody formulations and on finite element descriptions are presented and the issue of more realistic human body models, rather than dummy models, is brought up. The important issue of biofidelity of dummies and numerical models is thoroughly discussed and directions for future use of a biofidelity classification are also given.

The evaluation of highly complex nonlinear response of vehicle structures, which boils down to the assessment of the energy absorbing mechanisms of structural components has been a dominant feature in structural crashworthiness. In Part IV, the dynamic inelastic response of structures and structural components is described in view of the strain material properties under impact loading. The accurate and efficient analysis tools with increasing complexity, ranging from simplified 1D and 2D models embedding analytical solutions or experimental results, to sophisticated finite element procedures are introduced to model complex structural systems. The design of buses for crashworthiness and rollover protection exemplifies the application of these principles.

In Part V, recent trends and advances of the finite element analysis in crash simulation of motor vehicle and occupant are presented. The modeling and the use of different material models for vehicle structural components, restrain systems, air bags, crash barriers and dummies are addressed in the framework of the finite element methodology. The feasibility of the integrated simulation of full scale vehicle models, restrain systems and occupants is then demonstrated and discussed.

Many of the most popular analysis and design procedures used in crashworthiness of vehicles and occupants are based in multibody methodologies. In Part VI, the fundamentals of multibody dynamic methodology are reviewed regarding its applicability to crashworthiness and impact analysis. The important aspect of modeling and validating the contact-impact forces, including material and geometrical properties of the surfaces in contact, are addressed here. Finally, different formulations, with increasing complexity, suitable for the simulation of vehicle and occupant impact, are presented. The emerging use of optimization procedures in the crashworthiness design is considered in the framework of multibody dynamics.

In addition to the general assessment of the different aspects in crashworthiness, major applications of developments to date in aeronautics are presented and analyzed in Part VII of this book. A balanced presentation of the different issues in the design and simulation of aircraft and helicopter crashworthiness, together with its applicational aspects will help the reader to apprise the different design goals and methodologies and the efforts that are in progress to improve the present status of the art. Issues related with crash protection of aircraft occupants, failure of materials used in the aeronautics industry and adequacy of the simulation methods to specific problems of this area are addressed here.

The editors are indebted to the Co-Director of the Institute, Prof. Wlodek Abramowicz, and to the members of the Organizing Committee, Prof. Norman Jones, Dr. David Viano and Dr. Jac Wismans for their valuable suggestions and advise in the organization of the ASI. We extend our recognition to all lecturers and participants in the Institute for their active participation in the discussions and contributing presentations. Through Mr. João Sousa we would like to acknowledge the professional work of the Torralta staff for the excellent ambiance provided during the Institute. Our appreciation is also due for the collaboration and efforts of Ms. Andrea Freitas and Mr. Amândio Rebelo, from the IDMEC staff, that contributed to the smooth running of the ASI. The preparation and running of the Institute and the efficient editing work of this book is greatly due to the competence of Mrs. Sandra Castelo whose dedication is highly appreciated. Finally, without the support of the NATO Science Committee the Institute would not have been possible. Their support is gratefully acknowledged.

Lisbon, October 1996

Jorge Ambrósio
Manuel Pereira
Fernando Pina da Silva

LIST OF PARTICIPANTS

Lecturers:

Wlodek Abramowicz
Impact Design Inc., ul. 1 Maja 18, Michalowice 05-816, Poland

Jorge Ambrósio
Instituto de Engenharia Mecânica, I.S.T., Av. Rovisco Pais 1, 1096 Lisboa Codex, Portugal

Pierluigi Ardoino
FIAT-Auto, Product Engineering Safety Center, Via G. Gozzano 2, 10043 Orbassano (TO), Italy

Faris Bandak
National Highway Traffic Safety Administration, 400 Seventh St, SW Washington, D.C. 20590, U.S.A.

Dominique Cesari
INRETS 109 Avenue Salvador Allende 69675 Bron Cedex, France

Eberhard Haug
R/D - Engineering System International, SA 20 Rue Saarinen, Silic 270 94578 Rungis, France

Ian Hill
Guy's Hospital, Department of Forensic Medicine University of London, London Bridge, SE1 9RT, U.K.

Norman Jones
The University of Liverpool, Department of Mechanical Engineering Liverpool, L69 3BX , U.K.

Tom Khalil
Safety Center, General Motors Research and Development Center, 30200 Mound Road, 1-11 Warren, Michigan 48090-9010, U.S.A.

Chris Kindervater
Deutsche Forshungsanstalt für Luft Raumfahrt, Pfaffenwaldring 38-40, D-7000 Stuttgart 80, Germany

Albert King
Wayne State University, Bioengineering Center 818 W. Hancock, Detroit, MI 48202, U.S.A.

Hamid Lankarani
Wichita State University, Department of Mechanical Engineering, Wichita, KS 67260-0035, U.S.A.

Murray Mackay
University of Birmingham, Accident Research Unit Edgbaston Birmingham B15 2TT, U.K.

Matyas Matolscy
Ikarus Vehicle Manufacturing, Margit u. 114 1165 Budapest, Hungary

Priya Prasad
Ford Motor Company, Advanced Engineering Center, 20000 Rotunda - P.O. Box 2053, Mail Drop 66 Dearborn, MI 48121-2053, U.S.A.

Claude Tarriere
Renault, Department of Biomedical de l'Automobile, 132 Rue des Suisses, 92000 Nanterre, France

David Viano
General Motors Research Laboratories, Biomedical Science Department, 30500 Mound Road, Box 905 Warren, Michigan 48090-9055, U.S.A.

T. Wierzbicki
Massachussets Institute of Technology, 77 Massachusetts Av., Cambridge, MA 02139, U.S.A.

Jac Wismans
TNO Road Vehicles Research Inst., Schoemakerstraat 97, PO Box 6033 2600 JA Delft, The Netherlands

Participants:

João Abrantes
Faculdade de Motricidade Humana, Laboratório de Biomecânica, 1499 Lisboa, Portugal

Amein Alsuezi
George Washington University, P.O. Box 6019 McLean, VA 22106, U.S.A.

Massimiliano Avalle
Politecnico de Torino, C. So. Duca Abruzzi, 24-10129 Torino, Italy

Joaquim Infante Barbosa
Escola Náutica Infante D. Henrique, Av. Bonneville Franco, 2780 Paço d'Arcos, Portugal

Sergio Bianco
Centro Ricerche FIAT, Strada Torino 50, I-10043 Orbassano - TO, Italy

Soo-Won Chae
Hongik University, 72-1 Sangsu-Dong Mapo-Ku, Seoul, Korea

E. C. Chirwa
Bolton Institute Deane Road, School of Engineering, Bolton - BL3 5AB, U.K.

Dominique Cornette
Université de Valenciennes, LAMIH/LGM, B. P. 311, 59304 Valenciennes Cedex, France

Jeff Crandall
University of Virginia, Automobile Safety Labor, 1011 Linden Av., Charlottesville, VA 22902, U.S.A.

Humberto Cunha
Instituto Superior Técnico, Av. Rovisco Pais 1, 1096 Lisboa Codex, Portugal

Michael Damsgaard
Institute Mechanical Engng, Aalborg University, Pontoppidanstraede 101, DK-9220 Aalborg, Denmark

João Dias
Instituto Superior Técnico, Av. Rovisco Pais 1, 1096 Lisboa Codex, Portugal

Nima Edjtemai
DYNALIS, Rue Lavoisier, BP 24 91710 Vert-Le-Petit, France

Jan Gierej
Technical University of Warsaw, Akient 6, 01-937 Warsaw 4, Poland

Carlos Gomes
IDMEC-Pólo FEUP, Rua dos Bragas, 4099 Porto Codex, Portugal

João Gonçalves
Instituto de Engenharia Mecânica, I.S.T., Av. Rovisco Pais 1, 1096 Lisboa Codex, Portugal

Frank Graf
152 Court St., Portsmouth, New Hampshire 03801-4416, U.S.A.

John Hansen
Technical University of Denmark, Dept. of Solid Mechanics, Building 404, DK-2800 Lyngby, Denmark

Sherman Henson
Ford Motor Co., Fairlane Plaza South, 330 Town Center Drive, Suite 500, Dearborn, MI 48126, U.S.A.

Steven Hooper
National Institute for Aviation Research, Wichita State University Wichita, KS 67260-0093, U.S.A.

Ricky Houtris
MASCOTECH Engineering Europe Ltd., Canewdon House Locks Hill, Rochford, Essex SS4 1BB, U.K.

Mircea Ieremia
Technical University of Civil Engineering, Bd. Lacul Tei 124 OP 38, 72302 Bucharest, Romania

Szymon Imiekowski
Polish Academy of Sciences - IFTR, Swietokrzyska 21, 00-049 Warsaw, Poland

Mideya Innami
Toyota Motor Europe, Hoge Wei 33 - A, 1930 Zaventem, Belgium

Shin-You Kang
Kangwon National University, Dept. Precision Mechanical Engng., 192-1 Hyoja 2-Dong, ChunCheon Kangwon-Do 200-701, Korea

Dimitrios Karamanlidis
University of Rhode Island, Department of Civil and Environmental Engng., Bliss Hall, 1 Lippitt Road Kingston, RI 02881-0805, U.S.A.

Dusan Kecman
Cranfield Impact Center, Wharley End., Cranfield Bedford, MK43 0AL, U.K.

Noboru Kikuchi
University of Michigan, Mechanical Engng and Applied Mechanics Dept. Ann Harbor, MI 48109, USA

Agnes Kim
Ford Motor Company, Advanced Engineering Center, 20000 Rotunda P.O. Box 253, MD #66, Dearborn, MI 48121, U.S.A.

Heung-Soo Kim
Department of Mechanical Design and Production Engng.,College of Engineering, Seoul National Univ., Shilim-dong sa 56-1, Seoul 151-742, Korea

Heon Young Kim
Kangwon National University, Department of Mechanical Engineering, 192-1 Hyoja 2 - Dong, Chuncheon, Kangwon Do, 200-701, Korea

Ki Taek Kim
SAIT - Samsung Advanced Institute of Technology, CAE Team Dept. Supercomputer Applications, SAIT P.O. Box 111 Suwon, Kyung Ki-Do, 440-600, Korea

Yuichi Kitagawa
Nissan Motor Co., Ltd, Vehicle Res. Lab., Ikego 2-18-7-104, Zuchi-City 249, Japan

Matthias Kröger
Institut für Mechanik, Universität Hannover, Applestrasse 11 A, 30167 Hannover, Germany

Hans-Peter Lang
Robert Bosch GmbH, FV/SLM 3, Postfach 300240, 70442 Stuttgart, Germany

J. Andre Lavoie
Virginia Polytechnic Institute & State University, Engineering Science and Mechanics Department, Blacksburg, Virginia 24061-0219, U.S.A.

Anthony Lawson
Cranfield University, Cranfield Impact Center, Cranfield Bedfordshire, MK43 0AL, U.K.

Eugene Lee
Samsung Motors Inc., Technology Center, 493 Banwol-Ri, Taean-Eup Hwasung-Kun, Kyungki-Do 445-970, Korea

J. Lescheticky
BMW AG, EK-20, D-80788 München, Germany

Eric Markiewicz
Université de Valenciennes, LAMIH/LGM, B. P. 311, 59304 Valenciennes Cedex, France

Nigel Mills
Univ. of Birmingham, School of Metallurgy and Materials, PO Box B63 Birmingham B15 2TT, U.K.

Parviz Nikravesh
University of Arizona, Department of. Aerospace and Mechanical Engng., Tucson, AZ 85721 , U.S.A.

Stanislaw Opalinski
Polish State Railways, Railway Research Center, ul. Chlopickiego 50, 04-275 Warszawa, Poland

A. Otubushin
A & AE & TS, Loughborough University, Ashby Road, Loughborough Leics., U.K.

Sándor Vincze-Pap
AUTÓKUT, Csóka u. 9, 1115 Budapest, Hungary

Shin-Hee Park
Seoul National University, Dept. Mech. Design and Prod. Eng, San 56-1, Shilim-Dong, Seoul, Korea

Jerzy Pawlus
Polish State Railways, Railway Research Center, Ul. Chlopickiego 50, 04-275 Warszawa, Poland

Niels Pedersen
Technical University of Denmark, Dept. of Solid Mechanics, Building 404, DK-2800 Lyngby, Denmark

Pauli Pedersen
Technical University of Denmark, Solid Mechanics Dept., Building 404, DK-2800 Lyngby, Denmark

Manuel Seabra Pereira
Instituto de Engenharia Mecânica, I.S.T., Av. Rovisco Pais 1, 1096 Lisboa Codex, Portugal

Michiel van Ratingen
TNO Road Vehicle Research Inst, Schoemakerstrasse 97, PO Box 6033, 2600 JA Delft, The Netherlands

Peter Ravn
Technical University of Denmark, Dept. of Solid Mechanics, Building 404, DK-2800 Lyngby, Denmark

Juan Viaño Rey
University of Santiago de Compostela, Departamento de Matemática Aplicada, 15706 Santiago de Compostela, Spain

Majid Sadeghi
Cranfield Impact Center Wharley End., Cranfield, Bedford MK43 0AL, U.K.

Christian Schuster
University of Duisburg, Lotharstr. 1, 47058 Duisburg, Germany

Michael Sheh
CRAY Research, Inc. 1902 Grayslake Dr., Rochester Hills, MI 48306, U.S.A

Fernando Pina da Silva
Instituto Superior Técnico, Av. Rovisco Pais 1, 1096 Lisboa Codex, Portugal

Miguel Silva
Instituto de Engenharia Mecânica, I.S.T., Av. Rovisco Pais 1, 1096 Lisboa Codex, Portugal

Bo Cerup Simonsen
Technical University of Denmark, Department of MAOE, Building 101, E DK-2800 Lyngby, Denmark

Carlos Mota Soares
Instituto Superior Técnico, Av. Rovisco Pais 1, 1096 Lisboa Codex, Portugal

M.C. Tofan
Transilvania University of Brasov, Mechanics and Strength of Materials Department, Bd. Eroilor 29, 2200 Brasov, Romania

Paul Tomlison
Dalphi-Metal España, S.A., C/ Martires Concepcionistas 3, 28006 Madrid, Spain

Júlio Machado Viana
Universidade do Minho, Dept. de Eng. de Polímeros, Campus Azurem, 4800 Guimarães, Portugal

Rade Vienjevic
Cranfield University, College of Aeronautics, Cranfield, Bedfordshire, MK43 0AL, U.K.

Jaime Viso
Dalphi-Metal España S.A., C/ Martires Concepcionistas 3, 28006 Madrid, Spain

Jerky Wicher
Institute of Vehicles, 84 Narbutta Str,. 02-524 Warsaw, Poland

PART I
Impact Biomechanics

OCCUPANT KINEMATICS AND IMPACT BIOMECHANICS

ALBERT I. KING
Wayne State University
Bioengineering Center
Detroit, Michigan, USA

Abstract

This paper describes the kinematics of the driver and the front seat passenger of an automobile that is involved in four crash configurations; namely that of frontal, side and rear impact, and in rollovers. In each case, the associated biomechanical injury mechanisms are discussed involving body parts from head to foot. The effects of various restraint systems are considered, including their potential for causing injuries that are different from those they were designed to prevent. It is concluded that the present restraints are doing a good job for frontal impact but more needs to be done in side impact protection, for rearend crashes and in rollovers.

1. Introduction

The use of restraint systems to protect automotive occupants in the event of a crash is a form of environmental control which is often more effective than behavioral control because to err is human. The suddenness of the crash and its severity precludes the use of muscular forces to overcome the inertia of the body. However, the design of restraint systems and the avoidance of injury secondary to the system require knowledge of occupant kinematics and impact biomechanics. This paper describes the major features of the kinematics of an occupant in frontal, side, rearend and rollover crashes and the principal mechanisms of injury resulting from either the non-use or use of the common types of restraint systems, namely, belt restraints and airbags. Each of the four crash modes is described separately and for each mode, the kinematics of the occupant relative to the vehicle are discussed along with the principal injuries for the unrestrained occupant and those due to the use of three restraint systems: the three point belt, the airbag and the combination of belts with airbag.

J. A. C. Ambrósio et al. (eds.),
Crashworthiness of Transportation Systems: Structural Impact and Occupant Protection, 3–23.

2. Frontal Impact

This crash mode is still the most common and there are still many fatalities resulting from the failure to use the available restraints. With the introduction of the airbag in all US cars by the 1997 model year, fatalities rates and injuries due to frontal impact should be on a gradual decline. Evans [1] states categorically that there is an enormous body of scientific literature accumulated over three decades which provides a robust, consistent, coherent and clear picture that belt restraint systems reduce the risk of injury and fatality as well as the severity of the injury. The reason for unrestrained occupants sustaining severe injuries in frontal impacts is, of course, due to the fact that they continue to move at the pre-impact speed within the car which is coming to a rapid stop. They then impact the interior surfaces of the vehicle which are not as forgiving as a belt or an airbag. Belts and airbags on the other hand provide the necessary ride down to decrease the velocity of the occupant relative to the vehicle and thus reduce the g-level of impact. The ride-down concept is explained in greater detail by Eppinger [2] and by King and Yang [3].

2.1 THE UNRESTRAINED OCCUPANT

The kinematics for the unrestrained driver are quite different from those of the front seat passenger. Any description of their kinematics must take this into account since the driver interacts with the steering system first while the passenger impacts the instrument panel and windshield directly. Similarly, the injury picture can be quite different.

2.1.1 *The Unrestrained Driver*

The first thing an unrestrained driver encounters when there is frontal crash is the steering system and the instrument panel. The seated torso slides forward without flexing, causing the chest and /or the abdomen to strike the steering wheel. The precise location of the impact on the body is unpredictable because the orientation of the steering column can change during impact with the possibility of a head impact as well. The torso wraps itself around the wheel as the steering column collapses, absorbing some of the energy of impact. Most steering wheels will be deformed by the torso and, in severe impacts, the rim is bent backward to expose the hub which is only about 150 mm (6 inches) in diameter. Severe chest injuries can result from the chest impaling itself onto the hub. As for the lower extremities, the knees usually run into the instrument panel while the feet are braced against the brake pedal and/or the floor board. This sets up a complex loading pattern on the bones and joints of the lower extremities. Impact of the head with the windshield follows with the head in slight flexion at the time of contact with the glass. Because of the high penetration resistance windshield, the head is usually retained within the vehicle but it makes the familiar stellate pattern on the windshield.. The head of taller drivers may hit the roof rail above the windshield. The body then rebounds into the seat, after sustaining multiple injuries.

If the impact is slightly angled to the left, the head of the left sided driver can hit the left A-pillar. The impact scenario is quite variable because perfect head-on impacts are the exception rather than the rule. Recently, there has been much interest in offset frontal collisions in which the left half or the driver side of two opposing vehicles interact. The impact severity in terms of g-levels experienced by the vehicles may not be as severe as a full faced impact but there is usually much more intrusion of the driver side compartment, including the footwell area.

Fatalities are usually the result of brain injuries due to head impact and aortic ruptures due to chest impact with the steering wheel hub. Head impact with the A-pillar or the roof rail is responsible for the fatal brain injuries. The new US Federal Motor Vehicle Safety Standard (FMVSS) 201 for head impact will help reduce these fatalities. However, the standard does not account for rotational acceleration and more effective head injury criterion is needed for this standard and the associated standard FMVSS 208. As for the catastrophic injury of aortic rupture, it is well known to emergency physicians and various mechanisms have been proposed by Viano [4]. However, there has been no verification of these mechanisms [5]. Experimental duplication of this injury was demonstrated by Roberts et al [6] in live but anesthetized dogs and in both live and postmortem dogs by Nusholtz et al [7], but it could not be reproduced in cadaveric subjects, based on our own unpublished results. Severe or life-threatening injuries to the heart, lungs, and solid abdominal organs are also seen. Again, the reproduction of these soft tissue injuries is difficult in cadavers. Moreover, there have not been a large number of cadaveric studies on unrestrained occupants. Disabling injuries to the bones and joints of the lower extremities are also not uncommon. The tibia can be fractured by bending or axial compression while the femur is generally fractured in bending due to knee loads generated by impact with the instrument panel. The knee can form a pocket in the panel and can sustain a large shear load in either the lateral or vertical direction or both, relative to the normal seated position, subjecting the femur to asymmetrical bending. Although the current US FMVSS 208 calls for a compressive load limit of 10 kN on the femur, there is no evidence that femurs would actually fail in compression in a frontal impact. The foot, ankle, knee and hip are also at risk. If there is footwell intrusion, a variety of ankle injuries are seen. The right ankle can be subjected to extreme dorsiflexion of over 45 degrees [8], causing rupture of the ankle ligaments and malleolar fractures. The same injuries can occur due to eversion or inversion of the ankle [9]. Knee injuries can be a splitting fracture of the femoral condyles caused by the patella, patellar fractures, damage to the cartilage covering the patella and the femoral condyles and to the subchondral bone [10] or rupture of the posterior cruciate ligament (PCL) [11], due to initial impact of the tibial tuberosity with the instrument panel. It is imperative that the instrument panel slope forward from the top down so as to avoid contact with the tibial tuberosity.

There is no question that the unrestrained driver is at high risk for severe injuries in a frontal crash. Human muscular response is just not fast enough nor strong enough to resist the enormous inertial forces experienced by the body.

2.1.2 *The Unrestrained Front Seat Passenger*

In addition to the injuries described above, it is necessary to discuss a severe or perhaps fatal injury risk to the front seat passenger caused by head impact with the windshield. Because of the slope in the windshield, the head and neck are placed in extension as well as in compression [12]. The injury involves the second cervical vertebra which can be fractured in much the same way as in judicial hanging. It is now known as traumatic spondylolisthesis of the axis (C2). Chest injuries are due to impact of the torso with the dash and there is a similar risk of knee and femoral injury due to knee impact with the dash. Tibial and ankle injuries are also seen, particularly in slightly angled frontal impacts [13].

2.1.3 *Occupants Restrained by a Lap/Shoulder Belt*

When front seat occupants are restrained by a lap/shoulder or three-point belt, the severity of injury drops dramatically for the same frontal crash severity. Patrick and Andersson [14] estimated that the survivable barrier equivalent (BEV) speed of the Volvo 140 series cars should be in the range of 72 km/h (45 mph) if there was no intrusion of the occupant compartment. In terms of the kinematics of the belted occupant, the body slides forward a slight distance till the lapbelt becomes taut. Then the head and upper torso begin to flex as the shoulder belt loads up. After the torso stops rotating forward, the head and neck continue to flex until it either reaches the end of its forced range of motion or it hits an object in front of the occupant, such as the rim or hub of the steering wheel. In terms of the injuries sustained by the belted occupant, we can discuss both occupants under the same heading because belt-induced injuries sustained by the driver and passenger are similar. There is a risk of rib and sternum fractures due to loading from the shoulder belt, which in severe cases can cause injury to the abdominal organs, such as the liver [15]. Lower abdominal injuries can occur due to an improperly worn lapbelt. If it is not on tightly enough, there is a risk of submarining in which the torso slides down the seat and the lapbelt rides up above the pelvis, causing injury to the intestines [16]. Experimental evidence provided by Levine et al [17] show that the pre-impact locking of the knee joint through muscular action can reduce the incidence of submarining and thus reduce the risk of abdominal injuries.

The major difference in the injury pattern between the driver and passenger is the risk of head injury due to impact of the head with the steering wheel for the driver [18] Because of the fact that entire torso is restrained to ride down with the car, the only free segments are the head and neck and the upper extremities. For the passenger, flexion injuries to the neck without head contact were found to be rare [19] but the injuries can be severe [20]. There is, of course, the risk of head contact with the steering wheel for the driver. Although the severity of this head injury is minor compared to that sustained by an unrestrained driver, there is nevertheless a potential for brain damage, justifying the use of a driver-side airbag.

Huelke et al [21] also discussed a thoracolumbar injury among belted occupants. They were found to sustain a wedge type fracture of the vertebral body of lower thoracic and upper lumbar vertebrae. This injury was attributed to the slumping posture of the occupant. There have also been reports of these injuries in laboratory studies in which whole-body testing of belted cadavers were carried out. These include the work of Patrick and Levine [22] and Leung et al [23]. No quantitative explanation of the mechanism of injury was provided by these authors. However, the biomechanical basis for this injury was provided by Begeman et al [24] but it was a result that was reported before its time and was not widely accepted. The study by Begeman et al [24] was initiated by an interesting result of applying the 2-D vertical acceleration spinal model of Prasad [25] to a horizontal crash. While working on his dissertation, Prasad found that an axial compressive load was predicted by the model even though the input acceleration to the body was in the horizontal plane. He did not include this result in his thesis but after a thorough check of the model, it was concluded that this was a case in which the counter-intuitive predictions of a model should be verified experimentally. The experiments were done and large seat pan loads were measured when fully restrained cadavers were subjected to a -Gx acceleration (a horizontal deceleration). Interestingly enough, the same seat pan loads were not present when an anthropomorphic dummy was used. Care was taken to subtract the vertical component of the lapbelt load so that the measured seatpan load was the net load due to the seated cadaver. The only cause of this load was found to be the kyphotic curvature of the thoracic spine. When the upper torso is held back by the shoulder harness, the thoracic spine tends to straighten out and in the process pushes down on the thoracolumbar spine and up on the head and neck. This is why no seatpan load was measured when the subject was a dummy which has a straight and rigid spine. This phenomenon was found to occur in our other tests, including seatpan loads measured during tests using human subjects.

There can still be lower extremity injuries among belted occupants. Knee contact cannot be completely avoided and foot and ankle injuries are dependent in part on the extent of footwell intrusion. In any case, there is a high probability of surviving a frontal crash when the occupants are properly restrained by the three-point belt. Fatalities occurring among these restrained occupants are due mainly to intrusion of the passenger compartment by another vehicle [26], such as, in a vehicular underride crash of a passenger car into the rearend of truck with long overhang.

2.1.4 *Occupants Protected by an Airbag*

The airbag was first introduced by General Motors in 1000 1973 Chevrolet Impalas and subsequently in a total of 11,321 GM cars between the 1973-76 model years. Mertz [27] analyzed the results of 216 deployments and found that the overall results were quite good if the occupants were not out of position. Production airbags were introduced by Mercedes in 1987 and all new cars will be so equipped in 1997 in the United States. From the manufacturers' viewpoint, the principal purpose of these airbags is to provide supplemental protection to a belted occupant. In particular, it is meant to eliminate the risk of head injury to the driver. However, under FMVSS 208, the test procedure calls for each vehicle to meet the injury criteria using an unbelted dummy, based on the assumption that a high percentage of US front seat occupants will continue to remain unrestrained. To meet this standard, it is necessary to install a relatively "aggressive" airbag which can come to full deployment or pressure rapidly to withstand the full inertial load of the occupant. The kinematics of an unbelted occupant in an airbag equipped car start off very much like that of any unbelted occupant. The whole body slides forward as the vehicle comes to a rapid stop. However, at the same time, the airbag is deploying and, in most cases, the occupant impacts a fully deployed bag and avoids being severely injured. In a severe crash, however, the lower extremities are not protected from impact with the dash and the intruding footwell. If the occupant is restrained by a three point belt, the major effect of the airbag is to reduce the amount of head and neck flexion and to reduce belt loading on the chest and pelvis. There is no question that the airbag has the potential of saving many lives and of reducing injury severity dramatically, particularly for the unrestrained occupants. However, there are side effects which can also be quite severe. The most dramatic side effect is the fatal or near fatal injury it can cause to occupants who are too close to the airbag at the time of deployment. These so-called "out of position" (OOP) occupants can sustain severe neck and/or chest injuries due to impact with the rapidly deploying bag. King and Yang [3] have provided a biomechanical explanation of the hazards of being too close to a deploying airbag. Reports of such incidents can be found in the clinical literature such as the case of a neck injury described by Traynelis and Gold [28]. The mechanism of neck injury was first described by Cheng et al [29] who reported upper cervical injuries in cadaveric subjects when the bag gets under the chin of the test subject. The estimated traction loads on the neck were in the range of 10 kN. Short drivers are also at risk because the impact of the bag material on the chest as it is deploying can cause the heart to go into arrhythmia or ventricular fibrillation and damage the lung tissue. It is also important to ensure that children do not stand in front of the passenger seat or that rearward facing child seats not be placed in that seat. Mertz et al [30] have demonstrated the hazards of the OOP child standing in front of an airbag. It should be noted that most of these severe injuries would not occur if the front seat occupant is fully restrained by a three-point belt.

In severe frontal crashes, the lower extremities are still at risk, for both the belted and unbelted occupant but more so for the unrestrained occupant. There is anecdotal evidence that these severe lower extremity injuries are now occurring because these cases would have been fatalities had the airbag not been available. Minor to moderate injuries to the upper extremities and the face can occur as a result of contact with the air bag or its cover. Such injuries include dislocated thumbs and other forearm fractures [19], ocular injuries [31, 32] and facial abrasions and burns [33].

3. Side Impact

Approximately 8,000 US car occupants are killed annually from side impact. This crash can take two forms. Relatively young drivers in their 20's and 30's are more frequently involved in single-vehicle side impacts with a fixed object, such as a tree or pole. On the other hand, relatively older drivers above the age of 50 are more at risk in two vehicle crashes at intersections [34]. One can speculate that the involvement of older drivers in these crashes is due in part to their lack of ability to judge the speed of on-coming traffic at an intersection. The most frequent injuries are to the head, neck and chest [35]. There are also abdominal and pelvic injuries which can be quite severe. These injuries are most likely to occur to the near side occupant and when there is inward buckling of the side door. The use of belt restraints have been shown to be helpful because there is a frontal deceleration component in any side impact and belts help keep the occupants in place, especially when multiple occupants on a bench seat are involved. They can also prevent head injury to the driver in a far side impact, when the driver is the sole occupant of the vehicle [36].

3.1 OCCUPANT KINEMATICS IN A SIDE IMPACT

In a side impact, the near-side occupant initially has no lateral relative velocity with respect to the side wall of the car. He/she is first impacted by the intruding side door and then by the vehicle itself if the struck vehicle is impacted hard enough. Typical velocity time histories of the door and of the vehicles involved are shown in Figure 1. The stationary occupant is struck by the door when it is at or near its first peak door velocity.

The second peak can cause additional injury if it is high enough. The extent of the intrusion is a function of the speed of the striking vehicles, its bumper height and the strength of the door of the target vehicle. If the side window is made of tempered glass, there is usually no head impact with the window because it is shattered before the head reaches the plane of the window. However, the head can strike the B-pillar if the seat is positioned rearward or if the pillar is more to the front, such as in a four-door vehicle. The head can also strike the hood of the bullet car because the hood is designed to fold up in the form of an inverted V when the front of the car is impacted. The torso is subjected to a lateral impact from the intruding door and the location of the impact is dependent on the manner in which the door buckles inward.

3.2 MAJOR INJURIES TO THE NEAR-SIDE OCCUPANT

As mentioned above, head injury is responsible for 45% of side impact fatalities. A recent finite element model by Zhou et al [37] was used to simulate such an impact.The head impacted the hood of the striking vehicle at 64 km/h (40 mph) and the result was

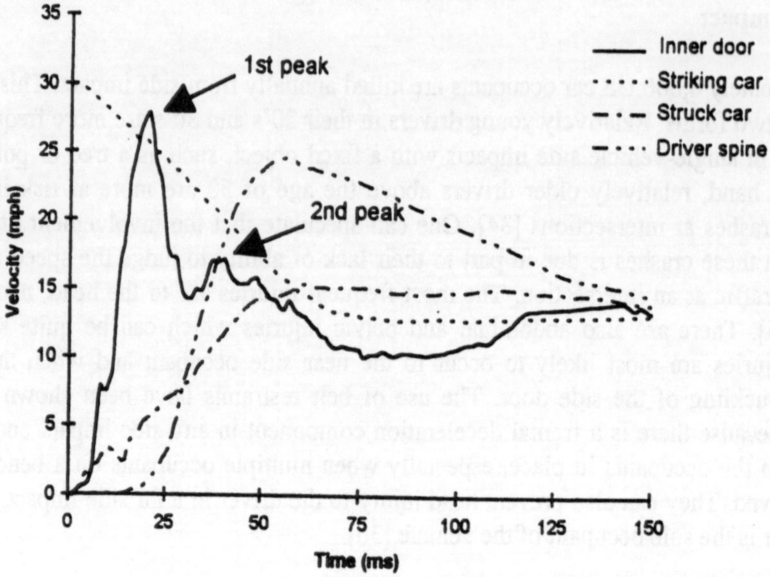

Figure 1. A typical door velocity profile in a side impact

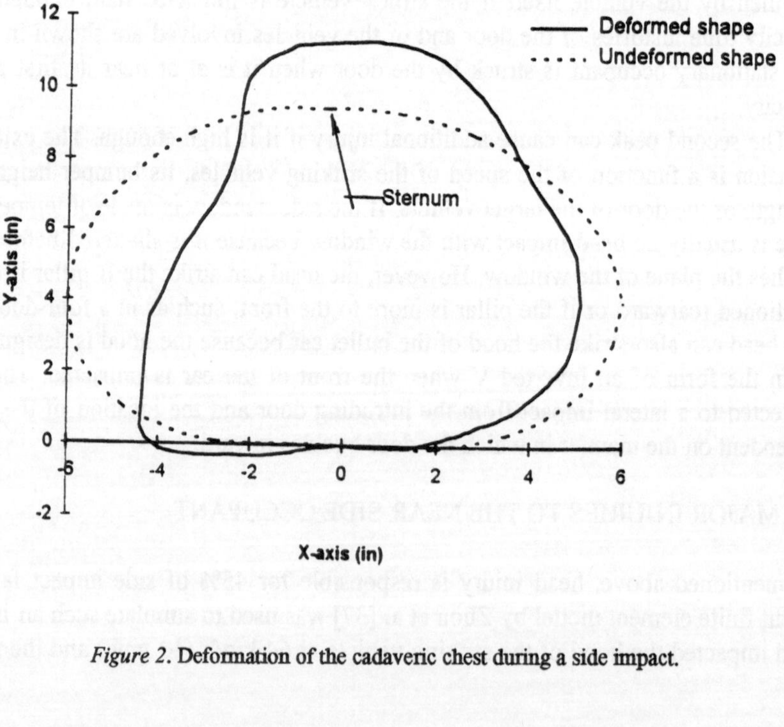

Figure 2. Deformation of the cadaveric chest during a side impact.

high shear strains in the corpus callosum and the brain stem. The results were comparable to autopsy findings of diffuse axonal injury in these areas of the brain. The other body regions susceptible to severe injury is the thorax, abdomen and pelvis. Using a chest band which can define the dynamic shape of the thorax during a crash, Figure 2 shows the deformed and undeformed shape of a cadaveric chest subjected to a side impact of about 32 km/h (20 mph). The extreme change in shape caused multiple rib fractures in almost every specimen tested. Life threatening soft tissue cadaveric injuries included aortic rupture and liver lacerations. As for the pelvis, the most common fractures are those of the pubic rami. In some cases, there is dislocation of the sacroiliac joint, depending on the impact point on the pelvis. If the initial impact is located at the level of the greater trochanter of the femur, pubic rami fractures are the result. To fracture the ileum or dislocate the sacroiliac joint, the impact must be directed at the wing of the ileum. Fracture of the acetabulum has also been reported.

3.3 SIDE IMPACT INJURY CRITERIA

The new FMVSS 214 was implemented in 1994 and total compliance by all manufacturers of cars sold in the US becomes effective with the 1997 model year. The acceleration based criteria are aimed at protecting the thorax and the pelvis. For the thorax, the criterion is called the Thoracic Trauma Index (TTI) and it is average of the peak accelerations measured on the impacted side of the thorax and on the spine of a government specified Side Impact Dummy (SID). This criterion was proposed by Eppinger et al [38] following an extensive series of cadaveric experiments conducted by various laboratories under government contract. The US automotive industry proposed a competing criterion termed the Viscous Criterion or V*C. It was first proposed by Viano and Lau [39] following a series of experiments on anesthetized animals. This criterion is the product of the chest wall velocity and the percent compression of chest in the lateral direction. Subsequent cadaveric testing by Cavanaugh et al [40] showed that the V*C criterion was a more sensitive measure of injury to the chest than TTI. In fact, Cavanaugh et al [41] proposed a third criterion called the Average Spine Acceleration which appears to work better than peak accelerations and better than TTI in the SID dummy [42]. The automotive industry also developed a more human like side impact dummy called the BIOSID. Its thoracic response is closer to that of the human than the SID. The principal difference between these two dummies is the more human like response of the BIOSID chest in comparison with that of the SID. The ribs of the BIOSID are long and slender and the chest can deflect up to 100 mm while the SID chest is very rigid and does not deform when impacted from the side. It simply rotates out of the way. In Europe, the safety community is recommending the V*C criterion and/or the compression criterion based on the response of a third side impact dummy, the EUROSID. The adoption of two different criteria based on two different side impact dummies for the same impact situation is a dilemma for all manufacturers that sell cars on both continents. It is certainly illogical to design cars to meet two different standards and to have to test every car twice to protect the same human occupant.

3.4 BIOMECHANICS OF SIDE IMPACT PROTECTION

The studies by Cavanaugh et al [40-42] referred to in Section 3.3 above compared the effectiveness of various forms of padding which can be used to protect the thorax. Based on the injuries sustained by the cadaveric subjects, it was found that the rib cage is very sensitive to the stiffness of the padding and the padding which produced the lowest value of TTI in the SID caused life-threatening injuries in the cadaver. It was found that padding for side impact protection should have a nominal crush strength of 15 psi with an actual crush strength of about 10 psi or 67 kPa. The material used was a paper honeycomb which had a flat response to load after a higher initial peak load to initiate the crushing. The optimal thickness was 100 mm (4 in). Plastic foams with the same response characteristics can be found to replace the paper honeycomb for use in the side doors of automobiles but the required thickness of 100 mm is not acceptable because the total width of the car would be decreased by 200 mm. If a thinner pad is used, it should have a stiffening response to keep the foam from bottoming out and to keep large loads from being generated. Alternately, side impact airbags can be used to achieve the desired stiffness and 100-mm thickness.

3.5 MATHEMATICAL MODELING OF SIDE IMPACT

The use of computer models to simulate impact events is becoming more popular as the cost of testing skyrockets. Gross motion simulators, such as the CAL 3D program or the MADYMO program, are used by auto manufacturers to study the response of the SID or BIOSID to side impact. However, since both of these dummies are not human-like, there is no assurance that padding and other designs based on simulations of dummy response will be safe for the human occupant. Huang et al [43] have developed a side impact MADYMO model which simulated the response of the human occupant, based on the data reported by Cavanaugh et al [40] and by Viano et al [44]. It is recommended that all side impact protective features be checked by using this model because it has been validated against two different series of cadaveric tests and because it can be used to predict both injury criteria, TTI and V*C. A parametric study was carried out using this model. It was found that engagement of the shoulder with the side door was a desirable characteristic because it would reduce the load on the thorax. Further, the model also predicted that strengthening the side door excessively is not necessary because with a non-deforming door, the side impact would accelerate the entire vehicle to a significant speed and the side door would impact the stationary occupant anyway. Datasets of this MADYMO model are available from Wayne State University, at no charge.

4. Rearend Impact

Although this impact mode is generally not life-threatening and fatalities are rare, the subject is of considerable interest because of the idiopathic nature of the neck pain that can plague many of the victims for a long time and because there is a high frequency of these minor impacts which can mysteriously cause so much suffering and disability among the victims. Headrests were mandated by NHTSA for cars built in the US after 1968. Their efficacy in reducing the number of patients with whiplash injuries has not been proven [45].

4.1 OCCUPANT KINEMATICS IN A REAREND COLLISION

Rearend collisions occur most frequently at intersections where the struck vehicle is stopped or slowing down to stop for a red traffic signal. Taking the case of a stopped vehicle which is accelerated forward by the striking vehicle, the occupant in the struck vehicle is initially stationary relative to an observer on the sidewalk. As the vehicle moves forward, the seat back engages the occupant who then deforms the seatback to generate a force to accelerate the occupant forward with the car while his/her head and neck remain stationary for a short time. To cause the head to move forward a shear force is generated at each level of the cervical spine until it reaches the occiput where it can begin to move the head forward to keep it attached to the neck and the rest of the body. The inertia of the head and neck results in their forward motion being delayed and if the head is not in contact with the headrest at the time of impact, its inertia will cause it to extend the neck and to rotate rearward. In a severe crash, hyperextension can occur if there is no headrest and the seat back only reaches the height of the shoulders. Thus, the most visible mechanical response of the neck is its hyperextension, followed by a rebound in flexion. The occupant also tends to ramp up the seat back as it is bent rearward by the inertial load of the occupant. Again, in severe impacts, the seat back can fail causing the occupant to end up in the rear compartment of the vehicle. However, deformation of the seatback was found to be more efficient than the headrest in protecting the neck [46], particularly for AIS 1 type injuries [47].

4.2 NECK INJURIES DUE TO REAREND IMPACTS

For minor rearend impact, the most common type of injury is to the soft tissues of the neck. There is complaint of neck pain but usually there is no immediate objective evidence of injury when the patient is seen in the emergency room, such as hemorrhage or obvious soft tissue tears. The usual diagnosis is cervical strain which is quite meaningless in terms of the tissues that were really injured. However, in more severe cases, there can be fractures of the anterior-superior aspect of the vertebral body or the separation of the anterior aspect of the disc from the vertebral endplate. If the patient continues to have pain, more sophisticated testing can reveal a herniated or bulging cervical disc. In the opinion of the author, such findings are coincidental and are the result of a pre-existing disc condition which may have been asymptomatic prior to the crash or the pain was not constant or severe enough to warrant a visit to the doctor. The complaint of chronic pain in the back of the neck, however, is not always a case of malingering but the cause of this pain as it relates to the forces applied to the neck and to the motion of the head and neck is still a subject of lively debate.

4.3 RECENT BIOMECHANICAL RESEARCH

Current biomechanical research interest in neck injuries due to rearend impacts is perhaps motivated by a large number of insurance claims and court cases against insurance companies and automobile manufacturers. Although some of the claims are for cervical disc herniation and bulging, there is an underlying problem of chronic pain, the cause of which has not been determined. The simple explanation of myalgia or muscle pain is not compatible with the kinematics described above, that of hyperextension of the neck. Since most of the pain is in the back of the neck and most of the muscles are in the back of the neck, there is no biomechanical basis for causing injury to these muscles when they are placed in compression. Furthermore, minor frontal crashes of equivalent severity cause stretching of these extensor muscles but do not cause a chronic pain problem. Because the most obvious response of the head and neck is hyperextension, most of the research in the past was focused on head rotation and neck extension [48, 49]. More recent work by Matsushita et al [50] provided evidence of neck compression in volunteer subjects, prior to the rearward motion of the head. The data were obtained by the use of high speed x-ray cinematography and the test speeds were from 2.5 to 4.9 km/h (1.6 to 3.0 mph).

4.4 NEUROPHYSIOLOGICAL BASES FOR SPINAL PAIN

Based on a decade long study by Cavanaugh et al [51] on the causes of low back pain, the research team at Wayne State University has formulated a neurophysiological basis for low back pain which is applicable to all intervertebral joints. The necessary conditions for the sensing of pain are the presence of nociceptors or pain sensing nerve endings in the tissue being considered and an adequate amount of deformation of the same tissue. The deformation must be relatively large because pain fibers normally have a high threshold and do not go off until the tissue is stretched beyond its normal range. These studies also showed that the capsules of the lumbar facet joints of New Zealand white rabbits are richly innervated with nerve endings which respond to pressure, stretch and spinal loading and satisfy all of the properties of a pain fiber [52, 53]. Ozaktay et al [54] found that such fibers are present in the human lumbar facet capsule. Earlier biomechanical studies have shown that the human lumbar facet capsule underwent a considerable amount of stretch when the facets were required to support large vertical loads [55]. It was also found that the threshold of these pain fibers in rabbits were lowered when the tissue was inflamed [56]. In fact, the pain fibers would go off without any mechanical stimulus and their rate of firing would increase when stimulated. The inflammatory condition was produced to simulate conditions which can be present during soft tissue injury.

These findings can be applied to the cervical spine because Bogduk [57] found that the human cervical facet capsule is innervated by sensory fibers from the cervical nerve root. Furthermore, McLain [58] described different types of mechanoreceptors, including pain sensing free nerve endings in the capsules of cervical facet joints from three human subjects. This anatomical evidence is bolstered by clinical studies conducted by Bogduk and Marsland [59], Aprill and Bogduk [60] and Lord et al [61]. Patients with chronic neck pain following whiplash were given facet injections of local anesthetics and 82 out of 128 patients experienced a relief of their pain. In another study by Barnsley et al [62] in which two different anesthetics were used, it was concluded that at least 54% of whiplash victims were suffering from facet pain.

4.5 A HYPOTHESIS FOR NECK PAIN FOLLOWING A REAREND CRASH

A hypothesis is proposed to explain the idiopathic nature of cervical pain secondary to a rearend collision. It begins with the study by Begeman et al [24] which indicated that the lumbar spine of an occupant restrained by a three-point belt was placed under compression in a frontal crash. This load was due to the straightening of the thoracic spine. In a rearend collision, the seat back would apply a posteroanterior load to straighten the thoracic spine causing a compressive lumbar and cervical load. The latter would occur before significant hyperextension begins. There is also a tendency for the torso to ramp up the seat back, resulting in a second compressive load on the cervical spine. It is hypothesized that these compressive loads loosen the ligaments around the intervertebral joints and decreases their shear resistance. At the same time, a shear force is developed across these joints due to the inertia of the head. This shear would cause the cervical facet capsules to be stretched and perhaps torn, causing the nociceptors to go off. In minor crashes with very little damage to the bumpers of the vehicles involved, it is not likely that the capsules would show a visible tear but the tissue can be stretched. Any report of immediate neck pain is generally attributable to a pre-existing degenerative condition of the cervical spine. For most people, the inflammation sets in several hours to a day later and the pain would then be apparent after inflammation has set in to lower the pain threshold. This hypothesis is consistent with the anatomy of the cervical spine. In particular, at the levels of C4-5 and C5-6, the facet surfaces are notably flat and the upper vertebra is more prone to slide rearward on the lower one more than it would slide forward. Testing of this hypothesis is underway and preliminary results in cadaveric tests indicate stretching of the facet capsules at the beginning of the rear impact.

4.6 PREVENTION OF NECK INJURIES

If the above hypothesis is valid, the method whereby neck pain can be prevented is to design a headrest which will push the head forward in concert with the seat back pushing the torso forward so that no shear is developed in the cervical intervertebral joints. The principle is relatively simple but the actual design of such a headrest can be quite involved because of the fact that the back of the torso has a different stiffness than the back of the head and the stiffness of the headrest needs to compensate for this difference. There are, of course, many other issues in the prevention of neck injuries due to a rearend collision. These are discussed in Viano and Gargan [45].

5. Rollover Crashes

With the advent of vans and recreational vehicles and increasing speed of travel, there has been an increase in rollover crashes. Some 10,000 motor vehicle occupant fatalities occur annually in the US [63] along with relatively severe central nervous system injuries, including quadriplegia and paraplegia. The primary cause of these fatalities and injuries is ejection from the vehicle, partial or total. It seems that ejection can take place through almost any opening in the vehicle, including the sun roof. Improved door latches have not prevented ejection through the side windows. Most of the ejected occupants are unrestrained by a lap belt and the single most effective countermeasure to ejection is the use of the lap-shoulder belt system. In rollovers with multiple rolls, there is usually a certain amount of roof crush which is often blamed as the cause of neck injuries to occupants who were not ejected. The validity of this claim is still being debated and needs to be assessed in relation to the kinematics of the occupant in a rollover crash [64].

5.1..OCCUPANT KINEMATICS IN A ROLLOVER CRASH

Recent experimental research by Bahling et al [65] involving the use of belted dummies in full scale rollover tests, showed that the restrained dummy did not roll in phase with the vehicle but assumed positions consistent with the centripetal acceleration generated by the roll. Its kinematics could also be affected by the presence of another occupant. Neck injury can be caused by head impact with almost any surface of the car, including not only the roof but also the roof rails, the ground and other occupants. This was based on measured neck loads in excess of 2000 N. These authors also showed that peak neck loads were measured well before the roof crush reached its maximum value and that the same force could occur with or without roof crush. However, field data suggest that there is no correlation between roof strength and neck injuries [66].

5.2 ROLLOVER CRASH INJURIES AND COUNTERMEASURES

The prevailing notion is that fatalities and severe injuries in a rollover are due mainly to ejection of the occupant. Contact of the head and neck with the hard road surface generally results in fatalities or catastrophic injuries. There is also evidence that the use of the belt restraint significantly reduced the number of severe to life threatening injuries to all body regions except the head [67]. Huelke et al [68] also concluded that the fatality rate was 4% among belted occupants and 10% among unbelted ones. Similarly, the belted occupants had fewer severe injuries than the unbelted occupants.

The obvious countermeasure is to prevent ejection by the use of the belt restraint system. However, the belts do not prevent partial ejection during which there can also be head to ground contact. In this case, other means of environmental control are needed to help reduce the incidence of rollovers, such as roadway design, including the design of highway lane partitions and of curbs.

6. Discussion

6.1 FRONTAL IMPACT

Attempts at protecting the occupant from severe injuries incurred in a frontal impact have been largely successful. The combination of a three-point belt system and a supplemental airbag restraint system should result in a further decrease in fatality and injury rates. The major remaining problems are intrusion of the occupant compartment and out of position front seat occupants. Current research interest in the effects of an offset frontal crash should be expanded into a more generalized study of the maintenance of the integrity of the occupant compartment for impacts from different directions. It is not clear if a continual increase in the number of airbags within the vehicle is the only solution to this problem.

The introduction of "smart" airbags which can control its rate of deployment based on occupant weight and crash severity is perhaps a first step in the use of faster computers to adjust its deployment aggressivity to the conditions at impact. Technology is perhaps available to prevent an airbag from deploying, if a rear-facing child seat is in the front passenger seat. Reliance on human intervention to turn the airbag on and off should only be a stop-gap measure. The severe neck and chest injury problems associated with out of position occupants still require more study. In the area of biomechanical response and tolerance, the airbag has generated interest in studying the biomechanical characteristics of the upper extremity, from the shoulder down to the fingers. Impact response and tolerance data for the upper extremity are presently unavailable.

6.2 SIDE IMPACT

The most crucial problem that needs to be resolved in side impact is a universal agreement on the best injury criterion to be used and the most acceptable side impact dummy to be adopted. The excessive burden of having to design and test all vehicles to meet two widely different side impact safety standards is illogical and extremely wasteful. As consumers, we should recognize that the cost of testing and development is eventually passed onto us. Harmonization of this standard should be a top priority among government agencies of all nations involved in the setting of these standards and it appears that the issue is not technical but rather one of national and/or personal ego.

In the implementation of design strategies to protect the occupant from side impact injuries, the design engineer should recognize the shortcomings of the currently available side impact dummies and in the criteria contained in FMVSS 214. Because of these uncertainties, it behooves the engineer to verify that any design which meets the relevant side impact safety standard is also safe for the human occupant. One method of doing so is to use computer models of the **human occupant** to determine the suitability of the design .The major issue is the stiffness of the padding or airbag to be used. Data provided by Cavanaugh et al [40] call for a soft padding to protect the elderly occupant whereas the SID prefers a stiff padding.

6.3 REAREND IMPACT

The root cause of the so-called "whiplash syndrome" needs to be found before any preventive measures can be effectively instituted and before it can be treated properly. The hypothesis proposed in this paper is based on the results of two separate areas of research; namely, spinal biomechanics and neurophysiology of the intervertebral joint. No such hypothesis has been proposed before because it is indeed unusual to have this combined expertise and knowledge in a single laboratory. In any case, testing of this hypothesis will provide a high probability of establishing the cause of whiplash induced neck pain and will lead to the intelligent design of headrests and other seat components.

The other problem with this syndrome is the unknown proportion of malingerers among the victims who are alleging injury to their cervical spine. It is suspected that in the US, the proportion of malingerers may be high because of the litigious nature of the US society. If so, the injury statistics are not valid and we may be chasing a non-existent problem when the impact is very minor. Furthermore, in these minor impacts, a complaint of pain does not automatically imply injury because in degenerated spines, the pain threshold is lowered and pain can be caused without an associated injury.

6.4 ROLLOVERS

The best way to avoid injury in a rollover crash is still to use all available restraints. There is no question that it is safer to remain with the vehicle than to exit it during a rollover. One proposed mechanism to prevent ejection of unbelted occupants and partial ejection of belted occupants is to use high penetration resistant (HPR) or laminated glass in the side and rear windows. Currently, HPR glass is used in the windshield only to prevent penetration of the head through the windshield and to eliminate the disfiguring lacerations of the face. Tempered glass is used in the other windows because of its strength and the non-lacerative nature of the glass fragments when it breaks. In fact, tempered glass shatters into small pieces when it breaks and because of this, the opening becomes a pathway for ejection. Arguments in favor of the use of HPR glass in side glazing include the retention of the occupant in a rollover even if the glass is fractured and the facilitation of the use of an airbag to protect the head in a side impact. There are, however, issues of durability and head injury and other hazards associated with the use of HPR glass in a side window [69]. The use of HPR glass in rear windows and rear hatches is not practical because of the inability of the glass to withstand torsion [69].

For non-ejected occupants, the ability to predict injury severity and, in particular, the probability of sustaining a catastrophic head or neck injury is limited because of the large number of variables needed to characterize a rollover. Among them are vehicle type and size, number of quarter turns executed, and the extent of roof and other structural deformation. Perhaps, to avoid injury, the best countermeasure is to minimize the incidence of rollover.

7. Conclusions

The occupant of an airbag equipped automobile should be able to survive most crashes if all available restraints are used. In severe crashes, the ability to walk away from the crash cannot be assured because the philosophy of the design is to protect most occupants from life-threatening injuries and to keep the cost of the safety features reasonably low.

Human tolerance to head impact needs to be revisited to make the current standards on head injury (FMVSS 201 and 208) more relevant to real world impacts. That is, the standard should consider both linear and angular acceleration of the head. At present, linear acceleration is the primary determinant of head injury. One of the proposed methods to change the standard is to use a finite element model of the brain to compute injury parameters that take into account not only the input accelerations but also the location of the impact on the head.

Many neck injury problems remain. Prevention of catastrophic neck injuries which result in paralysis is still not within reach and there is only a glimmer of hope that the whiplash problem can be finally understood.

There is very little known regarding the biomechanical response and tolerance of the upper extremity to impact. Additional research is needed for the intelligent design of side impact protection devices, such as side impact airbags..

Reduction of injury to the lower extremities is still necessary to provide the restrained occupant with more protection. The minimization of occupant compartment intrusion in frontal impacts, especially in offset frontal impacts, appears to be an effective countermeasure.

The prevention of rollover crashes is perhaps more important than countermeasures which are designed to prevent ejection or protect the occupant from roof crush.

8. Acknowledgments

The assistance of Dr. John Cavanaugh and Dr. King Yang for their input and advice is greatly appreciated. This work was supported, in part, by CDC Research Program Project Grant No. R49/CCR 505435-06

9. References

1. Evans, L (1995) How we know safety belts reduce injury and fatality risk, in Issues in Automotive Safety Technology, SAE Publication SP-1072, SAE Paper No. 950241.
2. Eppinger, RH (1993) Occupant Restraint Systems, in *Accidental Injury - Biomechanics and Prevention*, Chap. 8, Ed. by AM Nahum and JW Melvin, Springer-Verlag, New York.
3. King, AI and Yang, KH (1995) Research in Biomechanics of Occupant Protection, *J. Trauma*, **38**, 570-576.
4. Viano, DC (1983) Biomechanics of nonpenetrating aortic trauma: A review, Proc. 27th Stapp Conf., SAE Paper No. 831608, pp. 109-114.
5. Dischinger, PC, Cowley, RA, Shankar, BS and Smialek, JE (1988) The incidence of ruptured aorta among vehicular fatalities, 32nd Annual Proc., Assoc. for the Advancement of Automotive Med., pp. 15-23.
6. Roberts, VL, Moffatt, RC and Berkas, RM (1965) Blunt trauma to the thorax - Mechanism of vascular injuries, Proc. 9th Stapp Conf., Edited by Cragun, MK, pp. 3-12.
7. Nusholtz, GS, Kaiker, PS and Bosio, AC and Kirsh, M (1985) Thoracic response to frontal impact, Proc. 29th Stapp Conf., SAE Paper No. 851721, pp. 17-48.
8. Begeman, PC and Prasad, P (1990) Human ankle impact response in dorsiflexion, Proc. 34th Stapp Conf., SAE Paper No. 902308, pp. 39-54.
9. Begeman, PC, Levine, RS, Balakrishnan, P and King, AI (1993) Dynamic ankle response in inversion and eversion, Proc. 37th Stapp Conf., SAE Paper No. 933115, pp. 83-93.
10. Haut, RC and Atkinson, PJ (1995) Insult to the human cadaver patellofemoral joint: Effects of age on fracture tolerance and occult injury, Proc. 39th Stapp Conf., SAE Paper No. 952729, pp. 281-294.
11. Viano, DC and Haut, RC (1980) Biomechanics and kinematics of posterior cruciate ligament trauma, Proc. 26th Annual Orthopedic Research Soc., p. 19.
12. Garfin, SR and Rothman, RH (1983) Traumatic spondylolisthesis of the axis (Hangman's fracture), in *The Cervical Spine*, Ed. by RB Bailey, et al, JB Lippincott Co., Philadelphia, pp. 223-232.
13. Parenteau, CS, Viano, DC, Lovsund, P. and Tingvall, C (1995) Foot-ankle injuries: Influence of crash location, seating position and age, 39th Annual Proc Assoc. for the Advancement of Automotive Medicine, pp. 177-192.

14. Patrick LM and Andersson, A (1974) Three-point harness accident and laboratory data comparison, Proc. 18th Stapp Conf., SAE Paper No. 741181, pp., 201-282.60.Aprill, C and Bogduk, N (1992) the prevalence of cervical zygaphophysial joint pain: A first approximation. Spine, **17**, 744-747.
15. Schmidt, G, Kallieris, D, Barz, J and Mattern, R (1974) Results of 49 cadaver tests simulating frontal collision of front seat passengers, Proc 18th Stapp Conf., SAE Paper No. 741182, pp. 283-291.
16. Miller, MA (1989) The biomechanical response of the lower abdomen to belt restraint loading, *J. Trauma*, **29**, 1571-1584.
17. Levine, RS, Patrick, LM, Begeman, PC and King, AI (1978) Effect of quadriceps function on submarining, 22nd Annual Proc., American Assoc. for Automotive Med., Vol. 1, pp. 319-329.
18. Petty, SPA and Fen, MA (1985) A modified steering wheel to reduce facial injuries and an associated test procedure, Proc. 10th International Technical Conf. on Exp. Safety Veh., pp. 342-347.
19. Huelke, DF, Moore, JL, Compton, TW, Samuels, J and Levine, RS (1995) Upper extremity injuries related to airbag deployments, *J. Trauma*, **38**, 482-488.
20. Huelke, DF, Mackay, GM, Morris, A and Bradford, M (1992) Cervical fractures and fracture-dislocations without head impacts sustained by restrained occupants, 36th Annual Proc., Assoc. for the Advancement of Automotive Med., pp. 1-23.
21. Huelke, DF, Mackay, GM and Morris, A (1995) Vertebral column injuries and lap-shoulder belts, *J. Trauma*, **38**, 547-556.
22. Patrick, LM and Levine, RS (1975) Injury to unembalmed belted cadavers in simulated collisions, Proc. 19th Stapp Conf., SAE Paper No. 751144, pp. 79-115.
23. Leung, YC, Tarriere, C, Lestrelin, D, Got, C,Guillon, F, Patel, A and Hureau, J (1982) Submarining injuries of 3 pt. belted occupants in frontal collisions - Description, mechanisms and protection, Proc. 26th Stapp Conf., SAE Paper No. 821158, pp. 173-201.
24. Begeman, PC, King AI, Prasad, P (1973) Spinal loads resulting from -Gx acceleration, Proc. 17th Stapp Conf.., SAE Paper No. 730977, pp. 343-360.
25. Prasad, P (1973) The dynamic response of the spine during +Gz acceleration, Ph.D. Dissertation, Wayne State University.
26. Mackay, GM, Cheng, L, Smith, M and Parkin, S (1990) Restrained front seat car occupant fatalities - The nature and circumstances of their injuries, 34th Annual Proc., Assoc. for the Advancement of Automotive Med., pp. 139-161.
27. Mertz, HJ (1988) Restraint performance of the 1973-76 GM air cushion restraint system, in *Automatic Occupant Protection Systems,* SAE Special Publication SP-736, SAE Paper No. 880400, pp. 61-71.
28. Traynelis, VC and Gold, M (1993) Cervical spine injury in an air bag-equipped vehicle, *J. Spinal Disorders*, **1**, 60-61.
29. Cheng, R, Yang, KH, Levine, RS, King, AI and Morgan, R (1982) Injuries to the cervical spine caused by a distributed frontal load to the chest, Proc. 26th Stapp Conf., SAE Paper No. 821155, pp. 1-40.
30. Mertz, HJ, Driscoll, GD, Lenox, JB, Nyquist, GW and DA Weber (1982) Responses of animals exposed to deployment of various passenger restraint system concepts for a variety of collision severities and animal positions, Proc. 9th International Conf. on Exp. Safety Veh., pp. 352-368.
31. Gault, JA, Vichnin, MC, Jaeger, EA and Jeffers, JB (1995) Ocular injuries associated with eyeglass wear and airbag inflation, *J. Trauma*, **38**, 494-497.
32. Sastry, SM, Paul, BK, Bain, L, et al (1993) Ocular trauma among major trauma victims, *J. Trauma*, **34**, 223.
33. Reed, MAP, Schneider, LW and Burney, RE (1992) Investigation of airbag-induced skin abrasions, Proc. 36th Stapp Conf., SAE, Paper No. 922510, pp. 1-12.
34. Viano, DC, Culver, CC, Evans, L, Frick, M and Scott, R (1989) Involvement of older drivers in multi-vehicle side impact crashes, Proc. 12th International Technical Conf. on Exp. Safety Veh., pp. 699-705.

35. FMVSS 214, 49 CFR Part 571 (1990) FMVSS No. 214, Side Impact Protection, Federal Register, Docket No. 88-06, Notice 8, RIN 2127-AB86, Vol 55 (210), Oct. 30, 1990.
36. Mackay, GM, Parkin, S, Hill, J and Munns, JAR (1991) Restrained occupants of the non-struck side in lateral collisions, 35th Annual Proc., Assoc. for the Advancement of Automotive Med., pp. 119-132.
37. Zhou, C, Khalil, TB and King, AI (1995) A new model comparing impact responses of the homogeneous and inhomogeneous human brain, Proc. 39th Stapp Conf., SAE Paper No. 952714, pp. 121-137.
38. Eppinger, RH, Marcus, JH, Morgan, RM (1984) Development of dummy and injury index for NHTSA's thoracic side impact protection research program, SAE Paper No. 840885, Government/Industry Meeting and Exposition, Washington, DC.
39. Viano, DC and Lau, IV (1985) Thoracic impact: A viscous tolerance criterion, Proc. 10th International Conf. on Exp. Safety Veh., pp. 104-114.
40. Cavanaugh, JM, Walilko, TJ, Malholtra, A, Zhu, Y, King, AI (1990) Biomechanical response and injury tolerance of the thorax in twelve sled side impacts, Proc. 34th Stapp Conf., SAE Paper No. 902307. pp. 23-38.
41. Cavanaugh, JM, Zhu, Y, Huang, Y and King, AI (1993) Deflection responses of the shoulder and thorax in lateral impact, Proc. 37th Stapp Conf., SAE Paper No. 933127, pp. 199-221.
42. Cavanaugh, JM, Walilko, TJ, Walbridge, A, Huang, Y and King, AI (1994) An evaluation of TTI and ASA in SID side impact sled tests, Proc. 38th Stapp Conf., SAE Paper No. 942225, pp. 293-308.
43. Huang, Y, King, AI and Cavanaugh, JM (1994) A MADYMO model of near-side human occupants in side impact, *J. Biomech. Engg.*, **116**, 228-235.
44. Viano, DC, Lau, IV, Asbury, C, King, AI and Begeman, P (1989) Biomechanics of the human chest, abdomen and pelvis in lateral impact, *Accid, Anal. & Prev.*, **21**, 553-574.
45. Viano, DC and Gargan, MF (1995) Headrest position during normal driving: Implications to neck injury risks in rear crashes, 39th Annual Proc., Assoc. for the Advancement of Automotive Med., pp. 215-229.
46. Foret-Bruno, JY, Tarriere, C, Le Coz, JY, Got, C and Guillon, F (1990) Risk of cervical lesions in real-world and simulated collisions, 34th Annual Proc., Assoc. for the Advancement of Automotive Med., pp. 373-389.
47. Parkin, S, Mackay, GM, Hassan, AM and Graham, R (1995) Rear-end collisions and seat performance - To yield or not to yield, 39th Annual Proc., Assoc. for the Advancement of Automotive Med., pp. 231-244.
48. Wickstrom, J, Martinez, JL, Rodriguez, R, and Haines, DM (1970) Hyperextension and hyperflexion injuries to the head and neck of primates, in *Neckache and Backache,* Edited by Gurdjian, ES and Thomas LM, Charles C. Thomas, Publisher, Springfield, IL, pp. 108-119.
49. Mertz, HJ and Patrick, LM (1967) Investigation of the kinematics and kinetics of whiplash, Proc. 11th Stapp Conf., SAE Paper No. 670919, pp. 267-317.
50. Matsushita. T. Sato, TB, Hirabayashi, K, Fujimura, S Asazuma, T and Takatori, T (1994) X-ray study of the human neck motion due to head inertia loading, Proc. 38th Stapp Proc., SAE Paper No. 942208, pp. 55-64.
51. Cavanaugh, JM, Ozaktay, AC, Yamashita, T and King AI (1996) Lumbar facet pain: Biomechanics, neuroanatomy and neurophysiology. *J. Biomech*. (In press).
52. Yamashita, T, Cavanaugh, JM, El-Bohy, AA, Getchell, TV and King, AI (1990) Mechanosensitive afferent units in the lumbar facet joint, *J Bone Joint Surgery*, **72A**, 865-870.
53. Avramov, AI, Cavanaugh, JM, Ozaktay, AC, Getchell, TV, King, AI (1992) Effects of controlled mechanical loading on Group II, III and IV afferents from the lumbar facet joint and surrounding tissues: An *in vitro* study. J. *Bone Joint Surg*., **74A**, 1464-1471.
54. Ozaktay, AC, Yamashita, T, Cavanaugh, JM and King AI (1991) Fine nerve fibers and endings in the fibrous capsule of the lumbar facet joint, Proc. 37th Annual Meeting, Orthopaedic Res. Soc., p. 353.

55. King, AI and Cavanaugh, JM (1996) Neurophysiologic basis for low back pain, in *The Lumbar Spine*, Edited by Wiesel, SW et al, WB Saunders Co., Philadelphia, Chapter 3, pp. 74-85.
56. Ozaktay, AC, Cavanaugh, JM, Blagoev, D, Getchell, TV and King, AI (1994) Effects of carrageenan induced inflammation in rabbit lumbar facet joint capsule and adjacent tissue, *Neuroscience Res*, **20**, 355-364.
57. Bogduk N (1982) The clinical anatomy of the cervical dorsal rami. *Spine*, **7**, 319-330
58. Mclain, RF (1994) Mechanoreceptors endings in human cervical facet joints, *Spine*, **19**, 495-501,
59. Bogduk, N and Marsland, A (1988) The cervical zygapophysial joints as a source of neck pain, Spine, **13**, 610-617.
61. Lord, S, Barnsley, L and Bogduk, N (1993) Cervical zygapophysial pain in whiplash, in *Spine: Cervical Flexion-Extension/Whiplash Injuries*, 7, 355-372.
62. Barnsley, L, Lord, SM, Wallis, BJ and Bogduk, N (1995) The presence of cervical zygapophysial joint pain after whiplash, *Spine*, **20**, 20-26.
63. Cohen, D, Digges, K and Nichols, RH (1989) Rollover crashworthiness classification and severity indices, Proc. 12th International Technical Conf. on Exp Safety Veh., pp. 477-488.
64. Thurman, DJ, Burnett, CL, Beaudoin, DE, Jeppson, L and Sniezek, JE (1993) Risk factors and mechanisms of occurrence in motor-vehicle-related spinal cord injuries: Utah, 37th Annual Proc Assoc. for the Advancement of Automotive Med., pp. 201-208.
65. Bahling, GS, Bundorf, RT, Kaspzyk, GS, Moffatt, EA, Orlowski, KF and Stocke, JE (1990) Rollover and drop tests - The influence of roof strength on injury mechanics using belted dummies, Proc 34th Stapp Conf., Paper No. 902314, pp. 101-112.
66. Moffatt, EA and Padmanaban, J (1995) The relationship between vehicle roof strength and occupant injury in rollover crash data, 39th Annual Proc., Assoc. for the Advancement of Automotive Med., pp. 245-267.
67. Huelke, DF, Lawson, TE, Scott, R and Marsh, JC (1977) The effectiveness of belt systems in frontal and rollover crashes, SAE Paper No. 770148.
68. Huelke, DF, Lawson, TE, and Marsh, JC (1977) Injuries, restraints and vehicle factors in rollover car crashes, *Accid. Anal. and Prev.*, **9**, 93-107.
69. Patrick, LM (1995) Glazing for motor vehicles - 1995, Proc. 39th Stapp Conf., SAE Paper No. 952717, pp. 161-172.

INJURY MECHANISMS AND BIOFIDELITY OF DUMMIES

D. C. VIANO[1] and A. I. KING[2]
[1]*Research &Development Center*
General Motors Corporation
Warren, MI, U.S.A. 48090-9055
[2]*Bioengineering Department*
Wayne State University
Detroit, MI, U.S.A. 48202

Abstract

The principal aim of impact biomechanics is the prevention of injury through environmental modification, such as the provision of an airbag for automotive occupants to protect them during a frontal crash. To achieve this aim effectively, it is necessary that workers in the field have a clear understanding of the mechanisms of injury, be able to describe the mechanical response of the tissues involved, have a basic understanding of human tolerance to impact, and be in possession of tools which can be used as human surrogates to assess a particular injury. The tools can be in the form of anthopometric test dummies or mathmatical models for simulation of impact events and evaluation of injury risks. This article addresses the biomechanics of head and neck, thorocolumbar spine, and chest and abdomen. It provides basic information on injury mechanisms, biomechanical responses, human tolerances and surrogates to evaluate safety systems for injury prevention.

1. Head and Neck

1.1. INJURY MECHANISMS

1.1.1. *Head Injury Mechanisms*

It is postulated that the relative motion of the brain surface with respect to the rough inner surface of the skull results in surface contusions on the inferior surfaces of the frontal and temporal lobes and the tearing of bridging veins between the brain and the dura mater, the principal membrane protecting the brain beneath the skull. The irregular geometry and surface of intracranial bones and membranes contributes to deformation of the brain upon severe head impact that can result in injury. Gennarelli et al. [1] have found that rotational acceleration of the head can cause a diffuse injury

J. A. C. Ambrósio et al. (eds.),
Crashworthiness of Transportation Systems: Structural Impact and Occupant Protection, 25–51.

to the white matter of the brain in animal models, as evidenced by retraction balls developing along the axons of injured nerves. This injury was described by Strich [2] as diffuse axonal injury (DAI) in the white matter of autopsied human brains. Other researchers, including Lighthall et al. [3] have been able to cause DAI in the brain of a ferret by the application of direct impact to the brain without an associated head angular acceleration. Adams et al. [4] indicated that DAI is the most important factor in severe head injury, as it is irreversible and leads to incapacitation and dementia. It is postulated that DAI occurs as a result of the mechanical insult but cannot be detected by staining techniques at autopsy unless the patient survives the injury for at least several hours.

Among the other theories of brain injury due to blunt impact, are changes in intracranial pressure and the development of shear strains in the brain. Positive pressure increases are found in the brain behind the site of impact on the skull. Rapid acceleration of the head, in-bending of the skull and the propagation of a compressive pressure wave are proposed as mechanisms for the generation of intracranial compression which causes local contusion of the brain tissue. At the contrecoup site, there is an opposite response in the form of a negative pressure pulse which also causes bruising. It is not clear as to whether the injury is due to the negative pressure itself (tensile loading) or to a cavitation phenomenon similar to that seen on the surfaces of propellers of ships (compression loading). The pressure differential across the brain necessarily results in a pressure gradient which can give rise to shear strains developing within the deep structures of the brain. Furthermore, when the head is impacted, it not only translates but also rotates about the neck and in reaction to the orientation of the impact load.

1.1.2. *Neck Injury Mechanisms*

Injuries to the upper cervical spine, particularly at the atlanto-occipital joint, are considered to be more serious and life-threatening than those at the lower level. The atlanto-occipital joint can be dislocated either by an axial torsional load or a shear force applied in the anteroposterior direction or vice versa. A large compression force can cause the arches of C_1 to fracture, breaking it up into two to four sections. The odontoid process of C_2 is also a vulnerable area.

Hyperflexion of the neck is a common cause of odontoid fractures, and a large percentage of these injuries are related to automotive crashes of largely unrestrained occupants [5]. Fractures through the pars interarticularis of C_2, commonly known as "hangman's" fractures in automotive collisions, are the result of a combined axial compression and extension (rearward bending) of the cervical spine. Impact of the forehead and face of unrestrained occupants with the windshield can result in this injury. Garfin and Rothman [6] discussed this injury in relation to hanging and traced the history of this mode of execution. It was estimated by a British judiciary committee that the energy required to cause a hangman's fracture was 1,708 N.m (1,260 ft-lb).

In automotive crashes, the loading on the neck due to head contact force is usually a combination of an axial or shear load with bending. Bending loads are almost always present, and the degree of axial or shear force is dependent upon the location and direction of the contact force. For impacts near the crown of the head, compressive forces predominate. If the impact is principally in the transverse plane, there is less compression and more shear. Bending can occur in any direction because impacts can come from any angle around the head. The following injury modes are considered the most prominent: tension-flexion, tension-extension, compression-flexion and compression-extension in the midsagittal plane, and lateral bending.

1.1.3. *Tension-Flexion Injuries*

Forces resulting from inertial loading of the head-neck system can result in flexion of the cervical spine while it is being subjected to a tensile force. In experiments with restrained subjects in forward deceleration, Thomas and Jessop [7] reported atlanto-occipital separation and C_1-C_2 separation occurring in subhuman primates at 120 g. Similar injuries in human cadavers were found at 34 to 38 g by Cheng et al. [8], who used a preinflated airbag for the thorax but let the head and neck to rotate over the bag.

1.1.4. *Tension-Extension Injuries*

The most common injury due to combined tension and extension of the cervical spine is the "whiplash" syndrome. However, a large majority of such injuries involve the soft tissues of the neck and the pain is believed to reside in the joint capsules of the articular facets of the cervical vertebrae [9]. In severe cases, tear drop fractures of the anterior-superior aspect of the vertebral body can occur. Alternately, separation of the anterior aspect of the disc from the vertebral endplate is known to occur. More severe injuries occur when the chin impacts the instrument panel or when the forehead impacts the windshield. In both cases, the head rotates rearward and applies a tensile and bending load on the neck. In the case of windshield impact by the forehead, hangman's fracture of C_2 can occur. Garfin and Rothman [6] suggested that it is caused by spinal extension combined with compression on the lamina of C_2, causing the pars to fracture.

1.1.5. *Compression-Flexion Injuries*

When force is applied to the posterior-superior quadrant of the head or when a crown impact occurs while the head is in flexion, the neck is subjected to a combined load of axial compression and forward bending. Anterior wedge fractures of vertebral bodies are commonly seen, but with increased load, burst fractures and fracture dislocations of the facets can result. The latter two conditions are unstable and tend to disrupt or injure the spinal cord, the extent of the injury depends on the penetration of the vertebral body or its fragments into the spinal canal. Recent experiments by Pintar et al. [10, 11] indicate that burst fractures of lower cervical vertebrae can be reproduced in cadavers by a crown impact to a flexed cervical spine. Nightingale et al. [12] showed that fracture dislocations of the cervical spine occur very early in the impact event (within the first 10 ms) and that the subsequent motion of the head or bending of the cervical spine cannot be used as a reliable indicator of the mechanism of injury.

1.1.6. *Compression-Extension Injuries*
Frontal impacts to the head with the neck in extension will cause compression-extension injuries. These involve the fracture of one or more spinous processes and possibly, symmetrical lesions of the pedicles, facets and laminae. If there is a fracture-dislocation, the inferior facet of the upper vertebra is displaced posteriorly and upward and appears to be more horizontal than normal on x-ray.

1.1.7 *Injuries Involving Lateral Bending*
If the applied force or inertial load on the head has a significant component out of the midsagittal plane, the neck will be subjected to lateral or oblique bending along with axial and shear loading. The injuries characteristic of this type of bending are lateral wedge fractures of the vertebral body and fractures to the posterior elements on one side of the vertebral column.

Whenever there is lateral or oblique bending, there is the possibility of twisting the neck. The associated torsional loads may be responsible for unilateral facet dislocations or unilateral locked facets [13]. However, pure torsional loads on the neck are rarely encountered in automotive accidents. It was shown by Wismans and Spenny [14] that, in a purely lateral impact, the head rotated axially about the cervical axis while it translated laterally and vertically and rotated about an antero-posterior axis. These responses were obtained from lateral impact tests performed by the Naval Biodynamics Laboratory on human subjects who were fully restrained at and below the shoulders.

1.2. MECHANICAL RESPONSES

1.2.1. *Mechanical Response of the Head*
Many cadaver studies on blunt head impact have been carried out over the past 50 years. The head was impacted by rigid and padded surfaces and by impactors of varying shapes to simulate flat surfaces and knobs encountered in the automotive environment. In general, the impact responses were described in terms of head acceleration or impact force. Both of these responses are dependent on a variety of factors, including the inertial properties of the head and surface impacted by the head. In this section, the inertial properties of the head will be described and response data for head impact against a flat rigid surface will be provided. It should be noted that while response data against surfaces with a variety of shapes and stiffness are of interest, the only generally applicable and reproducible data are those of impacts to flat rigid surfaces.

1.2.2. *Inertial Properties of the Head*
There are several sources of data on the inertial properties of the human head. Data by Walker et al. [15], Hodgson and Thomas [16, 17] were analyzed by Hubbard and McLeod [18] who found that 31 heads had dimensions which were close to those of the average male and head mass. The average value is 4.54 kg and a specific gravity

of 1.097. The data of Reynolds et al. [19], Beier et al. [20], McConville et al. [21] and Robbins [22] provide information on the average mass moment of inertia of the head. They are $I_{xx} = 22.0$, $I_{yy} = 24.2$, and $I_{zz} = 15.9 \times 10^{-3}$ kgm^2.

1.2.3. *Cranial Impact Response*

Impact response of the head against a flat rigid surface was obtained by Hodgson and Thomas [17, 23] who performed a series of drop tests using embalmed cadavers. The

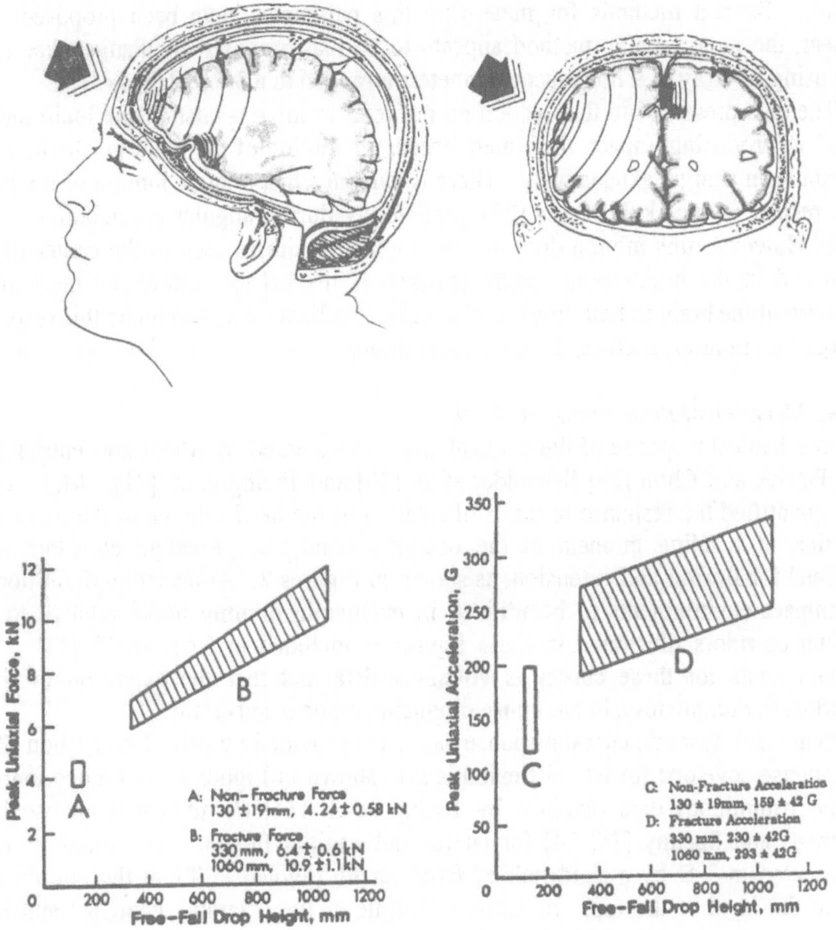

Figure 1. Blunt Impact of the head accelerates the skull. Injury of the brain can be caused as its motion lags that of the skull and strains develops in brain tissue and blood vessels. Experimental data are available on the peak force and acceleration related to skull fracture in free-fall, head drop tests.

responses in terms of peak force and acceleration are shown in Figure 1 as a function of an equivalent free-fall drop height. Details of adjustments made to the original data are described by Prasad et al. [24]. The data were for frontal, lateral and occipital impact directions although there was a large scatter in the peak values. For this reason, the data were pooled and individual data points were not shown.

The difficulty with acceleration measurements in head impact is twofold. The head is not a rigid body and accelerometers cannot be mounted at the cg of the head.The cg acceleration can be computed if head angular acceleration is measured but the variation in skull stiffness cannot be corrected for easily. That is why measurement of head angular acceleration is recommended in future head impact studies. Several methods for measuring this parameter have been proposed. At present, the most reliable method appears to be that proposed by Padgaonkar et al. [25] using an array of 9 linear accelerometers arranged in a 3-2-2-2 cluster.

The data presented in this section do not refer to the response of the brain during an injury producing impact. For intact heads, the motion of the brain inside the skull has not been studied exhaustively. There is evidence that relative motion of the brain with respect to the skull occurs [26], particularly during angular acceleration of the head. However, this motion does not fully explain injuries seen in the center of the brain and in the brain stem. More research is needed to explore the mechanical response of the brain to both linear and angular acceleration and to relate this response to observed injuries, such as, diffuse axonal injury.

1.2.4. *Mechanical Response of the Neck*

The mechanical response of the cervical spine was studied by Mertz and Patrick [27, 28], Patrick and Chou [29] Schneider et al. [30] and Ewing et al. [31]. Mertz et al. [32] quantified the response in terms of rotation of the head relative to the torso as a function of bending moment at the occipital condyles. Loading corridors were obtained for flexion and extension, as shown in Figures 2. An exacting definition of the impact environments to be utilized in evaluating dummy necks relative to the loading corridors illustrated in these figures is included in SAE J1460 [33]. The primary basis for these curves is volunteer data and that the extension of these corridors to dummy tests in the injury producing range is uncertain.

Static and dynamic lateral response data were provided by Patrick and Chou [29]. A response envelope for lateral flexion is also shown in Figure 2. A limited amount of the voluminous data obtained by Ewing et al. [31] (6 runs) was analyzed by Wismans and Spenny [14, 34] for lateral and sagittal flexion. The rotations were represented in 3-D by a rigid link of fixed length pivoted at T1 at the bottom and within the head at the top. In terms of torque at the occipital condyles and head rotation, the results fell within the earlier corridors for forward and lateral flexion.

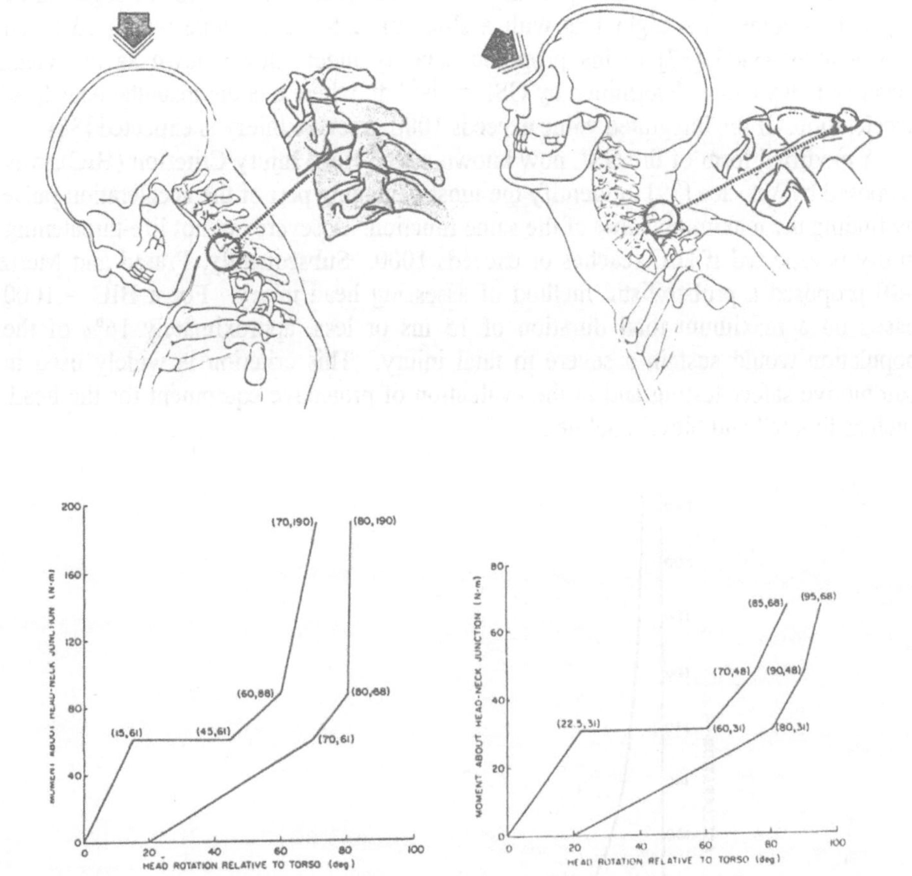

Figure 2. **Downward impact on the head can flex (forward bending) or extend (rearward bending) the neck with the potential for fracture-dislocation of the vertebrae and damage to the spinal cord. Moment-rotation corridors are available for neck flexion (left curve) and extension (right curve).**

1.3. INJURY TOLERANCES

1.3.1. *Regional Tolerance of the Head*

The most commonly measured parameter during head impact is acceleration. It is therefore natural to express human tolerance to injury in terms of head acceleration. The first known tolerance criterion is the Wayne State Tolerance Curve proposed by Lissner et al. [35] and modified by Patrick et al. [36] with the addition of animal and volunteer data to the original cadaver responses. The modified curve is shown in Figure 3.

The head can withstand higher accelerations for shorter durations and any exposure above the curve is injurious. When this curve is plotted on logarithmic paper, it becomes a straight line with a slope of -2.5. This slope was used as an exponent by Gadd [37] in his proposed severity index, now known as the Gadd Severity Index and is determined by GSI = $\int a^{2.5}\, dt$, where a is the instantaneous head acceleration. If the integrated value exceeds 1000, a severe injury is expected [38].

A modified form of the GSI, now known as the Head Injury Criterion (HIC), was proposed by Versace [39] to identify the most damaging part of the acceleration pulse by finding the maximum value of the same function. A severe but not life-threatening injury is espected if HIC reaches or exceeds 1000. Subsequently, Prasad and Mertz [40] proposed a probabilistic method of assessing head injury. For a HIC = 1000 based on a maximum total duration of 15 ms or less, approximately 16% of the population would sustain a severe to fatal injury. This criterion is widely used in automotive safety testing and in the evaluation of protective equipment for the head, such as football and bicycle helmets.

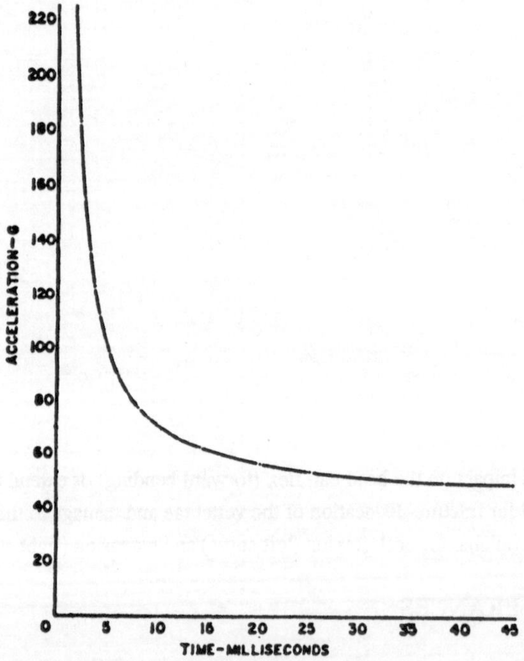

Figure 3. Wayne State Tolerance Curve for head injury

However, there is another school of thought which believes in the injurious potential of angular acceleration causing cerebral contusion of the brain surface and rupture of the parasagittal bridging veins between the brain and the dura mater. A proposed limit for angular acceleration is 4500 rad/s^2, based on a mathematical model developed by Lowenhielm [41]. This limit has not received universal acceptance. Many other criteria have been proposed but HIC is the current criterion for Federal Motor Vehicle Safety Standard (FMVSS) 214 and attempts to replace it have so far been unsuccessful.

1.3.2. *Regional Tolerance of the Neck*

Currently there are no tolerance values for the neck for the various injury modes. This is not due to a lack of data but rather to the many injury mechanisms and several levels of injury severity, ranging from life-threatening injuries to the spinal cord to minor soft tissue injuries which cannot be identified on radiographic or magnetic scans. It is likely that a combined criterion of axial load and bending moment about one or more axes will be adopted for injury assessment purposes.

1.4. HUMAN SURROGATES

1.4.1. *Experimental Surrogates*

The most effective experimental surrogate for impact biomechanics research is the unembalmed cadaver. This is also true for the head and neck, despite the lack of muscle tone because the duration of impact is usually too short for the muscles to respond adequately. It is true, however, that muscle pre-tensioning in the neck may have to be added under certain circumstances, since it influences head and neck kinematics. Similarly, for the brain, the cadaver brain cannot develop DAI and the mechanical properties of brain change rapidly after death. If the pathophysiology of central nervous system is to be studied, the ideal surrogate is an animal brain. Currently, the rat is frequently used as the animal of choice and there is some work in progress using the mini-pig.

1.4.2. *Injury Assessment Tools*

The response and tolerance data acquired from cadaver studies have been used to design humanlike surrogates, known as anthropomorphic test devices (ATD). These surrogates are required to have biofidelity, the ability to simulate the essential characteristics of the human response. They also need to provide physical measurements that are representative of human injury, and are designed to be repeatable and reproducible. The current frontal impact dummy is the Hybrid III which is humanlike in many of its responses, including that of the head and neck. The head consists of an aluminum headform covered by an appropriately designed vinyl skin to yield humanlike acceleration responses for frontal and lateral impacts against a flat rigid surface. Two-dimensional physical models of the brain have been proposed by Margulies et al. [42] using a silicon gel in which pre-inscribed grid lines would deform under angular acceleration. No injury criterion is associated with this gel model.

The dummy neck was designed to yield responses in flexion and extension which would fit within the corridors shown in Figure 2. The principal function of the dummy neck is to place the head in the approximate position of a human head in the same impact involving a human occupant.

1.4.3. *Computer Models*

Models of head impact first appeared over 50 years ago [43]. Extensive reviews of such models were made by King and Chou [44], and Hardy et al. [45]. The use of the finite element method (FEM) to simulate the various components of the head appears to be the most effective and popular means of modeling brain response. Despite the large number of nodes and elements used, the models are still not detailed enough to predict displacement and distortion of the brain and the location of DAI development following a given impact. The research is also hampered by the limited amount of animal DAI data currently available to validate model responses.

Many neck and spinal models have been developed over the past four decades. A recent paper by Kleinberger [46] provides a brief review of models. However, the method of choice for modeling the response of the neck is the finite element method, principally because of the complex geometry of the vertebral components and the interaction of several different materials. A fully validated model for impact response is still not available.

2. Thoracolumbar Spine

Although injuries to the bony portion of the thoracolumbar spine are rare in automotive crashes, paraplegia can result if the spinal cord is involved and complaints of low back pain are a common claim. Impact spinal biomechanics attempts to explain the various injury mechanisms and provides data on human response and tolerance. Surrogates in the form of a dummy spine or a computer model are developed based on the results of these biomechanical studies.

Historically, the study of thoracolumbar spinal injuries were motivated by the pilot ejection problem. The vertebral body of the lower thoracic and upper lumbar spine tend to sustain anterior wedge fractures when military pilots eject from disabled jet aircraft. The vertical acceleration necessary to effect a successful ejection is about 20 g. The mechanism of this injury was studied by Ewing et al. [47] and it was determined that the wedge fractures were due to a combined compressive load and a forward flexion moment. As a result of this research, the mechanism of support of the thoracolumbar spine was also determined by Prasad et al. [48]. The load through the spine is transmitted from one level to another via the intervertebral disc and the articular facets. In hyperextension, the inferior tips of the superior facets bottom out onto the lamina of the vertebra below and create a load path which relieves the vertebral body of some of the compressive load it carries due either to vertical acceleration or muscular compression [49].

Since research on the impact biomechanics of the spine was originally motivated by the pilot ejection problem, most of the work is related to the effects of caudocephalad acceleration. It has led to a better understanding of the effects of compressive loading on the spine, mechanism of load transmission, and injury. In fact, it has contributed to the search for causes of low back pain by identifying the facets as load bearing elements of the spine. Automotive related injuries to the thoracolumbar spine are rarely encountered. However, injuries involving fracture-dislocation of an intervertebral joint can result in paraplegia. Two types of such injuries were identified: Chance fractures in improperly lap-belted rear seat occupants and spinal fractures due to the shoulder belt. In terms of soft tissue injuries, there is no causal relationship between a single impact to the spine and subsequently diagnosed disc ruptures, if the impact does not result in spinal fractures.

2.1. INJURY MECHANISMS

Of the several major categories of thoracolumbar injuries, the mechanism of injury usually involves an applied force accompanied by a bending or twisting moment. This combined loading causes not only anterior wedge fractures but also dislocation and fracture-dislocation of the vertebrae, rotational injuries, Chance fractures and hyperextension injuries. The only injury in which a moment may not be involved is burst fracture of the vertebral body. Oxland [50] was able to reproduce these fractures in the laboratory by dropping a weight onto a spinal specimen. However, the study did not determine if the fracture occurred only if spine was slightly flexed and the facets were non load bearing.

In the automotive setting, the Chance [51] fracture is a belt induced injury which can occur to rear seat occupants who are usually restrained by a lap-belt over the abdomen or slouching in the seat. Because of the flat angle of the rear belt relative to the horizontal plane, the lap-belt tends to ride up above the iliac crest in a horizontal crash. The abdominal organs are compressed by the belt which then bears against the spine as the torso flexes. This causes the supraspinous and interspinous ligaments to rupture and the vertebral body to split horizontally, starting along its posterior aspect. The spinal cord is stretched and paraplegia can result.

Another automotive-related belt induced injury that may be on the increase is anterior wedge fracture of the thoracolumbar lumbar vertebral body, similar to that seen in pilots who eject from disabled aircraft. In this case, the shoulder harness imposes a large load across the torso of an occupant in a severe frontal crash, causing the curved thoracic spine to straighten out. As a result, the a compressive force is generated in the thoracolumbar spine to push the head and neck upward and the rest of the torso downward [52]. This load was measured on the seat pan in tests on three-point belted cadavers and volunteers but was not observed when an anthropomorphic test device (dummy) was used. In a survey of injury patterns in Monroe County of New York state, States et al. [53] found a large increase (69% thoracic and 80% lumbar) in minor to moderate spinal injuries following the seatbelt use law in the state.

In terms of soft tissue injuries, the complaint of low back pain is often associated with a diagnosis of disc rupture. The incident provoking such symptoms can range from a minor rearend impact to a very severe frontal crash. However, predominant findings in the literature indicate that disc rupture is a slow degenerative process and that an extremely violent single loading event is needed to cause the nucleus pulposus to extrude from the disc, in association with bony fracture. The disc does not herniate like a balloon and minor events which result in pain may not be due a defect in the disc. There are many sources of back pain, such as in the capsules of facet joints, and a causal relationship between an impact and a rupture usually does not exist. A more detailed discussion is provided by King [54].

2.2. MECHANICAL RESPONSE

Much information exists on the static mechanical response of a functional spinal unit (two vertebrae and a disc) but very much less is available regarding the response of the entire thoracic or lumbar spine. In terms of dynamic response in an impact, much of the work on the intact spine was done in relation to the pilot ejection problem. Thus, the early data were concerned with spinal response to a caudocephalad or $+G_Z$ acceleration. The first known whole-body cadaveric experiments were performed by Lissner et al. on a vertical accelerator housed in an elevator shaft of the School of Medicine at Wayne State University. The response data originally took the form of vertebral body strain as a function of time [55].

Subsequently, spinal force-time histories were produced by Prasad et al. [48]. However, because the development of a surrogate spine was not a primary aim, response data were obtained primarily on the mechanisms of load transmission and injury. A limited amount of moment and rotation response data was acquired from volunteer subjects in a frontal crash or $-G_X$ environment. Cheng et al. [56] used 7 young subjects, 4 male and 3 female, to obtain kinematic and kinetic response at various levels during spinal flexion. For example, the reported data include the linear and angular acceleration of T1, the relative position of T1 with respect to T12 and the pelvis and hip moment as a function of the rotation of T1 relative to the pelvis. The data were also compared with cadaver data obtained previously by Mital et al. [57].

2.3. HUMAN TOLERANCES

Tolerance data for the spine are available from several sources for a variety of loading conditions. These conditions range from static loading of individual vertebrae to dynamic loading of the whole body in impact tests using human cadavers. While data on the strength of the intact spine under dynamic loading are of particular importance in impact biomechanics, data sources for strength of individual vertebrae and functional spinal units are provided. Yamada [58] provides a summary of results by Sonoda [59] for tensile, compressive and torsional strength characteristics of cervical, thoracic and lumbar vertebrae and the intervertebral discs. The data include variation in strength along the column and its dependence on age. Myklebust et al. [60] applied quasi-static compressive loads to individual bodies, portions of the excised spine with ligaments intact and torsos of complete cadavers. Isolated thoracic and lumbar vertebral bodies were loaded in compression between parallel plates in the superior-inferior direction. Excised spinal segments and intact cadavers were also compressed to yield fracture loads which were lower than the loads borne by the individaul bodies. The level of shear force and bending moment that may have been present was not measured.

As early as 1959, human tolerance limits to impact acceleration from several directions were described by Eiband [61]. For caudocephalad acceleration, the available data from humans and animals indicates a 20 g limit for ejection seats, which is used by the U.S. Army Air Force. The limit assumed a trapezoidal acceleration pulse acting on a subject fully restrained by a military restraint system. Limits for frontal crash, rearend crash and downward acceleration were also proposed. For impact durations of less than 100 ms, a 40 g limit was proposed for well restrained seated individuals.

2.4. HUMAN SURROGATES

The thoracolumbar spine in dummies is not humanlike. The spine is a rigid steel box section and the lumbar spine is a relatively rigid curved rubber cylinder which has a steel cable running through its center. An attempt was made by Schneider et al. [62] to include a joint in the middle of the thoracic column to simulate the kyphotic curve. This represents an improvement but the entire column is too rigid to represent the human spine or injury of an individual vertebrae.

An alternative is to use mathematical or computer models of the spine to predict forces and moments under various loading conditions. There are many mathematical models of the spine but most of them deal with quasi-static loading. Models simulating ejection seat impacts were developed originally by Latham [63]. The history of models was reviewed by King and Chou [44]. Prasad and King [64] developed a two-dimensional discrete parameter model which was subjected to a validation process against cadaveric data. It was this model which first predicted spine loads due to the shoulder belt in a frontal crash. Finite element models of the thoracolumbar spine simulating impact events do not seem to have appeared in the literature as yet, primarily because injuries are rare in frontal crashes and the pilot ejection problem is considered by the military as a mature technology, not requiring further research.

3. Chest and Abdomen

Chest and abdomen injury is related to energy delivered by an impacting object as well as its shape, stiffness, point of body contact and orientation. Non-penetrating injury occurs when the body is struck by a blunt object, such as a vehicle instrument panel or side interior, at moderate velocity, and force is distributed over a relatively large area of the body. Impact energy can be absorbed by padding or other crushable materials, which allows the surfaces in contact to deform, extending the duration of impact and reducing loads. The torso has a viscoelastic behavior; load developed by the chest and abdomen increases with the speed of impact as internal structures resist deformation [65].

The biomechanical response of the body has three components: (1) inertial resistance by acceleration of body masses, (2) elastic resistance by compression of stiff structures and tissues, and (3) viscous resistance by rate-dependent properties of the body and its tissues. For low impact speeds, the elastic stiffness is critical to protection from crush injuries; whereas, for high rates of body deformation, the inertial and viscous properties determine the force developed to limit deformation. In all cases, the risk of skeletal and internal organ injury relates to energy transfer. At the highest speeds of impact, such as with light-weight projectiles and for blast waves, the viscous properties are the essential factor in human protection.

In any impact situation, the inertial resistance of body masses and the elastic and viscous properties of soft tissues combine to develop dynamic load, resist deformation, and prevent injury. Impact force produces acceleration and deformation of the body, and each relates to injury when the soft tissues are deformed beyond their recoverable limit. In most situations, the viscoelastic properties of the body protect vitals organs by absorbing energy and developing high forces which resistance body deformation during impact.

Deformation beyond a recoverable limit is the general injury mechanism in blunt chest and abdominal impact. This mechanism relates to compression or strain, defined as the change in dimension over the original thickness of the body, tissue or organ. For example, the typical anteroposterior thickness of the chest is about 22.2 cm (8.74"), so 20% compression involves 4.44 cm (1.75") deflection of the chest. Even in this condition, the primary types of strain that produce tissue damage are tensile and shear strain, which can fracture ribs and lacerate, rupture and avulse vessels. A third type is compressive strain which produces crush injury.

3.1. INJURY MECHANISMS

A mechanism of injury is a description of the mechanical changes that result in anatomical or functional damage. An understanding of how and when injury occurs provides the basis for determining appropriate safety systems which are compatible with the response and tolerance of the body [66]. The primary mechanism of chest and abdomen injury is direct compression of the body at high rates of loading. This causes deformation and stretching of internal organs and vessels. When compression of the torso exceeds the ribcage tolerance, fractures occur and internal organs and vessels can be contused or ruptured. In some chest impacts, however, internal injury occurs without skeletal damage. This can happen particularly during high-speed loading. It is due to the viscous or rate-sensitive nature of human tissue responses. Thus, biomechanical responses differ for low and high speed impact.

When organs or vessels are loaded slowly, the input energy is absorbed gradually through deformation which is resisted by elastic properties and pressure build-up in tissue. When loaded rapidly, reaction force is proportional to the speed of tissue deformation as the viscous properties of the body resist deformation and provide a natural protection from impact. However, there is also a considerable inertial component to the reaction force. In this case, the body develops high internal pressure and injuries can occur before the ribs deflect much. The ability of an organ or other biological system to absorb impact energy without compression failure is called the viscous tolerance. Internal organs and vessels can also be torn from attachment points during torso impact or high-level acceleration causing rapid motion of the body.

If an artery is stretched beyond its tensile strength, the tissue will tear. Organs and vessels can be stretched in different ways, which result in different types of injury. Motion of the heart during chest compression stretches the aorta along its axis from points of tethering in the body. This elongation generally leads to a transverse laceration when the strain limit is exceeded. In contrast, an increase in vascular pressure dilates the vessel and produces biaxial strain which is larger in the transverse than axial direction. If pressure rises beyond the vessel's limit, it will burst. For severe impacts, intra-aortic pressure exceeds 500-1000 mmHg, which is a significant, non-physiologic level, but is tolerable for short durations. When laceration occurs, the predominant mode of aortic failure is axial so the combined effects of stretch and internal pressure contribute to injury. Chest impact also compresses the ribcage causing tensile strain on the outer surface of the ribs. As compression increases, the risk of rib fracture increases. In both cases, the mechanism of injury is tissue deformation.

The abdomen is more vulnerable to injury than the chest, because there is little bony structure below the ribcage to protect internal organs in front and lateral impact. Blunt impact of the upper abdomen can compress and injure solid organs, such as the liver and kidneys, before significant whole-body motion or acceleration occurs. In the liver, compression increases the intrahepatic pressure and generates tensile or shear strains. If the tissue is sufficiently deformed, laceration of the major hepatic vessels can result in hemoperitoneum. Abdominal deformation also causes lobes of the liver to move relative to each other, stretching and shearing the vascular attachment at the hilar region.

For blunt impact, the primary factors that determine the type and severity of injury are the amount of body area over which the impact energy is spread, and the speed and stiffness of the impacting object. Effective restraints, occupant safety systems, and protective equipment not only spread impact energy over the strongest body structures but also reduce contact velocity between the body and the impacted surface or object. The design of protective systems is aided by an understanding of injury mechanisms, quantification of human tolerance levels and development of numerical relationships between measurable engineering parameter, such as force, acceleration or deformation, and injury. These relationships are called injury criteria. The following discussion outlines the biomechanics of torso impact, the unique viscoelastic properties of the body, and human tolerance to impact injury.

3.2. HUMAN TOLERANCES

3.2.1. *Acceleration Injury*

Stapp [67] conducted a series of rocket-sled experiments that demonstrated the effectiveness of belt-restraint systems in achieving high tolerance to long-duration, whole-body acceleration. This improved the protection of military personnel exposed to rapid but sustained acceleration. The experiments enabled Eiband [61] to demonstrate that the tolerance to whole-body acceleration increased as the exposure duration decreased. This led to Figure 4, which links human tolerance and acceleration for exposures of 2-1000 ms in duration. The tolerance data is based on average sled acceleration rather than the acceleration of the volunteer subject, which would be higher due to compliance of the restraint system used in the tests. Even with this limitation, the data provide useful early guidelines for the development of crash restraint systems for military and civilian personnel [68]. Analysis of the data also indicated that rate of onset affected acceleration tolerance, since high peaks could be tolerated if reached over a greater period of time.

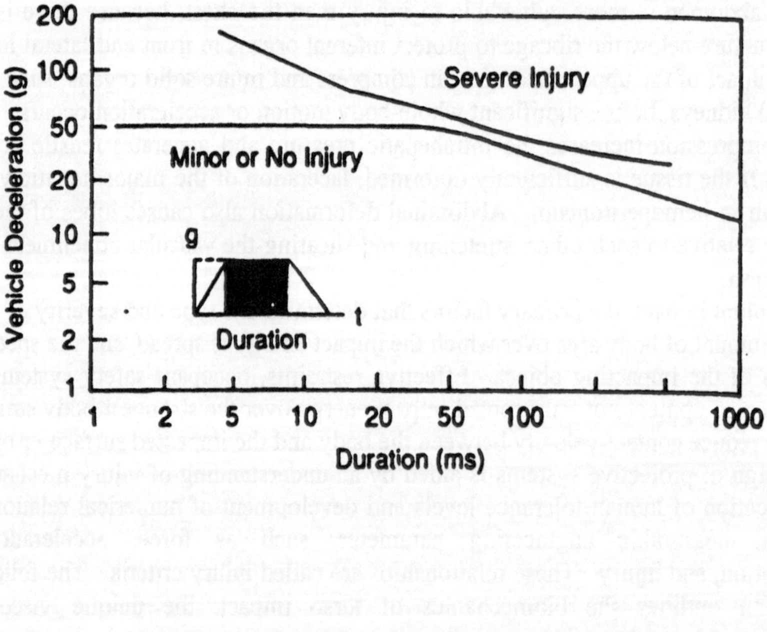

Figure 4. Whole-body tolerance to vehicle acceleration

More recent tests of side impact injury have led to other acceleration formulas for chest injury tolerance. Morgan et al. [69] evaluated rigid, side-wall cadaver tests and developed TTI, a thoracic trauma index, which is the average rib and spine acceleration of the chest. TTI limits human tolerance to 85-90g in vehicle crash tests. Somewhat better injury assessment has been achieved by Cavanaugh [70] using average spinal acceleration (ASA), which is obtained by integrating the thoracic spinal acceleration and determining the average slope. This term relates to the rate of momentum transfer to the body during side impact, and a value of 30g is proposed. In most cases, the torso can withstand 60-80g whole-body acceleration by a well-distributed load.

3.2.2. *Force Injury*

The basis for whole-body tolerance is Newton's second law of motion: acceleration of a rigid mass is proportional to the force acting on it, or the well-known F = ma. Although the human body is not a rigid mass, a well distributed restraint system allows the torso to respond as though it were fairly rigid if loads are applied through the shoulder and pelvis. The greater the acceleration, the greater the force acting on the body and the greater the risk of injury. For a high-speed frontal crash, a restrained occupant can experience 60g acceleration (60 times the force of gravity). For a body mass of 76 kg, the inertial load is 44.7kN (10,000 lb) and is tolerable if distributed over strong skeletal elements.

The ability to withstand high acceleration for short durations implies that tolerance is related to momentum transfer, because an equivalent change in velocity can be achieved by increasing the acceleration and decreasing its duration, as $\Delta V = a\Delta t$. The implication for occupant protection systems is that the risk of injury can be decreased if the crash deceleration is extended over a greater period of time. For occupant restraint in 25 ms (0.025s), a velocity change of 14.7 m/s (32.7 mph) occurs with 60g whole-body acceleration. This duration of deceleration can be achieved by the use of crushable vehicle structures and occupant restraints.

In the 1960s, automotive safety engineers were seeking information on the tolerance of the chest for localized impact loads from energy absorbing instrument panels and steering systems. This was prior to the widespread use of safety belts. The goal was to limit impact force to a tolerable level by using crushable materials and structures. By using the highest practical level of crush force, safety was extended to the greatest severity of vehicle crashes. GM Research and Wayne State University collaborated on the development of a crash sled facility which was used to simulate progressively more severe frontal impacts. Embalmed human cadavers were exposed to head, chest and knee impact on 15 cm (6") diameter load cells until bone fracture was recorded on Xray.

The tests by Patrick [71] demonstrated that blunt chest loading of 3.3 kN (740 lb) could be tolerated with minimal risk of serious injury. This is a pressure of 187 kPa. Subsequent experiments by Gadd and Patrick [72] demonstrated that tolerance was as high as 8.0 kN (1800 lb) if the load was distributed over the shoulders and chest by a properly designed steering wheel and column. More recent side impact tests show that the torso can tolerate similar forces as in frontal impacts and that shoulder loading is an important load-path. However, the loads are a conservative threshold of injury.

3.2.3. *Compression Injury*

Study of high-speed films from cadaver impacts shows that whole-body acceleration does not adequately describe torso impact biomechanics. Tolerance of the chest and abdomen must consider body deformation. Force acting on the body generates two simultaneous responses: (1) compression of the compliant structures of the torso, and (2) acceleration of body masses. The previously neglected mechanism of injury was compression, which causes the sternum to displace toward the spine as ribs bend and possibly fracture. Acceleration and force, per se, are not sufficient indicators of impact response and tolerance because they cannot discriminate between the two underlying responses. Numerous studies have shown that acceleration is less related to injury than compression.

The importance of chest deformation was confirmed by Kroell [73, 74] in a series of blunt thoracic impacts of unembalmed cadavers. Both peak spinal acceleration and impact force were poorer injury predictors than the maximum compression of the chest, as measured by the percent change in the anteroposterior thickness of the body. A relationship between injury risk and compression involves the concept of energy stored by elastic deformation of the body. Stored energy (E_S) by a spring representing the ribcage and soft tissues is related to the displacement integral of force: $E_S = \int F dx$. Force in a spring is proportional to deformation: $F = kx$, where k is a spring constant. Stored energy is $E_S = k\int x dx = 0.5kx^2$. Over a reasonable range, stored energy is proportional to deformation or compression, so $E_S \approx C$.

Tests with human volunteers showed that compression up to 20% during moderately long duration loading produced no detectable injury and was fully reversible. Cadaver impacts at levels of compression greater than 20% showed an increase in rib fractures and internal organ injury as the compression increased up to 40%. The original tolerance for chest deflection was set at 8.8 cm (3.5") for moderate but recoverable injury. This represents 39% compression. However, at this level of compression, multiple rib fractures and a range of serious injury can occur so a more conservative tolerance of 32% is used to avert the possibility of flail chest. This reduces the risk of direct loading on the heart, lungs and internal organs by a loss of the protective function of the ribcage.

3.2.4. *Viscous Injury*

The velocity of body deformation is determined by the rate of loading and is an important factor in high-speed, non-penetrating injury. For example, when a fluid-filled organ is compressed slowly much of the applied energy can be absorbed through tissue deformation without damage. When loaded rapidly, however, the organ cannot deform fast enough and rupture may occur without significant change in shape, even though the load on the organ has increased substantially over the level occurring in the slow loading condition.

Research on soft-tissue injury has made it increasingly evident that the body is not merely an elastic structure, but rather is viscoelastic for impacts causing body deformation velocities greater than 3 m/s. For lower speeds of deformation, such as in slow crushing loads or for a belt-restrained occupant in a frontal crash, the compression of tissue is the determining factor in skeletal and internal organ injury. For higher speeds of deformation, such as occupant loading by the door in a side impact or for an unrestrained occupant or pedestrian, maximum compression does not adequately address the viscous and inertial properties of the torso, nor the time of greatest injury risk. In these conditions, the tolerance to compression is progressively lower as the speed of deformation increases, and the velocity of deformation becomes an equally important factor.

Insight on a rate-dependent injury mechanism came from over twenty years of research by Jonsson and Clemedson [75] on high-speed impact simulating blast-wave exposures. The studies confirmed that tolerable compression inversely varied with the velocity of impact. The concept was further studied in relation to the abdomen by Lau and Viano [76] for frontal impacts in the range of 5-20 m/s (10-45 mph). The liver was the target organ. Using a maximum compression of 16%, the severity of injury increased with the speed of loading, including serious mutilation of the lobes and major vessels in the highest speed impacts. While the compression was well within limits of volunteer loading at low speeds (<3 m/s body deformation velocity), the exposure produced critical injury at higher speeds. Subsequent tests on other animals and target organs verified an inter-relationship between body compression, deformation velocity and injury.

The previous observations led Viano and Lau [77, 78] to propose a viscous injury mechanism for soft biological tissues. The viscous response (VC) is defined as the product of velocity of deformation (V) and compression (C), which is a time-varying function in an impact. The parameter has physical meaning to absorbed energy (E_a) by a viscous dashpot under impact loading. Energy is related to the displacement integral of force: $E_a = \int F dx$, and force in a dashpot is proportional to the velocity of deformation: $F = cV$, where c is a dashpot parameter. Absorbed energy is: $E_a = c\int V dx$, or a time integral by substitution: $E_a = c\int V^2 dt$. The integrand is composed of two responses, so: $E_a = c(\int d(Vx) - \int ax dt)$, where a is acceleration across the dashpot. The first term is the viscous response and the second an inertial term related to the deceleration of fluid set in motion. Absorbed energy is given by: $E_a = c(Vx - \int ax dt)$, or $E_a \approx VC$. The viscous response is proportional to absorbed energy during the rapid phase of impact loading prior to peak compression.

Subsequent tests by Lau and Viano [79] verified that serious injury occurred at the time of peak VC, which is much earlier than the peak compression. For blunt chest impact, peak VC occurs 15-20 ms earlier than maximum compression. Rib fractures also occur progressively with chest compression, as early as 9-14 ms in an impact requiring 30 ms to reach peak compression. Peak VC occurs at the initiation of rib injury in cadaver tests. Upper-abdominal injury by steering wheel contact also relates to viscous loading. Lau et al. [80] showed that limiting the viscous response by a self-aligning steering wheel reduced the risk of liver injury, as does force limiting an armrest in side impacts. Animal tests have also shown that VC is a good predictor of

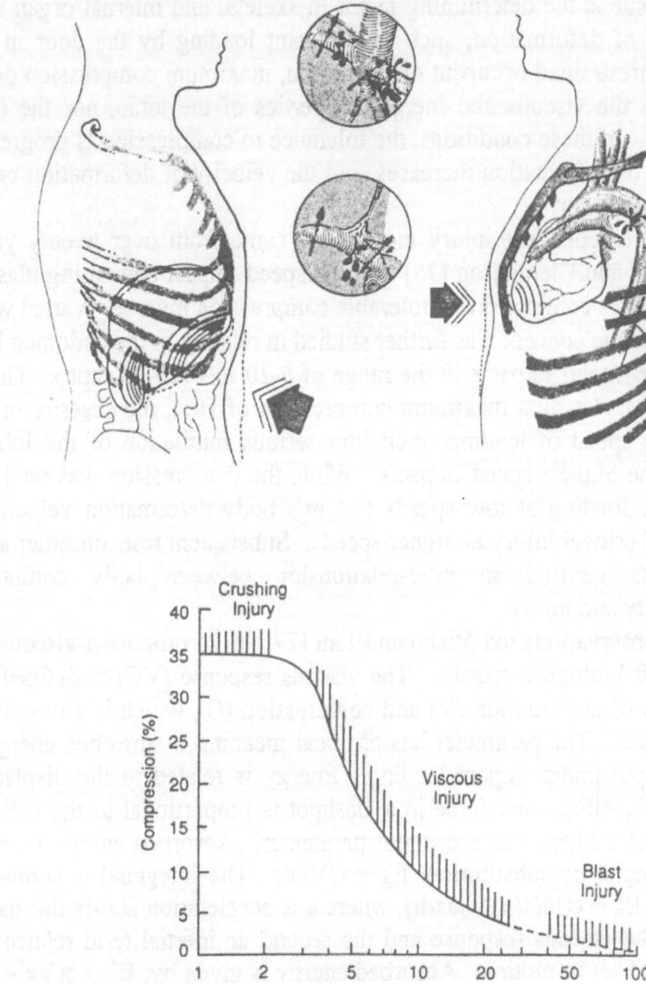

Figure 5. **Loading the abdomen directly compresses soft tissues. Chest impact compresses the rib cage and displaces internal organs with the potential to cause injury. Human tolerance is related to crushing injure by compression at C_{max} = 35%, viscous injury by rate-dependent deformation at VC_{max} = 1.0 m/s, and a blast injury.**

functional injury to heart and respiratory systems. In these experiments, Stein [81] et al. found that the severity of cardiac arrhythmia and traumatic apnea was related to VC. This situation is important to baseball impact protection of children, Viano et al. [82], and in the design of non-penetrating bullets and bullet-proof protective vests.

Figure 5 summarizes torso injury mechanisms associated with impact deformation. For low speeds of deformation, the limiting factor is the risk of crush injury from high compression of the body (C). This occurs at about 35-40% depending on the contact area and orientation of loading. For deformation speeds above 3 m/s, a similar level of injury risk is determined by the peak viscous response (VC). In a particular situation, there is a potential for injury related to compression or viscous responses; either or both mechanism can occur during impact. At extreme rates of loading, such as in a blast wave exposure, injury occurs with less than 10-15% compression by high energy transfer to viscous elements of the body.

3.3. BIOMECHANICAL RESPONSES

The reaction force developed by the chest varies with the velocity of impact, so biomechanics is best characterized by the force-deflection response of the torso The dynamic compliance is related to viscous, inertial and elastic properties of the body.

There is an initial rise in force which is related to inertial responses as the sternal mass is rapidly accelerated to the impact speed. This is followed by a plateau in force which is related to the viscous response and is rate-dependent, and a superimposed stiffness component related to chest compression. Melvin [83] analyzed frontal biomechanics and modeled the force-deflection response as an initial stiffness $k = 0.26 + 0.60(V-1.3)$ and a plateau force $F = 1.0 + 0.75(V-3.7)$, where k is in kN/cm, F is in kN and the velocity of impact V is in m/s. The force F reasonably approximates the plateau level for lateral chest and abdominal impact, but the initial stiffness is lower at $F = 0.12(V-1.2)$ for side loading.

The Hybrid III dummy reported on by Foster [84] was the first to demonstrate humanlike chest responses typical of the biomechanical data for frontal impacts. Rouhana [85] used other information to develop a frangible abdomen, useful in predicting the injury risk of lap-belt submarining. More recent work by Schneider [62] has led to a new prototype frontal impact dummy. Lateral impact tests of cadavers against a rigid wall and blunt pendulum have led to new side impact dummies, such as the Eurosid and Biosid [86], based on side impact response data [87-89].

3.4. INJURY RISK ASSESSMENT

Over years of study, tolerance levels have been established for most responses to assess injury from chest and abdomen impact. Reviews by Cavanaugh [90] and Rouhana [91] provide current tolerance information. For the chest, the compression tolerance is 32% and Viscous tolerance is 1.0 m/s in sternal impact. These are single level thresholds, which are commonly used to evaluate safety systems. The implication is that for biomechanical responses below tolerance, there is no injury, and for responses above tolerance, there is injury [92]. An additional factor is biomechanical response scaling for individuals of different size and weight. The commonly accepted procedure involves equal stress and velocity, which enabled Mertz et al. [93] to predict injury tolerances and biomechanical responses for different size dummies.

Injury risk assessment is frequently used. It evaluates the probability of injury as a function of biomechanical response. A Logist function relates injury probability p to a biomechanical response x by $p(x) = [1 + \exp(\alpha - \beta x)]^{-1}$ where α and β are parameters derived from statistical analysis of biomechanical data. This function provides a sigmoidal relationship with three distinct regions. For low biomechanical response levels, there is a low probability of injury. Similarly, for very high levels, the risk asymptotes to 100%. The transition region between the two extremes involves risk which is proportional to the biomechanical response. A sigmoidal function is typical of human tolerance because it represents the distribution in weak through strong subjects in a population exposed to impact [94]. Table 1 summarizes available parameters for injury risk assessment.

Table 1: Injury Probability Function Coefficients

Body Region	$ED_{25\%}$	α	β	χ^2	p	R
			Frontal Impact			
Head (skull fracture)						
HIC	560	2.048	0.0017	12.7	0.000	0.38
A	160 g	2.859	0.0112	6.0	0.015	0.23
Chest (AIS 4+)						
VC	1.0 m/s	11.42	11.56	25.6	0.000	0.68
C	34%	10.49	0.277	15.9	0.000	0.52
Femur (fracture)						
F	6.0 kN	7.590	0.0011	4.9	0.028	0.39
			Lateral Impact			
Chest (AIS 4+)						
VC	1.5 m/s	10.02	6.08	13.7	0.000	0.77
C	38%	31.22	0.79	13.5	0.000	0.76
Abdomen (AIS 4+)						
VC	2.0 m/s	8.64	3.81	6.1	0.013	0.60
C	47%	16.29	0.35	4.6	0.032	0.48
Pelvis (pubic ramus fracture)						
C	27%	84.02	3.07	11.5	0.001	0.91

3.5. ANTHROPOMORPHIC TEST DEVICES

Anthropomorphic test devices, or dummies, are mechanical analogs of the human body which are routinely used to evaluate the effectiveness of restraint systems, protective clothing, safety devices and automotive designs in preventing injury.

Dummies are designed to simulate the size, shape, mass, stiffness and energy absorption of the human body during impact. The most sophisticated frontal dummy is the Hybrid III. The family of Hybrid III dummies includes the 50th percentile and 95th percentile adult male, 5th percentile adult female, 3 and 6 year old child, and a range of infants for child safety seat testing. There are also a range of side impact dummies representing small to mid-size adults.

Each dummy mimics the trajectory, acceleration and impact deformation experienced by a human of that size during crash deceleration or impact. The body of biomechanical information has been used to define the humanlike response and tolerance of each member of the dummy family to impact, and the mechanical dummies realistically approximate the human response for a range of impact severities from non-injury through serious/fatal injury. The dummies are durable and repeatable in response, and are well-accepted in the industry and by governments worldwide. A recent summary by Mertz [86] provides background information and technical details on the available frontal and side impact dummies. Various dummies and special instrumentation are commercially available from First Technology Safety Systems, TNO, Vector Research, Denton and Endevco just to name a few companies. In addition, special instrumentation is available to study specific body region responses, including the abdomen, lower legs and ankles, and even the protection of women during pregnancy.

References

1. Gennarelli, T.A. (1983). Head injuries in man and experimental animals: Clinical aspects. *Acta neurochirurgica Suppl.* 32:1-13.
2. Strich, S.J. (1961). Shearing of nerve fibres as a cause of brain damage due to head injury. *The Lancet.* 2:443-448.
3. Lighthall, J.W., Goshgarian, H.G., Pinderski, C.R. (1990). Characterization of axonal injury produced by controlled cortical impact. *J. Neurotrauma.* 7(2):65-76.
4. Adams, J.H., Doyle, D., Graham, D.I., Lawrence, A.E., and McLellan, D.R. (1986). Gliding contusions in nonmissile head injury in humans. *Arch. Pathol. Lab. Med.* 110:485-488.
5. Pierce, D.A. and Barr, J.S. (1983). Fractures and dislocations at the base of the skull and upper spine. In The Cervical Spine. ed. R.W. Baily, p. 196-206. Lippincott, Philadelphia, PA.
6. Garfin, S.R. and Rothman, R.H. (1983). Traumatic spondylolisthesis of the axis (Hangman's fracture). The Cervical Spine. ed. R.W. Baily, p. 223-232. Lippincott, Philadelphia, PA.
7. Thomas, D.J. and Jessop, M.E. (1983). Experimental head and neck injury. Impact Injury of the Head and Spine. pp. 177-217. Ed. C.L. Ewing et al. Charles C. Thomas, Springfield, IL.
8. Cheng, R., Yang, K.H., Levine, R.S., King, A.I., and Morgan, R. (1982). Injuries to the cervical spine caused by a distributed frontal load to the chest. Proc. 26th Stapp Car Crash Conf. p. 1-40.
9. Lord, S., Barnsley, L., and Bogduk, N. (1993). Cervical zygapophyseal joint pain in whiplash. In Cervical Flexion-Extension/Whiplash Injuries. ed. R.W. Teasell and A.P. Shapiro, p. 355-372. Hanley & Belfus, Inc., Philadelphia, PA.
10. Pintar, F.A., Yoganandan, N., Sances, A. Jr., Reinartz, J., Harris, G.M., and Larson, S.J. (1989). Kinematic and anatomical analysis of the human cervical spinal column under axial loading. Proc. 33rd Stapp Car Crash Conf. p. 191-214.
11. Pintar, F.A., Sances, A. Jr., Yoganandan, N., Reinartz, J, Maiman, D., Suh, J.K., Unger, G. (1990). Biodynamics of the total human cadaveric spine. Proc. 34th Stapp Car Crash Conf. p. 55-72.

12. Nightingale, R.W., McElhaney, J.H., Best, T.M., Richardson, W.J. and Myers, B.S. (1993). Proc. 39th Meeting Orthopedic Res. Soc. p. 233.
13. Moffat, E.A., Siegel, A.W., and Huelke, D.F. (1978). The biomechanics of automotive cervical fractures. Proc. 22nd Conf. of Am. Assoc. for Automotive Med. p. 151-168.
14. Wismans, J. and Spenny, D.H. (1983). Performance requirements for mechanical necks in lateral flexion. Proc. 27th Stapp Car Crash Conf. p. 137-148.
15. Walker, L.B. Jr., Harris, E.H., and Pontius, U.R. (1973). Mass, volume, center of mass, and mass moment of inertia of head and neck of human body. Proc. 17th Stapp Car Crash Conf. p. 525-537.
16. Hodgson, V.R. and Thomas, L.M. (1971). Comparison of head acceleration injury indices in cadaver skull fracture. Proc. 15th Stapp Car Crash Conf. p. 190-206.
17. Hodgson, V.R. and Thomas, L.M. (1973). Breaking strength of the human skull versus impact surface curvature. Wayne State University, Detroit, MI.
18. Hubbard, R.P. and McLeod, D.G. (1974). Definition and development of a crash dummy head. Proc. 18th Stapp Car Crash Conf. p. 599-628.
19. Reynolds, H.M., Clauser, C.E., McConville, J., Chandler, R and Young, J.W. (1975). Mass distribution properties of the male cadaver. SAE Paper No. 750424, Society of Automotive Engineers, Warrendale, PA.
20. Beier, G., Schuller, E., Schuck, M., Ewing, C., Becker, E., and Thomas, D. (1980). Center of gravity and moments of inertia of human head. Proc. 5th International Conf. on the Biokinetics of Impacts. p. 218-228.
21. McConville, J.T., Churchill, T.D., Kaleps, I., Clauser, C.E., and Cuzzi, J. (1980). Anthropometric relationships of body and body segment moments of inertia. AMRL-TR-8-119. Aerospace Medical Research Lab, Wright-Patterson AFB, OH.
22. Robbins, D.H. (1983). Development of anthropometrically based design specifications for an advanced adult anthropomorphic dummy family. Volume 2: Anthropometric specifications for a mid-sized male dummy. Report No. UMTRI 83-53-2. Univ. of Mich. Transportation Res. Inst., Ann Arbor.
23. Hodgson, V.R. and Thomas, L.M. (1975). Head impact response. Vehicle Research Inst., Soc. Automotive Engrs., Warrendale, PA.
24. Prasad, P., Melvin, J.W., Huelke, D.F., King, A.I., and Nyquist, G.W. (1985). Head. In Review of Biomechanical Impact Response and Injury in the Automotive Environment - Phase 1 Task B Report: Advanced Anthropomorphic Test Device Development Program. DOT Report No. DOT HS 807042, Univ. of Mich., Ann Arbor, MI.
25. Padgaonkar, A.J., Krieger, K.W., and King, A.I. (1975). Measurement of angular acceleration of a rigid body using linear accelerometers. *J. Appl. Mech.* 42:552-556.
26. Nuscholtz, G., Lux, P., Kaiker, P., and Janicki, M.A. (1984). Head impact response - Skull deformation and angular accelerations. Proc. 28th Stapp Car Crash Conf. p. 41-74.
27. Mertz, H.J. and Patrick, L.M. (1967). Investigation of the kinematics and kinetics of whiplash. Proc. 11th Stapp Car Crash Conf. p. 267-317.
28. Mertz, H.J. and Patrick, L.M. (1971). Strength and response of the human neck.Proc. 15th Stapp Car Crash Conf. p. 207-255.
29. Patrick, L.M. and Chou, C. (1976). Response of the human neck in flexion, extension, and lateral flexion. Vehicle Res. Inst. Report No. VRI-7-3. Soc. of Automotive Engrs., Warrendale, PA.
30. Schneider, L.W., Foust, D.R., Bowman, B.M., Snyder, R.G., Chaffin, D.B., Abdelnour, T.A. and Baum, J.K. (1975). Biomechanical properties of the human neck in lateral flexion. Proc. 19th Stapp Car Crash Conf. p. 455-486.
31. Ewing, C.L., Thomas, D.J., Lustick, L, Muzzy, W.H., Willems, G.C. and Majewski, P. (1978). Effect of initial position on the human head and neck response to +Y impact acceleration. Proc. 22nd. Stapp Car Crash Conf. p. 101-138.
32. Mertz, H.J., Neathery, R.F., and Culver, C.C. (1973). Performance requirements and

characteristics of mechanical necks. In Human Impact Response. Measurement and Simulation. p. 263-288, ed. W.F. King and H.J. Mertz. Plenum Press, New York, NY.

33. Society of Automotive Engineers, Human Mechanical Response Task Force. (1985). Human mechanical response characteristics. SAE J1460. Society of Automotive Engrs., Warrendale, PA.
34. Wismans, J. and Spenny, D.H. (1984). Head-neck response in frontal flexion. Proc. 28th Stapp Car Crash Conf. p. 161-171.
35. Lissner, H.R., Lebow, M., and Evans, F.G. (1960). Experimental studies on the relation between acceleration and intracranial pressure changes in man. *Surg. Gynecol. Obstet.* 111:329-338.
36. Patrick LM, Kroell CK, Mertz HJ, (1965). "Forces on the Human Body in Simulated Crashes," 9th Stapp Car Crash Conference, SAE, pp 237-260, Society of Automotive Engineers, Warrendale, PA.
37. Gadd, C.W. (1961). Criteria for injury potential. In Impact Acceleration Stress Symposium. Nat. Res. Council Publication No. 977. p. 141-144. Nat. Acad. Sci., Washington, DC
38. Patrick, L.M., Lissner, H.R., and Gurdjian, E.S. (1965). Survival by design - Head protection. Proc. 7th Stapp Car Crash Conf. p. 483-499.
39. Versace, J. (1970). A review of the severity index. Proc. 15th Stapp Car Crash Conf. p. 771-796.
40 Prasad, P. and Mertz, H.J. (1985). The Position of the United States Delegation to the ISO Working Group 6 on the Use of HIC in the Automotive Environment. SAE Paper No. 851246. Soc. of Automotive Engrs., Warrendale, PA.
41. Lowenhielm, P. (1975). Mathematical simulation of gliding contusions. *J. Biomech.* 8:351-356.
42. Margulies, S.S., Thibault, L.E., and Gennarelli, T.A. (1990). Physical model simulation of brain injury in the primate. *J. Biomech.* 23:823-836.
43. Holbourn, A.H.S. (1943). Mechanics of head injury. *The Lancet.* 2:438-441.
44. King, AI; Chou, CC: (1976). Mathematical modelling, simulation and experimental testing of biomechanical systems crash response. *J. Biomech.*, 9:301-317.
45. Hardy, W.N., Khalil, T.B., and King, A.I. (1994). Literature review of head injury biomechanics. Int. *J. Impact Engg.* (In Press).
46. Kleinberger, M. (1993). Application of finite element techniques to the study of cervical spine mechanics. Proc. 37th Stapp Car Crash Conf. p. 261-272.
47. Ewing, CL; King, AI; Prasad, P: (1972). Structural consideration of the human vertebral column under +Gz impact acceleration. *J. of Aircraft*, 9:84-90.
48. Prasad, P; King, AI: (1974). An experimentally validated dynamic model of the spine. *J. Appl. Mech.*, 41:546-550.
49. El-Bohy, AA; Yang, KH; King, AI (1989): Experimental verification of facet load transmission by direct measurement of facet/lamina contact pressure. *J. Biomech.* 22:931-941.
50 Oxland, TR: (1992). Burst fractures of the human thoracolumbar spine: A biomechanical investigation. Ph.Dissertation, Yale University.
51. Chance, GO (1948): Note on a type of flexion fracture of the spine. Br. J. Radiol., 21:452-453.
52. Begeman, PC; King, AI; Prasad, P (1973): Spinal loads resulting from -Gx acceleration. Proc. 17th Stapp Conf., SAE Paper No. 730977, pp. 343-360.
53. States, JD; Annechiarico, RP; Good, RG; Lieou, J; Andrews, M; Cushman, L; Ingersoll, G: (1989). A time comparison study of the New York State safety belt use law utilizing hospital admission and police accident report information. 33rd Annual Proc. Assoc. Advancement Automotive Med., pp. 265-281.
54. King, AI: (1993). Injury to the thoraco-lumbar spine and pelvis. Accidental Injury: Biomechanics and Prevention, Ed. A Nahum and J Melvin, Springer-Verlag, New York, pp. 429 459.
55. Evans, FG; Lissner, HR; Patrick, LM: (1962). Acceleration-induced strains in the intact vertebral column. *J. Appl. Physiol.* 17:405-409.
56. Cheng, R; Mital, NK; Levine, RS; King, AI: (1979) Biodynamics of the living human spine during -Gx impact acceleration. Proc. 23rd Stapp Conf., SAE Paper No. 791027, pp. 721-763.

57. Mital, NK; Cheng, R; Levine, RS; King, AI: (1978). Dynamic characteristics of the human spine during -Gx acceleration. Proc. 22nd Stapp Conf., SAE Paper No. 780889, pp. 139-161.
58. Yamada, H: (1970). Strength of Biological Materials. Ed. by FG Evans. Williams and Wilkins, Baltimore, pp. 75-80.
59. Sonoda, T: (1962). Studies on the strength for compression, tension and torsion of the human vertebral column. J. Kyoto Prefectural U. of Med., Med. Soc., 71:659-702.
60. Myklebust, J; Sances, A; Maiman, D; Pintar, F; Chilbert, M; Rauschning, W; Larson, S; Cusick, J; Ewing, C; Thomas, D; Saltzberg, B: (1983). Experimental spinal trauma studies in the human and monkey cadaver. Proc. 27th Stapp Conf., pp. 149-161.
61. Eiband AM, (1959) "Human Tolerance to Rapidly Applied Acceleration. A Survey of the Literature." National Aeronautics and Space Administration, Washington DC, NASA Memo No. 5-19-59E.
62. Schneider, LW; Haffner, MP; Eppinger, RH; Salloum, MJ, Beebe, MS; Rouhana, SW; King, AI; Hardy, WN; Neathery, RF:(1992). Development of an advanced ATD thorax system for improved injury assessment in frontal crash environments. Proc. 36th Stapp Conf., SAE Paper No. 922520, pp. 129-155.
63. Latham, F: (1957). A study in body ballistics: Seat ejection. *Proc. Royal Soc. (B)*, 147:121-139.
64. Prasad, P; King, AI; Ewing; CL:(1974). The role of articular facets during +Gz acceleration. *J. Appl. Mech.*, 41:321-326.
65. Melvin JW, King AI, Alem NM, (1988). "AATD System Technical Characteristics, Design Concepts, and Trauma Assessment Criteria," AATD task E-F Final Report, DOT-HS-807-224, US Department of Transportation, National Highway Traffic Safety Administration, Washington, DC.
66. Viano DC, King, AI, et al. (1989). "Injury Biomechanics Research: An Essential Element in the Prevention of Trauma," *Journal of Biomechanics* 22:403-417.
67. Stapp JP, (1970). "Voluntary Human Tolerance Levels," In Impact Injury and Crash Protection, Gurdjian ES, Lange WA, Patrick LM, Thomas LM, Editors, pp 308-349, Charles C Thomas, Springfield, IL.
68. Mertz HJ, Gadd CW, (1971). "Thoracic Tolerance to Whole-Body Deceleration," 15th Stapp Car Crash Conference, pp 135-157, SAE Paper No. 710852, Society of Automotive Engineers, Warrendale, PA.
69. Morgan RM, Marcus JH, Eppinger RH. (1986). "Side Impact-The Biofidelity of NHTSA's Proposed ATD and Efficacy of TTI," 30th Stapp Car Crash Conference, pp 27-40, SAE Paper No. 861877, Society of Automotive Engineers, Warrendale PA.
70. Cavanaugh JM, et al. (1993) "Injury and Response of the Thorax in Side Impact Cadaveric Tests," 37th Stapp Car Crash Conference, pp 199-222, SAE Paper No. 933127, Society of Automotive Engineers, Warrendale, PA.
71. Patrick LM, Mertz HJ, Kroell CK, (1967). Cadaver Knee, Chest, and Head Impact Loads," 11th Stapp Car Crash Conference, pp 168-182, SAE Paper No. 670913, Society of Automotive Engineers, Warrendale, PA.
72. Gadd CW, Patrick LM, (1968). "Systems Versus Laboratory Impact Tests for Estimating Injury Hazards," SAE Paper No. 680053, Society of Automotive Engineers, Warrendale, PA.
73. Kroell CK, Schneider DC, Nahum AM, (1971). "Impact Tolerance and Response to the Human Thorax," 15th Stapp Car Crash Conference, pp 84-134, SAE Paper No. 710851, Society of Automotive Engineers, Warrendale, PA.
74. Kroell CK, Schneider DC, Nahum AM, (1974). "Impact Tolerance and Response to the Human Thorax II," 18th Stapp Car Crash Conference, pp 383-457, SAE Paper No. 741187, Society of Automotive Engineers, Warrendale, PA.
75. Jonsson A, Clemedson CJ, et al. (1979). "Dynamic Factors Influencing the Production of Lung Injury in Rabbits Subjected to Blunt Chest Wall Impact," *Aviation, Space and Environmental Medicine* 50:325-337.

76. Lau IV, Viano DC, (1981). "Influence of Impact Velocity on the Severity of Nonpenetrating Hepatic Injury," *Journal Trauma* 21(2):115-123.
77. Viano DC, Lau IV, (1988). "A Viscous Tolerance Criterion for Soft Tissue Injury Assessment, *Journal of Biomechanics* 21:387-399.
78. Lau IV, Viano DC, (1986). "The Viscous Criterion-Bases and Application of an Injury Severity Index for Soft Tissue," 30th Stapp Car Crash Conference, pp 123-142, SAE Paper No. 861882, Society of Automotive Engineers, Warrendale, PA.
79. Lau IV, Viano DC, (1988). "How and When Blunt Injury Occurs: Implications to Frontal and Side Impact Protection." 32nd Stapp Car Crash Conference, pp. 81-100, SAE Paper No. 881714, Society of Automotive Engineers, Warrendale, PA.
80. Lau IV, Horsch JD, et al. (1987). "Biomechanics of Liver Injury by Steering Wheel Loading," *Journal of Trauma* 27:225-237.
81. Stein PD, Sabbah HN, et al. (1982). "Response of the Heart to Nonpenetrating Cardiac Trauma." *J Trauma* 22(5):364-373.
82. Viano DC, Andrzejak DV, Polley TZ, King AI, (1992). "Mechanism of Fatal Chest Injury by Baseball Impact: Development of an Experimental Model," *Clinical Journal of Sport Medicine* 2:166-171.
83. Melvin JW, Weber K, Editors, (1988). "Review of Biomechanical Response and Injury in the Automotive Environment," AATD Task B Final Report, DOT-HS-807-224, US Department of Transportation, National Highway Traffic Safety Administration, Washington DC.
84. Foster JK, Kortge JO, Wolanin MJ,(1977) "Hybrid III-A Biomechanically-Based Crash Test Dummy." Stapp Car Crash Conference, pp 975-1014, SAE Paper No. 770938, Society of Automotive Engineers, Warrendale, PA.
85. Rouhana SW, et al. (1989). "Assessing Submarining and Abdominal Injury Risk in the Hybrid III Family of Dummies," 33rd Stapp Car Crash Conference, pp. 257-279, SAE Paper No. 892440, Society of Automotive Engineers, Warrendale, PA.
86. Mertz HJ, (1993). "Anthropomorphic Test Devices," In Accidental Injury: Biomechanics and Prevention, Nahum AM, Melvin JW, Editors, pp 66-84, Springer-Verlag, New York.
87. Viano DC, "Biomechanical Responses and Injuries in Blunt Lateral Impact," 33rd Stapp Car Crash Conference, pp. 113-142, SAE Paper No. 892432, Society of Automotive Engineers, Warrendale, PA, 1989.
88. Viano DC, (1987). "Evaluation of the Benefit of Energy-Absorbing Materials for Side Impact Protection," 31st Stapp Car Crash Conference, pp. 185-224, SAE Paper No. 872213, Society of Automotive Engineers, Warrendale, PA.
89. King AI, (1984). "Regional Tolerance to Impact Acceleration," In SP-622, Society of Automotive Engineers, Warrendale, PA.
90. Cavanaugh JM, (1993) "The Biomechanics of Thoracic Trauma," In Accidental Injury: Biomechanics and Prevention, Nahum AM, Melvin JW, Editors, pp 362-391, Springer-Verlag, New York.
91. Rouhana SW, (1993). "Biomechanics of Abdominal Trauma," In Accidental Injury: Biomechanics and Prevention, Nahum AM, Melvin JW, Editors, pp 391-428, Springer-Verlag, New York.
92. Society of Automotive Engineers,(1986). Human Tolerance to Impact Conditions as Related to Motor Vehicle Design, SAE J885, Society of Automotive Engineers, Warrendale, PA.
93. Mertz HJ, Irwin A, et al. (1989). "Size, Weight and Biomechanical Impact Response Requirements for Adult Size Small Female and Large Male Dummies," SAE Paper No. 890756, Society of Automotive Engineers, Warrendale, PA.
94. Viano DC, (1988). "Cause and Control of Automotive Trauma," *Bulletin of the New York Academy of Medicine, Second Series* 64:376-421..

BIOMECHANICS OF IMPACT TRAUMATIC BRAIN INJURY

FARIS A. BANDAK
National Highway Traffic Safety Administration
US Dept. of Transportation, Washington, DC, and
School of Engineering and Applied Science
The George Washington University, Washington, DC, U.S.A.

Abstract

Brain injury is a major cause of death and disability. A better understanding of the biomechanics of the brain under loading conditions that lead to injury can reduce mortality and morbidity. During normal activity, simple body movements such as twisting or jumping can cause the brain to deform within the cranial cavity. External head loading can be severe even during <u>intentional</u> body movements such as heading a football (soccer in the US), punching a boxer, or experiencing a head collision in US football. At large enough magnitudes, these head loadings can produce irreversible effects in the neural, axonal, and vascular structures of the brain. Mechanisms of tissue damage as well as links with clinical observations have not been completely characterized. Macro-mechanical and micro-mechanical characterization of head loading is a prerequisite to the determination of brain damage mechanisms.

Characterization of head loading occurring in motor vehicle crashes is a subject of continuing research. This loading can be of the contact or non-contact type and produces complex brain response states with a wide range injury severity. Analysis of head impact in motor vehicle crashes has recently become more amenable to computational methods. Structural models of the head, particularly finite element models, have acquired a greater degree of sophistication and are accounting for anatomical detail and representing more phenomena. Coupled fluid-structure response is increasingly being built into established finite element applications. Techniques for the treatment of boundary conditions associated with the complex interactions between the various components of the brain are areas of developing research. Material response characterization still remains a difficult challenge since such studies require a living brain.

The objective of this chapter is to describe the head-brain complex, with emphasis on its mechanical characteristics, and to give a brief exposition of some computational biomechanics techniques used in the study of the mechanisms of traumatic brain injury occurring under motor vehicle crash conditions.

J. A. C. Ambrósio et al. (eds.),
Crashworthiness of Transportation Systems: Structural Impact and Occupant Protection, 53–93.

1. Introduction

It is estimated that in the United States alone Traumatic Brain Injury (TBI) is responsible for over 50 thousand fatalities and nearly 1 million injuries each year [1]. Motor vehicle crashes are responsible for nearly one-half of those injuries [1]. The total societal cost of head injury in the US was estimated for 1985 at 37 billion dollars [2]. The majority of research resources dedicated to TBI research have historically been allocated to treatment and rehabilitation. Relatively slight resources have been allocated to research in the prevention of TBI.

TBI takes many forms and has many levels of severity. These are classified using several well-established systems. The most widely known is the International Classification of Diseases system developed by the World Health Organization. The system consists of three types of codes referred to as 1) ICD9 codes, also called N-codes, which classify injury diagnoses such as skull fracture, brain stem contusion etc., 2) E codes, which classify the external cause of injury such as motor vehicle crash, and 3) V codes, which is a set of supplementary codes classifying such things as physical examinations. A second well-established injury classification system is the Abbreviated Injury Scale (AIS). It was developed by the Association for the Advancement of Automotive Medicine as an indicator to associate an injury with its consequent level of threat to life. This system describes injuries for each body region as for instance skeletal system, nervous system, internal organs, etc. A special set of codes has been devised for brain injury classifying loss of consciousness. The AIS scale classifies injury severity using values from 1 to 6 where AIS values of 1 and 2 are designated minor injuries, 3,4 and 5 are serious, severe, and critical, and 6 is considered untreatable usually fatal.

The most widely used classification system particular to brain injury is the Glasgow Coma Scale (GCS) [3]. This system essentially associates levels of impaired consciousness with a numerical value ranging between 3 and 15, based on the sum of three numbers related to eye opening, verbal response and best motor response. The lowest GCS value associated with the worst outcome. A GCS≤ 8 is associated with severe brain injury. Several other systems such as the Trauma Score, Revised Trauma Score, Triss, and Ascot, have been developed for the classification of trauma. The AIS is used to classify motor vehicle crash TBI data collected by the US government through a system called the National Accident Sampling System Crashworthiness Data System (NASS-CDS). The data is based on police reported crashes of passenger cars and light trucks in which at least one of the vehicles is towed from the scene due to damage from the crash. NASS-CDS data indicates that crash victims experience multiple injuries with TBI constituting some of the most severe injuries. A large percentage of those hospitalized after head injury experience a less severe form of brain injury called mild traumatic brain injury (MTBI) [4]. Significant data on MTBI is collected by the sports medicine and biomechanics communities. This is potentially invaluable data in the study of transportation related MTBI that may not present as the acute, more visible, forms of severe TBI. A proposed classification distinguishing levels of MTBI has been proposed by Ommaya [5]. This classification is sensitive to the types of MTBI that evolve into severe

TBI. Ommaya suggested three levels with the first involving patients that recover fully after three months; the second with a more incomplete recovery involving slowly diminishing neurological deficits; and third a class of mild brain injuries that involve significant deterioration resulting in permanent severe impairment or death. Ommaya refers to the last category as TADD, which stands for "talk and die or deteriorate", a syndrome first identified by Schneider [6] and classified by Reilly and co-workers [7]. This injury will be discussed in a later section.

Injuries common to sports and transportation are not confined to MTBI. According to Cantu [8], sports responsible for the majority of catastrophic head injuries include US football, gymnastics, ice hockey, and wrestling. The modes of injury usually involve an impact to the head such as the use of the head to tackle, in US football, or a missed dismount in gymnastics. Other sports posing serious, though less frequent, threat to the head include pole vault events, equestrian sports, and more prominently boxing. Sports head injury and transport injury share more obvious common ground in areas such as motorcycle and automobile racing.

Although severe TBI is prevalent to some degree in sports, it is the less severe MTBI that has the highest incidence. US football for example, involves the participation of more than 1.5 million athletes and is the single most responsible sport for MTBI with an incidence of approximately 250,000 cases per year [9]. The types of head injuries most responsible for sports fatalities are those associated with intracranial hemorrhage [10] with subdural hematoma accounting for most of those deaths. These types of injuries include intracerebral hematoma, involving severe loading of the brain, and epidural hematoma occurring as a result of some sort of blood vessel breaching that places often significant amounts of blood between the inner surface of the skull and the outer surface of the dura mater. This results in brain herniation where otherwise the brain may not have been injured. This level of severity for this injury is usually associated with vessel rupture as for example the middle meningeal artery caused by temporal skull fractures crossing the meningeal groove. A discussion of these types of injuries and their mechanical categorization will be given in sections to follow.

2. Anatomical Characteristics of the Head-Brain Complex

Quantitative data on the macroscopic and microscopic morphology of the head-brain complex is essential for medical and mechanical assessment of injury mechanisms. Such data is of great use in mechanics when it contains information on shapes, numbers, and constituent makeup of the components of the head-brain complex that are significant participants in the injury process and any consequent functional deficits. This section describes the components of the human head-brain complex starting with the extracranial structure, followed by the skull, and finally the intracranial contents. Only those anatomical components that are relevant to the mechanical treatment of head-brain complex will be emphasized.

2.1. THE CRANIAL STRUCTURE

A partial cross section of the upper portion of the head showing the scalp, skull, and the boundary with the cerebrum is given in Figure 1a. The figure shows the multi-layered scalp covering the skull and having a thickness of about 1 to 1.5 cm. The scalp is a set of connected layers composed of 1) skin with hair covering, 2) subcutaneous fibro-adepose connective tissue that adheres tightly to both the skin above and the epicranium (aponeurosis) below, 3) a layer of loose areolar tissue, and 4) the pericranium which is a tough vascular layer covering the cranial bones. The aponeurosis is a broad tendinous thin sheet that serves as the origin or insertion point of skeletal muscle. It is essentially tendon fibers spread over the large area of the cranial bones.

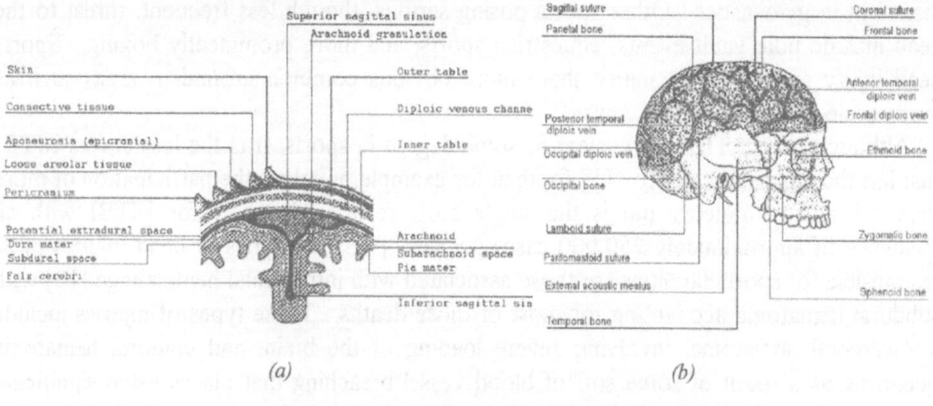

Figure 1: **(a) Layers of the cranium (modified from references 11 and 12) and (b) Sutures and diploic veins (modified from reference 11)**

The skull consists of a complicated set of bones and bone connections that make up the neurocranial and the facial regions. Both of these regions play a vital role in the protection of the brain from mechanically induced injury. The facial region acts as a shock absorber residing outside the cavity containing the brain while the neurocranial region encapsulates the brain and also provides protection through energy distribution and dissipation. The cranial cavity is made up of eight bones (Figure 1b); frontal bone, 2 parietal bones, 2 temporal bones, occipital bone, sphenoid bone, and ethmoid bone. They are divided by immovable joints called sutures. Sutures are evolutionary joints, or intertwining edges, that allow relative movement of the cranial bone plates in the early years of growth but continue to stiffen until they become solid allowing no relative movement at the final stages of growth. This is essentially an ossification process of the suture joints from a ligamentous to bony structure. The frontal bone forms the forehead and the two parietal bones form the greater portion of the sides and roof of the cranial cavity. The frontal and parietal bones meet along the coronal suture, and the two parietal bones meet along the mid-line sagittal suture as shown in Figure 1b. The internal surfaces of the bones contain many protrusions and depressions that accommodate the blood vessels supplying the outer covering of the brain called the dura mater. The two temporal bones form the inferior lateral aspects of the cranium and parts of the cranial floor. The occipital bone forms the posterior part of the base of the cranium that is a nonuniform surface with protrusions and openings. The sphenoid bone is one such protrusion that lies at the middle part of the

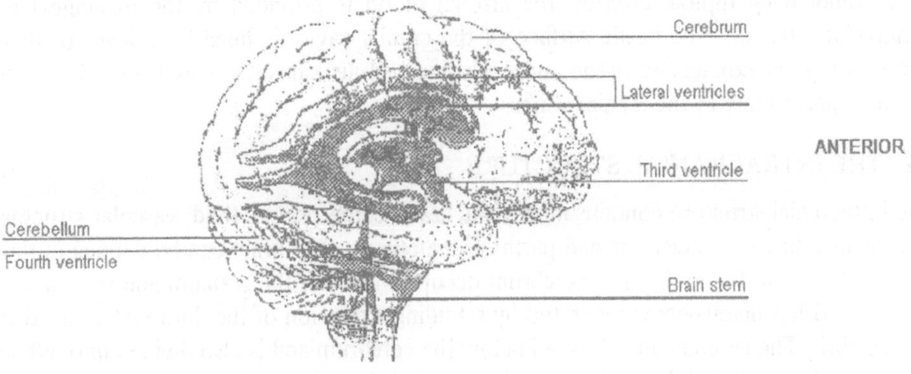

Figure 2: **Three dimensional view of the brain and the ventricular structure (modified from reference 13)**

base of the skull and articulates with all the other cranial bones, holding them together. It is in a location of significant interest from the point of view of injurious interactions with the brain. The ethmoid bone is a light, sponge-like bone located on the mid-line in the anterior part of the cranial floor medial to the orbits.

The neurocranial bone plates consist of three layers referred to as the outer table, diploe, and inner table. The diploe, consists of trabecular bone, and is located between the other two that are made up of compact bone. This arrangement exhibits great variability

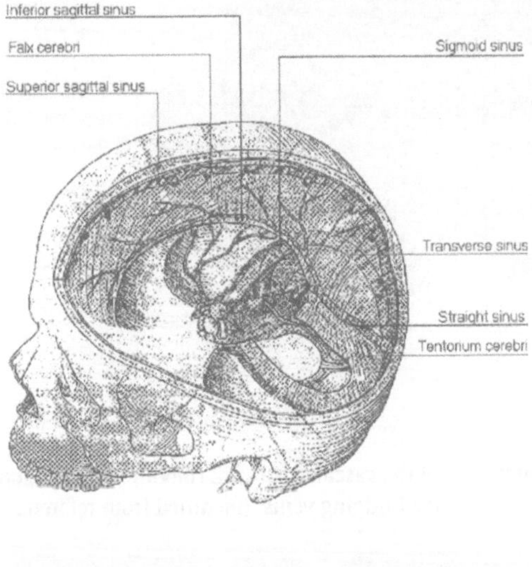

Figure 3: **Membranous partition and sinus structure (modified from reference 14)**

in the relative thickness of the layers. Figure 1a shows a cross section of these three layers and Figure 1b shows the outer table removed revealing the diploic venous channels. There are four paired (left and right) diploic veins 1) the frontal diploic vein opens into the supraorbital vein at the supraorbital notch, 2) the anterior temporal diploic vein opens into the sphenoparietal sinus, and finally 3) the posterior temporal and the occipital diploic veins both open into the transverse sinus and they may open into surface veins. There are no accompanying diploic arteries, the arterial blood is provided by the meningeal and pericranial arteries. The inside surface of the cranial cavity is lined by a layer of dense fibrous irregular connective tissue and enclosing venous sinuses. This layer adheres for the most part tightly to the cranial bones.

2.2. THE INTRACRANIAL STRUCTURE

The intracranial structure contains mainly the brain, cerebrospinal fluid, vascular structure, and the membranous coverings and partition structures. The brain can be defined in three parts as shown in Figure 2. The cerebrum occupies most of the cranium and is composed of right and left hemispheres separated by a folding extension of the dura mater called the falx cerebri. The cerebellum is located below the cerebrum and is also divided into left and right parts by another folding dural partition called the falx cerebelli. The cerebrum and the cerebellum are separated by another fold of the dura mater called the tentorium cerebelli (FIG. 3). The junction between the folds of the dura and the inner surface of the cranium forms some of the venous sinuses. These sinuses receive blood drained from the brain and also reabsorb cerebrospinal fluid in regions known as arachnoid granulations (FIG. 1). Large accumulations of these granulations are visible alongside the superior sagittal sinus.

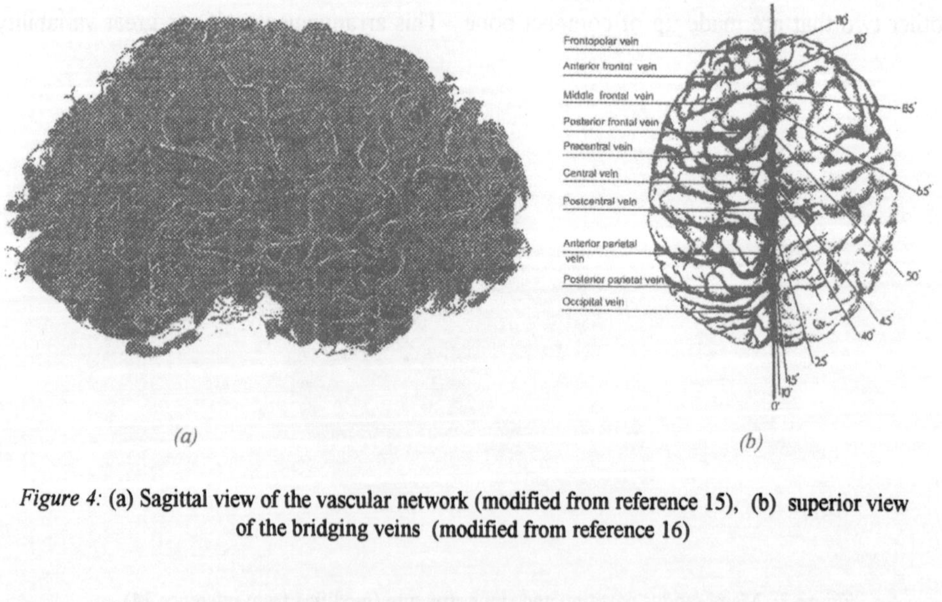

Figure 4: (a) Sagittal view of the vascular network (modified from reference 15), (b) superior view of the bridging veins (modified from reference 16)

The next layer below the dura mater is the arachnoid (FIG. 1a). This is a delicate avascular membrane having a spider web arrangement of delicate collagen fibers that surrounds the brain and envelopes the outer surface of the brain. It is bounded from above by the subdural space and from below by the pia mater. The pia mater (FIG. 1a) is a thin transparent connective tissue layer that adheres to the surface of the brain and the spinal cord. It consists of interlacing bundles of collagen fibers and some fine elastic fibers and contains many blood vessels. It also continues as a sheath around the many small vessels that penetrate into the brain. The space between the arachnoid and the pia mater is called the subarachnoidal space which is traversed by the arteries of the brain and cranial nerves and contains cerebrospinal fluid. The cerebrospinal fluid (CSF) continuously circulates through the subarachnoid space around the brain and spinal cord and through cavities within the brain. These are four cavities called ventricles (FIG. 2); two lateral ventricles, one in each of the cerebral hemispheres; the third ventricle is a vertical slit at the mid-line below the lateral ventricles and the right and left halves of the thalamus, with the thalamus forming the lateral walls of the third ventricle; and the fourth ventricle lies between the brain stem and the cerebellum.

The mean weight of the brain, as reported by several authors [17], is approximately 1400 grams for males and 1300 grams for females. Although there is great variability in the brain's dimensions and weights, some average values accounting as much as possible, for gender dimorphism and age differences are given in Table 1 [17]. As a percentage of total head volume the cortical

TABLE 1: Some dimensions and weights for the male and female brain

TABLE 1: Some dimensions and weights for the male and female brain

	Mean Weight (g)	Mean Sagittal Length (cm)	Mean Transverse Width (cm)	Mean Volume (cm^3)
Male Brain	1449 (20-30 yr.) 1344 (70-80 yr.)	16.7 --	14.0 / Coronal 9.3 --	1329 --
Cerebrum	1200			1160
Cerebellum	151	6.0	11.0	145
Brain Stem (3 parts)	29	2.5/2.5/1.0	3.6/1.8/1.0	
Cerebral Ventricles				9
Female Brain	1309 (20-30 yr.) 1213 (70-80 yr.)			1186
Cerebral Cortex	1068			1033
Cerebellum	135			130
Brain Stem		2.4/ /	3.3/ /	

	Mean Brain Weight (g)	
Age	**Male**	**Female**
1 Year	944.7	872.0
2 Years	1025.0	960.8
3 Years	1108.1	1040.2
4 Years	1330.1	1138.7
5 Years	1263.0	1220.9
6 Years	1359.1	1264.5
10 Years	1408.3	1284.4
18 Years	1444.5	1228.4

white matter is about 5%, cortical gray is 10%, cortical sulcal CSF is 1.5%, ventricular CSF is 1.25% [18]. Some of the effects of age on the brain are changes in shape and cortical gray to white matter ratio. The brain becomes more sagittaly elliptical (i.e. its length increases and its height decreases) and the ratio of cortical gray to white matter diminishes from about 3 at birth to about 1.5 at 30 years of age [18].

2.3. THE CEREBROVASCULAR STRUCTURE

The brain receives arterial blood and expels venous blood through a complex vascular network that is an integral part of the brain's mechanical structure (FIG. 4a). The main supply of arterial blood comes from the vertebral and internal carotid arteries. The vertebral arteries enter the posterior cranial fossa through the foramen magnum passing upwards and forwards uniting at the mid-line to form the basilar artery. Branches to the brain stem and cerebellum arise from the vertebral and basilar arteries before the latter divides at the upper border of the pons to form the left and right posterior cerebral arteries. Before supplying the posterior part of the cerebral hemispheres, each of these vessels gives rise to a posterior communicating artery which passes forwards to form part of the cerebral arterial circle by anastomosing with the internal carotid artery. Blood vessels that enter brain tissue pass along the surface of the brain (FIG. 4a), and as they penetrate inward, they are surrounded by a loose-fitting layer of pia mater. The venous drainage of the cranial cavity and its contents through the cranial venous sinuses (FIG. 3), which are endothelium-lined spaces, usually between the dura and the endosteum of the cranial bones. Most of this drainage goes to the superior and inferior sagittal sinuses before being delivered into the straight, transverse, and sigmoid sinuses. From there, drainage continues from the sigmoid sinuses into the internal jugular veins. The superior sagittal sinus is a major unpaired sinus that lies in the attached margin of the falx cerebri. It receives numerous superficial cerebral veins and has many accumulations of arachnoid granulations. The sinus usually turns laterally where the falx joins the tentorium cerebelli, and to the right in the attached margin of the tentorium, where it becomes the right transverse sinus. The inferior sagittal sinus lies in the free margin of the falx cerebri and runs posteriorly to join the straight sinus in the mid-line of the tentorium cerebelli. The straight sinus lies in the mid-line of the tentorium cerebelli. It receives blood from both the inferior sagittal sinus and the great cerebral vein draining deeper parts of the brain. The straight sinus turns to the left to become the left transverse sinus where a communication between the straight sinus and the superior sagittal sinus may be found and is called the confluence of sinuses. Each sigmoid sinus lie in an S-shaped groove in a part of the temporal bone and in the occipital bone. The groove carries the sinus downward to the posterior part of the jugular foramen where it becomes the internal jugular vein.

3. Categorization and Measures of Closed-Head Traumatic Brain Injury

Injury to the head can mean damage to the scalp, skull, and/or the intracranial contents. Closed-head injury can be defined as injury where the dural membrane remains intact. Serious head injuries, as defined by their relative threat to life, are those generally associated with the brain. Injuries to the brain have, in recent times, been referred to as traumatic brain injuries distinguishing them from the designation as head injuries. A particular class of TBI consists of any type of traumatic damage affecting brain function and resulting from non-penetrating mechanical head loading of the contact or non-contact type. Non-contact TBI-producing mechanical loading is generally an acceleration of the head transmitted through the neck as a result of overall body motion. Closed head TBI resulting from non-penetrating head impacts can be categorized, on a mechanical basis, as *diffuse* or *focal*. The first refers to bulk mechanical effects associated with axonal, neural, micro-vascular, and brain swelling injuries. The second is of the type of injuries that occur in localized regions of the brain subjected to tensile (rupturing) or compressive (contusive) stresses. A sub-category of focal injuries resulting from bulk mechanical loading is primarily dependent on movements of the brain relative to the cranial cavity. These include subdural hematomas and brainstem injuries. A brief description of a proposed mechanical categorization of impact brain injuries will be described below.

3.1. MECHANICAL CATEGORIZATION

3.1.1 *Diffuse Brain Injury*

Diffuse brain injury involves damage to the neural, axonal, and micro-vascular structures and damage resulting in diffuse brain swelling. This class of injuries is usually a consequence of distributed head loading conditions that generally induce relatively low-energy damage affecting substantial volumes as opposed to high energy damage affecting small, localized, volumes of the brain. A frequently occurring result of blunt head impact is an injury to the axonal structure referred to as Diffuse Axonal Injury (DAI) [19-21]. DAI is a distribution of focal lesions in the axonal components of the neural structure and thus the term diffuse. Holbourn [22] and then Ommaya and co-workers [23-25] investigated brain injury with the hypothesis that isochoric brain deformations resulting from combined rotational and translational accelerations of the head producing DAI.

Although the delicate axonal structure is more vulnerable to diffuse mechanical damage than the vascular structure, there is evidence of microvascular injury also consisting of a distribution of microlesions. Maxwell and co-workers [26] showed evidence of microvascular lesions in mechanically traumatized specimens used for the study of DAI. They observed morphological changes in the microvascular structure up to several hours after the DAI inducing accelerations were applied. It is not clear whether these morphological changes are direct result of the mechanical forces or are a physiological consequence of other resulting damage in the brain such as alteration of the endothelial membrane as has been hypothesized [26]. Maxwell and co-workers suggested that the microvascular response to brain insults occurs in two ways. A relatively rapid and localized swelling disrupting the blood-brain barrier and a more diffuse disruption of the endothelial structure occurring later in the injury cycle. Blumbergs [27] indicated that diffuse microvascular ruptures can occur in the corpus callosum without the presence of axonal injury.

Another form of diffuse brain injury is diffuse brain swelling which is an increase in cerebral volume. Brain swelling is distinguished from brain edema in that the latter also involves about 1-3% increase in brain water content [28]. The mechanisms for brain swelling as a result of brain trauma have not been clearly explained. Ommaya suggested that brain swelling results from vasospasm-induced arterial engorgement. This is based on his hypothesis that the mechanical loading causing venous constriction blocks the arterial blood contributing to brain swelling and the consequent ischemic response observed with such injuries.

3.1.2 *Focal Brain Injury*

Focal brain injuries result from relatively localized responses of the brain to loading. These include cranial and extracranial injuries affecting the bone and soft tissue components of the non-brain portion of the head. For instance a direct impact to the head can cause tissue damage to the scalp along with fracture damage to the skull. The type of loading determines whether the scalp-skull-dura laminate will experience 1) a depressed-type, localized shear-dominated fracture, 2) a linear-type, bending-dominated fracture, or 3) comminuted-type, large rapid stretch and shear dominated fracture. Cranial vault fractures are differentiated from fractures of the base of the skull by the types of brain injury that can result with each. In general the location and type of fracture is important mostly for its injurious effects on the brain. For instance basilar skull fracture can cause direct brain stem damage and temporal skull fracture crossing the meningeal groove can cause dural vessel ruptures resulting in extradural bleeding affecting the brain in a less direct mechanical way.

Bleeding that occurs between the inner surface of the skull and the outer surface of the dura mater (extradural) is referred to as epidural hematoma. This injury occurs as the primary lesion in about 6% of patients with severe closed head injuries [29]. Although it is commonly associated with skull fracture, it can result from other common focal damage such as torn dural arteries or, in some cases, torn venous sinuses. The effect of this injury on the brain can be compared by the degree of resulting unconsciousness. Unlike subdural hematoma which is almost always accompanied by immediate unconsciousness, only about one third patients with epidural hematomas experience unconsciousness, with one third experiencing no unconsciousness, and the last third mixed [30]. Approximately 91% of epidural hematoma in adults and 75% in children is associated with a skull fracture [31].

Another mode of brain damage that is also related to skull deformation with or without concomitant skull fracture or hemorrhage is focal brain contusion. Certain contusions result from the forces associated with direct contact between the brain and the solid and fluid material surrounding it and is usually in the form of cortical bleeding reaching some depth into the brain. Contusions resulting from any impact direction but lateral have been observed to occur mostly in the frontal and temporal regions of the brain [5]. They have been postulated to be a consequence of the anatomical characteristics associated with the sphenoidal region affecting the frontotemporal region of the brain. Lowenheilm [32] hypothesized that coronal accelerations can cause a type of lesion termed by Lindenberg and Freytag [33] as gliding contusions. They observed these lesions to be associated with subcortical traumatic hemorrhages in the upper region of the brain near the parasagittal sinus. Lowenheilm [32] reported that this type of lesion occurs in 25% of motor vehicle crash fatalities.

Relative Motion Brain Injury. A class of focal brain injuries results from somewhat tangential motion of the brain surface relative to the interior surface of the cranium. This motion can produce some of the focal contusions discussed above or can result in blood vessel rupture. A common vascular rupture injury is associated with the parasagittal bridging veins (FIG. 4b). These are a prominent set of vessels that cross the subdural region somewhat radially at several points on either side of the parasagittal sinus. Along with the CSF layer and the trabecula connecting the arachnoid with the dura, the bridging veins affect brain movement. They experience a tethering force when the brain moves tangentially relative to the dura along a direction vector in the mid-sagittal plane. This motion has been postulated to be mechanically sufficient to cause bridging vein rupture [32] that result in subdural hematoma. Subdural is the most frequent type of hematoma occurring as the primary lesion in approximately 24% of patients with severe closed head injuries [29]. Intracerebral hematomas, typically located in the frontal and temporal lobes, can also occur as a result of brain motion and are primary lesion in 10% of the severe closed head injuries [29].

3.2. TOLERANCE OF THE BRAIN TO MECHANICAL LOADS

3.2.1 *Classical Head Injury Criterion*

The term injury criterion is commonly used to refer to a quantity that relates injury potential with an associated mechanical action. Usually, such criteria are statistically based because of their intended range of application and the great variability in biological systems. The current criterion for the assessment of potential for motor vehicle head injury is the Head Injury Criterion (HIC) as specified in US Federal Motor Vehicle Safety Standard 208. This criterion has a historical basis in the work of Gadd [34] who used the Wayne State Tolerance Curve (WSTC) to develop what eventually became known as the Gadd severity index GSI [35]. The WSTC is based on the resultant translational head acceleration. It evolved from the early work of Gurdjian and co-workers (1955) who used the clinically observed prevalence of concomitant concussions in skull fracture cases (80% of all concussion cases also had linear skull fractures [36]) to relate cadaver impacts to brain injury. Gurdjian and co-workers concluded that by measuring the tolerance of the skull to fracture loads one is effectively inferring the tolerance to brain injury. Lissner and co-workers [37] later developed a relationship between the magnitude of the translational anterior-posterior acceleration and the load duration that became known as the WSTC.

The current HIC is based on the amplitude and duration characteristics of the resultant translational head accelerations. It relates the probability of head injury according to:

$$HIC = \max\left[\frac{1}{(t_2 - t_1)}\int_{t_1}^{t_2} a(t)dt\right]^{2.5} (t_2 - t_1) \qquad (1)$$

where $a(t)$ is the resultant translational acceleration at center of gravity of the head, $(t_2 - t_1)$ is the time interval for which the value of HIC is maximum. The suitability of the Values of the HIC in the assessment of the risk of head injury has been discussed by Prasad and Mertz [38]. More detail on the HIC is given in references [37] and [38].

There have been other measures developed to focus on such mechanisms as a basis for their formulation. Stalnaker and co-workers [39] proposed the Mean Strain Criterion. It is based on the assumption that strain is a basic mechanism for brain injury. It is basically the calculated displacement between two masses (coupled by a spring and a dashpot with stiffness, damping, and mass characteristics approximating those of the brain) under head impact conditions, averaged over the linear dimension of the brain, thus the mean strain. They calibrated the strain values against cadaver and living primate studies subjected to impacts causing a range of injury severity. This measure has not been shown to be sensitive to rotational loadings known to be responsible for generating significant strains in the brain. The next section will describe a new strain measure that is based in finite element calculations and can be related to strain damage in the brain.

3.2.2 *Mechanical Measures of Brain Damage*

Cumulative Strain Damage Measure. A mechanical measure to evaluate strain-related brain damage has been proposed [40] as a tool for the evaluation of deformation-related brain injuries resulting from head impacts. It is postulated that DAI is associated with the cumulative volume of the brain matter experiencing tensile strains over a critical level sometime during the impact. The severity of strain related injury may be associated with the magnitude and its extent with the volume of strain exceeding a particular level. The proposed measure, referred to as the *Cumulative Strain Damage Measure* (CSDM), monitors the accumulation of strain damage by calculating the volume fractions of the brain experiencing strain levels greater than various specified levels under dynamic loading. The measure, as implemented in finite element routines, is based on the maximum principal strain calculated from a strain tensor obtained by integration of the rate of deformation tensor, **D,** effectively giving the natural strain. The infinitesimal natural strain increment, in terms of the rate of deformation tensor, **D**, is given by $d\varepsilon_{ij} = D_{ij}dt$ where

$$D_{ij} = \frac{1}{2}\left(\frac{\partial v_i}{\partial x_j} + \frac{\partial v_j}{\partial x_i}\right). \tag{2}$$

are the components of **D**. Note that the velocity gradients in the above equation are with respect to the current position coordinates x_i. This is integrated for infinitesimal increments in particle position and time, to obtain the natural strain tensor.

$$\int_{t_0}^{t} d\varepsilon_{ij} = \int_{t_0}^{t} D_{ij}\, dt\,, \tag{3}$$

The integration of the rate of deformation tensor in LSDYNA3D [41], introduced above, is performed according to

$$D_{ij}^{n+1} = D_{ij}^{n} + Q_{ij}^{n} + \dot{D}_{ij}^{n+\frac{1}{2}} \Delta t^{n+\frac{1}{2}} \tag{4}$$

where Q_{ij} are the components of the rotational transformation of D_{ij} from the configuration at t^n to that at t^{n+1} and is defined by

$$Q_{ij}^{n+1} = \left\{ D_{ik}^{n} W_{kj}^{n+\frac{1}{2}} + D_{jk}^{n} W_{ki}^{n+\frac{1}{2}} \right\} \Delta t^{n+\frac{1}{2}} \tag{5}$$

where **W** is the spin tensor. This integration in the finite element program has been shown to be accurate to about 30% strain [41].

This strain measure is used to obtain the strain tensor components from which the maximum principal strains for the CSDM are calculated. At each time increment the volume of all the elements that have experienced a strain above prescribed threshold values is calculated based on the measure described above. The affected volume monotonically increases in time during conditions where regions of the brain are undergoing tensile stretching deformations. It remains constant (does not decrease) for all other deformation conditions such as those occurring during compression or unloading. The cumulative nature of this measure means that the end state of a calculation represents the strain damage that may be related to DAI associated with a particular loading regime up to that point in time. The final state can be related to DAI strain damage occurring during the whole event. The spatial distribution of affected volumes of brain matter exceeding various levels of strain can also be examined. The spatial distributions can be viewed sequentially, versus time, to evaluate nucleation and growth of damage sites aiding in the determination of the regions of the brain that may be more vulnerable than others under particular loading conditions. Additionally, the time history evolution of damage in the brain as a whole can be monitored for various choices of strain levels. The CSDM has been evaluated using an approximate human brain model and a pig model devised to simulate DAI experiments. Results from these applications will be discussed in a later section.

4. Measurement of Head Loading

Measurement of head impact loads in a car crash is an important aspect of head injury assessment. In this section we will discuss two techniques for measuring motor vehicle crash head loading. One of these is an established technique that measures head accelerations and has been used in crash test dummies. The second is a new, still experimental, technique for measuring the spatiotemporal pressure distribution occurring under contact head impact. The next two sections describe those two techniques and their uses for measuring head impact loading relevant to the assessment of brain injury.

4.1. ACCELERATION MEASUREMENTS

A technique using 9 accelerometers to measure head accelerations in motor vehicle crash tests is in current use [42]. The accelerometers are arranged in a 3-2-2-2 configuration where three accelerometers are at the CG aligned along each principal body axis, and two accelerometers on each principal arm, with sense axes oriented normal to the arm (FIG. 5). DiMasi [43] developed a procedure to utilize these crash head acceleration measurements in computer calculations. This is done by computing the generalized six degree-of-freedom angular and translational velocities, relative to inertial coordinates,

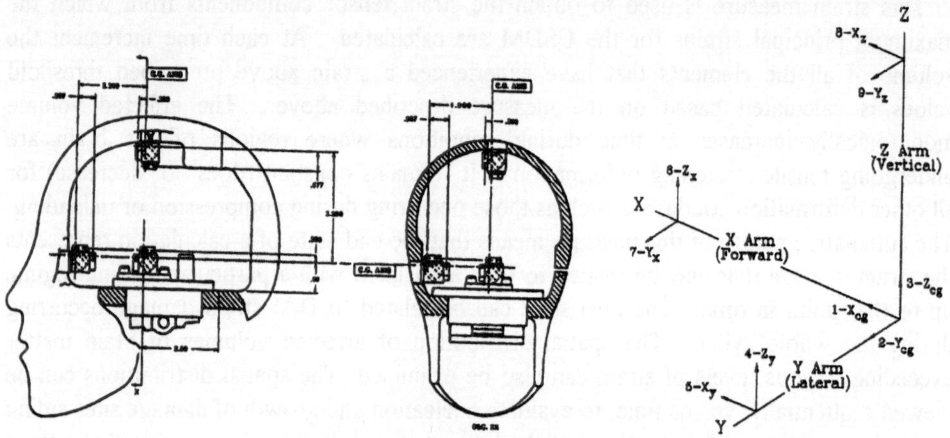

Figure 5: Orientation of accelerometers in the 3-2-2-2 arrangement in a crash dummy headpart

occurring at the center of gravity of the rigid dummy headpart instrumented with a 3-2-2-2 accelerometer arrangement. In a non-rotating system, a simple direction cosine matrix may be used to transform vector quantities from body to fixed coordinates. However, the transformation of vector quantities from the rotating, decelerating (non-inertial) dummy headpart requires considering changes due to both the rotating body coordinate system as well as time dependent vector variations within the body coordinate system. The transformation to fixed coordinates can be expressed as the summed transformation of the vector time-rate of change in the body coordinate system plus the cross product of the body angular velocity vector and the vector quantity in the body coordinate system. If changes in orientation between body and fixed coordinates are limited to very small incremental angular displacements (e.g., 1 milliradian or less), then both components may be related to the fixed coordinate system using a common transformation matrix. The orientation of the body coordinate system is computed and continually updated relative to fixed coordinates based on incremental changes in the body angular velocity vector. At the end of each time step, a new transformation matrix is computed based on the previous position, along with small incremental changes in body coordinate system orientation. This data is subsequently used to update the position of the new body coordinate system and recompute the direction cosine matrix for transforming of vector quantities from body to fixed coordinates. A compensation algorithm, which corrects for cross products of angular velocity and centripetal accelerations, resulting from the non-centroidal locations of accelerometers at the CG and on each arm, is also implemented. Typical NHTSA crash test head acceleration data is digitized at 8,000-10,000 samples/sec virtually assuring small incremental headpart angular displacements making this updated coordinate tracking algorithm suitable for this data. In the event the 1 milliradian limit on angular displacements is exceeded in a particular increment, the time step is sub-divided to assure a 1 milliradian maximum (vector) coordinate system angular displacement. The technique has been validated against data from several crash tests and results have been successfully compared [43] with corresponding film data. Additional data files containing related kinematic quantities in body vs. fixed coordinates are also produced, and a single output file containing the three translational and three angular velocities at the headpart CG is also generated for direct use in finite element analysis. The resulting generalized six-degree-of-freedom inertial loadings reflect all dummy headpart impacts as well as dummy neck reaction forces during the event.

4.2. CONTACT PRESSURE MEASUREMENTS

The spatiotemporal distribution of the pressure resulting from head impact provides measurable parameters, essential in the assessment of experimental head impact results, as well as providing accurate loading information in computational analysis. These measurements are very useful in the analysis of skull fracture but more importantly in the accurate evaluation of the local input energy affecting the brain. Several methods, using MetNet or Fuji film for example, have been used in the past to measure the contact pressure distribution resulting on the head from impact.

We investigated the spatiotemporal distribution of contact pressure using a commercially available pressure measuring device developed by TekScan, Incorporated [44]. The TekScan sensor is a thin, flexible, plastic sheet imprinted with pressure sensitive ink grid arranged in a rectangular array. The sensor is connected to an interface board in an Intel-based computer. Each pressure sensitive location is sampled in sequence by the hardware and stored for later display. While the individual pressure sensors can respond in a few microseconds, that system can only sample through the array once every 10 milliseconds. This sampling rate is inadequate for the impact duration times associated with motor vehicle crash head impacts. We developed performance specifications for a new system that can be used in the evaluation of head impacts. The new TekScan system can sample an array of pressure sensors on a 0.005 in flexible printed circuit every millisecond. The system can continuously sample up to 1000 frames per second and can measure pressures up to 1200 psi. Each sensor has a 14 by 14 grid of 0.2 in by 0.2 in sensing elements with each 14 by 14 data array constituting a data frame.

Figure 6 shows the handle, the connecting cable, and the grid patterns on the two sides of the TekScan sensor.

The new TekScan system was used as part of a test apparatus for measuring the

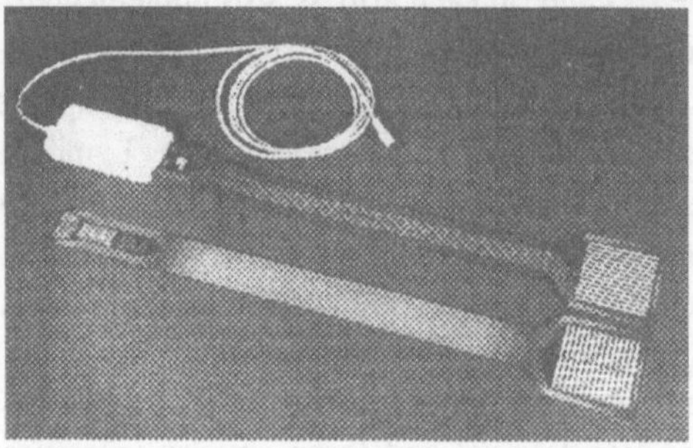

Figure 6: TekScan 9500 sensor: Front and back view, handle and cable

dynamic loading under the vertical drop of a sphere onto various impact partners. In addition to the spatiotemporal pressure measurement by the sensor, the force time history was measured with force ring transducer sandwiched between two 1/2 inch aluminum plates. The bottom plate was firmly fixed to prevent movement during impact and the top plate was overlaid with a neoprene sheet (with three thickness values of 0.25 in, 1.0 in, and 2.0 in) and used as the impact target. The TekScan sensor was evaluated under three different test configurations. In the first configuration, the sensor was placed directly against a rigid surface and covered with a 60 durometer neoprene sheet; in the second it is placed on top of a neoprene sheet. For these arrangements the load was delivered either by the free fall impact of a weighted sphere or through the compression of the spherical cap by the compression tester. In the third configuration the load was transferred from the compression tester through powder in a cylinder to the sensor. The powder provides a means to more uniformly distribute the pressure over the covered area. The setup for the static and drop impact test is shown in Figure 7a.

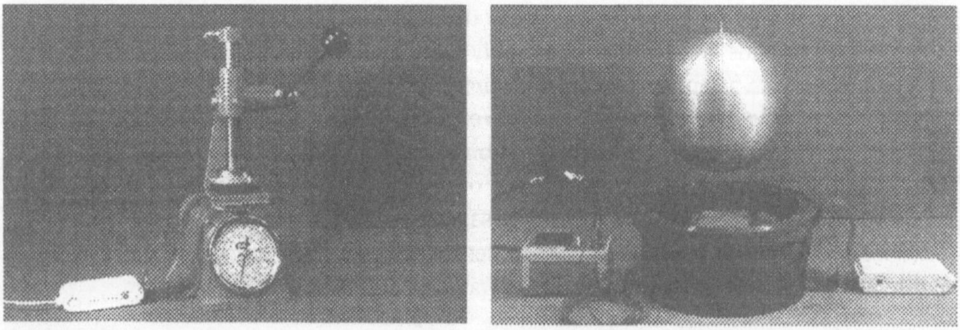

(a) (b)

Figure 7: Apparatus for testing the modified TekScan device (a) static test (b) dynamic test

Static and dynamic loading tests were conducted to assess the performance of the device. The TekScan sensor had to be statically loaded several times (up to 10) before it produces a repeatable output thereafter, the total load can be reproduced within ±5%. The sensor is quite durable provided that it is not stretched beyond its material strength. If the sensor receives too much lateral stress, it will wrinkle and the individual pressure sensors will fail. Such stretching can occur when the material is placed on a highly compliant surface or when it is covered with a compliant material that is undergoing large lateral stretching. Static and dynamic calibrations differ, perhaps because of a rate dependence of the sensor material. Once calibrated for a particular dynamic condition, there is little variation between sensors and impact strength. The total loading variation with time found by integrating the individual sensors agrees well with the force transducer output, as long as the impact has a duration of several milliseconds. If the impact is over a shorter period of time, the finite sampling rate of the device cannot produce a correct instantaneous distribution of pressure. Several signal processing techniques to reconstruct the true temporal distribution have proven promising. When calibrated dynamically and corrected for finite sampling rates, the TekScan instrument can be used to measure the load distribution under rapid impact conditions including some nonspherical impact partners.

We have been able to demonstrate that the TekScan system, modified to be suitable for head impact applications, is a viable tool for spatiotemporal impact pressure distribution measurement. This device is intended for use in an experimental head impact study to support finite element model development.

5. Computational Biomechanics and Modelling of Head Impact

In this section we will discuss two aspects of head impact biomechanics for their effects on the response of the brain. The first is the response of the skull where we consider finite element development, cadaver impact data, and physical model experiments. The second, along the same lines, is the fluid response of the subarachnoidal cerebrospinal fluid layer. The last section gives an application of finite element modelling in the assessment of brain injury. This example concentrates on strain related damage in the brain and highlights the dependence of such calculation on results from the first two sections.

5.1. APPROXIMATE HEAD AND BRAIN GEOMETRY

The modelling detail of the anatomical structure of the head-brain complex for finite element analysis depends on the level of output information required. Certain anatomical features play greater roles than others in the overall response of the brain. Realistic computational models that represent adequate structural and material features, internal constraints, and proper boundary conditions, have been developed by several workers as will be discussed later this section. These models have in part incorporated most of the features essential for adequate mechanical response using approximate brain geometry.

CT imaging techniques produce sequences of rectangular, two-dimensional cross-sectional slices of image pixels that are typically 1 by 1 millimeters in area. A sequence of slices can be taken at intervals as close as 1 millimeter apart and can be viewed as a three-dimensional rectangular array of image values or voxels. Voxel intensity has values ranging between 0 and 4095 and are directly proportional to the material mass density. We developed a head model [40] using CT scan data consisting of slices two millimeters apart and a pixel resolution of 0.9 millimeters (FIG. 8 a,b). The data was in ACR-NEMA data file format [45]

A very good source of geometrical information on the anatomy of the head is medical imaging. Vander Vorst and co-workers [45] eveloped a technique for obtaining geometrical information from medical imaging for the construction of finite element models. Several types of medical imaging techniques are in use today. X-ray Computed Tomography (CT) depends on the material density and provides good overall segmentation of bone and soft tissues. Magnetic Resonance Imaging (MRI) depends on proton spin and provides better visualization of soft tissue. In this section we will discuss the use of CT data in the construction of finite element models.

The geometry of the skull was obtained by detecting the layered surfaces of the skull from the inside out. Rays are extended from a point chosen in the interior of the skull. The intersection of these rays with the first occurrence of "bone" [47,48] generates a mesh of the interior surface of the cranium. The foramen magnum is automatically generated as a "hole" in the cranial mesh by specifying the lower vertical bound of the head in the CT scan. The outside surface of the cranium is given by the first occurrence of "not bone" along the ray extending from the inside surface of the cranium. Finally, the end of the ray occurs when the outside of the skin is reached. Each of these three sets of points forms a surface. The surfaces were smoothed to remove irregularities in the radial position introduced by artifacts in the CT scan. The face and scalp parts were modelled using hexahedral continuum elements that conform to and extend out from the cranial elements. Each element was assigned a mass that corresponds to the integral of the CT densities in the brick. We are utilizing the resulting model in a combined computational and experimental approach to characterize the deformational response of the skull to impact loading as an important step to the characterization of the input loadings to the brain. This approach involves a series of impact experiments on cadaveric specimens, where complementary, subject-specific, anatomically realistic mathematical models of the head are constructed from CT imaging data and checked against experiment allowing for a one-to-one comparison with experiments.

5.2. SIMULATION OF HEAD IMPACT

Several finite element models of the head have been developed by King and Chou [49], Shugar [50], Khalil and Hubbard [51], Hosey and Liu [52], Ward [53], Ruan and co-workers, [54], DiMasi [55], and Bandak [56]. In most of these models the skull was simulated using shell elements. Only Khalil and Hubbard [51], considered the scalp and none considered the sutures. Shugar [50] and Khalil and Hubbard [51], considered multi-layered skull elements to resolve the inner and outer tables and the diploe but using material properties to match single shell properties. Most of the models used approximate skull geometry with none representing individual human data such as realistic skull thickness variations relevant to the study of skull fracture. All of the material models were linear elastic.

In Ward's 3-D model [53], the geometry of the brain was found by tracing various anatomical planes without any assumption of sagittal symmetry. The skull was modelled using 4-node shell elements with thickness close to anatomical values. The material was assumed to be linearly elastic. Rotational and translational accelerations were used as loading inputs to the model. The brain was modelled using 8-node brick elements. To allow relative motion, the effective compressibility of the material was varied through the choice of Poisson's ratio ranging from 0.48 to 0.499 depending on the acceleration pulse duration. The values were calibrated to agree with brain pressure data.

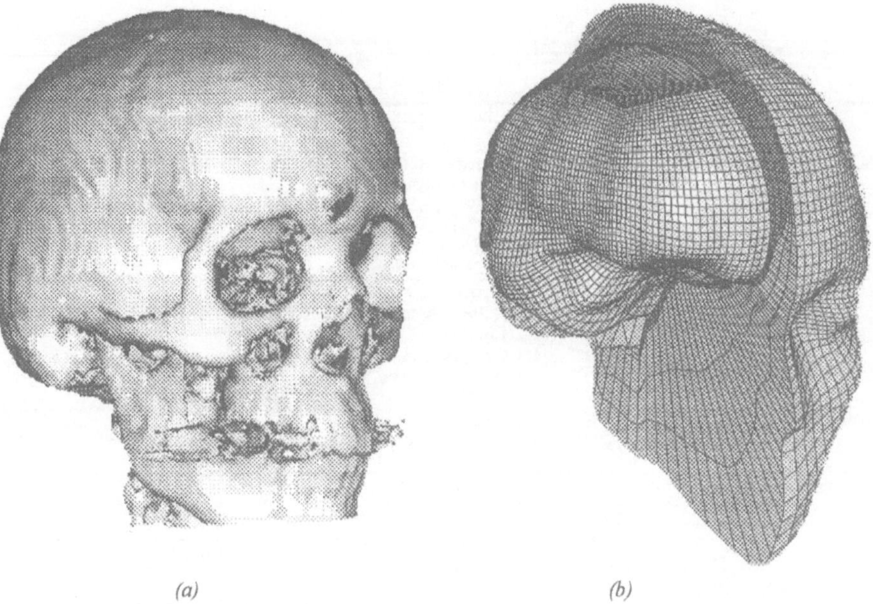

Figure 8: Reconstructed CT image of the skull (b) Imaging-based finite element of the head

Shugar's [50] three dimensional model assumed sagittal symmetry and utilized geometry from a plastic replica of the skull that was sliced in sagittal planes. The skull was resolved into the three layers and was modelled using brick elements not shell elements. The brain was modelled as a homogeneous structure with nearly incompressible properties. The brain and skull were separated by elements representing the CSF in order to allow the brain to move within the skull. Material properties were not given.

Hosey and Liu [52] used the brain geometry of Ward and Thompson [57], skull elements similar to Hardy and Marcal [58], and a structural formulation of the cervical column similar to Pontius [59]. They assumed symmetry across the midsagittal plane. They modelled the skull as a single layer of thin shell elements for frontal, parietal bones and the superior portions of the occipital and temporal bones. The inferior portions of the temporal bones, the jaw, and the brain were modelled using brick elements. The brain was separated from the skull by a single layer of brick elements representing the cerebrospinal fluid (CSF). The brain stem was modelled through the foramen magnum where it was constrained only by the CSF elements. This boundary condition was intended to allow brain motion through the foramen magnum to approximate volume changes produced by skull deformations. In one form or another, most of these models used material constants close to those in Table 2.

TABLE 2: Material constants for the Skull, Dura Mater, Falx Cerebri, and Brain

SKULL			
	Density	**Elastic Modulus**	**Poisson's Ratio**
McElhaney et al [60]	**dry mass** 1425 kg/m^3 (1.35 e-04 lb/in^3)	**Radial Compression** (2.4E+10 dyn/cm^2) 3.5E+05 psi **Tangential Compression** (5.55E+10 dyn/cm^2) 8.1E+05 psi	**Radial Compression** 0.19 **Tangential Compression** 0.22
McElhaney et al [61]		**Tensile** (1.2e+11 to 1.99e+11 dyn/cm^2) 1.8E+06 to 2.9E+06 psi **Compressive** (1.4E+10 dyn/cm^2) 2E+05 psi	
Wood et al [62]		**Tensile** (1.03E+11 to 2.1E+11 dyn/cm^2) 1.5E+06 to 3..2E+06 psi	
Simikin et al [63]		**Bending** (6E+11 dyn/cm^2) 9E+06 psi	
Rose et al [64]	1.94 g/cm^3 (1.8 e-04 lb/in^3)	**Bulk Modulus** 1.6E+11 dyn/cm^2 (2.32E+06 psi)	0.3

DURA/FALX		
	Strain Rate	**Elastic Modulus**
McElhaney et al [60]		6000-8000 psi
McElhaney et al [61]	Tensile	
	0.0666 sec^{-1}	6027 psi
	0.666 sec^{-1}	6430 psi
	6.66 sec^{-1}	8799 psi

BRAIN					
	Density	Elastic Modulus	Bulk Modulus	Shear Modulus	Poisson's Ratio
Ommaya [65]		0.8E+05 dynes/cm^2 to 1.5E+05 dynes/cm^2 (1.2 - 2) psi			
Fallenstein [66]				Dynamic Elastic Modulus G' 6.0E+03 dyn/cm^2 to 11E+03 dyn/cm^2 (0.1 - 0.16) psi Dynamic Loss Modulus G" 3.0E+03 dyn/cm^2 to 6.5E+03 dyn/cm^2 (0.05 - 0.11) psi	
Goldsmith [67]		(7.8e+05 dynes/cm^2) (10.4 psi)	(4.32 dynes/cm^2) 300000 psi		
Rose et al [64]	1.05 g/cm^3 (9.8 e-05 lb/in^3)		2.1E+10 dyn/cm^2 (3.05E+05 psi)		
Firoozbakhsh et al [68]	(1 g/cm^3) 9.38 e-05 lb/in^3			(1.92e+06 dyn/cm^2) 32.5 psi	
Margulies et al [69]	1.06 g/cm^3 (9.9 e-05 lb/in^3)	4.0E+04 Pa (6 psi)	2.0E+09 Pa (2.9E+05 psi)	1.38E+04 Pa (2 psi)	0.5

In the next section we will discuss the use of finite element models with cadaver test data and physical modelling to study skull fracture response.

5.2.1. *Impact Skull Response*

The response of the skull is a fundamental link in the understanding of the effect of contact impact loading on the brain. An evaluation of existing skull impact data was conducted using a spherical physical model and finite element analyses. The physical model was designed to replicate the drop test skull fracture experiments of Hodgson and Thomas [70]. Both the physical model and cadaver experiments were simulated using a finite element model of a sphere, similar to Khalil and Hubbard [51], and an anatomically detailed finite element model of the head developed by Bandak and co-workers [40]. The evaluation was conducted as a part of a study to determine skull fracture response of the head and the effects of surface shape, hardness, and impact location. A brief discussion of this study is given below with the report of the detailed study given by Vander Vorst and co-workers [71].

Spherical Physical Model. Drop tests were conducted to determine the spherical physical model's ability to replicate the dynamic quantities obtained in previous cadaver experiments. The model test apparatus, shown in FIG. 9, was devised and fabricated to record the transient force, acceleration, interfacial pressure, and rebound height for drops of the model onto various impact partners. The head was modelled using an aluminum

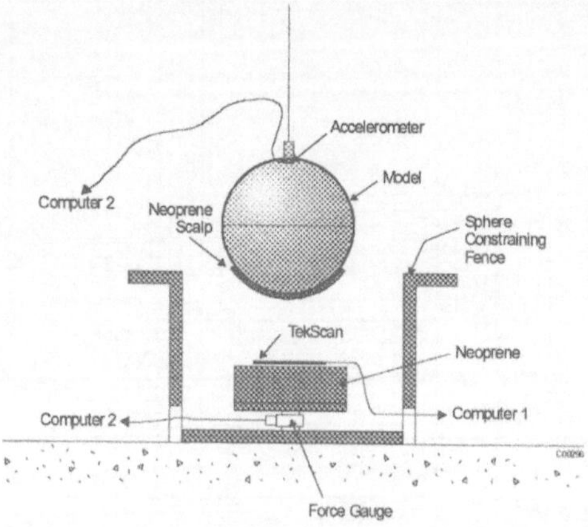

Figure 9: Skull impact physical model test apparatus

sphere that is 0.125 in. thick and has a 7 in. inside diameter. The sphere was filled with a mixture of rice and BB's for the exact model weights. Two layers of 1/8 in. thick neoprene sheets were used to simulate the skin. The impact partners include the following configurations: rigid flat surface, 2 in. thick durometer 60 and durometer 90 flat neoprene sheets, 1 in. radius durometer 60 and durometer 90 neoprene cylinders, and 1/8 in, 5/16 in, ½ in, and 1 in. radius rigid cylinders. At least four drops were made for each case. Acceleration of the impact sphere was measured by a single axis Endevco Model 7250A-2 and Model 7250A-10 Isotron piezoelectric accelerometer. The accelerometers were mounted at the inner surface of the apex of the sphere directly under the mounting stud of the pull string. Dynamic impact force was measured by a PCB Model 203A quartz force ring gauge. The gauge was sandwiched between two 1/2 in. thick aluminum plates. The bottom plate was fixed to prevent movement during impact. The impact partners were mounted on the top plate. The force and acceleration data were recorded at a rate of 50 samples per millisecond using the National Instrument LabVIEW system. Pressure distribution was measured, by the modified TekScan described earlier, for drops onto neoprene targets. The rebound height of the sphere was read from a Hi-8 video recording of the drop test using the frame-by-frame playback feature of the recorder. The measured data was processed to give key parameters, including HIC. The matrix of physical model drop tests, coinciding with Hodgson's experiments, is given in Table 3. Each condition was repeated several times and individual tests that deviated by more than two standard

TABLE 3. Matrix of Cases for Sphere Drop Tests

Impact Partner	Category Name	Sphere Weight	10 in.	12 in.	15 in.	18 in.	20 in.	24 in.	25 in.	30 in.	36 in.	44 in.	48 in.
Flat Rigid	fr	10 lb	✔						✔	✔			
Flat Durometer 90✦	f90	10 lb		✔ ✪	✔			✔ ✪			✔ ✪	✔	✪
Flat Durometer 60✦	f60	10 lb	✔	✔ ✪	✔		✔	✔ ✪		✔	✔ ✪	✔*	✔ ✪
1" Radius Rigid Cylinder	1cr	10 lb	✔		✔	✔	✔			✔			
1" Durometer 90 Cylinder✦	1c90	10 lb		✔	✔			✔			✔		✔
1" Durometer 60 Cylinder✦	1c60	10 lb	✔				✔			✔			✔
1/2" Rigid Cylinder	1/2cr	10 lb			✔								
5/16" Rigid Cylinder	5/16cr	10 lb	✔		✔		✔						
1/8" Rigid Cylinder	1/8cr	10 lb		✔	✔	✔		✔					
3" Radius Rigid Hemisphere	3h4												
5" Radius Hemisphere	5hr												
8" Radius Hemisphere	8h4												
Flat Durometer 90✦	f90.8	8 lb		✔				✔					✔
Flat Durometer 90✦	f90.12	12 lb		✔				✔					✔
✦ TekScan Used													
* 50 inch Drop													
✪ Rebound Observed													

deviations from a linear regression of change of velocity, Δv, during impact vs. drop height for a given impact partner were not used in subsequent statistical analyses. In addition to the Hodgson and Thomas data [70], we included the data of Got and co-workers [72] and McIntosh and co-workers [73] who also investigated the relationship between cadaver skull fracture and HIC. The data of McIntosh included only a small set of data under a limited impact conditions. Our analysis of the published data of Hodgson and Thomas, indicated that their calculated values of HIC may have been based on suspect acceleration data and that the device they used for measuring impact area may have influenced the acceleration. Hodgson's reported impact speeds were consistent with idealized free fall speed calculations, however, the velocity changes calculated from the acceleration curves yielded physically impossible values for some test cases. Additionally, their calculated HIC values for head fracture cases cannot be used in a statistical analysis since HIC values change with fracture. Figure 10a shows the effects of skull fracture on

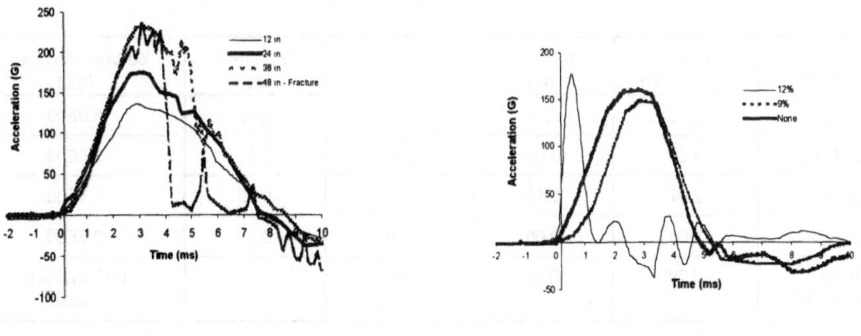

(a) *(b)*

Figure 10: Effects of (a) fracture and (b) MetNet on head impact acceleration measurements.

the measured accelerations. Another influence on the accelerations measured in those experiments is the Met Net between the head and the impact partner (FIG. 10b). The total force for the physical model drops was measured and compared with the total force integrated from the TekScan pressure measurements. Figure 11a compares these measured values with the force time-history reported by Hodgson. The skull fracture data of Hodgson and Thomas supports the conjecture that the occurrence of fracture decreases with increasing impact area.

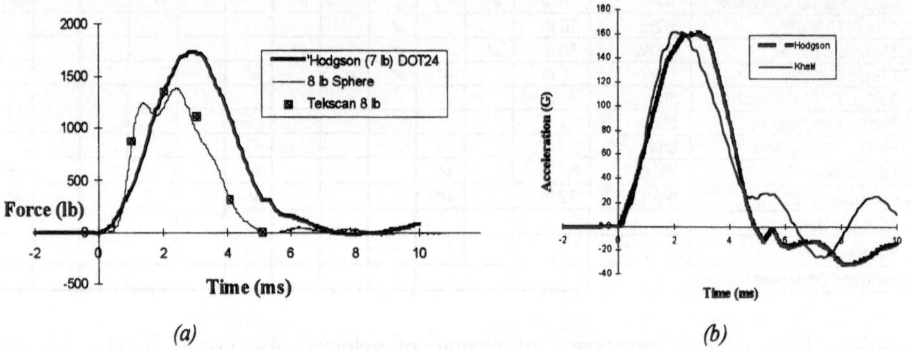

Figure 11: (a) Comparison of forces from TekScan pressure data with force ring measurements and Hodgson's data (b) comparison of finite element spherical model acceleration results with the Hodgson and Thomas data.

The reconstructed dynamics of these tests provide us with a more complete dynamic description of the contact impact that can be used to determine what aspects of the interaction are most important to head dynamics and injury.

Spherical and Anatomical Finite Element Models. A finite element model, approximating the skull as a sphere, similar to the model of Khalil and Hubbard [51], was used to study some of the characteristics of the Hodgson and Thomas experiments. Table 4 describes the parameters for the spherical model. This model simulated the brain as a filled cavity

TABLE 4. Spherical Finite Element Model of the head

Material	Outer Radius (in)	Thickness (in)	Specific Density	Poisson's Ratio	Young's Modulus (psi)
Scalp	3.27	0.24	1.2	0.4	5.0E03
Outer Table	3.03	0.06	3.0	0.35	2.6E06
Diploe	2.97	0.12	1.74	0.05	1.1E05
Inner Table	2.85	0.06	3.0	0.35	2.6E06
Brain Fluid model	2.79	Filled	1.0 weight		Bulk Modulus 3.2E05

making up a layer with a width of 0.3 in. The density was adjusted to give the correct overall head mass. The Shore durometer 60, and durometer 90, neoprene impact partners were modelled using linear elastic materials with specific gravity 1.03, Poisson's ratio, 0.49, and calibrated elasticity of 10,000 psi and 2,000 psi respectively. The elasticity of each neoprene material was determined by calibrating an acceleration curve from a finite element calculations against selected results of Hodgson and Thomas. For tests that gave a nearly elastic rebound, the acceleration traces of Hodgson and Thomas agree with the finite element analyses. (Figure 11b). The spherical model was used to calculate HIC values for several cases, given in Table 5, corresponding to Hodgson and Thomas' tests.

Additional calculations were conducted using an imaging-based anatomical finite element model of the skull (Figure 8a) [40] to investigate the differences between the idealized spherical model and a more geometrically accurate model. The same impact partner material properties were used in both the spherical and anatomical models. HIC and maximum principal strain were calculated using the spherical finite element model. The anatomically based finite element model gives slightly larger strains, which are less evenly distributed, than the spherical model. The maximum principal strain for the anatomical model is 0.20% compared with 0.17% for the spherical model. Figures 12a, b show the distribution of maximum principal strain for the spherical and anatomical models respectively. The case selected for comparison was a 36 inch drop onto a flat, two inch section of durometer 90 neoprene. In this analysis, the threshold value of maximum principal strain for skull fracture is about 0.12%. The maximum principal strain reported is

TABLE 5. Cases for Finite Element Analysis.

Type	Dot Number	Drop Number	Drop Height(in)	Partner Geometry	Partner Type	MetNet
3	26	2	24	flat	duro90	9
3	26	3	36	flat	duro90	9
4	24	1	12	flat	duro90	no
4	24	2	24	flat	duro90	no
6	30	1	24	flat	duro60	9
6	30	2	36	flat	duro60	9
6	30	3	48	flat	duro60	9
7	17	7	30	flat	duro60	no
11	27	1	12	cyl_1	duro90	9
11	27	2	24	cyl_1	duro90	9
11	28	1	24	cyl_1	duro90	9
21	102	0	10	flat	rigid	no

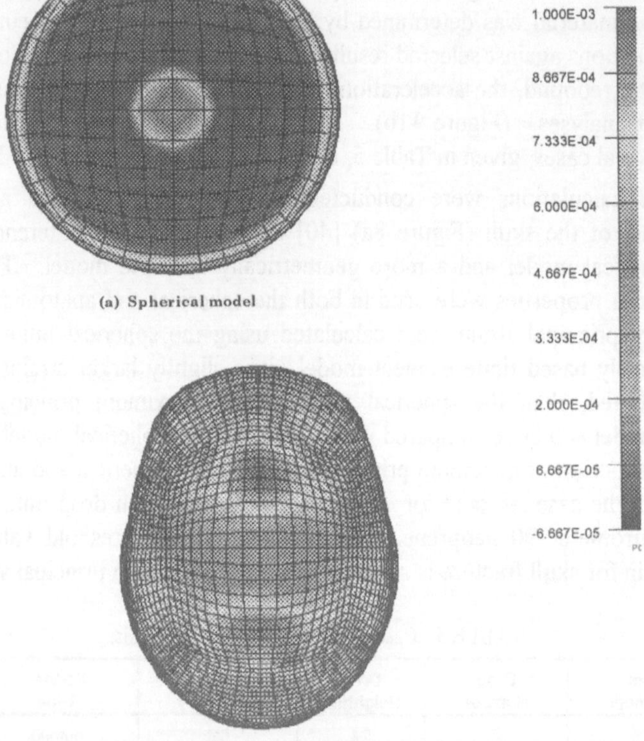

Figure 12: Principal strain distributions for impacts of the spherical and anatomical finite element models

the largest value appearing anywhere in the inner or outer cranial tables. Wood [62] showed that cranial bone fails at a strain level of 0.5% in tension. A possible explanation for this difference is the uniformity of the spherical model of the head as opposed to the human skull.

These results indicate that computational methods can reasonably reproduce the dynamics of contact impact to the head, as measured by forces and accelerations. In addition, the computed mechanical response of the skull leads to stress concentrations that are indicative, and eventually predictive, of skull fracture. More data and finite element model development are needed, of course, to establish a final correlate of head injury that accounts for population variability.

5.2.2. *Impact Fluid Response*
Head impact has a direct effect on the brain even without the presence of skull fracture. As we discussed earlier, the behaviour of the CSF surrounding the brain is an important factor in the determination of loading effects on the brain. The mechanical impact response of the human head, and the consequent strain field in the brain, are affected by the presence of the subarachnoidal CSF. The CSF layer influences load transmission, mitigation, and shear relief during dynamic interactions occurring between the brain and the inner surface of the cranial cavity. It, therefore, substantially affects the pressure gradients occurring in the loaded brain and the relative movement between the brain and the dura mater.

Denny-Brown and Russell [74] observed both positive and negative pressures in the brain as a result of rapid head acceleration. Gross [75] postulated that as the magnitudes of the negative pressures increase cavitation would occur and bubbles would form and subsequently collapse to cause damage the brain. Kopecky and Ripperger [76] showed experimentally and computationally that cavitation can occur as a result of a 150g acceleration. Analysis of the CSF response has, in the past, been limited to fluid-filled shells with the brain taken as a nearly incompressible homogeneous fluid and the skull as a thin spherical or ellipsoidal shell [76-80]. We conducted an experimental and computational study to understand some of the response features of the subarachnoidal CSF in the impact process and to gain insight into modelling the interface between the skull and the brain. Some results are given in the following sections with more detail given in Bandak and co-workers al [81] and Chan and co-workers [82].

Physical Model. A series of tests was conducted using an experimental setup to study the impact response of the subarachnoidal CSF [81] under loading conditions relevant to actual head impact TBI cases. The condition of cavitation under severe impact conditions was analyzed to determine its potential in producing the observed coup and contre-coup injuries. The effects of some of the brain properties influencing CSF flow and brainstem motion out of the foramen magnum region was also studied. Also, the slight density difference between the brain and the CSF was evaluated for its contribution to acceleration-driven relative motion between the brain and the dura. The physical model consisted of a fluid-filled spherical shell instrumented to measure the internal pressure response under various impact loadings (FIG. 13a). The skull part was made of two hollow, Plexiglas hemispheres bolted together at an equatorial flange. The top of the sphere had a 2 in. opening with a fiber-reinforced diaphragm for local impact load application. The inside of the sphere was filled with a balloon that, itself, was filled with a variety of materials. The outside of the balloon was surrounded with water to simulate the CSF. Four pressure gauges were located at polar locations of 45, 135, 180, and 225 deg on the same vertical plane (FIG. 13a). The model was tested under two impact conditions. The first was focal impact on a stationary model through a controlled drop of a weight onto the neoprene diaphragm (FIG. 13c) using a pulley arrangement to control the accuracy of the drop height and the impact point on the diaphragm. The second was a pendulum impact on the model's nearly rigid equatorial flange where the model was suspended to allow freedom to move (FIG. 13b). Both test configurations were confirmed for repeatability. The effects of the flow through the simulated foramen magnum opening and the brain simulant material properties on the pressure were monitored using the drop

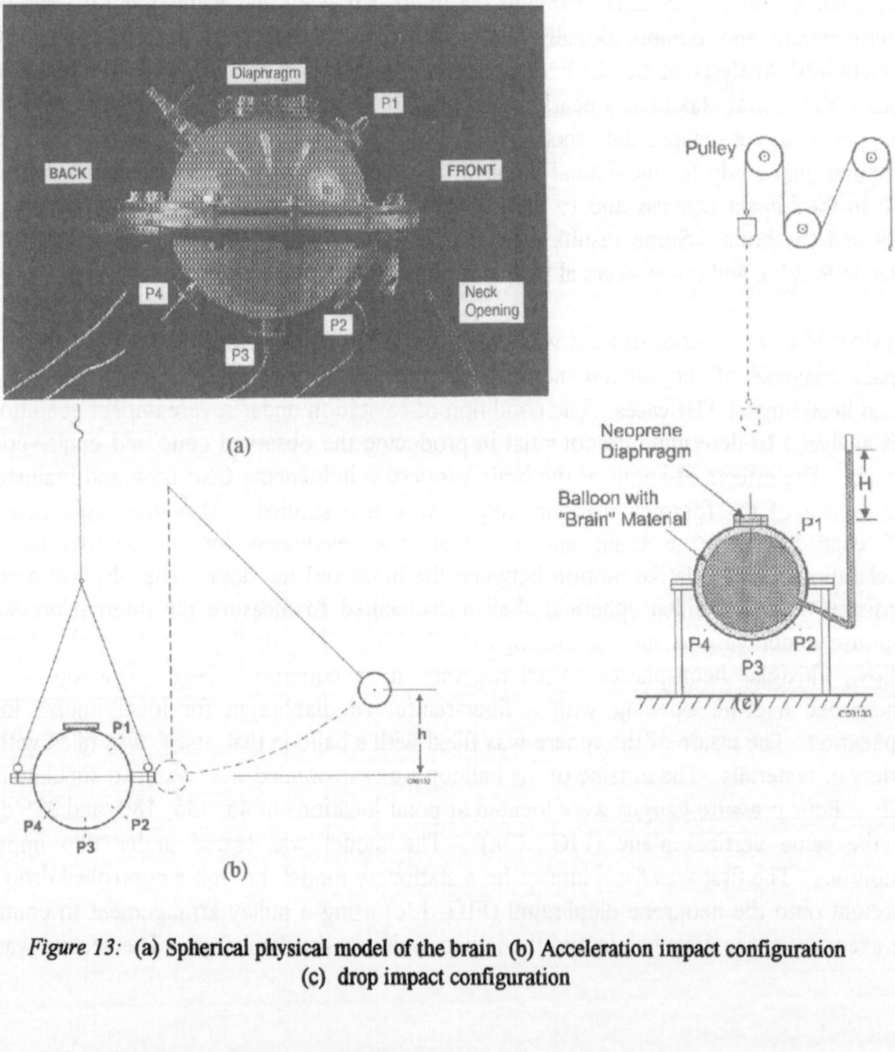

Figure 13: **(a) Spherical physical model of the brain (b) Acceleration impact configuration (c) drop impact configuration**

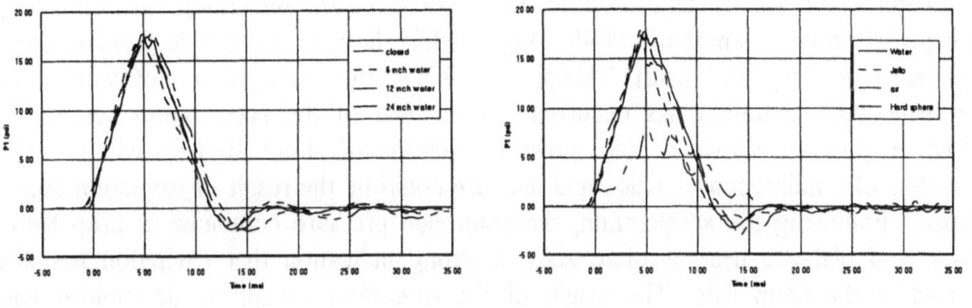

(a) *(b)*

Figure 14: **Effect of (a) fluid exit boundary condition and (b) compressibility on pressure in the brain.**

test configuration. As the exit pressure, through the opening, is reduced, the flow out of the balloon is increased and the internal pressure is relieved (FIG. 14a). Reductions of positive peak pressures of about 10% were observed for the conditions tested. Several brain simulant materials were tested including water (base case), gelatin, and water with an air cavity (a 2.5 inch diameter air balloon resulting in an overall void content of 3.1%) and a solid sphere. The response of the solid and water-filled balloon were nearly identical. The gelatin-filled model showed a faster reduction of pressure following the peak which might indicate a direct interaction between the diaphragm and the balloon. Pressure measurements indicated that the air cavity, though small by volume, has significant effects on the pressure magnitude. Figure 14b shows a 50% reduction in the pressure magnitude with indications of high frequency oscillatory behaviour in the time response. The oscillations are probably due to the compressibility of the air cavity creating an entirely different internal flow pattern where the volume intrusion of the diaphragm was mostly absorbed by the decrease in air cavity volume. All four gauges locations showed similar pressure oscillations that were out of phase with one another, indicating that there is considerable local flow variation. Figure 15 shows the pressure time response to

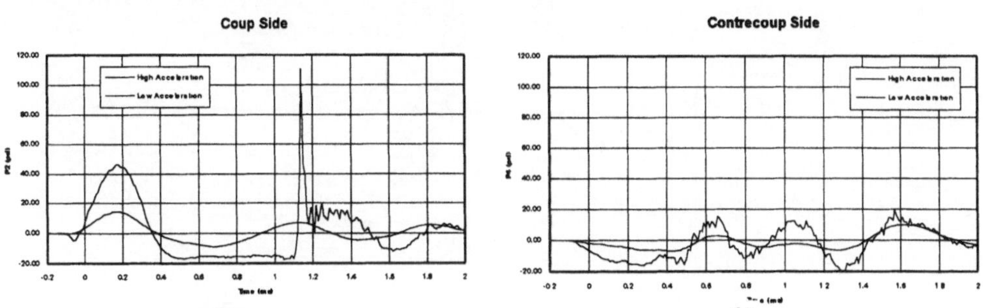

(a) (b)

Figure 15: **Impact coup and contre-coup pressure-time histories for high and low accelerations indicating cavitation at both sides under high acceleration**

acceleration where the acceleration phase lasted for about 0.5 ms during which the coup side experienced an overpressure of 40-50 psig, while the contrecoup side pressure drops to near absolute zero (-14.7 psig). This pressure is near the vapor pressure of water and it is probable that cavitation has occurred. At the end of the acceleration period, the contrecoup pressure shows a rapid pressure increase of about 30 psig within a few hundredths of a millisecond. This rapid increase could be the result of cavitation bubble collapse. Following the acceleration, the coup side pressure continues to drop below ambient and plateaus near absolute zero, a strong indication that cavitation has also occurred at the coup side. The origin of the cavitation cannot be determined with certainty, but it is plausible that it arises from the flow that occurs from the coup to contrecoup as a result of the initial acceleration. The measurements indicate that the cavitation state is maintained for over 0.5 ms. When the flow finally reverses and the bubble collapse takes place, the overpressure spike reaches 100-120 psig. These pressure levels indicate the possibility of significant brain tissue damage if they occur in the brain.

The violent fluid behaviour observed in these tests is related to the magnitude of the acceleration impact. At low acceleration, the pressures oscillate in time and are approximately out of phase, as would be expected, on the two sides. At no time did the pressures approach absolute zero. When the acceleration produces cavitation, the character of the pressure variation changes dramatically and the cavitation collapse spikes are clearly visible. These results suggest that there is a critical acceleration at which cavitation and its subsequent damage appear.

Computational Model. Computations were conducted to simulate the physical model experiments described in the previous section. A code that solves the full transient by an 8

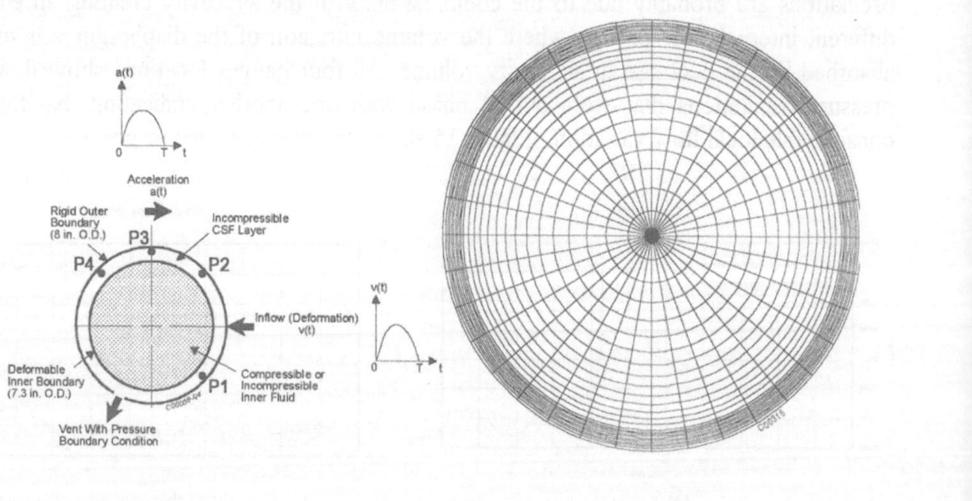

(a) (b)

Figure 16: Computational model schematic (a) and mesh (b) for the spherical brain model in. diameter rigid outer cylinder, and the brain is located concentrically inside with 7.3 in. diameter analogous to the experiment. A layer of fluid, simulating the CSF, is placed between the skull and the brain. The outer cylinder has a neck opening with a pressure boundary condition. The brain and the CSF are modelled as two distinct fluids with a sharp interface.

The model was used to study the effects of brain motion resulting from density differences between the brain and the CSF under a translational acceleration loading. Even though researchers agree that the brain and the CSF have about the same density as water, the exact values are still not well defined. It is somewhat accepted that the brain is slightly heavier than the CSF [76]. For our calculation, the brain was taken to be 15% heavier than the CSF [76], and the CSF was taken to be water. The kinematic viscosity of the brain was estimated to be 0.009 m^2/s [83], which is 10,000 times that of water. The model was loaded using a sinusoidal acceleration pulse with a nominal amplitude of 150g and a duration of 4 ms. A baseline case was first calculated with the brain and the CSF having the same water density and viscosity and subject to the nominal acceleration loading. No brain movement was observed. With the brain to CSF density ratio being 1, the coup and contrecoup pressures are symmetric about the brain mid-plane, indicating that the linear pressure gradient. The pressure difference across the brain increases with acceleration and reaches the maximum peak values of ±21.5 psi at 2 ms. The contrecoup pressure actually drops below the vapor pressure at 0.9 ms, and cavitation (if allowed) should occur. The base calculation was repeated by setting the brain density 15% heavier than the CSF, and the kinematic viscosity of 0.009 m^2/s was prescribed for the brain simulant. The acceleration causes the brain to move forward and squeeze the CSF fluid backward with flow velocity reaching 1.6 ft/s. The 15% heavier brain increases the peak coup pressure by 24% to 26 psig when compared to the equal density base case. However, the rearward tension caused by the brain's forward motion does not proportionally reduce the contrecoup minimum pressure that only drops by 7% to -23 psig at 2 ms. This uneven pressure change may be partially due to the CSF flow effect. Since the brain moves forward, the CSF has to flow backward, but it must also stagnate at the contrecoup point and may prevent the pressure to drop proportionally to the coup pressure increase. Overall, the heavier brain increases the peak pressure gradient across the brain by about 16%. The previous calculation was repeated with a 2 ms acceleration pulse, with the peak remaining at 150 g. Other than the pulse width being shorter, the asymmetric coup-contre-coup pressure pattern is similar to that of the long pulse. Both the short and long pulse results show slight movement of the brain after the end of the acceleration, as indicated by the pressure oscillations.

Another set of calculations were performed to simulate the equatorial pendulum impact tests described above. A case with a pulse duration of about 1 ms was chosen where the water was used for both the brain and the CSF. The structural deformation of the skull was approximated as a velocity inflow condition at the skull boundary over a 20 deg segment centered at the impact site. The inflow pulse follows the sinusoidal acceleration

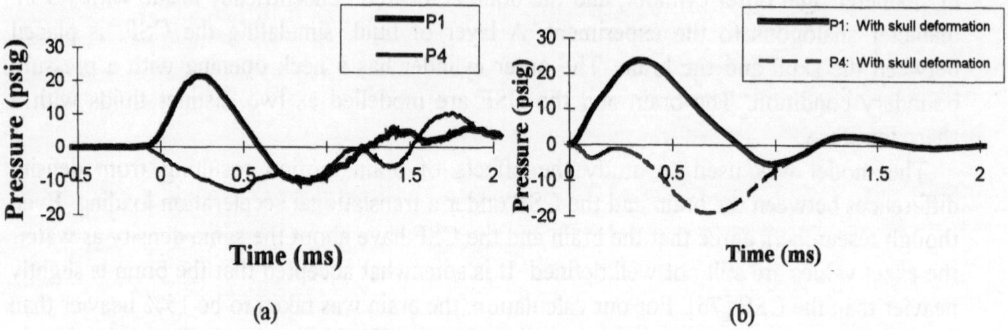

Figure 17: (a) Physical model coup and contre-coup pressure time histories (b) computed values allowing for skull deformations

pulse, with a peak value of -1 ft/s. This corresponds to a structural deformation of about 0.006 in. over 1 ms. As in the experiment, CSF pressures at four locations were monitored with P1 and P2 on the coup side, P3 on the mid-plane, and P4 on the contrecoup side (FIG. 16a). The calculated coup-contrecoup pressure asymmetry between P1 and P4 is confirmed by the data and is due to the structural deformation effect. The deformation of the impact region causes the coup pressure at P1 to rise faster than the contrecoup pressure drop at P4. Toward the end of the pulse at about 0.8 s, the structural snap back of the impact site creates a fluid tension, causing P1 to drop below zero. The calculated results with skull deformation show good qualitative agreement with the data (FIG. 17). The calculated peak values and timing at P1 and P4 agree fairly well with the measurements, even though the calculation was based on a planar two-dimensional model. The recorded oscillations beyond 1 ms probably came from the Plexiglas structural ringing excited by the loading and was not modelled. The favorable data comparison suggests that a realistic loading should include both acceleration and structural deformation, and a fluid-structure coupling at the CSF-skull boundary should be developed for future fluid dynamics modelling.

5.2.3. *Modelling of Strain Damage Response*

A three dimensional model approximating the human brain was used to investigate the development of the Cumulative Strain Damage Measure (CSDM) under a range of kinematical conditions [56]. The model was subjected to non-contact impulsive loadings to study the relative damaging potential for various combinations of rotational and translational accelerations.

Time dependent angular velocities were applied to the model's rigid skull about two axes of rotation. One axis passes through the center of gravity of the head model and the other passes through a point 30.48 cm (12 inches) below its center of gravity. Anterior-posterior and medial-lateral rotations were imposed about each of the two axes. The types of induced motions were chosen to produce various levels of rotational accelerations and to represent an upper bound on the combined translational and rotational acceleration conditions that might be expected under automotive crash restraint system forces. They were also chosen to isolate deformation modes that may be responsible for inducing the damaging strains associated with diffuse axonal injury.

The evolution of the CSDM value for a case of coronal rotations with peak angular velocity of 35 rad/sec and duration of 0.03 seconds, about a point 30.48 cm (12 inches) below the head's center of gravity was evaluated. The accumulation of strain was observed to occur along the periphery of the brain with concentrations near the surface of the falx structure. It was also observed that the lateral deformation of the falx structure contributes to the rotational motion of the brain. In fact it is the brain that is resisting the rotational motion of the falx as the brain lags behind the velocity imposed on the rest of the head. The difference in compressibility of the dura mater relative to the brain allows for the deformations that produce a slight volume change in the cranial cavity. The main accumulation of strain occurs at the outer radii and is more pronounced in the regions where the translational acceleration provides a tensile component combined with the rotational shear to release the brain-dura connection condition under the threshold prescribed. The same loading applied to the head in the anterior-posterior (A-P) direction produces CSDM vs. time values that rise much faster than the coronal case because of the more pronounced rotational motions. This is to be expected for this model since the A-P rotational motions are not constrained by the falx structure as the coronal rotations are. These motions are not as severe for the midbrain region. There is no indication that the artificial fixity in the model at the midbrain region below the corpus callosum is inducing excessive shear strain and thus contributing to the volume damage growth. These rotations may induce, in actual cases, tensile loadings on the brain stem causing it to move in and out of the foramen magnum. An observation common to all loading cases in this study (36 cases) is the increase in the CSDM value with increasing angular velocity. There is also a tendency for the CSDM to increase with decreasing duration although that is not the case for the non-centroidal A-P rotation at the 35 rad/sec amplitude of the angular velocity. There is little change in the CSDM values between the centroidal and the non-centroidal cases possessing the same magnitude and duration. This is the case even though the resultant accelerations for two such cases are different.

The CSDM was compared with the head injury criterion. Preliminary results indicate that resultant acceleration alone may not be sensitive enough for detecting this type brain injury. The centroidal rotation cases resulted in one extreme, that is significant strain damage for low or non-existent HIC value, while the translational cases showed little strain damage for very high values of HIC. The combined rotational and translational cases resulted in damage values similar to those for rotation only providing further support to the conclusions of Holbourn [22] and Ommaya [84].

The analysis above represented gave approximate results indicating the need for an anatomically more detailed model of the brain. Figure 18a shows the cerebrum, cerebellum and brain stem parts of a model under development for this purpose. This model along with verification of the CSDM against experimental data will serve as a tool to study strain damage in the brain.

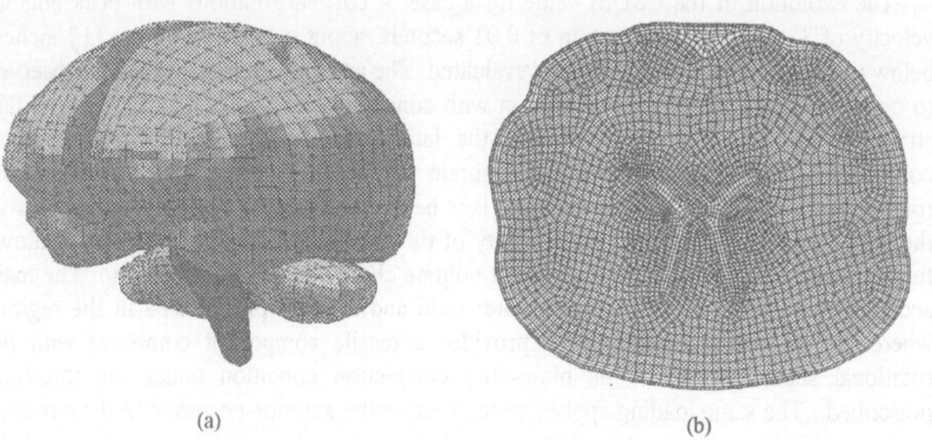

Figure 18: (a) Three dimensional finite element model of the human brain and (b) two dimensional finite element model of the mid-coronal slice of the minipig brain

An initial evaluation of the CSDM against existing experimental data on diffuse axonal injury has been conducted using a two-dimensional plane strain finite element model of a mid-coronal slice of the minipig head. The cross sectional geometry of the slice was obtained from photographs of actual test subjects [85] and from anatomical data. Figure 18b shows the model representing the brain, ventricular structure, cerebrospinal fluid simulant, and the sulci. The material properties were taken to be those typically used for

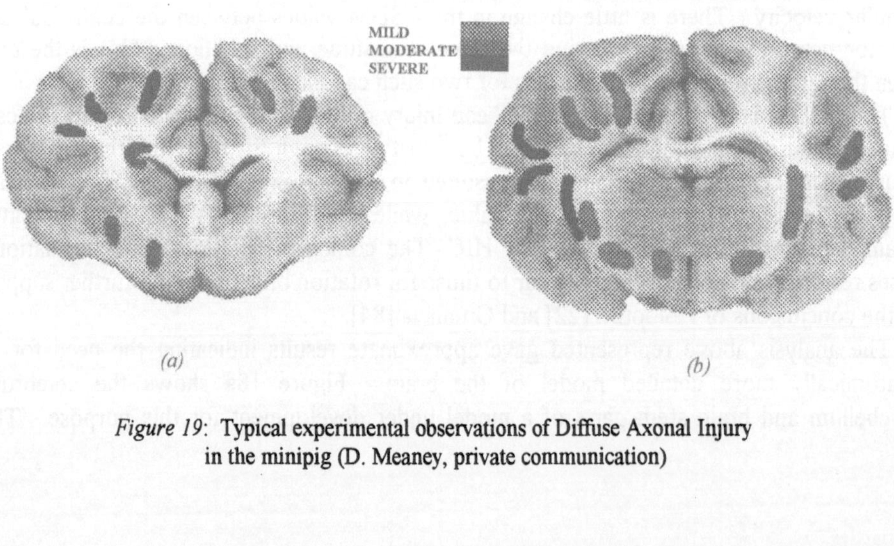

Figure 19: Typical experimental observations of Diffuse Axonal Injury in the minipig (D. Meaney, private communication)

the human brain. Results from the simulations indicate that that the localization of the strain damage as measured by the CSDM value occurs in regions where DAI is observed experimentally. Figure 19 shows a typical distribution of DAI lesions occurring at two coronal locations of the minipig brain. Several experimental cases were simulated giving early indications that the CSDM values are consistent with observed DAI. A more comprehensive analysis of these experimental cases using a three dimensional version of the 2D minipig model will provide a link between computational analysis and observed injury. Such models will help identify and develop measures of TBI that account for other types of injuries occurring under head impact.

6. Acknowledgments

The author wishes to thank M. Vander Vorst, J. Yu, P. Chan, J. Stuhmiller, A. Alsuezy, A. Sherif, and F. DiMasi for their contributions. Thanks are also extended to R. Eppinger, A. Ommaya and D. Meaney for valuable discussions. This work was done under the auspices of the National Highway Traffic Safety Administration's Associate Administration for Research and Development.

7. References

1. Waxweiler, R.J., Thurman, D., Sniezek, J., Sosin, D., and O'Neil, J. (1996) Monitoring the Impact of Traumatic Brain Injury: A Review and Update, in F.A. Bandak, R.H. Eppinger, and A.K. Ommaya (Eds.), *Traumatic Brain Injury: Bioscience and Mechanics*, Mary Ann Liebert, Inc.: Larchmont, NY. pp. 1-8.

2. Max, W., Mackenzie, E.J., and Rice, D.P. (1991) Costs and consequences, *J. Head Trauma Rehabilitation* **6**(2): pp. 76-91.

3. Jennett, B. and Teasdale, G. (1974) Assessment of coma and impaired consciousness, *Lancet* **2**: pp. 81-84.

4. Kraus, J. and Nourjah, P. (1988) The epidemiology of mild uncomplicated brain injury, *Journal of Trauma* **28**: pp. 1637-1643.

5. Ommaya, A.K. (1996) Head Injury Mechanisms and Concept of Preventive Management: A Review and Critical Synthesis, in F.A. Bandak, R.H. Eppinger, and A.K. Ommaya (Eds.), *Traumatic Brain Injury: Bioscience and Mechanics*, Mary Ann Liebert, Inc.: Larchmont, NY. pp. 19-38.

6. Schnieder, R.C. (1973) *Head and Neck Injuries in Football. Mechanisms, Treatment, and Prevention.* Williams and Wilkins, Baltimore.

7. Reilly, P.L., Graham, D.I., and Adams, J.H. (1975) Patients with Head Injury Who Talk and Die, *Lancet* : pp. 375-377.

8. Cantu, R.C. and Mueller, F. (1990) Catastrophic Spine Injury in Football 1977-1989, *J. Spinal Dis.*

9. Gerberich, S.G., Priest, J.D., and Boen, J.R. (1983) Concussion incidence and severity in secondary school varsity football players, *Am. J. Public Health* **73**: pp. 1370-1375.

10. Cantu, R.C. (1991) Minor head injuries in sports, *Adolescent Medicine: State of the Art Reviews* **2**.

11. Agur A. M. R. and Lee M. J., (1995) *Grant's Atlas of Anatomy*, Williams and Wilkins, Baltimore.

12. Akesson, E. J., Loeb, J. A., and Wilson-Pavwel L. (1990) *Thompson's Core Textbook of Anatomy*, J. B. Lippincott, Philadelphia.

13. Tortora, G. J. (1993) *Principles of Human Anatomy*, Harper and Row, New York.

14. Hall-Craggs, E. C. B. (1995) *Anatomy as a Basis for Clinical Medicine*, Williams and Wilkins, Baltimore.

15. McMinn R. M. H. and Hutchings R. T. (1984) *Color Atlas of Human Anatomy*, Year Book Medical Publishers.

16. Carter L., Spetzler R. F., and Hamilton M. G. (1994) *Neurovascular Surgery*, McGraw Hill.

17. Blinkov, S.M. and Glezer, I.I. (1968) *The Human Brain in Figures and Tables*. Plenum Press. New York.

18. Pfefferbaum A., Mathalon, D., Sullivan, E., Rawles, M., Zipursky. R., and Lim, K. (1994) A Quantitative Magnetic Resonance Imaging Study of Changes in Brain Morphology From Infancy to Late Adulthood, *Archives of Neurology*, **51**, 9, 874-887.

19. Gennarelli, T.A., Thibault, L.E., Adams, J.H., Graham, D.I., Thompson, C.J., and Marcincin, R.P. (1982) Diffuse Axonal Injury and Traumatic Coma in the Primate, *Annals of Neurology* **12**(6): pp. 564-574.

20. Gennarelli, T.A., Thibault, L.E., Tomei, G., Wiser, R., Graham, D., and Adams, E. (1987) Directional Dependence of Axonal Brain Injury due to Centroidal and Non-Centroidal Acceleration, in S.H. Backaitis (Ed.), *Biomechanics of Impact Injury and Tolerances of the Head-Neck Complex*, Society of Automotive Engineers, Inc. pp. 595-599.

21. Ross, D.T., Meaney, D.F., Sabol, M.K., Smith, D.H., and Gennarelli, T.A. (1994) Distribution of Forebrain Diffuse Axonal Injury Following Inertial Closed Head Injury in Miniature Swine, *Experimental Neurology* **126**(2): pp. 291-9.

22. Holbourn, A.S. (1943) Mechanics of head injuries, *Lancet* **2**: pp. 438-441.

23. Ommaya, A.K., Hirsch, A.E., and Flamm, E.S. (1966) Cerebral concussion in the monkey: An experimental model, *Science* **153**: pp. 211-212.

24. Ommaya, A.K., Fass, F., and Yarnell, A.R. (1968) Whiplash injury and brain damage: An experimental study, *Journal of American Medical Association* **204**: pp. 285-289.

25. Ommaya, A.K. and Corrao, P. (1969) Pathologic biomechanics of central nervous system injury in head impact and whiplash trauma, in K.M. Brinkhous (Ed.), *Accident Pathology*, US Government Printing Office.

26. Maxwell, W.L., Irvine, A., Adams, J.H., Graham, D.I., and Gennarelli, T.A. (1988) Response of cerebral microvasculature to brain injury, *J. Pathology* **155**: pp. 327-335.

27. Blumbergs, P.C., Scott, G., Manavis, J., Wainwright, H., Simpson, D.A., and McLean, A.J. (1996) Topography of Axonal Injury by Amyloid Precursor Protein an dthe Sector Scoring Method in Mild and Severe Closed Head Injury, in F.A. Bandak, R.H. Eppinger, and A.K. Ommaya (Eds.), *Traumatic Brain Injury: Bioscience and Mechanics*, Mary Ann Liebert, Inc. pp. 61-68.

28. Zauner, A. and Bullock, R. (1996) The Role of Excitatory Amino Acids in Severe Trauma: Opportunities for Therapy: A Review, in F.A. Bandak, R.H. Eppinger, and A.K. Ommaya (Eds.), *Traumatic Brain Injury: Bioscience and Mechanics*, Mary Ann Liebert, Inc. pp. 97-104.

29. Foulkes, M., Eisenberg, H.M., and Jane, J.A. (1991) The traumatic coma databank: design methods, and baseline characteristics, *Journal of Neurosurgery* **75(Suppl.)**: pp. s8-s13.

30. Jamieson, K.G. and Yelland, J.D.N. (1968) Extradural hematoma: report of 167 cases, *Journal of Neurosurgery* **29**: pp. 13-23.

31. Jennett, B., Teasdale, G., and Galbraith, S. (1977) Severe head injuries in three countries, *J. Neurol. Neurosurg. Psych.* **40**: pp. 291-298.

32. Lowenheilm, P. (1974) Dynamic properties of the parasagittal veins, *Z. Rechtsmedizin* **74**: pp. 55-62.

33. Lindenberg, R., and Freytag, E.(1960) A mechanism of Cerebral Contusions: a pathalogical-Anatomic Study, Arch. Pathology, **69**, pp. 440-469.

34. Gadd, C.W. (1961) *Criteria for Injury Potential*, National Research Council Publication No. 977 (pp.141-144). National Academy of Sciences.

35. Gadd, C.W. (1966) Use of wieghted impulse criterion for estimating injury hazard. in the Proceedings of the *10th Stapp Car Crash Conference*. pp. 164-174.

36. Melvin, J.W., Lighthall, J.W., and Ueno, K. (1993) Brain Injury Biomechanics, in A.M. Nahum and J.W. Melvin (Eds.), *Accidental Injury: Biomechanics and Prevention*, Springer-Verlag: New York. pp. 268-291.

37. Lissner, H. R., Lebow, M. Evans, F. G. (1960) Experimental Studies on the Relation Between Acceleration and Intracranial Pressure Changes in Man, *Surg. Gynecol. Obstet.*, **111**, pp. 329 338. et al (1960)

38. Prasad, P., and Mertz, H. J. (1985) The Position of the United States Delegation to the ISO Working Group 6 on the Use of HIC in Automotive Environment, SAE Paper No. 851246.

39. Stalnaker, R.L., Low, T.C., and Lin, A.C. (1987) Translational injury criteria and its correlation with head injury in the sub-human primate. in *Internaltional Conf. on Biomecahnics of Impact*. Birmingham. pp. 223-238.

40. Bandak, F.A., Vander Vorst, M.J., Stuhmiller, L.M., Mlakar, P.F., Chilton, W.E., and Stuhmiller, J.H. (1996a) An Imaging-Based Computational and Experimental Study of Skull Fracture: Finite Element Model Development, *Journal of Neurotrauma* **12**(4): pp. 679-687.

41. Hallquist, J. (1994) *LSDYNA3D Theoretical Manual,* Report 1018, revision 3. Livermore Software Technology Corp.

42. Padgaonkar, A.J., Kreiger, K.W., and King, A.I. (1975) Measurement of Angular Acceleration of a Rigid Body Using Linear Accelerometers, *J. of Applied Mechanics*.

43. DiMasi, F.P. (1995) *Transformation of Nine-Accelerometer Package (NAP) Data for Replicating Headpart Kinematics and Dynamic Loading,* DOT HS 808-282. US/DOT/Volpe Center.

44. Yu, J. H.-Y. , Ho, K. H.-H. , and Stuhmiller, J. H. (1995) Evaluation of a Device for Measuring Load Distribution During Impact, JAYCOR Report J95-2837-00/018.

45. Vander Vorst, M. J., Chilton, W. E., Stuhmiller, J. H., and Stuhmiller, L. M. (1995) Automated Finite Element Modeling of the Skull, JAYCOR Report J93-2837-00/019.

46. Nielsen, E.A. (1991) *A-D-F FORMAT, Interfacing to ACR-NEMA,* Department of Radiation Medicine, Loma Linda University School of Medicine. Loma Linda, CA.

47. Billet, F.P.J., Schmitt, W.G.H., and Gay, B. (1992) Computed tomography in traumatology with special regard to the advanced of three-dimensional display, *Archives of Orthopaedic and Trauma Surgery* **111**: pp. 131-137.

48. Stytz, M.R., Frieder, G., and Frieder, O. (1991) *Three-Dimensional Medical Imaging: Algorithms and Computer Systems, ACM Computing Surveys,* Association for Computing Machinery. New York, NY.

49. King, A.I. and Chou, C.C. (1976) Mathematical Modelling, Simulation and Experimental Testing of Biomechanical System Crash Response, *Journal of Biomechanics* **9**: pp. 301-317.

50. Shugar, T.A. (1977) *A Finite Element Head Injury Model,* DOT HS 803-211. NTIS.

51. Khalil, T.B. and Hubbard, R.P. (1977) Parametric Study of Head Response by Finite Element Modelling, *Journal of Biomechanics* **10**: pp. 119-132.

52. Hosey, R.R. and Liu, Y.K. (1982) *A Homeomorphic Finite Element Model of the Human Head and Neck in Finite Elements in Biomechanics*, ed. R.H. Gallagher, *et al.* John Wiley & Sons, Ltd. Chichester.

53. Ward, C.C. (1983) Finite Element Modelling of the Head and Neck in Impact Injury of the Head and Spine, in C.L. Ewing, *et al.* (Eds.), *Impact Injury of the Head and Spine*, Charles C. Thomas Publishers: Springfield, IL. pp. 421-474.

54. Ruan, J.S., Khalil, T.B., and King, A.I. (1991) Human Head Dynamic Response to Side Impact by Finite Element Modelling, *Journal of Biomechanical Engineering* **113**: pp. 276-283.

55. DiMasi, F.P., Eppinger, R.H., Gabler, H.C., and Marcus, J. (1991) Simulated Head Impacts with Upper Interior Structures Using Rigid and Anatomic Brain Models, in R. Strombotne (Ed.), *Auto and Traffic Safety*.

56. Bandak, F.A. and Eppinger, R.H. (1994) A three dimensional Finite Element Analysis of the Human Brain Under Combined Rotational and Translational Accelerations, in the Proceedings of the *38th Stapp Car Crash Conference*. pp. 145-163.

57. Ward, C.C. and Thompson, R.B. (1975) The development of a detailed finite element brain model. in the Proceedings of the *19th Stapp Car Crash Conference*. Society of Automotive Engineers, Inc. pp. 641-674.

58. Hardy, C.H. and Marcal, P.V. (1971) *Elastic Analysis of a Skull*, ONR Report No. N00014-0007/8.

59. Pontius, U.R. (1975) The Effects of the Cervical Neuromusculature on the Dynamics of Whiplash, Tulane University: New Orleans, LA.

60. McElhaney, J.H., Fogle, J.L., Melvin, J.W., Hanes, R.R., Roberts, V.L., and Alem, N.M. (1970) Mechanical Properties of Cranial Bone, *Journal of Biomechanics* **3**: pp. 495-511.

61. McElhaney, J.H., Melvin, J.W., Roberts, V.L., and Portnoy, H.D. (1973) Dynamic characteristics of the tissues of the head, in R.M. Kenedi (Ed.), *Perspectives in Biomedical Engineering*. pp. 215-222.

62. Wood, J.L. (1971) Dynamic Response of Human Cranial Bone, *Journal of Biomechanics* **4**: pp. 1-12.

63. Simikin, A. and Robin, G. (1973) The Mechanical Testing of Bone in Bending, *Journal of Biomechanics*, **9**.

64. Rose, J.L., Gordon, S.L., and Moskowitz, G. (1974) Dynamic Photoelastic Model Analysis of Impact to the Human Skull, *Journal of Biomechanics* 7: pp. 193-199.

65. Ommaya, A.K. (1967) Mechanical Properties of Tissues of the Nervous System, *Journal of Biomechanics* **1**: pp. 127-138.

66. Fallenstein, G.T. and Melvin, J.W. (1969) Dynamic Mechanical Properties of Human Brain Tissues, *Journal of Biomechanics* **2**: pp. 217-226.

67. Goldsmith, W. (1972) Biomechanics of Head Injury, in Y.C. Fun g, N. Perrone, and M. Anliker (Eds.), *Biomechanics: It's Foundation and Objectives*, Prentice Hall, Inc.: Englewood Cliffs, N.J.

68. Firoozbakhsh, K.K. and Desilva, C.N. (1975) A Model Of Brain Shear under Impulsive Torsional Loads, *Journal of Biomechanics* **8**: pp. 65-73.

69. Margulies, S.S. and Thibault, L.E. (1989) An Analytical Model of Traumatic Diffuse Brain Injury, *Journal of Biomechanical Engineering* **111**: pp. 241-249.

70. Hodgson, V.R. and Thomas, L.M. (1973) *Breaking Strength of the Human Skull vs. Impact Surface Curvature*, DOT HS 801-002. National Technical Information Services.

71. Vander Vorst, M. J., Yu, J. H.-Y., and Stuhmiller, J. H. (1996) Analysis of Skull Fracture Data, JAYCOR Report J96-2997-02/001.

72. Got, C., Patel, A., Fayon, A. Tarriere, C. and Walfisch, G. (1984) Results of Head Impacts on Cadavers--The Various Data Obtained and their Relationships to Some Physical Parameters, Proceedings of the *22nd STAPP Car Crash Conference*, pp. 57-99.

73. McIntosh, A.S., Kallieris, D. Mattern, R. and Miltner, E. (1993) Head and Neck Injury Resulting from Low Velocity Direct Impact, Proceedings of the *37th STAPP Car Crash Conference*, pp. 43-57.

74. Denny-Brown, D. and Russell, W.R. (1941) Experimental Cerebral Concussion, *Brain* **64**(9): pp. 93-164.

75. Gross, A.G. (1958) A New Theory on the Dynamics of Concussion and Brain Injury, *Journal of Neurosurgery* **15**: pp. 548-561.

76. Kopecky, J.A. and Ripperger, E.A. (1969) Closed Brain Injuries: An Engineering Analysis, *Journal of Biomechanics* **2**: pp. 29-34.

77. Engin, A.E. and Liu, Y.K. (1970) Axisymmetric Response of a Fluid-Filled Spherical Shell in Free Vibrations, *Journal of Biomechanics* **3**: pp. 11-12.

78. Merchant, H.C. and Crispino, A.J. (1974) A Dynamic Analysis of An Elastic Model of the Human Head, *Journal of Biomechanics* **7**: pp. 295-301.

79. Chan, H.S. and Liu, Y.K. (1974) The Asymmetric Response of a Fluid-Filled Spherical Shell-A Mathematical Simulation of a Glancing Blow to the Head, *Journal of Biomechanics* **7**: pp. 43-59.

80. Talhouni, O. and Dimaggino, F. (1975) Dynamic Response of a Fluid-Filled Spheroidal Shell-An Improved Model for Studying Head Injury, *Journal of Biomechanics* **8**: pp. 219-228.

81. Bandak, F.A., Yu, J.H.Y., Chan, P.C., and Stuhmiller, J.H. (1996c) Influence of the subarachnoidal cerebrospinal fluid on the impact response of the brain I: An experimental and computational study, *In preparation* .

82. Chan, P.C., Yu, J.H.Y., Stuhmiller, J.H., and Bandak, F.A. (1996) Influence of the subarachnoidal cerebrospinal fluid on the impact response of the brain II: An experimental and computational study, *In preparation* .

83. Chan, R.K.-C., Masiello, P.J., and Srikantiah, G.S. (1986) PORTHOS - A computer code for solving general three-dimensional, time-dependent two-fluid equations. in *ASME 86-WA/NE-3, Winter Annual Meeting*. Anaheim, CA.

84. Ljung, C. (1975) A Model for Brain Deformation Due to Rotation of the Skull, *Journal of Biomechanics* **8**: pp. 263-274.

85. Ommaya, A.K., Thibault, L.E., and Bandak, F.A. (1994) Mechanisms of impact head injury, *Int. J. Impact Engineering* **15**(4): pp. 535-560.

86. Meaney, D., private communication.

FORENSIC ANALYSIS AND DATA FOR ROAD USERS

IAN R. HILL
Consultant in Forensic Medicine
Department of Forensic Medicine
United Medical and Dental Schools of Guy's and St Thomas's Hospitals
Guy's Hospital, London SE1 9RT, U.K.

1. Introduction

Improving transport safety has not been easy. Considerable effort was needed to overcome prejudice and entrenched views before the existing achievements could be made. Thus despite the evidence of marked reductions in facilities and serious injuries in car crashes, it took eleven attempts to get seat belt laws through the British Parliament [1]. Many more years were to pass before the legislation was extended to the rear seats of cars, and bus and coach passengers are largely unprotected by seat belts. Many spurious arguments have been used with great facility by the antagonists, ranging from infringements of civil liberties to erroneous claims of dangers, to counter evidence of the benefits achievable by safety measures. These arguments must be remembered and considered in the context of their first use, for derivatives will be deployed in the future.

Although there have been great improvements in safety design the toll of accidents is still unacceptably high. However, potent reductions may not necessarily be achieved by safety design alone, for it is to accident prevention that we must look for the greatest benefits. Nevertheless there is still room for better design, but because its potential for improvement is not so great as has hitherto been the case, raw accident data alone will not be sufficient to prove the fact. Diminishing returns and economic complaints about what is perceived as the imposition of "state of the art" safety design will mitigate against claimed advantages [2, 3]. Consequently the data used will have to be comprehensive and cautiously interpreted if it is to find favour. Biomechanics has to take into account the pathophysiological consequences of a physical effect, if its reactions are to be understood. This may mean that new tools have to be used if we are to understand how accidents cause their effects upon the body and how they can be overcome. This paper discusses the background to this idea and places it in the context of current arguments for a greater understanding of trauma.

J. A. C. Ambrósio et al. (eds.),
Crashworthiness of Transportation Systems: Structural Impact and Occupant Protection, 95–111.

2. Characteristics of Injuries

Injuries occur when stimuli exceed bodily resistance, causing reactions which may range from altered physiology to structural damage and their various biochemical reactions. For the most part, the primary insult in transport accidents is mechanical trauma. Fires do, of course, occur. Cold and exposure may also be factors which have to be considered. Apart from fuel, chemical spillages are not commonplace.

Whilst mechanical impulses are the initiators of injury in transport accidents, and the immediate effects are localised to the region involved, they are not isolated. Much has been said and written about the transmission of forces within the body. The arguments will not be rehearsed. Suffice it to say that injury may be remote from the point of contact. This will, like the primary insult, be associated with localised and systemic effects; much of the nature of these is outside the scope of this paper, other than to say that they are of considerable relevance to the outcome. It is also important to note that a mechanical force can produce a pathophysiological reaction which can be sufficiently profound to cause death by itself. Thus accelerations or decelerations can cause cardiac dysrrhythmias which, in a susceptible individual, may cause death without there being any obvious mechanical injury, the only signs being those of heart failure. Here the problem is one of interpretation and the diagnosis of causes from their effects.

Insofar as physical injuries are concerned, if their causes are to be fully understood, and their consequences appreciated, then a variety of observations have to be made. These are important, not just as a catalogue of observations, but because they indicate their causes. The main features recorded for mechanical injuries are listed in Table 1.

TABLE 1. Essential features of mechanical injuries

(i) The nature of the injury - laceration
- incised
- bruise, graze, etc
- fracture(s)

(ii) The anatomical location
(iii) The dimensions
(iv) Its severity
(v) The presence of foreign material
(vi) Timing

Whilst this is a simplified list, it does provide a basic idea of the way in which injury recording should be approached if a comprehensive study is to be made. Inevitably there will be constraints upon the realisation of the ideal. A patient's clinical condition may dictate that resuscitation takes precedence and in carrying this out trace evidence may be lost and some of the features of the injuries may be altered. Thus the geometry of a wound may be changed by suturing needed to stop bleeding. This is not an unusual problem, especially in situations where there are penetrating injuries. Obviously no criticism can be reasonably levelled in such circumstances. It is, however, less easy to be sanguine when pressure of work is used as an excuse for a lack of attention to detail.

By themselves, comprehensive catalogues of injuries are useful in that they can help to determine causes, but care has to be exercised in attributing causation. The geometry of injuries may appear different when seen compared with when they were caused because of factors such as positional change, other than any deliberate new act. Thus a circular injury caused when the arm is flexed may become oval when it is extended.

Trace evidence is valuable because it may help to identify contacts. Locard's principle states that every contact leaves a trace. This may be revealed as an injury which has the characteristics of the impactor causing it. However, care has to be exercised because different objects may cause similar injuries. Thus a knife, a piece of glass or a sharp piece of metal can cause injuries which may be indistinguishable from one another. This led to difficulties in the interpretation of injuries in the Kegworth aircraft accident in England in 1989. The overhead bins had collapsed during the crash sequence, spilling their contents, but because of the complex kinematics of the accident, no single injury could be unequivocally attributed to this spillage. In contrast there is unequivocal evidence of lower limb injury in cars due to the rigidity of the parcel shelf in some models.

Trace evidence can be of considerable help in identifying contacts. The transfer of fibres, paint fragments, glass and other objects places the individual and the environment in contact with one another. In so doing it helps to answer questions of causation. Such evidence may be of profound significance, putting that person in a particular seat during the crash sequence. In so doing it may help in resolving questions about who was driving a crashed car when disputes occur. As a corollary of this, questions of injury causation and thus the effectiveness of safety design may be resolved.

This leads to another important feature of the forensic investigation of accidents. Long and detailed descriptions of injuries tell only part of the story. As was stated above, the differential diagnosis of causation may not be practicable purely by looking at the characteristics of an injury, because different objects may produce similar looking wounds. A systematic analysis has to be carried out, which includes factors such as the circumstances of the accident, a study of the wreckage and the site, as well as the medical evidence. To this must be added evidence gathered by other specialists, such as engineers, chemists and others whose expert opinion may help to resolve the issues.

Explanations are never easy in transport accidents and so attention to detail and accuracy are very important. Confusing lacerations with incised wounds, for example, can lead to errors. The former are caused by blunt trauma and the latter by sharp cutting objects. In this context it has to be remembered that more than one impactor can cause similar looking injuries. The timing of injuries is also important. Those which antedate the accident may be irrelevant, unless they cause deterioration in performance leading to accident causation. Injuries occurring after the accident sequence are of importance. The subject of injury timing is fraught with difficulty, because of the lack of finite parameters which can be used. To take a simple example, bruises change colour over time, but these changes can take place at various rates. Medico-legally this may be of great significance and our limitations in expressing finite opinions is a significant disadvantage. Occasionally, though, it is possible to provide some idea of the sequence of causation.

A man was killed in a road traffic accident. A lorry pulled out in front of him. His wife said that he braked fiercely, but to no avail. He died from a subarachnoid haemorrhage. The insurers claimed that this was a natural sequel to his severe hypertension and that he had failed to stop because he was unconscious before the crash. There was a large bruise on the sole of his foot, consistent with the application of the brakes, indicating that he was, in fact, alive and conscious before the impact.

Injury severity is much debated and, although much has been done to try and answer the questions raised, the answers provided are debatable. Part of the difficulty lies in the problems generated by attempts to resolve the injury caused with the amount of energy used and the amount of energy needed. A sharp instrument like a knife will easily penetrate the skin, very little force being needed. Once it has passed through the skin then it can easily enter the viscera. This does not mean that in such an instance only the minimum level of force needed was actually used; much more may actually have been used, without any appreciable changes in the features of the wound. Thus, whilst the available injury scoring methods, such as the AIS, have been of considerable value in accident injury research, some caution has to be exercised in interpreting any results, especially those relating to experimental findings. Experimental results obtained from tests carried out on isolated body parts may, because they are devoid of physiological inputs, bear little relevance to the behaviour of living tissues.

Injury severity is important and although there are difficulties in resolving the various issues, attempts should be made to assess it wherever possible. In carrying out such assessments, all of the features of each individual case should be analysed. Values which do not take such factors into account may be misleading.

3. Biological Difficulties

The foregoing outlines the basic forensic needs for the successful analysis of injuries; however, it must be remembered that this is a very much simplified summary, outlining an approach but not detailing the methods involved. It is not unusual to find that injuries may be superimposed upon one another. Similarly there may be widely different responses to the same injury. These and a range of other factors contrive to make for further analytical difficulties. Whilst it is undoubtedly true to say that biological structures must obey the laws of physics and chemistry, the ways in which they do show considerable variation, and as a consequence their reactions may not always be accurately predicted. Engineering tests are readily repeatable and, provided that there are no impurities within the samples, and no alterations in the test procedures, the results will be the same every time the test is carried out. When we deal with biological materials this may not necessarily be the case, and it is not unknown for wide variations to be found.

Of course not all of these variations have any biological significance, for they fall within what is commonly referred to as the normal range. What we cannot necessarily guarantee, though, is that when we get some variation in response to trauma, this will of necessity be part of the normal response.

There is a paradox here, which is both helpful in understanding trauma, and unhelpful in making the differential diagnosis of causation difficult. A fracture of the tibia sustained whilst playing football evokes just the same pathophysiological responses as it does when caused by a road traffic accident. This raises one of the least understandable issues of biomechanical research, and that is the apparent reluctance to take account of evidence acquired from other aspects of the study of trauma. Moreover, the felony is compounded by the attention paid to experiments carried out on isolated parts of the human body, and the uncritical translation of information gained from them to what happens in real accidents and their aftermath. Their weakness resides in the lack of any physiological response. Field studies of accidents have shown that it takes up to four times more force to break the femur in a living patient than was predicted by laboratory tests [4]. This does not mean that all laboratory experiments are meaningless. What has to be done is that the results thus obtained have to be treated with caution, as indicators of what might possibly be the case, rather than placing undue reliance upon them concerning actual structural strengths. Studies have shown that patients exhibit great variation in their responses to impact trauma [5]. A number of factors cause these differences, some of which are specific and others which are less easily defined. They include diurnal variations, the age and health of the individual, as well as their sex. These may be summarised by saying that the younger and fitter a person is, the better they will do in response to an injury, and the response will vary depending upon the time of day it is.

Tolerance to injury and inevitably to its effects, has been studied greatly. Summaries of the literature reveal that there are controversies and hiatuses in our knowledge [6, 7]. Perhaps even more alarming to the doctor, this has led to arguments that medicine as a representative of the biological sciences is not a proper science [8]. From the point of view of accident - injury research, this has important implications because so much of the study depends upon the comparability of different series.

It is often very difficult to say just how a particular patient will do. Medical history is replete with examples of miracle cures and miraculous escapes. Usually there is a valid explanation; sometimes though the science fails us and we are unable to say why someone lived and the other person died. If the survivor has very little injury and the fatality has very severe injuries, then there is little difficulty in explaining the outcome, at least on an individual basis; but it might not be so easy when a series of such cases are considered because it is probable that there will be those who die from treatable causes, and vice versa. Sometimes there is an easy explanation.

An elderly man sustained minor soft tissue injuries to the side of the face and the scalp in a road traffic accident. There was a little bruising to the left side of the body and he had a fracture of the mid shaft of the left tibia and fibula. He died three days after the accident. The accident was a low velocity impact in which there was very little structural deformation. The car had been hit on the passenger side by another car, which was travelling slowly. The man who died had been sitting in the passenger seat. His injuries were eminently survivable, but he died because he had very severe cardiovascular disease. Obviously the combination of the injuries and parlous health were too much for him to survive.

The advent of various scoring methods for injury severity has greatly improved our capacity for analysis. Their derivatives and additions have greatly improved predictability and helped to remove some of the most glaring deficiencies. Nevertheless, and perhaps predictably, there are many impediments to achieving any measure of success [6, 9]. Inevitably some of these result from the failure to appreciate that other kinds of trauma can help to fill the gaps in our knowledge.

After an accident, patients are taken to hospital and then if necessary they are admitted to the ward via the Accident and Emergency department and/or the operating theatre. As well as those who have been injured in road traffic accidents, there will be patients who have acquired their injuries in other ways. These add to the volume of information which can be used for comparisons. They also provide useful information on the effects of trauma and the outcome. Much of their analytical value lies in the fact that they help to provide a more statistically valid database than would be provided by analyses of road traffic accidents alone.

Although there are world-wide many road traffic accidents, they are not all carefully studied. Moreover, such is the breadth of individual variation that we need as much information as possible if we are to fully understand the effects of trauma. Recent information, for example, has shown that recovery from head injury is better if patients are well perfused and well oxygenated from an early stage [10].

The pathophysiological responses to trauma are complex, and imperfectly understood. It seems likely that they may be the ultimate factor dictating whether or not an individual survives their injuries. A detailed discussion of the factors involved is outside the scope of this paper. Suffice it here to look at a very simplified example of what this means. After an accident, patients may bleed; this loss of blood volume produces hypovolaemic shock. In an attempt to control this, and to ensure that the vital centres receive an adequate supply of blood and are therefore well oxygenated, there is peripheral shutdown. Those areas of the body which are better able to survive, albeit with poor function, lose much of their blood supply until such times as the shock is relieved. This is a useful defence mechanism, but it does have disadvantages. If the shocked state is prolonged there may not be any recovery. The smaller blood vessels which have been shut down may not re-open when the circulating blood volume is restored, thus irreversible shock has developed.

This concept is a difficult one, involving as it does complex biochemical reactions. It is one of many such factors which affect the outcome in an accident. There are others.

A hang glider pilot carried out a manoeuvre on a test flight. The aircraft plunged into the earth and he received a number of survivable injuries. He died shortly after arriving at hospital. At autopsy it was shown that he had a severe myocarditis. This had been present for some time, although he clearly had no knowledge of his illness, ascribing his malaise and weight loss to sadness at losing his longstanding girlfriend. Here the medical condition explained the cause of the accident. The myocardium was so compromised that the extra forces generated in the manoeuvre overcame his cardiac functioning and caused loss of consciousness. Also, because his heart was diseased he was unable to withstand the pathophysiological demands of moderate trauma, and died.

For the doctor, the ability to predict outcome is extremely important, this being one of the first questions relatives and friends ask. It is also one of the most difficult to answer, because of the unpredictability of human response. It is also a question which is increasingly being asked by those who fund health care. Here, the inference is that, if survival is not guaranteed, is the expenditure of finite resources reasonable? Insofar as accident - injury research is concerned and, with it, patient care, the answer is unequivocally yes; but it would help the argument if there were sound information to back up the opinion.

Many patients attend hospital after an accident and their injuries are treated. They are then allowed to go home and seek further care as an out-patient. Their injuries and the causes are not necessarily well researched. In London, road traffic accidents are only subjected to detailed police analysis if there is a fatality. For the most part, this is not unreasonable, because there are so many that it would be prohibitively expensive to carry out a full study of every single accident. Nevertheless there are many serious injury accidents which are insufficiently investigated. These accidents, and the responses of patients in lesser crashes and other types of accident, are the people who could help to fill the gaps in our knowledge about the effects and the outcome of accidents. It should be remembered that, when we talk about outcome, we are not just talking about living or dying; the issue is far more complex and revolves around the loss of any function which may result from trauma.

Alcohol's association with accidents has been well studied, and yet there are still some arguments about its effects. There are those who say that AIS grade for AIS grade, those who have been drinking will do worse than those who have not, and there are others who hold the opposing view. There is something to be said in favour of the former opinion [11]. However, not surprisingly, when we deal with medicine the answer is not straightforward. It may be that what we are really talking about when we say that the drinkers fare worse, is that we are looking at people who, in effect, abuse alcohol and it is not those who may have overindulged or even just have been drinking who develop post-traumatic complications after an accident.

Regular heavy drinkers may develop a variety of health problems, which may complicate the response to injury. Liver damage can result in prolongation of the bleeding time due to decreased production of clotting factors; these are made in the liver, thus they may suffer greater blood loss than someone who does not abuse alcohol. Also, if there is liver damage such as alcoholic steatosis, then other metabolic responses may be impaired. Moreover, fatty liver itself can cause death, and this may occur in an injured person. Thus it may not be the alcohol itself which causes them to do worse.

Whilst alcohol has been widely studied in road traffic accidents as a cause thereof, its wider implications have not been subjected to such careful scrutiny, at least in this context. Part of the problem lies in the fact that there are no certain tests which will unequivocally identify the heavy drinker and, added to this, there is the failure to comprehensively assimilate the knowledge gained by studying patients who abuse alcohol into studies of the responses of trauma victims.

Another factor which, in all probability, has a marked effect upon accident causation and outcome, is the effect of drugs. Many people take a variety of medicines legitimately, and others abuse drugs. These can have effects upon bodily functions which may profoundly affect outcome in an accident. Someone who has ischaemic heart disease may be taking medication, and they have an accident. If they die, then the death may be ascribed solely to the accident; but in reality this may be only a part of the answer, for their capacity to respond to the pathophysiological effects of the trauma is impaired. If their death is solely attributed to the accident then our picture of the way in which people respond to a particular level of injury severity is biased. Admittedly there is some allowance made for this, as has been alluded to earlier, when we acknowledged that older people do not respond so well to injury as does youth, but this is only part of the answer, like so many other features of accident - injury research. Ischaemic heart disease is not only the province of the elderly; moreover, can we always be absolutely certain that it is the disease itself that is responsible for the poor outcome, what effect does any medication have? In short, we do not know, largely because we do not look at the problem. Those studies which have been done on drugs and accidents have centred on accident causation and have not looked upon their later effects.

Causation of accidents is obviously extremely important, and as prevention is better than cure, emphasis upon this was reasonable. However, we do not need to know what happens to patients who have taken various drugs when they have an accident. This ought not to be a difficult technical problem, but it does raise some ethical issues which have to be resolved. It could be argued, for example, that it was reasonable to do a drug screen and alcohol estimation on all victims of accidents, so that those looking after them would have as full a clinical picture of their patient as is possible. Withdrawal symptoms from illicit drugs, adverse reactions to prescribed medication and complications due to non-administration of maintenance doses of regular medication, could adversely affect prognosis. Civil libertarians and others might argue differently, but if the test is in the patients' best interests, then it seems that there can be no logic in their view.

People who die of trauma in the United Kingdom will almost certainly have an autopsy. It would only be in the most unusual circumstances that they did not. This presents an opportunity for the pathologist to carry out a full investigation, including a toxicological analysis. If paramedics and others have put up intravenous infusions, and especially if blood transfusions have been given, then pre-transfusion specimens must be sought. These tests have to be paid for, and some authorities are not keen to do so. Also, the information gathered may not necessarily be comprehensive and it may not be analysed centrally, being confined to the depths of local archives . A large-scale study of drugs of abuse in traffic accident victims is being carried out in England and Wales, but common drugs are ignored.

This is another feature of a problem which has been much debated, and that is the inadequate reporting of accidents. This has been much debated recently. Police reports do not always equate with hospital findings and the statistics produced by insurance companies [12, 13]. Contemporary medical practice is moving towards the concept of meta analysis as a means of providing adequate information. Large studies can iron out some deficiencies, but even they may be biased, thereby frustrating the wider aims of the study.

Injury severity analysis has a long history, and in recent years it has been developed into a sophisticated analytical tool, which has been widely used in accident - injury research [14, 15, 16, 17, 18]. Not surprisingly, the evolution of these scales has not been a problematical. The major difficulties seem to have involved observer bias.

Injury scoring is often done by people who do not treat the patients. By itself, this is not necessarily a disadvantage, because it creates uniformity, the effects of observer bias and subjectivity being reduced. The latter have caused some problems in the past. There is, however, one source of difficulty, and that is the inadequacy of clinical descriptions. Marked differences have been found between the injuries documented in casualty departments and in in-patient notes [19]. This is not surprising and is not necessarily culpable, though it is frustrating. Casualty departments are places where emergency treatment is carried out, thereafter the patient is transferred for definitive therapy. Consequently the emphasis in casualty will be in saving life, and any record will refer only to the brief descriptions of major injuries. The wider forensic needs will have to await a more propitious time. Inevitably some forensic evidence will be lost. Those working in casualty cannot be expected to concern themselves all the time with the preservation of evidence when the medical problems facing them assume an overriding importance. Nevertheless, it is frustrating to find that a thoracotomy incision passes through a bullet or stab wound, thus destroying its characteristics.

Hospital notes are notoriously illegible and irretrievable, which adds to the frustration of those who code the injuries. What is needed is a system in which comprehensive, legible notes are provided for researchers.

The advantages of having dedicated coders is that, although they may manifest their own observer bias, this should be a constant feature, and its degree ought to be minimised. What cannot be easily verified is the comparability between centres. There may be some advantages to blind studies in which samples of cases are sent out for analysis by a body devoted to quality control in injury scoring. Regular samples of identical cases should be sent to all centres engaged in injury scoring. All of those thus involved would be required to carry out the scoring on all of the cases. Their performance could then be compared against an agreed standard, and the standard obtained by their centre.

Accuracy and reproducibility are of paramount importance if the manufacturers of vehicles and the legislators are to be convinced that they need to make changes, which might be expensive. Over the years the AIS system has evolved and there have been many changes, some of which have had a profound effect. It is likely, therefore, that it may not be practicable to compare a series carried out fifteen years ago with one being completed now. These changes have resulted from the observations made during the application of the AIS to accidents [20].

Predicting outcome has arguably been the most difficult feature of injury scoring, and has generated fierce debate at meetings. It has to be remembered that, "the AIS is an injury coding system and not an outcome or results system" [21]. Various statistical tools have been developed using the AIS system to try and develop an index of prediction. These have met with limited success, and this highlights what could be argued was an important shortcoming in the AIS system. This should not be regarded as a criticism of the system, as such; rather it illustrates the complexity of the problem with which we are having to deal, and which has been highlighted here.

Amongst other systems which have been developed is ASCOT. This uses anatomical, physiological, Glasgow coma scoring and AIS data in an attempt to obviate the shortcomings of ISS data. TRISS is another of these systems. Both have been well documented in the literature, and their details will not be discussed further here. Suffice it to say that there is not universal agreement about their usefulness [22, 23].

This debate, and the range of different measures, is perhaps symptomatic of the problem. Outcome is not easily predicted and, whilst we may be better at it than hitherto, the absolute criteria of acceptability, namely accuracy and reproducibility, are not yet achievable.

4. Crash Dynamics

Although this is a topic which is largely outside the scope of a pathologist, some mention of its principles is needed, because it is, with crashworthiness, an important arbiter in survivability. Also pathologists and other medical practitioners need to be aware of the mechanisms of injury if they are to understand causation and prevention.

Crash testing and analyses of accidents in which witness statements have played a part have provided the basic knowledge of how vehicles react during an accident. Crashes are not simple events - they usually involve complex movements which are potentially injurious. Abrupt movements, with or without changes in accelerations and decelerations can be dangerous to any occupants, inducing pathophysiological reactions which cause tissue damage.

Classifying accidents is not easy, because there are so many variations; nevertheless, it has been found acceptable to categorise them into broadly based groups, such as frontal, side and rear. From the point of view of injury causation, the important feature in all types of accident is that the impact forces are greatest at the point of impact. Experiments with older aircraft have shown that there can be a forty-seven per cent attenuation of impact forces in the region of the engines. This is because in that type of aeroplane, where the engine was at the front, it absorbed the impact forces due to the induced structural deformation. Also the rate of deceleration at the centre of the fuselage was a function of the structural characteristics of the airframe, and not the speed. The duration of the impact forces, which is an important factor in injury causation, is affected by speed. In these types of aircraft the decelerations produced had a spiky pattern, which is of considerable importance, predisposing to increased severity and incidence of injuries.

These observations and those of others have helped to define the way in which vehicles behave in a crash [24, 25, 26]. In so far as injury causation is concerned, the features outlined in Table 2 are a summary of the most important factors.

TABLE 2. Principle factors in impact behaviour affecting injury causation

Crumpling of the structure
Upward rotation of the floor
Rotation and shearing of the structure
Break up of the structure

Obviously these can be subdivided into a range of other elements, but in the final analysis, it is these which produce the environment which predispose to injury causation.

TABLE 3. Factors affecting the outcome in an accident (Partyka 1979)

1. At the accident level
   (a) The number of vehicles involved
   (b) Urbanisation
2. At the vehicle level
   (a) The change of velocity at impact
   (b) The direction of the impacting force
   (c) The damage area
   (d) The size of the damage
3. At the occupant level
   (a) Seating position
   (b) Age
   (c) Sex
   (d) Ejection
   (c) Restraint use

There has been much debate in the literature on motor vehicle accidents about the terminology and the way in which impacts should be described. There seems to be some general agreement now, and frontal impacts are said to be 12 o'clock, rear 6 o'clock and the sides 9 and 3 o'clock respectively. The intermediate spaces can be filled in as required. Rollover accidents are treated separately. The potential variations are so great that compromises have had to be made [27, 28]. Partyka's work [28] has allowed us to amplify Table 2, so that we can more accurately assess the potential outcome (Table 3).

To a large extent, these are self-explanatory, their detailed analysis having been widely studied in the literature. From the point of view of this discussion, what we are concerned with is the complexity of the situation. Clearly if we are to explain how an accident victim acquired their injuries then this can only be satisfactorily done if there is a multidisciplinary approach. Pathologists and clinicians cannot, by reviewing their findings alone, give a satisfactory opinion on injury causation. Rollover accidents are a good example of this. In the majority of cases the speeds involved are higher. Taken alone this could be regarded as the reason why the injuries were more severe than might perhaps have been expected in view of the vehicular damage recorded. In these accidents there is a prolongation of the period during which energy is being dissipated, and consequently there is a longer time in which the occupants are exposed to danger. In some ways this is similar to the accident in which there are multiple impacts by a single vehicle, without the rotational vector.

5. Discussion

Accident - injury research is a relatively new discipline, despite the fact that many people have expressed an interest in injury causation for a number of years. It is also a subject which has been comparatively underfunded [9]. Whilst it is true that the motor manufacturers and the aircraft industry have prosecuted various areas of research, spawning many valuable studies, their interest has been largely confined to those aspects which affect their activities. This is neither surprising nor is it unreasonable, and it has advanced our knowledge greatly. Arguably the greatest deficiency in all of this research is that it has tended to concern itself with the anatomical injury, neglecting what may be the ultimate arbiter or survivability, the pathophysiological response.

If a vehicle manufacturer finds that a particular model tends to cause a specific injury, such as a fractured femur in a frontal impact, whereas other similar models do not, then obviously he will want to know why. This approach has had a marked influence on some of the studies. The finding has been verified, the reasons why it occurs established and the design has been altered, so that no further fractures of the femur occur in those circumstances. To a certain degree this is reasonable, that is until the consequences of the design change are examined.

Theoretically, designers could make vehicles immensely strong, so that they would not break up on impact. This would mean that there were no more injuries due to contact with broken pieces of motor car. However, the consequences of such a move are that there could be many more injuries due to dynamic overshoot. Instead of absorbing the impact forces, the vehicle would transmit them all to the occupants.

It may well be that by removing the cause of a particular injury safety is improved, but if the only way we approach the problem is to look at a part of the body in isolation, then there can be no guarantee that we will improve survivability. The human body has to be treated as a whole.

Crashworthiness can only be achieved if the information upon which it is based is sound. The remarkable advances which have been made have come about because those involved paid careful attention to detail. They were, however, rather more fortunate than today's researchers, because they had a somewhat simpler set of problems. Unrestrained individuals in a motor car were obviously more likely to sustain serious injury and to die than those who were restrained. Admittedly, there were those who could and would not see the evidence, just as there are still those who do not wholly accept the dangers of alcohol in accident causation. Thus they had an uphill struggle to get their ideas accepted. It is precisely the fact that there was so much resistance to safety improvements that causes so much concern now, especially in those instances where scarce resources are devoted to analysing isolated injuries.

In the past there were great potential gains; it may be that there are still equally large gains to be made, but the evidence would suggest otherwise. Falling accident rates have reduced the toll, although it is still alarmingly high, and so there is or should be a reduced potential for improvement. In those countries where there is still a very high toll on the roads, large reductions could be made simply by adopting those safety measures found to be effective in other countries. Thus governments and other grant-awarding authorities may not feel quite so inclined to fund research programmes which look at what may be seen as proposals which can only bring in a small return. This is obviously a shortsighted view, but it seems that there is some reason for believing that it holds currency.

Forensically, we need to be able to predict an outcome from a particular course of action. We also need to be able to say how an injury was caused and what force was used. Unfortunately we often fail to answer these questions satisfactorily, because the database is too small and it is inadequate. Despite all of the research which has been done over many years, we still have an inadequate knowledge about tolerance. This is because research has tended to look at the anatomical factors involved, forgetting that structure and function are dependent upon physiology. There is no doubt that anatomy plays a very important part in the response to trauma. The maxillae are an excellent example of this, being very ingeniously designed to withstand normal physiological forces. Thus the antero-posterior strength of the maxilla is very low. The bone is designed to transmit force up through a set of buttresses, spreading it over a large area. This happens when we chew: the forces pass up these buttresses to the base of the skull, where they are dissipated. Similarly, the lower jaw is designed to break just below the tempero-mandibular joint, so that force is not transmitted up through the base of the skull causing an even more serious injury.

These ingenious features are only part of the story. They are the part which we understand best and which it is so much easier to research. Altered physiology is more difficult to study, particularly as we do not know all of the biochemical pathways involved. Those who argue that it is not important to know if the induced cerebral hypoxia is due to a jolt caused by an intermediate rate of onset of G or decreased blood supply as a result of shock, or some other factor, do the subject a disservice. The intermediate rate of onset of G force may be amenable to modification, but the other causes may not respond to the same measures. Consequently, if such an observation was used as an argument for changing a design, in the hopes of improving safety, it would fail, and credibility would be lost.

Outcome is obviously a very important factor. Medically, it has a slightly different meaning from that implied by others and it is of considerable interest at the moment, because of a variety of conflicts in patient care. Many of the arguments involved are outside the present scope. However, they do have a direct bearing upon the present topic. Moreover, the context in which they lie is controversial, especially in those areas where there are conflicting demands for limited funds. The relevance to this discussion lies in the fact that outcome is an arbiter of the success of safety design.

Regionalisation of trauma care has been shown to be effective. In one American study, it was found that seventy-three per cent of non-central nervous system deaths were preventable. Within a year of setting up a regional trauma centre, this was reduced to nine per cent [29, 30]. Despite results such as these, there are still those who argue that trauma centres are not needed. Their greatest wrath is reserved for some of the accident after-care schemes, especially those using helicopter evacuation. As in so many other areas, opinions can be based more upon anecdote than they are on careful scientific evaluation. There has even been a suggestion that they may be ego-driven [31].

Whenever results are being assessed, care has to be exercised to ensure that like is being compared with like. It is easy to show that there may be eight per cent of preventable deaths in a non-trauma hospital, against a two per cent incidence in a trauma hospital [32], but we have to be sure that we are looking at the same thing. It would be easy to say that the degree of improvement in outcome did not necessarily justify the expense. Improvements in care at the non-trauma hospital could lead to a reduction in the preventable death rate. If this could be achieved, then the obvious question is, why has it not been done? The answer probably is that it could not be done, because the difference does not represent a true analysis of the patients seen at trauma centres. They get referrals and some of these will inevitably be people who are very seriously injured. Their sample is biased [33]. Admittedly, this ought not to affect their preventable death rate, but again it depends upon their expectations.

One example of the difficulties found when looking at outcome statistics, is the depth of the investigation. Accident - injury research is so complicated that it can only be done successfully by a multi-disciplinary team which has comprehensive data. One early study showed a fifty per cent excess preventable death rate in a matched study of road traffic accidents in a rural area, compared with a non-rural area. A repeat of the study some years later found that, despite a greater overall severity of injury in the rural accident victims, the preventable death rate was higher in the non-rural environment [34]. At first, this seems to raise very serious questions about patient management. A careful analysis revealed that the autopsy rates were not comparable and the information yield from the rural area was incomplete.

It is a great pity that some of the information which is readily available from other kinds of trauma is ignored by students of transport accidents. General principles which have been shown to have stood the test of time, such as the effects of gender, age and pre-existing disease on the outcome from trauma, seem to be largely ignored. It is certainly true to say that some workers have looked at these problems; but, for the most part, the studies have tended to be a static approach, there being little or no attempt to establish why the effect is exerted. It is a curious fact that many who have carried out biodynamic studies of accidents, and who have carried out experiments, seem to be more interested in establishing what happens, rather than searching for the reasons why something has occurred.

Simple epidemiological studies do have their place. Thus it is relevant to know that motor cyclists who are above the limit for driving are more likely not to be wearing helmets than those who have not been drinking, and that the non-helmet wearers show significantly greater incidences of head and facial injuries [35]. This type of information is only of limited value, despite the fact that it allows us to identify those who take unnecessary risks.

Forensically, such a study has much to commend it, because it identifies the cause and the effect, and to some extent it establishes reasons, but not the underlying factors. In the context in which it was carried out it was valuable. The problem with many such studies is that they confirm what we already know. Thus one must ask, is the day of the epidemiologically based study over? Some might argue that they are useful for identifying so-called accident black spots, but questions such as these can be easily answered by looking at the basic police accident statistics. It is not part of the research base's remit to use its sparse resources thus. Rather, it should concentrate on identifying and understanding the complex biochemical and physiological responses to trauma.

It has been well established that the greatest improvements in traffic safety have accrued from social changes, such as the approach to drinking and driving. Seat belts and other improvements in crashworthiness, road design and speed limitations have all added their weight to the corpus of improvement. Future improvements will be slender by comparison, but it is quite possible that their effects may be very dramatic.

Current methods of resuscitation mean that people who, some years ago, would have died at the scene of an accident, are now living. A number of these people have such severe brain damage that they develop a persistent vegetative state. There is some evidence to suggest that early good perfusion may reduce the incidence of this very distressing predicament, by ensuring that the blood vessels remain open. Thus when the initial phases of shock have been, to some extent, ameliorated and there is some restoration of oxygenation, the brain receives some blood and does not suffer from irreversible shutdown. It is in areas like this where there is so much advantage to be gained.

Inevitably this means that the collection of information for study has to be beyond reproach, and this cannot be claimed at the present time. Insofar as injury description is concerned, the basic principles have long been established. Bruises, grazes, lacerations, incised wounds and fractures should all be differentiated according to their nature and their type. Their sizes, direction and anatomical location should all be recorded, as should all of the other noticeable details. To this information, knowledge about the clinical management and course should be added, together with autopsy, toxicological and histological findings. This is a counsel of perfection which unfortunately will not be realisable, except in closely confined research circles. The constraints of time and cost mitigate against such comprehensive studies in the normal course of events. Also, autopsy studies are not always carried out, even in those jurisdictions where autopsy rates are routinely high.

It is, perhaps, worthwhile questioning the continued crash testing of vehicles by government departments. The United Kingdom's Department of Transport has spent six million pounds on this in the past four years. It may be that this is an entirely justifiable expense, but in an era when research funds are short, and in the context of crash-injury research - where the fundamental questions about survivability are biochemical and physiological - this would perhaps seem to be the more realistic approach. It is, perhaps, symptomatic of the conflicts and of the paucity of resources, that neglect of such fundamental studies should evoke such strong emotions. It is hard to escape the feeling that, as far as governments are concerned, the newsworthy response is the one which counts, not necessarily the one which really needs attention.

6. Conclusions

Great improvements in transport safety have been achieved by the considerable efforts of a variety of workers from a wide range of specialities. To a certain extent, and because of the scope for improvements, their success has created a paradox. Further research will be difficult to justify because the scope for improvement is much reduced. However, many people are still dying and suffering injuries, some of which will result in impaired function, thereby causing a drain on resources.

If any significant inroads are to be made into this residuum of death and injury, then greater attention will have to be paid to the wider forensic features of accidents and the way in which they are caused. Implicit in this is a comprehensive analysis of accidents, including the effects of drugs and disease as well as alcohol not just as an accident causation, but also on injury causation. Moreover, if we are to improve our understanding of the effects of trauma, then any studies should be more widely ranging. They should take into account information gained by studying other types of accident, as well as road traffic accidents.

Also full cognisance should be taken of pathophysiological responses to trauma, for these may be the ultimate arbiters of survivability.

References

1. Hill I R, (1989, *Maxillofacial Injuries*, PhD Thesis, University of London
2. Anon (Dec 1995), European Transport Safety Council, *Safety Monitor*
3. Morton, Sir A (Dec 1995), Westminster Lecture, Parliamentary Advisory Council for Transport Safety
4. Anon (1988), Personal Communication
5. Cesari, D et al (1979), Evaluation of human intolerance in frontal car impacts, *23rd Stapp Car Crash Conference*, Society of Automotive Engineers, New York
6. Hill, I R (1984), *Aerospace Pathology*, MD Thesis, University of Cambridge
7. Lau, I V et al (1992), *The Viscous Criterion: Traffic Injuries, Causes, Consequences and Costs*, Association for the Advancement of Automotive Medicine, London
8. Hacking, I (1983:3), *Representing and Intervening*, Cambridge University Press,
9. Hill, I R (1993), In: *The Pathology of Trauma*, Ed Mason JK, Edward Arnold, Ch 3
10. Goode, A W (1994), Personal communication
11. Evans, L (1991), *Traffic Safety and the Driver*, Van Norstrand Reinhold, New York
12. Hutchinson, T P (1987), *Road Accident Statistics*, Rumsby Scientific Publishing, Australia
13. Barancik, J I et al (1985), Discrepancy in vehicular crash injury reporting, *Accid Anal & Prev* **17**, 147
14. Baker, S P et al (1985), The injury severity score, *J Trauma* **9**, 196
15. Bull, J P (1977), Measurement of injury severity, *Injury* **9**, 184
16. MacKay, G M (1979), Personal communication
17. Kruse, T et al (1979), *AIS as a Measure of Injury Related Incapacitation*, IVth IRCOBI Conference, Goteborg, Free University Amsterdam
18. Jorgenson, K et al (1979), *Description of 3225 Victims of Road Traffic Accident Trauma*, IVth IRCOBI Conference, Goteborg, Free University Amsterdam
19. Mackenzie, E J et al (1985), Rating injury severity using emergency sheets versus in-patient charts, *J Traum* **25**, 984
20. Copes, W S et al (1981), A comparison of AIS scales 1980 and 1985, *J Traum* **28**, 78
21. Petrucelli, E et al (1981), The abbreviated injury scale, *Accid Anal & Prev* **12**, 25
22. Champion, H R et al (1990), A new characterisation of injury severity, *J Trauma* **30**, 539
23. Yates, D W (1990), Scoring systems for trauma, *BM J* **301**, 1090
24. Fryer, D I (1965), In: *A Textbook of Aviation Physiology*, Ed Gillies J A, Pergammon, Oxford
25. Snyder, R G (1975), *Crashworthiness investigation of general aviation accidents*, Society of Automotive Engineers, New York
26. Snyder, R G (1978), *General Aviation Crash Survivability*, Society of Automotive Engineers, New York
27. Hardy, J I G et al (1973), *Field Accident Damage as a Basis for Crash Tests*, IRCOBI, Amsterdam
28. Partyka, S C (1979), *Fatal Accidents in the First 15 Months of the National Crash Severity Study*, AAAM proceedings, Louisville, Kentucky
29. West, J G et al (1979), Systems of trauma care - a study of two countries, *Arch Surg* **114**, 455
30. West, J G et al (1983), Impact ofregionalisation: the Orange County experience, *Arch Surg* **118**, 740
31. Moore, E E (1995), Trauma systems, trauma centers and trauma surgeons: opportunities in managed competition, *J Trauma* **39**, 1
32. Sauaia, A et al (1995), Epidemiology of trauma deaths: a reassessment, *J Trauma* **38**, 185
33. Waller, J A et al (1995), Trauma center-related biases in injury research, *J Trauma* **38**, 325
34. Chen, B et al (1995), Geographic variation in preventable deaths from motor vehicle crashes, *J Trauma* **38**, 228
35. Johnson, R M et al (1995), Craniofacial trauma in injured motorcyclists: the impact of helmet usage, *J Trauma* **38**, 876

PART II
Road Data, Compatibility Issues and Testing

A REVIEW OF THE BIOMECHANICS OF IMPACTS IN ROAD ACCIDENTS

MURRAY MACKAY
Professor of Transport Safety,
University of Birmingham,
Birmingham, United Kingdom.

Abstract

The real world characteristics of road accidents relevant to biomechanical considerations will be outlined. A brief introduction to the Abbreviated Injury Scale will be given, and then the frequencies and severities of the major collision types will be presented in terms of velocity change for various injury levels. Crash severity distributions for restrained and unrestrained conditions will be analysed. Distributions of injuries by anatomical regions will be illustrated for car occupants in the main types of collisions; frontal, lateral, rear and rollover. The limits of protection of current restraint systems will be illustrated. Gender, age, height, weight and sitting positions will be analysed to show the wide biomechanical response of the actual population exposed in collisions. The consequences of these real world variations for the optimisation of seat belts, using intelligent restraint system technology, will be presented.

1. Early Knowledge on the Biomechanics of Impacts

An implicit recognition of impact biomechanics exists in all of us and is engendered from an early age. Soft, deformable structures are benign; sharp, rigid objects, like kitchen knives, deserve caution. From the earliest of times, the underlying principles of impact biomechanics have been recognized: structures designed to maximize trauma are hard and concentrate loads, such as spears and clubs, while conversely, shields and armour absorb and distribute loads and protect vulnerable parts of the anatomy. Hippocrates, writing around 400 BC, noted that for head injuries:

Of those who are wounded in the parts about the bone or in the bone itself, by a fall, he who falls from a very high place upon a very hard and blunt object is in most danger of sustaining a fracture and contusion of the bone, and of having it depressed from its natural position. Whereas, he that falls upon more level ground, and upon a softer object is likely to suffer less injury in the bone, or it may not be injured at all.

Like most subjects, impact biomechanics have evolved from early observations of natural phenomena, through an experimental period to a theoretical framework that outlines general laws and precepts. Hugh de Haven is normally credited with the first insights into human tolerance of crash loads. During World War I, he was involved in a

J. A. C. Ambrósio et al. (eds.),
Crashworthiness of Transportation Systems: Structural Impact and Occupant Protection, 115–138.

midair collision. While convalescing, he realised that his survival was due to the maintenance of the integrity of his cockpit that, together with a safety harness, protected him from the localized contacts and catastrophic injuries that killed the other pilot. He also observed that his own serious abdominal injuries related to the buckle of his harness, causing a severe internal haemorrhage with laceration of the liver. Crash-protective design, rather than capricious good fortune, had ensured his survival.

In 1942 De Haven analysed the circumstances of eight people who fell from considerable heights, seven of whom survived. Speeds at impact ranged from 37-59 mph and the objects struck consisted of fences, a wooden roof, soft ground, and in two cases the bonnets of cars. Decelerations at impact were estimated and the groundwork for whole body tolerance was laid. Subsequently Snyder et al (1977) developed this approach into a most useful methodology for obtaining such data.

In 1941, Sir Hugh Cairns published a paper on fatalities occurring among Army dispatch riders. He showed that for those who wore helmets the head injuries were relatively mild. Later work by Cairns (1946) showed that following the compulsory wearing of helmets by Army motorcyclists there was a progressive fall in the death rate. He also noted that most blows on helmets were to the front and side rather than to the crown. The consequences of that observation have been reflected 30 years later in the evolution of the jet-style and full-face helmets of today and the demise of the cradle suspension inside the helmet.

John Lane in Australia in 1942 noted that aircraft should be certified in two ways: they should be both airworthy and crashworthy, and so the term 'crashworthiness' was born, but its application to automobile design did not begin until some 20 years later.

The experimental period of biomechanics got under way after World War II, with cadaveric studies by Gurdjian (1945) examining head injury in the main, and volunteer studies conducted by Stapp (1951). The great contribution by Stapp was to show that the primary forces acting in the majority of car collisions are entirely survivable if the packaging of the human frame is satisfactory. He showed that accelerations of 30 G for up to 0.5 s were entirely tolerable with only reversible soft tissue injuries occurring. At 45 G, signs of concussion and retinal haemorrhage begin to show. These accelerations were measured on the seat of the dynamic sledge. The accelerations experienced by the head itself were much greater.

In the 1950s, Severy and Mathewson (1954) were developing the techniques of experimental crash testing with instrumented dummies and high-speed film analysis. By the mid-l960s, a body of knowledge had developed that gave insights into the general frequencies of traffic collisions and injuries, some understanding of the actual mechanisms that generate the injuries, and some means whereby the forces and accelerations applied to car occupants could be modified. What was largely missing was accurate information on the tolerance of the actual human frame to specific impact loadings, ideas of the likely benefits that could be obtained from practical changes in car design, and what the penalties in design terms would be.

In parallel with the increasing amount of experimental work on both cadavers and animals, studies of real-world trauma continues. An elegant paper by Sheldon (1960), entitled 'On the natural history of falls in old age', showed how the routine observations of a practicing clinician could lead to new insights into the aetiology and mechanisms of injury, particularly for long-bone fractures in the elderly.

Pioneering work on seat belts was conducted in Sweden where, by 1960, some 50 per cent of private cars had belts fitted. The appropriate elongation and geometrical characteristics of belt systems were evaluated experimentally by Aldman (1962), who demonstrated the importance of correct anatomical positioning and dynamic properties appropriate to the deformation and geometrical characteristics of specific car designs.

The subject of biomechanics in relation to car-occupant crash protection grew rapidly in the 1950s and 1960s and became institutionalized with an extraordinarily important legislative act in the USA. As a result of government hearings that illustrated the great potential of crash protective design, in 1966 the National Highway Safety Bureau was created by act of Congress and it initiated a set of standards controlling the performance of cars in terms of their crashworthiness. The effects of those standards have reverberated through the automotive world ever since. They have been copied, modified, and adopted by almost every country with a significant car population and they have changed car design from a free market, styling dominated activity to one in which certification, or passing the standards, with all the attendant engineering problems, is of prime importance in the priorities of car manufacturers.

The scientific basis of these first crash performance standards was not well founded; many of the requirements were informed guesswork only. With the benefit of hindsight, quite extraordinarily few major mistakes were made, but what has also become clear is that the subject is a more complex one. The real world of collisions contains many surprises. Common sense has the most curious property of being more correct retrospectively than prospectively, and a major gap was left as the subject became more under the control of government and industry and away from the individual efforts of the early workers.

The problem was one of evaluation. As an example, consider door-latch design. Early work by the Ford Motor Company and by Garrett (1962) had shown, on very unrepresentative samples of collisions, that it was statistically better to remain in a car during a collision than to be thrown out. Most people were thrown out because doors opened during the crash and the concept of an antiburst door latch developed. Thus a problem was identified, a solution proposed, and the legislative machine went into action. Standards were written specifying the longitudinal and lateral strengths of door latches. The car industry, virtually worldwide, redesigned door latches to meet the new requirements, the new cars were built and sold, and the problem of ejection was then supposed to go away. It was left to the individual workers such as Huelke et al (1963) and Gissane and Bull (1962) actually to look at the real world of collisions and attempt to find out if these new rules were in fact producing the benefits for which they were conceived. Systematic feedback into the legislative process did not then and still does not exist.

There is a parallel to be drawn here with the drug industry. Before a new drug is marketed, it is subjected to rigorous trials, and then its performance in use is monitored for effectiveness and possible side-effects. In contrast, the introduction of crash-protective vehicle-design measures is relatively haphazard. In most countries until recently there has been no systematic evaluation of design and legislative changes.

In the case of antiburst door-latch design, evaluation studies in Britain on 400 crashes showed that antiburst designs reduced door-opening rates in collisions from 30 per cent to 15 per cent (Gissane and Bull, 1962). Overall, car occupant deaths were reduced by 5 per cent. Doors still continue to open in collisions, however, and the detailed crash-investigation studies by Mackay showed that doors were opening in collisions in ways not covered by the regulations.

2. General Collision Frequencies

Most motorized countries have basic crash-injury recording systems, almost always based on police reports. Systematic data linkage between hospitals records and police systems does not exist anywhere on a large scale, although for fatalities the system of the coroner's courts allows adequate pathological data to be linked with crash information in a few countries such as Australia and Sweden.

Serious under-reporting of minor injuries and some classes of serious injuries, especially to cyclists, motorcyclists and pedestrians occurs in most countries using a police-based system. Of cyclists treated in hospital 50 per cent to 80 per cent never appear in the police records in the UK.

With these provisos, Table 1 gives the range for the main classes of road user fatalities in motorized countries in Europe and the USA. The UK, for example, has a high proportion of pedestrian deaths, at 36 per cent of the total, compared with 16 per cent in the USA.

TABLE 1. Categories of fatalities

Class	Range (%)
Car and light truck occupants	38-66
Pedestrians	16-36
Motorcyclists	10-20
Pedalcyclists	2-5
Large truck and bus occupants	2-4

Table 2 gives the ranges of the main types of collisions in which car occupants are killed. Although frontal impacts are the majority of collision types, side impacts are important, particularly in countries with high levels of seat belt use because seat belts primarily reduce deaths in frontal crashes. Rollovers, rear-end collisions and multiple collisions, which are mainly side impacts followed by rollover, also contribute significantly to the total.

TABLE 2. Collision types for car occupant fatalities

Type	Range (%)
Frontal impact	50-60
Side impact	22-35
Rollover	8-15
Rear	3-5
Multiple	3-6

These general frequencies obviously have a bearing on the overall effectiveness of crashworthiness design which are now considered for the main types of collisions.

2.1. MECHANISMS OF INJURY IN FRONTAL COLLISIONS

In a standard frontal collision represented by a 30 mile/h (50 km/h) impact with a flat rigid barrier, most modern cars will have an overall stopping distance of approximately 2 feet (60 cm). Such a collision is similar to two cars of equal weight colliding with a closing speed of 60 mile/h (100 km/h). In the latter case the actual stopping distance is usually somewhat greater than the 2 feet in the rigid barrier test, and would be about 30 inches (76cm) because of mutual penetration of the two front structures. An approximate rule of thumb is that an inch of crush equates to 1 mile/h within the middle range of crash severities.

A stopping distance of 2 feet means that from 30 mile/h, the undeformed parts of the car, such as the passenger compartment, have an average deceleration of 15 G. In practice the decelerations are not uniform and the peak values are normally in the range of 18 G to 28 G.

An unrestrained occupant however experiences forces of much higher orders of magnitude. Consider the driver. As the collision develops, the passenger compartment starts to slow down but the driver continues to move forward at his initial speed, closing up the interior space between him and the forward structures. Loads are applied to him sequentially through the feet from the toeboard, to the knees from the instrument panel, to the chest from the steering wheel and to the head from the windscreen or the windscreen frame in the sun visor region.

The knee contacts occur early during this crash phase so that their localized contact velocity is low because the instrument panel is still moving forward at a significant velocity, around 20 mile/h (33 km/h). Thus the knees have a low localized impact of some 10 mile/h (l6km/h) and then they 'ride-down' on the remaining crush of the front of the car.

The chest and head contacts occur later in the crash sequence so that their localized contact velocities are higher. Indeed for most cars the head initially is normally about 2 feet (60 cm) from the windscreen or header so that the car has already come to rest by the time the head has a specific impact, and that occurs at the full initial velocity of the car. A head striking a modern laminated windscreen at 30 mile/h (50 km/h) will deform the interlayer of the glass approximately 15 cm causing it to bulge outward. Such a condition produces an average deceleration on the head of 60 G with a peak value of about 90 G. If, however, the head strikes the windscreen frame, then the stopping distance will be very much shorter, perhaps only 2 cm, being the amount which the metal of the windscreen frame deforms. That produces decelerations on the head in the range of 500 G.

Thus, no simple relationships exist between the overall crash severity and the specific forces experienced by the unrestrained driver. For a passenger the sequence of loads changes. The foot and knee contacts occur, and then the head strikes the forward structures before the chest hits the facia. That can result in major loads to the neck.

The main patterns of injuries for unrestrained occupants are thus to the knee femur-hip complex, the chest and the head. The lower limb injuries arise from compressive loads up from the foot and from the knee striking the lower instrument panel. Depending on the localized nature of that knee contact there may be specific fractures to the patella or a knee-joint injury, or the loads may produce injury remote from that contact with fractures to the shaft or neck of the femur or dislocation of the hip joint. Modern designs of instrument panel have diminished the localized injuries by providing several inches of ride-down for the knee contact.

The interaction of the chest with the steering assembly is complex. Regulations require a limit on rearward motion of the hub of the steering wheel of 5 inches (12.5 cm) measured horizontally. In addition a limit of 2500 lb (1130 kg) is set for the permissible load on the chest measured in a 15 mile/h (25 km/h) bodyblock test. There are various design solutions to these requirements usually involving some telescoping or deforming element in the shaft or hub of the steering wheel (Appel and Wustemann, 1986).

The onset of serious chest injury from in-depth crash investigations for both steering wheel contacts and from seat belt loads appears to relate to multiple rib fractures. That is associated with significant pulmonary dysfunction and risks to the great vessels in the chest. However, for other loading conditions, notably from an airbag, from an instrument panel, or from a door panel in a lateral collision, there is evidence to show that rib fracture is not an appropriate threshold for assessing injury risk. This is because the thorax has a visco-elastic response and thus exhibits different characteristics at high rates of loading. Viano and Lau (1985) have proposed a viscous criterion (VC) for thoracic loading to allow for this rate dependency. At low rates of loading, deflection, leading to rib fracture at the limiting condition, is a satisfactory parameter to measure. At high rates of load, however, intrathoracic injuries occur at relatively low deflections of the rib cage. This leads to the need for more complex dummies to assess injury risk experimentally, and it raises doubts about the appropriateness of current regulatory procedures (Cesari, 1993).

The most frequent and most serious injuries occurring to unrestrained occupants in frontal impacts are to the head. Head injuries represent a complex set of trauma to the skull, the brain and the face occurring singly and in combinations. The mechanisms of brain injury in particular show a spectrum of conditions of extraordinary complexity. Brain injuries can occur with and without skull fracture, they can be focal or diffuse in nature, they can be predominantly vascular, with epidural or subdural haemorrhage or rupture of the bridging veins, or there can be diffuse axonal damage throughout the entire brain mass. Gennarelli et al (1982) have shown how various combinations of linear and angular accelerations of varying duration lead to either focal or diffuse injuries or combinations of both. The brain exhibits different responses to motions and accelerations in the coronal and sagittal planes, with severe diffuse axonal injury (DAI) occurring primarily in lateral impacts where the head moves in the coronal plane.

It is against this background that the effectiveness of crashworthy design must be considered. At present, head-injury risk is assessed experimentally using a dummy in which the resultant linear acceleration, measured at the centre of gravity of the head, is the only criterion. That acceleration is the basic input into the calculation of the head injury criterion (HIC) which is a time-weighted function of acceleration. Historically an HIC of 1000 is taken as the limit of head-injury tolerance and that value is written into regulatory procedures worldwide.

The recent work of Gennarelli et al (1982) and Viano (1988) questions the validity of the HIC as an acceptable measure of head injury risk. Ultimately, brain injury is probably most directly related to local shear strains occurring within the brain. The patterns of shear strain are a function of the direction, magnitude and duration of the applied accelerations and they cannot be equated simply to one linear peak value (Viano, 1988).

Modern interior car design has produced yielding structures in most areas where head contacts occur. Those structures can be either of contoured metal which deforms appropriately or of plastic padding materials. One consequence is that skull fractures and related focal injuries have been reduced but diffuse trauma appears now as a relatively more important type. Indeed padding can under certain conditions of a glancing blow to the head enhance injury risk by increasing angular accelerations (Careme, 1990). In general terms however the aims of the designer are to provide the largest amount of ride-down distance for occupant contacts consistent with all the other functions of the interior of the car.

2.1.1. *The Restrained Occupant in the Frontal Collision*

The fundamental benefit of a seat belt in a frontal crash is to prevent or diminish those specific interior contacts described above. Soon after the car starts to decelerate, the occupant moves forward enough to load the seat belt. The webbing stretches, allowing some forward motion yet preventing head and chest contacts, so that ideally the occupants stopping distance is actually greater than the frontal crush of the car. If the belt allows a foot of forward motion then the occupant could theoretically have a ride-down distance of 3 feet when the front structure of the car crushes 2 feet. In practice that condition is never achieved because time, and hence distance, is required for the seat belt load to build up, for the 30 mile/h collision the forces on a restrained occupant's chest can be brought down to entirely tolerable levels, and chest contact with the steering wheel and instrument panel are prevented.

For most interior designs, a restrained occupant will still have knee contacts with the forward structures. There is clearly a trade-off between a very stiff seat-belt system limiting forward motion but generating high seat belt loads, and a softer system which gives more forward motion. Optimization of restraint design is exceedingly complex in this context because of the interaction of several variables, some of which are not well researched. Crash-severity distributions, human-tolerance variation, the occupant's initial sitting position, the occupant's size and weight and the presence or absence of pre-impact braking, all have a bearing on restraint performance.

In the collision, the head flexes forwards until the chin touches the sternum. Consequently the forward motion of the head of the restrained occupant is about 18 to 28 inches (46 to 72 cm) in a 30 mile/h collision. For convenient use, the steering wheel is some 16 inches (40cm) from the chest, and thus a head contact with the steering wheel is inevitable for the driver in a serious collision.

It is possible to diminish somewhat the amount of forward motion allowed by a conventional seat belt if a pre-loading device is used. A sensor detects crash level decelerations at the front of the car. It activates a device which applies a load to the belt, thus coupling thc occupant more efficiently to the car as it decelerates. Such devices may reduce head forward motion by approximately 2 inches (5 cm) as well as reducing seat-belt loads by about 10 per cent (Mitzkus and Eyrainer, 1984).

The preferred solution for the restrained driver having head and face contact with the steering wheel is to have a supplementary airbag. In a severe collision, the airbag deploys from the hub of the steering wheel and cushions the head as it arcs forwards and downward (Grosch, 1985).

Seat belts themselves can cause injuries. In the limiting condition, rib and sternal fractures will occur. Abdominal injuries are also a feature of three-point seat belts if the lap section is incorrectly positioned above the iliac spines of the pelvis. A phenomenon called 'submarining' can occur when the pelvis rotates out from under the lap section during the collision, allowing the lap belt to load the soft organs of the abdomen. Submarining risk is diminished by careful placement of the lower anchorages of the seat belt and designing the seat cushion to limit the downward and forward motion of the buttocks (Rouhona et al., 1989).

2.2. OCCUPANT KINEMATICS IN ANGLED COLLISIONS

An occupant in a collision always tends to move toward the position from which the principal crash force is applied. Thus, as a lateral component begins to act during a frontal collision so the occupant's trajectory alters. A crash coming from the 10 or 11 o'clock direction (12 o'clock being head-on) results in a front-seat occupant having a path angled across the interior so that the head, for example, would strike either the left-side A pillar in the case of an occupant sitting on the left or the rearview mirror in the case of an occupant sitting on the right.

Great care needs to be exercised in the reconstruction of angled and intersection collisions in terms of assessing occupant trajectories. For example, in a crash in which two cars are on perpendicular paths moving at equal speeds when they strike, the resulting occupant trajectories will be roughly at 45° across the compartment. If the actual contact between the two cars was such that the front of one struck the side of the other, then the pattern of the damage may give a misleading impression of the occupant's path during the crash phase.

Conventional three point seat belts have been shown to be very effective in angled collisions. Experimental studies have shown that the diagonal section effectively restrains the torso up to angles of about 45°. Even when the chest begins to come out from under the diagonal section, the lap restraint acts effectively in diminishing lateral head excursion (Herbert, 1976). In this context, dummies do not give a particularly accurate representation of the actual human response because of poorbiofidelity of the shoulder linkage.

2.3. OCCUPANT KINEMATICS IN SIDE IMPACTS

Collisions in which occupants receive injuries from the side structures of the car constitute almost a quarter of all crashes for fatal and serious injuries. For the occupant on the struck side, loads are applied directly by the door at the levels of the pelvis and chest. The armrest may produce a localized loading of the lateral abdomen.

In the case of a car-to-car impact, the direct loading of the occupant occurs from shoulder level downward, but the head may well flex laterally through the side window to strike the bullet car. With a lorry, the head is more exposed to severe direct loads directly on the front of the striking vehicle and in the case of trees or poles where there is intrusion of the side of the car from door sill to roof level the head is particularly exposed to a direct contact on that intruding object (Mackay, 1983).

Because of the small space between the occupant and the bullet car, it is entirely possible for the occupant to receive a velocity change greater than that experienced by the car in which he is sitting. If the side structure provides negligible resistance to the bullet car before the occupant is contacted, then it is possible for the occupant to receive a velocity change that approaches the absolute impact speed of the bullet car.

That velocity change can be diminished somewhat by increasing the rigidity of the side structure of the target car, but there are severe practical constraints on that approach. A better solution can be achieved by accepting a somewhat higher velocity change on the occupant, and by positioning padding as close as possible to him or her, providing a large ride-down distance. Thus, the magnitude of the applied forces can be diminished.

For the occupant on the struck side in a side impact, a belt has negligible benefit. For an occupant on the non-struck side, however, a seat belt will retain the pelvis close to the initial sitting position, diminish the chances of head contacts across the compartment, and also reduce the interactions with an adjacent occupant, an important mechanism of injury in severe lateral collisions (Mackay, 1983).

2.4. MECHANISMS OF INJURY IN REAR END COLLISIONS

If a car is hit from behind, it is accelerated forward. The back of the seats therefore act against the inertial mass of the occupants. If the seat back does not deflect excessively, then the occupant is accelerated forward with the car. In the absence of any head restraints, the head tends to remain in its initial position in space and consequently it lags behind the thorax. As a result, the neck is extended over the top of the seat back and put into tension. If severe enough, that motion can generate hyperextensive injuries of the neck, which in extreme cases can cause vertebral fractures and damage to the spinal cord. Most car seats have backs that flex elastically, and as a consequence, after loading the seat in a rearward direction the occupant is then projected forward and under some conditions can experience severe contacts with the forward structures. Those conditions for a restrained occupant can produce a forward flexion of the neck after the rearward extension, and it is that sequence of extension and flexion that is often described as a 'whiplash'.

The purpose of head restraints is to limit the initial hyperextension by supporting the head at the same time as the back is being loaded. To do that effectively the head restraint must not only be strong enough but it must be high and close to the head to prevent relative motion developing. In design terms, this presents obvious conflicts with comfort and vision requirements (Thomas et al., 1982).

2.5. OCCUPANT KINEMATICS IN ROLLOVER CRASHES

Rollover crashes are the most random as far as the motion and contacts of the occupant are concerned. Almost every rollover is unique in that small variations in terrain and in vehicle characteristics produce vastly different vehicle kinematics. The great majority of rollovers, over 90 per cent, occur off the roadway and there is the added complexity of striking objects while the car is in an unusual attitude.

On the other hand, field studies show that, providing an occupant is not ejected and the car does not strike any rigid objects, then rollovers are one of the least injurious of crash types. This is because the kinetic energy of the car is being dissipated in small amounts over a large distance.

There is much confusion in the literature over any supposed relationship between the amount of roof deformation that occurs in rollovers and the severity of injuries to occupants. Some studies have shown a statistical relationship, but that does not imply a direct, causal link. The greater the applied load, the greater the amount of roof crush. The greater the applied load, the more severe will be the contacts made by an occupant with the interior. A stronger roof would diminish the roof deformation but would leave the forces applied to an unrestrained occupant undiminished (Strother et al., 1984).

Another area of uncertainty in rollovers is the benefit obtained from use of a seat belt. Field studies indicate quite significant benefits in general terms. Experimental studies using dummies tend to be unsatisfactory because of the poor biofidelity of the shoulder and torso of current dummies. In comparison with the human frame, the torso of current dummies is far too rigid and the shoulder linkage is not adequately reproduced. As a consequence, the dummy does not replicate properly the way in which real people actually move during rollovers. Current crash-test dummies are only satisfactory in the limited condition of a frontal collision or a side impact.

2.6. THE KINEMATICS OF CYCLISTS AND MOTORCYCLISTS IN COLLISIONS

Crashes involving cyclists and motorcyclists can be divided into two general classes. First, there are the single-vehicle events where a loss of control occurs. The rider parts company from his machine and then slides and rolls along the road surface receiving a number of blows that often generate substantial angular forces because of the oblique nature of those impacts.

The second general class of collisions involving cyclists and motorcyclists occurs when there is a vehicle-to-vehicle collision followed by the subsequent passage of the rider along the road surface as a result of any residual velocity that he or she may have. In this case, the closing speeds in the first collision are often high, and the rider is projected from the machine to receive his or her most injurious contacts from the other vehicle. There is a tremendous variation in the properties of the actual structures that are contacted under these conditions. The rider's head may strike almost any part of the car above the wrist rail in a motorcycle-to-car collision, while his or her legs will contact the bumpers or the side panels according to the orientation of the vehicles at impact. Often the legs will receive substantial localized impacts but the rider will maintain a significant velocity, so that he or she travels over the car to have a second series of contacts with the ground and any roadside objects that may be in the rider's path.

Superficially these conditions may seem to be so random that little good can be achieved by attempting to introduce crashworthiness concepts into vehicle exterior design or motorcycle design. However, careful epidemiological studies of the circumstances of pedestrian collisions have shown that great benefits can be obtained by improved design of car exteriors (Ashton and MacKay, 1983). Many of those changes, relating to more compliant bumpers, recessed windshield edges, soft hoods and grills, plus improved external geometry, will benefit cyclists and motorcyclists as well as pedestrians.

2.7. PEDESTRIAN KINEMATICS AND MECHANISMS OF INJURY

In global terms, pedestrians are numerically the largest single group of road users killed in traffic collisions. Although no satisfactory statistics are collected in most Asian, African, or South American countries, there are enough sources of data to suggest that traffic fatalities worldwide total some 500 000 annually and at least half of these are pedestrians (Trinca et al., 1989).

Until the late 1960s, the main sources of information on pedestrian trauma were based either on police or hospital data files, but those sources were intrinsically limited in giving insight into the precise relationships between injuries to pedestrians, the relevant components of vehicles, and the associated kinematics. Several research workers, therefore, in the USA, Australia, and the UK conducted detailed at-the-scene studies within 30 min of the occurrence of the collision, when the cars would be examined in situ and the environmental circumstances and vehicle damage evaluated in detail (Huelke and Davis, 1969; Ryan and McLean, 1966; Ashton et al., 1977).

Field studies in general show that in some 8 per cent of cases of pedestrian casualty collisions, either more than one pedestrian or more than one vehicle is involved. The great majority of pedestrians are injured in single-vehicle and single-pedestrian impacts. In most Western countries, studies show that some three-quarters of pedestrians are struck by cars, the remainder mainly by lorries (14 per cent), motorcycles (5 per cent), and buses (4 per cent). The involvement rate on a vehicle base varies greatly. Buses, for example, are six times more likely, and trucks twice as likely, to be involved in a pedestrian collision than is a private car on a yearly basis. A better measure in terms of vehicle design is the relative risk of striking a pedestrian throughout the total operating life of the vehicle (Trinca et al., 1989).

On that basis, a bus is some 12 times more likely to injure a pedestrian than is a car. Other similar high-risk situations can be detected; for example, the average Manhattan taxi has a one in eight chance of injuring a pedestrian in 2 years; taxis in other large cities, such as London and Paris, may well have equivalent involvement rates. The chance of being killed also varies with the type of vehicle; only 3 per cent of pedestrians struck by cars or taxis are killed, while for lorries the percentage killed rises from 4 per cent for vehicles of less than 1.5 tons to 13 per cent for those over 4.5 tons in weight.

Ashton et al (1977) reporting on at-the-scene studies in Birmingham in the UK, noted that pedestrians were most frequently struck by the front of the vehicle, but in an asymmetrical manner. With left hand driving in the UK, most pedestrians are struck on their right sides by the left front corner of the car as soon as the pedestrian has moved into the road rather than when he or she is over half way across.

The most frequent type of pedestrian collision is that in which the pedestrian is struck by the front of a car. The initial contacts are from the bumper, which strikes the leg, and from the leading edge of the bonnet, which strikes the thigh or pelvis. The exact location of these contacts depends on the relative heights of the pedestrian and of the parts of the vehicle. The pedestrian then rotates about the leading edge until the head, shoulders, and chest strike the bonnet, the windscreen or its frame. At high impact speeds, the pedestrian rotates about a second contact of the head or shoulders, and the legs may strike the roof. By this time, the casualty is travelling at approximately the speed of the car.

Should there have been little or no braking during the crash phase it is possible that, at high speeds, the pedestrian may pass over the top of the car. If, however the vehicle is being braked - and this is the most frequent condition - the car slows down faster than does the pedestrian who thus continues to move forward and lands on the road in front of the vehicle The person will finally come to rest after sliding and rolling along the road surface.

The extent and severity of the contacts on the vehicle depends on its speed and on the relative heights of the pedestrian and front structure. The initial contact results in the pedestrian being pushed forward and at the same time rotated about his or hercentre of gravity. He/she is thus not 'run over' but 'run under' by the front of the car.

There are thus two phases to a pedestrian collision. The first phase consists of multiple contacts with the car; the second phase is the rolling and sliding motion that occurs on the road surface. Research has shown that injury severity is strongly associated with impact speed for the first phase, but for the ground contacts there is no dependence on speed and those injuries are generally minor.

The overall relationship of vehicle exterior shape to pedestrian injury represents one of the major areas of biomechanical research at the present time. Enough evidence exists to suggest that, in practice, different car profiles do present different risks of injury. Perhaps the most striking study that illustrates this point is one by McLean (1972) who compared two groups of pedestrians, the first being struck by Volkswagen Beetles and the second by Cadillacs. The samples were standardized as far as possible with regard to other relevant variables. He concluded that there would be a 30 per cent reduction in pedestrian fatalities in the USA if all cars that struck pedestrians had Volkswagen Beetle fronts.

At present, however, neither experimental studies nor field investigation projects are sufficiently far advanced for the ideal vehicle exterior to be specified. What may appear to be an optimum design in terms of contour and resilience for one impact speed and one particular height of pedestrian may well produce particularly unfavourable impact conditions in different circumstances.

One unfortunate development from the pedestrian's point of view has been the promulgation of the bumper standard in the USA. This is not a safety standard at all, it is an economic standard aimed at reducing repair costs by standardizing bumper heights and specifying strength requirements in a 5 mile/h impact. The standard has resulted in very rigid beams across the front of cars at 70 inches (50 cm) above the ground. This corresponds to adult knee height and results in a greater incidence of disabling injuries to the knee joint, involving damage to the articulating surfaces of the joint and ligamentous injuries, than does a bumper set at a lower height (Kaiser, 1991).

3. The Optimization of Car Occupant Protection by Advanced Restraint Technologies

Current seat belts have been shown to be very effective in diminishing the frequency and severity of injuries to car occupants. So much so that high levels of seat belt use are a prime aim of all national transport safety policies in motorized countries. The limitations of the protective abilities of current seat belts have been well documented in many analyses of both field accident data and experimental studies [Bacon, 1989].

Real world accident studies have identified five categories of limitations to the performance of current seat belts. These are:

1) Head and face contacts with the steering wheel by restrained drivers [Rogers et al, 1992] - It is inherent in the kinematics of a restrained occupant that, in a severe collision at a velocity change of around 50 km/hr, the head will arc forwards and downwards, having a horizontal translation of some 60 to 70 cms (see Figure 1). If a normal steering wheel position is superimposed on such a trajectory, the head and face necessarily will strike the steering wheel. Such contacts usually produce AIS 1 to 3 injuries and are best addressed with the supplementary airbag systems becoming common throughout the new vehicle fleet.

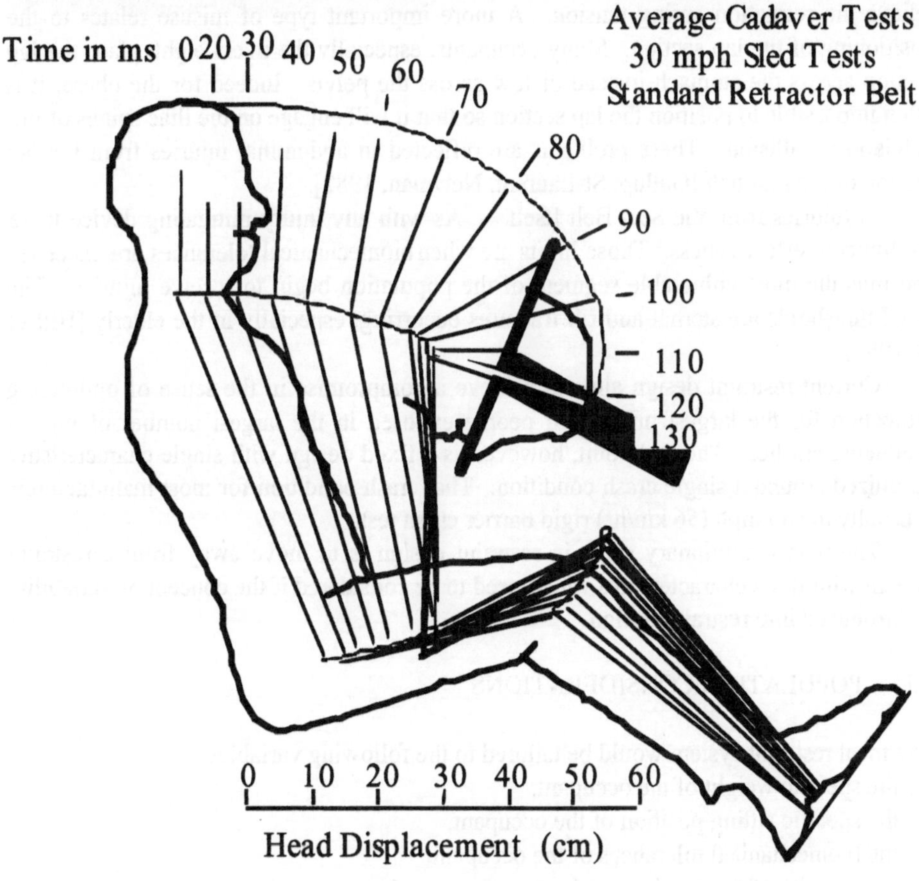

Figure 1. Seat belt excursion

2) Intrusion of Forward Structures - A seat belt requires a zone ahead of the occupant so that the occupant can be decelerated by the compliance of the restraint system. If intrusion compromises that space, then specific localized contacts can occur. The injury risk from such contacts may well be small if they are occurring with structures which have been engineered appropriately. Indeed, in the ultimate condition, it is better for the occupant to be decelerated not just by the seat belt alone but through a

combination of belt loads and contact loads. Those contact loads are through the feet at the firewall, through the knees into the lower dash and through the airbag and belt at chest level. In severe collisions, however, major intrusions are destroying the passenger compartment so that exterior objects are actually striking the occupants. This is a feature of restrained fatalities in frontal impacts [Mackay et al, 1990].

3) Rear Loading - Correctly restrained front seat occupants can receive injuries from unrestrained occupants, luggage or animals from the rear seats. Such events contribute to some 5% of restrained front seat fatalities [Griffiths et al, 1976].

4) Misuse of the Seat Belt - Seat belts must be positioned correctly on the human frame to work effectively. Dejeammes (1993) in a survey of belt use in France found that some 1.6% of front seat occupants had the shoulder belt under the arm or behind the back whilst some 3.3% had introduced slack because of the use of some clip or peg to relieve the retraction spring tension. A more important type of misuse relates to the positioning of the lap section. Many occupants, especially the overweight, place the lap section across the stomach instead of low across the pelvis. Indeed for the obese, it is often impossible to position the lap section so that it will engage on the iliac spines of the pelvis in a collision. These problems are reflected in abdominal injuries from the lap section of the seat belt [Gallup, St-Laurent, Newman, 1982].

5) Injuries from the Seat Belt Itself - As with any injury mitigating device there are limits to effectiveness. Those limits are when biomechanical tolerances are exceeded and thus the most vulnerable segment of the population begin to receive injuries. The usual thresholds are sternal and rib fractures occurring, especially in the elderly [Hill et al, 1992].

Current restraint design aims to achieve a compromise in the sense of optimizing protection for the largest number of people exposed in the largest number of injury-producing crashes. The end point, however, is a fixed design with single characteristics optimized around a single crash condition. That crash condition for most manufacturers is usually the 35 mph (56 km/hr) rigid barrier crash test.

The next evolutionary stage in restraint design is to move away from a restraint system with fixed characteristics which need to be considered if the concept of variability is introduced into restraint design.

3.1. POPULATION CONSIDERATIONS

The ideal restraint system would be tailored to the following variables:

- the specific weight of the occupant,
- the specific sitting position of the occupant,
- the biomechanical tolerances of the occupant,
- the severity of the specific crash which is occurring,
- the chances of specific passenger compartment intrusion occurring which might compromise restraint performance
- might compromise restraint performance,
- the specifics of the compartment geometry and crush properties of the car.

3.2. ANTHROPOMETRIC CONSIDERATIONS

Current dummies and modeling cover the 5th percentile female to 95th percentile male range. Assuming for simplicity that males and females are exposed equally and that there are few males smaller than the 5th percentile female or females larger than the 95th percentile male, these conventional limits put 2.5% (1 in 40) of the small population and 2.5% of the larger population beyond those limits; 5% or 1 in 20 overall.

Table 3 gives the 1% and 99% ranges for height, sitting height and weight. These data show what would be required if the design parameters were extended to cover this wider range, so that only 1 in 50 of car occupants would be outside the design parameters [Society of Actuaries, 1979].

Table 3. Population ranges for height, sitting and weight

Adult	Height ins/cm	Sitting Height ins/cm	Weight lbs/kg
1%ile female	57/145	28/72	82/37
5%ile female	59/150	29/75	90/41
95%ile male	73/185	37/93	225/102
99%ile male	75/190	38/96	236/107

More importantly, it is implicitly assumed in current designs that height (or sitting height) and hence sitting position are colinear with the weight of the occupant. In fact, there are data available to suggest that the relationship between height and weight are rather complex. For example, the body mass index (i.e., the ratio of weight in kilograms to height in meters squared) varies to a greater degree in women than in men, and particularly at the 75th percentile and above, women have higher BMIs than men. In addition, the prevalence of overweight increases with age, more with females than males [Williamson, 1993].

Therefore to optimize a restraint system it would appear appropriate that sitting position and body weight should be assessed independently if variability is to be introduced into restraint design.

3.3 POPULATION CHARACTERISTICS BY POSITION IN THE CAR

European data show that some 80% of drivers in injury-producing collisions are male, whilst some 65% of front seat passengers are female [Bull and Mackay, 1978]. Approximately one-third of rear passengers are children of 10 years of age or under [Huelke, 1987]. These simple frequencies suggest that restraint characteristics should not necessarily be the same for all sitting positions in the car.

3.4 SITTING POSITIONS

Current design is predicated on the positions established for the three conventional dummies. Observational studies by Parkin et al (1993) have demonstrated that there are substantial differences between those three positions and an actual population of drivers. Passive observations of drivers in the traffic stream have been made using video recording techniques, and drivers classified by sex and general age groups of young (35 years), middle (36-55 years) and elderly (56 years and older). Make and model of car were recorded and measurements made of the following distances:

- nasion to steering wheel upper rim and hub,
- top of head to side roof rail,
- back of head to head restraint, horizontally and vertically,
- shoulder in relation to 'B' pillar.

Such techniques allow thousands of observations to be made quickly and therefore population contours can be drawn. Figure 2 illustrates how particularly for the 5th percentile female population the actual sitting position is significantly closer than that of the 5th percentile dummy, by some 9.2cm. The 5th percentile, small female population sits some 38cm (15 inches) or closer to the hub of the steering wheel.

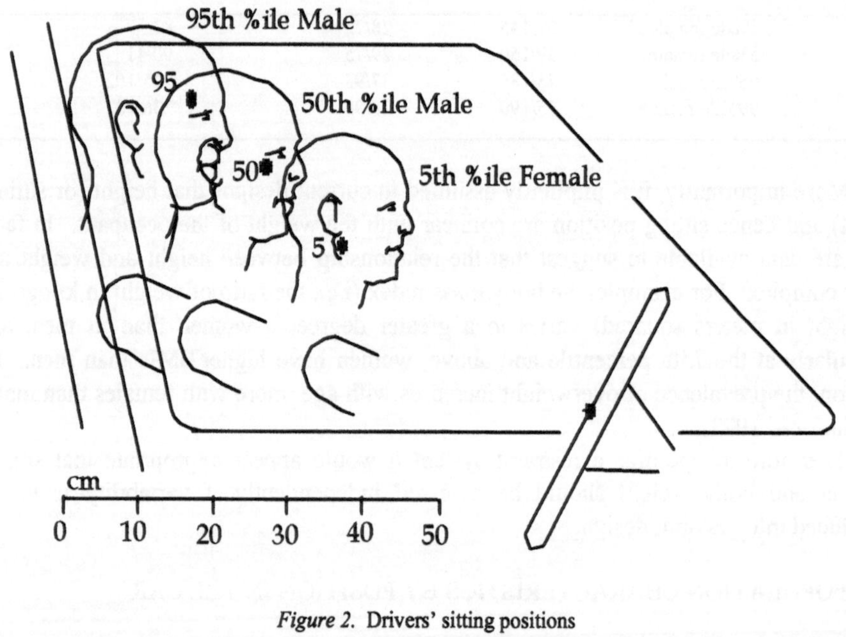

Figure 2. Drivers' sitting positions

3.5 BIOMECHANICAL VARIATION

An extensive literature exists concerning human response to impact forces, mostly conducted in an experimental context. A general conclusion from that body of knowledge is that for almost any parameter, there is a variation of at least a factor of 3 for the healthy population exposed to impact trauma in traffic collisions [McElhaney, Roberts, Hilyard, 1976]. That variation applies to variables which are relatively well researched such as the mechanical properties of bone strength, cartilage, ligamentous tissues and skin. It is likely to be even greater when applied to gross anatomical regions such as the thigh in compression, the thoracic cage, the neck or the brain.

How such variability is demonstrated in populations of collisions is less well understood. Data from a ten year in-depth study of European crashes for restrained front seat occupants are given in Figures 3 and 4. The methodology of that work has been described elsewhere [Mackay et al, 1985].

Figure 3 illustrates the effect of age on injury outcome in terms of the frequency of AIS 2 and greater injuries for three age groups. The 60+ age group especially shows greater vulnerability than the younger groups. As a broad generalization one may conclude that for the same injury severity, the younger age groups must have a velocity change of some 10 km/hr more than the elderly. The effect is more marked if a more severe injury level is chosen. Figure 4 illustrates the cumulative frequencies for the three age groups for injuries of AIS 4 and greater.

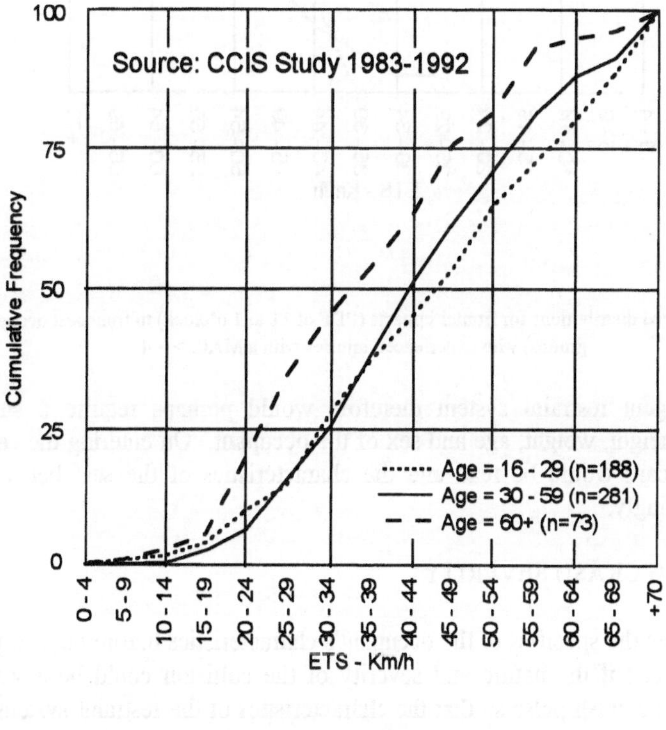

Figure 3. **Crash speed distributions for frontal impacts (PDF of 11 to 1 o'clock) to drivers (by age groups) who experienced injuries with a MAIS > = 2**

Figure 5 shows similar frequency curves for crash severity by sex of occupant. Thus at a velocity change of 48 km/hr (30 mph), some 2/3 of male and some 80% of female AIS 2+ injuries have occurred. As a starting point, therefore, as well as specific body weight and sitting position, a combination of age, sex and biomechanical variation could be developed as a predictor of the tolerance of a specific person within the population range.

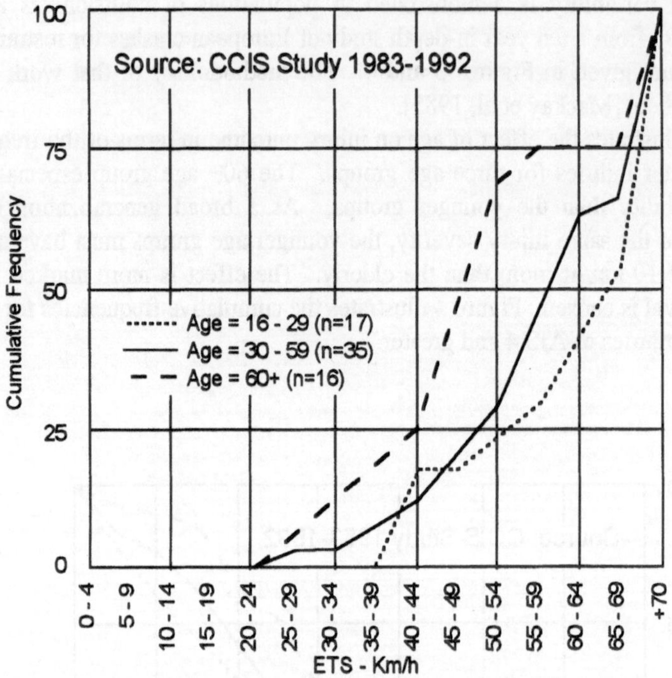

Figure 4. **Crash speed distributions for frontal impacts (PDF of 11 to 1 o'clock) to front seat occupants (by age groups) who experienced injuries with a MAIS >= 4**

An intelligent restraint system therefore would perhaps require a smart card, specifying the height, weight, age and sex of the occupant. On entering the card for the first time, the card would be read and the characteristics of the seat belt and airbag adjusted accordingly.

3.6. SENSING CRASH SEVERITY

Besides assessing the specifics of the occupant's characteristics before impact, protection could be enhanced if the nature and severity of the collision could be assessed early enough during the crash pulse so that the characteristics of the restraint system could be modified. That would require, for example, sensors to discriminate between distributed versus concentrated impacts, and between, for example, three levels of collision severity such as less than 30 km/hr, 30 to 50 km/hr, and greater than 50 km/hr. In addition, conceptually one might have an array of sensors which would detect the early development of compartment intrusion. Such electronic data could then instruct the restraint system to change its characteristics early enough during the crash phase to alter the characteristics of the restraint and thus the loads on and forward excursion of the occupant.

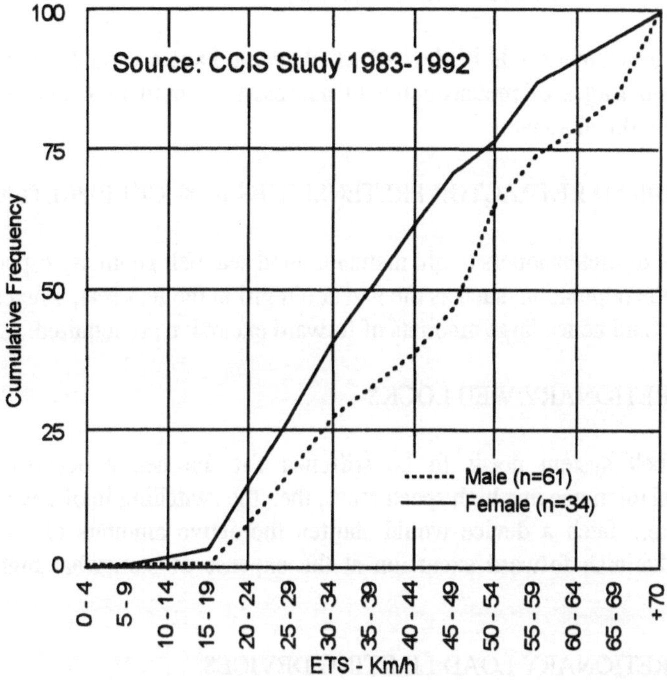

Figure 5. **Crash speed distributions for frontal impacts to front seat passengers (by sex) who experienced injuries with a MAIS >= 2**

3.7 VARIABLE RESTRAINT CHARACTERISTICS

The advantages of a variable restraint system are illustrated by considering some examples. A front seat passenger, 70 years of age and female, weighing 45 kg sitting well back, in a 30 km/hr frontal collision with no intrusion, would be best protected by a relatively soft restraint system which would maximize the ride-down distance and minimize the seat belt loads. That would require a low pretensioning force, a long elongation belt characteristic provided by load limiters and a soft airbag.

Such a system is very different from what would be required by a 25 year old, 100 kg male, sitting close to the steering wheel in a 70 km/hr offset frontal collision. He would need a very stiff seat belt, an early deploying stiff airbag and a large amount of pretensioning load.

Consider thirdly a 9 year old girl, weighing 30 kg sitting in a rear seat in a 56 km/hr frontal impact. Maximizing her ride-down distance and minimizing the seat belt loads would require low pretensioning loads and a very soft belt system, but one which would still have a biomechanically satisfactory geometry at the forward limit of excursion. Possible techniques for introducing variability into restraint design are now discussed.

3.8. VARIABLE PRETENSIONING FORCE

A retractor pretensioner might be devised which would have a variable stroking distance or perhaps two stages of pretensioning to address the population and crash severity requirements outlined above.

3.9. COMBINED RETRACTOR PRETENSIONER & BUCKLE PRETENSIONER

Such a system of pretensioners might maintain good seat belt geometry especially for the small end of the population, such as the 9 year old girl in the rear seat, when soft restraint characteristics and hence large amounts of forward excursion are required.

3.10 DISCRETIONARY WEB LOCKS

If the seat belt system needs to be stiffened for the heavy occupant with high biomechanical tolerance in a high speed crash, then the switching in of a web lock would be appropriate. Such a device would shorten the active amounts of webbing being loaded and diminish forward excursion at the expense of somewhat higher seat belt loads.

3.11. DISCRETIONARY LOAD LIMITING DEVICES

One way of providing for biomechanical variability would be to have a load limiting mechanism which would be calibrated for the specifics of the occupant's age, sex and weight. Such a device could also be adjusted according to transient sitting position. Belt loads would be limited at the expense of increased forward excursion.

3.12. VARIABLE SITTING POSITIONS

Ultrasonic, infrared or other techniques of sensing might be used to monitor continuously the head position of each occupant. Such information could be used at a minimum to provide a warning that an occupant was sitting too far forward and in particular too close to the steering wheel. At a more advanced level it could be used to tune the seat belt and airbag characteristics to be optimized for that occupant in that specific position by adjusting the other restraint variables.

3.13. VARIABLE AIRBAG FIRING THRESHOLD

The need for an airbag varies according to seated positions in the car and the characteristics and sitting position of the occupant. For most drivers in most sitting positions a supplementary steering wheel airbag becomes desirable in crash severities above 30 km/hr [Rogers et al, 1992]. For a front seat passenger however, particularly one who is towards the top end of the biomechanical tolerance spectrum and sitting well back, an airbag at 30 km/hr is unnecessary. For a child sitting a long way forward in such a crash, it might also be disadvantageous. Hence specific sensing techniques at a minimum could discriminate between the presence or absence of a passenger, and at the next level assess the need for the airbag to inflate or not.

3.14. VARIABLE AIRBAG CHARACTERISTICS

In response to the sensing data about the occupant's characteristics and transient sitting position, and the accelerometer data about the nature and severity of the collision which is occurring, the airbag properties could be varied. Specifically, gas volume and inflation rate could be changed. Compressed gas systems instead of chemical gas generators have the potential for providing those characteristics by having time-based adjustable inflation ports. This requires very advanced sensing and control systems but these aims could well be addressed through future research and development.

4. Other Crash Configurations

The discussion so far has focused on frontal collisions which constitute some 50% to 65% of injury producing collisions in most traffic environments. Lateral, rear and rollover crashes also suggest opportunities for optimizing protection through intelligent restraint systems.

4.1. LATERAL COLLISIONS

The technology is now developing for side impact airbags with two versions becoming available on 1995 model-year passenger cars. The observational data of Parkin et al (1993) have illustrated the range of driver sitting positions which reflect the requirements of side impact airbag geometry to cover both the door and the B pillar. Because a significant part of the population, tall males, choose to sit as far rearward as possible, in a side impact in many four door vehicles the thorax would be loaded by the B pillar rather than the door.

A practical issue is the nature and position of the sensor for a side impact. Because of the extremely short time available for sensing, around 5 milliseconds, a simple switch system is appropriate [Haland, 1991]. An analysis of a representative sample of AIS 3 plus lateral collisions has demonstrated that if a switch sensor is located in the lower rear quadrant of the front door then approximately 90% of all such side impacts would be sensed appropriately. A set of several sensors would be required to address the remaining few collisions, whilst rear seat occupant protection would also be addressed in large part by a sensor in the same position in the front door as is appropriate for front seat occupants [Hassan et al, 1994].

4.2. REAR IMPACTS

Occupant protection in rear end collisions is addressed largely through the appropriate load deflection characteristics of seat backs and the provision of correctly positioned head restraints. The real world data of Parkin et al (1993) demonstrates that head restraints are frequently positioned both too low and too far to the rear of the occupant's actual head position. The head position sensors discussed above could also be used for adjusting automatically both the vertical and horizontal position of the head restraint. Such a technology is relatively simple but the costs and reliability, as well as acceptability by the driving population, present serious practical problems.

4.3. ROLLOVER ACCIDENTS

Actual mechanisms of injury in rollover accidents have been well researched byBahling et al (1990) for occupants in current seat belts. Conceptually one can suggest that a buckle pretensioner might have some benefits in rollover circumstances by diminishing the relative vertical motion of an occupant. However, in rollovers current dummies do not have the appropriate soft tissue or thoracic and lumbar spine response characteristics, in comparison to the human frame. The basic clearance of currentbodyshell design and packaging limit intrinsically the ability of any restraint system to modify the nature of any roof contacts under the forces of actual rollover circumstances even with no roof deformation taking place. Raising current roof lines leads to many undesirable consequences. Nevertheless it would be of interest to explore occupant kinematics in rollovers using more realistic techniques with volunteer and cadaver subjects in the context of buckle pretensioners and the requirements of a sensor to detect incipient rollover.

5. Conclusions

The above has outlined in conceptual form some of the issues which need to be addressed in advancing from today's seat belts and airbags towards some form of intelligent restraint system. Of fundamental importance is to recognize the population issues of size, sitting position, biomechanical variation and changing crash exposures. Beyond these issues lies a larger amount of challenging research and development to actually produce the sensors and hardware to provide variability in a seat belt and airbag system. Proximity sensing has its advocates, and if radar techniques could actually discriminate an impending collision from a near miss or a passing object, then the provision of say 500 milliseconds warning would alter many of the restraint issues reviewed in this paper. However, the basic premise remains; the next generation of restraints must change from having single fixed characteristics towards variable ones which recognize the real world population variables of weight, sitting position, biomechanical tolerance and crash exposure.

6. Summary

This review has touched superficially oh the various mechanisms of injuries to the main classes of road-user casualties, and illustrated some of the desirable vehicle design characteristics which can minimize road trauma. There is still much to be gained by better vehicle design. For occupants, enhanced performance from restraint systems, supplementary airbags, better structural integrity and better compatibility between cars and lorries are obvious areas where known solutions are waiting to be applied.

For pedestrians there is an extensive literature outlining the characteristics of friendly vehicle exteriors; many current models of cars are exhibiting some of these properties but better agreement on an optimum specification for the car's exterior is still required.

For motorcyclists there is the promise of improved leg protection from energy-absorbing fairings although substantial technical debate still surrounds these proposals. For the clinician, a more accurate appreciation of the likely injuries which occur in given collision circumstances can help in diagnosis and treatment. It is for these reasons that some understanding of vehicle design and crashworthiness is worthwhile.

7. References

1. Aldman B. (1962) Biodynamic studies of impact protection. *Acta Physiol.* Scand. 56, 192.
2. Appel H. and Wustemann J. (1986) Risk order of injury-causing car parts in various types of car accidents. *Int. J. Vehicle Design* 7, 232.
3. Ashton S. J and Mackay G. M. (1983) Benefits from changes in vehicle exterior design.*Soc. Auto. Eng. Society of Automotive Engineers Technical paper* number 830626, 255.
4. Ashton S. J., Pedder J. B and Mackay G. M. (1977) Pedestrian injuries and the car exterior.*Soc. Auto. Eng.* 770092.
5. Bacon, D.G.C. (1989) The Effect of Restraint Design and Seat Position on the Crash Trajectory of the Hybrid III Dummy, *12th International Technical Conference on Experimental Safety Vehicles*, 12:451-457, Gōteborg, Sweden,
6. Bahling, G.S., Bundorf R.J., Kaspzyk G.E., Moffatt E.A., Orlowski K.R and Stocke J.E. (1990) Rollover and Drop Tests. The Influence of Roof Strength on Injury Mechanisms Using Belted Dummies, *Proceedings 34th Stapp Car Crash_Conference,* 34:101-112, Society of Automotive Engineers, Warrendale, PA.
7. Bull J.B and Mackay G.M. (1978) Some Characteristics of Collisions, the Population of Car Occupant Casualties and Their Relevance to Performance Testing,*Proceedings IRCOBI 3rd Conference*, pages 13-26, Lyon, France, September.
8. Cairns H. (1941) Head injuries in motorcyclists. *Br. Med. J.* No. 4213, 465.
9. Cairns H. (1946) Crash helmets. *Br. Med. J.* No. 4470, 322.
10. Careme L. M. M. (1990) Biomechanics of head injury in frontal impacts. *Society of Automotive Engineers Technical paper* number 900541.
11. Cesari D. (1983) A review of injury mechanisms tolerance data and protection criteria in side impact accidents. *Proc. Seminar Biomechanics Impacts,* EEC Eur 8939, 138.
12. DeHaven H. (1942) Mechanical analysis of survival in falls from heights of fifty to one hundred and fifty feet. *War Med.* 2, 586.
13. Dejeammes M., Alauzer A and Trauchessec R. (1990) Comfort of Passive Safety Devices in Cars: Methodology of a Long-Term Follow-Up Survey, *SAE Paper* No.905199, Society of Automotive Engineers, Warrendale, PA.
14. Gallup B.M., St. Laurent A.M and Newman J.A. (1982) Abdominal Injuries to Restrained Front Seat Occupants in Frontal Collisions, *AAAM Proceedings*, 26:131-145.
15. Garrett J. W. (1962) An evaluation of door lock effectiveness. *Proc. 5th Stapp Conf.* 20.
16. Gennarelli T. A., Thibault L. E and Adams I H. (1982) Diffuseaxonal injury and traumatic coma in the primate. *Am. Neurol.* 12, 564.
17. Gissane W. and Bull J. P. (1962) Injures from road accidents. *Practitioner* 1988, 489.
18. Griffiths D., Hayes M., Gloyns P.F., Rattenbury S and Mackay M. (1976) Car Occupant Fatalities and the Effects of Future Safety Legislation. *Proceedings, 20th Stapp Car Crash Conference,* 20:335-388, Society of Automotive Engineers, Warrendale, PA,
19. Grosch L. (1985) Injury criteria for combined restraint systems. *Proc. 10th Exp. Safety Veh. Conf.* NHTSA 338.

20. Gurdjian E. S. (1945) Experimental studies on the mechanism of head injury.*Res. Bull. Assoc. Nerv. Ment. Dis.* 24, 48.
21. Haland Y and Pipkorn B. The Protective Effect of Airbags and Padding in Side Impacts - Evaluation of a New Subsystem Test Method. (1991) *13th International Technical Conference on Experimental Safety Vehicles*, 13:523-533, Paris.
22. Hassan A., Morris A and Mackay M. (1991) The Best Place for a Side Impact Airbag Sensor. *Proc. AAAM/IRCOBI Conference on Advances in Occupant Restraint Technologies*, Lyon, France, September.
23. Herbert D.C. (1976) Occupant Head Space in Passenger Cars. *TARU Report* 1/176. Sydney.
24. Hill J.R., Mackay M., Morris A.P., Smith M.T and Little S. (1992) Car Occupant Injury Patterns with Special Reference to Chest and Abdominal Injuries caused by Seat Belt Loading,*Proc. IRCOBI Annual Conference*, pp. 357-372, Verona, Italy, September.
25. Huelke D. F and Davis R. A. (1969) A study of pedestrian fatalities. *Univ. Mich.* HSRI 9.
26. Huelke D. F., Gikos P. W and Hendrix R. C. (1963) Patterns of injury in fatal automobile accidents*Proc. 6th Stapp Conf. 44.*
27. Huelke D.F., Ostrom M., Mackay M and Morris A. (1993) Thoracic and Lumbar Spine Injuries and the Lap/Shoulder Belt. *Proceedings SAE Congress*, SAE930640, Society of Automotive Engineers, Warrendale, PA.
28. Huelke D.F. (1987) The Rear Seat Occupant in Car Crashes, *AmericanAssn for Automotive Medicine Journal* 9:21-24.
29. Kaiser J. (1991) The biomechanics of knee injuries*[Thesis]*. Chalmers Tech. Univ. Goteborg, Sweden. June.
30. Mackay G. M. (1969) Some collision aspects of British road accidents. *Auto. Eng.* 59, 500.
31. Mackay G. M. (1983) Characteristics of lateral collisions. *Proc. Seminar Biomechanics of Impacts.* EEC Eur 8939:11,1.
32. Mackay G.M., Cheng L., Smith M and Parkin S. (1990) Restrained Front Seat Car Occupant Fatalities. *AAAM Proceedings* 34:139-162, October 1990.
33. Mackay M., Ashton S., Galer M and Thomas P. (1985) Methodology ofIn-Depth Studies of Car Crashes in Britain. Proceedings, Accident Investigation Methodologies, SP159, pages 365-390, *SAE Paper* 850556, Society of Automotive Engineers, Warrendale, PA.
34. Mackay, G. M. (1977) Belted occupants in frontal crashes. *Proc. 6th Int. Conf. IAATM* 351.
35. McElhaney J.H., Roberts V.L and Hilyard J.F. (1976) *Handbook of Human Tolerance,* Japanese Automobile Research Institute, Tokyo.
36. McLean A. J. (1972) Car shape and pedestrian injury.*Proc. Symposium Road Safety.* Dept. Transport. Canberra. 179.
37. Mitzkus J. E. and Eyrainer H. (1984) Three-point belt improvements for increasing occupant protection. *Soc. Auto. Eng.* P141, 245.
38. Parkin S., Mackay M and Cooper A. (1993) How Drivers Sit in Cars.*Proc. AAAM*37:375-388, November.
39. Rogers S., Hill J and Mackay M. (1992) Maxillofacial Injuries Following Steering Wheel Contacts by Drivers Using Seat Belts. *Brit. J. Oral & Maxillofacial Surgery,* 30:24-30, 1992.
40. Rouhona S. W., Horsch J. D and Kroell C. K. (1989) Assessment of lap-shoulder belt restraint performance in laboratory testing. *Proc. 33rd Stapp Conf.* P227, 43.
41. Ryan G. A and McLean A. J. (1966) Pedestrian survival. *Proc. 9th Stapp Conf.* 321.
42. Severy D. M and Mathewson J. H. (1954) Automobile-barrier impacts. *Natl. Res. Council Pub.* 334. 39.
43. Sheldon J. H. (1960) On the natural history of falls in old age. *Br. Med. J.* 10, 1685.
44. Snyder R. G., Foust D. R and Dowman B. M. (1977) Study of impact tolerance through free-fall investigation. *Univ. Mich. HSRI Rep.* 77. Society of Actuaries Build and Blood Pressure Study, London, 1979.
45. Stapp, J. P. (1951) Human exposure to linear acceleration. *Aero. Med. Lab. Air Force Report 5912, 2.*
46. Strother C., Smith G. C., James M. B and Warner C. Y. (1984) Injury and intrusion in side impacts and rollovers. *Soc. Auto. Eng.* P141, 317.
47. Thomas C., Faverjon G., Hartemann F., Tarriere C., Patel A and Got A. (1982) Protection against rear-end accidents. *Proc. 7th IRCOBI Conf.*
48. Trinca G. W., Campbell B. J., Haight F., Johnston 1. R., Knight P., Mackay G. M., McLean A. J and Petrucelli E. (1989) Reducing traffic injury - a global challenge. *Royal Aust. Col. Surg.*
49. Viano D. C and Lau I, V. (1985) Thoracic impact a viscous tolerance criterion.*Proc. 10th Exp. Safety Veh. Conf.* NHTSA 104.
50. Viano, D. C. (1988) Biomechanics of head injury - toward a theory linking head dynamic motion brain tissue deformation and neural trauma. *Society of Automotive Engineers.* Technical paper number 881708.
51. Williamson D.F. (1993) Descriptive Epidemiology of Body Weight and Weight Change in U.S. Adults. *Ann Intern Med.* Oct 1; 119(7 Pt 2):646-9.

COMPATIBILITY ISSUES AND VULNERABLE USERS

CLAUDE H.E. TARRIERE
Head of the Automobile Biomedical Department
Renault S.A.
132 rue des Suisses
92000 Nanterre - France

Abstract

Compatibility issues; *1.Car-to-car:* the ratio of the masses of the cars governs the level of risk for the occupants of the two cars involved in a fronto-frontal crash $\Delta V_1 / \Delta V_2 = M_1 / M_2$. Because the heterogeneity of the car masses is a given parameter of the real world, the only solution for improvement is to modify the stiffness of the cars' structures. The strength limit of the lightest car (before the deformation of the passenger compartment) constitutes the limit of the force acceptable for the heaviest cars; *2.Car-to-truck:* obviously the maximal incompatibility is associated to the car-to-truck front end impact. Avoiding the « underrun » of the car under the front of the truck and dissipating the impact energy are the two main countermeasures that can reduce the injury risk for the car occupants. Both are related to the truck because in this case the behaviour of the car structure itself is negligible. Vulnerable users: pedestrians could be considered as another case of user incompatibility when impacted by a vehicle. Countermeasures are more easily developed for impact avoidance than for car-pedestrian impact compatibility. Child occupants are also vulnerable users. Not only because they have some weaknesses at neck level for example but also because they have no autonomy and their protection depends entirely on the care given by the parents. Each child age group requires a specific protection system : rearward facing devices up to 3 or even 4 years old, booster cushions as a complement of the 3 points belt between 3 and 10 years old. Such requirements will be explained according to age, location in the car, belt type (3 and 2 point belts). The roles of ISOFIX or alternative approach are discussed as potential improvements for the near future.

J. A. C. Ambrósio et al. (eds.),
Crashworthiness of Transportation Systems: Structural Impact and Occupant Protection, 139–171.

1. Compatibility issues[1]

1.1. INTRODUCTION

Since an acceptable level of protection is obtained (or planed to be so) for the main crash modes and especially in frontal impact, another major item comes as a priority for the near future the compatibility issues.

Ignoring the compatibility requirements between cars of different masses when looking for an improvement of the car behaviour in frontal impact could be counter-productive if resulting in heaviest car stiffer and stiffer, still aggravating the unacceptable existing incompatibility between heaviest and lightest cars.

Another issue is car-to-truck compatibility. It has also to be considered as a high priority.

The compatibility between vehicles is not a new idea. Compatibility and aggressivity were introduced by European researchers several decades ago (1) to (6) both in frontal and side impacts analysing the aggressivity in terms of masses, stiffness and architecture and the structural compatibility between cars in a mass ratio of 1 to 2 was shown as feasible to realise in 1972 (3). All the seventies were marked by intense activity on this topic (1) to (17).

But the feasibility of a structural compatibility is not sufficient. To reach an equal protection effectiveness in both vehicles very powerful restraint systems are needed. And it is only in the present decade they will come available with smart optimised belts and air-bags systems (18) (19) (20).

While not a new idea, car manufacturers and governmental agencies have initiated renewed efforts to study compatibility as a mean of reducing crash related injuries below the best level already possible by more traditional crashworthiness approach ... (18 to 43).

« ... pursuing an optimal crashworthy design without regard to its collision partners can lead to very aggressive vehicles. Design modifications which minimise injuries in one vehicle may actually accentuate injury levels in the collision partner » (31).

1.2. ACCIDENTOLOGICAL ISSUES

1.2.1. *The Weight of the Compatibility*

The analysis of the French accidentology allows to evaluate the weight of the compatibility in terms of victim numbers.

For a total of 6,000 fatalities inside the car, it is an amount of 2,285 that we have to consider as relevant of the compatibility domain.

The whole analysis is described in the Table 1. On the second line appears what would be the number of fatalities in case of 100 % seat belt wearing. And 2,285 represents 54 % of these 4,200 belted occupants : 30 % between cars (car-to-car) and 24 % between car-to-truck.

---

[1] With the help of D. BELLOT, head of Safety Engineering Department

TABLE 1. Occupants taken into account in compatibility car-to-car

	Fatalities	Seriously injured
Total in France all obstacles[1]	6,000	18,000
Corresponding value[2]	4,200	16,000
Car-to-car head-on collisions	700	5,750
Car-to-car lateral collisions	in lateral: 520 in side impact[3]: 385	in lateral: 1,740 in side impact[3]: 920
Car-to-car rear-end collisions	30	240
Car-to-truck head-on collision	500	850
Frontal car against the truck[4]	140	620
Frontal truck against car lateral	350	320
Frontal truck against car rear	45	110
Total car-to-car	1250	7730
Total car-to-truck	1035	1900
TOTAL	2285	16630

[1] all crash configurations
[2] if belt wearing = 100 %
[3] with direct intrusion against the occupant
[4] lateral or rear areas

Compatibility = 54 % fatalities in which 30 % car-to-car
24 % car-to-truck

This Table 1 also ranks the compatibility issue according to the collision modes:

- Fronto-frontal car-to-car and car-to-truck come first : 1,200 fatalities
- Fronto-lateral car-to-car and car-to-truck follow : 870 fatalities

The severely injured victims (AIS 3+ or AIS $\geq$ 3) distribution is analysed according to the same rational resulting in the same priority ranking. One fact clearly appears : the car-to-truck incompatibility induces a considerably more unfavourable ratio fatalities/ severe injuries than the car-to-car incompatibility (1035 / 1900 and 1250/7730) : in the existing road situation few people survive to a car-to-truck collision.

The key question comes now : how many fatalities and severe injuries could be avoided if all the national car fleet was « car-to-car and car-to-truck compatible » ?

The analysis shows that 815 fatalities among 2,285 would be the saving potential : 50 % in car-to-car, 50 % in car-to-truck.

Similarly 4,760 severe injuries would be avoided among 16,630 : 80 % in car-to-car, 20 % in car-to-truck (Table 2).

Types of the main hypothesis used for these calculation are now summarised. For example in fronto-frontal car-to-car collisions it is considered that :

- all the fatalities become severe injured for a closing speed lower than 112 km/h,
- fatalities for people lower than 35 years old become severe injured for a closing speed between 112 and 130 km/h,
- all the severe injured become moderate injured for closing speed lower than 112 km/h,
- between 112 and 130 km/h closing speed, the number of fatalities becoming severe injured equals the number of severe injured becoming moderate injured.

TABLE 2. Potential gains expected thanks to compatibility countermeasures

	Total in France		Potential Gain	
	Fatalities	Severely injured	Fatalities	Severely injured
Car-to-car head-on collisions	700	5,750	350	3,450
Car-to-car lateral collisions	520 (385)[1]	1,740	70	460
Car-to-car rear-end collisions	30	240	0	0
Car-to-truck head-on collisions	500	850	225 (100)[2]	500
Frontal car against the truck (lateral or rear areas)	140	620	80	250
Frontal truck against car lateral	350	320	80	80
Frontal truck against car rear	45	110	10	20
	2,285	16,630	815 (420)[3]	4,760[3]

[1] in which (.) just behind the door panel

[2] in which (.) front underrun device without absorption energy system

[3] in which (.) car-to-car

Reaching such results would suppose that all possible protection technologies would be generalised to the whole national fleet. The car structure would be optimised for such closing speeds and also the restraint systems including smart belt and airbags.

The structural compatibility would satisfy the requirements described in the next chapter.

1.2.2. *Influence of Car Weights on Occupant Injury Severity and Fatality in Head-on Collisions*

Statistical studies aiming to analyse the influence of car weights on the occupant injury risk are available at least for USA, France, Germany (22) (24) (25) (31) (38). The last Fifteenth International Technical Conference on the Enhanced Safety of Vehicles, Melbourne, Australia, 13-16 May 1996 has devoted a full session to the « Vehicle Aggressively and Compatibility for Occupant Protection » still increasing the amount of available data (20) (31 to 43).

One of the most recent and comprehensive statistical study looks at injury severity and fatalities for belted drivers involved in collisions between two cars having a known mass. It allows an in-depth evaluation of risk as a function of the mass of the impacting cars for the drivers in both involved cars, as well as the overall severity of the collisions (38).

Sample Analysis. The computerised accident database of the French Gendarmerie (French police force) was used to carry out this study. These files contain information on approximately 3,000,000 road users (car occupants, pedestrians, two wheeled vehicles, heavy trucks) who were involved in accidents resulting in injury which occurred outside of cities of over 5000 inhabitants between 1978 and 1995.

We selected 446,498 records concerning car drivers in collisions involving two and only two cars as well as records on single car accidents involving no pedestrians or two wheeled vehicles. In order to be selected, the car's registered weight as approved by theDépartement des Mines (division of the Department of Transportation) had to be known.

In the first part of our study, we exclusively looked at the injury severity and fatalities for drivers involved in head-on collisions which took place outside of intersections. This sample was made up of 41,688 head-on collisions (83,376 drivers).

In the second part, in which we compare the relative risk to drivers in all two car collisions (inside and outside of intersections), we applied the two methods described in the abstract to our sample of 253,836 drivers. A comparison of likelihood of fatality by weight class will also be made for single car accidents involving a rigid obstacle or other obstacles.

As well we will use a second sample from the accident database mentioned above which includes both pedestrians involved in accidents with cars of known mass as well as pedestrians killed.

Table 3 gives a breakdown of the sample sizes for the different accident types involving pedestrians and driver.

TABLE 3. Breakdown of driver and pedestrian sample size used in this study by accident type and severity of impacts

	involved	fatalities
sub sample drivers	446498	24624
car-to-car all impact points	253836	6626
single car accidents against rigid obstacles	97520	11735
single cars accidents others	95142	6263
head-on car-to-car collisions (outside intersections)	83376	3090
sub sample pedestrians	56481	7406

Head-on Car-to-Car Collision. The sample of 83,376 drivers involved in head-on collisions includes 20,999 seriously injured and 3,090 fatalities.

Gravity and Mortality Rate for Drivers as a Function of Their Car's Mass. We can see, by looking at the 4 mass categories in «Figures 1. and 2.», that the gravity rate (number of seriously injured and fatalities/number of drivers involved) and mortality rate (number of fatalities/number of drivers involved) for drivers increase significantly when the drivers involved are in the car that weighs less. The mortality rate recorded for cars of less than 850 kg is double that for cars of over 1200 kg and the gravity is multiplied by 1.5.

Gravity and Mortality Rate for the Driver as a Function of His/Her Car's Mass and that of the Other Impacting Car. Tables 4, 5 and 6 give the number of drivers by gravity and by the cars' masses.

From these three graphs we can calculate the mortality and gravity rate in order to see the difference in risk to the drivers based on whether they were in a light car in a collision against a car of equal or greater mass.

The results given in TABLES 7 and 8, which are graphed in « Figures 3. and 4. », are compatible with impact physics in terms of the risks involved in accidents between cars of the extreme upper and lower mass range, and represent the results most often published in many articles.

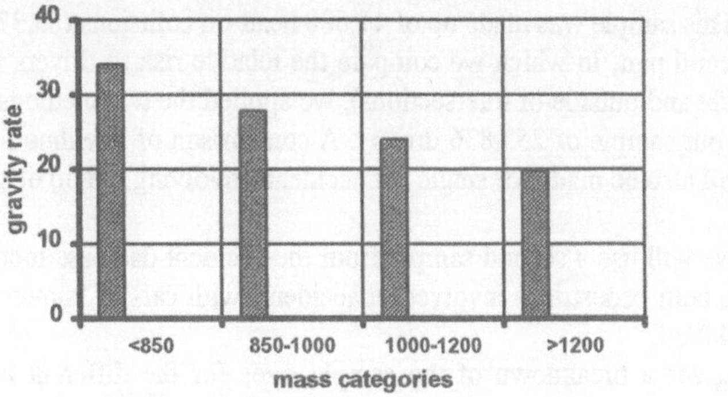

Figure 1. **Driver gravity rate as a function of 4 mass categories in head-on car-to-car collisions**

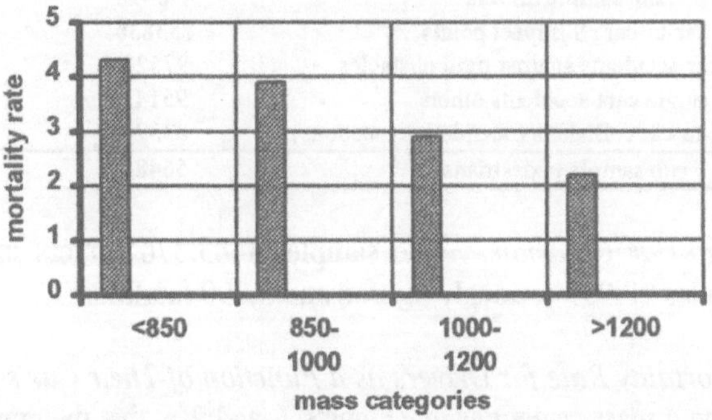

Figure 2. **Driver mortality rate as a function of 4 mass categories in head-on car-to-car collisions**

TABLE 4. Number of Drivers Involved in Cars of Mass M1 in Collisions with Cars of Mass M2

	cars M1	cars M2				
		m1	m2	m3	m4	total
<850 kg	m1	13134	11827	4515	3652	33128
850-1000 kg	m2	11827	10728	4034	3150	29739
1000-1200 kg	m3	4515	4034	1732	1183	11464
>1200 kg	m4	3652	3150	1183	1060	9045
	total	33128	29739	11464	9045	83376

TABLE 5. Number of Drivers Severely Injured and Fatalities in Cars of Mass M1 in Collisions with Cars of Mass M2

	cars M1	cars M2				
		m1	m2	m3	m4	total
<850 kg	m1	3780	4115	1789	1588	11272
850-1000 kg	m2	2495	3180	1376	1231	8282
1000-1200 kg	m3	791	1044	484	418	2737
>1200 kg	m4	525	671	284	318	1798
	total	7591	9010	3933	3555	24089

TABLE 6. Number of Driver Fatalities in Cars of Mass M1 in Collisions with Cars of Mass M2

	cars M1	cars M2				
		m1	m2	m3	m4	total
<850 kg	m1	353	514	276	275	1418
850-1000 kg	m2	243	432	226	249	1150
1000-1200 kg	m3	79	108	72	68	327
>1200 kg	m4	41	65	30	59	195
	total	716	1119	604	651	3090

TABLE 7. Driver Gravity Rate

	m1	m2	m3	m4
m1	28.7	34.8	39.6	43.4
m2	21.1	29.6	34.1	39.1
m3	17.5	25.9	27.9	35.3
m4	14.3	21.3	24.0	30.0

TABLE 8. Driver Fatality Rate

	m1	m2	m3	m4
m1	2.69	4.35	6.11	7.53
m2	2.05	4.03	5.60	7.90
m3	1.75	2.68	4.16	5.75
m4	1.12	2.06	2.54	5.57

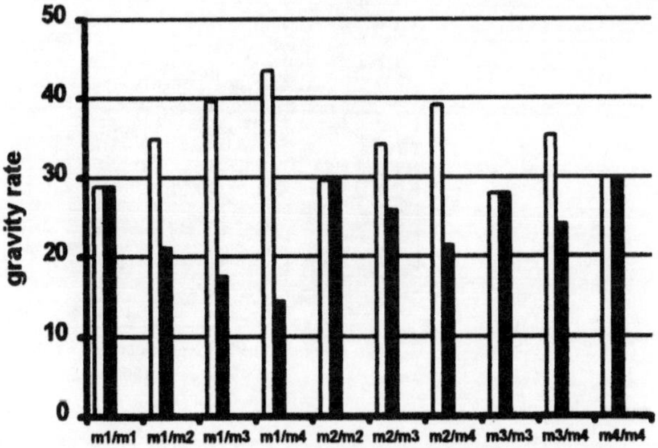

Figure 3. Driver gravity rate by mass category

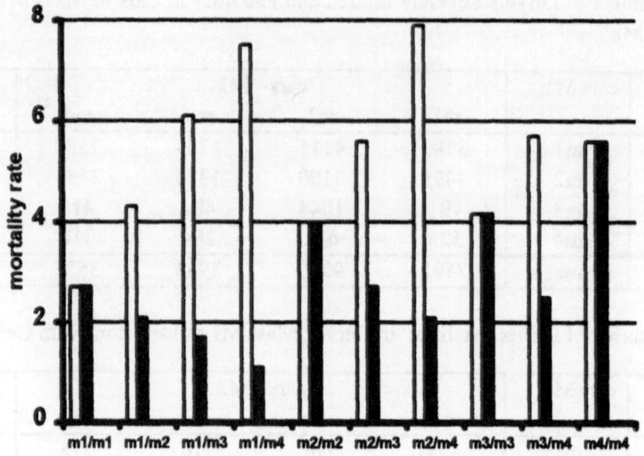

Figure 4. Driver mortality rate by mass category

Hence, we can see that mortality for the driver of a car in the lightest range (<850 kg) in a head-on collision with a car in the highest mass range (>1200 kg) is 7 times higher than that of the driver of the bigger car involved (see Figure 4). The gravity rate for that lighter car driver is also approximately 3 times higher.

But in the case of a collision between two cars of equal mass the average mortality rate for both drivers steadily increases as the mass of the impacting cars increases, becoming twice as high (2.69 for m1 and 5.57 for m4) for accidents between two cars in the upper mass range.

However, as far as the overall gravity is concerned, the rate is practically stable for all mass categories in head-on collisions involving two cars of equal mass.

Are these results, so clearly shown in « Figures 5. and 6. », perhaps due to higher speeds in collisions involving two heavy cars or perhaps to other factors not taken into account ?

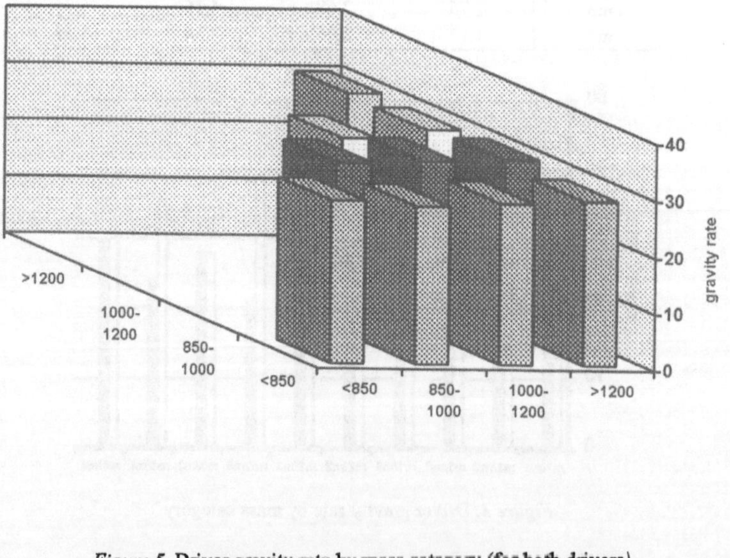

Figure 5. Driver gravity rate by mass category (for both drivers)

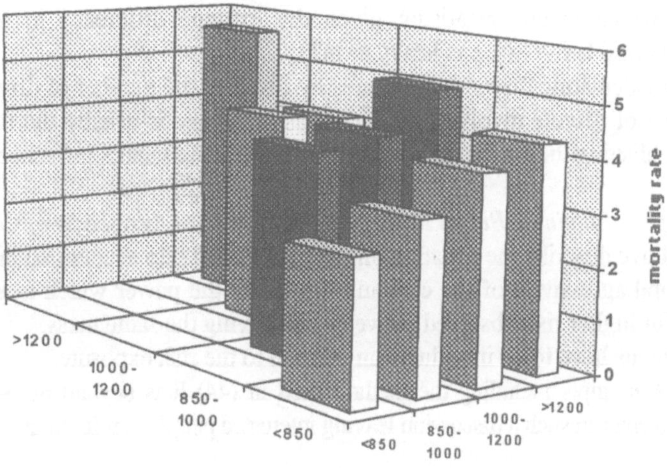

Figure 6. Driver fatality rate by mass category (for both drivers)

In the last part of our paper we will try to find an explanation for these results, but first we will look at the risks to drivers in other types of accidents and use the method put forward by Evans to calculate exposure to risk.

Car-to-car Collisions, All Impact Point. In order to obtain the largest sample possible and to verify the trend observed in head-on collisions, we selected all collisions in which two and only two cars of known mass were involved.

These mass categories are of course different from those used byEvans due to the fact that cars in our country are substantially lighter on the average than American cars. However, our aim here is only to compare the results obtained using the two methods for the same number of mass categories.

As we had a sample of 253,836 drivers involved, of which there were 6,626 fatalities, we sub-divided all the impacting cars into six mass categories, the lightest being less than 750 kg and the heaviest over 1300 kg.

Drivers Involved and Driver Fatalities as a Function of the Mass Categories of the Impacting Cars. Tables 9 and 10 below give the distribution of driver fatalities and drivers involved as a function of the 6 mass categories of the impacting cars. Using the figures from these tables we were able to calculate the mortality rate that is found in Table 11.

Once again, in the case of collisions involving two cars of equal mass, the mortality rate for both drivers increases steadily as the mass of the two impacting cars increases, being 2.6 times higher for the highest mass category than for the lowest.

For collisions involving a car from each of the two extreme mass categories (upper and lower) we confirm obviously that the mortality rate is twenty times higher for the smaller car, which is not surprising given the extreme difference in mass of the impacting cars (m6 being twice as heavy as m1).

Directly calculating driver fatality rates by dividing the number of driver fatalities by the number of drivers involved for each mass category is usually the most reliable means of establishing these rates as long as no other biasing factors influence results.

1.2.3. *Attempt to Eliminate Biases Related to the Structural Agressivity of Cars*

The results above describe the situation on the road taking into account all biases related to the structural agressivity of the cars and especially the power which is a part of the explanation for higher risk observed between cars having the same mass.

Adjustments have to be introduced in relation to the risk exposure.

Different attempts including the method used in (44) It is beyond the scope of this presentation to enter in such a discussion leaving interested people to refer to the source (38).

TABLE 9. Number of Drivers Involved in Cars of Mass M1 in Collisions with Cars of Mass M2

	cars M1			cars	M2			
		m1	m2	m3	m4	m5	m6	total
<750	m1	7216	9974	11114	7933	4647	2281	43165
750-850	m2	9974	14942	15703	11211	6285	3130	61245
850-950	m3	11114	15703	16682	11733	6582	2998	64812
950-1100	m4	7933	11211	11733	8804	4533	2155	46369
1100-1300	m5	4647	6285	6582	4533	2632	1189	25868
>1300	m6	2281	3130	2998	2155	1189	624	12377
	total	43165	61245	64812	46369	25868	12377	253836

TABLE 10. Number of Driver Fatalities in Cars of Mass M1 in Collisions with Cars of Mass M2

	cars M1			cars	M2			
		m1	m2	m3	m4	m5	m6	total
<750	m1	101	220	388	335	266	142	1452
750-850	m2	112	308	418	438	310	168	1754
850-950	m3	111	271	439	422	334	205	1782
950-1100	m4	70	191	245	244	168	117	1035
1100-1300	m5	22	77	121	89	86	52	447
>1300	m6	7	23	40	30	33	23	156
	total	423	1090	1651	1558	1197	707	6626

TABLE 11. Driver Fatality Rate

	m1	m2	m3	m4	m5	m6
m1	1.4	2.21	3.49	4.22	5.72	6.23
m2	1.12	2.06	2.66	3.91	4.93	5.37
m3	1.00	1.73	2.63	3.60	5.07	6.84
m4	0.88	1.70	2.09	2.77	3.71	5.43
m5	0.47	1.23	1.84	1.96	3.27	4.37
m6	0.31	0.73	1.33	1.39	2.78	3.69

The conclusion may be given : « When two cars of equal mass collide, the higher they are in the mass range, the greater the number of occupant fatalities. This observation, which also applies when looking at all car-to-car collisions or at single car accidents, is quite simply due to higher average impact speeds when heavier cars are involved. This slightly higher speed is also observable when pedestrians are hit by heavier cars. We can estimate the average speed at impact to be between 5 to 10 km/h higher for heavier cars based on our Accidentology Laboratory database. These extra km/h are enough to result in the increased risk to drivers. »

1.2.4. *The Side Impact Compatibility: The Influence of the Architecture*

It is well established since 1979 that the relative height of the main structural elements on impacted and impacting cars governs the severity of the injuries in side impact (27)

Another architectural feature plays a very significant effect : the seating height in the impacted car. The risk of fatal injury is about two times less when the seating height increases from 480 mm to 630 mm (« Figure 7. »). These values correspond to the height of the « hip point » above the ground level.

After the correction of the age and mass bias a strong linear correlation appears between the mortality rate and the seating height, still stronger when all side impacts are considered mixing car-to-car collisions either out of as well as in cross section than only in cross section area. The difference comes of higher closing speeds when side impact happens in open road or highway following a loss of control for the impacted car.

Mention is given on « Figure 7. » of the French monospace « Espace » for which the mortality rate is zero. No fatality has been yet reported in France in side impact for this specific car where the seating height reached 722 mm. The size of the « Espace » subsample (77 cases) is still to small to be associated to the correlation calculation.

The same analysis done for the gravity rate (fatal and severe injuries) does not allow to show the same results. If the thoracic and abdominal injury risks decrease when the seating height increases (explaining the lower fatality rate), the pelvis injury risk increases and neutralises the favourable effect of the high seating on the torso. This « pelvic effect » is less true for a car such as « Espace » which still appears in a more favourable situation (« Figure 8. »).

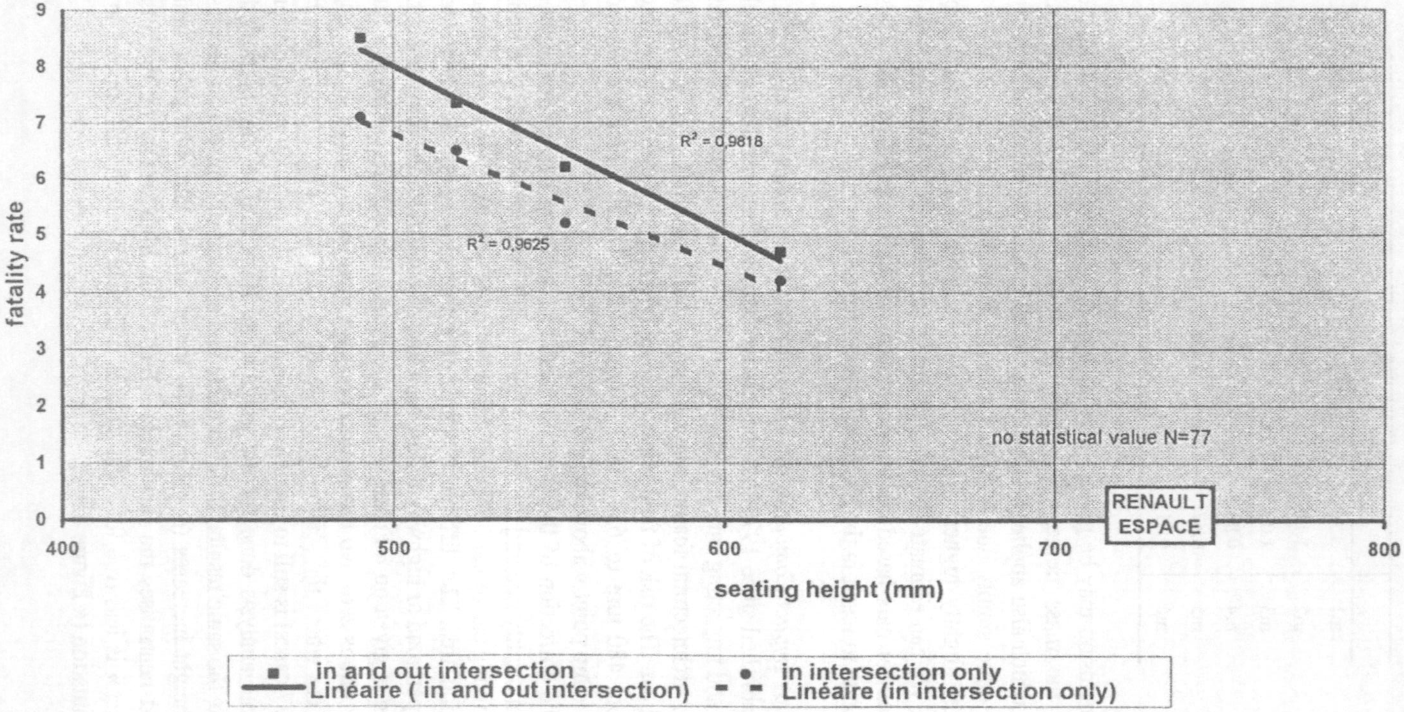

Figure 7. **Fatality rate for front seat occupants involved in side impacts in relation with the seating height (car-to-car with age and mass bias corrected)**

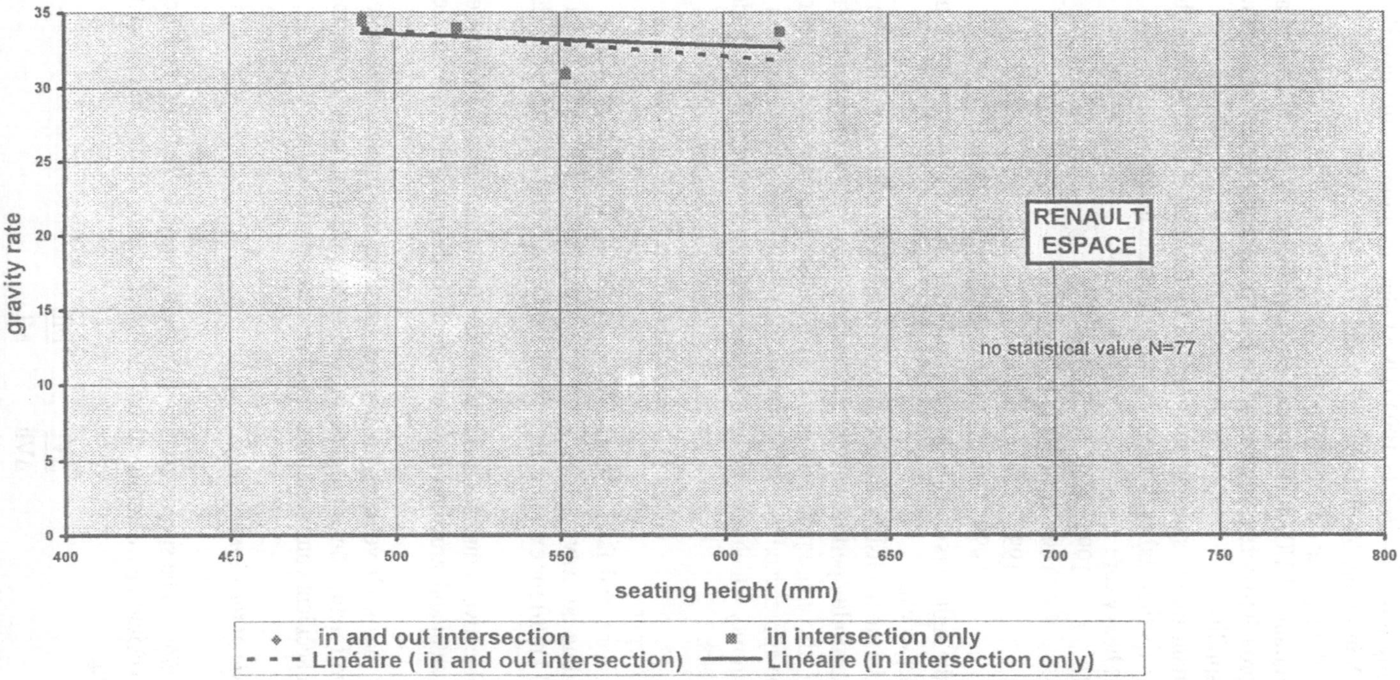

Figure 8. **Gravity rate for front seat occupants involved in side impacts in relation with the seating height (car-to-car with age and mass bias corrected)**

1.3. EXPERIMENTAL APPROACH FOR THE COMPATIBILITY

A NHTSA working group created in 1993 (USA) aims to minimise the global number of injured for each introduction of a new car model.

A BAST working group (Germany) tests a car against itself in fronto-lateral crash.

A work programme EUCAR (European car manufacturers) has undertaken a global experimental approach to define what could be European requirements to reach a global car-to-car compatibility.

Very important for the spontaneous progress by the market of the compatibility is the launching of experimental evaluations by some German media :

light against heavy car :

ADAC:	1994 - Golf / D.B. class S
AMS:	1995 - Corsa / D.B. class S
	1995 - Polo / Audi A8
	1995 - Calcul D.B. : vision A / D.B. class E

It appears the light cars become better and better and an important work needs to be done on the heavy cars.

As the results are published, these Ratings constitute an efficient pressure to take in account the compatibility performance in the design of new cars at least for the smallest. For the heaviest, the consequence is not clear because if agressivity is a poor social value, it could be appreciated by a large part of their usual customers asbullbars are for an increasing minority which doe not take care of other users especially the pedestrians or the two wheelers.

1.3.1. *General Target*

- To avoid severe and fatal injuries (AIS ≤ 3) for the occupant of a light vehicle when colliding an heavy vehicle for a given mass ratio [1.5] and a given closing speed [100 km/h] for example. In a second step, ratio and closing speed should be higher.
- Such an objective induces a serie of precise protection criteria, a very efficient restraint system and structural requirements for both light and heavy vehicles.

It is clear that such vehicles do not yet exist on the market. The objective is ambitious and will induce huge progress in road safety, the same cars having to be also safe in the other accident configurations, fix obstacles especially.

1.3.2. *The Physic of the Compatibility*

Mass Compatibility. The speed variation (ΔV) will always be higher for the occupants of the lightest vehicle and determined by the inverse ratio of the masses of the two vehicles colliding :

$$\frac{\Delta Vl}{\Delta Vh} = \frac{Mh}{Ml} \qquad (1)$$

where ΔVl and Ml are speed variation and mass for the light vehicle, ΔVh and Mh are those for the heavy vehicle

and $$\Delta V1 = Mh / (Mh + Ml) \;.\; C.S. \qquad (2)$$

$$\Delta Vh = Ml / (Mh + Ml) \;.\; C.S. \qquad (3)$$

where C.S. = Closing speed

Stiffness Compatibility. At each impact instant, both vehicles at their interface oppose their forces to avoid to be destroyed (collapse force). The weaker takes the deformation.

Precisely these strengths depends on the mass of the vehicle and on the total deformation. Generally these strengths are higher for the heaviest car and the deformations more important on the lightest.

The influence of the stiffness itself clearly appears when comparing the deformations of two cars having the same mass (« Figure 9. »).

Architectural Compatibility. To limit the complex interactions between frontal structures and to have in car-to-car a crash behaviour similar to a rigid barrier test one, the architects have to give the structural elements (longitudinal, transversal structures, engine ...) compatible between the two cars.

This is possible inside a same car manufacturer serie but how to get such a compatibility between different makes and models ? It appears that a very strong transversal member behind the bumper would be the mean to reach this universal architectural compatibility.

The basic rules.

Compatibility in terms of energy. Light and heavy cars will be compatibles if the absorbed energy during a car-to-car collision is the same than against a rigid wall. This absorbed energy in car-to-car collision is similar to their masses-ratio.

$$\text{Light Absorbed Energy} = Ml / (Ml + Mh) \times \text{Total Absorbed Energy} \qquad (4)$$

$$\text{Heavy Absorbed Energy} = Mh / (Ml + Mh) \times \text{Total Absorbed Energy} \qquad (5)$$

and $$\text{Total Absorbed Energy} = 0{,}5\,(Ml + Mh) / (Ml + Mh)\; C.S.\,.\,2 \qquad (6)$$

Compatibility in terms of Habitacle Mean Acceleration. Light and heavy vehicles will be said compatible if the mean habitacle deceleration for the two vehicles in car-to-car is not more severe than in crash against a rigid wall.

Rigid Barrier Test : Specific Energy = l acceleration x l displacement
h acceleration x h displacement

$$\frac{\text{h displacement}}{\text{l displacement}} = \frac{\text{l acceleration}}{\text{h acceleration}} \qquad (7)$$

Car-to-car Collision

$$\frac{h\,\Delta V}{l\,\Delta V} = \frac{\text{h acceleration}}{\text{l acceleration}} = \frac{l\,m}{h\,m} \qquad (8)$$

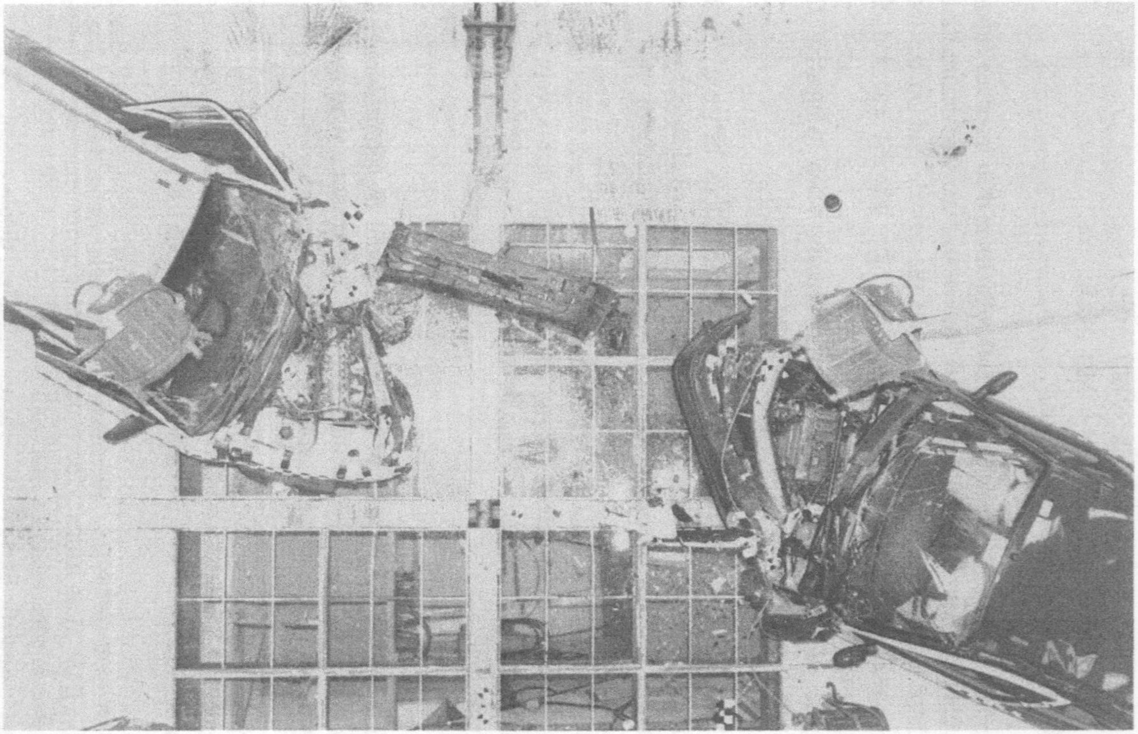

Figure 9. Mégane an illustration of the influence of the stiffness.
It is a most realistic confrontation with the rival, up to now considered as the best in the crashes held in Germany. It is a car-to-car head-on test collision with a closing speed of 112 km/h in anasymetrical crash with 50 % front overlap only for both cars. On this view from above, the two bonnets have been removed in order to bettervisualize the structure deformations. Mégane, situated on the right hand side, shows the passenger compartment without any deformation. For the « rival », situated on the left hand side, the deformation of the left door highlights the passenger compartment intrusion.

Figure 10. **Renault numerical simulation at work between a light and a heavy car. The interactions between front-wheels and engines are clearly demonstrated (here at time 60.ms)**

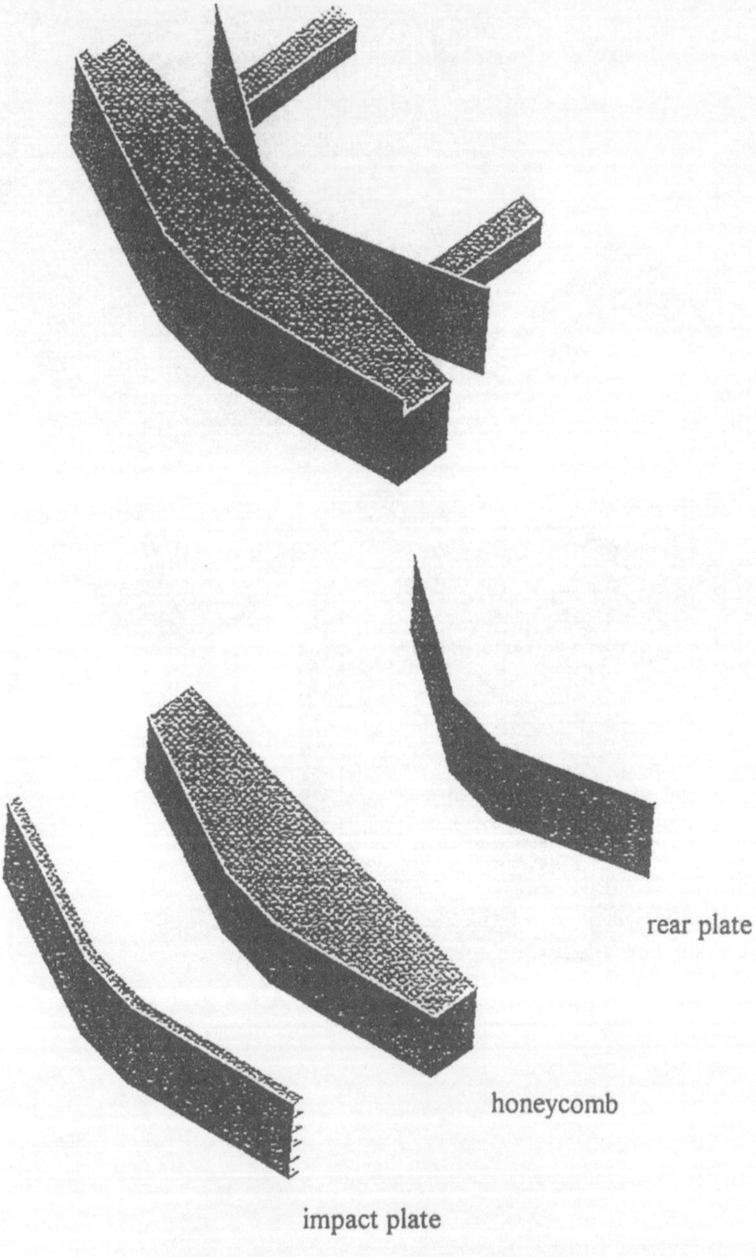

Figure 11. **Design Concept Derived From Simulation**

RENAULT at work on compatibility. The methodology uses a numerical simulation and the experimental tests allow the necessary adjustments and check up (« Figure 10. »).

The structural compatibility associated to high performance in the restraint system demonstrate the feasibility of such an ambitious objective.

A cooperation with Zurich University, another pioneer in small compatible car since their first demonstration given in 1991, contributes to establish the feasibility through a successful experimental validation (30).

Car-to-truck Obviously the maximal incompatibility is associated to the car-to-truck front end impact. Avoiding the « underrun » of the car under the front of the truck and dissipating the impact energy are the two main countermeasures that can reduce the injury risk for the car occupants. Both are related to the truck because in this case the behaviour of the car structure itself is negligible.

Such approaches are conducted at pre-regulation level but could be decided soon by the European authorities.

A more recent study (44) described another concept of a front-end design for heavy trucks. « The new bumper design involves an energy absorbing honeycomb block, covered by an impact surface which swivels upon impact, thus deflecting the car away from the path of the truck. » (« Figure 11. ») (45).

2. Vulnerable Users: The Children

2.1. INTRODUCTION - A BETTER CHILD PROTECTION : A PRIORITY ISSUE IN EUROPE

Each year, 1,000 children are killed on European roads, and 80,000 are injured. These figures are especially unbearable in that the great majority of these victims could be avoided if very significant progress would take place in the design of appropriate child restraint systems. The problem is worldwide. Even if some differences appear in some countries. To improve the situation, the design of reliable dummies, the development of measurement techniques and corresponding injury criteria are first priorities.

2.2. STATE OF THE ART AND DEGREE OF INNOVATION

2.2.1. *Weakness of the Evaluation Tools*

Protection systems for child car passengers used in the European Union countries must comply with ECE regulation 44 which aims at ensuring a good safety level. From the accident protection viewpoint, this regulation consists mainly of a frontal impact test on impact simulator during which it is verified that the protection criteria deduced from physical measurements performed on dummies do not exceed the specified limit values. These criteria cover kinematics, which must remain within a certain range, while an accelerometric criterion covers the thoracic segment only.

The analysis of accidents involving children reveals that Child Restraint Systems (CRS) in compliance with European regulations give highly contrasted levels of protection in real-world accidents. There are many reasons for this :

- Lack of biofidelity of dummies (dynamic behaviour too different from that of children).
  Example: The 3-year and 6-year dummies cannot detect the risk of submarining and can therefore not distinguish between a poor booster because not equipped of belt guides from a good booster equipped with belt guides ; however only the last one can prevent abdominal risk.
- Insufficient biomechanical knowledge to select the relevant physical parameters and determine their limit values measured on dummies for the various body areas most exposed to high risks of injuries.

2.2.2. *Unsustained Cervical Risk in Young Children (age less than 2 to 3 years old)*

Injuries to the cervical column for children less than two to three years old happen in real accidents. Even if there are still few known cases (a few cases per year in the countries of the European Union), they should be assigned greater importance than that bases on the statistics because, on the one hand, when they do not cause death they result in tetraplegia or paraplegia, which represent a permanent major handicap for life. On the other hand, these very severe injuries generally occur in accidents of medium violence in which usually the other occupants of the vehicle are unharmed or not severely injured.It may be thought that the real frequency of neck injuries is even higher and that numerous cases are unrecorded due to lack of an autopsy following, accidents in which the child died.

Comparative analyses between Europe, North America and Australia seem to show that, for seats with 5-point or 6-point harness, when the system is connected to the vehicle by means of a top tether strap such injuries do not occur ; likewise, no cases of injuries of this type were encountered when a rearward-facing device was used.

The development of more efficient devices requires first obtaining a good understanding of injury mechanisms involved ; insofar as concerns injuries to the cervical column, it must be elucidated whether their occurrence is associated or not with head contact and, if so, what are the characteristics of such contact and the spine motion generated by it. It is also essential to specify the age above which this cervical vulnerability is attenuated or disappears. This could be made possible by accident research analyses.

2.2.3. *Risk of Abdominal and Lumbar Spine Injuries*

In general, the interaction between the restraint system and the body of a young child is not well known, particularly as regards the lap belt part of the restraint, since as it grows the child's pelvis shape changes. The risk of submarining phenomenon cannot be ascertained by the present dummies. Indeed it is acknowledge that their pelvis, abdomen design are quite unrealistic.

Analysis of real-world accidents has shown that shield or table shield restraint systems provided satisfactory protection but only for a limited ages range. These results must be confirmed and analysis should be carried out on the optimum dimensions of such systems relative to children's dimensional characteristics, so as to optimise protection by preventing the risk of slippage while ensuring acceptable comfort. This is especially important in that these devices are the only ones which can provide good protection for the central rear position equipped with a lap belt only.

2.2.4. *Extent of Risk of Severe Injury in Side Impact*
Child protection systems are designed and certified for operation in frontal impact. Now, more than a quarter of car occupant fatalities and severe injuries are sustained in side impact. A new regulation concerning car performance in side impact should come into application in the coming months in Europe and Australia (as it is already the case in USA), but it does not concern protection systems for child passengers, a field in which nothing has been done as yet. Some running studies have important implications for the development of enhanced-safety child protection systems ; it is bound to have consequences for the methods used to evaluate these systems and for designing the dummies required for said evaluation.

2.2.5. *Conditions of Standard Certification Test and Real Conditions*
On a more secondary level, differences of performance are observed for a given CRS, depending on whether it is evaluated in the Directive conditions or in the specific environment of a given vehicle.

2.3. SCIENTIFIC OBJECTIVES OF EXISTING RESEARCH PROGRAMMES

Protection of child car passengers in accidents requires the use of appropriate devices (seats, boosters, car beds, etc.) specially designed for the purpose. The characteristics of each category of device must be compatible with the impact tolerance of the groups of children liable to use the device, in the conditions applicable in real-world accidents.

Unlike adults, children's impact tolerance cannot be determined directly by experiment on human bodies. The most appropriate method for understanding the biomechanics of the injuries to child car passengers is to perform reconstruction of well documented real-world crashes, during which physical parameters are measured on fully instrumented dummies, and to compare the values measured with the corresponding injuries sustained in the real-world accident chosen for the reconstitution.

This knowledge is absolutely essential to be able to optimise the design of protection systems for child car passengers. It requires two necessary and complementary MAIN TASKS :

- I : to make a detailed analysis of road accidents,
- II : to make experimental reconstructions of some of these accident cases carefully selected.

Several organisations are at work on the same line :

2.3.1. *The International Task Force on Child Restraint Systems (ITFCRS).*
It is in cooperation between 14 partners from 8 countries (Australia, Canada, France, Germany, the Netherlands, Sweden, U.K., USA). The work started in 1990 (46) to (49).

2.3.2. *The specific ISO ad hoc group « Children in Cars - Side Impact Studies »*
Aiming to define a test procedure to assess the CRS protection in side impact.

2.3.3. *The CREST programme (Child REstraint-system STandards)*
Is a European cooperation between 12 organisations (universities, car manufacturers, public laboratories).

For 2.3.3., the aim and methodology are the same that for ITFCRS. Based on in detailed accident cases where restrained children are involved, both programmes have to establish a relationship between injuries observed in the real-world of accident and the measurement obtained during experimental reconstructions with fully instrumented dummies. These biomechanical data will allow to establish future regulation for CRS similar to the current practices used to assess the adult protection.

The work to do represents a huge sum of efforts :

. accident analysis (2.3.1.) (2.3.2.) and (2.3.3.)
. accident reconstruction with dummies (2.3.1.) and (2.3.3.),
. creation of a new dummies generation with their instrumentations (2.3.3.),
. creation of new test procedures (2.3.2.) and (2.3.3.).

The summary of CREST is given in ANNEX 1.

2.4. MISUSE - ISOFIX

It is usual in several countries (USA, European states) to observe that 50 % of child injuries are due to a false utilisation of the CRS. This misuse covers numerous practices and some very basic:

. CRS not correctly fixed to the car : partial fixation, fixation with large slack, no fixation at-all or fixation on a wrong way, rearward-facing used in forward-facing.
. CRS correctly fixed to the car but the child not using the straps into the child seat.
. CRS not fitted to the child age, etc.

There is an explanation for this poor situation : the parents' task is really difficult and needs a true expertise ; so many different CRS, different fixations, different requirements according to the child age and the location inside the car (front seat, rear-side seat and middle rear) ...

So a new concept is under study : to have for all countries, all cars, all seat locations, all CRS, the same simple and easy CRS fitting to the car (50).

ISOFIX is the name given to this project. After several years of work, the product is not yet available. They are at least two challengers : a solution with 4 fixations : two on the rear, two on the front of the normal car seat or bench (« Figure 12 ») another solution keeping the two rear fixations and having, instead of the two front fixations, a device allowing to get a firm coupling between the CRS and the car seat by a strong compression of the foam. This compression device aims to avoid slack and rotation of the CRS during a crash.

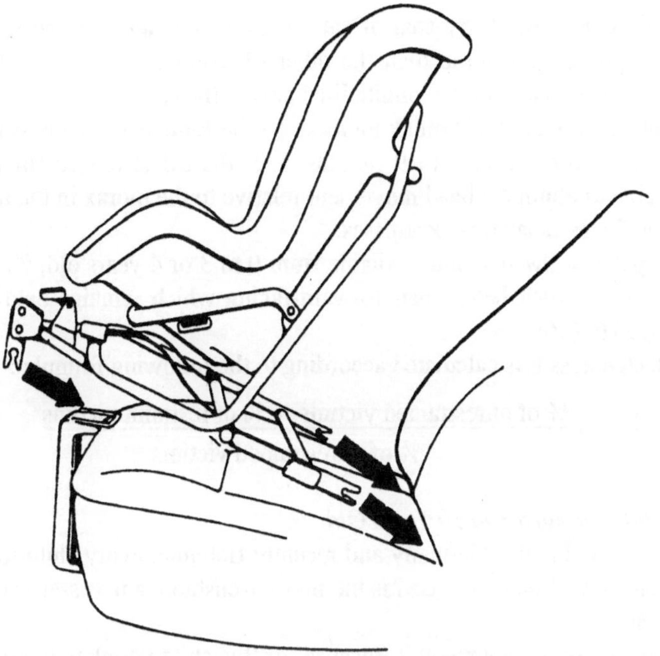

Figure 12. ISOFIX Type 3 (UNIFIX)

2.5. C.R.S. IN USE IN EUROPE

2.5.1. *For children aged 0 to 3 years old*

The most common device is the classic forward-facing child seat with harness fastened to the car with the adult 3 point-belt. It is a configuration said « forward-facing » having severe limitations :

- **This device is not usable for children before they get the seating posture (not before about 6 months).**
- **This device is normally fastened to the car by the 3-point belt with a slack allowing too more rotational movements and too often head impact during frontal crashes.**
- **The effectiveness is limited by the head injury and neck injury risks induced by the possible head impact against the back of the front seat.**
- **For the youngest users between 6 and 24 months there is also a neck injury risk only induced by the forward-facing posture explained by the inertia mechanism and more difficult to be tolerated for this youngest child population.**
- **This device is normally fastened to the car by the 3-point belt with a slack allowing too more rotational movements and too often head impact during frontal crashes.**
- **The effectiveness is limited by the head injury and neck injury risks induced by the possible head impact against the back of the front seat.**
- **For the youngest users between 6 and 24 months there is also a neck injury risk only induced by the forward-facing posture explained by the inertia mechanism and more difficult to be tolerated for this youngest child population.**

Before 6 months, the carry-coat or car-bed is a well fitted device for the infants. However, it requires more room than the other (2 seats) because it has to be located transversally, perpendicular to the longitudinal axis of the car.

The challenger, knowing a quick increase on the European market is therearward-facing seat for children from 0 to 9 months. It is the safest device for the youngest because it is able to eliminate head movement relative to the thorax in the main phase of the impact and the associated neck stresses.

Very popular in Sweden and available from 0 to 3 or 4 years old, its effectiveness reaches 90 % (51) much better than forward-facing which remains below 55 % (see TABLE 12) (52) to (56).

The effectiveness E is calculated according to the following formula :

$$E = \frac{\text{\% of unrestrained victims - \% of restraint victims}}{\text{\% of unrestrained victims}} \quad (9)$$

2.5.2. *For Children Aged 3 to 10 Years Old*

In countries such as France, Germany and recently Belgium, every child restraint has to be specific. The most frequently used is the booster-cushion, a necessary complement of the 3-point belt.

Why a necessary complement ? Because for this child population, the size and the shape of the pelvis do not allow to safely keep the lap belt on the pelvis.

The consequence is a very high abdominal and lumbar spine injury risk. The pelvic part of the 3 point-belt will always unfortunately tend to slip upward the iliac crests and penetrate into the abdomen. The force applied can be up to several hundred deca-newtons. It is easy to understand the induced injuries : liver, spleen and intestinal ruptures, luxation or fracture of lumbar vertebraes.

The only prevention is to avoid the belt wearing without a well designed booster-cushion. Such a booster provides lap-belt guides (one for each side) positioning the strap flat at the base of the thighs of children ages 3-4 and over (56).

Misuses of booster cushion :

. to use boosters unable to modify the lap-belt trajectory (booster without strap-guides or strap-guides fitted for older children),
. to use a booster with a lap belt only,
. the same result is obtained when a child passes the thoracic strap of the 3-point belt behind his torso and the child dies of brain and neck injuries in case of a frontal impact.

A booster-cushion sometimes includes a seat back and head-restraint to provide more comfort and sleep posture.

There is an obvious drawback for the middle rear seat when it is still too often equipped with a lap belt. The only solution in this case is the shield device. Some are very popular in Germany : the Römer-Peggy from 2 to 4 years old, the Römer Vario from 3 to 6 years old. « Figure 13. » illustrates the selection of the best C.R.S. done by a car-manufacturer as an example of whole set necessary to cover the needs for all ages, all seat location, all car-models.

TABLE 12. Children - France : Injury Distribution by Body Area and Restraint System

1629 children involved in 1090 accidents in 4 months of 1922-1993

	Harness seat (N=316)		Booster - Seat Belt (2-point / 3-point) (N=624)		Unrestrained (N=613)	
	E = 45%	E = 54%	E = 8%	E = 23%		
	AIS 1+ (N=112)	AIS 2+ (N=30)	AIS 1+ (N=316)	AIS 2+ (N=92)	AIS 1+ (N=332)	AIS 2+ (N=109)
HEAD + NECK + THORAX	78.7	58.3	62.5	45.4	64	54.6
	E = +33%	E = 51%	E = 10%	E = 36%		
ABDOMEN + PELVIS	1.5	2.8	13.7[2]	15.1	6.8[3]	8.6
	E = 89%	E = 86%	E = -87%	E = -38%		

All Restraint Systems (N=992)		Unrestrained (N=613)		Severe Accidents PVM 1990[1] All Restraint Systems AIS 1+	
AIS 1+ (N=442)	AIS 2+ (N=125)	AIS 1+ (N=332)	AIS 2+ (N=109)	All configu-rations (N=41)	Frontal (N=29)
50.4	38.7	51.9	47.4	54%	52%
4.5	5.2	3.9	0.7	29%	21%
8.3	4.5	8.2	6.6	24%	31%
7.5	16.1	10.1	15.1	5%	3%
9.3	9	4.7	5.9	17%	24%
1.7	3.2	2.1	2.6	5%	9%
12.8	19.4	16.1	20.4	7%	10%
A	C	B	D		E

(PVM = Fatal Accident Police Reports)

$$E = \text{Efficiency} = \frac{\%\text{ of unrestrained injury victims} - \%\text{ of restrained injury victims}}{\%\text{ of unrestrained injury victims}}$$

[1] 41 children involved in severe accidents in which at least one adult or child occupant was killed.
[2] The corresponding percentage is 10.8 for the USA (restrained children aged 6 to 12 years - FARS).
[3] The corresponding percentage is 2.6 for the USA (restrained children aged 6 to 12 years - FARS).

Source : C. TARRIERE and C. GOT (December 1994)

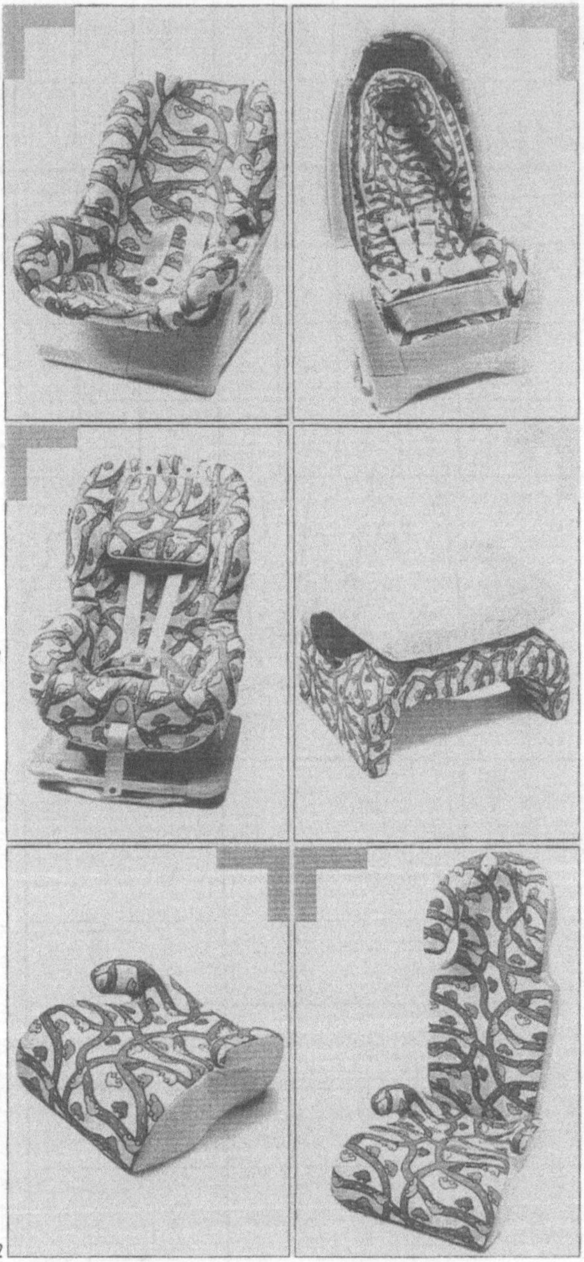

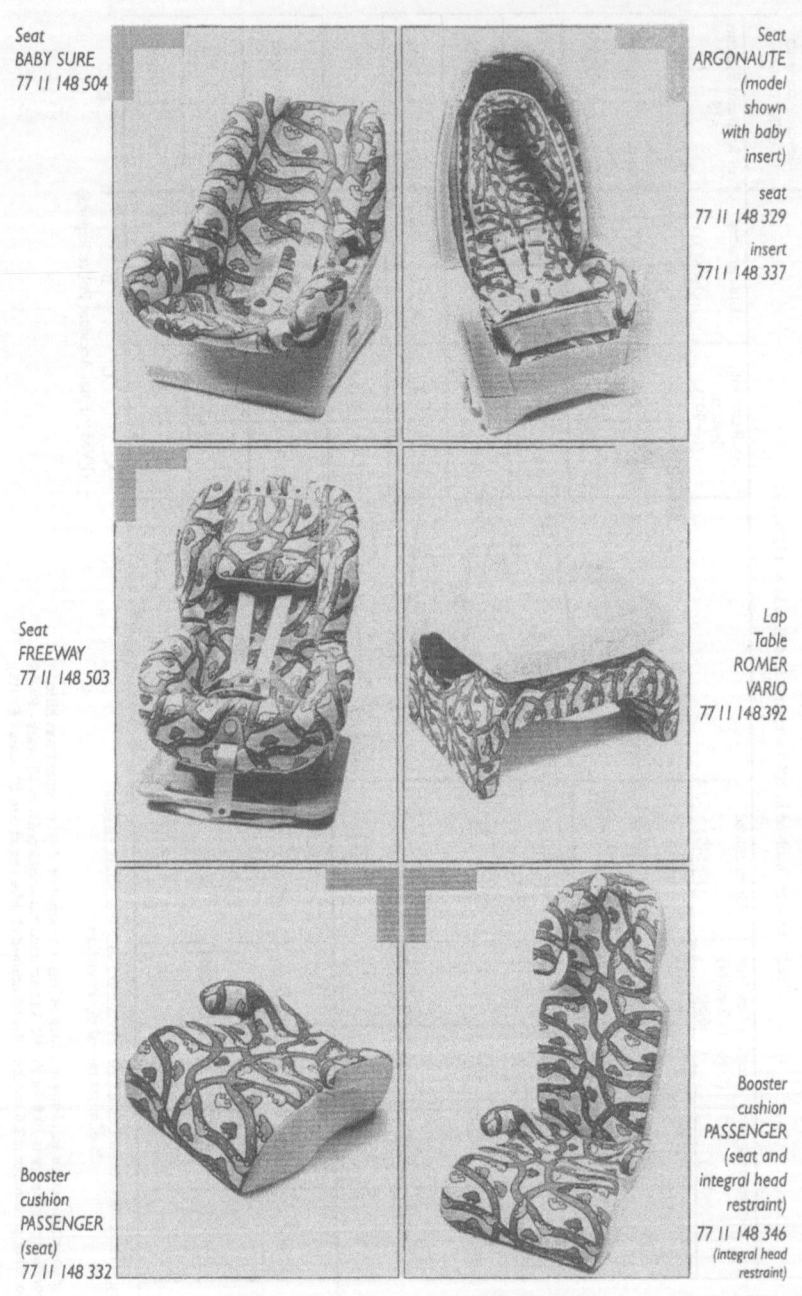

Figure 13. **Children seats and boosters**

2.6. CONCLUSIONS AND PROSPECTS

The effectiveness of the existing C.R.S. is variable but on the whole positive, comprises between 25 to 55 %.

Due to the poor knowledge relative to the child tolerance to impact, the existing regulation are very insufficient.

Research programmes are in progress to acquire the missing knowledge to improve a new child dummy family and to define new criteria found onbiomechanical data both for frontal and lateral impacts.

For the present time, progress in the child protection have to be based on the expert recommendations drawn from the analysis of the CRS behaviour in real-world accidents.

The priorities to improvements are clearly designed :

- to increase the percentage specific of Child Restraint Systems usage from 0 to 12 years old,
- to promote rearward-facing Child Restraint Systems between 0 to 3-4 years old,
- to promote the booster-cushion between 3 and 12 years old.

Simultaneously, the UN/ECE Regulation 44 has to be improved to increase Child Restraint Effectiveness.

E.C. planning to transfer UN/ECE Regulation 44 in a directive, it would be an opportunity to improve the regulation 44 on the basis of the most recent available knowledge.

The need of information is obvious to help parents to select the best CRS on the existing market. National Administrations, automobile industry, private organisations such as consumer associations, public and private health services have to contribute to this information.

3. Vulnerable Users: The Pedestrians

3.1. INTRODUCTION

The number of pedestrian victims in road accidents very regularly decreases since 15 years. « Between 1979 and 1992, the number of fatalities was reduced by 62.7 % and the number of serious injuries decreased by 45.2 % despite growing road traffic figures. These good results were obtained even without a regulation on pedestrian protection ». It is an extract of a report issue by ACEA (57).

Despite these achievements, a further reduction of pedestrian injuries is obviously desirable and therefore appropriate measures should be taken to achieve this aim.

Research in this area indicates that benefits might be obtained from three approaches

- improvement of the infrastructure,
- action to modify the behaviour of pedestrian themselves,
- changes to the geometry and compliance of the frontal structures of cars.

3.2. IMPROVEMENT OF THE INFRASTRUCTURE

The number of pedestrians fatalities varies widely throughout the countries of Europe. Nevertheless, the designs of passenger cars are virtually identical in these countries. The differences in pedestrian fatalities rates must therefore be due to other factors such as :

- improved traffic management,
- separation of pedestrian from other traffic,
- traffic lights controlling pedestrians crossing,
- speed reduction inside cities.

3.3. ACTION TO MODIFY THE BEHAVIOUR OF PEDESTRIANS THEMSELVES

It is well known that over 50 % of pedestrian fatalities involve blood alcohol levels exceeding 0.5 g/l.

Alcohol use is one of the factors of relatively high « at fault » finding against pedestrians.

In a recent Australian survey, overall 74 % of pedestrians involved in fatal crashes were primarily responsible for the crash and a further 8 % were partially responsible (males 79 % ; females 62 %) (58).

These figures in themselves are a persuasive argument for continuing attempts to change pedestrian behaviour and to study in more detail the involvement of alcohol in pedestrian crashes.

Behaviour change on a large scale is a difficult and time consuming task justifying greater reinforcement of laws and controls.

Education is also useful for the drivers. Approximately half of Australian drivers attempted to brake or swerve or both when confronted with a likely pedestrian collision. It has been established than the ABS braking obtains its better efficiency in pedestrians avoidance (59) (60).

3.4. CHANGES TO THE GEOMETRY AND COMPLIANCE OF THE FRONTAL STRUCTURE OF CARS.

Another approach to reduce the severity of car to pedestrian accidents is by making changes to the frontal structure of cars. This is a difficult task to achieve. Test procedures have been developed, particularly that of EEVC using impactors simulating the main body parts : head, pelvis, legs.

This approach called « sub-system tests » was preferred taking into account the difficulty to control a global approach with pedestrian dummies adult and child.

3.4.1. *EEVC Sub-System Tests Recently Reviewed (61)*

The leg form to bumper test was developed by INRETS in France. An instrumented mechanical leg is propelled against the front of a car to a given speed (35 or 40 km/h). The deformation of the knee is controlled by two deformable elements providing the relevant force/deformation characteristics in bending and shearing. An accelerometer is fixed on the lower leg just below the knee. Three protection criteria are proposed :

. bending : 15°
. shearing : 6 mm
. tibia acceleration : 150 g

The aim of the bonnet leading edge is to protect adult pedestrian against femur fracture and child pedestrian against pelvic-abdominal injuries.

The protection of the head is assessed by a sub system head form (child head form and an adult head form respectively 2.5 and 4.8 kg) equipped with a triaxial accelerometer. A HIC limit of 1000 is the proposed protection criteria for all the tests. The impact speed is 40 km/h with impact angle of 50° for the child head form and 65° for the adult one. These proposals are still in discussion (61).

3.4.2. *Cost Benefit Analysis*

Several European institutions conducted cost benefit studies concerning the implementation of the procedures (2) to (5).

The main results are indicated in TABLE 13. The differences are related to the value chosen for the cost of the modifications but also to the determination of the number of pedestrians accidents for the years after 2000. All studies use the trend of last years and project this variation in the future but this projection is not homogeneous between the teams.

The conclusion of the recent review of independent expert is the following : « The work done by EEVC to propose sub system tests for pedestrians protection assessment is very promising. However, at its present stage of development, it seems that additional work to validate the method and possibly to amend it is needed » (61).

ACEA has undertaken a programme of assessment of seven vehicles representative of the European car fleet according to the EEVC procedure ; the results are not yet available.

TABLE 13. Cost benefit of the EEVC pedestrian impact requirements

Study	Cost (Ecus)	Benefits (Ecus)	Cost/benefit ratio
TRL	11 to 19	-	1:4.3 to 1:7.5
BAST	-	25 to 34	-
SWOV	-	49.6	-
ACEA	-	4.7	57:1

Simultaneously, EUCAR has initiated a project, which will be finalised in 1999, to analyse the consequences of vehicles design modifications aimed at meeting the requirements laid down for pedestrian friendly vehicles.

References

1. Chillon G. (1971) "The Importance of Vehicle Agressiveness in the Case of a Transversal Impact", Proceedings of 1st International ESV Conference , January 25-27, 1971.
2. Appel H. (1971) "Optimal Deformation Characteristics of Front, Rear and Side Structure of Motor Vehicles in Mixed Traffic", Report on the Second International Conference on ESV, 1971.
3. Ventre P. (1972) "Homogeneous Safety Amid Heteregeneous Car Population", Report on the Third International Technical Conference on ESV, Washington, May 30 thru June 2, 1972.
4. C. Berlioz C. (1972) "Comparison of the Aggressiveness of Different Vehicles and the Safety They Afford", Report on the Third International Conference on ESV, Washington, May 30thru June 2, 1972.
5. Hamon J. (1972) "Rôle de l'agressivité des Structures en Collision Latérale", Proceedings 14th FISITA Congress Inst. Mech. Engrs., London, June 1972.
6. Ventre P. (1973) "Proposal for Test Evaluation of Compatibility Between Very Different Passenger Cars", Report of the Fourth International Technical Conference on Experimental Safety Vehicle, Kyoto, Japan, March 13-16, 1973.
7. Kossar J.M. (1974) "Big and Little Car Compatibility", Report on the Fifth International Technical Conference on ESV, London, June 4-7, 1974.
8. Chandler E. (1973) "Car-to-Car Compatibility", Proceedings of 4th International ESV Conference, Kyoto, Japan, March 13-16, 1973.
9. Seiffert U. (1974) "Compatibility on the Road", Report on the Fifth International Technical Conference on ESV, London, June 4-7, 1974.
10. Ventre P. (1974) "Compatibility Between Vehicles in Frontal and Semi-Frontal Collisions", Report on the Fifth International Technical Conference on ESV, London, June 4-7, 1974.
11. Ventre P. (1974) "A method to Approach the Problems of Vehicle Compatibility", Report on the Third International Congress on Automotive Safety, San Francisco, July 15-17, 1974.
12. Marumo N., Aya N., Takahashi K. and Nousho H. (1974) "Compatibility Between Different-Sized Vehicles on Crash Survivability", Report on the Third International Congress on Automotive Safety, San Francisco, July 15-17, 1974.
13. Reidelbach W. and Schmid W. (1974) "An Attempt to Define and Evaluate Vehicle Compatibility", Report on the Third International Congress on Automotive Safety, San Francisco, July 15-17, 1974.
14. Seiffert U.W., Hamilton J. and Boersch F. (1974) "Compatibility of Traffic Participants", Report on the Third International Congress on Automotive Safety, San Francisco, July 15-17, 1974.
15. Schmidt R. (1976) "Realistic Compatibility Concepts and Associated Testing", Report on the Sixth International Technical Conference on ESV, Washington (D.C.), October 12-15, 1976.
16. Schimkat H. (1979) "A concept of Increasing Compatibility of Passenger Cars", Report on the Seventh International Technical Conference on ESV, Paris, June 5-8, 1979.
17. Fabian G.J. (1980) "Compatibility in the Calspan Research Safety Vehicle", Report on the Eight International Technical Conference on ESV, Wolfsburg, October 21-24, 1980.
18. Ernst G., Bruhning E., Glaeser K.P. and Schmid M. (1991) "Compatibility Problems of Small and Large Passenger Cars in Head-on Collisions", 13th International Technical Conference on Experimental Safety Vehicle, Paris, 1991.
19. Tarrière C., Morvan Y., Steyer C. and Bellot D. (1994) « Accident Research and Experimental Data Useful for an Understanding of the Influence of Car Structural Incompatibility on the Risk of Accident Injury », 14th International Technical Conference on the Enhanced Safety of Vehicles, Munich, May 23-26, 1994.
20. Maurer D., Weyersberg A. Blaßer S., Müller R. Böttcher C., Thompson G. andGraffe M. (1996) « Safety Concepts for Very Small Vehicles - Example : OPEL MAXX », 15th International Technical Conference on the Enhanced Safety of Vehicles, Melbourne, May 13-16, 1996, paper nb. 96-S4-0-13.
21. Seiffert U. (1994) "So Sicher Wie Die Grossen" (Aussi sûres que les grosses"), Auto Motor und Sport, Veran staltang Die Besten Autos 1994, Mittwoch, February 9, 1994.
22. Thomas C., Faverjon G., Henri C., Le Coz J.Y., Got C. and Patel A. (1990) "The Problem of Compatibility in Car-to-Car Collisions", 34th Annual Proceedings AAAM, 1990.
23. Tarrière C., Thomas C. and Trosseille X. (1991) "Frontal Impact Protection Requires a Whole Safety System Integration", Proceedings of the 13th ESV Conference, Paris, 1991.

24. Chelimsky E. (1991) "Automobile Weight and Safety", GAO -Before the Subcommittee on Consumer Committee on Commerce, Science and Transportation, United States Senat (1991).
25. Fontaine H. and Gourlet Y. (1992) « Agressivité des Véhicules et Protection de leurs Occupants", R.T.S., 1992.
26. Fontaine H. and Gourlet Y. (1994) "Sécurité des Véhicules et de leurs Conducteurs", Rapport INRETS n° 175, Février 1994.
27. Hartemann F., Foret-Bruno J.Y., Thomas C., Tarrière C., Got C. and Patel A. (1979) « Influence of Mass Ratio and Structural Compatibility on the Severity of Injuries Sustained by the Near Side Occupants in Car-to-Car Side Collisions », 23rd Stapp Car Crash Conference, San Diego, October 1979, Paper n° 79, 1010.
28. Richter B., Jahn N., Sinnhuber R., Stender Ch., Zobel R. and Zogalla G. (1994) « How Safe can Lightweight Cars Be ? An Analytical Study of the Limits of Passive Safety », 14th International Technical Conference on the Enhanced Safety of Vehicles, Munich, May 23-26, 1994.
29. Neilson I.D. (1994) « Improved Protection Through Greater Compatibility Between Road Vehicles», 14th International Technical Conference on the Enhanced Safety of Vehicles, Munich, May 23-26, 1994.
30. Neiderer, Kaser, Walz, Brunner, Faerber (1994) « Compatibility Considerations for Low Mass Rigid Belt Vehicles », 14th International Technical Conference on the Enhanced Safety of Vehicles, Munich, May 23-26, 1994.
31. Hollowell W. and Gabler H. (1996) « NHTSA's Vehicle Agressivity and Compatibility Research Program », 15th International Technical Conference on the Enhanced Safety of Vehicles, Melbourne, May 13-16, 1996, paper nb. 96-S4-0-01.
32. Akiyama A., Ngatomi K. and Kobayashi T. (1996) « Bumper Structure for Pedestrian Protection », 15th International Technical Conference on the Enhanced Safety of Vehicles, Melbourne, May 13-16, 1996, paper 96-S4-0-02.
33. Neilson I. (1996) « Achieving Compatibility at Impact », 15th International Technical Conference on the Enhanced Safety of Vehicles, Melbourne, May 13-16, 1996, paper 96-S4-0-03.
34. Shearlaw A., Thomas P. and Williams A. (1996) « Vehicle to Vehicle Compatibility in Real World Accidents », 15th International Technical Conference on the Enhanced Safety of Vehicles, Melbourne, May 13-16, 1996, paper nb. 96-S4-0-04.
35. Hobbs A., Williams A. and Coleman D. (1996) « Compatibility of Cars in Frontal and Side Impacts », 15th International Technical Conference on the Enhanced Safety of Vehicles, Melbourne, May 13-16, 1996, paper nb. 96-S4-0-05.
36. Schimmelpfennig K. (1996) « Gliding Zone, A New Approach to Increase Passive Safety for Passenger Cars », 15th International Technical Conference on the Enhanced Safety of Vehicles, Melbourne, May 13-16, 1996, paper nb. 96-S4-0-06.
37. Bloch J. and Biard R. (1996) « In-Depth Analysis of Offset Frontal Crash Tests to Introduce External Aggressivity Criteria », 15th International Technical Conference on the Enhanced Safety of Vehicles, Melbourne, May 13-16, 1996, paper nb. 96-S4-0-07.
38. Foret-Bruno J.Y., Morvan Y., Le Coz J.Y. and Tarrière C. (1996) « Influence of Car Weights on Driver Injury Severity and Fatalities in Head-On Collisions», 15th International Technical Conference on the Enhanced Safety of Vehicles, Melbourne, May 13-16, 1996, papernb. 96-S4-0-08.
39. Prasad P. (1996) « Effect of Frontal Impact Regulations on Vehicle-to-Vehicle Compatibility », 15th International Technical Conference on the Enhanced Safety of Vehicles, Melbourne, May 13-16, 1996, paper nb. 96-S4-0-09.
40. Mizuno K. and Yonezawa H. (1996) «The Effects of Vehicle Properties in a Head-On Collision », 15th International Technical Conference on the Enhanced Safety of Vehicles, Melbourne, May 13-16, 1996, paper nb. 96-S4-W-16.
41. Zuppichini F. (1996) « Vehicle Crash Compatibility : Some considerations Upon Real Cases », 15th International Technical Conference on the Enhanced Safety of Vehicles, Melbourne, May 13-16, 1996, paper nb. 96-S4-W-17.
42. Pipkorn B., Ouchterlony B. and Edvardsson M. (1996) « A Model for Evaluating and Improving Vehicle Performance in Car-to-Car Side Impacts », 15th International Technical Conference on the Enhanced Safety of Vehicles, Melbourne, May 13-16, 1996, paper nb. 96-S4-W-19.
43. Schöneburg R., Bush D., Cakmak M. and Zobel R. (1996) « Evaluation of Crash Compatibility of Vehicles with the Aid of Finite Element Analysis », 15th International Technical Conference on the Enhanced Safety of Vehicles, Melbourne, May 13-16, 1996, paper nb. 96-S4-W-25.
44. Evans, Wasielewski (1987) « Serious or Fatal Driver Injury Rate Versus Car Mass in Head-on Crashes Between Cars of Similar Mass, Accident Analysis and Prevention, vol 19, N° 2, p 119.

45. Mendis K and Mani A. (1996) « Concept to Reduce Heavy Truck Agressivity in Truck-to-Car Collisions », 15th International Technical Conference on the Enhanced Safety of Vehicles, Melbourne, May 13-16, 1996, paper nb. 96-S4-0-14.
46. Tarrière C., Trosseille X. and Carlsson G. (1991) «Initial Conclusion of an International Task Force on Child Restraining Systems», Proceedings of 13rd ESV Conference, November 4-7, 1991, Paris, France.
47. Trosseille X. and Tarrière C., (1993) «Neck Injury Criteria for Children from Real Crash Reconstructions», Child Occupant Protection Symposium, San Antonio, Texas,U.S.A., November 7-8, 1993.
48. Weber K., Dalmotas D. and Hendrick B. (1993) « Investigation of Dummy Response and Restraint Configuration Factors Associated with Upper Spinal Cord Injury in a Forward-Facing Child Restraint», Child Occupant Protection Symposium, San Antonio, Texas,U.S.A., November 7-8, 1993.
49. Brun-Cassan F., Page M., Pincemaille Y., Kallieris D. and Tarrière C. (1993) « Comparative Study of Restrained Child Dummies and Cadavers in Experimental Crashes », 37thStapp Car Crash Conference, San Antonio, USA, November 7-8, 1993.
50. Turbell T., Lowne R., Lundell B. and Tingwall C. (1993) «ISOFIX - A New Concept of Installing Child Restraints in Cars», Child Occupant Protection Symposium, San Antonio, Texas,U.S.A., November 7-8, 1993.
51. Carlson G. Norin H and Ysander L. (1989) « Rearward facing child seats - the safest car restraint for children ? » 33rd AAAM Conference, October 2-4, 1989 (Baltimore, Maryland).
52. Vallée H., Cailleret M.C., Coltat J.C. et al. (1991) « Child Casualties in Fatal Car Crashes », Proceedings of the 13th ESV Conference, November 4-7, Paris, 1991.
53. Got C. and Cuny S. (1994) « Analyse Accidentologique de la Protection des Enfants dans les Véhicules Automobiles en France ». In : Observatoire national interministériel de sécurité routière. Les dispositifs de retenue pour enfants en 1994 - Utilisation etefficacité. La documentationfrançaise, 1994.
54. Tarrière C. (1994) « La recherche et la sécurité des enfants en voiture. La protection actuelle n'est pas satisfaisante. Comment l'améliorer ? » Communicationaux Journées Parisiennes de Pédiatrie 1994. Paris : Médecine-Sciences-Flammarion, 1994.
55. Tarrière C. (1995) « Sécurité des Enfants en Voiture - Nouveaux Dispositifs de Retenue », Journal de Pédiatrie et de Puériculture, Vol VIII - N° 2, 65-128, 1995.
56. Tarrière C. (1995), « Children are not Miniature Adults »,Bertil Aldman Lecture, 1995 International IRCOBI Conference, September 13-15, 1995, Brunnen, Switzerland.
57. ACEA White Paper on Road Safety, to be published in 1996.
58. McFadden M. (1996) »Pedestrian Safety », 15th International Technical Conference on the Enhanced Safety of Vehicles, Melbourne, May 13-16, 1996, paper nb. 96-S7-W-20.
59. Evans L. (1995) »ABS and Relative Crash Risk Under Different Roadway, Weather, and Other Conditions », SAE Technical paper n° 950353.
60. Kahane C.J., (1994) « Preliminary Evaluation of the Effectiveness of Antilock Brake Systems for Passenger Cars », NHTSA Technical Report, December 1994.
61. Cesari D., Fontaine H., Lassare S. (1996) « The Validity of the Proposed European Pedestrian Protection Procedure and Its Expected Benefits », 15th International Technical Conference on the Enhanced Safety of Vehicles, Melbourne, May 13-16, 1996, paper nb. 96-S4-W-25.

ANNEX 1

CHILD RESTRAINT SYSTEMS FOR CARS

Starting date: 1 January 1996
Contract no: SMT4 CT95 2019
Total costs: 4 282 656 ECU
Duration: 54 months
Proposal no: PL-95-2098
EC Funding: 2 390 000 ECU

SUMMARY

Child restraint systems (CRS's) for cars are intended to protect children in the case of a car accident. Unfortunately their effectiveness is still too low: in the range 30-50 % when it would be expected to be about 70-80 %. Consequently, there are high mortality and morbidity rates even in those European countries, such as France and Germany, which have stringent regulations and where the use of child restraint systems is mandatory from birth to 10 or 12 years of age. The low effectiveness of child restraint systems can partly be explained for the youngest passengers by their greater cervical vulnerability and for the oldest (from 3 to 12 years old) by the morphological immaturity of the pelvis. There is a need to improve the design of instrumented child dummies used to test child restraint systems so that they better simulate the dynamic characteristics of children.
This project aims ultimately to improve the effectiveness of child restrainst systems. New test procedures for determining the effectiveness of child restraint systems using instrumented child dummies will be developed. Data will be collected from accident investigations and from crash tests in order to determine the physical parameters (forces and accelerations on the child) which correspond to the various child injury mechanisms. Hence, limits will be prescribed under which injuries can be avoided. The new test procedures that are developed will be applied to prototype innovative child restraints systems in order to demonstrate the feasibility for industry to meet more stringent requirements for child protection.
All accident configurations which are found to involve child victims will be studied in the project. It is expected that side impacts will be particularly important as the few statistics currently available indicate that they account for a high rate of severe and fatal injuries to children.

COORDINATOR

Dr. Claude TARRIERE/
Mrs. Charlette POIRIER
Renault
Département Biomédical de l'Automobile
132, rue des Suisses
FR - 92000 Nanterre

Tel: +33 1 47773554
Fax: +33 1 47773636

MAIN PARTNERS

- Fiat Auto SpA, IT
- Institut National de Recherche sur le Transport et la Sécurité, FR
- GIE PSA Peugeot Citroën, FR
- Nederlandse Organisatie voor Toegepast Natuurwetenschappelijk Onderzoek, NL
- Technische Universität Berlin, DE
- Loughborough University of Technology, UK
- Bundesanstalt für Strassenwesen, DE
- Swedish Road and Transport Research Institute, SE
- Research Institute for Consumer Electronics, UK
- Verbond der Schadenversicheren e.V., DE
- Medical University Hannover, DE

ADVANCED RESTRAINT SYSTEMS FOR OCCUPANT PROTECTION

DOMINIQUE CESARI
INRETS
109 Avenue S. Allende
69500 BRON FRANCE

Abstract

Restraint systems are a major contributor to the protection of car occupants in traffic accidents. They are used for more than 30 years, but they have evolved during these three decades. Analysis of restraint systems effectiveness shows that the three points seat belt is very effective but the combination of belt plus airbag pushes up the limits of protection. The airbag and belt technologies are described, and the recent works to develop smart restraint systems which can adapt to the passenger to be protected and to the crash conditions are analysed. In the last part the further steps for more intelligent restraint systems for frontal and side impact protection are discussed.

1. Introduction - Traffic Accidents Injuries : a Worldwide Problem

Every year in the European Union, more than 50 000 people are killed in 1.5 million traffic accidents. More than half a million of Europeans are every year admitted to an hospital as a consequence of these accidents, and the associated cost of these injuries is around 70 billion ecus.

Even if the number of traffic accident victims has been divided by a factor greater than 2 during the last twenty years, traffic safety remains a priority in all developped countries. Two directions of actions can be developed to improve the situation : the first one deals with active (or primary) safety, and the second one with passive (secondary) safety; the effectiveness of restraint systems are related to the latter.

J. A. C. Ambrósio et al. (eds.),
Crashworthiness of Transportation Systems: Structural Impact and Occupant Protection, 173–187.

2. The needs for Occupant Protection Systems

An accident is a transitory phenomenon during which high magnitude forces can apply in a short time. In a frontal impact accident, the car is stopped by the deformation of the structures ; this velocity change in a very short time applies forces to the occupant. If we consider a collision of a car at 50 km/h impact speed against a flat rigid barrier, the car is stopped within approximately 60 cm.

If the occupant is unrestrained, the velocity change of 50 km/h in 60 cm corresponds to an average deceleration of more than 15 g. As the car begins to decelerate, the occupant is put forward keeping the car initial speed until he contacts with body parts the interior of the passenger compartement. For a specific impact speed, the severity of the occupant impact is related to the deformation capabilities of the impacted structure and of the involved body part. This explains that for example if the head hit the steering wheel or the A pillar, the head acceleration may be over 150 g, which is 10 times the average car deceleration. As indicated on figure 1 the velocity change is not the same for all body components : it is lower for the feet than for the head which is far from car structures and which then sustain a velocity change equal to the car impact speed [1].

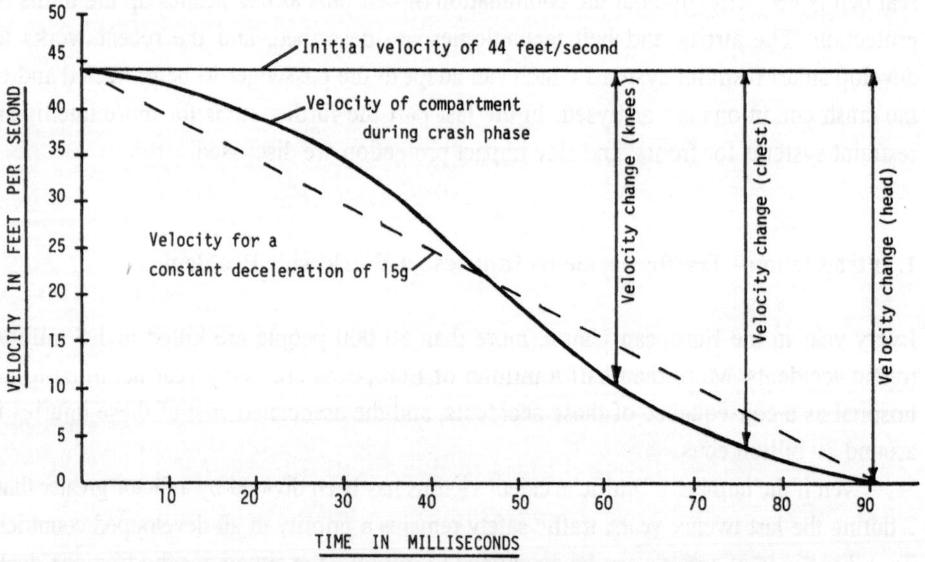

Figure 1. Velocity versus Time for a 48 km/h Impact with 60 cm of crush [1]

To avoid this very severe secondary impact, it is necessary to keep the occupant away from any rigid car structure : this is the role of restraint systems. However the restraint system will create contact forces on the body whose value will depend on the contact surface on the body (contact pressure) and on the deformation capability of the restraint device, the general goal being that as the same time the restraint system avoid contacts with rigid car compartment components, it generates forces loading the body at levels below the tolerance to injuries.

The body to restraint system contact forces will also be influenced by the body mass and mass distribution. Moreover the human tolerance to a specific impact injury varies from one person to another according to his age and other individual characteristics, and this tolerance is very variable from a body segment to another.

3. Effectiveness of Frontal Impact Protection Systems

The 3 points safety belt (lap and shoulder belt) restraining the occupant through the thorax and the pelvis is considered as the most effective protection device for a very low production cost. However the safety belt has its own limitations, and to push forward the limits of protection, the airbag was invented and developed. It is recognized now especially in Europe that safety belt and airbag do not compete, but on the opposite the combination of both improves the protection.

3.1. EFFECTIVENESS OF THREE POINTS BELT

The effectiveness of the safety belt has been analysed widely through accident analysis and experimental research. All these studies conclude in a high effectiveness of the safety belt compared to the unrestrained situation.

This problem of safety belt effectiveness was carefully analysed by L. Evans [2]. He demonstrated that the effectiveness, which is the percentage of fatally injured unrestrained occupants which would have been saved if they were restrained is greatly influenced by accident conditions, as a protection device is designed to have its highest effectiveness in a predetermined crash condition.

Using the FARS (Fatal Accident Reporting System), he proposed a value of about 40 % for the effectiveness of the 3 points belt. The details of 3 points belt effectiveness for different impact directions are included in table 1. This table shows a higher effectiveness for drivers than for right front passengers. According to Evans, half of the effectiveness of the belt is related to ejection avoidance.

TABLE 1. Effectiveness of 3 points belt in accidents

Accident Type	Driver	Right Front Pass.
Pure Frontal	43%±8%	39%±9%
Frontal Right	41%±18%	30%±20%
Side Right	39%±15%	27%±19%
Rear	49%±14%	45%±20%
Side Left	27%±17%	19%±20%
Frontal Left	38%±15%	23%±20%
Rollover	59%±10%	46%±15%
Total	42%±4%	39%±4%

3.2. EFFECTIVENESS OF AIR BAG

The determination of airbag effectiveness was made more recently and is still in progress, especially in Europe.

In the USA, Evans applying the same technique as for safety belts concluded that the effectiveness of the airbag in reducing fatalities is equal to 18 %±4 % for the drivers and 13 %±4 % for the right front passenger, which is less than half of the safety belt effectiveness.

The airbag affects all severities of accidents, and Malliaris [3] using NASS and FARS files proposes a distribution according to the injuries severity for different restraint devices, as indicated in table 2.

TABLE 2. Distribution of injuries severity for different restraint conditions [3]

Severity	Belt	Airbag	Belt+Airbag	Unrestrained
MAIS 1	35% to 37%	33% to 50%	36% to 43%	43% to 46%
MAIS 2	5% to 5.6%	6% to 12%	5% to 8%	10.5% to 11.6%
MAIS 3	1.6% to 1.8%	3.3% to 6.2%	0.9% to 1.6%	5.2% to 5.8%
MAIS 4	0.4% to 0.6%	0.9% to 1.9%	0.2% to 0.6%	2.2% to 2.5%
MAIS 5	0.3% to 0.4%	0.4% to 1.3%	0.2% to 0.5%	1.7% to 2%
Fatal	0.3% to 0.4%	0.5% to 1.3%	0.2% to 0.5%	1.6% to 1.8%

Analysis of table 2 indicates that the combination of belt and air bag is more efficient than the belt alone especially for severe accidents.

In a recent study [4] based on the analysis of accidents involving front airbag-equipped cars it is indicated for the eight collisions in which the airbag did not deploy (because the impact severity was below the trigger level), none of the occupant sustained injuries of AIS greater than 2. This study also indicates that the protective effect of the airbag associated with a 3 points belt is well established for the driver even for high severity accidents.

3.3. LIMITATIONS OF AIRBAG EFFECTIVENESS

The use of air bag is associated with a specific injury typology, which corresponds mainly to slight injuries (for example minor face burns, or arm wounds), but the airbag may be injury producing if the occupant is in contact with or very close to the cover at the time of the inflation. If the occupant is « out of position », he may be hit during the inflation process, and then injured by the contact with the bag moving at a very high speed. This probem can concern either unrestrained adults or children. Especially if the airbag deploys in a frontal impact in a car with a reaward facing child seat, the airbag will accelerate severely in the rear direction the child seat and produce very severe or fatal injuries to the child.

4. Restraint systems Technology

Seat belt as well as air bag may be designed with different technological solutions, and especially the seat belt has evolved from the first ones to the most recent during about 30 years.

4.1. SEAT BELT TECHNOLOGY

The first generation of three points belts produced in the late sixties had fixed anchorage points and were manually adjustable. This seat belt was efficient when correctly ajusted, but loosed a part of its protective effect when worn loose, which was very common. From the mid seventy, the new seat belts include a belt retractor : the shoulder strap is rolled on a spool activated by a spring aimed to wind the belt on the spool. If an accident happens, the spool is locked by the motion of a small steel ball, and another device based on the action of centrifugal force locks the spool if it turns too fast (or if the strap is pulled out too quickly). Retractor belt allows the belt to be always thighten, always in contact with the body; however because a certain length of the strap is stored on the retractor spool, when it is locked on its axis, a few centimeters of the belt may get out when body inertia forces pull on it.

By comparison with a static 3 points belt, the retractor does not improve the protection level, but has two main advantages : it ensures that the belt is correctly worn, and it increases the wearing rate as a retractor belt is much easier to wear.

Belt geometry is very important to provide an optimised restraint capability. Two evolutions dealing with the improvement of belt geometry were made. The first one concerns the adjustable shoulder anchorage point. By moving up or down the anchorage point on the B pillar it is possible to optimise the position of the shoulder belt on the thorax for different occupant sizes and different seat adjustments. The second one was the fixation of the buckle (and sometimes the low external anchorage) on the seat structure (instead of the car floor); this feature gives the same position of the buckle relative to the body for different seat adjustments.

Accidents studies have showed the limitations of the protection of the belt which may induce specific injuries, especially on elderly occupants. To limit the loads applied to the body two types of devices were proposed : a belt force limiter and a belt pretensionner.

<u>Process</u>

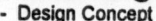

- Determine Customer Requirements - Spring Definition
  - 100-150 mm @ 500-700 lb.
  - Design Concept

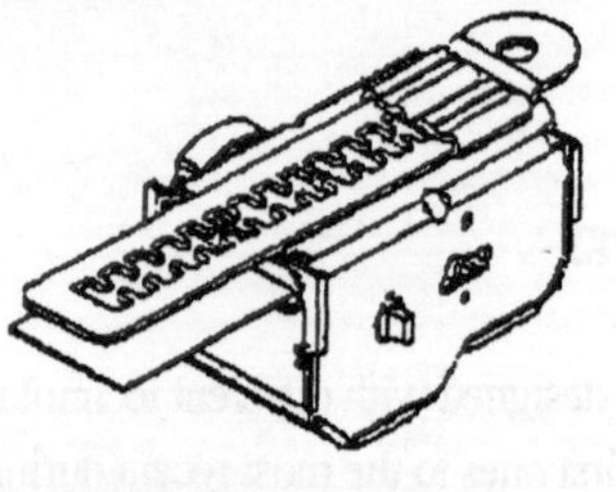

  - NCAP Evaluation Results

	Chest (g)	Shoulder Belt Load (#)	Star Rating
Base Line	54	1400	*** ****
EM System	44	1100	*****

Figure 2. Energy Management Retractor [5]

A first belt force limiter was used on static belts. When the belt force reached a limit value, the rupture of a fabric belt « by-pass » limited the force exerted on the thorax and gave a few centimeters of extra length of the shoulder belt ; this device has the inconvenience of providing an extended kinematic to the occupant, and then favours impacts of the body with components inside the passenger compartment. New recent developments of belt force limiters were made ; they incorporate an energy dissipative mechanism such as deformable belt anchors, deformable buckle attachement or load compensating retractor [5]. Figure 2 is an example of a load compensating retractor. These devices would work correctly if the car is also fitted with an airbag : then the force limiter keeps the shoulder force below the force injury threshold, and the airbag avoids direct contacts with hard components.

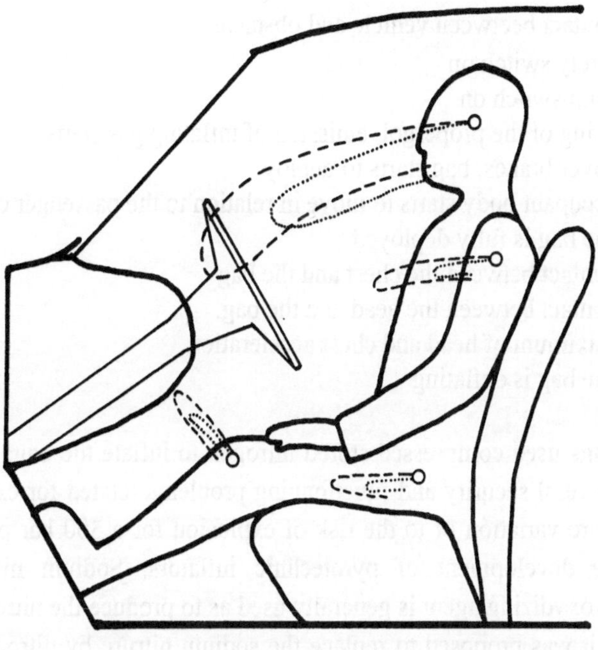

Figure 3. Trajectory of a Restrained Occupant with a Retractor Belt and with a Pretensioner at 48 km/h impact speed [1]

Belt pretensioner is aimed to diminish the motion of the occupant in relation to the car by pulling on the belt if a crash occurs. This device limits the occupant kinematic as shown on figure 3 [1]. Moreover because it gives a better coupling between the car and the occupants it avoids the second impact inside the restraint system and then decreases the belt force.

4.2. AIRBAG TECHNOLOGY

An airbag is constituted of three main components : the sensor, the gas generator and the bag. The sensor is aimed to activate the airbag if the accident occurs under certain conditions (acceleration, speed, direction of impact...). The gas generator delivers in a very short time the gas necessary to inflate the bag. The inflated bag would friendly stopp the occupant during his motion.

The sensor analyses the crash and activates the airbag. Hereafter is a typical chronology of events for a frontal collision of an airbag equipped car [6].

00 ms : Contact beetween vehicle and obstacle
10 ms : Safety switch on
23 ms : Main switch on
25 ms : Firing of the propergol, emission of inflating gas starts
28 ms : Cover brakes, bag starts to deploy
40 ms : Occupant body starts to move in relation to the passenger compartment
55 ms : The bag is fully deployed
60 ms : Contact between the chest and the bag
70 ms : Contact between the head and the bag.
95 ms : Maximum of head and chest acceleration
150 ms : The bag is deflating

The first airbags used compressed stored nitrogen to inflate the bag; however this technique raised several security and functionning problems related for example to the effect of temperature variation or to the risk of explosion for a 300 bar pressure tank. This justified the development of pyrotechnic inflators. Sodium nitrure (NaN_3) associated with an oxydizing agent is generally used as to produce the nitrogen inflating the bag. Recently it was proposed to replace the sodium nitrure by nitrocellulose; this allows to reduce the size of the inflator, and to solve the problem of gaz toxicity, but the gaz is generated at high temperature (up to 3000° C) and then it is necessary to cool it during the airbag inflation. As some problems remain with pyrotechnic inflators, there is a come back to pressure stored gas or hybrid systems [7].

The component ensuring the protection is the inflated bag. The maximum speed of a deploying bag is between 160 and 340 km/h, which explains injuries occuring if the bag hits an occupant during the deployment ; however even for a small driver seating close to the steering wheel, the contact speed is lower than the maximum speed value [8]. The use of internal straps controlling the inflation may reduce its agressivity to the occupant.

The bag has vents allowing it to deflate slowly, otherwise the bag would be too elastic and the body may rebound on it; their number, size and positions are important to optimize the airbag behaviour and to avoid hands or face burns.

5. Smart Restraint Systems

Considering the limits of present restraint systems and the improvement of the knowledge in the field of human impact biomechanics, a new concept, called « Smart (or Intelligent) Restraint System » was developed. In principle it consists in designing restraints systems whose characteristics adapt to the seating posture, anthropometry, age and accident conditions.

5.1. SMART RESTRAINT FOR FRONTAL PROTECTION

The SMART™ airbag [9] was developed to adapt its characteristics to the crash situation with the logic indicated on figure 4. It includes a variable inflator with three levels of inflation (level 1, level 2, and level 1+2) and a possible delay of the air bag firing.

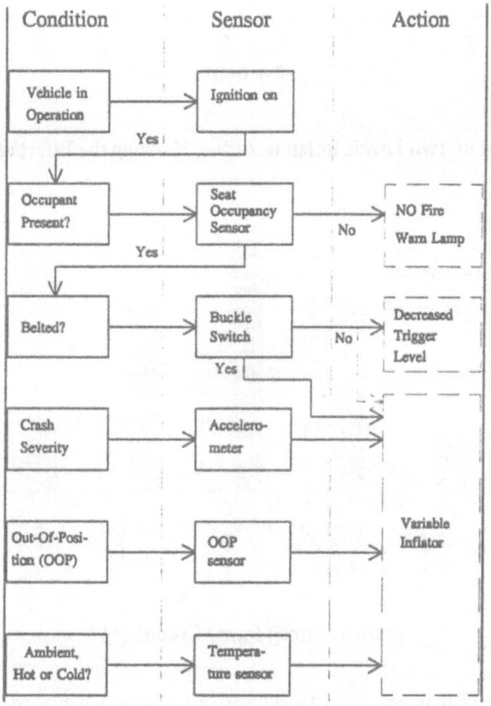

Figure 4. Logic for Airbag Triggering [9]

Figure 5 shows the effects of various delays on the balistic performance of the multi level gas generation source. This approach allows also to compensate for temperature variation

Using a smart airbag with a 350 kPa output instead of a 475 kPa normal airbag in sled tests with an out of position 5th percentile female dummy shows a drastic decrease of all injury predicting parameters as indicated on figure 6.

To determine the characteristics of the different levels of an adjustable inflator S. Andrews uses an equivalent energy approach [10]. This approach shows that a 4 levels system allows to cover adult occupants in the mass range of 45 to 100 kg and sitting at a distance of 0.2 to 0.7 m within ± 31 % of the optimized energy, as indicated in figure 7.

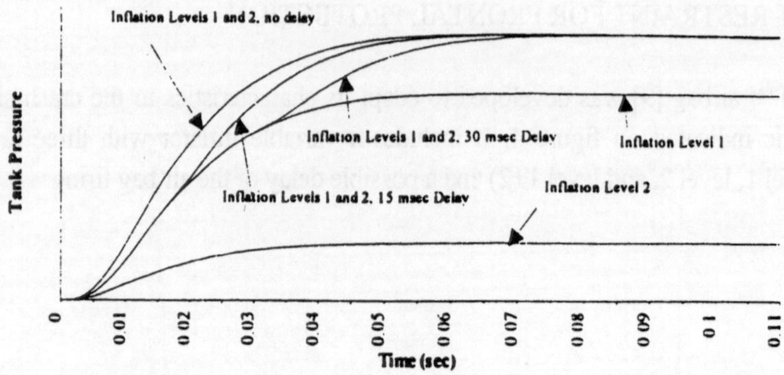

Figure 5. Combination of Two Levels Inflation Airbag Showing the Effect of Various Delays [9]

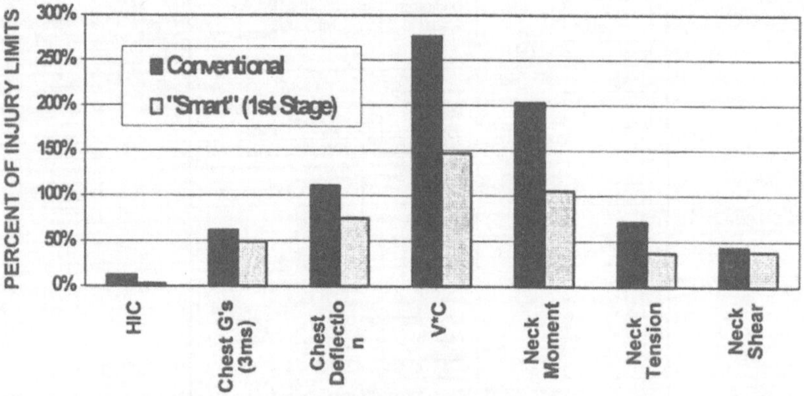

Figure 6. Airbag Injury Criteria [11]

To obtain the four levels it is propsed to design an airbag with a two levels inflator and to have adjustable vents as indicated hereunder :

Level 1 : Low Inflation, High Vents
Level 2 : Low Inflation, Low Vents
Level 3 : High Inflation, High Vents
Level 4 : High Inflation, Low Vents

As the seat belt interacts early with the occupant and the airbag latter it seems interesting to optimise the specific characteristics of each system to design the best performing restraint system. As stated by H. J. Miller [12] an ideal restraint system performance is one which results in a linear velocity time history and therefore constant acceleration of the occupant. According to the occupant size, interior space available and crash severity it is then possible to determine the characteristics of an optimised restraint system. Figure 8 shows a typical and an ideal (constant force) force displacement histories which are used to determine the effect on different restraint conditions.

Mass (Kg)	Position Relative to Deployed Restraint (m) 0.2	0.3	0.4	0.5	0.6	0.7
100	-24%	14%	-9%	14%	-1%	15%
95	-28%	9%	-13%	9%	-6%	9%
90	-31%	3%	-18%	3%	-11%	4%
85	31%	-3%	30%	-3%	17%	-2%
80	23%	-9%	22%	-9%	10%	-8%
75	15%	-14%	14%	-14%	3%	-14%
70	8%	-20%	7%	-20%	-4%	12%
65	0%	-26%	-1%	24%	-11%	4%
60	-8%	-31%	-9%	14%	-18%	-4%
55	-15%	27%	-16%	5%	26%	-12%
50	-23%	15%	-24%	-5%	14%	-20%
45	-31%	4%	-31%	-14%	3%	20%
40	-38%	-8%	23%	-24%	-9%	7%
35	-46%	-19%	8%	-33%	-20%	-7%
30	-54%	-31%	-8%	15%	-31%	-20%
25	-62%	-42%	-23%	-4%	15%	-33%
20	-69%	-54%	-38%	-23%	-8%	8%

Figure 7. Energy Level Variance from Class Mean Versus Occupant Mass and Position [10]

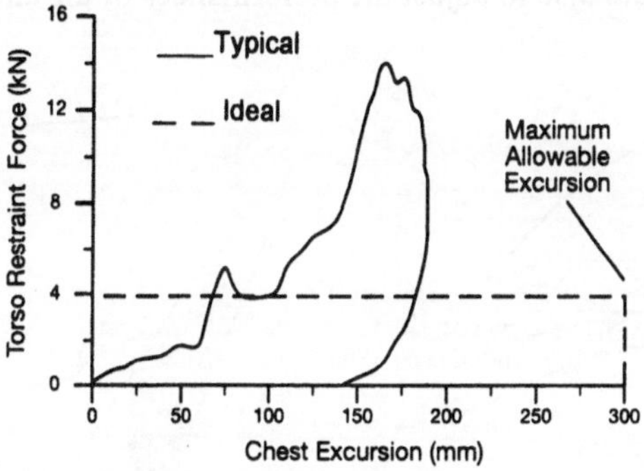

Figure 8. Thorax Force Displacement Plots for Typical and Constant Force Restraint Systems [12]

In a pragmatic approach it is proposed to optimise the protection provided by the seat belt airbag combination by reducing coupling of combined forces of belt and bag on the sternum [5]. This may be obtained by applying a load phase shift and by reducing the peak applied load. This would result in the reduction of the chest acceleration as shown on figure 9.

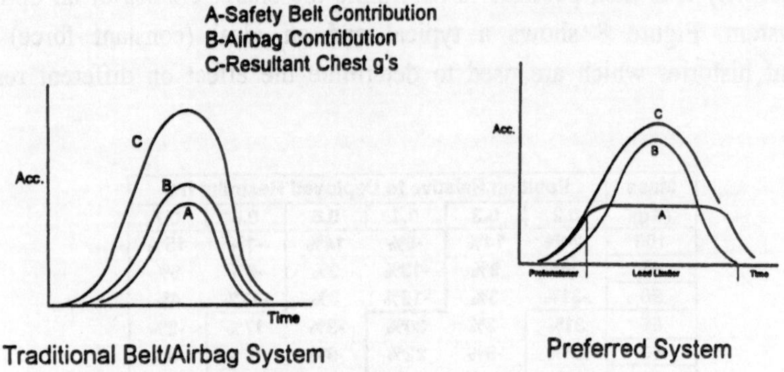

Figure 9. Chest Acceleration for Two Restraint Conditions [5]

Intelligent restraint systems would be also able to detect the presence of a child restrained in a rearward facing child seat. Beside the use of labels and pictograms to warn the car users not to put a rearward facing child seat on the front passenger seat, several technical designs are proposed to detect the presence of a child seat and how it is used. Figures 10a and 10b show an example of a configuration using electronic « tags » to detect the child seat and then possibly desactivate the passenger air bag [10]. Figure 11 shows a more general device using an infrared sensor and human presence detectors in the right front passenger seat; this device may not only used to activate or desactivate the airbag, but also to adjust the performances of the airbag.

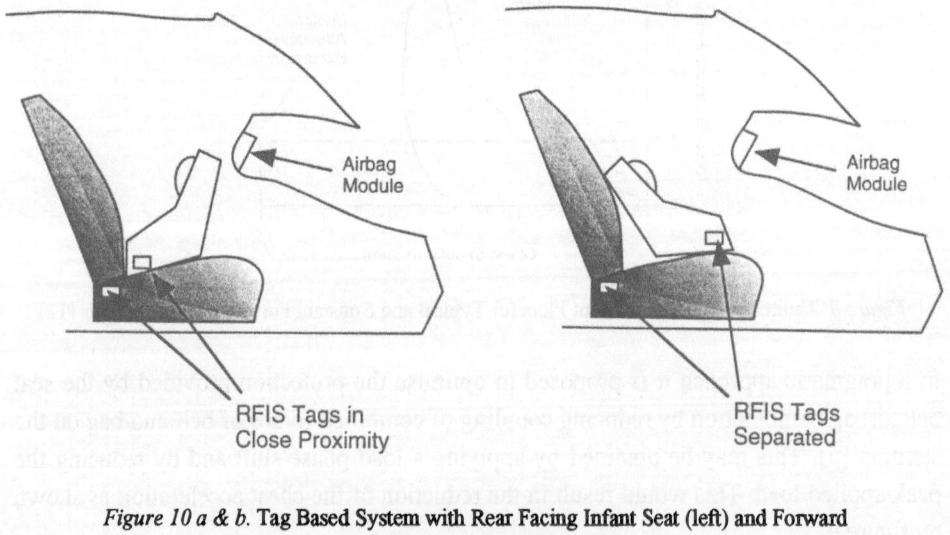

Figure 10 a & b. Tag Based System with Rear Facing Infant Seat (left) and Forward Facing Infant Seat (right) [10]

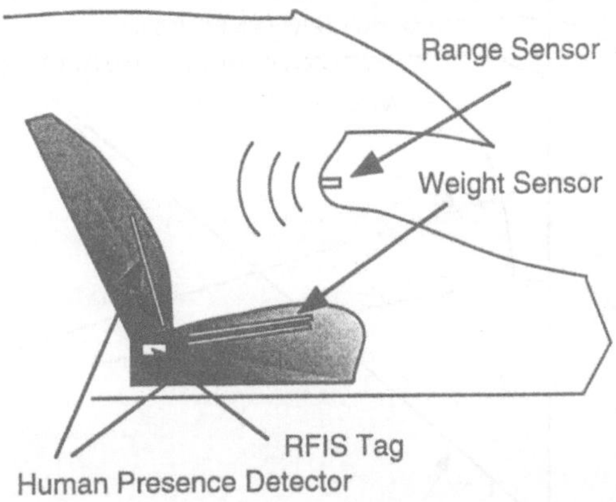

Figure 11. **Typical Occupant and RFIS Sensing System** [10]

5.2. SPECIFIC DEVELOPMENTS FOR SIDE IMPACT PROTECTION

New regulation to be applied in Europe by October 1998 will require car design modifications to fulfill the requirements. The protection of car occupants in side impact is a difficult matter as, unlike in frontal accidents, the distance to control the occupant kinematic is very short, ant then the available time to activate a restraint system is also very short. Based on injury mechanisms in side impact, the improvement of protection in side impact needs to combine crashworthiness modifications to decrease the effects of intrusion, and the use of internal deformable structures to damp the contact the occupant in the internal side of the door [13]. For the latter either a padding or an airbag may be used, and it is easy to use a padding for lower body protection (pelvis), but the protection of the thorax and of the head can be afforded with an airbag, as it can match the tolerance and do not affect the passenger compartment habitability.

Figure 12 clearly illustrates how short is the time available to trigger a side impact airbag [14]. Consequently side impact airbags are designed to be fully deployed within 10 to 15 ms. Figures 13a and 13b show two examples of side impact thoracic air bags, the first one fitted in the seat back [15], and the second one on the door panel [16].

Several developments are made to improve the head protection [17] ; the main proposed solutions are an airbag integrated in the headrest deploying between the window and the side of the head, an airbag fitted on the side roof rail deploying down, or a tighten curtain moving quickly down from the top of the door. There is some attempts to ensure protection of the head and of the thorax with a unique airbag, however it seems difficult to achieve because of the short available time to inflate a large airbag.

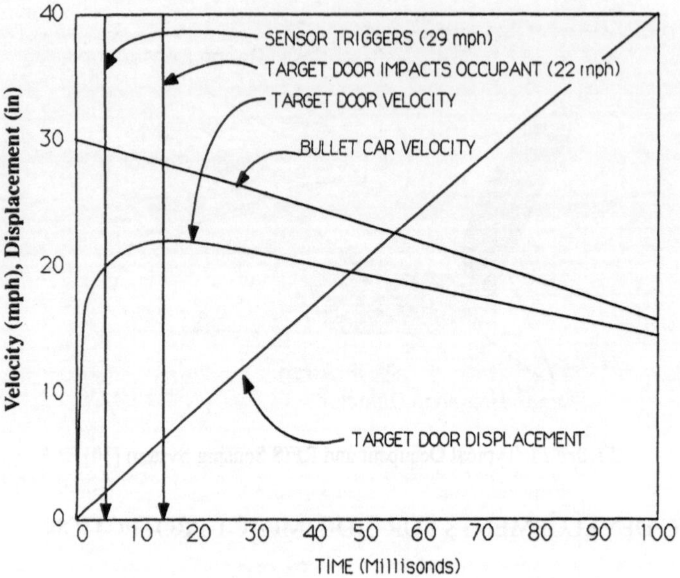

Figure 12. Time Events in Side Impact Airbag Trigger [14]

6. Conclusion

Restraint systems are greatly contributing to the protection of car occupants in accidents, especially in frontal an side collisions. Their effectiveness is fully proved, but we are now facing new challenges to optimise the protection to the specific characteristics of the person to be protected and to the accident conditions, instead of adjusting the restraint system behaviour to standard 50th percentile dummy response : this is the concept of smart restraint systems. Recent developments show that, for frontal impact, the most promising system is a combination of a 3 points belt with a pretensioner and a force limiter and a multistage airbag which adapts its inflation conditions. The problem of side impact is more difficult, as the time available to blow the airbag is very short, and then limits the possibility of designing an adjustable system.

7. References

1. Mackay, M. (1984) An Historical Perspective on Impact Biomechanics and Some Basic Kinematics, *The Biomechanics of Impact Trauma* Elsevier Science Publishers, Amsterdam
2. Evans, L. (1987) Occupant Protection Device Effectiveness, Some Conceptual Considerations, *Journal of Safety Research* **18**, 137-144

3. Malliaris, A. C.,Digges, K. H., DeBois, J. H. (1995) Evaluation of Airbag Field Performance, *SAE Technical Paper # 950869* Detroit
4. Langwieder, K., Hummel, T. (1996) Experience with Airbag Equipped Cars in Real-Life Accidents in Germany *15th International Technical Conference on the Enhanced Safety of Vehicles,* paper #96-S1-O-04, Washington
5. Bernat, A. R. (1995) « SMART » Safety Belts for Injury Reduction, *39th A.A.A.M. Annual Conference Proceedings,* 567-576 Des Plaines
6. Page, Y., Saint Ourens, A. (1996) Que Savons-nous de l'Efficacité des Sacs Gonflables? *Observatoire National Interministeriel de Sécurité Routière,* Paris
7. Mahon, G., Whiston, B. (1996) BREED Temperature Compensated Stored Gas Inflator : The Only True Greeen Solution *15th International Technical Conference on the Enhanced Safety of Vehicles,* paper #96-S1-W-23, Washington
8. Mateja, J. (1995) Are Bigger Cars Safer?, *Traffic Safety Journal,* March/April issue
9. Smith, B. W. (1994) SMART™ Airbag System, *The 14th International Technical Conference on Enhanced Safety of Vehicles,* paper #94-S4-O-11, Washington
10. Andrews, S. (1995) Occupant Sensing in Smart Restraint Systems, *39th A.A.A.M. Annual Conference Proceedings,* 543-555, Des Plaines
11. Fredin, S. R; (1995) Injury Reduction Potential for « SMART » Airbags, *39th A.A.A.M. Annual Conference Proceedings,* 557-566, Des Plaines
12. Miller, H. J. (1995) Injury Reduction with Smart Restraint Systems, *39th A.A.A.M. Annual Conference Proceedings,* 527-541, Des Plaines
13. Cesari, D., Ramet, M., Herry-Martin, D. (1978) Injury Mechanisms in Side Impact,*22nd Stapp Car Crash Conference,* 429-448, Detroit
14. Breed, D. S., Sanders, W. T., Castelli, V. (1994) Sensing Side Impacts, *The 14th International Technical Conference on Enhanced Safety of Vehicles,* paper #94-S6-O-12, Washington
15. Pilhall, S., Korner, J., Ouchterlony, B. (1994) SIPSBAG - A New Seat-Mounted Side IMPACT Airbag System, *The 14th International Technical Conference on Enhanced Safety of Vehicles,* paper #94-S6-O-13, Washington
16. Haland, Y. (1994) Sensor for a Side Airbag : Evaluation by a New Subsystem Test Method, *The 14th International Technical Conference on Enhanced Safety of Vehicles,* paper #94-S6-W-26, Washington
17. Kompass, K., Haberl, J., Messner, G. (1996) Field Study on the Potential Benefit of Different Side Airbag Systems *15th International Technical Conference on the Enhanced Safety of Vehicles,* paper #96-S6-O-01, Washington

CAR CRASH AND SAFETY TESTING

PIERLUIGI L. ARDOINO
Fiat Auto, Product Engineering, Safety Center Manager,
Via G. Gozzano 2, 10043 Orbassano, Turin, Italy

1. Introduction

Initially, passive safety was conceived as the preservation of the integrity of the car's interior compartment in a collision, to be ensured at the design stage, so as to offer a satisfactory degree of protection to the occupants.

In the late '60s, this way of thinking gave rise to a set of standards aimed at the preservation of a series of geometrical characteristics when the vehicle components are subjected to predetermined dynamic or static loads. Most of the passive safety standards in force nowadays are of this kind.

For example, one requirement is that in head-on collision against a fixed barrier the car will undergo a backward and upward displacement of the steering wheel not to exceed 12.7 cm. Another requirement concerns intrusions, which must not exceed fixed values for loads applied to the sides or roof of the car.

In the 70's, this approach reached its apex in the U.S.A., when the Experimental Safety Vehicle (ESV) project specifications were drawn up. Test on the ESV cars, showed the inadequacy of the geometrical approach and pointed to the need to assess vehicle safety by biomechanical criteria. According to this new approach, passive safety means to ensure that the stresses to which car occupants are exposed in an accident will not exceed the limits of human tolerance.

2. Procedure for Formulation of the Standards of a Biomechanical Type

The logic of the biomechanical approach is illustrated in fig.1.

We distinguish three essential steps, in terms of knowledge and activities, leading to the formulation of biomechanical standards:

- background accident analysis, as a means to identify priority interventions and to check the effectiveness of the standards;
- the definition of testing conditions, so as to be able to subject the interior compartment to the same stresses as are produced by the actual accidents that we wish to simulate;
- the definition of performance criteria expressed in biomechanical terms.

J. A. C. Ambrósio et al. (eds.),
Crashworthiness of Transportation Systems: Structural Impact and Occupant Protection, 189–205.

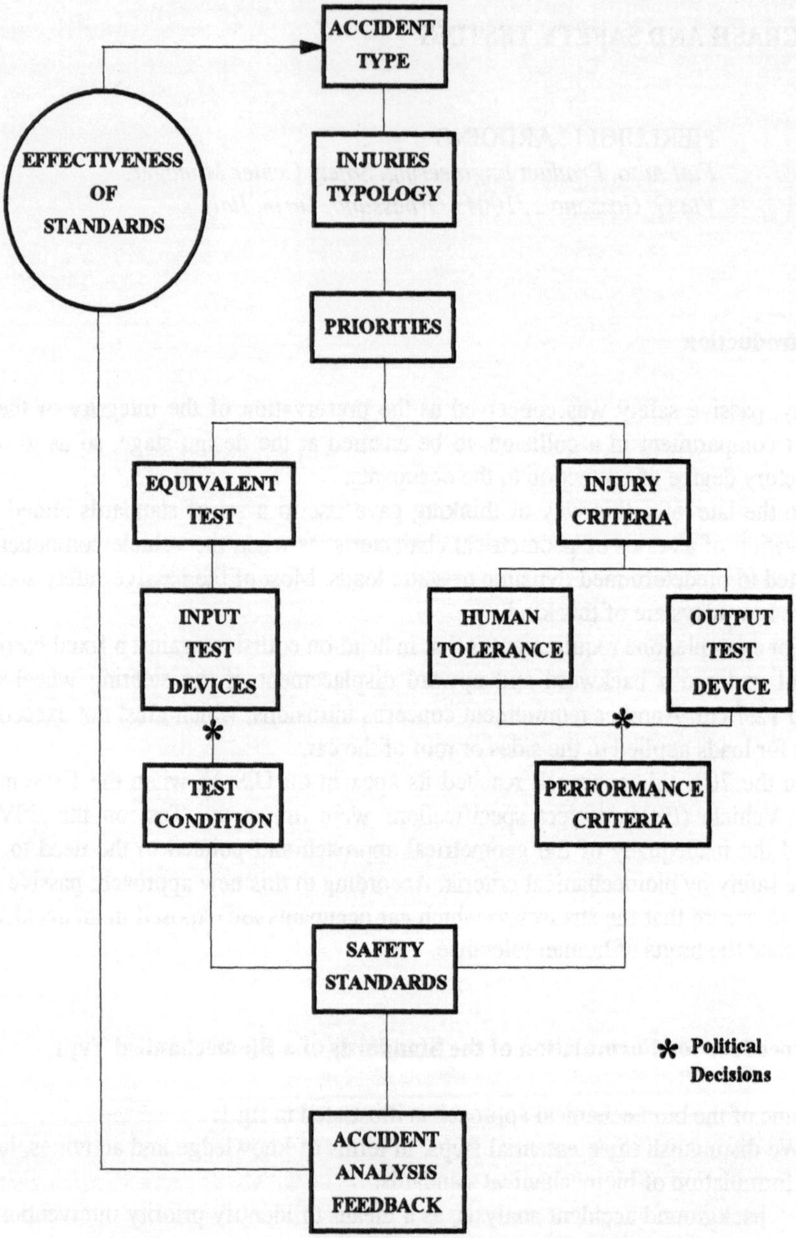

Figure 1 . Biomechanical Approach.

2.1 BACKGROUND ACCIDENT ANALYSIS

The basic knowledge has to be gathered from a through analysis of accidents, to be carried out by multidisciplinary teams. The different types of accident should be analysed in order to determine their priorities, their complexity and the resulting injuries to the various parts of the human body.

2.2 TESTING CONDITIONS

A definition of an "equivalent test" with respect to the type of actual accident to be reproduced is of fundamental importance. In general, this means defining a barrier or a particular obstacle, as well as its area and the angle of impact, which amounts to defining the "Input Test Devices". However, it is up to the Political Authorities to determine the severity of the testing conditions (collision speed, weight and size of the barrier, and so on), since they must decide on the limits within which to protect road users. This decision making stage is marked out by an asterisk (*) in the diagram.

2.3 PERFORMANCE CRITERIA EXPRESSED IN BIOMECHANICAL TERMS

Having determined the body segments to be protected for each accident (accident analysis priorities) it is necessary to define, for each part of the body, those measurable physical parameters (injury criteria) which are correlated with the severity of an injury (for example, if an increase in chest deflection results in an increase in the number of rib fractures, then chest deflection should be viewed as an injury criterion for the chest).

Once the injury criteria have been established, the next step is to determine what values the human body may whitstand (human tolerance levels); at the same time, it is also necessary to work out an "anthropomorphic device of compliance" or -"Output Test Device"- able to measure the injury criteria during the collisions.

It is important to point out that the definition of acceptable values ("Performance Criteria") is the responsibility of the Political Authorities, that must establish which part of the population at risk they intend to protect and the injury levels they are willing to accept for the established " Testing Conditions".

The moment of decision, as has already been done for the testing conditions, it is marked out by an asterisk (*) in the diagram.

2.4 SAFETY STANDARDS

The "Testing Conditions" together with the "Performance Criteria" are the essential part of a standard based on biomechanical criteria. Two full scale tests based on biomechanical criteria for the protection of car occupants in head-on and side collisions have been developed in Europe so far (fig. 2-3-4-5-6). They shall be applied according to the following schedule:

- 1/10/1998 to all new vehicle models;
- 1/10/2003 to all new registered vehicles.

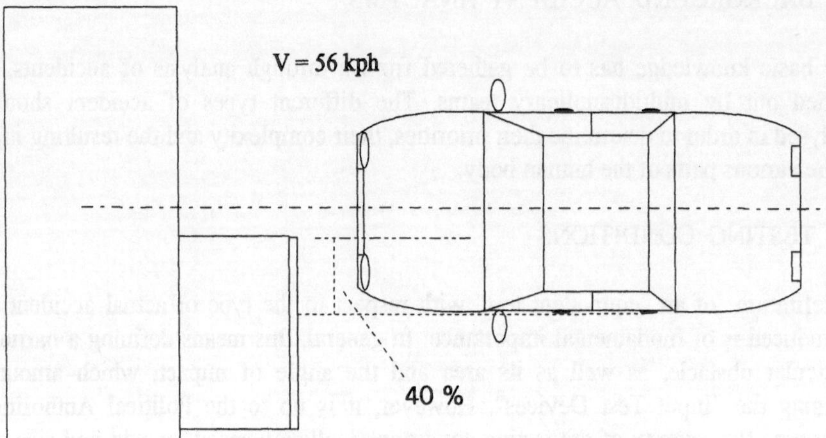

- Moving vehicle against stationery deformable barrier
- Main block stiffness 50 PSI
- Bumper stiffness 250 PSI
- Slotted bumper facia
- Barrier ground clearance : 200 mm
- Overlaps : 40 % on the driver side
- 2 HYBRID III 50 % on front seats
- Test speed : 56 Kph

Figure 2. **Test Conditions for the Frontal Impact.**

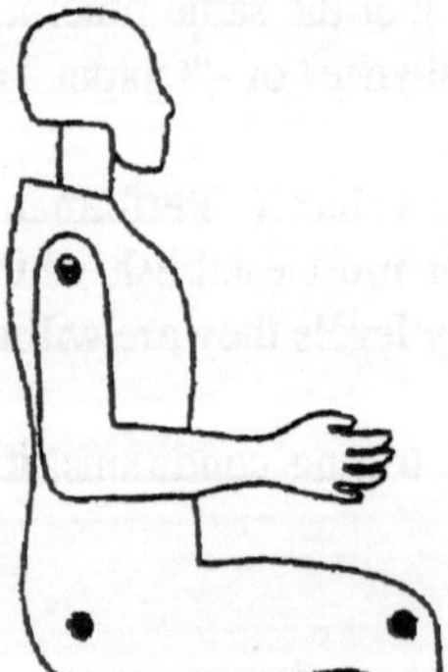

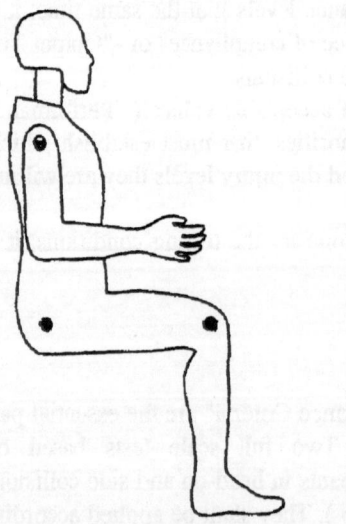

	Hybrid
Head	HIC < 1000 (36 ms)
	a < 80 g x Δt < 3 ms
Neck	Extension : Mt < 57 Nm
	Tension[1]: (Fz x Δt)
	Shear[1]: (Fx x Δt)
	a < 60 g x Δt < 3 ms
Chest	Chest deflection < 50 mm
	VC < 1 m / sec
Femur	(F x Δt)[1]
knee	Sliding Knee < 15 mm
Tibia	Fz < 8 KN
	(Tibia Index) T.I. < 1.3

[1] according to the fig. 4 performance limits

Figure 3. **Dummy's Performance Criteria - Frontal Impact (Europe).**

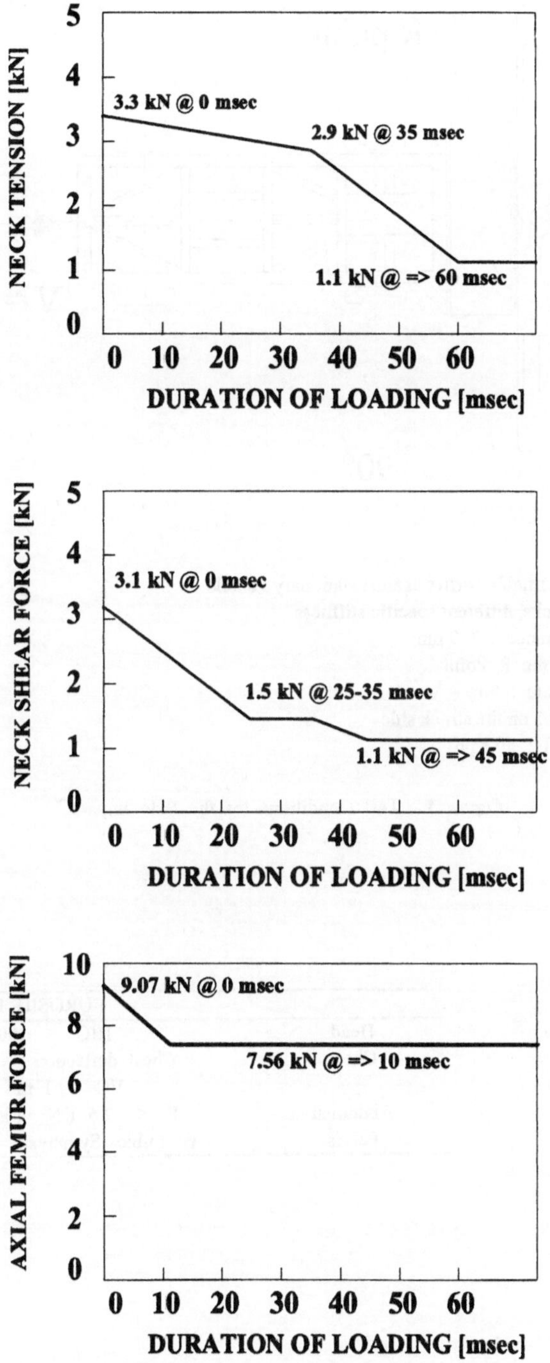

Figure 4. Neck and Femur Performance Limits.

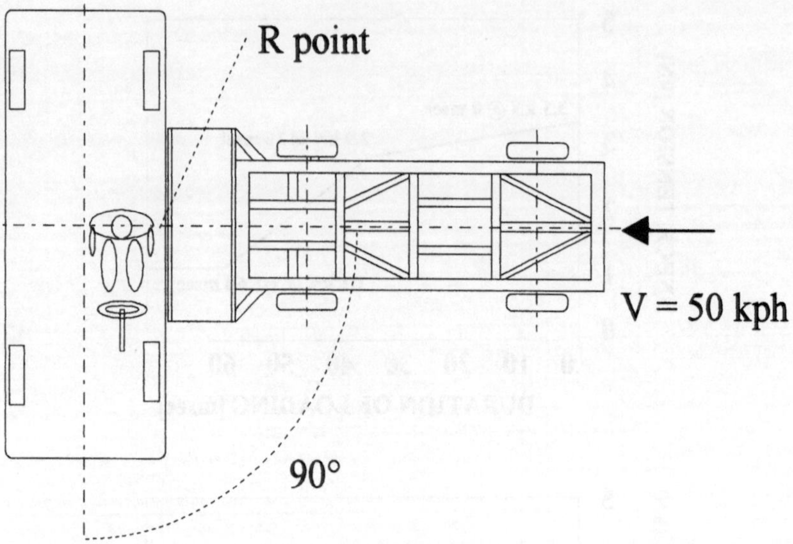

- Moving deformable barrier against stationary vehicle
- 6 barrier blocks, different specific stiffness
- Ground clearance : 300 mm
- Centered on the R Point
- Collision angle : 90°
- 1 EUROSID 1 on the struck side
- Barrier speed : 50 Kph

Figure 5. Test Conditions for the Side Impact.

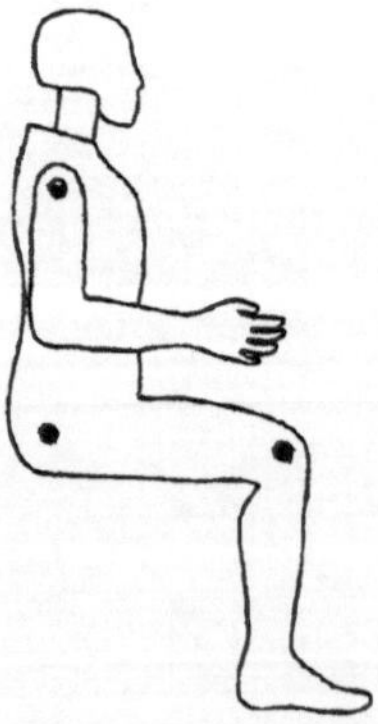

	EUROSID 1
Head	HIC < 1000
Chest	Chest deflection < 42 mm
	VC < 1 m / sec
Abdomen	F < 2,5 KN (internally)
Pelvis	F Pubic Synphysis < 6 KN

Figure 6. Dummy Performance Criteria - Side Impact (Europe).

3. Main Shortcomings of the Fully Experimental Biomechanical Approach

Whilst the superiority of the biomechanical approach is undeniable, nevertheless, in the assessment of overall vehicle safety, a full-scale biomechanical test may serve as a reliable final check, but it is not a suitable tool for the development of a new car model.

It is simply impossible, for example, to ask an engineer to design the front end of a car in such a way that, in a crash against a wall, the driver's HIC value will not exceed the 1000 threshold. The biomechanical standard must be translated into a set of component performance ratings to be checked through specially developed tests, which often are just internal company standards.

As for FIAT, nowadays this process involves the utilisation of nearly 60 different types of test and some 20 different computer simulation models. Furthermore, the biomechanical performance capabilities of a newly developed model are evaluated through nearly 20 different kinds of full-scale tests, in order to obtain the best safety performance balance for the entire range of real accidents.

This process, applied to the development of one of our latest models, resulted in a global testing programme involving approximately:

70	Car crash tests
80	Crash simulations tests (sled tests)
250	Component tests
50	Computer simulations

(These figures do not include the tests performed by subsystem and component suppliers).

Obviously, the trend in present day engineering is to limit the testing activities to the mere checking of fully developed products; then the difference between the number of tests performed and the number of test types is a good indicator of the expected reduction in time to market.

4. A Proposal to Improve the Crashworthiness of a New Car Model

In addition to complying with internal and external safety standards, it is of paramount importance to define the safety tasks of a new car model; to establish on which parts, components and subsystems we have to concentrate our efforts in order to improve the vehicle's crashworthiness.

In this connection, we have developed a methodology aimed at evaluating, in biomechanical terms, the industrially feasible countermeasures applicable to the new model. As an example, we may mention an analysis performed a few years ago for a model which is now on the market and which obtained a high score under the most popular European Crashworthiness Rating Systems.

At present, the procedure remains the same. What is changing with time is the industrial availability of new safety countermeasures, technical capabilities and safety knowledge.

The first step of the procedure, as described above, is the definition of accident type priorities.

At the very start of the process, we made an initial decision as to the severity of the injuries and the population at risk involved. In view of the fact that seat-belts are now compulsory throughout Europe, we decided to focus our efforts on:

- the reduction of severe injuries and casualities,
- the safety of the restrained occupants.

From an examination of accident statistics (see fig. 7) we find that the first priority is the head-on collision, the second is side impact and the third is the roll-over situation.

	RESTRAINED OCCUPANTS					
	Severe Injuries (%)		Fatalities (%)			
	S	F	S	USA	UK	F
FRONTAL	64	69	59	50	45	51
LATERAL	17	20	25	32	36	35
Rear	3	3	1	4	11	2
ROLLOVER	16	3	15	12	8	4
Others	0	5	0	2	0	8
Total	100	100	100	100	100	100

Figure 7. Accident Priority : Type of Impact.

4.1 HEAD-ON COLLISIONS

Injury distribution, as inferred from in depth accident analyses (see, for instance, figs. 8 and 9, concerning only drivers wearing a seat-belt in head-on collisions), makes it possible to identify the body segments most frequently injured, and the relative causes, i.e., the associated contacted areas, for both the driver and the front passenger. The upper limbs were excluded because of lack of informations about the causes. These data are crossed (see scheme and fig. 10 a,b,c,d) with the industrially available and feasible countermeasures.

At each intersection we may evaluate the effectiveness of the single countermeasure for a single type of injury. Through this kind of analysis we may define the categories of the most effective countermeasures applicable to vehicles in head-on collisions, which are :

(1) The use of safety belts;
(2) Limited average deceleration at different speeds and limited underdash intrusions;
(3) Energy absorbing steering wheel; availability of air bags for the driver and the front passenger;
(4) Pretensioners and optimized belt attachment geometry;
(5) Anti-submarining beams;
(6) Dashboard and car interior with friendly geometry and energy absorbing features.

During the vehicle engineering process, special attention was payed to the activities relating to these six groups of countermeasures.

BODY SEGMENTS	AIS					ΣAIS^3	
	1	2	3	4	>5		
Head	45	97	5	3	-	1148	
Face	207	19	5	-	-	+ 494	= 1642
Neck	81	3	1		-	132	
Chest	175	37	19	5	-		1304
Upper Limbs	91	47	17		-	926	
Spine	27	7	1		-	110	
Abdomen	33	4	5	3	1		517
Lower Limbs							
Upper Segment/Pelvis	183	37	34		-		1397
Lower Segment	127	49	14				897

Figure 8. Frontal Impact : Belted Drivers.

CONTACTED AREAS ASSOCIATED WITH INJURIES

	1	2	3	4	5	6	Total	%	Related Body Segments
Windscreen	83	28	2	4	2	4	123	13.9	H
Windscreen Flame	12	9	1	2	0	0	24	2.7	H
A - Pillar	26	14	5	1	1	1	48	5.4	H
Steering System	124	35	15	11	23	3	211	23.8	H - C
Dashboard. Top	8	6	0	0	0	0	14	1.6	-
Dashboard. Middle	9	1	0	0	0	0	10	1.1	-
Dashboard. Bottom	120	17	11	13	0	0	161	18.2	P - U.S.
Pedals	32	8	3	0	0	0	43	4.9	L.S.
Bulkhead	11	7	6	1	0	0	25	2.8	L.S.
Belt	123	24	11	23	23	0	204	23.1	A - C
Outside of Car	8	3	3	4	2	2	22	2.5	H
Total	556	152	57	59	51	10	885	100	

Legend : H = Head; C = Chest; A = Abdomen; P = Pelvis;
U.S. = Lower Limbs - Upper Segment; L.S. = Lower Limbs - Lower Segment.

Figure 9. Frontal Impact : Belted Drivers.

Body Segments and Causes of injury \ Countermeasures		Limited medium decelerations at different speeds ("compatible" car).	Limited deformations at the underdash driver's side structure.	Limited deformations at the underdash passenger's side structure.	Steering column kynematism limiting the trasmission of motion from steering box to steering wheel.	Energy-absorbing steering wheel to limit decelerations at the brain and damage to the facial bones.	Driver's air-bag.	Passenger's air-bag.	Dashboard far away from passenger.	Dashboard and car interior with friendly geometry and energy-absorbing features.	Use of safety belts.
HEAD / FACE	Impact against steering wheel	+	+	/	++	++	++	/	/	/	++
	Impact against dashboard	+	/	+	/	/	/	++	+	++	++
	Impact against windscreen pillars.	+	++	++	/	/	+	++	/	+	++
	Impact against front crossmember under roof panel	+	+	+	/	/	+	+	/	+	++
	Impact against windscreen	+	+	+	/	/	++	++	/	/	++
	Impact against outside objects due to ejection of occupants	/	/	/	/	/	/	/	/	/	++
CHEST	Impact against steering wheel	+	++	/	++	++	++	/	/	/	++
	Interaction with safety belt	/	/	/	/	/	/	/	/	/	-
	Injuries to front passengers by rear passengers	+	/	/	/	/	/	/	/	/	++
	Impact against outside objects due to ejection of occupants	/	/	/	/	/	/	/	/	/	++
		{	2	}{		3		}{	6	}{	1 }

++ Highly effective + Effective / No influence
- Negative effects ? Doubtful effects

Figure 10 a . Type of impact : Frontal Impact.

Body Segments and Causes of injury \ Countermeasures		Limited medium decelerations at different speeds ("compatible" car).	Limited deformations at the underdash driver's side structure.	Limited deformations at the underdash passenger's side structure.	Steering column kynematism limiting the trasmission of motion from steering box to steering wheel.	Energy-absorbing steering wheel to limit decelerations at the brain and damage to the facial bones.	Driver's air-bag.	Passenger's air-bag.	Dashboard far away from passenger.	Dashboard and car interior with friendly geometry and energy-absorbing features.	Use of safety belts.
ABDOMEN	Front passengers submarining	/	/	/	/	/	/	/	/	/	-
	Rear passengers submarining	/	/	/	/	/	/	/	/	/	-
	Impact against outside objects due to ejection of occupants	/	/	/	/	/	/	/	/	/	++
PELVIS	Forces coming from the thighs due to interaction with dashboard	+	+	+	/	/	/	/	+	+	++
	Impact against outside objects due to ejection of occupants	/	/	/	/	/	/	/	/	/	++
LOWER LIMBS	Impact against lower section of dashboard	+	+	+	/	/	/	/	+	+	++
	Interaction at front footboard level	/	/	/	/	/	/	/	/	/	/
	Forces coming from pedals	/	/	/	/	/	/	/	/	/	/
WHOLE BODY	Impact against outside objects due to ejection of occupants	/	/	/	/	/	/	/	/	/	++
		{	2	}	{	3		}	{	6 }	{ 1 }

++ Highly effective + Effective / No influence
- Negative effects ? Doubtful effects

Figure 10 b. Type of impact : Frontal Impact.

Body Segments and Ca} ses of injury	Countermeasures	Safety windscreen.	Pretensioners.	Height adjustments of front safety belts.	Optimised geometry of belt attachments.	Anti-submarining beam for front seats.	Anti-submarining beam for rear seats.	Shield in front seat backs.	Underdash without parts that might hurt the knees.	Limited deformations in the pedal unit area.	Limited deformations in the passenger's foot board area.	Generously-sized and torsionally stable pedals.	Height adjustable rear seat belts or adaptive safety belts.
HEAD / FACE	Impact against steering wheel	/	+	+	+	/	/	/	/	/	/	/	/
	Impact against dashboard	/	+	+	+	/	/	/	/	/	/	/	/
	Impact against windscreen pillars.	/	+	+	+	/	/	/	/	/	/	/	/
	Impact against front crossmember under roof panel	/	+	/	+	/	/	/	/	/	/	/	/
	Impact against windscreen	++	+	+	+	/	/	/	/	/	/	/	/
	Impact against outside objects due to ejection of occupants	+	/	/	/	/	/	/	/	/	/	/	/
CHEST	Impact against steering wheel	/	+	+	+	/	/	/	/	/	/	/	/
	Interaction with safety belt	/	+	++	++	+	++	/	/	/	/	/	++
	Injuries to front passengers by rear passengers	/	/	/	/	/	++	++	/	/	/	/	/
	Impact against outside objects due to ejection of occupants	+	/	/	/	/	/	/	/	/	/	/	/

{ 4 }{ 5 }

++ Highly effective   + Effective   / No influence
- Negative effects   ? Doubtful effects

Figure 10 c. Type of impact : Frontal Impact

Body Segments and Causes of injury		Safety windscreen.	Pretensioners.	Height adjustments of front safety belts.	Optimised geometry of belt attachments.	Anti-submarining beam for front seats.	Anti-submarining beam for rear seats.	Shield in front seat backs.	Underdash without parts that might hurt the knees.	Limited deformations in the pedal unit area.	Limited deformations in the passenger's foot board area.	Generously-sized and torsionally stable pedals.	Height adjustable rear seat belts or adaptive safety belts.
ABDOMEN	Front passengers submarining	/	+	/	+	++	/	/	/	/	/	/	/
	Rear passengers submarining	/	/	/	+	/	++	+	/	/	/	/	/
	Impact against outside objects due to ejection of occupants	+	/	/	/	/	/	/	/	/	/	/	/
PELVIS	Forces coming from the thighs due to interaction with dashboard	/	+	/	/	+	/	/	++	/	/	/	/
	Impact against outside objects due to ejection of occupants	+	/	/	/	/	/	/	/	/	/	/	/
LOWER LIMBS	Impact against lower section of dashboard	/	+	/	/	+	/	/	++	/	/	/	/
	Interaction at front footboard level	/	/	/	/	/	/	/	/	++	++	/	/
	Forces coming from pedals	/	/	/	/	/	/	/	/	++	/	+?	/
WHOLE BODY	Impact against outside objects due to ejection of occupants	+	/	/	/	/	/	/	/	/	/	/	/

{ 4 }{ 5 }

++ Highly effective   + Effective   / No influence
- Negative effects   ? Doubtful effects

Figure 10 d . Type of impact : Frontal Impact.

4.2 SIDE IMPACT

The same methodology has been applied to the side impact so as to define the injured body segments and the associated causes of injury for the struck side occupants. These data were crossed with the available countermeasures and the effectiveness of the latter was evaluated (fig 11 a,b). On this basis, we defined four categories grouping together the most effective countermeasures, and namely:

(1) Wearing the safety belts, which reduce the severity of injures caused by impacts between car occupants and risk of being thrown out of the car.
(2) To reduce intrusions account taken of the fact that it is important to create a favourable "deformation profile" (i.e. deformations at waist level should be more limited than deformations in the floor area).
(3) Increasing the flexural stiffness of the doors by fitting reinforcing beams, so as to reduce the risk of intrusion in impacts against concentrated objects (poles, motorcycles, car edges).
(4) Softening the interior and extending the armrest shape.

4.3 ROLLOVER SITUATION

Finally for the roll-over situation the groups of the most effective countermeasures (fig.12) has been defined as follows :

(1) Wearing the safety belts and increasing the restraining action in order to avoid ejection.
(2) Increasing the stability of the car minimizing the ratio between the height of the centre of gravity and the track.
(3) Making the interior friendly, with softer materials, no sharp edges and improved energy absorption capabilities of key components such as the steering wheel and dashboard.
(4) Increasing the structural resistance of the car roof.

5. Final Remarks

The purpose of this integrated biomechanical approach is not to optimize specific aspects in a "spectacular" full scale test. Its purpose is the true safety of the occupants over the widest possible range of real accidents in road traffic.

Body Segments and Causes of injury \ Countermeasures		Reduction of deformation at waist level.	Reduction of deformation at floor level.	Increased flexural rigidity of the doors.	Soft interiors without sharp edges.	Soft, extended armrest with restricted projection.	Side air-bag.	Use of safety belts.
HEAD / FACE	Impact of the head against the car interior.	+	-	+	+	/	+?	/
	Impact of the head against the impacting object.	++	-	+	/	/	+?	/
	Impact against outside objects due to ejection of the occupant.	/	/	/	/	/	/	++
CHEST	Impact of the chest against the car interior.	+	-	+	+	/	+?	/
	Impact of the chest against parts of the bodies of the other occupants.	+	/	+	/	/	+?	++
	Impact of the chest against outside objects due to ejection of the occupant.	/	/	/	/	/	/	++
		{	2 }	{ 3 }	{	4 }	?	{ 1 }

Legend fig. 11 a, b:
++ Highly effective + Effective / No influence
- Negative effects ? Doubtful effects

Figure 11 a. Type of Impact : Side Impact

Body Segments and Causes of injury \ Countermeasures		Reduction of deformation at waist level.	Reduction of deformation at floor level.	Increased flexural rigidity of the doors.	Soft interiors without sharp edges.	Soft, extended armrest with restricted projection.	Side air-bag.	Use of safety belts.
ABDOMEN	Impact of the abdomen against the car interior.	/	+	+	+	/	+?	?
	Impact of the abdomen against the door armrest	/	+	+	/	++	++?	/
	Impact of the abdomen against parts of the bodies of the other occupants.	/	/	+	/	/	+?	++
	Impact of the abdomen against outside objects due to ejection of the occupant.	/	/	/	/	/	/	++
PELVIS	Impact of the pelvis against the car interior.	-	+	+	+	/	+?	/
	Impact of the pelvis against parts of the bodies of the other occupants.	-	+	+	/	/	+?	++
	Impact of the pelvis against outside objects due to ejection of the occupant.	/	/	/	/	/	/	++
		{	2	}{ 3	}{	4 }	?	{ 1 }

Figure 11 b . Type of Impact : Side Impact.

Body Segments and Causes of injury \ Countermeasures		Structural resistance of the car roof panel and top (reduction of deformations).	Soft car interiors with no sharp edges.	Energy-absorbing steering wheel.	Energy-absorbing dashboard.	Safety windscreen.	Use of safety belts.	Pretensioners.	Car stability (height of centre of gravity / track ratio)
HEAD	Impact against structural parts, against the roof and against interior trim parts	+	+	/	/	+	+	+	+
	Impact against outside objects due to ejection of the occupants.	+	/	/	/	+	++	/	+
CHEST	Impact against structural internal parts and against interior trim parts	/	+	+	+	/	+	+	+
	Impact against outside objects due to ejection of the occupants.	+	/	/	/	+	++	/	+
		{ 4 }	{	3	}	{		1 }	{ 2 }

Legend fig. 12:
++ Highly effective   + Effective   / No influence
- Negative effects   ? Doubtful effects

Figure 12. Type of Impact : Roll Over.

PART III
Occupant Simulation Models

OCCUPANT SIMULATION MODELS:EXPERIMENT AND PRACTICE

PRIYA PRASAD
Biomechanics and Advanced Safety CAE
Ford Motor Company
Dearborn, Michigan, USA

Abstract

This paper is an overview of the evolution of math modeling as it pertains to occupant/restraint interactions. Specifically, math models involving lumped mass, finite element, and hybrid methods are examined. The process by which math models of both the dummy and restraints are developed, validated, and integrated into the vehicle design process is also presented. Finally, the progression of modeling impact responses of instrumented dummies to that of modeling actual human responses is motivated by way of example.

1. Introduction

With the advent of computers, mathematical modeling of crashes involving vehicles, occupants and pedestrians has been pursued over the last thirty-three years. Whereas early models were one- and two-dimensional in nature, the emergence of multi-body dynamics and non-linear finite element techniques, coupled with the development of supercomputers and workstations, has led to the current availability of three-dimensional rigid body and finite element models of vehicle structures, occupants, and pedestrians. These models have moved on from research tools in the hands of a few analysts to vehicle design tools utilized by many engineers involved in the development of new vehicles. In most automotive companies and their suppliers, these models are being used as necessary tools for design decisions leading to optimization of designs and shortening of product design time.

Whereas the ultimate goal of mathematical modeling is to eliminate the testing of prototype hardwares, testing has been a necessary part in the development of mathematical models. Most models require a substantial number of experiments to establish their validity. Comparisons between experimental and model results lead to further improvements in modeling techniques and test methods. Therefore, experimentation and modeling are in an iterative loop which facilitates further development of both technologies. However, with continual model validation exercises, enough confidence in the predictive capability of models will be gained, thereby leading to substantial reduction in hardware testing.

The objective of this paper is to briefly review the following: Current occupant/pedestrian simulation models, their utilization in vehicle design, methods for experimental verification, and their future involvement in the biomechanical field.

J. A. C. Ambrósio et al. (eds.),
Crashworthiness of Transportation Systems: Structural Impact and Occupant Protection, 209–219.

2. Brief Overview of Occupant Simulation Models

Almost thirty-three years have elapsed since McHenry (1) proposed one of the first mathematical simulation models to describe the dynamic response of a vehicle occupant involved in a collision. Since then, many more sophisticated models have been developed for simulating occupant kinematics in crashes. During the past decades, a great deal of emphasis has been placed on the use of mathematical models in research and development in the field of automotive safety.

A review of several "Gross-motion Simulators" was made by King and Chou (2) in 1975. One of the simulation codes reviewed, namely CAL3D, has gone through extensive use and development. A second code called MADYMO2D/3D has been developed by TNO in Holland in 1979. The basic features of MADYMO2D were described and reviewed by Prasad (3) in a benchmark study (4). Further developments in these gross-motion simulators and others such as SOMLA (5) as well as some one-dimensional and special purpose restraint system models were reported by Prasad and Chou (6, 7).

All the above reviews concentrated on mathematical models derived on the basis of rigid body dynamics. Although the SOMLA program had integrated an occupant model with a finite element seat model for light aircraft, use of finite element analysis in occupant simulation was limited in the past due to lack of both software and computational speed. Recent developments in finite element analysis and coupling of finite element analysis codes with rigid body dynamics codes have opened a new era in occupant dynamics simulation.

Finite element codes are now available in which the entire occupant or parts of the occupant can be modeled using finite elements, making it possible to model the occupant body deformations due to contact loadings more accurately than with rigid body dynamics codes. Although the majority of studies so far have concentrated on modeling dummies, the use of mathematical models to explore injury mechanisms and injury criteria is also being undertaken. These new developments of finite element dummy models and finite element human injury models have been reviewed extensively by Prasad and Chou (8). The above detailed reviews of simple to very detailed occupant/pedestrian models will not be repeated in this paper. However, some of the more useful developments are worth reporting.

Model occupant responses are sensitive to restraint system modeling. An early limitation in belt system modeling was that the belts were constrained to pass over several fixed points on the occupant torso. As a result, even in conditions where a belt system would slip off an occupant segment, the models could not predict such belt motions. Additionally, situations where occupant kinematics would dictate complete disengagement of the belt system, e.g. the upper torso belt in a rear impact or far side impact, could not be accurately predicted. These phenomenon can now be modeled by the new finite element belt models and an advanced belt routine incorporated in the ATB (9), as well as in the MADYMO codes.

Another useful development in restraint system modeling has been the introduction of finite element airbag models. Whereas earlier airbag models utilized the final shape of the airbag for volume and pressure predictions, and membrane effects were simulated with highly simplified assumptions, the current finite element airbag models can simulate the shape of the bag as it inflates and membrane effects are included in the formulation. Additional advances in bag folding and unfolding during inflation, have led to more realistic simulations. The Euler-Lagrange coupling in MSC-

Dytran (10) has led to much more realistic airbag deployment predictions in terms of fabric leading edge velocity and airbag/occupant contact forces. Application of this technology has been reported by Prasad and Laituri (11).

The development of hybrid finite element and rigid body modeling is leading towards better discretization of flexible and deformable elements in the human body, as well as vehicle structures (12). Coupling of finite element and rigid body dynamics has been utilized by Ruan and Prasad (13) to study the effect of various restraint systems on skull stress and brain strains (14, 15) and will be described later in this paper.

In summary, the field of occupant simulation is evolving rapidly as a direct result of new simulation tools and faster computers. Implementation of this growth into the design process will now be discussed.

3. Utilization of Models During Vehicle Design Process

Occupant responses in crashes are determined by vehicle structural responses, interior geometry, stiffnesses of occupant contact surfaces, and characteristics of the restraint system. In frontal impact, the structural responses of interest are the acceleration time history of the passenger compartment and the deformation of the passenger compartment. In side impacts, the side structure acceleration and velocities are important parameters. The interior geometry of the vehicle dictates the initial conditions of the occupant and distances that the occupant can travel within the vehicle without contacting vehicle components. Although the belt system and airbags are the commonly recognized restraint systems, the seat and all surfaces contacted by the occupant complement the restraint system. Therefore, the shape and stiffness of these contact surfaces are also important. Restraint system parameters that affect occupant responses are generally belt stiffness and geometry, airbag shape, vent sizes, and inflator characteristics like gas mass inflow rates and temperature. All of the above restraint and vehicle interior parameters combine as a system to affect occupant responses.

The challenge to the vehicle designer is to integrate all design parameters to control occupant responses in various modes of impact. Some of the impact modes are in safety regulations worldwide and are shown in Figure 1. Some regulations and requirements have strong opposing interactions. For example, the roof crush requirement demands stiff upper structures; the head impact requirement demands softer structures. The low speed damageability requirements demand stiffer structures; pedestrian requirements demand softer structures. The United States' FMVSS 208 passive restraint frontal impact standard require softer front ends; the proposed European offset crash standards require stiffer front ends. In general, safety standards have a tendency to increase vehicle mass, but lower weight designs are required by the Corporate Average Fuel Economy Standard in the US. To achieve the right balance of conflicting requirements while meeting occupant protection goals, mathematical models of the occupant and structure are essential tools during the development phases of the vehicle.

Early in the design process lumped mass, multi-body dynamic models reported by Chou, et al. (16), Prasad, et al. (1) Low, et al. (18), and Midoun et al., (12) are used to evaluate and guide the development of the basic architecture of the vehicle. As the design progresses, sizing of the vehicle components involved in crash are determined by non-linear beam analysis methods proposed by Mahmood and Paluzny (19, 20). At this stage, detailed sub-system finite element models are utilized to further guide the designs. As hardwares of sub-systems become available, dynamic tests are used to evaluate the adequacy of the sub-system designs. The structure and occupant models are updated to reflect the latest level of hardware designs, and further iterations of vehicle designs are performed with the models, until the next level of hardware is available. The transition from lumped mass models to detailed finite element models of the total vehicle is a gradual process involving several levels of modeling and design verification, with the final full-vehicle FE model representing the finished product.

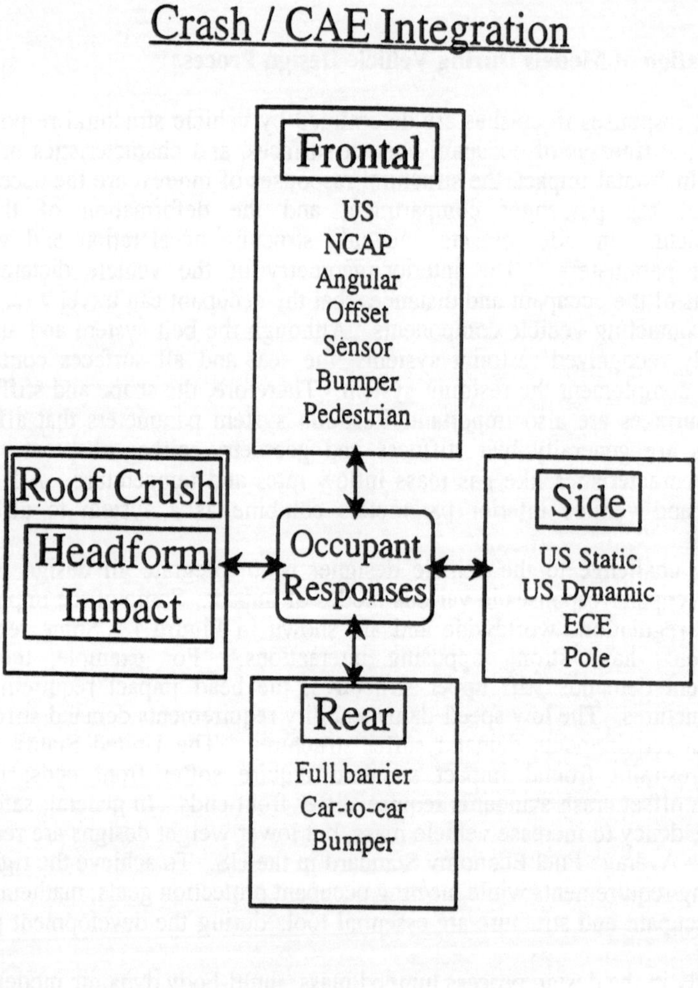

Figure 1. CAE integration in vehicle crash design

4. Experimental Validation Of Occupant Models

4.1 GENERAL PRINCIPLES

To validate models experimentally, some general principles followed by the author are worth discussing. A single test validation of any model, structural or occupant, may be misleading, since it is possible that the model parameters may be valid for only that test. As a result, validation should be conducted over various test severities that cover the range of expected severities in which the model may be utilized. This method of validation recognizes that model or test results may be non-linear in nature. In general, tests themselves may not be repeatable. As a result, several tests should be conducted at the same severity. In general, sled tests are more repeatable than vehicle crash tests. Therefore, wherever possible, occupant models should be validated against sled tests. The reproducibility of test results are also important since the results may vary due to test operator, test site, and dummy/vehicle differences. Therefore, validation of models may require many tests to determine the stability of test results. Assuming that repeatability and reproducibility of model results are not in question, the best that any model can predict is within the known variability of test results.

The "goodness" of a model in predicting test results can be entirely in the mind of the modeler. An experienced modeler can utilize a model for design purposes if the modeler can account for systematic differences between model and test results. Objective criteria for determining the "goodness" of a model do not currently exist, although an attempt to develop such objective criteria was made by a sub-committee of the SAE. This sub-committee was composed of occupant modelers from the automotive industry, code developers, and the academia. A proposed Validation Index was agreed upon in principle, but various issues dealing with the methods for comparing two time varying responses, the total number of required tests, and the region of validation were left unresolved. As a result the Validation Index was not finalized. The proposed Validation Index consisted of various levels of validation of a model, with higher levels implying higher quality of model predictions of a reference test event(s). This proposed Validation Index is shown in Table 1, hoping that the modeling community can further refine this Index.

TABLE 1. Summary of proposed validation index levels

Class	Characteristics
Level 0	No agreement between predictions of model and "reference event"
Level 1	Qualitative agreement: a) Trends of predicted parameters same b) Kinematics correspond qualitatively c) Contacts between the occupant and vehicle interior are the same in general
Level 2	HIC and similar indicators predicted by simulation are within 20% of those obtained in reference event
Level 3	Peak values of important occupant responses limited to a relative error of 20% (20% on vector magnitude, 11.31 deg on vector direction)
Level 4	Same as level 3 except 5%
Level 5	Timing of peaks of important vector responses limited to 5% relative difference
Level 6	All peaks and valleys in the duration of time-dependent predictions must match the reference event within 10%
Level 7	Same as level 6 except 5%
Level 8	1% relative error, point-by-point, over the durations of the reference and predicted events

4.2 DUMMY DATA GENERATION

As occupant models have developed over the last thirty-five years, some level of experimental validation studies have also followed. Depending on the geometrical discretization and linkage system describing the occupant, different types of data are required. The most extensive validation studies have been conducted for the test devices currently or previously utilized in crash testing of vehicles, e.g., the Hybrid II and the Hybrid III dummies for frontal impacts, and the SID and Eurosid dummies for side impacts. In general, the dummy segments lend themselves to be described as a collection of rigid links. The size, shape, and the inertial properties of the dummy linkages are required to describe the dummy. The joints and their elastic, viscous, and friction properties are further required to describe their interactions during relative movements. Commonly accepted test methods to obtain the above data have been developed. The dummy surfaces and their stiffness characteristics are also required to predict the interaction of the dummy with vehicle components and the restraint systems. Standardized test procedures for the determination of the dummy surface stiffness characteristics are not currently available since they may be dependent on the type/mode of crash being simulated. However, with the greater use of finite element techniques, it is expected that advanced foam models of the dummy surfaces will lead to generalized description of the dummy surfaces. An extensive study to develop a dataset for modeling to Hybrid II dummy, previously used in frontal crash testing worldwide, has been reported by Fleck, et al. (21). This report contains the test methods for determining the geometrical, inertial, and stiffness properties of the dummy. Whereas the Hybrid II dummy has now been superceeded by the Hybrid III dummy for frontal crash testing worldwide, the test methods to determine dummy modeling parameters have been used by the Armstrong Air Force Medical Research Laboratory to determine the modeling parameters for the Hybrid III dummy (22). Modeling parameters determined by the AAMRL are now in datasets for the ATB (22). A parallel study has been conducted by TNO in the Netherlands and has resulted in datasets for the 5th, 50th and the 95th percentile dummies for the MADYMO occupant simulation models (23). Datasets describing the side impact dummies are also available in MADYMO.

4.3 EXPERIMENTAL VALIDATION

The datasets currently available for describing the Hybrid III dummy have been utilized by various organizations to determine their predictive capability in frontal sled tests. A benchmark test series using a rigid seat and rigid floor pan was conducted by Prasad (24) in 1985 to determine the dynamic responses of the Hybrid II and the Hybrid III dummy in frontal sled tests. Three tests each were conducted at three severity levels. Three-point belt restraint systems of known stiffness characteristics were used. Occupant-to-vehicle contacts were limited to seat/pelvis and foot/toe board only. This arrangement dictated that the dummy responses were not influenced by other possible vehicle frontal component contacts. The exact geometry of the belt system and their initial contacts with the dummies were controlled in each test. The results of these tests were released to a sub-committee of the Society of Automotive Engineers for validation exercises of the ATB and the MADYMO occupant simulation models. These exercises have been reported by Obergefell, et al. (25), Wismans and Hermans (26) and Khatua, et al. (27) at the 1988 SAE Congress. The author's own simulations of these tests utilizing the MADYMO software and dummy joint characteristics measured at Ford are seen in Figure 2 and Table 2, and were discussed at the SAE Congress. The simulation results were well within test-to-test variability of important responses like the Head Injury Criteria (HIC), chest acceleration, chest deflection, and belt loads. The overall kinematics of the dummy were reproduced well, as measured by head displacements. The time varying responses were also within test corridors. Further improvements in the dummy datasets have followed as a result of further usage in different impact conditions.

Details of the development and experimental validation of the side impact dummy, SID, have been reported by Low and Prasad (29). In this series of validation exercises, repeat tests of the SID dummy against rigid load cells were conducted at three impact velocities - 10, 15, and 20 mph. The efficacy of the dummy model was further investigated in crash tests at three different velocities (17, 18). For model validation purposes, test methods to determine the stiffnesses of the side structure and occupant/side interactions were developed and are reported by Prasad, et al. (17) and Low, et al. (18). Other validation studies have been cited in Reference 8.

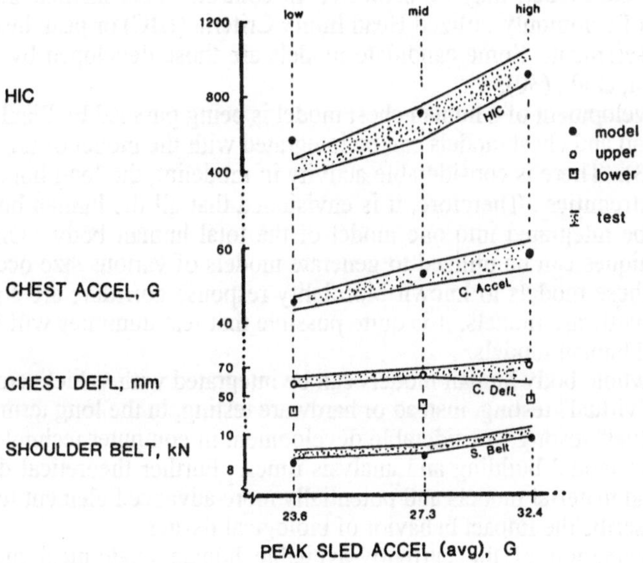

Figure 2. Test and model results of sled series

TABLE 2. Test and model results (Head)

	Head Displacement W.R.T. Seat Back (m)			
	Longitidinal		Vertical	
Peak Sled Accel G's	Test	Model	Test	Model
23.5-23.7	0.46-0.47	0.45	0.20-0.24	0.28
27.2-27.4	0.44-0.47	0.47	0.25-0.28	0.31
32.1-32.6	0.47-0.50	0.49	> 0.30	0.34

5. Biomehanical Modeling

With the available modeling technology, the logical progression is to advance from dummy modeling to modeling of the human itself. Modeling of various parts of the human body has been an essential activity in the field of biomechanics to gain better understanding of human response to impact and associated injury criteria. Rigid body dynamic models, as well as finite element models, have been developed for many years. A review of such biomechanical modeling activities has been presented by King and Chou (2) in 1975. Since 1975, substantial improvements in rigid body dynamic modeling and finite element analysis techniques have been made. These improvements have led to the development of detailed anatomical finite element models of the human body.

There is considerable activity worldwide in biomechanical modeling. Human skull and brain models are being developed and may be considered in future rulemaking. The goal is to measure the linear and angular accelerations of the dummy head in crash tests, and to utilize these measurements to excite a human skull/brain model. The model responses such as skull stresses and/or brain stresses and strains, predicted by the model, may be subjected to control. This method may lead to the replacement of commonly-utilized Head Injury Criteria (HIC) or peak head acceleration for injury assessment. Some candidate models are those developed by DiMasi, et al., (29) and Ruan, et al., (30).

The development of a human chest model is being pursued by Plank and Eppinger (31). The head and chest models can be integrated with the model of the spine (32) and the pelvis (33). There is considerable activity in modeling the long bones of the lower and upper extremities. Therefore, it is envisioned that all the human body anatomical models can be integrated into one model of the total human body. Once developed, scaling techniques can be utilized to generate models of various size occupants. Since "tuning" of these models to known biofidelity response corridors are somewhat easier than that of hardware models, it is quite possible that test dummies will be replaced by mathematical human models.

These whole body human models can be integrated with vehicle crash models and may lead to "virtual" testing, instead of hardware testing, in the long term. To get to the stage of "virtual" testing, considerable development in computer technology is required towards faster model building and analysis times. Further theoretical development in biomechanical material models and potentially more advanced element formulations are needed to describe the impact behavior of biological tissues.

The integration of the currently-available human anatomical models is in its infancy. These models are being utilized for hypothesis testing to gain better understanding of human responses to impact and injury mechanisms. One such example is discussed below.

Scientific interest in head injuries continues due to the high incidence of fatality rates from head impact and the related high treatment costs of these injuries. Head injury mitigation has been a high priority in the designs of current vehicles. Most regulations worldwide utilize the Head Injury Criterion (HIC) as the indicator of potential injuries to the skull and brain. The Head Injury Criterion is based on the Wayne State University Tolerance Curve first proposed by Patrick, et al. (34) in 1965, and later verified by Ono, et al. (35) in 1980. A shortcoming of the HIC identified by many researchers is that it is based on the measurement of only the resultant linear acceleration at the center of gravity of the head. It is well known that angular accelerations can also cause head injuries. As a result, the control of HIC may not be a necessary and sufficient condition to minimize head injuries. To examine the above hypothesis, it was decided to utilize a detailed finite element model of the skull and brain developed by Ruan, et al. (30). As a first step, the skull/brain model was coupled to a detailed rigid body model of the Hybrid III dummy through a multi-segment neck model. Short duration acceleration pulses ranging between 1 to 8 m secs were used to excite the skull. A constant skull tensile stress curve was generated by varying the average acceleration and time duration. It was found that a constant 100 MPa tensile stress curve corresponded to the skull fracture curve suggested by Ono, et al. (35). A constant brain shear strain curve representing 5% shear strain was also generated by model exercises. This curve also was in good agreement with proposed brain concussion tolerance curves of Patrick, et al. (34) and Ono, et al.(35). These modeling studies have been reported by Ruan and Prasad (13,15).

The coupled skull/brain and the Hybrid III dummy model was further used to simulate frontal impacts, with no head contact, airbag/head contact and head contacts with a soft, medium stiffness, and semi-rigid steering hub. It was found that in the cases of no head contact, airbag and soft hub contacts, the HIC's were nearly the same (553-665), angular accelerations were low (2700-3700 rad/S^2) and the maximum shear stress in the brain was below 12 kPa. With the medium stiffness head/hub contact, the maximum shear stress was 15 kPa and was associated with HIC of 1543 and angular acceleration of 19500 rad/S^2. The semi-rigid hub contact yielded a maximum shear stress of 22 kPa associated with a HIC of 2359, and an angular acceleration of 15,500 rad/S^2. It could be seen that whereas the HIC rated the increasing severity of head contact as evidenced by the shear stress in the brain, the angular acceleration did not. Based on the above simulations of vehicle frontal crashes, it appears that HIC rates the severity of head impact properly and reduction of HIC in these cases should be associated with the reduction of brain shear stresses.

6. Conclusions

Occupant simulation models have been discussed from several standpoints: 1) Their evolution from two to three dimensions, 2) their refinement as a consequence of the availability of finite element methods, computational advances, and extensive validation efforts, 3) their introduction into the automobile design process, and 4) their evolution into the field of biomechanics.

7. References

1. McHenry, R. R. (1963) Analysis of the Dynamics of Automobile Passenger restraint Systems, *Proc., 7th Stapp Car Crash Conference,* SAE, Warrendale, Pennsylvania.
2. King, A. I. and C. C. Chou (1975) Mathematical Modeling, Simulation and Experimental Testing of Biomechanical System Crash Response, *AIAA 11th Annual Meeting and Technical Display, also in: 1976 Journal of Biomechanics,* pp. 301-317.
3. Prasad, P. (1984) Overview of Major Occupant Simulation Models, SAE Paper No. 84055 in SAE P-146, *Mathematical Simulation of Occupant and Vehicle Kinematics.*

4. Prasad, P. (July 1985) Comparative Evaluation of the MVMA2D and the MADYMO2D Occupant Simulation Models with MADYMO - Test Comparisons, *Proc. of the Tenth International Conference on Experimental Safety Vehicles,* Oxford, England.
5. Laananen, D. H. (1986) Simulation of Passenger Response in Transport Aircraft Accidents, Symposium on Vehicle Crashworthiness Including Impact Biomechanics, *AMD - 79, BED - 1, ASME,* pp. 47-56.
6. Prasad, P. and C. C. Chou (1989) A Review of Mathematical Occupant Simulation Models, in Crashworthiness and Occupant Protection in Transportation Systems, *ASME, AMD - 106, BED - 13, pp. 96-112.*
7. Prasad, P. and C. C. Chou (1993) A Review of Mathematical Occupant Simulation Models, in Accidental Injury - *Biomechanics and Prevention, Chapter 6,* Edited by A. A. Nahum and J. W. Melvin, Springer - Verlag.
8. Prasad, P. and C. C. Chou (1995) Biodynamics Simulation Models, in Shock and Vibration Computer Programs, *Reviews and Summaries,* pp. 105-182, Edited by Walter and Barbara Pilkey, Booze Allen & Hamilton.
9. Fleck, J. T. (1991) Improvements in ATB/CVS Body Dynamics Model, *Proc. of the Thirteenth International Conference on Experimental Safety Vehicles,* Vol. II.
10. Buijk, A. J. and C. J. L. Flone (1991) Numerical Simulation of Deploying Airbags and Their Interaction with Vehicle Occupants, Report by the MacNeal-Schrindler Company, B. V., Groningenweg 6, 2803 PV Gonda, The Netherlands.
11. Prasad, P. and T. R. Laituri (May 13-16, 1996) Consideration for Belted FMVSS 208 Testing, Paper No. 96-53-0-03, *Proc. of the Fifteenth International Technical Conference on Enhanced Safety of Vehicles,* Melbourne, Australia.
12. Midoun, D. E., D. M. Johnson, M. D. Rao and P. Prasad (Sept. 21-23, 1993) Hybrid Modeling for Frontal Impact Into a Rigid Barrier, *Proc. of International Body Engineering Conference,* Detroit, Michigan.
13. Ruan, J. and P. Prasad (August 1995) Coupling of a Finite Element Human Head Model with a Lumped Parameter Hybrid III Dummy Model - Preliminary Results, *Journal of Neurotrauma, Vol. 12, No. 4,* Mary Ann Liebert, Inc. Publishers.
14. Ruan, J. and P. Prasad (1994) Head Injury Potential Assessment in Frontal Impact by Mathematical Modeling, *Proc. of the 37th Stapp Car Crash Conference,* SAE.
15. Ruan J. and P. Prasad (1995) Skull Isostress and Brain Isostrain Responses of a Finite Element Human Head Model, *Proc. of the 1995 Bioengineering Conference, ASME BED,* Vol. 29.
16. Chou, C. C., S. Neriya, T. C. Low and P. Prasad (1993)MADYMO2D/3D Vehicle Structural/ Occupant Simulation Models, *AMD* Vol. 169, *BED* Vol. 25, *Crashworthiness and Occupant Protection in Transportation Systems,* ASME.
17. Prasad, P., T. C. Low, G. G. Lim, C. C. Chou and S. Sundararajan (1991) Side Impact Modeling Using Quasi-Static Crush Data, SAE Paper No. 910601, SAE.
18. Low, T. C., P. Prasad, G. G. Lim, C. C. Chou and S. Sundararajan (1991) *A* MADYMO3D Side Impact Simulation Model, *Proc. Crashworthiness and Occupant Protection in Transportation Systems,* ASME Winter Annual Meeting, Atlanta Georgia.
19. Mahmood, H. F. and A. Puluzny (April 1984) Axial Collapse of Thin-Wall Cylindrical Column, *SAE Fifth International Conference on Vehicle Structural Mechanics,* SAE.
20. Mahmood, H. F. and A. Paluzny (Nov. 1981) Design of Thin-Walled Columns for Crash Energy Management - Their Strength and Mode of Collapse, *SAE Fourth International Conference on Vehicle Structural Mechanics,* SAE.
21. Fleck, J. T., F. E. Butler and S. L. Vogel (July 1974) An Improved Three Dimensional Computer Simulation of Motor Vehicle Crash Victims, Vol. I -IV, Report Nos. DOT-HS-801507, -508, -509, -510, *USDOT.*

22. Kaleps, I. and J. Whitestone, Hybrid III Geometrical and Inertial Properties, SAE Paper No. 880638, also in Hybrid III - *The first Human-Like Crash Test Dummy, SAE PT-44*, Warrendale, Pennsylvania.
23. Pilippens, M., J. Niebor and J. Wismans, An Advanced Database of the 50th Percentile Hybrid III Dummy, SAE paper No. 910813, also in SAE PT-44, SAE, Warrendale, Pennsylvania.
24. Prasad, P. (1990) Comparative Evaluation of the Dynamic Responses of the Hybrid III and the Hybrid III Dummies, SAE paper No. 902318, *Proc. of the 34th Stapp Car Crash Conference.*
25. Obergefell, L., I. Kaleps and S. Steele, Part 572 and Hybrid III Dummy Comparisons in Sled Test Simulations, SAE paper No. 880639, also in *SAE PT-44,* SAE Warrendale, Pennsylvania.
26. Wismans, J. and J. Hermans, MADYMO3D Simulations of Hybrid III Dummy Sled Tests, SAE paper No. 880645, also in *SAE PT-44,* SAE, Warrendale, Pennsylvania.
27. Khatua, T., L. Chang and R. Piziali, ATB Simulation of Hybrid III Dummy in Sled Tests, SAE paper No. 880646, also in *SAE PT 44,* SAE, Warrendale, Pennsylvania.
28. Low, T. C. and P. Prasad (1990) Dynamic Response and Mathematical Model of the Side Impact Dummy, *Proc. of the 34th Stapp Car Crash Conference,* SAE Warrendale, Pennsylvania.
29. DiMasi, F. P., R. Eppinger and J. Marcus (1991) 3-D Anatomic Brain Model for Relating Cortical Strains to Automobile Crash Loading, *Proc. 13th International Experimental Safety Vehicle Conference,* paper No. 91-58-0-11, DOT.
30. Ruan, J. S., T. B. Khalil and A. I. King, Finite Element Modeling of Direct Head Impact, *Proc. 37th Stapp Car Crash Conference,* SAE, Warrendale, Pennsylvania.
31. Plank, G. and R. Eppinger (1989) Computed Dynamic Response of the Human Thorax From a Finite Element Model, *Proc. 12th International Experimental Safety Vehicle Conference,* pp. 665-672, DOT.
32. Kleinberger, M., Application of Finite Element Techniques to the Study of Cervical Spine Mechanics, *Proc. 37th Stapp Car Crash Conference,* SAE paper No. 933131, SAE, Warrendale, Pennsylvania.
33. Renaudin, F., H. Guillemot, F. LaVaste, W. Skalii, F. LeSage and C. Pecheux, A 3-D Finite Element Model of Pelvis in Side Impact, *Proc. 37th Stapp Car Crash Conference,* SAE paper No. 933130, SAE, Warrendale, Pennsylvania.
34. Patrick, L. M., H. R. Lissner, E. S. Gurdjian (1965) Survival by Design - Head Protection, *Proc. 7th Stapp Car Crash Conference,* pp. 483-499.
35. Ono, K., A. Kikuchi, M. Nakamura, H. Kobayashi and N. Nakamura (1980) Human Head Tolerance to Sagittal Impact - Reliable Estimation Deduced From Experimental Head Injury Using Subhuman Primates and Human Cadaver Skull, *Proc. 24th Stapp Car Crash Conference,* SAE paper No. 801303.

MODELS IN INJURY BIOMECHANICS FOR IMPROVED PASSIVE VEHICLE SAFETY

JAC WISMANS
TNO Crash-Safety Research Centre
Delft, The Netherlands

Abstract

Thorough knowledge of the characteristics of the human body and its behaviour under extreme loading conditions is essential in order to prevent the serious consequences of road and other accidents. In order to study the human body response five type of models for the human body can be distinguished: human volunteers, human cadavers, living and dead animals, mechanical models (crash dummies) and mathematical models. These models briefly will be introduced. The main part of this paper will concentrate on mathematical models of the human body. The high standard current crash simulation technology has reached can be attributed mainly to three developments: (1) developments in the field of multi-body techniques (2) developments in the field of finite element techniques and (3) developments in methodology and technology dealing with the determination of input data and validation of models for crash dummies. A review of recent developments in this field will be presented, with special emphasis on a suitable general research methodology to be used for crash dummy model development and validation. Various types of models will be presented with their application fields, capabilities and limitations. Area's of future development will be identified. Most of the current human models in use for vehicle design and safety system optimisation represent crash dummies, as prescribed in vehicle safety regulations, rather than simulating the real human body. The challenge for the future is to develop mathematical models that offer a more realistic description of the human body than current crash dummies do.

1. Introduction

Every year 50 thousand people die in road accidents in Europe and many more are injured seriously. In order to protect the human body against injuries in case of extreme loading conditions, a clear insight in the ways injuries arise and into loads at which they occur is needed. Accident analysis studies and biomechanical research are carried out for this purpose. In injury biomechanics research several types of models for the human being are used to study the mechanical behaviour of the human body and to study injury

J. A. C. Ambrósio et al. (eds.),
***Crashworthiness of Transportation Systems: Structural Impact and Occupant Protection*, 221–236.**

mechanisms and tolerances. Also for the assessment of the protection offered by safety provisions, like seat belts and crushable vehicle structures, models for the human being are needed. In fact five types of human models can be distinguished: human volunteers, human cadavers, animals, mechanical models and mathematical models.

1.1 HUMAN VOLUNTEERS

For obvious reasons it is impossible to experiment with human beings fitted with instrumentation under injury producing conditions. Only in low severity tests, i.e. below the pain thresholds, sometimes human volunteers are used. Such tests, which are limited by rigid regulations and guidelines, can contribute to general knowledge on the human body non-injurious response. The results are very important for the development of mechanical or mathematical models. The human subjects are mostly young, well-trained, military volunteers. Their pain tolerance is usually much higher than that of the general population. Therefore the test results are not representative for females, children and elderly people.

An advantage of the use of human volunteers is that the effect of muscle tone and pre-bracing on the dynamical response can be studied. But this influence, which might be relatively large at low impact levels, cannot be simply extrapolated to higher impact levels.

The first well-documented human volunteer test was conducted in 1954 in the dessert of New Mexico at Holloman Air Force Base. Test subject was Colonel John Paul Stapp who sustained, without serious complaints, a velocity change of 1000 km/h during 1.4 sec on a rocket propelled sled device. The maximum deceleration was 40g, in other words 40 times the deceleration of gravity. For comparison: the average deceleration during a 50 km/h vehicle crash test against a brick wall is about 25g. And this deceleration usually doesn't take longer than 0.1 sec.

1.2 HUMAN CADAVERS

The primary research tool to evaluate injurious biomechanical response is a human cadaver (sometimes referred to as PMHS: post-mortem human subject). The anthropometry of a human cadavers is almost identical to the living human and material properties are often close to living tissue. This latter aspect however is strongly dependent on the preparation techniques applied and the time duration since death. Very often the lungs of the cadavers are inflated during the test and the blood pressure is partly restored by infusion.

Disadvantages of the use of human cadavers are the absence of muscle tone and that physiologic responses cannot be determined. Furthermore the age of cadavers is often high and since the mechanical strength of most tissue in the human body tends to decrease with age, the data obtained are not necessarily representative for the general population. Like for human volunteer tests also human cadaver tests are restricted by regulations and guidelines.

1.3 ANIMALS

Research using anaesthetized animals as human surrogates, is needed to obtain information on physiologic responses, in injury producing loading conditions, for specific body areas like the brains and the spinal cord. Furthermore tests with animals can provide insight in the differences between dead and living surrogates and as such provide the information for correct interpretation of human cadaver testing. Due to differences in size, shape and also structural differences between humans and animals, quantitative scaling and extrapolation of results of animal testing to the human is very difficult. The use of animals for injury biomechanics research is in most countries nowadays strongly restricted or not possible at all.

1.4 MECHANICAL MODELS

Mechanical models or crash dummies (sometimes also referred to as anthropomorphic test devices) normally consist of metal or plastic skeleton, including joints, covered by a flesh-simulating plastic or foam. They are constructed such that dimensions, masses and mass-distribution, and therefore the kinematics in a crash are humanlike (also called biofidelic). The dummy is fitted with instrumentation to measure accelerations, forces and deflections during the tests, that correlate with injury criteria for human beings. Dummies are often used in approval tests on vehicles or safety devices, in which the measured values should remain below certain (human tolerance) levels. Very important in this respect is a repeatable response of the dummy in identical tests.

Furthermore crash dummies sometimes are used for the reconstruction of real accidents in order to study the correlation between real world injuries and actual loading conditions.

Mechanical human models are not necessarily complex pieces of equipment. Three levels of complexity can be distinguished:

- complete (crash) dummies for the evaluation of vehicles and/or safety systems in full-scale tests
- body segment impactors for the evaluation of vehicle sub-systems like the body-block impactor consisting of the front of a dummy, used for steering-wheel testing.
- impactors for the evaluation of vehicle components like impactors simulating the head for padding testing.

Crash dummies have fixed sizes. In automotive tests a 50th *percentile* adult male dummy is most frequently used. In a frequency-distribution of a certain anthropometric measurement, 'one percentile' is a column with an area of 1% under the frequency-curve. So the 50th percentile value indicates the measurement value for which 50% of the population shows a lower value (and therefore also 50% a higher value). Therefore a 50th percentile measurement represents an 'average'. All these 50th percentile values are combined in one 'mid-sized' dummy, which then represents a 'defined average' adult male dummy. Occasionally two other sizes are used in vehicle crash tests as well, a 95th percentile (large) male and a 5th percentile (small) female. Furthermore different sized child dummies have been developed like the TNO series of child dummies consisting of a 9-month, 3-year, 6-year and a 10-year old child.

Most dummies for evaluation of vehicles or safety devices are used only in one specific impact mode; frontal or lateral. The most widely used dummy for frontal impacts is the Hybrid III dummy. The development of a new frontal impact dummy is underway in the US the so-called AATD (Advanced Anthropomorphic Test Device). The first prototype of this dummy is available in 1996.

For side impacts several dummies are available: the EUROSID-I developed in Europe and the SID and the BIOSID developed in the US. There is a large need for harmonization resulting in a single wordwide accepted side impact dummy (sometimes referred to as WORLDSID).

Dummies with a vertical performance capability are used in aircraft research concerned with ejection seat testing. For the evaluation of motorcycle safety recently a special motorcycle dummy has been developed. Attempts have been made to design 'omni-directional' dummies, however without much succes up to now.

1.5 MATHEMATICAL MODELS

This paper further deals with mathematical models of the human body. Together with a mathematical description of the environment (e.g. steering-wheel, dashboard, seat structure and the restraint systems) and the impact conditions, the model provides a numerical description of the crash event. The usage of computer simulations in the field of passive safety, like in many other engineering disciplines, has shown a strong increase in the past twenty years. This is partly due to the fast developments in computer hardware and software, but in the case of automotive safety, probably even more due to emphasis which has been placed on the development of reliable models describing the human body in an impact situation, as well as numerous validation studies which have been conducted using these models.
In the crash safety field mathematical models are applied in various areas including:

- human impact biomechanics research
- reconstruction of actual accidents
- design of vehicle structure, restraint systems, safety devices and roadside facilities

Furthermore computer simulations may become part of safety regulations, much like standard crash tests. But up to now usage of computer models for this purpose has been rather limited. One example is the bus rollover regulation in the United Kingdom [1] and a proposal to introduce a relative simple occupant model as part of a European side impact regulation, the so-called CTP testprocedure [2].

One of the first human body gross-motion simulation models was developed already 30 years ago by McHenry in the United States, who proposed a two-dimensional seven segment numerical model to describe the motion of a vehicle occupant in a collision event [3]. The results of this model were so encouraging that since then many, more sophisticated, models have been developed including MVMA-2D, CAL3D, ATB and MADYMO. For a comparison of a number of the basic features between these programs the reader is referred to reviews by Prasad [4] and Prasad and Chou [5].

By definition a model is a simplification of the reality. The amount of detail required in a model strongly depends on the objectives of the model. In general it can be stated that for human body models with an increasing complexity more input data is required. Usually additional assumptions have to be introduced since the current knowledge on human body mechanical properties is still rather limited. Taking also into account the increased CPU time required for more complicated models it is recommended to make models not more complicated than what is required for its type of application.

Due to the assumptions which have to be introduced in simple as well as the more complicated crash models, validation of the models under well defined experimental conditions is an essential part of the model development strategy.

The most important numerical techniques in crash analyses are the multi-body method and the finite element (FE) method. In this paper first these techniques briefly will be introduced. Multi-body models are the most widely used and most economical way for human body dynamic simulations. Human body models can be subdivided into models of crash dummies and models of the real human body. Well-validated models thus far, have been developed mainly for crash dummies. An overview of the modelling strategy for the development of crash dummy models together with a number of examples of dummy databases will be presented. Next the modelling of the real human body will be discussed which is much more complicated than crash dummy modelling, among others due to the lack of reliable biological material data. This paper concludes with an illustration of some typical applications in an automotive passive safety environment followed by a discussion on future developments.

2. Numerical techniques

Dependent on the nature of research, several types of crash computer programs have been developed, each with their own, but often overlapping, area of applicability. Most of the models are deterministic, that is, based upon measured or estimated parameter values, representing characteristics of the human body, safety devices, the vehicle and its surroundings and using well established physical laws, the outcome of the crash event is predicted. Although the various deterministic models may differ in many aspects, they are all dynamic models. They account for inertial effects by applying equations of motions for all movable parts and solve these equations by some integration scheme. The numerical approaches used for these programmes can be subdivided into multi-body techniques and finite element (FE) techniques. Furthermore, lumped mass models can be distinguished but they can be considered as a special case of multibody models. In a lumped mass model a system is represented by a number of rigid bodies connected by mass-less elements like springs and dampers. The most important difference with a multi-body model is the absence of kinematic joints. Furthermore, lumped mass models usually are one-dimensional while multi-body models are two- or three-dimensional.

Multi-body models are particularly suitable for simulation of spatial motions and forces in systems of bodies connected by joints. Most earlier multi-body programs for crash analyses only allowed rigid bodies connected by simple pin joints and ball-and-socket joints. But more recent developed software also allows flexible bodies and several other joint types like translational, cylindrical and planar joints and even user defined joints. Degrees of freedom in typical multi-body human body models usually vary from 5-15 in simple models up to 100 degrees or more in complicated systems.

In a FE model the object is represented by a complex system composed of relatively simple elements interconnected at a discrete number of nodal points. FE models allow the calculation of detailed deformations and stress and strain distributions. The number of degrees of freedom in a typical FE human body model is much larger: from 1000-10000 or more and require a much smaller time step for the explicit integration method due to which the required computer time (CPU) is much higher than in multi-body models.

The examples presented in this paper are obtained with the MADYMO program. MADYMO from its origin is a general multi-body program with a number of special features for crash analyses [6]. It has a 2D and 3D version. Since 1991 also FE capabilities have become available in MADYMO 3D allowing the user to perform fully integrated coupled multi-body/FE calculations also referred to as hybrid simulations [7]. Figure 1 illustrates the general structure of MADYMO.

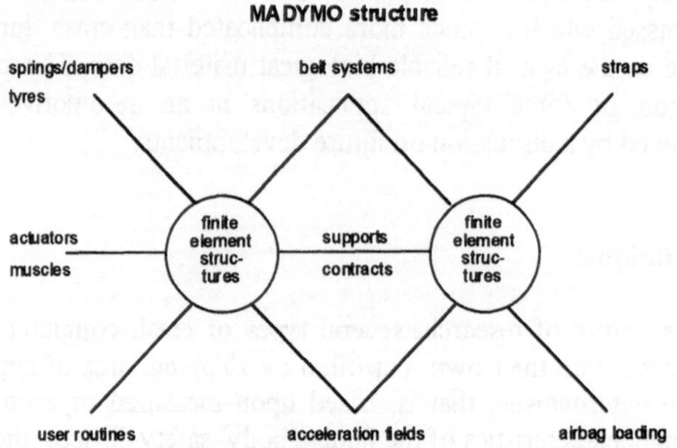

Figure 1. General structure of MADYMO 3D

In MADYMO special options for modelling restraint systems and other features of a vehicle are available. For contact interaction between multi-bodies and the environment special contact models have been developed. For the analysis of muscular activity, models for the active and passive behaviour of skeletal muscle have been developed.

In the FE part of MADYMO several element types can be used including brick, truss, beam, membrane and shell elements. Material models include elastic, elasto-plastic and visco-elastic models. Specific fabric material models have been implemented for the membrane elements for the simulation of airbag and seat-belt dynamics [8].

Crash simulation programs offer many output options including the parameters that are usually measured in a crash test. Examples of output parameters are: 3D body positions, velocities and accelerations, contact forces, belt forces, local stresses and strains (FE programs) and injury criteria. By means of interactive graphical post-processors the motions can be visualized. Particular the option of animation of the crash event appears to be very effective.

3. Multi-body models of crash dummies

Well-validated human body crash models presented in literature so far mainly have been developed for mechanical models of the human body, i.e. crash dummies, rather than for real human beings themselves. The rationale for this is that most model input data in case of crash dummy models can be measured relatively easy. Moreover, results of experiments with crash dummies often are available for model validation and if not, such experiments, unlike tests with biological models, readily can be carried out in many well equipped crash laboratories. Another reason for the emphasis on modelling crash dummies sofar is the need, particular from the design departments in the automotive industry, for well-validated design tools which can reduce the number of regulatory tests with crash dummies in order to shorten and optimize the development process of a new car model.

The first step in the modelling process of a crash dummy is the division of the dummy in segments and the specification of the parts that belong to each segment. The segments are selected by dividing the dummy into functional components. Each part of the dummy having significant mass and a flexible connection with other parts is considered as a segment. Dummy parts which do not show any relative motion are usually considered to be part of another segment except if load information is required at the interface between the two segments. In MADYMO a variety of joint types are available to link these segments Figure 2. In present dummy designs usually four types of kinematic connections between segments can be distinguished: revolute (or pin joints), translational joints, universal joints and ball and socket joints.

Furthermore often flexible structures are present in a dummy. Some are partly or completely made of rubber like the lumbar spine and neck. They are usually modelled by two ball and socket joints located in the centres of the end planes of these structures or by a 6-degree of freedom joint. Recently also a special restraint available within the MADYMO multi-body formulation has been applied for this purpose. Flexible parts like the ribs in the dummies are usually represented by a number of rigid bodies, but a flexible body approach may be more accurate and efficient [9].

If the general model set-up has been specified the geometrical parameters are measured. This includes the joint locations within the individual segments, the joint axes orientations and the outside surface geometry. These three-dimensional measurements are conducted at a disassembled dummy. Some of the geometrical joint data have to be determined in an indirect way since the requested joint data may not be directly accessible by the measuring device digitizing arm. Additional measurements of landmarks specifying segment local co-ordinate systems have to be conducted in order to express

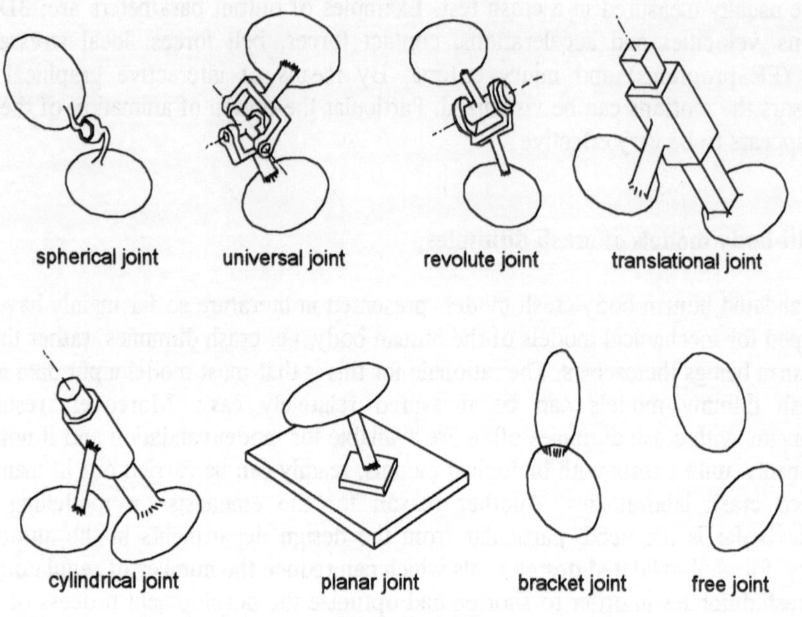

Figure 2. Joint types

the data in a common body-fixed base.

The outside surfaces of the dummy segments are usually represented by means of ellipsoids, but in the MADYMO program recently also arbitrary shaped surfaces can be specified for this purpose. The ellipsoids or arbitrary surfaces are used for visual presentation of the occupant kinematics as well as for the contact interactions between dummy segments and environment (e.g. the vehicle interior).

The next step is the measurement of the inertial properties. The mass, location of the centre of gravity, the principal moments of inertia and the orientation of the principal axes must be determined for each dummy segment. In addition the position of segment landmarks have to be determined in order to express the inertia data in the body-fixed coordinate system. The moments of inertia are determined with a torsional vibration table [10]. The object which is fixed in a box is measured in several positions in order to get the complete inertia tensor.

The stiffness of the connections (joints) between the different segments is one of the parameters having a major effect on the motions of the dummy segments in a crash environment. These joint resistive properties are determined using various static and dynamic test methods. In these tests the range of motion corresponding to a joint coordinate is determined as function of the externally applied load. If a joint has more than one degree of freedom, like in a ball and socket joint, for each degree of freedom separate measurements are conducted, keeping the others fixed. Since the actual joint resistance often will depend on the value of multiple joint coordinates, large test series may be required. In practice up to now this dependency on more than one degree of freedom is neglected and joints are tested with the other degrees of freedom fixed.

The last step is the specification of the surface compliance properties. Static as well as dynamic measurements with several penetrating surfaces must be performed at different locations on the dummy segments. The surface compliance is dependent on the skin covering thickness and density as well as the compliance of the underlying structure.

On the basis of these measurements a database of the dummy can be compiled now. After formulating a database, verification simulations are carried out to insure that the database adequately represents the complete dummy. For this purpose well controlled impactor tests and sled tests with the assembled dummy at different acceleration levels are used. If results are not completely satisfying further model refinements with corresponding input measurements may be required. A well validated computer model allows the user to apply the model for predictive simulations of events outside the range of validated simulations.

Figure 3 and Figure 4 illustrate validated multi-body dummy databases currently available with the MADYMO program, which include databases for the frontal Hybrid III dummy family (5th, 50th and 95th%), the TNO child dummies and the side impact dummies BIOSID, SID and EUROSID. Figure 5 shows MADYMO FE databases for the European pedestrian impactors representing, respectively an adult head, and the femur and knee joint part of the leg.

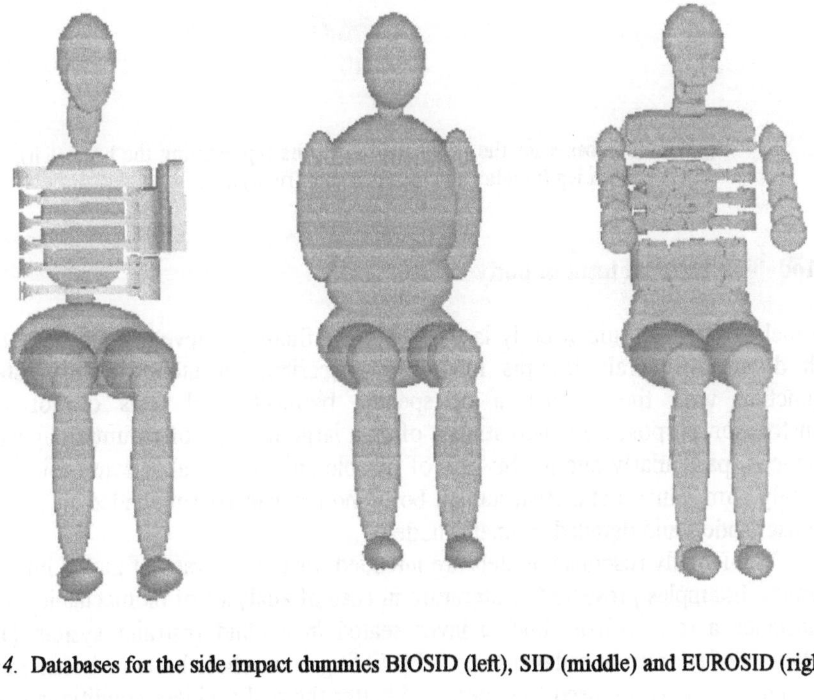

Figure 4. Databases for the side impact dummies BIOSID (left), SID (middle) and EUROSID (right)

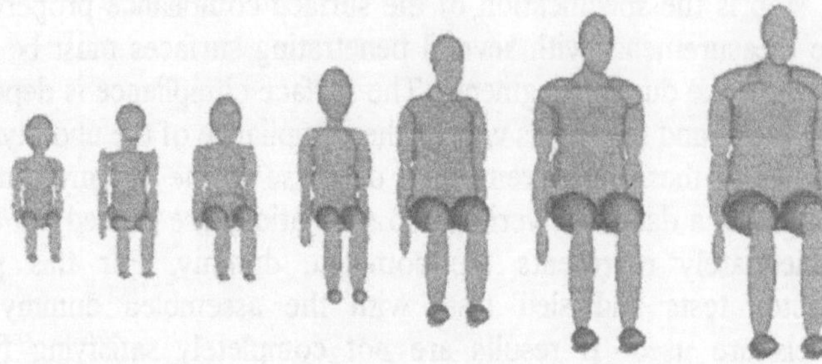

Figure 4. **TNO child dummies and the 5th%, 50th% and 95th% Hybrid III dummy family**

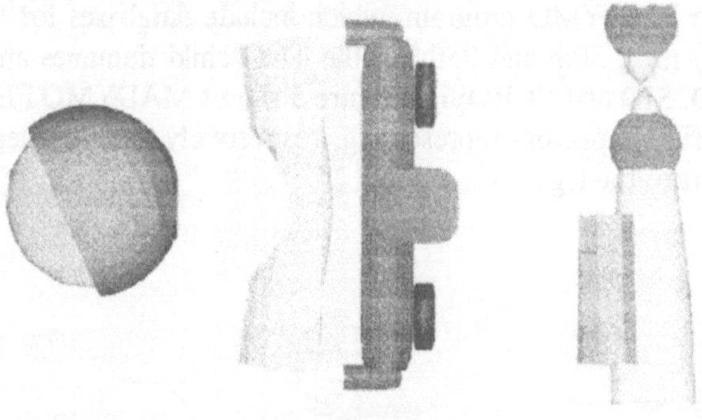

Figure 5. **Databases for the pedestrian impactors representing the head (left), upper leg (middle) and the knee joint (right) area**

4. Modelling the real human body

A model of the real human body is much more difficult to develop than a model of a crash dummy. Several attempts have been described in literature for instance in conjunction with the simulation of specific biomechanical tests or for accident reconstruction purposes. In such studies often a large number of assumptions had to be introduced, particularly due to the lack of reliable and statistical relevant joint property and body compliance data. Real human body models can be subdivided in whole body response models and detailed segment models.

Whole body response models are intended for the analyses of global human body dynamics. Examples presented in literature in case of analyses of biomechanical tests are for instance a test with a child cadaver seated in a child restraint system [11] and simulations of pedestrian cadaver tests [12]. In litigation whole body response models of the real human body are used to understand better the real accident conditions in relation to the resulting human body dynamics. For such simulations the human body geometry and mass distribution for MADYMO can be generated by a computer program called

GEBOD (GEnerator of BOdy Data), which produces geometric and inertial properties of human beings [13]. The program is developed by the Air Force Aerospace Medical Research Laboratory in Dayton USA. It generates a model up to 17 segments: head, neck, upper and lower arms, hands, thorax, abdomen, pelvis, upper and lower legs and feet using regression equations based on anthropometric surveys and stereophotometric data. Input for GEBOD are height and/or weight to generate male or female adult datasets and age, height and/or weight to generate child data sets. A user may also input up to 35 specific anthropometric measurements to achieve better estimations from the program.

Detailed body segment models consider the loading and deformation in specific anatomical structures and thereby directly relating external loading to injury mechanisms. Three examples of such models developed in MADYMO are illustrated in Figure 6. The first example is a model of the tibia in an impactor test [14]. The model consists of ten segments and allows the prediction of tibia fracture. The model was compared with lower extremity cadaver impacts. The second example is a model of the human kneejoint developed by Yang et al [15]. In this model femur and tibia are represented by multi-body elements connected by non-linear elements representing the ligaments. The femoral condyles are represented by ellipsoids and the tibial condyles by planes. Model predictions were compared in dynamic bending and shearing tests with lower extremity specimen experiments and a realistic response was reported for the knee bending angle and the shear displacement. The model allows to predict ligament failure. The third example is a more detailed model and concerns the human cervical spine [16]. In this model vertebra, intervertrebal disks, facet joints, ligaments, capsule and also passive and active muscle response are included. This model has been validated in detail for human volunteer tests in different impact directions. Simulations with this model revealed that the muscular response has a significant effect on the dynamical behaviour of the head-neck system. The model is intended to gain a better understanding of neck injury mechanisms under various loading conditions and to study neck injury prevention measures.

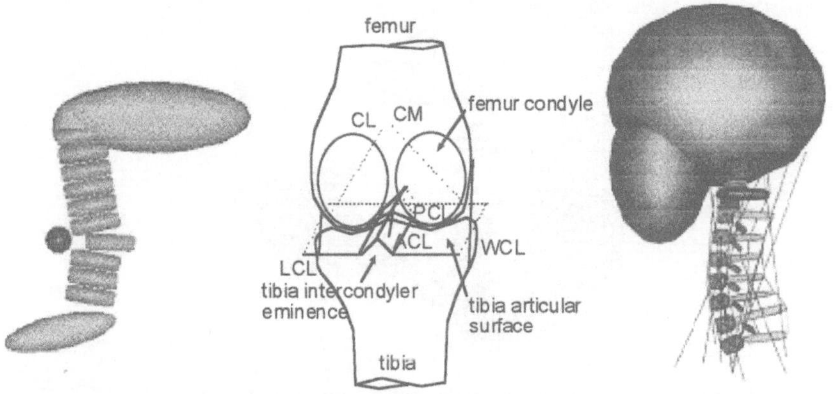

Figure 6. **Segment models of the real human body: tibia (left), kneejoint (middle) and neck (right)**

5. Examples of applications in automotive crash analyses

Simulations with crash dummy models have been carried out for a wide range of applications, many of them for which validation studies were conducted. An illustration of a number of typical applications in automotive passive safety are presented in Figure 7 - Figure 9. Figure 7 shows the simulation of unrestrained truck occupants and a cyclist impacted by a truck front using multi-body techniques only. Figure 8 shows hybrid simulations in which the airbag is represented by finite elements. The human body in these simulations and the motorcycle (including suspension) are modelled by multi-body elements. Also Figure 9 shows hybrid simulations in which the kneebolster and the seat, respectively are modelled by finite elements. Also in other fields of transport safety crash simulations have been applied. See [17] and [18] for some typical applications in the field of aircraft safety.

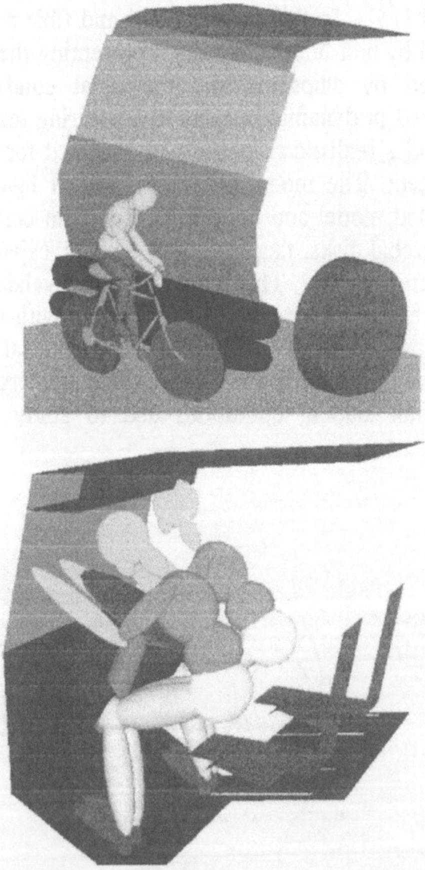

Figure 7. Applications using multibody techniques

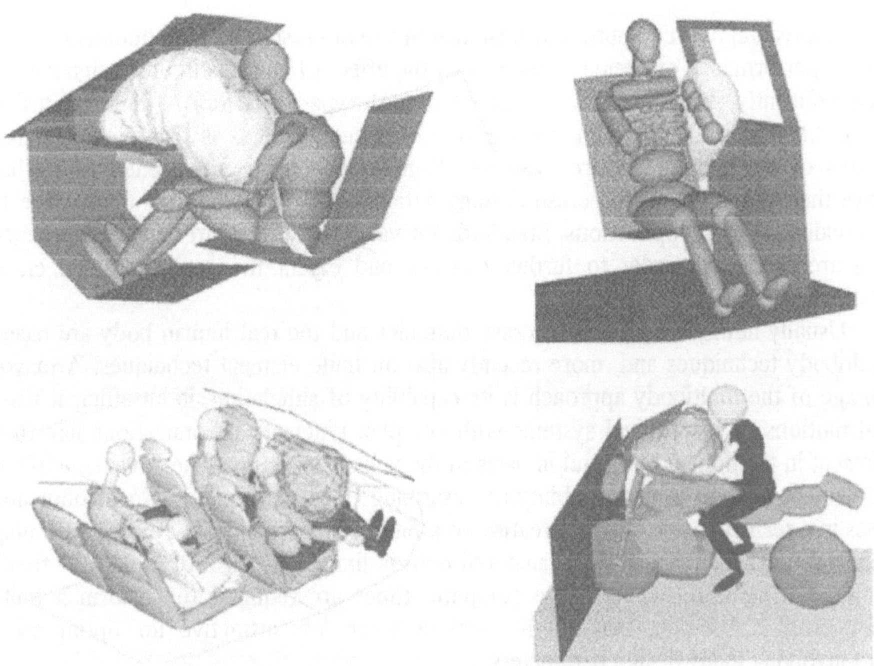

Figure 8. **Applications using the hybrid approach with finite elements for the airbags**

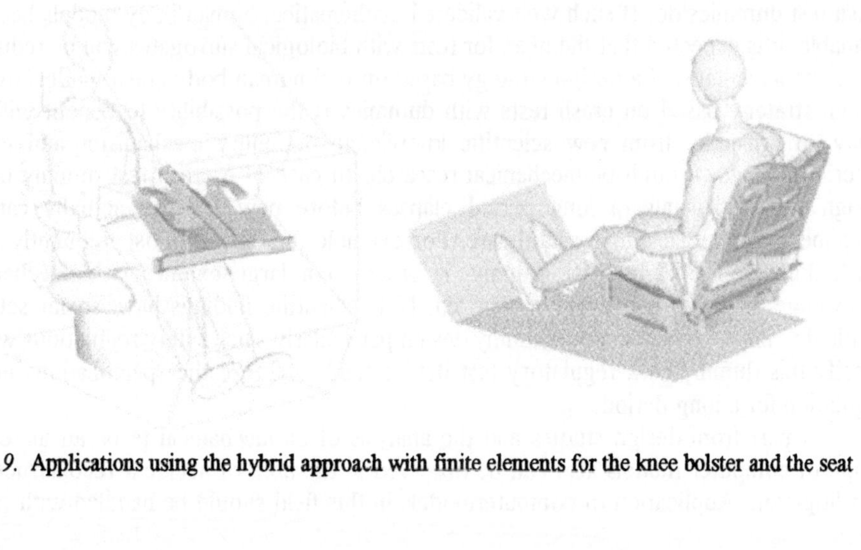

Figure 9. **Applications using the hybrid approach with finite elements for the knee bolster and the seat**

6. Discussion and conclusions

A general advantage of computer crash simulations over crash tests with dummies is that the safety performance of design concepts and the effect of changes in the design can be studied efficiently, sometimes even without a prototype to be build (concept design studies). An important condition for the usage of such models is that well-validated databases of crash dummies are available. Continuous efforts are needed to further improve the quality of existing crash dummy databases in order to allow their usage in even a wider range of applications. Standards for validation procedures and performance criteria are needed in order to further enhance and extent the applicability of crash simulations.

Usually numerical models of crash dummies and the real human body are based on multibody techniques and, more recently also on finite element techniques. A mayor advantage of the multibody approach is its capability of simulating, in an efficient way, spatial motions of mechanical systems with complex kinematical connections like they are present in the human body and in parts of the vehicle structure. The advantage of the finite element method is the capability of describing (local) structural deformations and stresses in a realistic way. But the creation of a finite element model is a time consuming job and the availability of realistic material data is limited, in case of biological tissue response. Furthermore usually large computer times are required to perform a finite element crash simulation, making the method much less attractive for optimization studies involving many design parameters.

Compared to crash dummy models the state of development of models of the real human body is still in its infancy. Real human body models allow e.g. the study of the effect of body size, posture as well as muscular activity. Up to now studies reporting model verification using human volunteer or human cadaver tests appear to be rather limited. A major step forward will have been made once mathematical models reach a stage where they offer a more realistic representation of the human body than current crash test dummies do. If such well-validated mathematical human body models become available, it is expected that the need for tests with biological surrogates will be reduced. A unique advantage of a design strategy based on real human body crash models over a design strategy based on crash tests with dummies is the possibility to benefit without delay, in principle, from new scientific knowledge on injury mechanisms and injury criteria obtained through biomechanical research. In case of a crash test dummy based design strategy usually a long period elapses before new findings actually can be implemented in crash dummy hardware. For example the current most frequently used crash dummy, the Hybrid III dummy, is based to a large extent on biomechanical knowledge of more than twenty years old. New scientific findings have so far seldom resulted in improvements in the dummy design particularly since safety regulations which specify this dummy as a regulatory test device tend to freeze the specifications in the regulation for a long period.

Apart from design studies and the analysis of biomechanical tests, an increased usage of computer models also can be observed in the area of accident reconstructions and litigation. Application of computermodels in this field should be handled with much care, due to the limited level of development of real human body models, the usually large number of unknown accident parameters and the lack of experimental data available for validation for the case under consideration. Development of a code of practice with guidelines for usage of models in this field is highly recommended.

Several areas can be identified in the field of human body crash simulations where further developments should take place. As far as crash dummies is concerned, in particular realistic models for the foam type structures (skin and damping material) are required. Area's of future developments in the field of real human body models include further improvements in the description of the non-linear dynamic behaviour of muscles (incl. neuro-muscular control), the modelling of complex human joints and the study of constitutive equations and parameters for biological materials (e.g. brain, skin). The complexity of human body modelling in this area is however such that significant progress only can be expected if the various (international) research groups work closely together according to a well-defined research plan.

This paper has concentrated in particular on the usage of multi-body technology. Further refinements of the models using FE techniques constitutes an increase in model complexity, however, with the advantage that detailed stress and strain analysis can be performed. Particular for detailed studies of injury mechanisms in specific body part this is a necessary and feasible approach. Boundary conditions for such segment models may be obtained from experiments or from results obtained through more global models. The usage of FE techniques coupled with multi-body techniques will allow the user to benefit from the capabilities of both approaches and will offer him the flexibility of merging more global multi-body models with, whenever needed, detailed representations for certain parts in his model.

7. References

1. Uniform provisions concerning the approval of large passenger vehicles with regard to the strength of their superstructure (1987), *E/ECE/324, E/ECE/TRANS/505,* Addendum 65: Regulation No. 66.
2. Bourdillon (1994), Program of comparison of side impact testing methods, paper 94-S6-O-15, *Proceedings of the 14th ESV Conference*, p. 1039-1057, 1994.
3. McHenry, R.R. (1963), Analysis of the dynamics of automobile passenger restraint systems, *Proc. 7th Stapp Car Crash Conference*, pp.207-249.
4. Prasad, P. (1984), An overview of major occupant simulation models, Mathematical simulation of occupant and vehicle kinematics, *SAE Publication P-146*, SAE paper no. 840855
5. Prasad, P. and Chou, C.C. (1989), A review of mathematical occupant simulation models, Crashworthiness and occupant protection in transportation systems, *Proceedings AMD-Vol. 106, BED-Vol. 13 of the Winter Annual Meeting of ASME*.
6. Lupker, H.A., de Coo, P.J.A., Nieboer, J.J. and J. Wismans (1991), Advances in MADYMO crash simulations, *SAE 910879, Proceedings SP-851,* Int. Congress and Exposition, Detroit, Society of Automotive Engineers Inc.
7. Lupker, H.A., Helleman, H.B., Fraterman, E. and Wismans, J. (1991), The MADYMO Finite Element Airbag Model, *Proceedings of the 13th International Technical Conference on Experimental Safety Vehicles*, Paris.
8. Fraterman, E. and H.A. Lupker (1993), Evaluation of belt modelling techniques, *SAE paper 930635, In: SP-947,* Int. Congress and Exposition, Detroit, Society of Automotive Engineers Inc.
9. Koppens, W.P., Lupker, H.A. and C.W. Rademaker (1993), Comparison of modelling techniques for

flexible dummy parts, *Proceedings of the 37nd Stapp Car Crash Conference*, SAE paper no. 933116, P-269.
10. Kaleps, I., J. Whitestone (1988), Hybrid III Geometrical and Inertial Propertie", *SAE Paper 880638*, International Congress and Exposition, Detroit.
11. Wismans J.S.H.M., Maltha J., Melvin J.W., Stalnaker R.L. (1979), Child restraint evaluation by experimental and mathematical simulation, *Proceedings of the 23th STAPP Conference*, SAE-791017.
12. Ishakawa H., Kajzer J., Schroeder G. (1993), Computer simulation of impact response of the human body in car-pedestrian accidents, *Proceedings of the 37th STAPP Car Crash Conference*, SAE-933129.
13. Baughman, L.D. (1983), Development of an interactive computer program to produce body description data. University of Dayton Research Institute, Ohio, USA, *Report nr. AFAMRL-TR-83-058*, NTIS doc. no. AD-A 133 720.
14. Wismans, J., R. Happee, J. Thunnissen and H.A. Lupker (1995), Modelling the lower extremities using multi-body techniques, *In: Proceedings of the International Conference on pelvic and lower axtremity injuries (PLEI)*, Washington.
15. Yang, J., Cavallero, C. and Bonnoit, J. (1995), Computer simulation of shearing and bending response of the knee joint to a lateral impact, SAE 95 2727, p. 251-264, *Proceedings of the 39th Stapp Conference*.
16. M. de Jager, A. Sauren, J. Thunnissen and J. Wismans (1994), A three-dimensional head-neck model: Validation for frontal and lateral impacts, In: *Proceedings of the 38th Stapp Car Crash Conference*, Fort Lauderdale, Florida, pp. 93-109, *SAE Paper 942211*, October 31-November 2.
17. Nieboer, J.J., J. Wismans and R. Verschut (1992), Occupant simulation as an aspect of flight safety research, In: *Agard Conference Proceedings 532, Aircraft accidents, Trends in Aerospace Medical Investigation Techniques*, Advisory Group for Aerospace Research and Development, NATO, pp 14.1-14.9.
18. Wismans, J. and Obergefell, L. (1996), Data Bases and Analytical Modelling, *Chapter 8 in AGARD Advisary Report Anthropomorphic Dummies for Crash and Escape system testing, AGARD/AMP/WG21.*

BIOMECHANICAL MODELS IN VEHICLE ACCIDENT SIMULATION

E. HAUG
ESI Group
20 rue Saarinen, Silic 270, 94578 Rungis-Cedex , France

Abstract

An overview on biomechanical modelling and its application in simulations of vehicle accidents is given. For the fabrication of biomechanical finite element models, acquisition of geometrical data, material properties and validation tests are needed. The utilization of biomechanical models is discussed and demonstrated on examples.

1. Introduction

1.1. PREVIOUS ESI GROUP WORK IN BIOMECHANICS

The present paper may be considered to follow the line of presentations given on PUCA'93 [1] and PUCA'94 [2]. In the first reference the authors discuss the numerical modelling techniques of mechanical occupant surrogates (dummies), using several "generations" of rigid, articulated rigid body, deformable articulated and finite element models of mechanical dummies. The paper ends with the outlook : "Towards the Human Surrogate ?", in which then available models and results of human and mammalian "parts" (head, knee, cervical spine) were presented.

In the second reference the authors discuss the needs for biomechanical tests and simulations by describing head injuries, neck injuries, thorax injuries, abdominal injuries, spine injuries, pelvis injuries and injuries of the extremities. Next, these authors give an overview on methods of testing, including the use of biofidelic mechanical dummies, cadavers, animals and live human volunteers. A following paragraph mentions data collection from police reports on automobile crashes, reports by accident investigation teams, hospital reports, autopsy reports and numerical autopsies. Finally, simulation techniques are discussed, starting with models of mechanical dummies, gradually evolving towards biomechanical finite element models, with the needs of biomechanical material modelling. Acquisition of geometry data, interface modelling and validation are also discussed. Perspectives announce the creation of a library of biomodels by which the prediction of injury and trauma of mammalian vehicle occupants can be done best by computer simulation, which may in this case be the only valid alternative.

The present paper comes at a time when biomechanical finite element simulations of accidental injuries are starting to be considered realistic means of obtaining refined information on occupant injuries by the research and development departments of car companies, i.e., when the subject is about to transit from the realm of academic research at universities and government laboratories to industry. Reasons for this progression are found in the continued general progression of numerical simulation techniques and software packages, the growing efficiency of computer hardware

J. A. C. Ambrósio et al. (eds.),
Crashworthiness of Transportation Systems: Structural Impact and Occupant Protection, 237–259.

platforms and code architecture that permit to envisage numerical models of sufficient detail and complexity, the growing databases and improved techniques for the fabrication of geometrical and material data and models, more accidentological and trauma data, and the general recognition that simulation is the only practical way to include and obtain mammalian occupant response in the crashworthy design of transport vehicles, constituting perhaps the ultimate consecration of this discipline.

The paper outlines ways to obtain geometrical data of mammalian parts. It then demonstrates how such information can be transformed into finite element models. The all important aspects of material data acquisition and modelling are addressed next. The paper then gives an overview on currently developed biomodels and simulation results, for the human head, neck, thorax, knee and ankle. Finally, a biomedical application on the impact of a small missile on the human eye, currently under investigation at Nihon ESI, and one on the operative repair of skin lesions, demonstrate the great potential of biomechanical modelling beyond standard transport vehicle accident simulation.

2. Geometry of Biomechanical Human Models

The acquisition of geometrical data of biomechanical human models can be a considerable task and requires novel techniques, where the establishment of a geometrical description as CAD data, clouds of points, etc., may be expensive and time consuming. Once established for a given part, the question arises whether this geometry is representative, e.g., for a 50-th percentile cross section of male or female, child or adult, young or old population of a given continent. Recently data bases are established that collect CAD geometries of the human body, for example, the BIOCAD data base fostered by the NHTSA organization of the United States.

Several techniques are utilized to establish human part geometries, such as

- computer tomography (CT) scans, where the geometry of slices of the parts is recorded, slice-by-slice, and digitalized ; finite element models can be obtained by connecting the digitalized slices with finite elements ; an example is shown in Figure 1, where CT scans of a human skull were transformed into a finite element model of the bony structure of a human skull without the lower jaw, consisting of 1342 solid elements in 31 slices [3] ;
- external laser scanning or similar techniques, leading to clouds of points that serve as a basis for elaborate CAD models of the outer surfaces of bone, ligaments, muscles, tendons, etc.; such models are available, e.g., by Viewpoint Datalabs [4]. A representative sample of human parts is collected in Figure 2;
- mechanical slicing, abrasion or laser cutting, where human parts are embedded in a supporting matrix, slices of 2 to 0.33 millimeter thickness are prepared or exposed at low temperature, photographed, then detailed by anatomists and ultimately digitized ; due to the identification of bone, ligament, muscle, tendon, skin, fat and other tissue materials, 3D models of the interesting constellations (e.g., bones + ligaments) can be prepared from the slice geometries ; Figure 3 shows the photograph of a typical slice ready for digitization (from the Visible Human Project, available on CD-ROM) ;
- copying from textbooks on the human anatomy (e.g., [5]) and anatomical atlases, which contain views of slices through human parts with corresponding explanations, which can be most helpful in the understanding of the overall and detailed human anatomy, Figure 4.

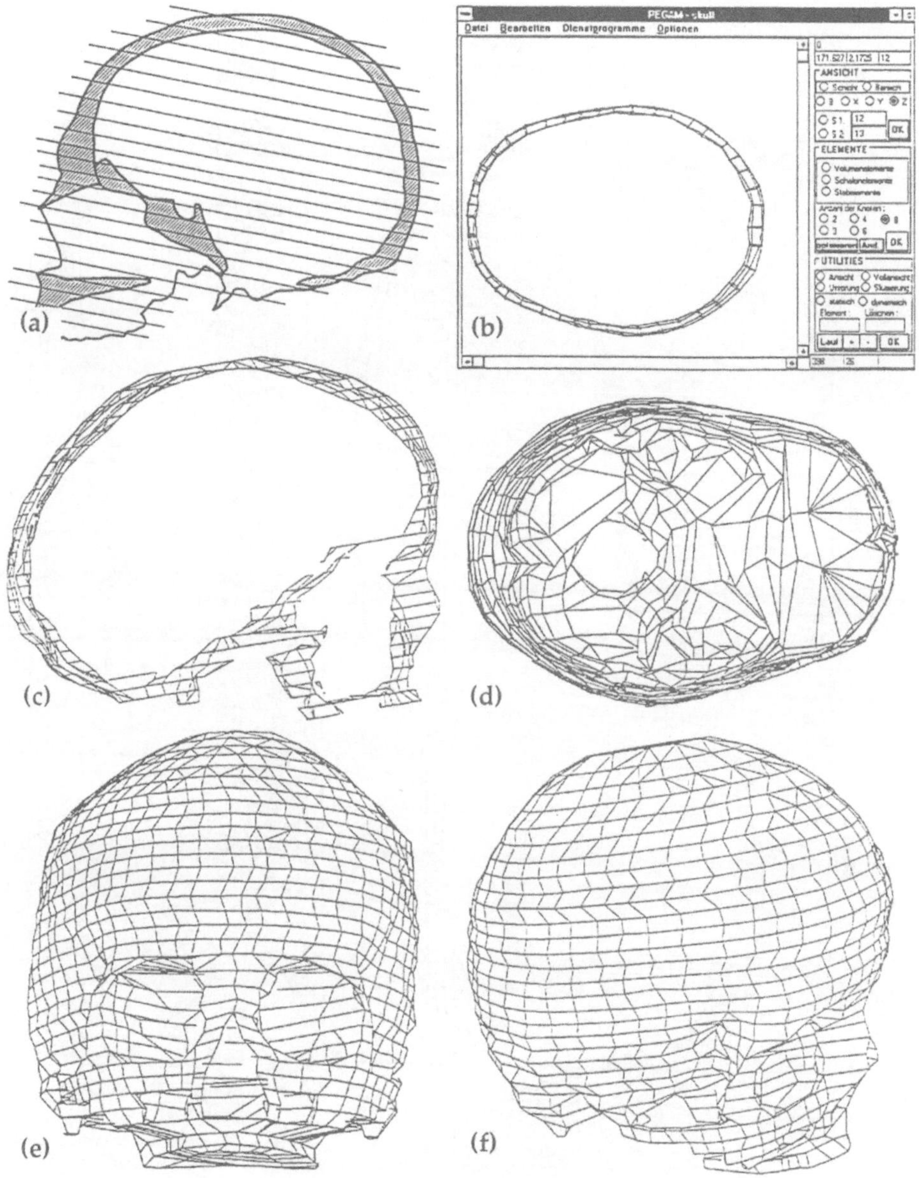

Figure 1. Computer Tomography (CT) scan based FE model of a human skull (1342 solids, 2874 nodes, 31 slices) : (a) parallel slice pattern, (b) digitalization of slices (TUB-PEGAM code), (c) mid-sagittal view, (d) inferior view, (e) frontal view, (f) laterial view (courtesy Technical University Berlin (TUB), Institute for Automotive Engineering, Prof. Appel) (after [3])

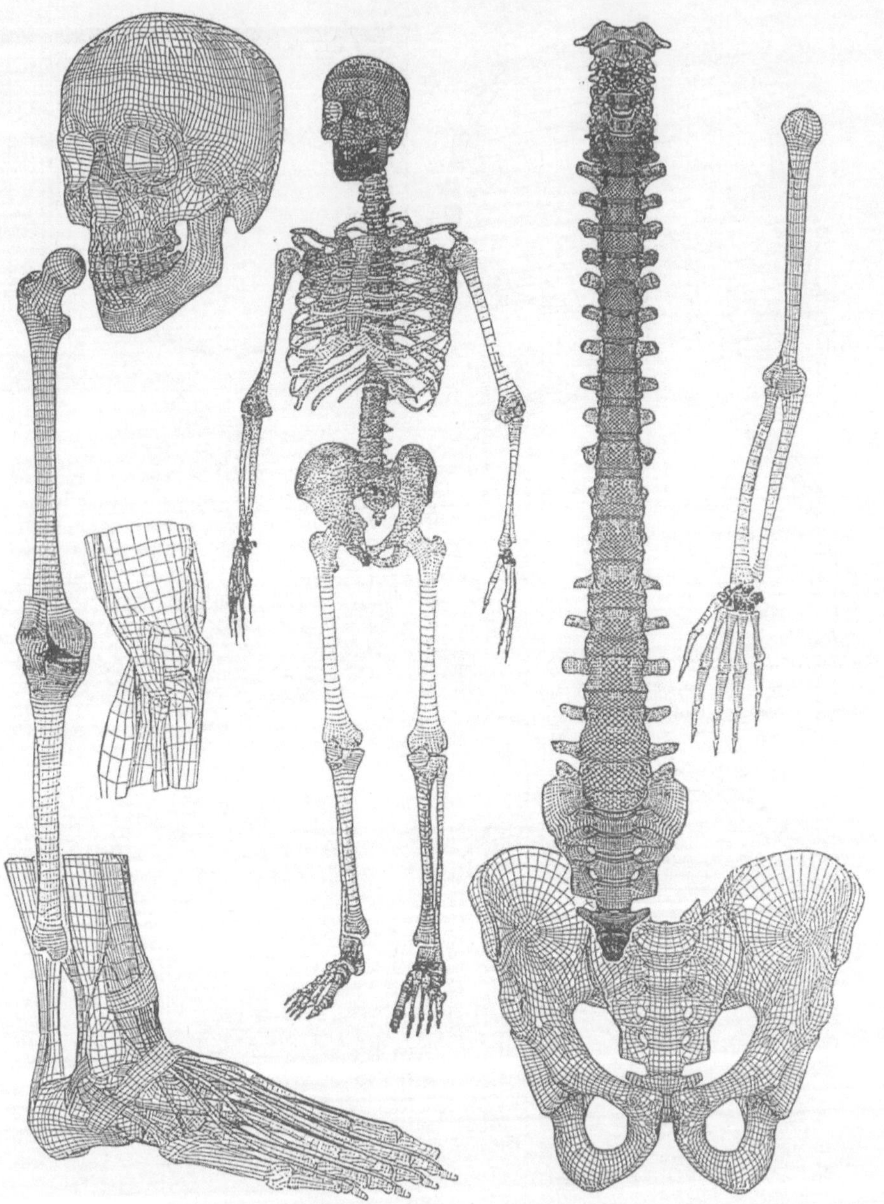

Figure 2. Surface discretizations of human parts (courtesy : Viewpoint Datalabs Int'l) (after [4])

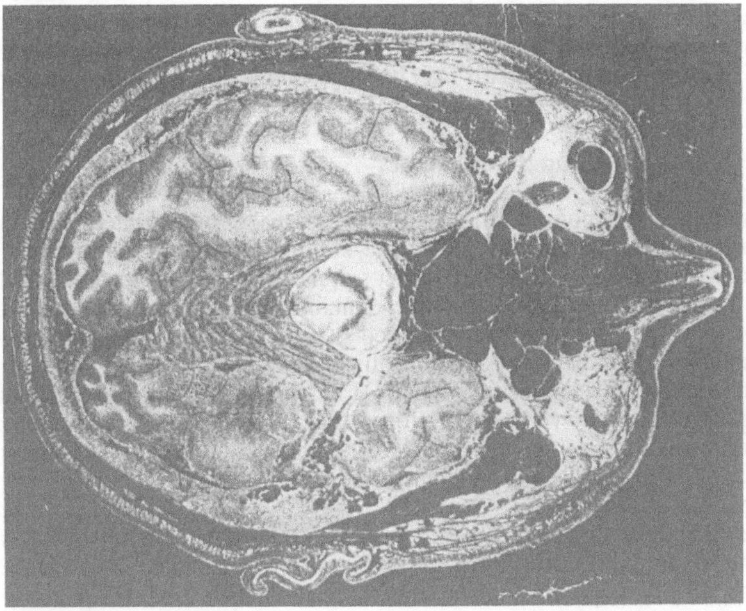

Figure 3. Exposed human slice (head) with anatomical details (from CD-ROM : Visible Human Project)

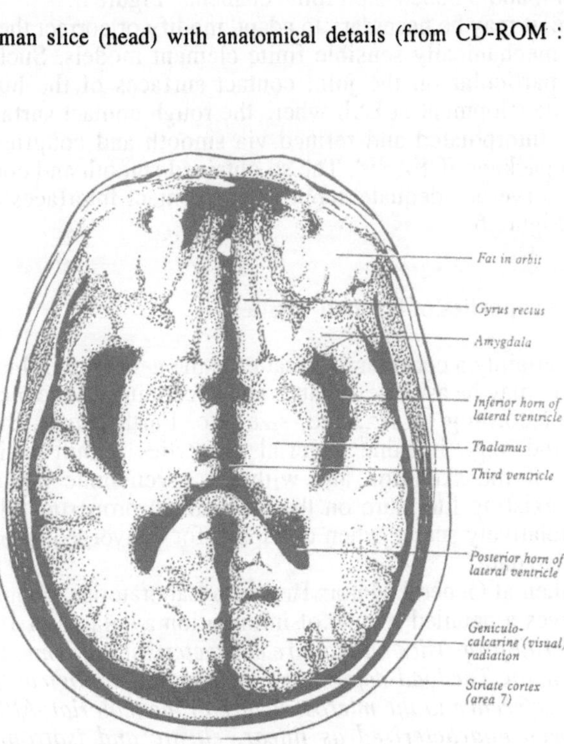

Figure 4. Human Slice (head) from texbook with annotations (from Grays Anatomy) (after [5])

All mentioned methods, being feasible to establish more and less detailed geometries of human parts, can do so at best only for a given individual. The biomechanical analyst is still confronted with the lack of average properties or even coherence of models. Different parts may be based on data from different individuals.

3. Finite Element Models in Biomechanics

It is generally agreed upon that the finite element discretization method is the best suited spatial discretization method for the simulation of impacts and accidental conditions on human parts. Dynamic impact simulations can then be performed using the explicit direct integration in the time domain. Quasi-static simulations can be carried out via implicit solvers or by damped slow dynamic explicit solutions where the static solution is approached asymptotically in time.

In connection with the above mentioned CT scan geometry acquisition method, for example, preprocessors have been developed that connect 2D sliced discretizations with 3D solid or surfacic finite elements, see Figure 1. Similarly, such preprocessors might be applied to data acquired from the process of mechanical slizing or from exposed sections of parts.

External laser scanning, e.g., can only lead to surface data, that must be meshed to accomodate triangular and quadrilateral surfacic finite elements. The surface CAD data furnished by Viewpoint Datalabs consists in IGES files containing 3 to n noded surface patches, with a majority of 4 and 3 node patches, that might directly serve as the geometry of 4 and 3 noded shell finite elements, Figure 5.

Furthermore, it may be necessary to adapt, modify or correct the original geometry data to obtain mechanically sensible finite element models. Such corrections have been made in particular on the joint contact surfaces of the human ankle model actually under development at ESI, where the rough contact surfaces of the original geometry were interpolated and refined via smooth and congruent surface splines using the CAD package IDEAS™. The so obtained smooth and congruent patches of bone can now serve as adequate supports for contact interfaces of the mechanical solver codes, Figure 6.

4. Mechanical Properties of Human Tissues

While there is certainly a considerable scatter in the geometry of human parts between individuals, there may be as large a scatter and uncertainty in the material properties of human tissue, according to sex, age, size, etc. Furthermore, there may be large differences found, e.g., in bone material properties within a given individual at different parts of the skeletton and within a given bone (spongy and compact portions). The existing literature on the mechanical properties of human tissues is abundant, but relatively scarce when one looks for converged, exact, comprehensive and representative data.

Viano [6], then at General Motors Research Laboratories, Biomechanical Science Department, gives a detailed review of information available in 1986 on *"biological structures, material properties and failure characteristics of bone, articular cartilage, ligament and tendon. The load-deformation response of biological tissues is presented with particular reference to the microstructure of the material. Although many of the tissues have been characterized as linear, elastic and isotropic materials, they actually have a more complicated response to load, which includes stiffening with increasing strain, inelastic yield, and strain rate sensitivity. Failure of compact and cancellous bone depends on the rate, type, and direction of loading. Soft biological tissues are visco-elastic and exhibit a higher load tolerance with an increasing rate of loading. The paper includes a discussion on the basic principles of biomechanics and emphasizes material properties and failure characteristics of biological tissues subjected to impact loading"*, from the Abstract of [6].

This author presents on more than 30 pages what should be known from an engineering point of view about biological tissues and what types of fibers (collagen), bulky tissue with visco-elastic properties, which can also consolidate (hyaline cartilage), and crystals in mineralized bone tissues (calcium), are responsible for the cohesion of the skeletton (ligaments), the attachment of muscles to the skeletton (tendons), the transmission of compression forces across joints (articular cartilage) and for maintaining the overall shape (bones). For each discussed material the paper describes its biological microstructure and composition, it discusses laboratory setups for material testing, it gives typical stress-strain samples and it outlines possible mathematical models to describe the measured properties up to rupture or fracture.

Other works (e.g. by Yamada [7]) contain global information about basic material properties, such as average Young's modulus or fracture strength that helps to obtain basic ideas about the mechanical properties of human tissues. Papers that give detailed data on the complete stress-strain behaviour, including rate effects, damage and rupture, as needed in modern computer simulation, are scarce. The numerical analyst is therefore constrained to use approximate, incoherent or incomplete data, or to order biomechanical material tests to be carried out by friendly laboratories. Even then is it clear that the information remains punctual and the relationship between "in vitro" (cadaver tests) and "in vivo" (life tests) remains uncertain in many cases.

Given all these predicaments, the reality of meaningful simulation of the mechanical behaviour of human tissue materials may seem far away. Promising head starts, nevertheless, are possible when one calibrates the applied material models from in vitro sample material tests from the same individual that furnished the parts for tests under selected accidental conditions. One can then be sure that the used material data correspond to the ones of the tested part and the tests are a meaningful basis for validating the numerical models. Given the same individual, the model can then be used to extrapolate its response to similar but different impact conditions, and valuable information can be obtained from parametric studies, e.g., for investigation of the influence of design changes of the car interior on head impacts or on ankle injuries. It will require a large amount of international collaboration, however, before accepted sets of "standard" values of biomechanical data will be available, such as the averaged "50-th percentile" male or female human.

One example of the resulting stress-strain curves obtained for a ligament in a human foot is given next. The tests were carried out by Professor Begeman at Wayne State University [8]. The tests are tensile tests on an isolated ligament where the bony attachments were preserved, all other connections between the two bones were eliminated, the bones were cast into medical cement and the cemented bones where gripped in a tensile test machine and tested at different load levels and deformation velocities. The resulting force-displacement curves, Figure 7, exhibit linear and nonlinear loading behaviour, visco-elastic hysteresis effects in the lower ranges of loading, irreversible fiber damage effects in the upper ranges and finally rupture. A creep test and a sustained vibration test were also carried out.

Knowing the cross sectional area of the tested ligament it is now possible to evaluate an average stress-strain behaviour from which the parameters of a suitable material model can be calibrated. Alternatively, the entire ligament may be modeled

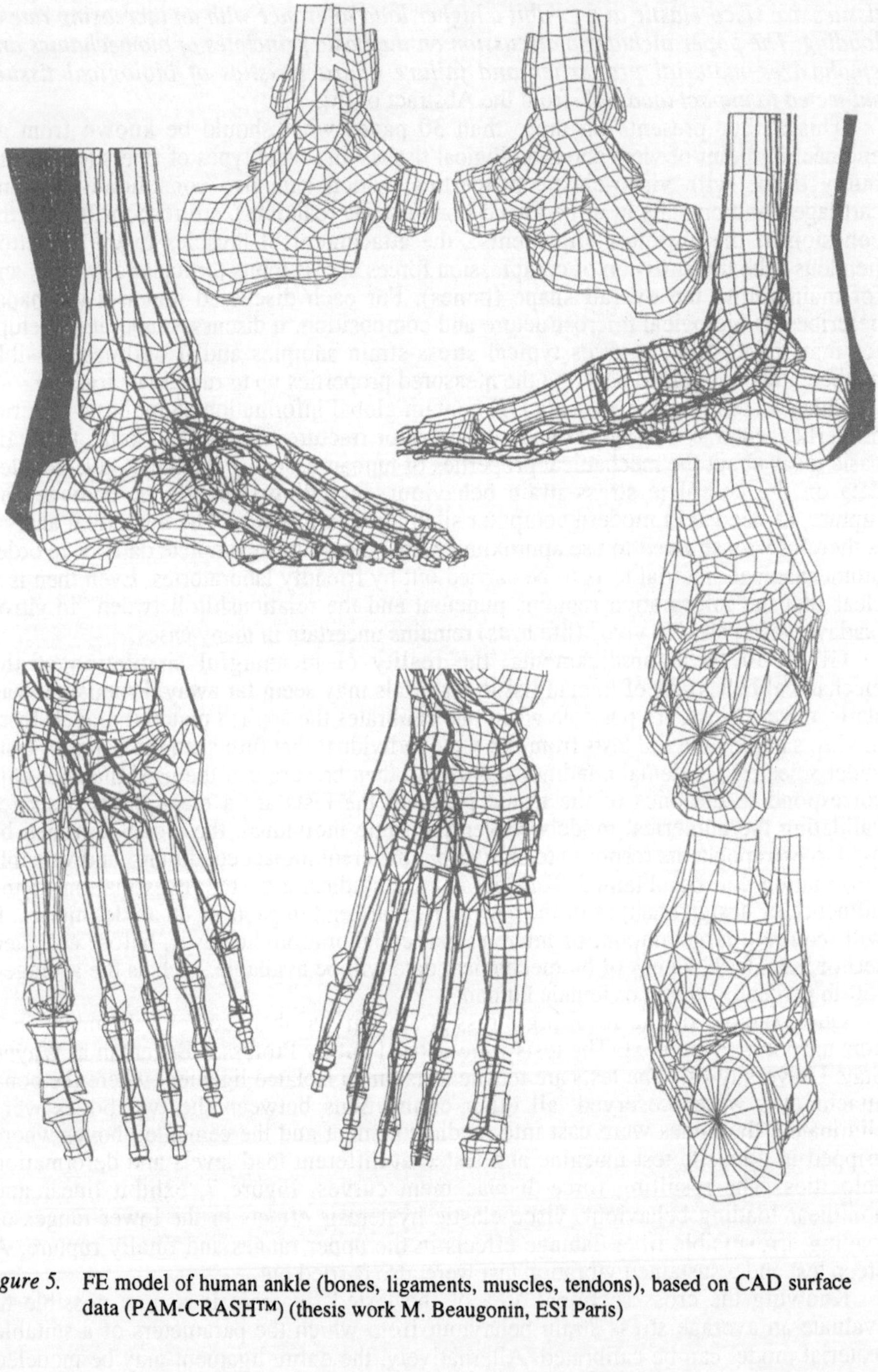

Figure 5. FE model of human ankle (bones, ligaments, muscles, tendons), based on CAD surface data (PAM-CRASH™) (thesis work M. Beaugonin, ESI Paris)

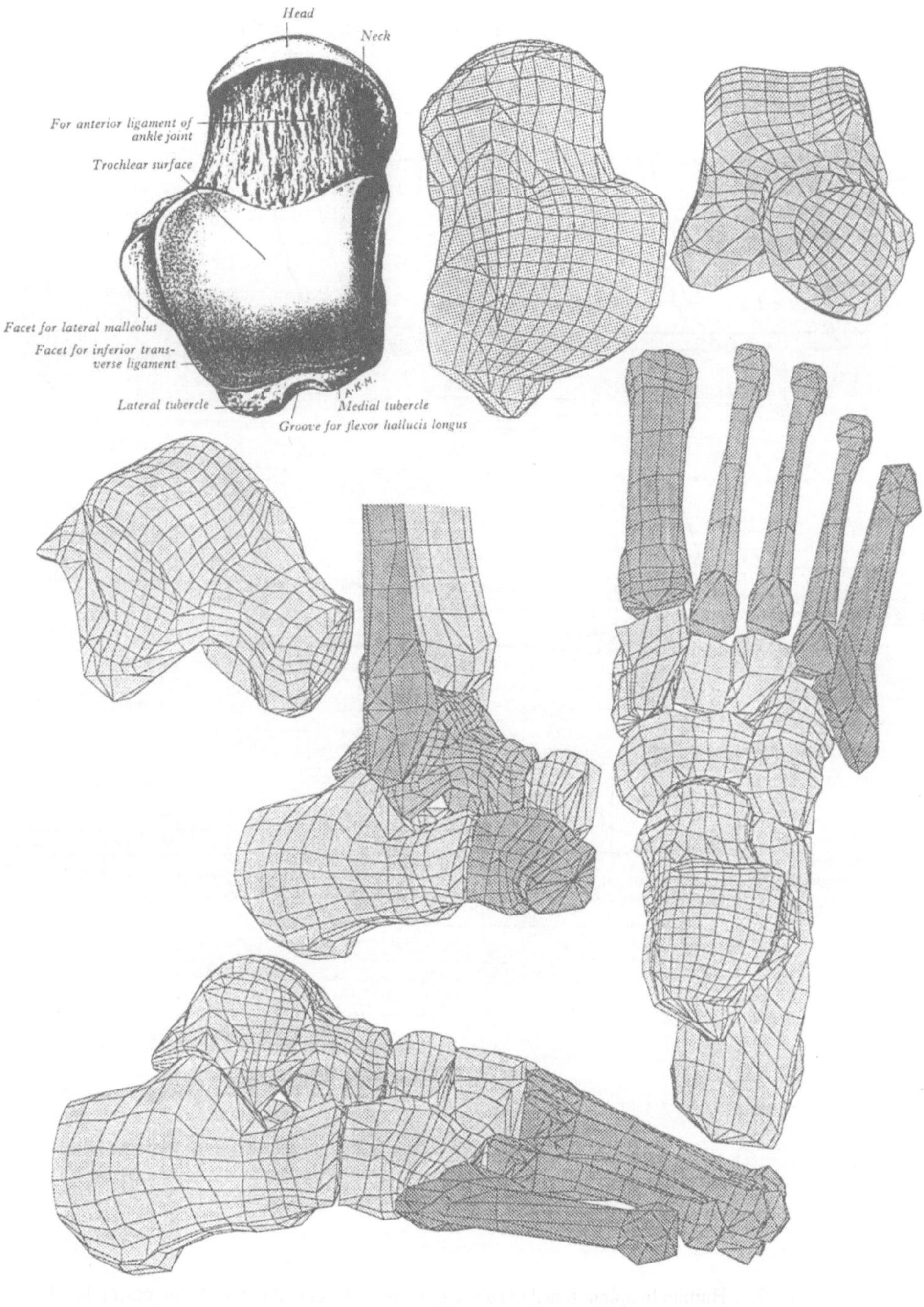

Figure 6. Improved joint surfaces in human ankle FE model (PAM-CRASH™), (inset from Gray's Anatomy) (thesis work M. Beaugonin, ESI Paris)

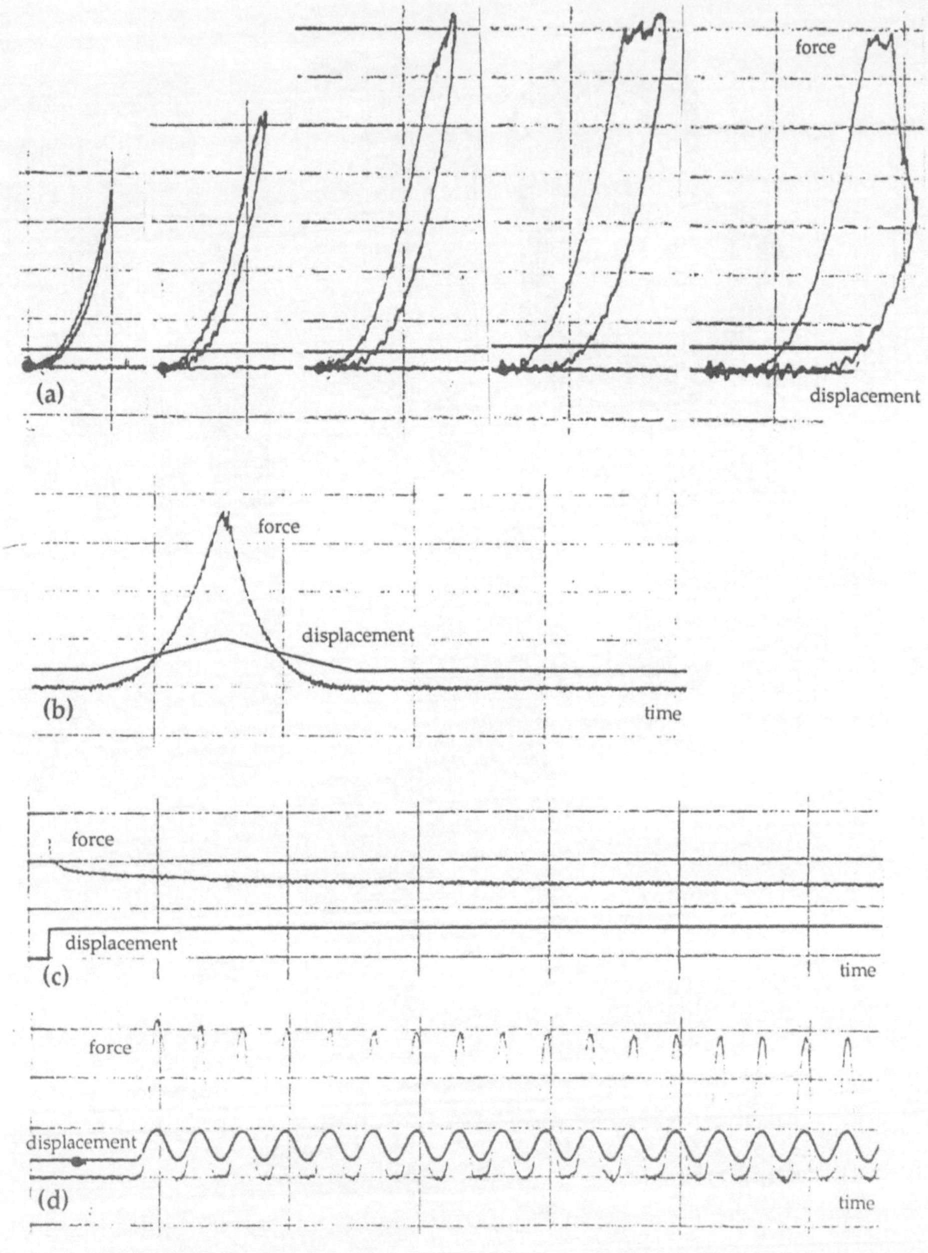

Figure 7. Human ligament tensile tests : (a) load-displacement curves at increasing level, (b) load time and displacement time curve, (c) creep test, (d) vibration test (courtesy : Wayne State University, Prof. Begeman)

by an equivalent nonlinear bar element, whereupon the experimental force-displacement relationships can be used without reduction. Proceeding this way enables the analyst to work with plausible data, where the behaviour of similar ligaments in the same part is extrapolated by adjusting to their respective cross sectional areas.

5. Currently Developed Biomodels using PAM-CRASH™

The remainder of this overview paper briefly indicates biomodels of human parts that are currently existing or under development, mostly using the PAM-CRASH™ software. The models are developed by collaborating institutes such as Wayne State University, INRETS Lyon, LBA Marseille, Technical University Berlin and by ESI Paris. For modeling details, material data and results the reader is invited to consult the cited references.

5.1. HUMAN HEAD MODEL

Figure 8 shows the evolution of a human head finite element model over a time span of about five years, References [9] to [13]. The latest model presents the most detail, Figure 9. Figure 10 presents results of impact tests obtained with this model. All models were developed and applied by WSU and recent models are used and jointly further developed by ESI in collaboration with WSU. Applications carried out by ESI concern the defense related study of skull and brain injuries and helmet protection and applications in airplane crash and car crash are under discussion with aircraft and car manufacturers. A possible application for response evaluation under strong electromagnetic wave sollicitation is also envisaged, where the mechanical properties of the biological tissues are replaced by their electrical properties.

5.2. HUMAN NECK MODEL

Figure 11 shows a finite element model of the human cervical spinal column that has been developed at the Technical University Berlin [14]. The model has been subjected to various loading conditions and some results are presented in Figure 12.

5.3. HUMAN THORAX MODEL

Figure 13 presents aspect and response under a lateral pendulum impact of a human thorax model that was developed at Wayne State University [15]. This model contains all important skeletton parts and the major organs are modeled in sufficient detail, so as to be able to predict internal injuries such as rupture of the aorta. The model was fabricated by Kevin Wang from measurements and from drawings of cross sections found in altases of the human anatomy.

5.4. HUMAN KNEE MODEL

Figure 14 shows the geometry of a human knee model prepared at INRETS Lyon [16], [17], that was subjected to lateral impacts typical in pedestrian impact with the bumpers of passenger cars, and it indicates the response of the model to an impact. Only the major skelettal and ligamental parts have been modeled, and the response of the model was reported to correlate well with test results.

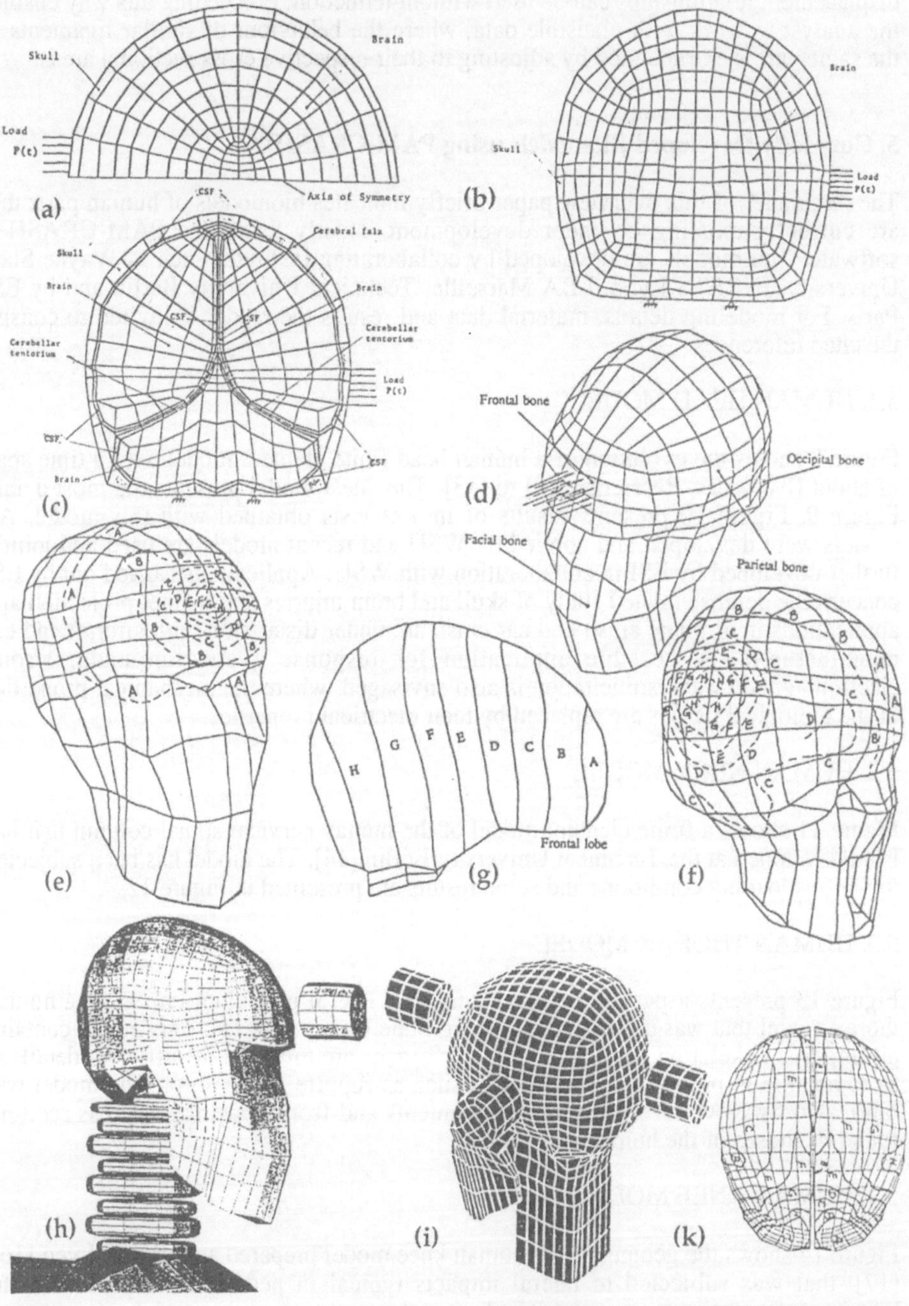

Figure 8. Evolution of human head FE models : (a) axisymmetric, (b) plane strain, (c) plane strain with anatomical details, (d) 3D model (1110 solids), (e) von Mises stresses in skull for frontal impact, (f) for rear impact, (g) pressure contours for frontal impact, (h) 3D FE head model with articulated neck and dummy (5456 solids, 1895 shells), (i) detailed 3D FE head model (22995 elements), with (k) brain shear distributions for frontal impact (Wayne State University, Bioengineering Center) (after [9] to [13])

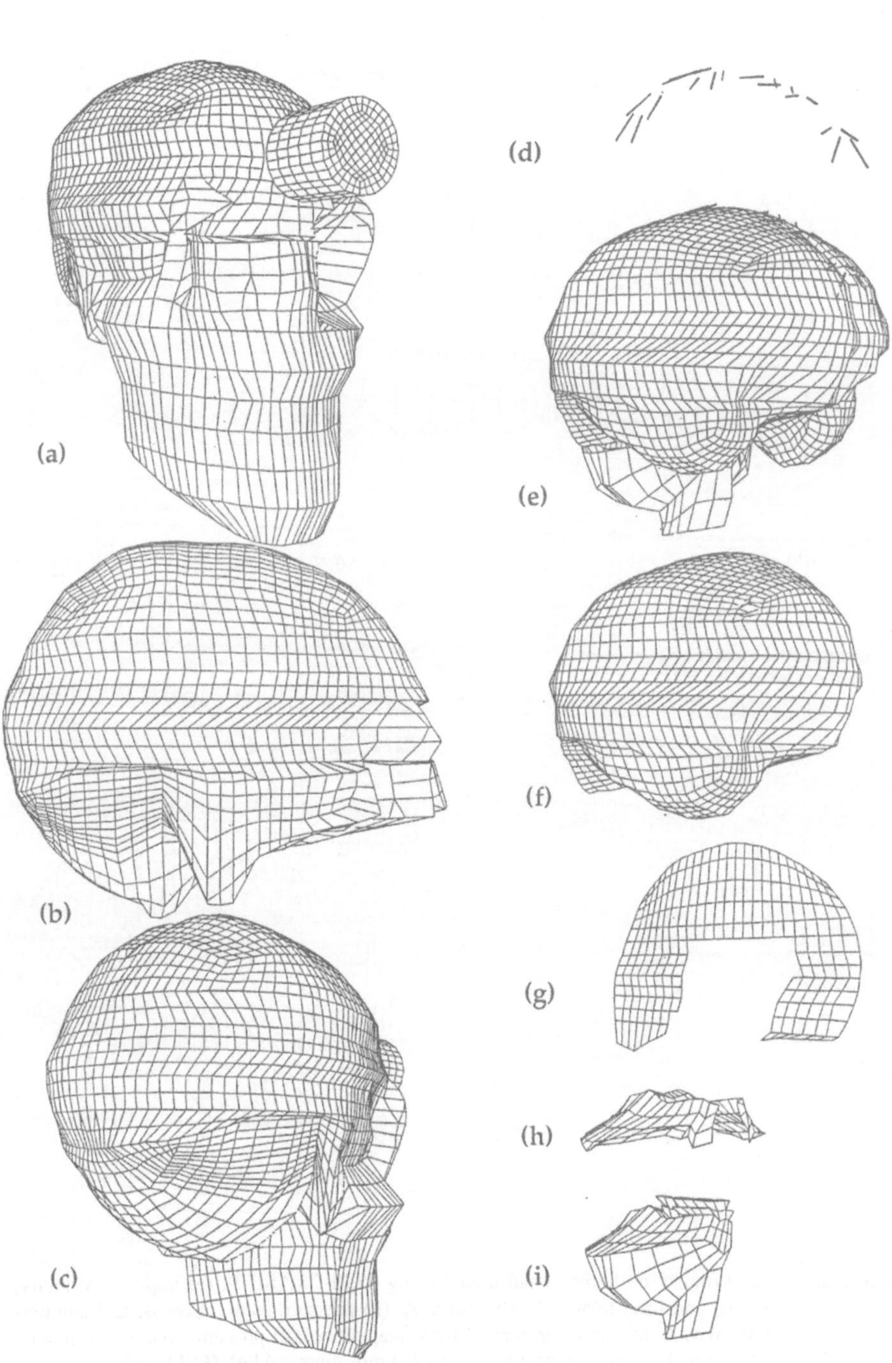

Figure 9. FE model of human head (15784 solids, 7609 shells, 20 beams, 22 materials), (a) frontal view with frontal impactor, (b) side view without facial bones, (c) rear view, (d) bridging veins, (e) pia mater and bridging viens, (f) gray and white matter (left), (g) falx cerebri, (h) tentorium cerebrelli, (i) brain stem (Wayne State University and ESI Paris)

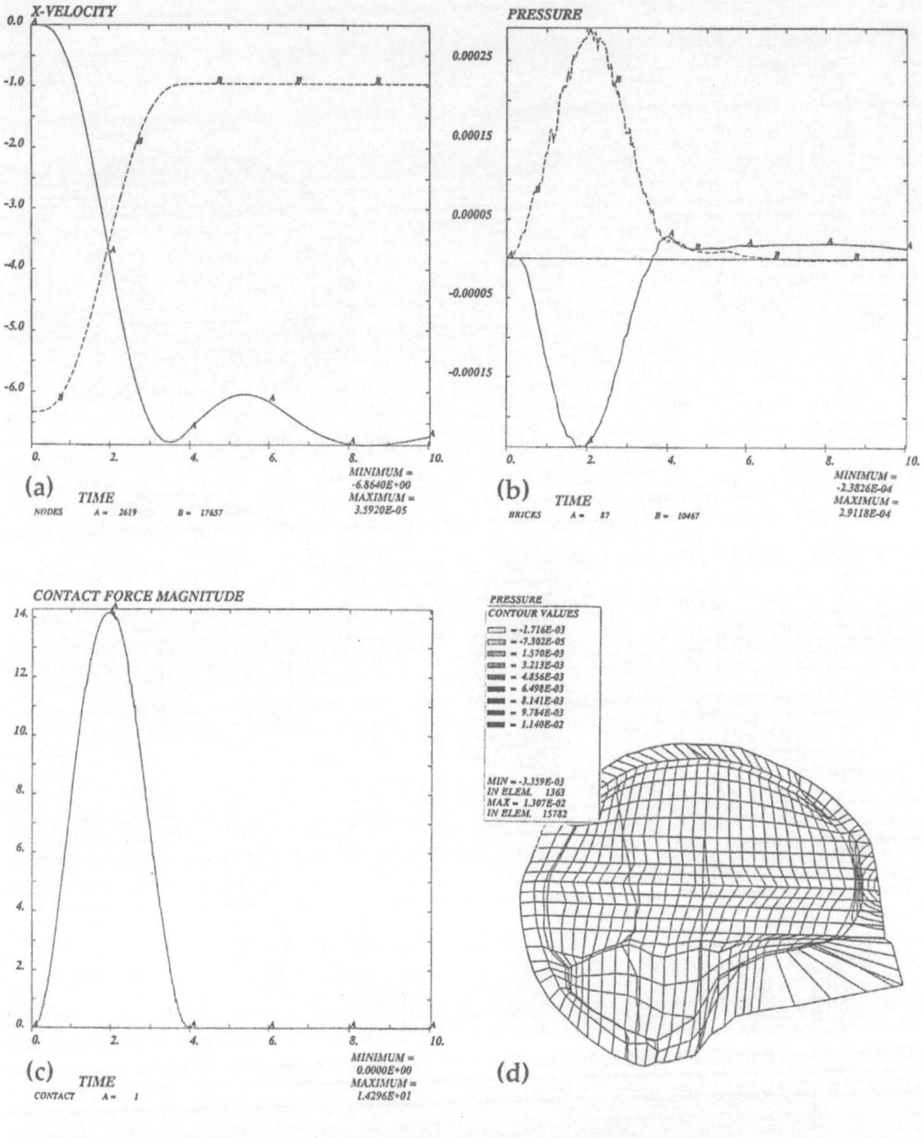

Figure 10. Frontal impact on human head model using PAM-CRASH™, (a) impactor velocity, curve B, and head node velocity, curve A, (b) pressure coup, curve B, and counter-coup, curve A, (c) impactor contact force magnitude, (d) pressure contours in medial plane (sagittal section) (impactor velocity 6.4 m/s, mass ~ 4 kg) (ESI Paris)

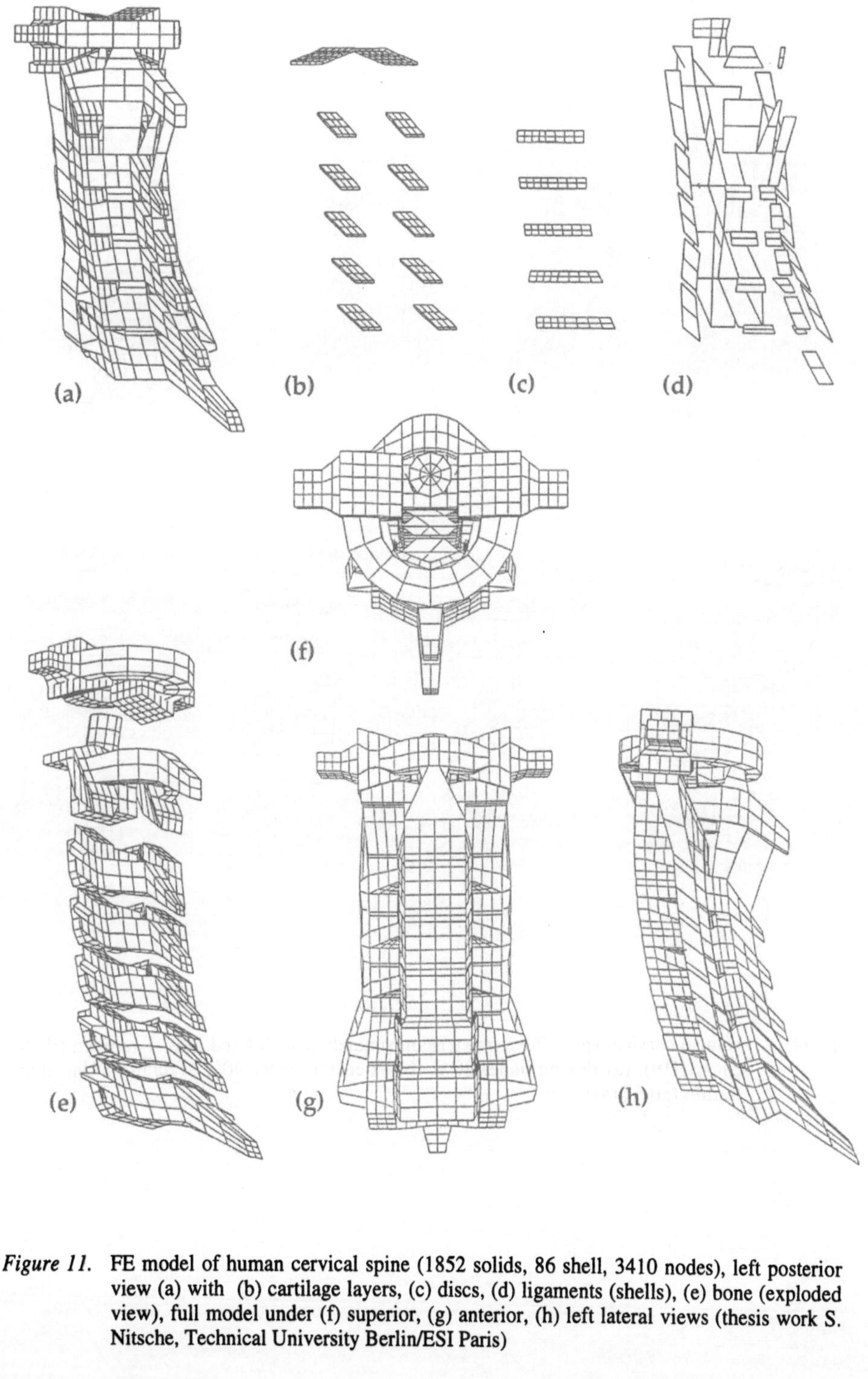

Figure 11. FE model of human cervical spine (1852 solids, 86 shell, 3410 nodes), left posterior view (a) with (b) cartilage layers, (c) discs, (d) ligaments (shells), (e) bone (exploded view), full model under (f) superior, (g) anterior, (h) left lateral views (thesis work S. Nitsche, Technical University Berlin/ESI Paris)

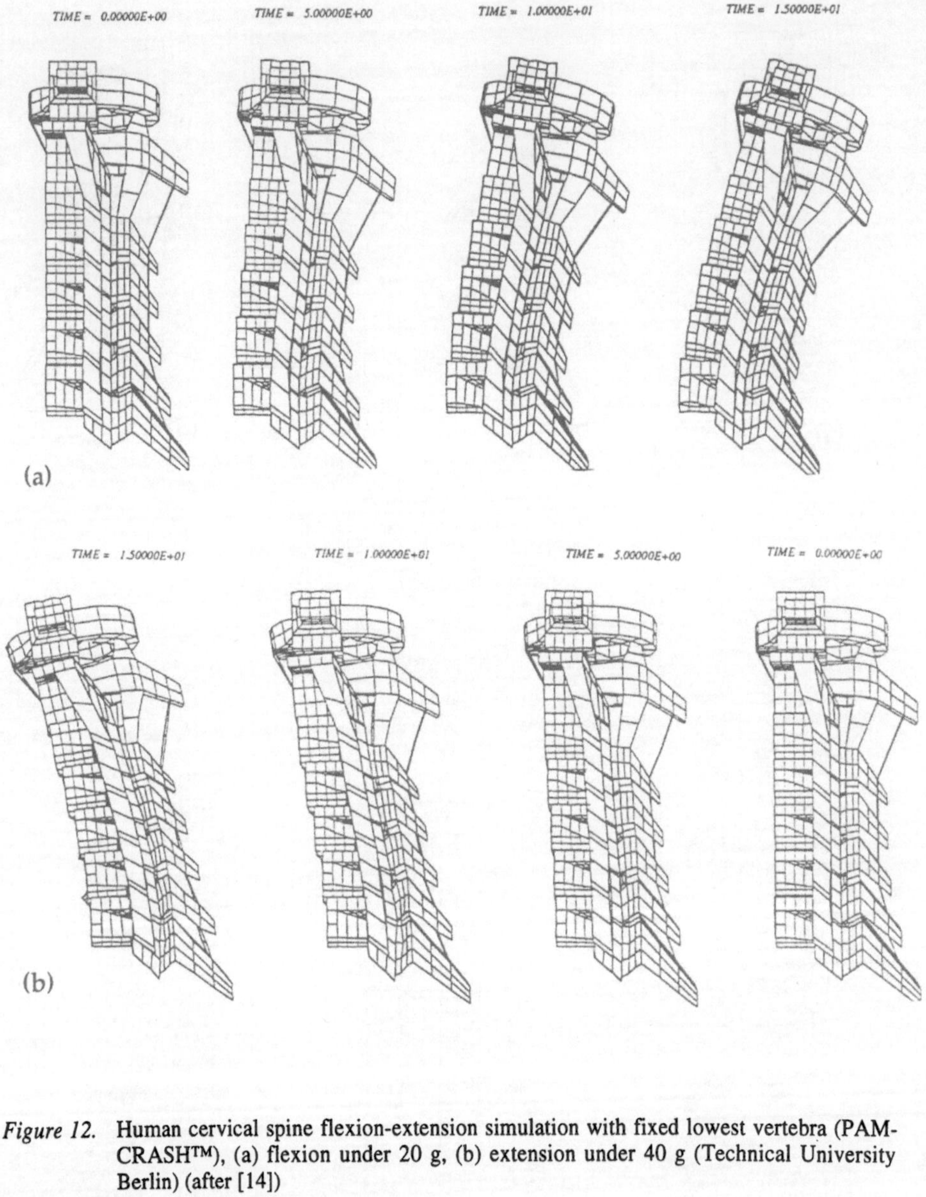

Figure 12. Human cervical spine flexion-extension simulation with fixed lowest vertebra (PAM-CRASH™), (a) flexion under 20 g, (b) extension under 40 g (Technical University Berlin) (after [14])

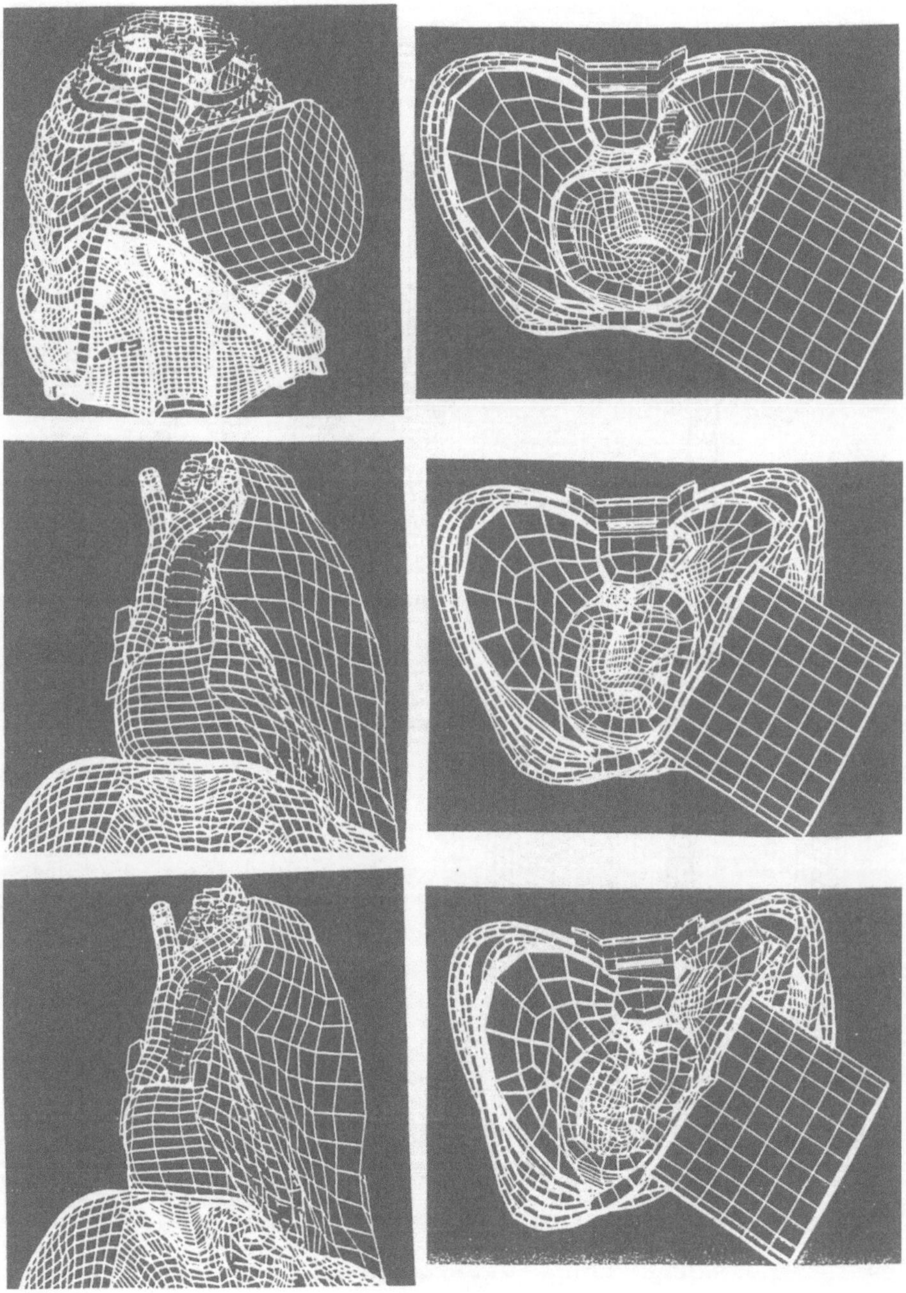

Figure 13. FE model of the human thorax under lateral impact (courtesy : Wayne State University)

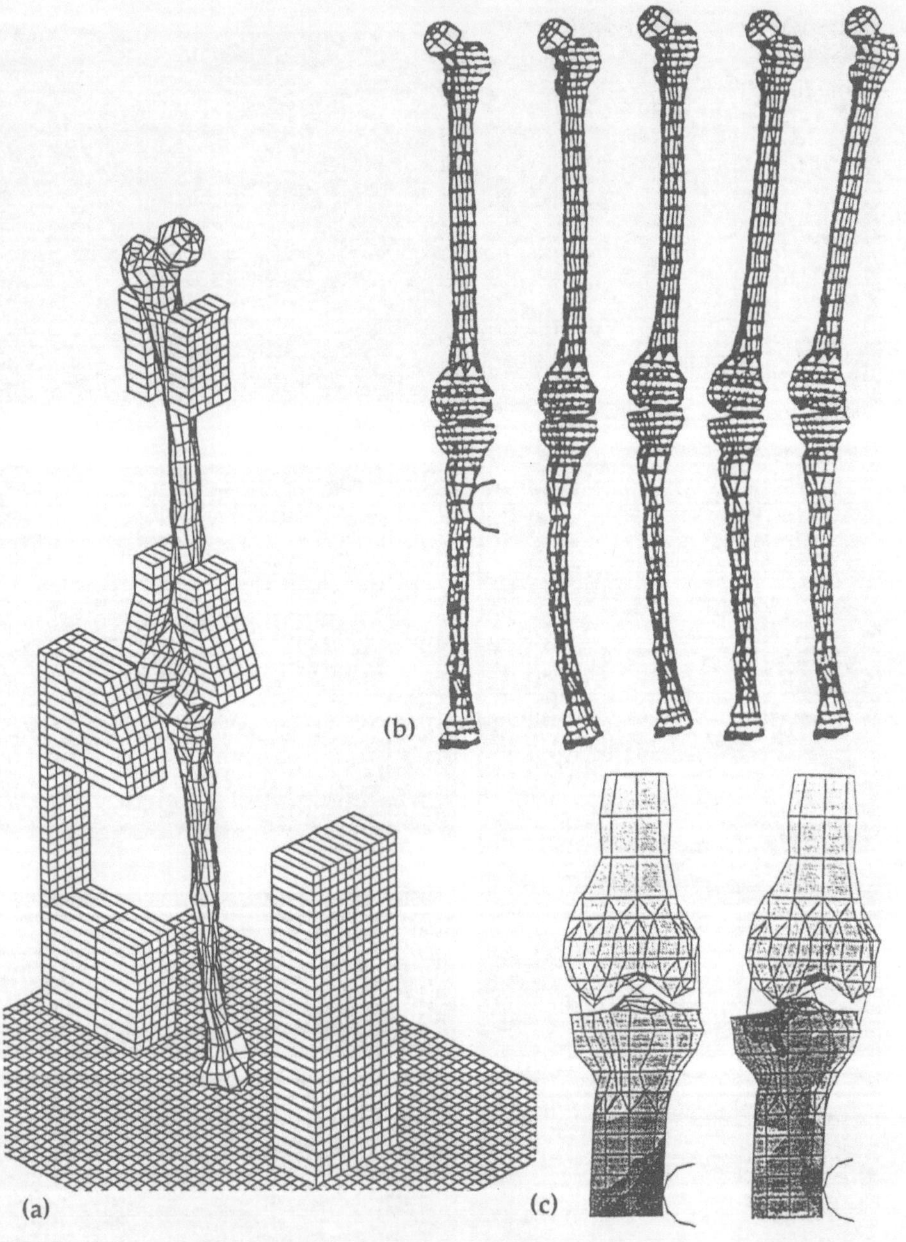

Figure 14. FE model of the human knee and leg under pedestrian impact with PAM-CRASH™ (1200 shells in the skeletton bones) (courtesy : INRETS Lyon, F. Bermond), (a) model and impactor setup, (b) simulation of an impact (after [17]), (c) details of relative motion between tibia and femur showing ligaments

5.5. HUMAN ANKLE MODEL

This model is actually under development at ESI Paris, where a Ph.D. thesis (M. Beaugonin) is under preparation under the general supervision of Dr. D. Cesari, INRETS Lyon and where Prof. Bonnoit, INRETS Marseille provides the facilities of his institute (LBA, Laboratory of Applied Biomechanics Hospital Nord) for carrying out the necessary disections, slicing operations, data acquisition and mechanical in vitro testing for geometries, biomaterials and model calibration and validation.

Figure 15 contains the major ingredients of the model including all relevant skeletton bones (rigid and deformable bodies), ligaments (shell and bar models), muscles and tendons (deformable). The geometry of the model was based on CAD IGES outer surface data of a human ankle, acquired from Viewpoint Datalabs. The mechanical properties of bones, ligaments, tendons and muscles were drawn from test results reported in the literature and from ligament test results provided by Wayne State University, Professor Begeman [8].

A recent Ph.D. thesis on the mathematical modeling and material data calibration on skeletton bone at the Technical University of Lausanne (ETL) was found helpful to confirm data and to understand the mechanical behaviour of bone [18]. Figure 15 also contains first simulation of dynamic tests of dorsiflexion carried out at WSU.

5.6. HUMAN EYE MODEL

At the time of writing Nihon ESI performs a study of an impact simulation of projectiles emanating from industrial machinery (grinders, cutters, etc.). Figure 16 shows the geometry and a schematic finite element model of the human eye, by which simulations of the impact of splinters on the surface of the eye will be carried out.

5.7. HUMAN SKIN MODELS

The excision of cutaneous lesions or malformations demands careful planning of their shape and direction, so that subsequent closure leads to minimal disturbance of the skin surface, taking into account the locally more or less pronounced nonlinearity and anisotropy of the skin. Figure 17 schematically shows a lens shaped excision before and after closure and the deformation of a grid as painted on the skin before the operation. The "ears" that appear near the tips of the excision hole after closure are due to local compressive stresses in the skin, and they should be minimized. The stress pattern is shown as von Mises stress contours.

6. Conclusion

A brief overview on problems and solutions related to the establishment of biomechanical models and computer simulations of accidental conditions using such models in transport industries and in national defense was given. It is felt that biomodeling and simulations are now on the verge to make their way into industry, where major industry research departments are actively pursuing this subject. Being involved in studies, code and compute model provision in transport vehicle crashworthiness and occupant protection, ESI Group recognizes the feasibility and the potential of biomodels for the near future and invests time and effort in the establishment of appropriate simulation knowhow, software and compute models.

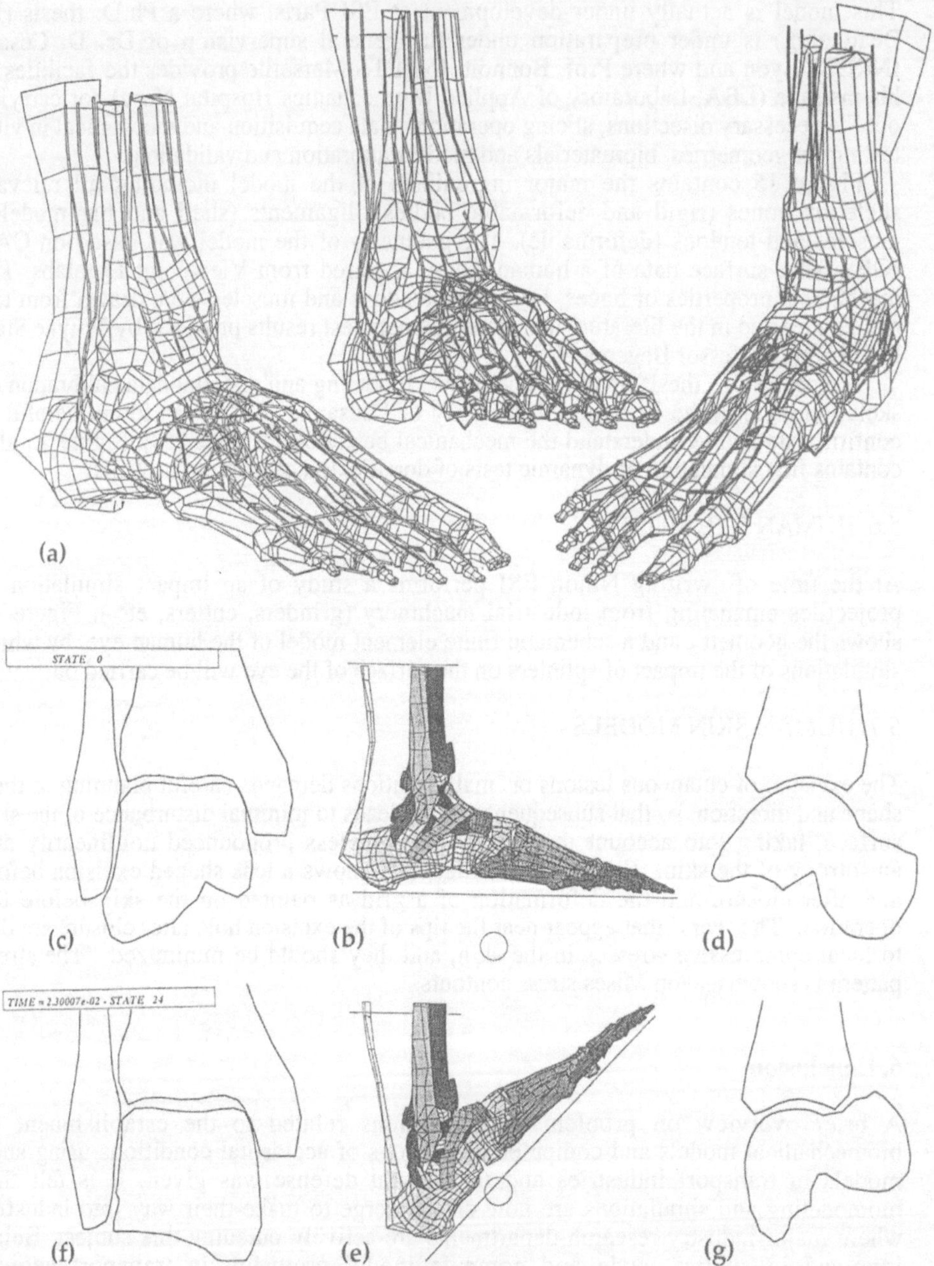

Figure 15. **Human Ankle FE model and dorsiflexion impact simulation (3670 shells, 412 beam/bar elements, 13 rigid bodies, 12 joints, 84 sliding interfaces, 46 sliprings), (a) different views with skeletton bones, ligaments, tendons, muscles and retinaculae, (b) initial position in dorsiflexion impact test simulation with coronal (c) and sagittal sections (d) through the ankle joint (tibia, fibula, talus), (e) final position at about 45° flexion angle with coronal (f) and sagittal (g) sections (thesis work M. Beaugonin, ESI Paris)**

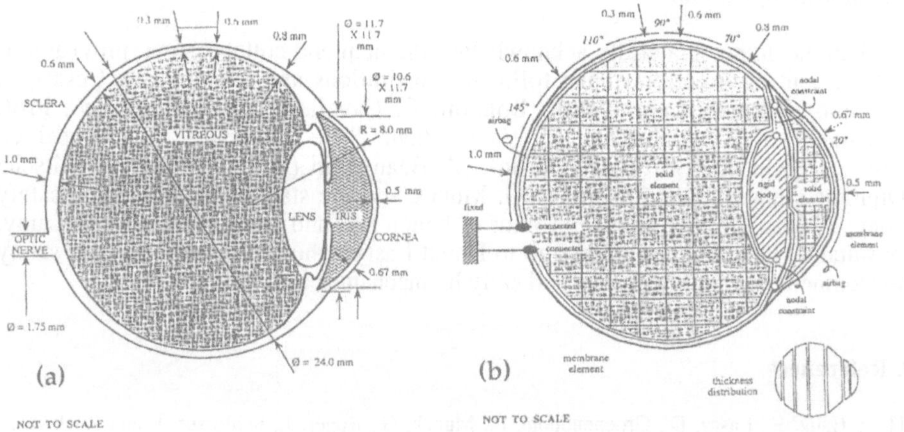

Figure 16. FE model of the human eye for projectile impact simulation from industrialmachinery with PAM-CRASH (Nihon ESI), (a) geometry, (b) FE model (schematic)

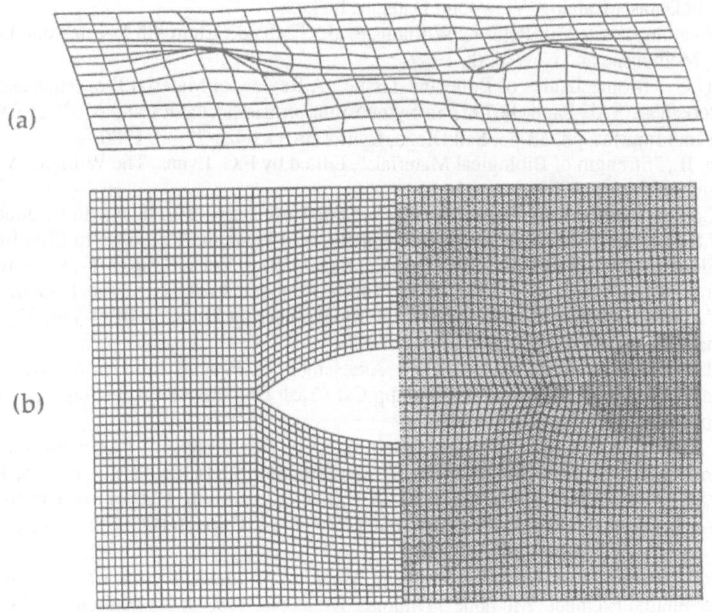

Figure 17. FE model of the human skin and simulation of the closure of a lens shaped excision, (a) deformation of a macro grid painted on the skin, (b) excision hole before (left) and after (right) closure with von Mises stress contours (right)

7. Acknowledgements

The humble author gratefully acknowledges the help and collaboration, provision of material and information of the following institutions and persons : D. Cesari, F. Bermond, INRETS Lyon ; Prof. Bonnoit, C. Cavallero, LBA Marseille ; Prof. Begeman, Prof. King, K. Yang, K. Wang, Wayne State University ; Prof. Appel, G. Krabbel, Technical University Berlin ; M. Beaugonin (Ph.D. student), S. Nitsche (Diploma student) at ESI Paris ; G. Munck and the staff of the ESI Paris safety group; T. Kisielewicz and Allen Chhor of Nihon ESI and Pacific ESI, and R. Lamy, consultant. Our particular thanks go to David Lasry, who set up our occupant safety development and analysis group and early biomechanics activities.

8. References

[1] Haug, E., Lasry, D., Groenenboom, P., Munck, G., Roger, J., Schlosser, J. and Rückert, J., "Finite Element Models of Dummies and Biomechanical Applications using PAM-CRASH", PAM-User's Conference in Asia ; PUCA'93, pp. 123-158, Shinyokohama, Japan, Nov. 1993.

[2] Kisielewicz, T.K. and Andoh, K., "Critical Issues in Biomechanical Tests and Simulations of Impact Injuries", PAM-User's Conference in Asia, PUCA'94, Shinyokohama, Japan, pp. 209-222, November 1994 and World Congress on Computational Mechanics, WCCM'94, Makuhari, Japan, August 1994.

[3] Krabbel, G., Nitsche, S. and Appel, H., "Development of an Anatomical 3D FE-Model of the Human Head Utilizing CT-Data", 14-th ESV Conference, Paper Nb. 94-S1-O-12, Munich, May 23-26, 1994.

[4] Viewpoint Dataset Catalog, Viewpoint Datalabs Int'l.

[5] Gray's Anatomy, edited by Peter L. Williams et al., 37-th ed., Churchill Livingstone Edinburg, London, Melbourne and New York, 1989.

[6] Viano, D.C., "Biomechanics of Bone and Tissue : A Review of Material Properties and Failure Characteristics", SAE Paper 861923 in Symposium on Biomechanics and Medical Aspects of Lower Limb Injuries, pp. 33-63, San Diego, California, October 29-30, 1986.

[7] Yamada, H., "Strength of Biological Materials", Edited by F.G. Evans, The Williams & Wilkins Company, Baltimore, 1970.

[8] Direct Communication Prof. Begeman, Wayne State University, Detroit, Michigan, June 1995.

[9] Ruan, J.S., Khalil, T.B. and King, A.I., "Human Head Dynamic Response to Side Impact by Finite Element Modeling", J. of Biomechanical Engineering, Vol 113, pp. 276-283, Aug. 1981.

[10] Ruan, J.S., Khalil, T.B. and King, A.I., "Finite Element Analysis of the Human Head to Impact", Winter Annual Meeting of ASME, Bioengineering Division BED-Vol. 22, pp. 249-252, Anaheim, California, November 8-13, 1992.

[11] Ruan, J.S. and Prasad, P., "Head Injury Assessment in Frontal Impacts by Mathematical Modeling" ; SAE Paper 942212, 38-th Stapp Car Crash Conference Proceedings, pp. 111-121, November 1994.

[12] Ruan, J.S., Khalil, T.B. and King, A.I., "Brain Injury in Direct Impact", Proceedings of the 4-th Injury Prevention Through Biomechanics Symposium, pp. 133-141 and BED-Vol. 28, 1994.

[13] Chun Zhou, Khalil, T.B. and King, A.I., "A Human Head Finite Element Model for Impact Injury Analysis", Proceedings of the 5-th Injury Prevention Through Biomechanics Symposium, pp. 137-147, Detroit, May 4 and 5, 1995.

[14] Rade, R., "Aufbau eines Dreidimensionalen Modells der Menschlichen Halswirbelsäule mit Hilfe der Finiten Elemente Methode", Diploma Thesis Nb. 1193 at the Institute for Automotive Engineering, Technical University Berlin, 1993.

[15] Wang, K.H.C. and Yang, K.H., "The Development of a Finite Element Human Thoracic Model", Proceedings of the 5-th Injury Prevention Through Biomechanics Symposium, pp. 129-135, Detroit, May 4 and 5, 1995.

[16] Bermond, F., Ramet, M., Bouquet, R. and Cesari, D., "A Finite Element Model of the Pedestrian Knee Joint in Lateral Impact", Proceedings of the 1993 International IRCOBI Conference on the Biomechanics of Impacts, pp. 117-129, Eindhoven, September 8-10, 1993.
[17] Bermond, F., Bouquet, R., Caire, Y., Ramet, M. and Maistre, O., "Modélisation Mathématique de la Jambe d'un Piéton lors de Chocs avec l'Avant d'une Voiture", XX-ème Congrès de la Société de Biomécanique, p. C-128, Lausanne, September 11 and 12, 1995.
[18] Zysset, P., "A Constitutive Law for Trabecular Bone", Thesis Nb. 1252 at Ecole Polytechnique Fédérale de Lausanne EPFL, Dept. de Génie Mécanique, Dr. A. Curnier, 1994.

PART IV
Structural Impact

DYNAMIC INELASTIC STRUCTURAL RESPONSE

NORMAN JONES
Impact Research Centre
Department of Mechanical Engineering
The University of Liverpool
Liverpool L69 3BX, U.K.

Abstract

The topic of structural impact is concerned with the response of structures subjected to dynamic loadings producing large inelastic strains and permanent deformations. This is relevant to safety calculations and energy absorption estimates for structural crashworthiness studies. The rigid-plastic method of analysis is introduced and used to examine the dynamic plastic response of a fully clamped beam struck by a rigid mass travelling with an initial impact velocity. This theoretical solution is used to assess the accuracy and range of validity of a quasi-static method of analysis which is often employed to simplify impact problems in the structural crashworthiness field. The dynamic axial crushing behaviour and energy absorbing capacity of a thin-walled tube is examined briefly. Also reported are the results of some recent studies on the influence of axial length on the response modes of various thin-walled sections subjected to axial impact loads. The chapter concludes with some brief comments on the large strain properties of materials under dynamic loads, the scaling of structures subjected to large impact loads and the dynamic inelastic failure of structures.

1. Introduction

Structural crashworthiness is concerned with the design of vehicles containing structural members and systems which are required to absorb the dynamic energies and loads arising during a collision or impact event, but in a controlled manner which minimises injury to any occupants [1-6]. The studies associated with the development of human injury criteria and the fundamental research into the dynamic response of structural members under large impact loads might be used for structural crashworthiness problems in a wide range of industries (e.g., aircraft, bus, car, marine and train).

Generally speaking, a designer, in the structural crashworthiness field, seeks the permanent deformations and the associated impact energy which must lie within specified limits of structural damage and human injury [4]. Clearly, the practical problems in this area are extremely complex. They involve not only dynamic effects, but often have

J. A. C. Ambrósio et al. (eds.),
Crashworthiness of Transportation Systems: Structural Impact and Occupant Protection, 263–290.

complicated structural geometries which undergo finite displacements and finite rotations with large inelastic strains, together with tearing and fracture of the material. Numerical schemes have been developed to examine some of the more important practical problems, but many are dogged by lack of experimental data [7].

Many papers have been written on the topic of structural crashworthiness [1,3, 5, 6], but considerable research work is still required, as noted, for example, in a recent railway accident enquiry [8]. Impact velocities in this field are often considered sufficiently slow so that the structural response can be taken as quasi-static even though material strain rate sensitivity effects may be important. The neglect of inertia simplifies a theoretical analysis, but considerable complexity still remains because large deflections and plastic strains may develop during the structural response. The axial crushing of thin-walled tubes is a simple and efficient way of absorbing impact energy and, therefore, is of particular interest in structural crashworthiness studies. The behaviour of thin-walled tubes with circular, square, rectangular, top-hat and double-hat cross-sections and subjected to axial loads has been of particular interest since the pioneering work of Pugsley on the axial impact of idealised railway coaches.

It is the aim of this article to equip the reader with an understanding of the impact behaviour of some simple structures. The dynamic response of the particular cases studied may be adequate to predict the response of various practical problems, particularly when recognising the lack of information on the impact loading characteristics, and the shortage of data on the properties of materials under dynamic loads. If simple methods of analysis are inadequate for a particular problem, then the understanding gained from simple calculations provides a foundation for a reader to make further progress with other solution strategies. In particular, this understanding and insight is indispensable for the efficient use and interpretation of numerical codes which play an increasingly important role in engineering design.

The rigid-plastic approximation, which has been developed for the static plastic behaviour of structures, is introduced in the next section. This method is also used to obtain the dynamic plastic response of structures. The static plastic collapse load is the largest possible external load which may act on a perfectly plastic structure according to the limit theorems of plasticity. Thus, a structure is not in static equilibrium for larger external loads, so that inertia forces are generated, and motion commences. This motion continues until all of the external energy has been consumed by internal plastic work. It is evident that the permanent displacements and response duration are of particular interest for dynamic loads.

The dynamic response of a fully clamped beam, which is struck by a rigid mass at the mid-span, is examined in Section 3. The energy consumed in the plastic deformation dominates the elastic energy, so that a rigid-plastic method of analysis, which neglects elastic effects, is suitable. However, in some practical situations, the inertia effects may be small so that the behaviour can be predicted using a quasi-static method of analysis. Thus, the quasi-static procedure is introduced in Section 4 and used to study the mass impact loading of the beam examined in Section 3. Particular attention is given to the accuracy of a quasi-static method of analysis compared with the exact theoretical solution and some guidelines are drawn on the likely range of validity.

The behaviour of a circular tube subjected to a dynamic axial load is studied in Section 5. This crushing load produces many axisymmetric convolutions or wrinkles, in the tube, and gives rise to a fluctuating resistance about a mean crushing force. This phenomenon is known as dynamic progressive buckling, because the deformations form progressively with time from one end of a tube. The transverse inertia forces in the wall of a tube are not significant, and are neglected, so that the mode of deformation is taken to be the same as for static loads. It transpires that considerable energy may be absorbed in an axially crushed tube before bottoming-out. Some experimental results for impact velocities up to 11.28 m/s on mild steel tubes are reported in References [9-11] together with relatively simple rigid, perfectly plastic analyses which are suitable for design use. A more recent study has been undertaken on the dynamic axial crushing of circular and square steel tubes having various cross-sectional dimensions and axial lengths [12]. The transition between initial global bending and progressive buckling was of particular interest in this work and is reported in Section 6. Another recently study [13] has examined the dynamic axial crushing of top hat and double-hat thin-walled sections made from mild steel and having several axial lengths. Some preliminary comments are reported in Section 7.

The topic of structural impact is a highly non-linear area. These non- linearities arise through the dynamic inelastic material properties, large strains and geometry changes which are exacerbated by the temporal dependence of all quantities. Although a considerable understanding exists already in this field, which contributes valuable information to present engineering design practice, there are, nevertheless, many outstanding issues requiring further study. In particular, some current studies are presented on dynamic material properties in Section 8 and recent work on relating the behaviour of small-scale models to the response of full-scale prototypes, or scaling, is discussed in Section 9. The dynamic inelastic failure of structures is introduced briefly in Section 10.

2. Rigid, Perfectly Plastic Method of Analysis

It was noted in the Introduction that this article is focused on the response of structures which are subjected to dynamic loads causing large inelastic strains and permanent deformations. In this circumstance, stress wave propagation effects may be considered independently of the structural behaviour because the associated time-scales differ, usually, by several orders of magnitude. Moreover, the external dynamic energy absorbed elastically is small compared with the total energy absorption through inelastic effects and, therefore, may be neglected. This leads directly to the idealisation of a rigid, perfectly plastic material for a ductile material, as shown in Figure 1.

A considerable body of literature has been published over the last 40 years on the behaviour of structures made from a rigid, perfectly plastic material and subjected to static loads [14-16]. For example, a plastic hinge would form at the mid-span of the simply supported beam in Figure 2(a) when subjected to a sufficiently large uniform static pressure, p, across the entire span, 2L. The beam would collapse under a pressure, p_c, which transforms it into a mechanism such as the one illustrated in Figure 2(b). Thus, equating the external work, $p_c(2L)(L\theta)/2$, to the internal energy consumed at the central hinge, $M_o(2\theta)$, gives the static collapse pressure

$$p_c = 2M_o/L^2 , \tag{1}$$

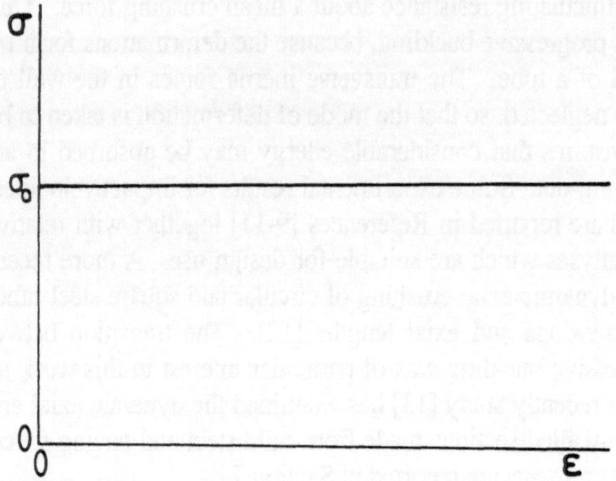

Figure 1. A rigid-perfectly plastic idealisation of the uniaxial stress-strain behaviour of a ductile material

where M_o is the plastic bending moment capacity of the beam cross-section and the transverse displacements in Figure 2(b) remain infinitesimally small.

A static analysis predicts that equation (1) gives the largest uniform pressure, which may be supported by the beam in Figure 2(a). A rigid, perfectly plastic beam remains rigid and undeformed for uniform pressures with $p < p_c$.

It is evident that a uniform pressure having a magnitude $p \geq p_c$ would cause the beam in Figure 2(a) to accelerate leading to large displacements and destruction. However, if the external pressure was active for a short time, τ, as indicated for the pressure pulse in Figure 3, then a finite amount of external work would be imparted to the beam while it was accelerating during the time τ. Energy would be absorbed by the central plastic hinge in the beam while it accelerates during the first phase of motion $0 \leq t \leq \tau$. In addition, energy would continue to be absorbed by the central plastic hinge during the second phase of motion, $\tau \leq t \leq T$, when the unloaded beam decelerates from the maximum transverse velocity at $t = \tau$ to a new permanently deformed equilibrium position at $t = T$ when motion ceases.

The literature which has been published on the theoretical analysis of rigid, perfectly plastic beams subjected to dynamic loads has incorporated all of the basic assumptions which have been developed for the static loading of beams [2, 17, 18]. However, it is necessary, of course, to obtain the time-dependence of the response from the governing equilibrium equations in order to predict the maximum permanent transverse displacement profile and the response duration which are of principal interest.

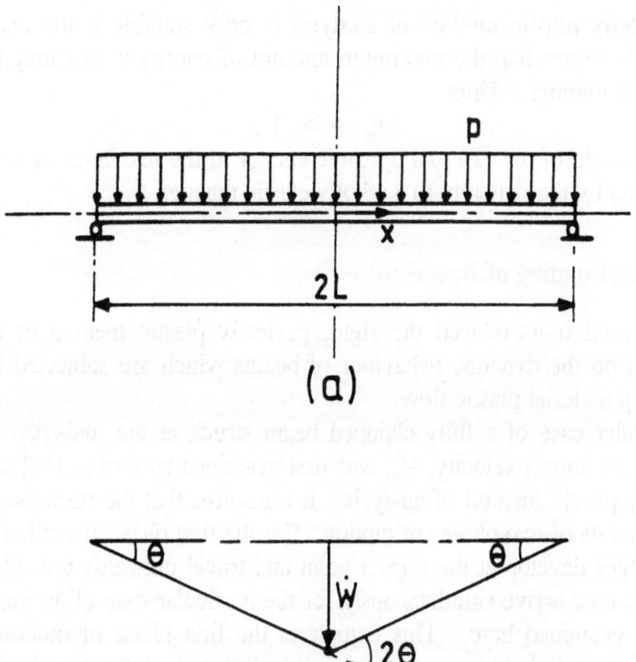

Figure 2 **(a) Simply supported beam subjected to a uniformly distributed pressure across the entire span; (b)Transverse velocity profile for a simply supported beam with a plastic hinge at the mid-span**

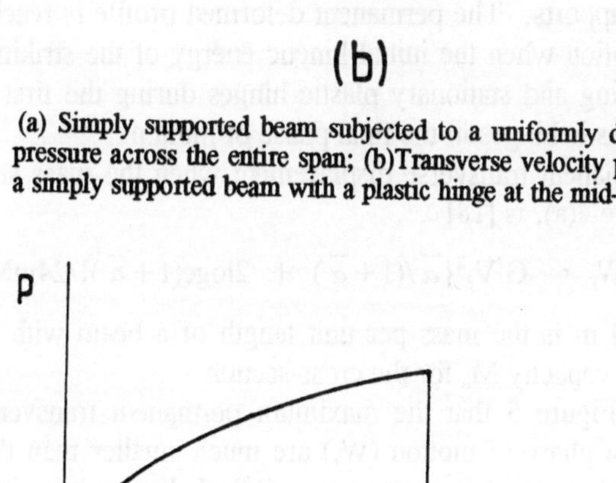

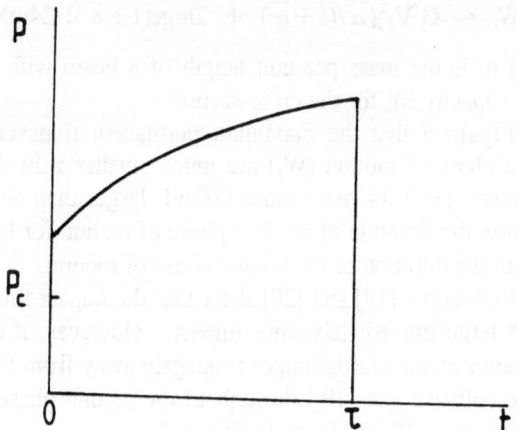

Figure 3. **Idealised dynamic pressure pulse**

A rigid, perfectly plastic method of analysis is only suitable if the external dynamic energy is much larger than the maximum amount of energy which may be absorbed in a wholly elastic manner. Thus,

$$E_r >> 1 , \tag{2}$$

where E_r is the ratio of the external dynamic energy to the maximum strain energy which may be absorbed by the structure in a wholly elastic manner.

3. Mass Impact Loading of Beams

The previous section introduced the rigid, perfectly plastic method of analysis. This section focuses on the dynamic behaviour of beams which are subjected to large impact loads producing material plastic flow.

The particular case of a fully clamped beam struck at the mid-span by a mass, G, travelling with an impact velocity, V_o, was first examined by Parkes [18] with the aid of a rigid, perfectly plastic method of analysis. It transpires that the response of the beam in Figure 4(a) consists of two phases of motion. For the first phase of motion in Figure 4(b), two plastic hinges develop at the impact point and travel outwards towards the respective supports where they arrive simultaneously for the particular case of an impact at the mid-span which is examined here. This completes the first phase of motion which is then followed by a second phase of motion in Figure 4(c) with stationary plastic hinges at the mid-span and at both supports. The permanent deformed profile is reached at the end of the second phase of motion when the initial kinetic energy of the striking mass has been absorbed by the travelling and stationary plastic hinges during the first phase of motion and by the stationary plastic hinges in the final phase of motion.

The maximum permanent transverse displacement when the mass strikes at the mid-span, as shown in Figure 4(a), is [18]

$$W_f = G^2V_o^2\{\bar{\alpha}/(1+\bar{\alpha}) + 2\mathrm{loge}(1+\bar{\alpha})\}/24mM_o , \tag{3}$$

where $\bar{\alpha} = mL/G$ and m is the mass per unit length of a beam with a span 2L and a plastic bending moment capacity M_o for the cross-section.

It is evident from Figure 5 that the maximum permanent transverse displacements acquired during the first phase of motion (W_1) are much smaller than those accumulated throughout the second phase (W_2) for mass ratios G/2mL larger than about 5 to 10. This observation occurs because the duration of the first phase of motion for large mass ratios is very short compared with the duration of the second phase of motion.

The calculations in References [19] and [20] show that the impact force underneath the striker (P) is infinitely* large immediately after impact. However, it decreases quickly with time for all mass ratios as the plastic hinges propagate away from the impact position and is close to the static collapse force (P_c) throughout the second phase phase of motion for mass ratios larger than about 10, as shown in Figure 6.

* In fact, the magnitude P cannot exceed the force $2Q_o = 2\tau_0$ BH for a beam with a rectangular cross-section when the transverse shear force is retained in the yield condition[2], unlike the theoretical study of Parkes [18]

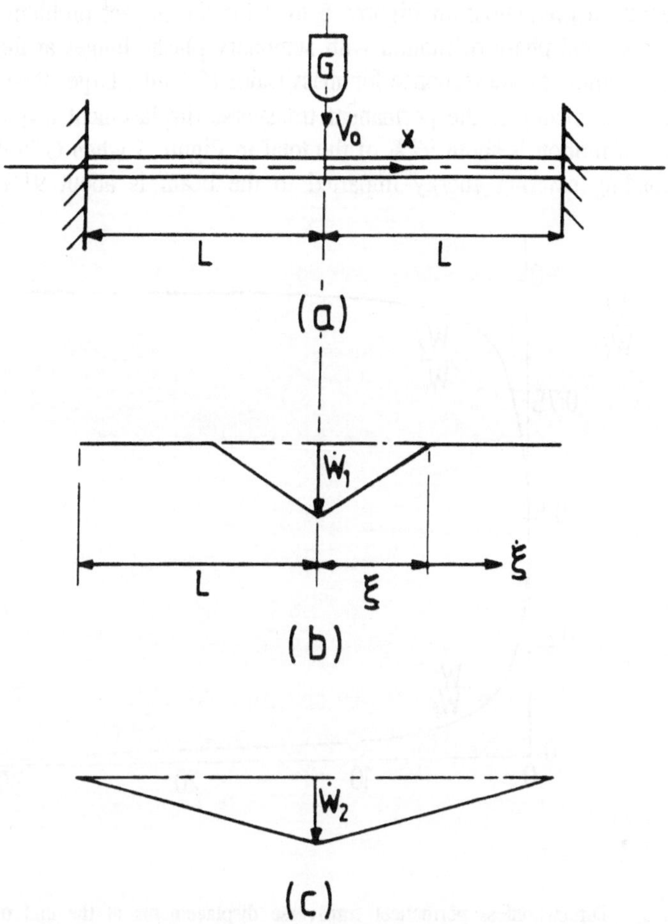

Figure 4 (a) A beam which is fully clamped across a span 2L, struck by a mass, G, travelling with an initial velocity, V_o. (b) Transverse velocity profile during the transient phase of motion with travelling plastic hinges at $x = \pm \xi$ and a stationary plastic hinge at the mid-span. (c) Transverse velocity profile during the final phase of motion with stationary plastic hinges at both supports and at the mid-span.

The external energy, which is imparted to the beam in Figure 4(a) during the second phase of motion ($E_e = \int PdW$), is shown in Figure 7. A significant amount of energy is imparted to a beam during the first phase of motion when the mass ratio is small. However, most of the external energy is imparted during the second phase of motion for large mass ratios notwithstanding the large impact forces immediately after impact.

The theoretical predictions in Figures 5 to 7 for the impact problem in Figure 4(a) show that the second phase of motion with stationary plastic hinges at the mid-span and both supports dominates the response for mass ratios (G/2mL) larger than about 10 (i.e., $\bar{\alpha} < 0.05$). For example, the permanent transverse displacement acquired during the second phase of motion is about 97% of the total in Figure 5 when G/2mL = 10, while the corresponding external energy imparted to the beam is about 91% of the initial

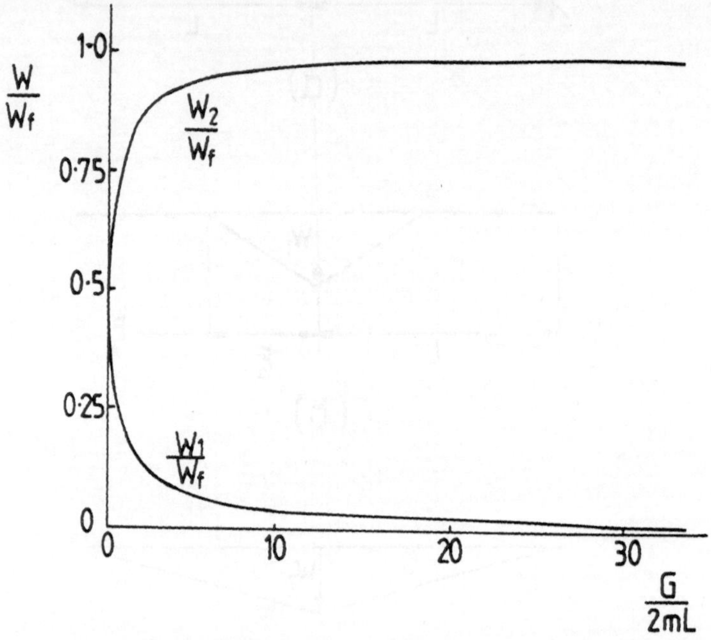

Figure 5. Dimensionless permanent transverse displacements at the end of the first phase of motion (W_1) and acquired during the second phase of motion (W_2) for the beam in Figure 4(a). W_f is the total permanent transverse displacement given by equation (3).

kinetic energy (K) from Figure 7. These observations suggest that the rigid-plastic analysis of many practical engineering problems may be simplified with the aid of a quasi-static method of analysis which disregards the transient or travelling plastic hinge phase of motion.

4. Quasi-static response

Inertia forces are not important in many practical engineering problems, as noted at the end of the last section. These particular problems may be idealised as quasi-static which is a considerable simplification and saving of computational effort compared with a full dynamic analysis. In this circumstance, the transverse displacement profiles of structures subjected to dynamic loads are similar to the corresponding ones which are developed for static loads so that no travelling plastic hinges or time-dependent plastic zones are

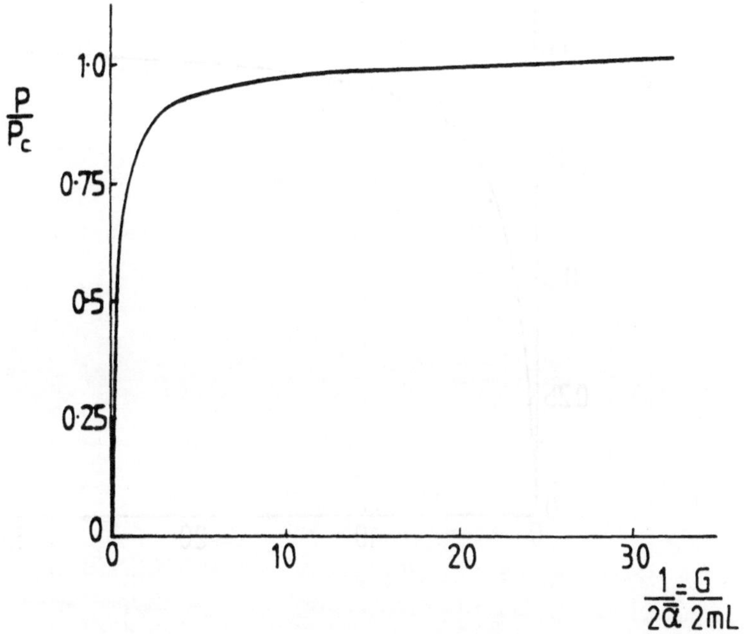

Figure 6. **Dimensionless force at the mid-span of the beam in Figure 4(a) during the second phase of motion. P_c is the static plastic collapse force given by equation (5).**

generated. Thus, for the mass impact loading case in Figure 4(a), a quasi-static estimate for the transverse displacement, W_q, at the impact location, is obtained by equating the initial kinetic energy, $K = GV_o^2/2$, to the work done by a static concentrated load P at the impact location, i.e.,

$$K = PW_q . \tag{4}$$

Designers require a rational criterion to decide whether or not a structural crashworthiness problem may be examined using quasi-static methods of analysis. It has been suggested, for example, that quasi-static methods of analysis could be used to examine the dynamic response of simple structures when the duration of an impact force exceeds the natural period of vibration of the structure. More recently, the accuracy of quasi-static methods of analysis has been studied in References [19] to [21].

A simple illustration of the quasi-static method of analysis is now given for the fully clamped beam struck by a mass at the mid-span as shown in Figure 4(a), and which was studied in the previous section.

Now, the static plastic collapse load for a fully clamped beam in Figure 8(a) subjected to concentrated force at the mid-span is [2]

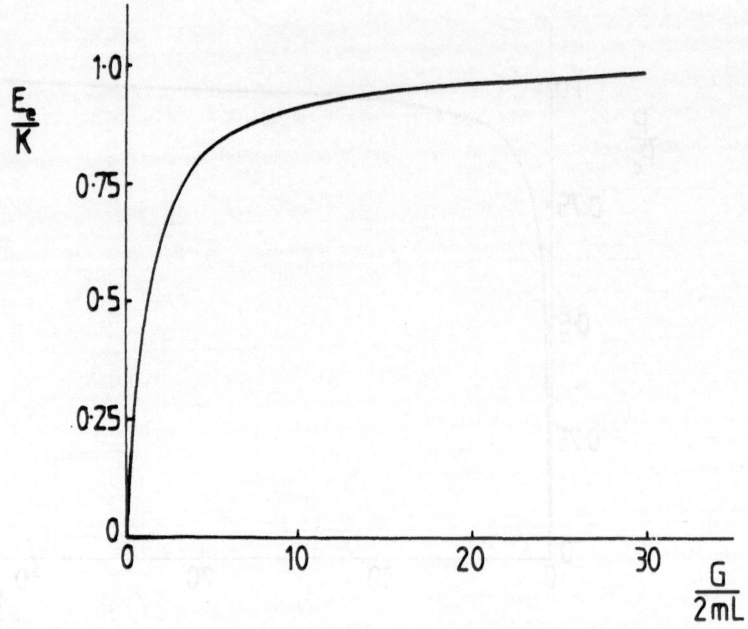

Figure 7. **Dimensionless external energy (E_e) imparted to the beam in Figure 4(a) during the second phase of motion. $K = GV_o^2/2$ is the initial kinetic energy of the mass G.**

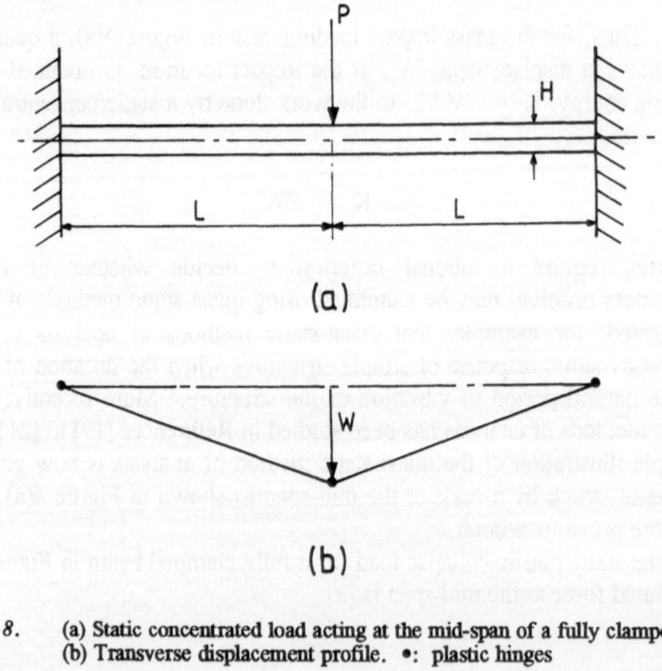

Figure 8. **(a) Static concentrated load acting at the mid-span of a fully clamped beam**
(b) Transverse displacement profile. •: plastic hinges

$$P_c = 4M_o/L \tag{5}$$

for the simple triangular displacement profile in Figure 8(b). The external work done by P_c, for a transverse displacement W_q underneath the concentrated load, is P_cW_q, which, for a quasi-static method of analysis, must equal the initial kinetic energy of the mass in Figure 4(a) ($K = GV_o^2/2$) according to equation (4). Thus,

$$W_q = GV_o^2L/8M_o \tag{6}$$

which is independent of the mass of a beam. Equation (3) reduces to equation (6) for large mass ratios (i.e., $<< 1$). The quasi-static theoretical prediction of equation (6) may also be obtained by equating the initial kinetic energy (K) to the energy absorbed by stationary plastic hinges which form at the mid-span and supports as in the second phase of motion in the theoretical analysis [18] leading to equation (3). In this circumstance, the travelling plastic hinge phase of motion in the Parkes solution [18] makes a negligible contribution to the theoretical predictions for W_f, as shown in Figure 5 for large mass ratios. In fact, it may be seen from Figure 6 that the reaction force (P) underneath the striker mass equals the static collapse pressure (P_c) given by equation (5) when $G/2mL >> 1$ throughout the second or modal phase of motion.

A comparison is presented in Figure 9 between the theoretical predictions of a complete dynamic analysis (equation (3)) and a quasi-static method (equation (6)). It is evident that the quasi-static procedure overpredicts the maximum permanent transverse displacements of a dynamic analysis by 6.6% when $G/2mL = 5$ and that this difference increases to only 10.8% for a striker which weighs three times the beam mass ($G/2mL = 3$). The quasi-static method of analysis overpredicts the maximum permanent transverse displacement because all of the initial kinetic energy is absorbed in three stationary plastic hinges (i.e., $W_1 = 0$, $W_2 = W_f$), whereas, in the complete dynamic analysis, the travelling plastic hinges cause plastic energy to be absorbed throughout the entire span as well as at stationary plastic hinges (i.e., $W_1 \neq 0$, $W_1 + W_2 = W_f$). This difference in the energy absorbing mechanisms is particularly significant when the striker masses are smaller than the total beam mass. In this case, the travelling plastic hinge phase of motion in the theoretical solution leading to equation (3) is important, as shown in Figure 5. In fact, for $G/2mL = 0.05$, almost all of the initial kinetic energy is absorbed during the first phase of motion with very little remaining to be absorbed during the second or modal phase, as indicated in Figure 7. It may be shown that the theoretical analysis [18] leading to equation (3) predicts that $W_1 \rightarrow W_f$ and $W_2 \rightarrow 0$ when $G/2mL \rightarrow 0$. Thus, it is clear that quasi-static methods of analysis would be wholly inappropriate when $G/2mL < 1$, approximately.

The percentage difference between equation (6) for the maximum permanent transverse displacements according to a quasi-static theoretical procedure (W_q) and the exact dynamic analysis (W_f) given by equation (3) for the problem in Figure 4(a) is defined as

$$e_w = (W_q - W_f)100/W_f\,. \tag{7}$$

The differences according to equation (7) between the quasi-static and exact theoretical predictions in section 3 for the beam problem in Figure 4(a) subject to striking masses, which produce infinitesimal displacements, are shown as vertical dotted lines in Figure 10.

The theoretical predictions in this section and the previous one were obtained for beams subjected to dynamic loads which produced only infinitesimal transverse

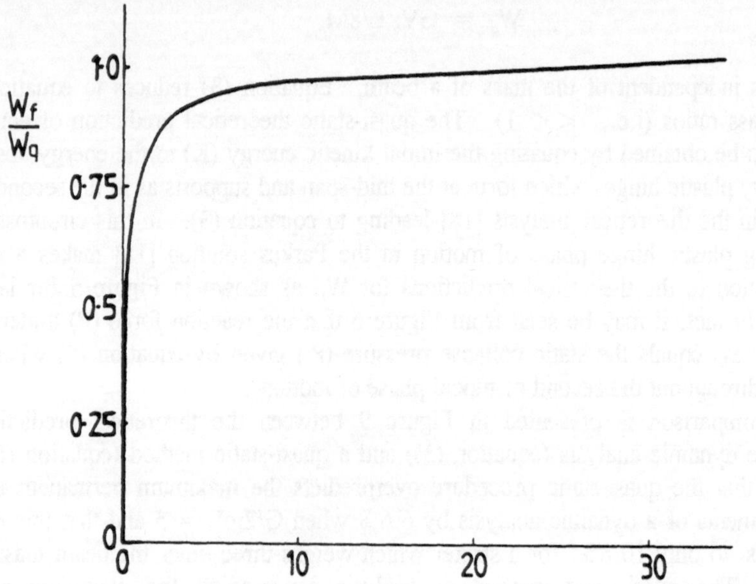

Figure 9. Ratio of the maximum permanent transverse displacements for a rigid, perfectly plastic beam in Figure 4(a) according to an exact dynamic analysis (W_f) (equation (3)) and a quasi-static analysis (W_q) (equation (6)).

displacements. However, it is well known that the influence of finite transverse displacements, or geometry changes, exercise a significant role during the response of fully clamped beams struck by masses which produce maximum permanent transverse displacements greater than about one-half of the beam thickness [2, 22]. The plastic yielding of a beam undergoing small transverse displacements is controlled by the bending moment with the transverse shear force customarily taken as a reaction force which does not contribute to plastic flow. As the transverse deflections increase, the influence of the bending moment and the transverse shear force diminishes, while the axial membrane force increases until the response approaches that of a plastic string for which plastic flow is controlled by the membrane force alone.

Reference [21] contains a quasi-static analysis for the mass impact problem in Figure 4(a) which caters for the non-linear effect of finite-displacements, or geometry changes.

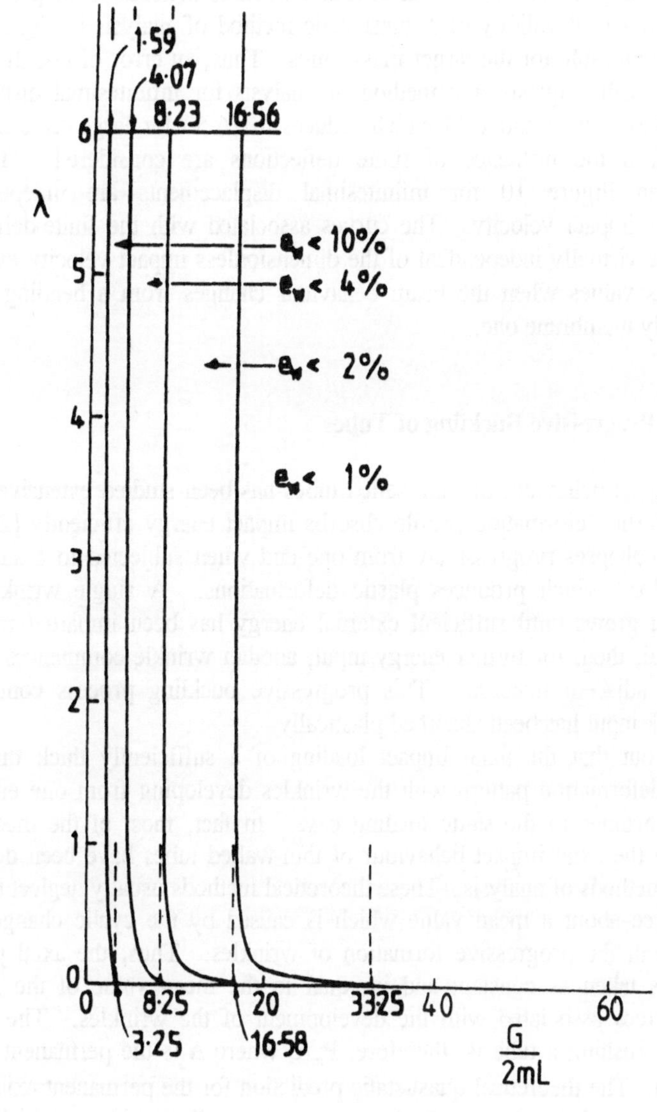

Figure 10. **Percentage difference between the theoretical predictions of a quasi-static analysis and a dynamic rigid-plastic analysis [21] for a beam impact problem in Figure 4(a).($\lambda = mV_o^2L^2/4M_oH$). ____: finite-displacement effects retained in analysis [21]. - - - -: infinitesimal displacements (equations (3) and (6))**

It is interesting to observe that the influence of finite-deflections, or geometry changes, expands the range of validity of a quasi-static method of analysis in Figure 10. This is particularly noticeable for the larger mass ratios. Thus, an error of less than one per cent is associated with a quasi-static method of analysis for infinitesimal displacements and mass ratios larger than about 33 which reduces to 16.56 for large dimensionless impact velocities when the influence of finite deflections are considered. The theoretical predictions in Figure 10 for infinitesimal displacements are independent of the dimensionless impact velocity. The curves associated with the finite-deflection case in Figure 10 are virtually independent of the dimensionless impact velocity except for small dimensionless values when the beam behaviour changes from a bending response to a predominantly membrane one.

5. Dynamic Progressive Buckling of Tubes

The axial impact behaviour of thin-walled tubes has been studied extensively because the associated plastic deformation profile absorbs impact energy efficiently [2,23]. A thin-walled tube collapses progressively from one end when subjected to a sufficiently large static axial load which produces plastic deformations. A single wrinkle, or buckle, develops and grows until sufficient external energy has been imparted to complete the formation and, then, for further energy input, another wrinkle commences to form in the immediately adjacent material. This progressive buckling process continues until all external work input has been absorbed plastically.

It turns out that the axial impact loading of a sufficiently thick thin-walled tube produces a deformation pattern with the wrinkles developing from one end in a similar progressive manner to the static loading case. In fact, most of the theoretical studies published on the axial impact behaviour of thin-walled tubes have been developed using quasi-static methods of analysis. These theoretical methods usually neglect the variation of the axial force about a mean value which is caused by the cyclic change in resistance associated with the progressive formation of wrinkles. Thus, the axial plastic collapse force (P_m) is taken as constant and is equal to the mean value of the maximum and minimum forces associated with the development of the wrinkles. The plastic energy absorbed in crushing a tube is, therefore, $P_m \Delta$, where Δ is the permanent axial crushing displacement. The theoretical quasi-static prediction for the permanent axial displacement (Δ_q) of a similar tube struck axially by a mass G travelling with an initial velocity V_o is obtained from the energy conservation relation

$$P_m \Delta_q = G V_o^2/2 . \tag{8}$$

This quasi-static method of analysis [9] has given good agreement with experimental results which have been obtained for the axial impact loading of both circular and square ductile metal tubes, as shown in Figure 11 for some circular steel tubes. The mass ratios (G/mass of tube) for the specimens in Figure 11 are large and range from 153 to 444, while the impact velocities are smaller than 10.4 m/s, approximately. The energy ratios defined by equation (2) are also large and range from 350 to 1200 for the specimens in Figure 11.

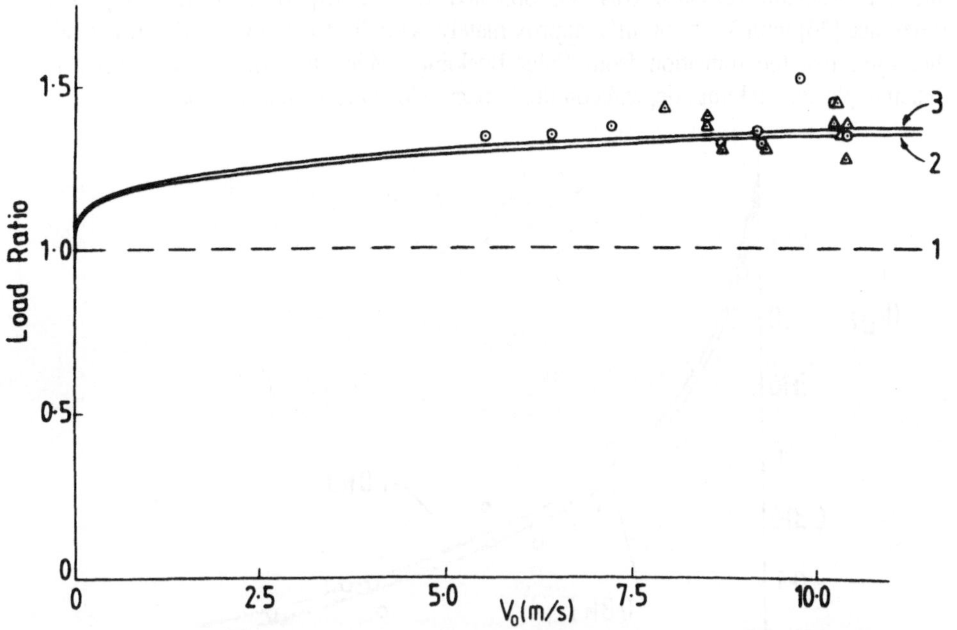

Figure 11. Ratio of dynamic axial crushing loads to static axial crushing loads (load ratio) for thin-walled cylindrical shells made from mild steel for impact velocities up to 10 m/s, approximately [2,9]. - - - - 1: theoretical predictions for a strain rate insensitive material, _____, 2,3: theoretical predictions for a strain rate sensitive material, O,_: experimental results [9].

Now, as the axial impact velocity is increased for circular tubes, the dynamic progressive (quasi-static) buckling discussed above gives way to a phenomenon known as dynamic plastic buckling [2, 24]. The wrinkles form and grow simultaneously throughout the entire length of a circular tube undergoing dynamic plastic buckling which is different from the deformation mode for the dynamic progressive buckling behaviour which is observed at the lower impact velocities and discussed above. The transverse inertia of the tube wall cannot be neglected in this case so that a complete dynamic analysis is required [2, 24, 25].

A recent experimental study [26] has examined the transition with increasing impact velocity from the static buckling mode (progressive or Euler) to dynamic plastic buckling for aluminium alloy circular tubes struck axially by rigid masses. It was found for the particular tubular specimens in Figure 12 that the quasi-static Euler buckling occurs for impact velocities up to about 12 m/s and that dynamic plastic buckling develops for velocities larger than 50 m/s, approximately. Mixed modes of dynamic progressive buckling and dynamic plastic buckling were observed for intermediate velocities. The mass ratios in these experiments ranged from over 500 for the quasi-static Euler buckling to about 30 for the transition to dynamic plastic buckling, as indicated in Figure 12. A mass ratio of 30 is well within the range of validity of the quasi-static methods of analysis, but a quasi-static response was not obtained for the experimental tests reported in Reference [26] with $V_o > 50$ m/s, approximately, as indicated in Figure 12. It is evident, therefore, that the transition from Euler buckling, which is a quasi-static response, to dynamic plastic buckling, depends on the impact velocity as well as the mass ratio.

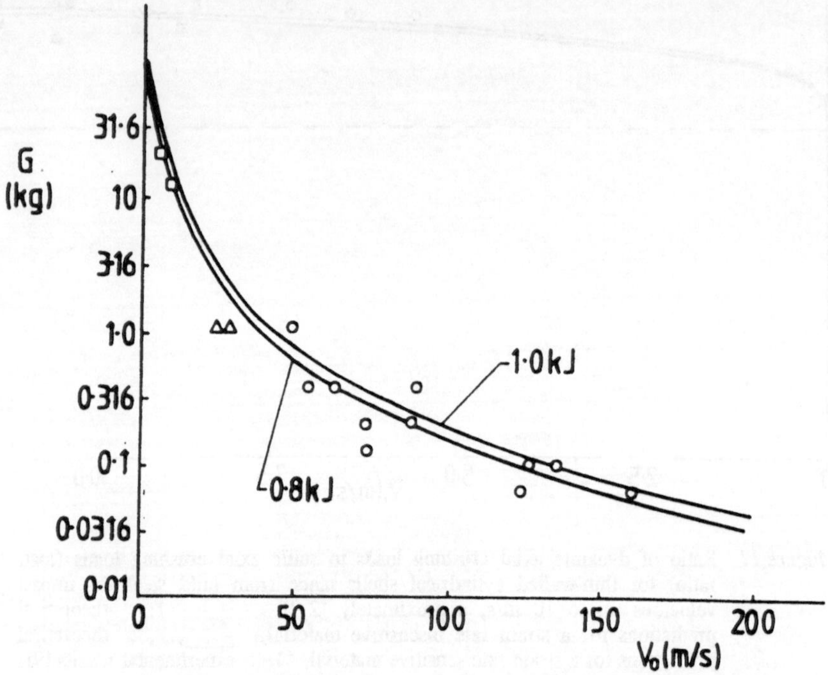

Figure 12. *The effect of striker mass and striker velocity on the failure mode of aluminium alloy 6061 T6 cylindrical tubes loaded axially (wall thickness = 1.65 mm, outside diameter = 25.4 mm, axial length = 101.6 mm, tube mass = 0.034 kg). Experimental results [26]: □: Euler (global) buckling; △: mixed dynamic progressive and dynamic plastic buckling, ○: dynamic plastic buckling.*

6. Transition Between Initial Global Bending and Progressive Buckling of Tubes

This section contains some preliminary results from a more detailed study to be published in Reference [12] of the transition behaviour from static and dynamic initial global bending of 'thick' circular and square mild steel tubes to a progressive buckling mode. Apart from a study on the static behaviour of aluminium circular tubes [27], no design guidelines appear to be available on the effect of the initial length of a tube on the subsequent behaviour.

The square columns in Reference [12] have six C/t ratios from 5.5 to 38, where C is the width of a side face and t is the wall thickness. The corresponding L/C ratios, where L is the axial length, range from 2.4 to 5.8 for C/t = 5.5 and from 6.1 to 33.5 for the square columns with C/t = 38. The circular tubes in Reference [12] have five D/t ratios from 9.5 to 48, where D is the mean diameter, and L/D ratios from 2.2 to 7.7 for the specimens with D/t = 9.5 and from 5.1 to 9.2 when D/t = 48. All 128 of these specimens enter the plastic range in the initial straight configuration and buckle plastically in the Euler or global mode in the sense that the lateral displacements commence to grow with a further increase of the axial compressive load.

These specimens are, therefore, regarded as 'thick' to distinguish them from 'thin' specimens which would buckle initially within the elastic range of the material.

However, many of the tubular specimens switch to a progressive buckling mode and, indeed, a visual inspection of most of these specimens after a test would lead one to conclude that progressive buckling had occurred throughout the entire deformation process. Nevertheless, other specimens continued to bend and collapsed with an Euler or global bending mode. The effect of lateral inertia in the dynamic crushing case acts so as to promote, or at least favour, a dynamic progressive buckling mode which may be maintained until near to the end of motion when the resisting force in some cases might be insufficient to prevent the development of a global bending mode.

$$(L/C)_{crit} = 2.48 \exp (0.041\ C/t) \quad (9)$$

Figure 13 summarises the static axial crushing tests on square tubes reported in Reference [12]. All of these tubes buckled initially with a plastic bending mode but the curved line in this figure which is given by marks the approximate boundary between square columns which continue to bend when they are long enough and the shorter ones which switch to progressive folding.

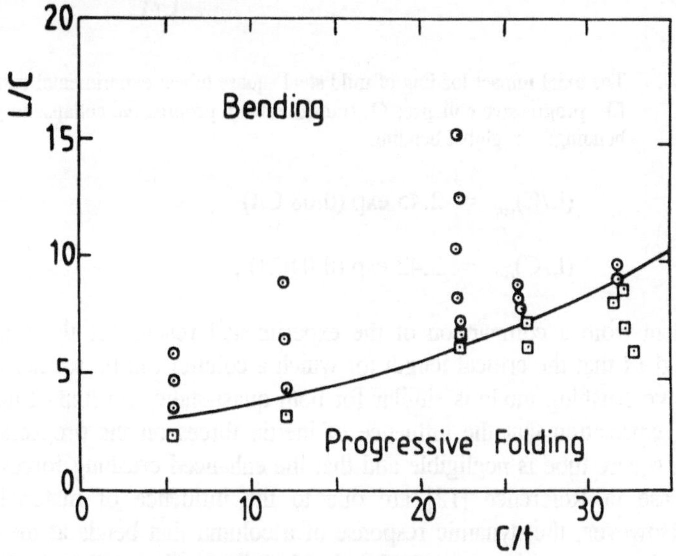

Figure 13. **The static axial loading of mild steel square tubes. ______: equation (9); experimental results [12]: □: progressive collapse; O: global bending**

The axial impact test results on square tubes are summarised in Figure 14, where, unlike the static crushing results in Figure 13, a transition region lies between those cases characterised as progressive folding and bending. Thus, the tubular test specimens lying within the transition region first buckle plastically in the Euler or global sense, as do all of the specimens in Figures 13 and 14, then switch to progressive collapse and eventually switch back to global bending. The upper and lower curves in Figure 14 are given by respectively.

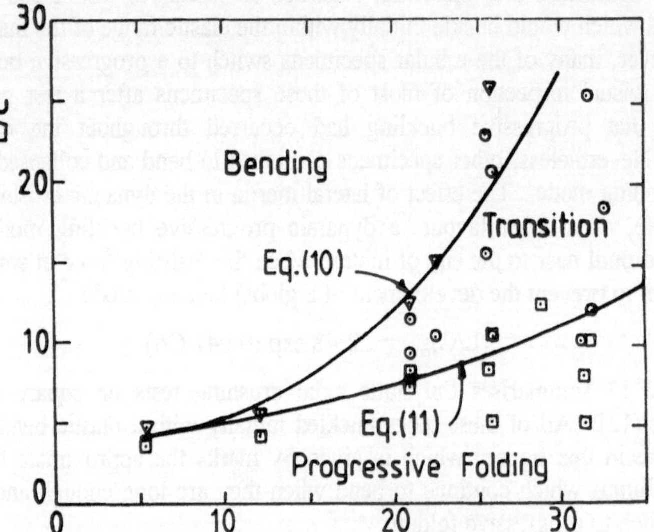

Figure 14. **The axial impact loading of mild steel square tubes; experimental results [12]; □: progressive collapse; O: transition from progressive collapse to global bending; _: global bending**

$$(L/C)_{crit} = 2.45 \exp (0.08\ C/t) \tag{10}$$

and

$$(L/C)_{crit} = 3.42 \exp (0.04 C/t)\ , \tag{11}$$

It is evident from a comparison of the experimental results for the square tubes in Figures 13 and 14 that the critical length for which a column can be squeezed completely in a progressive crushing mode is similar for both quasi-static and impact loadings. This confirms the expectation that the influence of inertia forces on the progressive crushing response of a square tube is negligible and that the enhanced crushing forces observed in the impact case in Reference [12] are due to the influence of material strain rate sensitivity. However, the dynamic response of a column that bends at the onset of the dynamic crushing process is, on the other hand, influenced significantly by the lateral inertia force. In the early stage of the bending deformation process, the lateral inertia force slows down the bending of a column and allows the formation of a few complete plastic lobes at the impacted end. As the crushing process progresses, however, the stabilising effect of the lateral inertia force diminishes gradually and a column eventually bends.

Some elementary theoretical considerations are presented in Reference [12] on the initiation of the collapse process and the maximum load-carrying capacity for tubes with square cross-sections and the transition from global bending to progressive buckling.

It was observed that the theoretical predictions of equations (12) and (29) in Reference [28] for the static and dynamic axial progressive buckling of square tubes gave excellent agreement with the experimental results for the progressively crushed columns in Reference [12] when the ultimate tensile stress is used for the characteristic stress. This lends further support to the use of these equations by designers provided the tubular specimens lie within the progressive crushing regions indicated in Figures 13 and 14.

7. Recent Studies on the Axial Crushing of Top-Hat and Double-Hat Structures

A comprehensive experimental programme on the behaviour of top-hat and double-hat structures subjected to axial static and axial impact loads has been undertaken recently in the Impact Research Centre at The University of Liverpool. All of the specimens were manufactured from mild steel but had various axial lengths to explore any changes in the buckling mode and four different flange widths with several joining methods. The experimental results for static axial loads are presented in Reference [13], while theoretical analyses are developed in Reference [29]. The experimental impact tests have been completed and will be reported in due course.

A number of quasi-static crush tests were carried out on top-hat, double-hat and square tube structures and are reported in Reference [13]. Various collapse modes have been identified and the relative energy absorbing characteristics of the mild steel structures has been assessed.

The width of the external flange altered the collapse stability for the various structures and by increasing the flange width from 10 mm to 25 mm, an increase in the regularity of the collapse mode was found. Generally speaking, an increase in the flange width produced an increase in the axial displacement and the peak loads but with a reduction in the associated mean load values. For the double-hat structures, an increase in the flange width increased all of the energy absorbing characteristics, except for 25 mm width flange specimens.

For top-hat structures, the length of the specimen was also found to influence the collapse performance. The shortest specimens deformed in an irregular manner, while a three-fold increase in the length produced a tendency towards global buckling. Increasing the specimen length increased both the amount of energy absorbed and the axial displacement, while a decrease in the maximum and the mean loads were observed.

Higher maximum and mean loads occurred in the double-hat specimens compared to the top-hat specimens, being approximately 40% and 15% higher, respectively. The lowest values were observed in the square tube specimens. For a given flange width, the top-hat specimens absorb more energy and undergo greater axial displacements than the double-hat specimens before an instability intervenes in the response. The various energy absorbing characteristics and post-test collapse profiles, suggest that the top-hat structures are better energy absorbing devices than the double-hat structures tested in Reference [13].

A theoretical method was developed in Reference [29] to describe the collapse behaviour of top-hat and double-hat structures under static axial loads. The method of analysis employed superfolding elements for a regular symmetric progressive collapse mode. This theoretical procedure has been used previously for the static and dynamic axial crushing of thin-walled square tubes, as reported in References [9] and [28]. It was observed that good agreement was obtained between the theoretical predictions for the mean crushing forces and the axial crush deformations and the corresponding experimental results on mild steel square columns. Thus, the same general method of analysis was adopted in Reference [29] to predict the behaviour of thin-walled structures having more complex cross-sections as encountered, for example, in the automobile industry.

For those test structures in which regular symmetric collapse occurred, the theoretical mean loads [29] compared favourably with those obtained experimentally [13], the best correlation being for the 20 mm and 25 mm wide flange top-hat specimens. Any instability in the collapse process, leading to an Euler buckling or to a catastrophic failure, cannot be predicted using the theory in Reference [29]. However, the theory in Reference [29] predicts more closely the collapse characteristics of top-hat and double-hat structures than any previously published theories, at least for the range of parameters examined in the reported experimental tests.

8. Dynamic Properties of Materials

Considerable experimental and theoretical effort has been expended over the last several decades into studying the strain rate sensitive behaviour of materials. Most of the tests have been conducted on materials subjected to uniaxial stress states producing relatively small strains only up to a few per cent because of the experimental difficulties which are encountered when conducting controllable and repeatable tests with large strains at high strain rates. However, many practical structural crashworthiness applications are concerned with dynamic loads which produce large plastic strains often up to rupture. This topic is discussed further in References [30] and [31] and the constitutive equation

$$\frac{\sigma_{do}}{\sigma_o} = 1 + \left\{ \frac{\dot{\varepsilon}}{B + F\varepsilon} \right\}^{1/q}, \quad \varepsilon_y \leq \varepsilon \leq \varepsilon_u \tag{12}$$

where

$$B = C - F\varepsilon_y \tag{13}$$

and

$$F = \frac{C_u - C}{\varepsilon_u - \varepsilon_y} \tag{14}$$

is proposed in order to cater for the variation of the strain rate effect on the flow stress with increasing strain. σ_{do} and σ_o are the respective dynamic and static flow stresses, is the uniaxial strain rate and ε_y and ε_u are the yield and ultimate strains, respectively. Equation (12) with $\varepsilon = \varepsilon_y$, reduces to the familiar form of the Cowper-Symonds relation

[2] when C and q are identified with the usual coefficients for small strains. Equation (12) with $\varepsilon = \varepsilon_u$ for large strains, again takes on the usual form of the Cowper-Symonds equation except that the coefficient C_u is evaluated from the strain rate sensitive properties at the ultimate tensile strength of the material.

The rupture strains of some materials also change with strain rate and for mild and stainless steels it was shown that the dynamic rupture strains are [30-32] where ε_{rs} is the static uniaxial rupture strain, as illustrated in Figure 15 for mild steel.

$$\varepsilon_r = \{1 + (\dot{\varepsilon}/C)^{1/q}\}^{-1}\, \varepsilon_{rs} \tag{15}$$

The values of the various constants in equations (12)-(15) are given in Reference [32] for mild steel and for equation (15) in Reference [31] for stainless steel.

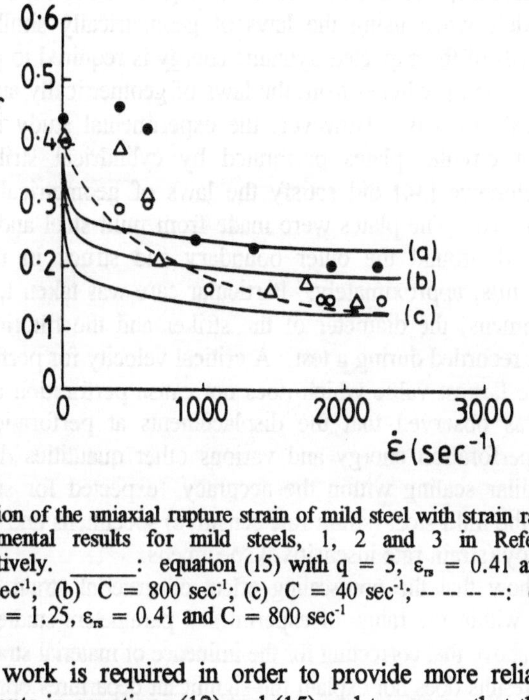

Figure 15. Variation of the uniaxial rupture strain of mild steel with strain rate O, ●, _: experimental results for mild steels, 1, 2 and 3 in Reference [59], respectively. _____: equation (15) with $q = 5$, $\varepsilon_{rs} = 0.41$ and (a) $C = 6340\ \text{sec}^{-1}$; (b) $C = 800\ \text{sec}^{-1}$; (c) $C = 40\ \text{sec}^{-1}$; -- -- -- --: equation (15) with $q = 1.25$, $\varepsilon_{rs} = 0.41$ and $C = 800\ \text{sec}^{-1}$

Further experimental work is required in order to provide more reliable data for the various material constants in equations (12) and (15) and to ensure that the forms of these equations are adequate for design purposes.

The development of numerical codes and theoretical work for the dynamic plastic behaviour of structures has run ahead of the available experimental data. Thus, many studies ignore the variation of rupture strain with strain rate as shown by equation (15). Moreover, very little information is available on these phenomena for biaxial and triaxial stress states. Thus, the usual method [2] of generalising uniaxial test data, with the attendant limitations, is employed when required for calculations.

9. Similitude

It may be shown that the experimental studies on plates [33,34], pipelines [35], and, possibly, shells [36], which are subjected to impact loads producing large ductile deformations without any rupture or tearing of the material, satisfy the elementary laws of geometrically similar scaling. This occurs despite the presence of gravitational and strain rate effects which do not satisfy the laws of geometrically similar scaling. However, the gravitational forces are small compared with the dynamic forces and the influence of material strain rate sensitivity, although significant in a particular test, does not play a major role when comparing geometrically similar test results over practical scale ranges [2].

It is evident from the experimental impact studies in References [37] to [39] that there is a strong measure of agreement between the results on quite different structural geometries subjected to dynamic loads causing large plastic deformations and cutting, rupture or tearing of the material. These studies show that, roughly speaking, the maximum permanent displacements are twice as large as those expected from tests on one-quarter scale models when using the laws of geometrically similar scaling. In other words, only one-half of the expected dynamic energy is required to produce the maximum permanent displacement predicted from the laws of geometrically similar scaling and tests on one-quarter scale models. However, the experimental study into the geometrically similar scaling of circular plates perforated by cylindrical strikers with blunt ends undertaken in Reference [34] did satisfy the laws of geometrically similar scaling, as indicated in Figure 16. The plates were made from mild steel and aluminium alloy and were fully clamped around the outer boundary and struck in the centre by strikers travelling up to 5 m/s, approximately. Particular care was taken to scale the dimensions of the plate specimens, the diameter of the striker and the filtering frequency for the experimental data recorded during a test. A critical velocity for perforation was defined as the average of the largest value which does not cause perforation and the smallest value that does. It was observed that the displacements at perforation, interfacial forces, response times, perforation energy and various other quantities did satisfy the laws of geometrically similar scaling within the accuracy expected for such tests over a scale range of four for the mild steel (strain rate sensitive) specimens and approximately five for the aluminium alloy (strain rate insensitive) specimens.

The results show that the non-scaling effect of material strain rate sensitivity is not important at least within the range of experimental parameters studied. This supports the observation made above that correcting for the influence of material strain rate sensitivity in the experimental test results does not explain the significant departures observed from the laws of geometrically similar scaling in Reference [37] and other references.

Calladine [40] studied the influence of typical axial load-axial displacement curves for struts, plates and other structures which when loaded axially support increasing load with displacement until a peak value is reached after which the load-carrying capacity decreases. Calladine observed that the non-scaling effect in structures with these load-displacement characteristics is exacerbated by the influence of material strain rate sensitivity. He suggested that this non-scaling effect could be removed by distorting the scaling law and using impact velocities which are proportional to the scale factor rather than remaining constant, as required by the laws of geometrically similar scaling.

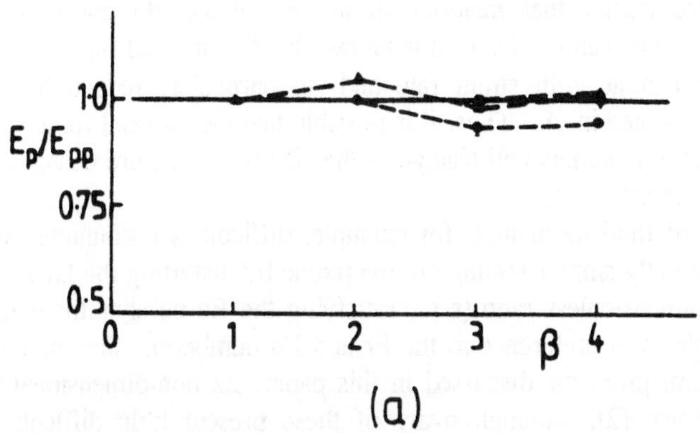

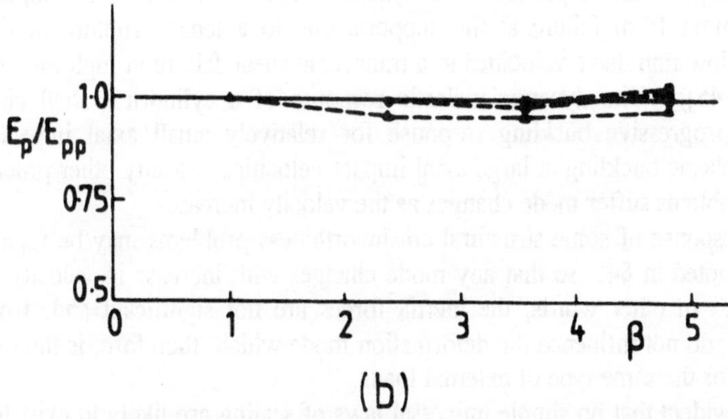

Figure 16. **Ratio of perforation energy (E_p) and the corresponding predicted value (E_{pp}) from a small-scale test versus the scale factor β (β > 1) for geometrically similar scaled impact perforation tests on circular plates. (a) mild steel, (b) aluminium alloy. Further details are given in Reference [34].**

However, it is noted in Reference [41] that using Calladine's particular scaling law would lead to difficulties in scaling the initial linear momentum and the initial kinetic energy. In addition, the nature of the response of some problems is dependent on the magnitude of the initial velocity as observed for the critical mode numbers of impulsively loaded spherical shells and for other structural impact problems.

It is well known that simply changing the size of a structure according to the laws of geometrically similar scaling can lead to a change in the failure mode. A large structure could therefore crack before plastic yielding, whereas the same material could display general plastic yielding before cracking in a small structure. Kendall [42] has illustrated this phenomenon both theoretically and experimentally for a glassy material loaded statically, as discussed further in References [2] and [41].

It was noted earlier that material strain rate effects do not satisfy the laws of geometrically similar scaling. Now, it is known that the uniaxial rupture strains of some materials change with strain rate and, in particular, reduce for mild steel and stainless steel, as noted in §8. Thus, it is possible that the uniaxial rupture strains of some strain rate sensitive materials will change as the size of a structure changes due to the lack of scaling of strain rate [43].

In the field of fluid mechanics, for example, difficulties encountered when using the laws of geometrically similar scaling are overcome by distorting the laws and focusing on one dominant dimensionless ratio (e.g., satisfying the Re number for certain Newtonian fluid flow problems in preference to the Fr and Eu numbers). For the class of dynamic inelastic structural problems discussed in this paper, 22 non-dimensional parameters are listed in Reference [2], although many of these present little difficulty in practice. Unfortunately, the structural mechanics community does not have enough experimental data to select the dominant dimensionless parameter and distort the laws of geometrically similar scaling with confidence. For example, changing the impact velocity could cause a mode change in the response. The dynamic inelastic failure of an impulsively loaded beam changes from failure at the supports due to a tensile rupture of the material at relatively low impulsive velocities to a transverse shear failure at high impulsive velocities ([32, 44, 45]). The dynamic inelastic response of a cylindrical shell changes from a dynamic progressive buckling response for relatively small axial impact velocities to dynamic plastic buckling at large axial impact velocities. Many other practical structural impact problems suffer mode changes as the velocity increases.

The response of some structural crashworthiness problems may be regarded as quasi-static, as noted in §4, so that any mode changes with increase in velocity would be less important. In other words, the inertia forces are not significant and, from a practical viewpoint, do not influence the deformation mode which, therefore, is the same as a static response for the same type of external loads.

It is evident that no simple universal laws of scaling are likely to exist for relating the permanent deformations and other features of small-scale models to the response of full-scale prototypes subjected to dynamic loads causing large inelastic deformations. The principal reason for this difficulty is that this class of problems is time-dependent with elastic loading, plastic loading, elastic unloading, plastic reloading, rupture, etc. occurring at different times with different scaling laws governing different phases of the response and indeed controlling the behaviour in different regions of a structure at a given time. One possible remedy to this situation is to follow the lead of the fluid mechanics community and generate additional experimental work on the dynamic inelastic behaviour of structures with a view to understanding the principal non-scaling phenomena and identifying the dimensionless parameters which must be satisfied and the associated range of validity.

10. Dynamic Inelastic Failure

Some comments are offered in this section on several current research topics on structural failure which are relevant to the field of structural crashworthiness.

The rigid-plastic method of analysis discussed in §2 may be used to predict the impact energy absorbed by a structure and the associated magnitude of the permanent deformations when assuming that the material has an unlimited ductility. No information is obtained, therefore, on the structural integrity which is important for structural crashworthiness calculations.

A survey on the dynamic inelastic failure of beams was presented in Reference [32]. The beams were made from ductile materials, which could be modelled as rigid, perfectly plastic, and they were subjected either to a uniformly distributed impulsive velocity, as an idealisation of an explosion, or to a mass impact to idealise a dropped object loading. Thus, large plastic strains could be produced and the possibility of material rupture was studied for sufficiently large dynamic loads. It was observed that the different dynamic loadings examined in Reference [32] may cause the development of different failure modes. The simplest failure modes were associated with a uniformly distributed impulsive velocity loading. A rigid-plastic method of analysis for this particular problem [2] shows that membrane forces as well as bending moments must be retained in the basic equations for the response of axially restrained beams subjected to large dynamic loads which cause transverse displacements exceeding the beam thickness, approximately. This is known as a Mode I response where the dynamic energy is absorbed plastically without material failure.

If the external impulse is severe enough then the large strains, which are developed at the supports of an axially restrained beam, would cause rupture of the material which is known as a Mode II failure [2, 44]. At still higher impulsive velocities, the influence of transverse shear forces dominates the response and failure is more localised and occurs due to excessive transverse shearing displacements (Mode III). It is observed in Reference [32] that despite the lower impact velocities of the mass impact case, the failure behaviour is much more complex than the impulsive loading case. The Mode II and III failure modes discussed above for an impulsive loading also occur for a mass impact loading but other more complex failure modes may also develop.

It is evident [32] that the dynamic inelastic rupture of beams and other structures is an extremely complex phenomenon and that there is a pressing need for the development of a reliable criterion which can be used in theoretical methods, numerical schemes and computer codes in order to predict the onset of structural failure due to material rupture in structural crashworthiness assessments.

In an attempt to obtain a universal failure criterion, which could be used for a large class of dynamic structural problems, an energy density failure criterion was introduced in Reference [46] and discussed further in Reference [45]. It is assumed that rupture occurs in a rigid-plastic structure when the absorption of plastic work(per unit volume) reaches a critical value. The numerical predictions of the energy density failure criterion in Reference [46] for the dimensionless impulses at the transitions between the failure modes compare favourably with the corresponding experimental results of Menkes and Opat [47] and the theoretical rigid, perfectly plastic predictions in Reference [44]. The critical energy density failure criterion was also used in References [48] and [49] to examine the failure of fully clamped beams subjected to mass impact loads and circular plates subjected to impulsive loads, respectively.

It is observed that the transverse shear forces in structural members play a much more important role for dynamic loads than they do for static ones, as noted above for the mode III failure of beams. In this circumstance, the transverse shear force must be incorporated in the plastic flow condition: usually it is retained in the equilibrium equations but not in the yield condition [2]. Structural failure at the impact location of structures struck by objects and at the supports or other hard points of dynamically loaded structures are often dominated by transverse shear effects [32, 50] so that theoretical rigid-plastic methods of analysis have been developed to explore this phenomenon [51-54] and experimental work is continuing [55-58].

11. Conclusions

This chapter has introduced several topics on structural impact which are relevant to the structural crashworthiness of air, land and marine transportation systems. A considerable amount is known already about this field, but some important areas do require further study. For example, the paucity of dynamic material properties at large plastic strains is noted since this is important for all methods of analysis including finite-element and other numerical schemes. The potentially significant departures from the elementary geometrically similar scaling laws would be relevant in any model testing and calibration studies of numerical analyses for the structural crashworthiness of larger transportation systems. These and other topics are being examined vigorously in many current research studies being conducted in the structural impact field.

12. Acknowledgements

The author is indebted to the Impact Research Centre at The University of Liverpool and, in particular, to Mrs. M. White for her secretarial assistance.

13. References

1. *Structural Crashworthiness* (1983) Jones, N. and Wierzbicki T. (eds.), Butterworths Press.
2. Jones, N. (1989) *Structural Impact*, Cambridge University Press.
3. *Structural Failure* (1989) Wierzbicki, T. and Jones, N. (eds.), Wiley, New York.
4. Jones, N. (1990) Structural Safety Due to Large Dynamic in *Advances in Engineering*, R. S. Agarwal, R. N. Mittal and G. S. Sekhon (eds.). Tata McGraw Hill, pp. 3-15.
5. *Structural Crashworthiness and Failure*, N. Jones and T. Wierzbicki (eds.), Elsevier Applied Science, London, 1993.
6. Johnson, W. and Mamalis, A. G. (1978) Crashworthiness of Vehicles, *Mechanical Engineering Publications*, London.
7. Jones, N. (1990) Some Trends in Structural Impact in *The Future of Structural Testing, Computational Mechanics Publications*, Southampton and McGraw Hill, New York, pp. 63-91.

8. Department of Transport (1989) Investigation into the Clapham Junction Railway Accident, HMSO, London.
9. Abramowicz, W. and Jones, N. (1986) Dynamic progressive buckling of circular and square tubes, *International Journal of Impact Engineering*, **4**(4), 243-270.
10. Abramowicz, W. and Jones, N. (1984) Dynamic axial crushing of square tubes, *International Journal of Impact Engineering*, **2**(2), 179-208
11. Abramowicz, W. and Jones, N. (1984) Dynamic axial crushing of circular tubes, *International Journal of Impact Engineering*, **2**(3), 263-281.
12. Abramowicz, W. and Jones, N. (1994) Transition from global bending to progressive buckling at the onset of plastic collapse of tubes, The University of Liverpool, Impact Research Centre Report No. ES/93/93.
13. White, M. D. and Jones, N. (1994) Experimental behaviour of the quasi-static axial crushing of top-hat, double-hat and square structures, The University of Liverpool, Impact Research Centre Report No. IRC/111/94, (Confidential).
14. Hodge, P. G. (1959) *Plastic Analysis of Structures*, McGraw Hill, New York.
15. Horne, M. R. (1971) *Plastic Theory of Structures*, The M.I.T. Press, Cambridge, U.S.A.
16. Martin, J. B. (1975) *Plasticity: Fundamentals and General Results*, The M.I.T. Press, Cambridge, U.S.A.
17. Symonds, P. S. (1954) Large plastic deformations of beams under blast type loading, *proc. of 2nd U.S. National Congress of Applied Mechanics*, 505-515.
18. Parkes, E. W. (1958) The permanent deformation of an encastré beam struck transversely at any point in its span, *proc. Inst. Civil Engineers*, **10**, 277-304.
19. Jones, N. (1995) Quasi-static analysis of structural impact damage, *Journal of Constructional Steel Research*, **33**, 151-177.
20. Jones, N. (1995) Structural impact, *Science Progress*, **78**(2), 89-118.
21. Shen, W. Q. and Jones, N. (1991), A comment on the low speed impact of a clamped beam by a heavy striker, *Mechanics, Structures and Machines*, **19**(4), 527-549.
22. Liu, J. and Jones, N. (1987) Experimental investigation of clamped beams struck transversely by a mass, *Int. J. Impact Engineering*, **6**(4), 303-335.
23. Johnson, W. and Reid, S. R. (1978) Metallic energy dissipating systems, *Applied Mechanics Reviews*, **31**, 277-288; (1986) **39**, 315-319.
24. Lindberg, H. E. and Florence, A. L. (1982) *Dynamic Pulse Buckling*, SRI Report to Defence Nuclear Agency and Martinus Nijhoff Publishers, Mass. (1987).
25. Jones, N. (1989) Recent studies on the dynamic plastic behaviour of structures, *Applied Mechanics Reviews*, **42**(4), 95-115.
26. Murase, K. and Jones, N. (1991) The transition from progressive plastic buckling to dynamic plastic buckling, Reports of the Faculty of Science and Technology, Meijo University Nagoya, Japan, **31**, 81-87.
27. Andrews, K. R. F., England, G. L. and Ghani, E. (1983) Classification of the axial collapse of cylindrical tubes under quasi-static loading, *International Journal of Mechanical Sciences*, **25**(9/10), 687-696.
28. Jones, N. and Abramowicz, W. (1985) Static and Dynamic Axial Crushing of Circular and Square Tubes in *Metal Forming and Impact Mechanics*, S. R. Reid (ed.), Pergamon Press, Oxford, pp. 225-247.
29. White, M. D., Jones, N. and Abramowicz, W. (1994). A theoretical analysis of the quasi-static axial crushing of top-hat and double-hat sections, The University of Liverpool, Impact Research Centre Report No. IRC/112/94 (Confidential).
30. Jones, N. (1989) Some Comments on the Modelling of Material Properties for Dynamic Structural Plasticity in *Mechanical Properties of Materials at High Rates of Strain*, J. Harding (ed.), Institute of Physics Conference Series No.102, Bristol, pp. 435-445.
31. Jones, N. (1993) Material Properties for Structural Impact Problems in *Advances in Materials and Their Applications*, P. Rama Rao (ed.), Wiley Eastern Ltd., pp. 151-163.
32. Jones, N. (1989) On the Dynamic Inelastic Failure of Beams in *Structural Failure*, T. Wierzbicki and N. Jones (eds.), John Wiley and Sons, New York, pp. 133-159.
33. Duffey, T. A., Cheresh, M. C. and Sutherland, S. H. (1984) Experimental verification of scaling laws for punch-impact-loaded structures, *International Journal of Impact Engineering*, **2**(1), 103-117.
34. Wen, H. M. and Jones N. (1993) Experimental investigations of the scaling laws for metal plates struck by large masses, *International Journal of Impact Engineering*, **13**(3). 485-505.

35. Jones, N., Birch, S. E., Birch, R. S., Zhu, L. and Brown, M. (1992) An experimental study on the lateral impact of fully clamped mild steel pipes", *Proceedings of the Institution of Mechanical Engineers*, **206**(E), 111-127.
36. Duffey, T. A. (1971) Scaling laws for fuel capsules subjected to blast, impact and thermal loading, SAE paper 719107, *Proceedings of the Intersociety Energy Conversion Engineering Conference*, 775-786.
37. Booth, E., Collier, D. and Miles, J. (1983) Impact Scalability of Plated Steel Structures, *Structural Crashworthiness*, N. Jones and T. Wierzbicki (eds.), Butterworths Publishers, London, pp. 136-174.
38. Jones, N. and Jouri, W. S. (1987) A study of plate tearing for ship collision and grounding damage, *Journal of Ship Research*, **31**(4), 253-268.
39. Jouri, W. S. and Jones, N. (1988) The impact behaviour of aluminium alloy and mild steel double-shear specimens, *International Journal of Mechanical Sciences*, **30**(3/4), 153-172.
40. Calladine, C. R. (1983) An Investigation of Impact Scaling Theory, *Structural Crashworthiness*, N. Jones and T. Wierzbicki (eds.), Butterworths Publishers, London, pp. 169-174.
41. Jones, N. (1984) Scaling of Inelastic Structures Loaded Dynamically in *Structural Impact and Crashworthiness, Volume 1*, Keynote Lectures, G. A. O. Davies (ed.), Elsevier Applied Science Publishers, London and New York, pp. 45-74.
42. Kendall, K. (1978) Complexities of compression failure, *Proceedings of the Royal Society*, London, **361(A)**, 245-263.
43. Jones, N. (1967) Influence of strain-hardening and strain-rate sensitivity on the permanent deformation of impulsively loaded rigid-plastic beams, *International Journal of Mechanical Sciences*, **9**, 777-796.
44. Jones, N. (1976) Plastic failure of ductile beams loaded dynamically, *Transactions of the ASME, Journal of Engineering for Industry*, **98(B1)**, 131-136.
45. Jones, N. and Shen, W. Q. (1993) Criteria for the Inelastic Rupture of Ductile Metal Beams Subjected to Large Dynamic Loads in *Structural Crashworthiness and Failure*, N. Jones and T. Wierzbicki (eds.), Elsevier Applied Science, London, pp. 95-130.
46. Shen, W. Q. and Jones, N. (1992) A failure criterion for beams under impulsive loading, *International Journal of Impact Engineering*, **12**(1), 101-121 and p. 329.
47. Menkes, S. B. and Opat, H. J. (1973) Broken beams, *Experimental Mechanics*, **13**, 480-486.
48. Shen, W. Q. and Jones, N. (1993) The dynamic plastic response and failure of a clamped beam struck transversely by a mass, *International Journal of Solids and Structures*, **30**(12), 1631-1648.
49. Shen, W. Q. and Jones (1993) Dynamic response and failure of fully clamped circular plates under impulsive loading, *International Journal of Impact Engineering*, **13**(2), 259-278.
50. Yu, J. and Jones, N. (1991) Further experimental investigations on the failure of clamped beams under impact loads, *International Journal of Solids and Structures*, **27**(9), 1113-1137.
51. Li, Q. M. and Jones, N. (1995) Blast loading of fully clamped beams with transverse shear effects, *Mechanics, Structures and Machines*, 23(1), 59-86.
52. Li, Q. M. and Jones, N. (1994) Blast Loading of Fully Clamped Circular Plates with Transverse Shear Effects, *International Journal of Solids and Structures*, **31**(14), 1861-1876.
53. Li, Q. M. and Jones, N. (1995) Blast loading of a 'short' cylindrical shell with transverse shear effects, *International Journal of Impact Engineering*, **16**(2), 331-353.
54. Liu, J. and Jones, N. (1996) Shear and bending response of a rigid-perfectly plastic circular plate struck transversely by a mass, *Mechanics, Structures and Machines* **24**(3), 359-386.
55. Teeling-Smith, R. G. and Nurick, G. N. (1991). The deformation and tearing of thin circular plates subjected to impulsive loads, *International Journal of Impact Engineering*, **11**(1), 77-91.
56. Olson, M. D., Nurick, G. N. and Fagnan, J. R. (1993) Deformation and rupture of blast loaded square-plates - predictions and experiments, *International Journal of Impact Engineering*, **13**(2), 279-291.
57. Nurick, G. N., Olson, M. D., Fagnan, J. R. and Levin, A. (1995) Deformation and tearing of blast-loaded stiffened square plates, *International Journal of Impact Engineering*, **16**(2), 273-291.
58. Nurick, G. N. and Shave, G. C. (1996) The deformation and tearing of thin square plates subjected to impulsive loads - an experimental study, *International Journal of Impact Engineering*, **18**(1), 99-116.
59. Kawata, K., Fukui, S., Seino, J. and Takada, N. (1968) Some analytical and experimental investigations on high velocity elongation of sheet materials by tensile shock, *IUTAM Symposium on Behaviour of Dense Media under High Dynamic Pressure*, Dunod, Paris, 313-323.

THE MACRO ELEMENT APPROACH IN CRASH CALCULATIONS

WLODEK ABRAMOWICZ
Institute of Fundamental Technological Research
Polish Academy of Sciences
Swietokrzyska 21, 00-049 Warsaw, Poland.

Abstract

The objective of this chapter is to present basic concepts of a consistent and mathematically tractable method of calculating large shape distortions in shells subjected to crash loading. The characteristic feature which distinguishes the present method from all other classical formulations in nonlinear mechanics is that trial deformation functions are postulated on the basis of experimental observations rather then on the basis of expected simplicity of integration schemes. The trial solutions are postulated as global space-time fields rather then local space fields which render solution in one configuration only. Such an approach provides for a natural and convenient means of continuously updating an actual configuration of the shell and therefore, lead to a global rather then incremental formulation of the problem. Another concept is consideration of local deforming regions with floating rather then fixed boundaries with stringent conditions of kinematic continuity at the boundary between neighboring elements. As a result of such a formulation the number of degrees of freedom is dramatically reduced without compromising accuracy of calculations.

1. Introduction

The plastic response of crashed structures involves a number of highly nonlinear phenomena such as: localization of plastic flow, interaction of local and global buckling modes, large deformations and tearing of material - to name just few of them. This inherent complexity of crash phenomena poses a serious challenge to the developers of computational tools. From the early days of crashworthiness the Finite Element (FE) approach is used most frequently in modeling crash events. Typically, a FE model for crash calculations requires thousands of elements and involves massive computing.

Unlike classical FE method which builds on relatively simple formulation of an individual element, heavy discretization and incremental formulation the Macro

J. A. C. Ambrósio et al. (eds.),
Crashworthiness of Transportation Systems: Structural Impact and Occupant Protection, 291–320.

Element approach uses large elements with specialized shape functions and global rather then incremental formulation.

The Macro Element approach is derived from the kinematic method of mechanics where kinematic variables like displacement, velocity or acceleration fields are postulated as a set of trial solutions in variational or extremal formulations of solid and structural mechanics. Typically the postulated kinematic fields are selected from a class of polynomials or Fourier series in order to simplify integration procedures which constitute an ever-present ingredient of any solution algorithm.

In the Macro Element approach we are postulating the kinematic fields on the basis of careful observations of an actual deformation process, Ref.[1] [6]. This method is especially appealing in the field of crushing mechanics where large deformations of structures are tangible and can be easily observed in laboratory experiments or during the inspection of vehicles involved in crash accidents. Furthermore, a typical, repeatable deformation patterns are observed in a large class of shell-like structures subjected to crush loading. This observation lead to the formulation of the Superfolding, Superbeam and Supertearing Elements which are now used successfully in large segments of the transportation industry worldwide.

In most cases a successful formulation of an individual Macro Element imposes strict limitations onto the type of a structure that can be modeled by means of such an element, as well as certain restrictions onto the admissible boundary and loading conditions. Therefore, the range of applicability of a single Macro Element is much more narrow then general formulations of engineering theories like e.g. plate theory or FE method. However, the advantage of such a restricted formulation of an element is its simplicity. Frequently solution to the crushing problem is obtained in a closed form while computer programs based on the Macro Element method are fast and do not require complex input data. Therefore, they are especially useful in the design of energy absorbing structures.

This chapter presents formulation of the Macro Element method and overviews corresponding solution procedures. The general method of the Macro Element approach is then applied to an elementary formulation of the Superfolding Element which is used successfully in the design and calculation of thin-walled prismatic members.

2. Formulation of the Macro Element Method

The purpose of this section is twofold. First, it overviews the general formulation of the governing problem of solid and structural mechanics with particular reference to the kinematical aspects of the general formalism. Second, by referring to general formulation it is shown how the formulation of Macro Element approach is build on first principles of structural mechanics by introducing internal kinematic constrains and by imposing restrictions onto the set of admissible boundary and loading conditions.

2.1. NONLINEAR ENGINEERING THEORIES - AN OVERVIEW OF GENERAL FORMULATION

In the continuum mechanics description of the deformation (motion) of a material element of a solid, structure or a representative part of a structure is based on two fundamental concepts, Ref. [7], [8].

- It is assumed that a certain region V of the Euclidean $\boldsymbol{R}^3$ space of spatial points, with a piecewise smooth boundary S represents a *reference configuration* of a real solid or structure. Points of V are denoted as $X \equiv (X_1, X_2, X_3)$, $X \in V$, and are referred to as material points. The set V is frequently referred to as the undeformed body in order to emphasize the fact that V constitutes a mathematical description of a real solid or structure.
- *Configurations* of the body are defined as invertible and continuously differentiable[1] mappings $x = \chi(X)$, $X \in V$. Each mapping $\chi(V)$ is referred to as a deformed body (with respect to V) provided $\chi(\cdot)$ is not an identity mapping. A non empty set of all admissible configurations is denoted as $K(V)$.

The *deformation* of a body in a time interval $[t_o, t_f]$ is defined as a one-parameter family of configurations $\chi(\cdot,\tau) \in K(V)$; $\tau \in [t_o, t_f]$, in such a way that for each $X \in V$ the function

$$\mathbf{x} = \chi(\mathbf{X}, \tau) \qquad \tau \in [t_o, t_f] \tag{1}$$

is continuous and has a continuous first- and second order time derivatives. Every $x = \chi(X,t)$ is called a position of the material point X, $X \in V$, at time instant t, $t \in [t_o, t_f]$, in the deformation $\chi(\cdot,\tau) \in K(V)$, $\tau \in [t_o, t_f]$. A region of Euclidean space corresponding to a *current configuration* of the body, $v \in K(V)$, is denoted as $v \equiv \chi(V,t)$.

In the continuum mechanics any invertible and continuously differentiable mapping $x = \chi(X)$, $X \in V$ can be taken as a configuration (deformed body). In structural mechanics problems the class of admissible configurations is typically limited to same special sub-classes by imposing the internal kinematic constrains. A specific example of such a constrained class of deformation is the Love-Kirchhoff shell theory in which it is assumed that the material fiber that is normal to the midsurface in one configuration remains normal in any other configuration. Another example is the problem of plane deformations. In the formulation of any problem in the field of structural mechanics it is assumed, as a rule, that the region V and the set $K(V)$ are known a priori.

---

[1] The term *continuously differentiable* which will be used frequently in this chapter should be understood, unless otherwise stated, as: *continuously differentiable as many times as required except at some surfaces, lines or points.*

Once the set of admissible configurations is defined a generic form of the nonlinear problem of structural mechanics is formulated by specifying the governing equations. These are:

- equations of equilibrium
- constitutive relations
- compatibility conditions
- kinematic boundary constraints, boundary loading and initial conditions

These equations constitute a set of governing relations for the unknown deformation $\chi(X,\tau)$; $\tau \in [t_o, t_f]$. The function $\chi(X,\tau)$ which constitutes the solution to the generic problem will be referred to as a fundamental solution and denoted as $\chi^o(X,\tau)$[2]. Likewise, all other fields corresponding to the fundamental solution $\chi^o(X,\tau)$ will be identified by superscript 'o'.

2.2. GENERAL FORMULATION OF THE MACRO ELEMENT METHOD

The general methods of structural mechanics (e.g. beam or shell theories) are formulated in such a way that the solution to the generic problem can be obtained for a wide class of reference configurations V (undeformed body), boundary and initial conditions and specific constitutive relations.

In contrast to general methods the Macro Element formulation is dedicated to narrow classes of structural elements or even to a single type of a structural element or its representative part. Consequently, sharp restrictions are imposed onto the admissible reference configuration V which, in the remaining of this chapter, will be identified with a reference configuration of a Macro Element. Typical examples of structural elements modeled by dedicated Macro Elements are: cylindrical tubes, circular rings or thin-walled prismatic members.

Furthermore, the class of boundary, loading and initial conditions, for which a given method is designed, is also restricted and in several cases only one type of boundary conditions is permitted. For example, a Macro Element that models axial crushing response of a cylindrical tube will typically require that one end of the tube is clamped while the opposite end moves with a constant velocity and remains parallel to the clamped end throughout the entire deformation. In terms of boundary conditions it means that the spatial position $x_u = \chi(X,\tau)$; $X \in S_u$, $\tau \in [t_o, t_f]$ of the surface $S_u \subset S$ on which displacements are prescribed is known for each time instant. Also, the velocity of each point on the surface S_u is known a priori (kinematic loading). When the boundary conditions are changed usually another type of Macro Element must be formulated.

Under such strict limitations the deformation of a structure (or its representative part), $x = \chi(X,\tau)$; $X \in V$; $\tau \in [t_o, t_f]$, resulting from a strictly defined loading and boundary conditions can be determined on the basis of careful experimental

---

[2] In the following considerations it is assumed that such a solution exists and is unique.

observations of an actual deformation pattern. As a rule, the deformation $\chi(X,\tau)$, is postulated with an accuracy to a vector of free scalar parameters $\beta=[\beta_1, \ldots, \beta_N]$ and thus constitutes a set of trial deformations.

$$\{\mathbf{x}^* \in K(V):\ \mathbf{x}^* = \chi(V,\beta,\tau);\ \ \tau \in \left[t_o, t_f\right];\ \ \beta \in \mathbf{R}^3\} \tag{2}$$

In order to avoid needless complexity it is further assumed that the vector β does not depend on time. All variables pertinent to trial deformations, Eq.2, will be denoted by a superscript '*'.

Once the class of admissible deformations is determined the general form of the Macro Element method is formulated in the following way. First, we note that due to limitations imposed onto the admissible set of deformations, Eq.2, the boundary, loading and initial conditions are automatically satisfied for each $t \in [t_0, t_f]$. Second, the kinematic variables i.e. the material and spatial velocity fields

$$\mathbf{v}^* = \mathbf{G}(\mathbf{X},\tau) \equiv \mathbf{g}[\mathbf{x}(\mathbf{X},\tau),\tau] \quad where \quad \mathbf{G}(\mathbf{X},\tau) \equiv \frac{\partial}{\partial \tau}\chi(\mathbf{X},\tau);\ \tau \in [t_o, t_f]$$

gradient of deformation, $F^* = \nabla x^*$, and its rate, $\dot{\mathbf{F}}^*$, are determined from the admissible deformation fields, Eq.2, by time and spatial differentiation, respectively. Therefore, the compatibility conditions are also automatically satisfied.

All that remains to be done is to solve the equations of equilibrium. In the remaining of this chapter we shall consider only quasi-static deformations. In this case the equilibrium of a Macro Element, V, at a time instant t, $t \in [t_0, t_f]$, can be conveniently expressed via the following form of the principle of virtual velocities (weak formulation)

$$\dot{E}^o_{ext} = \dot{E}^o_{\text{int}} \tag{3}$$

where $\dot{E}^o_{ext} = \int_{\partial v} \mathbf{T}\,\mathbf{v}\, d(\partial v)$ denotes the power input while $\dot{E}^o_{\text{int}} = \int_v \sigma\,\mathbf{d} dv$ is the stress power (rate of total internal dissipation in the case of plastic solids) corresponding to a fundamental solution. In the preceding expressions T denotes surface tractions, σ is the Cauchy stress tensor while d is the rate of deformation tensor. In the following calculations the spatial description is used consistently, so that, the integration in expressions for the total rate of internal and external work is performed over the current configuration $v \equiv \chi(V,t)$. In the subsequent step of the solution procedure a set of trial functions for the stress power is established.

$$\left\{\dot{E}^*_{\text{int}}(\beta,\tau):\ \ \dot{E}^*_{\text{int}} = \int_{v^*} \sigma^*\,\mathbf{d}^*\, dv^*;\ \ d_{ij} = \frac{1}{2}(\mathrm{v}^*_{i,j} + \mathrm{v}^*_{j,i})\right\} \tag{4}$$

Each trial solution corresponds to a trial deformation $x^* = \chi^*(X,\beta,\tau)$. Since deformations are finite the configurations $v \equiv \chi(V,\tau)$ and $v^* \equiv \chi^*(V,\beta,\tau)$; $\tau \in [t_0, t_f]$ may occupy different regions in space and therefore the limits of integration in Eq.'s 3 and 4 are, in general, different. The 'discrepancy' between the exact and approximate rates of internal dissipation is then expressed as a residual

$$\hat{R}(\beta,\tau)=\dot{E}^{*}_{\text{int}}(\beta,\tau)-\dot{E}^{o}_{\text{int}}(\tau);\quad \tau\in[t_o,t_f]$$

which is a function of time and a vector of free parameters β only. Since the set of admissible deformations is known a global residual, R(β), which describes a 'global accuracy' of each trial deformation $x^{*}\equiv\chi^{*}(X,\beta,\tau)$; $\tau\in[t_0, t_f]$, can be determined as

$$R(\beta)=\int_{t_o}^{t_f}\dot{E}^{*}_{\text{int}}(\beta,\tau)\,d\tau-\int_{t_o}^{t_f}\dot{E}^{o}_{\text{int}}(\tau)\,d\tau \qquad (5)$$

The global residual R(β) is a function of the vector β only. Consequently, the solution to the problem is reduced to the determination of an optimal vector of free parameters, β^{o}, which renders the global residual R(β) an absolute minimum and thus, corresponds to the best approximation in the entire space-time subdomain. Once the vector β^{o} is known the unknown external loading can be determined directly from Eq.3. A more detailed discussion of this problem is presented in the next section.

It should be noted at this point that the solution to the governing equation Eq.5 can be obtained directly by applying to the principle of virtual velocities, Eq.3, one of the collocation methods or incremental procedures based on the general variational formulations. The advantage of such an approach is that it holds for a wide class of constitutive relations. However, it does not answer the question whether the approximate solution renders an upper or lower bound to the fundamental solution. On the other hand, the minimum property plays an important role in more advanced applications of the Macro Element method concerned with transition of deformation modes and bifurcation of the equilibrium path.

A solution which constitutes a global upper bound to the fundamental solution can be obtained in the case of a perfectly plastic material and certain limited class of deformation processes. An appropriate procedure, based on the upper bound theorem of the Theory of Plasticity, is discussed in the next section.

3. Solution Procedures

This section shows how a solution to the governing relation of the Macro Element method, Eq.5, is constructed. First we shall identify a class of allowable deformation processes for which an upper bound solution can be found. Then, the general procedure is applied to the steady state and quasi-steady state processes. The simple minimum criterion for the mean crushing force is then derived for progressive axial crushing of tubular members. Finally, a generalization of the minimum criterion, valid for a wide class of constitutive relations, is suggested.

3.1. THE GENERAL SOLUTION PROCEDURE

In the energy-time space, *{E, τ}*, function

$$E^o_{\text{int}}(\tau) = \int_{t_o}^{\tau} \dot{E}^o_{\text{int}}(\tau)\, d\tau; \qquad \tau \in [t_o, t_f]$$

defines a single curve referred to as a fundamental energy trajectory. Likewise the set of trial functions

$$\left\{ E^*_{\text{int}}(\beta,\tau): \quad E^*_{\text{int}}(\beta,\tau) = \int_{t_o}^{\tau} \dot{E}^*_{\text{int}}(\beta,\tau)\, d\tau; \quad \tau \in [t_o, t_f],\ \beta \in \mathbf{R}^3 \right\}$$

define a set of curves referred to as trial trajectories. All trajectories start from a common origin, corresponding to the reference configuration $v_O = \chi(V, t_O)$, compare Fig.1 Typically, the reference configuration is identified with the initiation of the crushing process.

In the following we shall restrict the class of admissible constitutive laws to rigid/perfectly plastic isotropic materials with convex yield surface and the associated flow rule. In this case all trajectories are non-decreasing functions of time τ, or a time-like parameter, δ. In the vicinity of a reference configuration, $\tau = t_O$, the fundamental and trial trajectories can be expanded into a Taylor series

$$E^*_{\text{int}}(t_o + \Delta t) = E^*_{\text{int}}(t_o) + \dot{E}^*_{\text{int}}(t_o)\,\Delta t + O(\Delta t^2) \tag{6}$$

where the rate, $\dot{E}^*_{\text{int}}$, defines the slope of a given trajectory at the reference configuration. In the limit analysis of the classical Theory of Plasticity the upper bound theorem for infinitesimal deformations is proved, see e.g. Ref.[8]. This theorem states that out of all kinematically admissible velocity fields, defined over the unchanging reference configuration, $v_O = \chi(V, t_O)$, an actual velocity field minimizes the rate of energy dissipation.

In terms of energy trajectories defined in the energy space it means that the trial trajectory with the smallest slope at the reference configuration coincides with fundamental trajectory at least in the immediate vicinity of that configuration and with an accuracy to at least first order terms[3]. In other words the upper bound theorem defines certain property of *the state* of the system at the reference configuration and does not provide any definite clues as to *the ongoing deformation process*. In particular the principle of minimum work can not be derived from the upper bound theorem, at least in a general case. This conclusion is illustrated schematically in Fig.1 which

---

[3] It can be shown that for a stable deformation processes the curvature of the trajectory at the reference configuration is also positive, compare ref. [10].

shows that a trial trajectory which coincides with the fundamental trajectory at the reference configuration my intersect other trajectories later in the deformation process. Thus, the total expenditure of work predicted by the corresponding solution might be larger or smaller then an actual work input. Furthermore, a trajectory that is optimal within the immediate vicinity of a reference configuration may diverge from the fundamental solution and vice versa a trajectory which overestimates an incremental response may constitute a better approximation later in the process or may be an optimal approximation in an average sense, compare Fig.1.

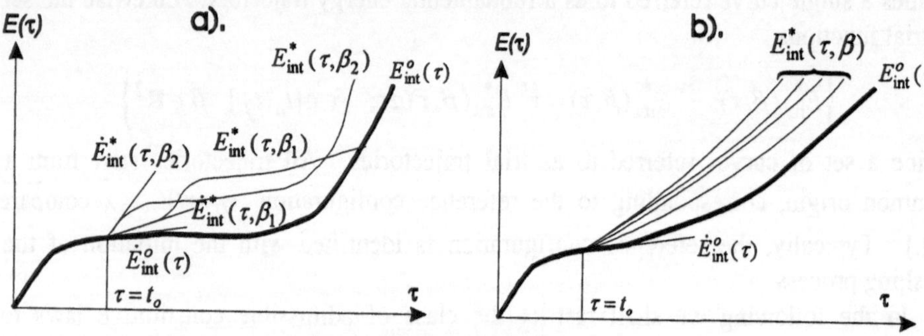

Figure 1 a). Fundamental energy trajectory, $E^{o}_{\rm int}(\tau)$, and trial trajectories, $E^{*}_{\rm int}(\beta,\tau)$, with a common origin corresponding to the reference configuration at t = to. b). A family of non-intersecting trial trajectories.

There are, however, certain sets of trial trajectories for which the upper bound theorem can be used as an effective tool in selecting optimal solution. These are the non-intersecting sets of trial trajectories illustrated schematically in Fig.1b. In this case a trajectory selected on the basis of a minimum slope condition at the reference configuration remains the best solution later in the deformation process. The necessary condition for a *local minimum* of the slope is

$$\left(\frac{\partial \dot{E}^{*}_{\rm int}(\tau,\beta)}{\beta}\right)_{\tau=t_o}=0$$

This condition constitutes a set of N algebraic equations for N unknown components of the optimal vector β^o. A *global* character of the minimum is usually demonstrated by referring to a particular form of the function $\dot{E}^{*}_{\rm int}$. In an abbreviated notation a set of necessary and sufficient conditions for the global minimum is denoted as

$$\beta^{o}=\min_{\beta}\left[\dot{E}^{*}_{\rm int}\right] \tag{7}$$

An example of non-intersecting trajectories is a set of straight lines with a common origin which describes trial solutions for steady-state deformations. Typical example of a steady-state deformation is propagation of a buckle in pipeline or flow of the material over a toroidal surface during plastic inversion of a cylinder. In steady deformation, all the process parameters are constant in time *at each spatial location*. Therefore, the rate of internal dissipation, calculated as a volume integral over the instantaneous configuration of the body is also constant in time and the corresponding energy trajectory is a straight line.

All kinematically admissible steady-state processes that originate from the same reference configuration also result in a set of straight lines and quite obviously the solution selected on the basis of the upper bound theorem, Eq.7 provides for the best solution in a global sense. So that, the optimal solution to a steady-state process, β^o, can be defined from either of the conditions

$$\beta^o = \min_{\beta}\left[\frac{\Delta E^*_{\text{int}}}{\Delta \tau}\right] \quad or \quad \beta^o = \min_{\beta}\left[\lim_{\tau\to 0}\frac{\Delta E^*_{\text{int}}}{\Delta \tau}\right] = \min_{\beta}\left[\dot{E}^*_{\text{int}}\right] \tag{8}$$

Another example of a deformation which can be solved by applying the upper bound theorem is the quasi-steady deformation process. This process is defined as a 'perturbed' steady-state deformation where the trial trajectory oscillates periodically around a straight line which represents a stationary process, Fig.2.

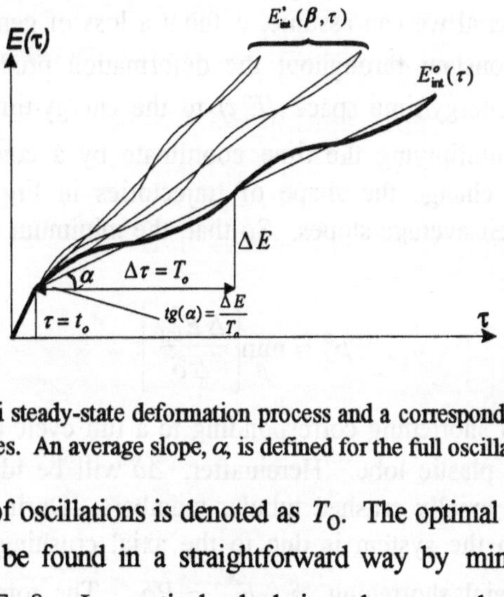

Figure 2 The quasi steady-state deformation process and a corresponding family of trial trajectories. An average slope, α, is defined for the full oscillation cycle, T_O.

In Fig.2 the period of oscillations is denoted as T_O. The optimal solution to the quasi-steady deformation can be found in a straightforward way by minimizing slope at the reference configuration, Eq.8$_1$. In practical calculations, however, the optimal solution β^o is usually determined from an average slope for the full cycle of oscillations

$$\beta^o = \min_{\beta}\left[\frac{\Delta E^*_{\text{int}}}{T_o}\right] \tag{9}$$

where ΔE is an increment of internal dissipation corresponding to the full cycle. Obviously, a number of other approximations based for example on weighted residuals method can be applied here. Quite surprisingly, however, this possibility seems not to have received due mention in the relevant literature.

3.2. SOLUTION FOR PROGRESSIVE CRUSHING OF TUBULAR MEMBERS

Progressive crushing of tubular members is a special case of the quasi-steady deformation. It have received a significant attention in the literature due to its applicability to energy absorbing devices. The crushing of tubular members is characterized by the presence of highly localized zones of plastic flow. Outside these zones the shell is assumed to be rigid. The rigid parts, however, can be subjected to an arbitrary rigid motion characterized by a rigid-body translation vector, δ, and rigid-body rotation vector θ. In the rigid-body dynamics, the external loads are the global cross-sectional forces, P, and moments, M. Therefore the rate of external work is, Ref.[4] and in general include six terms.

$$\dot{E}^{o}_{ext} = \mathbf{P}\dot{\delta} + \mathbf{M}\dot{\theta} \tag{10}$$

In the next section of this chapter progressive axial crushing of tubular members will be discussed as a representative example of the application of the Macro Element method. In the case of axial crushing it is convenient to use an actual axial shortening of the member, δ, as a time-like parameter of the process. Furthermore, for a perfectly/plastic material we can assume, without a loss of generality, that the rate of deformation $\dot{\delta}$ is constant throughout the deformation process. Consequently, the transition from the energy/time space $\{E, \tau\}$ to the energy/time-like parameter space $\{E, \delta\}$ is done by multiplying the time coordinate by a constant velocity $\dot{\delta}$. This transformation does change the shape of trajectories in Fig.2 but does not affect relations between their average slopes. So that, the minimum condition, Eq.9, can be rewritten in the form

$$\beta^{o} = \min_{\beta}\left[\frac{\Delta E^{*}_{\text{int}}}{\Delta\delta}\right] \tag{11}$$

where $\Delta\delta$ is an axial shortening corresponding to a full cycle of oscillations or to the creation of a single plastic lobe. Hereinafter, $\Delta\delta$ will be identified as an effective crushing distance in axially crushed tubular members. During an axial crushing the total power input to the system is due to the axial crushing force, P, acting on a conjugate rate of axial shortening $\dot{\delta}$, $\dot{E}^{*}_{ext} = P\dot{\delta}$. The total expenditure of work corresponding to a full cycle of oscillations $\Delta\delta = \delta(t_o + T_o) - \delta(t_o)$, is simply $E^{*}_{ext} = P_m\,\Delta\delta$ and equals to the total internal dissipation E^{*}_{int}. Substituting E^{*}_{ext} into Eq.11 finally predicts

$$\beta^{o} = \min_{\beta}\left[\frac{\Delta E^{*}_{\text{int}}}{\Delta\delta}\right] = \min_{\beta}\left[P_m\right] \tag{12}$$

The above result is simply the criterion of an absolute minimum of the mean crushing force, P_m, per single plastic lobe. This criterion was conjectured rather then proved by Alexander in 1960, Ref.[1], and has been used ever since in virtually all solutions concerned with progressive crushing although up to now the conjecture has never been substantiated by a rigorous proof.

3.4 GENERALIZATION OF THE GLOBAL MINIMUM CRITERION

In the preceding sections it has been shown that the solution to the progressive crushing problem can be obtained from the upper bound theorem in the case of a particular set of non-intersecting energy trajectories, Fig.1. The question remains, however, whether an actual deformation is governed by the requirement of an absolute minimum of the total expenditure of work. Deeper insight into this problem is obtained from the analysis of the straining history of a representative material point.

In Ref.[9] Hill has proved that there exist optimal paths of homogenous deformation between different states of finite strains for which the total expenditure of work is a global minimum. An optimal path is such that certain triad of material fibers is perpetually orthogonal, while the logarithms of their stretches vary monotonically in fixed ratio. In other words optimal paths correspond to a pure straining deformation which, moreover coincides with a path of proportional loading in the space of logarithmic strains

$$\frac{\log \lambda_1}{\dot{\varepsilon}_1} = \frac{\log \lambda_2}{\dot{\varepsilon}_2} = \frac{\log \lambda_3}{\dot{\varepsilon}_3} = t \tag{13}$$

In Eq.13 λ_i denotes i-th component of the principal stretch while $\dot{\varepsilon}_i$ is the rate of logarithmic strain, $\dot{\varepsilon}_i = \dot{\lambda}_i / \lambda_i$. In a plane-strain deformation of incompressible material the principal Eulerian strain-rates are always automatically in fixed ratio, namely {1, -1, 0}. Hence, the condition of proportional loading, Eq.13, is identically satisfied. Therefore, any (locally) plane pure straining deformation coincides with an absolute minimum of the total work expenditure (for given initial and final strains).

There is a strong evidence that in fact the above conditions are fulfilled in a majority of representative deformation mechanisms associated with local collapse of crushed shells. Indeed, the deformed pattern of thin-walled members can be assembled out of axisymmetric shells such as moving and/or stationary cylinders, cylindrical cones and toroids.. Moreover, it is observed that plastic deformations are negligible in one of the principal directions and therefore majority of computational models assume inextensibility in this direction. Such an axisymmetric deformations are 'locally plane' in planes that are perpendicular to the direction of inextensibility. Since during such deformations the same material fibers are principal throughout the crushing process both optimum conditions are satisfied. This implies that an actual deformation of crushed shells follows the path of minimal work expenditure.

Even more importantly the above conclusion remains valid for a strain-hardening materials as well as for a purely geometrical state variable, Ref.[9].

$$\int_0^t \sqrt{\dot{\varepsilon}_1^2 + \dot{\varepsilon}_2^2 + \dot{\varepsilon}_3^2}\ dt \tag{14}$$

The path integral in Eq.14 is frequently identified with an equivalent strain (or is proportional to such a strain measure). This corollary to the minimum condition has an important implication: if the requirement of an absolute minimum of work is an underlying law that governs crushing response of shells then elements made of highly strain-hardening as well as non-metallic materials should deform in a manner similar to elements made of plastic material. This supposition have a strong experimental evidence. It is observed that typical deformation patterns of shell-like structures made of sufficiently ductile materials can be approximated, with a reasonable accuracy, by segments of axisymmetric shells. This simple geometrical observation suggests that regardless of a material property crushing deformation of shells tends to follow an optimal path defined by Eq.12. Therefore the criterion of the minimum mean crushing force is also used in the case of shells made of strain hardening as well as non-metallic materials. The only limitation here seems to be a sufficient ductility of material, so that, large strains in localized deformation zones can be accommodated without the damage of material or rupture of a shell.

4. Formulation of the Superfolding Element

The crushing deformation of shell-like structures results from a local loss of stability and creation of the so-called local plastic fold or plastic wave. Once created, the plastic fold accommodates most of the plastic deformation in a shell. The local deformation process continues up to the point where local contacts prevent further deformation of an actual fold and encourage development of subsequent fold. Such a deformation process is referred to as a progressive crushing or progressive folding process.

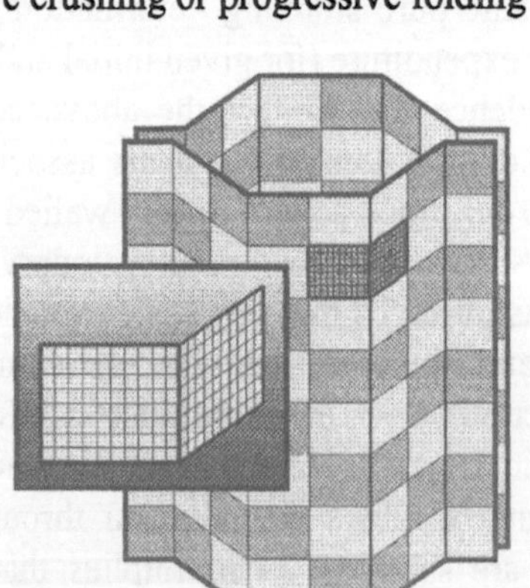

Figure 3 A checkerboard of SE in a spot welded hexagonal column illustrates the discretization procedure by using SE's. The insert reveals the 'macro-size' of a SE as compared to the standard FE mesh.

The progressive folding is a typical mode of deformation of prismatic columns subjected to the predominantly axial loading and is of primary interest to engineers involved in the calculation and design of energy absorbing structures. An interesting feature of the progressive folding is its 'geometrical similitude'. It has been observed that most of actual deformation patterns can be assembled from a single, typical folding lobe. The crushing behavior of such a lobe is modeled by a single Superfolding Element (SE). This section provides for an overview of underlying concepts of the Superfolding Element method. The presentation starts from general formulation of the SE used in computerized applications. Then, a detailed discussion of possible simplifications to the general solution are presented. All results of this section are valid for shells made of plastic isotropic strain hardening materials with the convex yield surface and an associated flow rule.

4.1. DISCRETIZATION OF A COLUMN INTO SUPERFOLDING ELEMENTS

In the initial, undeformed configuration, a SE represents the segment of a corner line of a prismatic column, refer to Fig.3. It is cut off from a column by a set of two parallel horizontal planes. The distance between planes, *2H*, equals the length of the plastic folding wave of the column. The vertical boundaries of a SE are defined by a set of two vertical planes equally distanced from the neighboring corners and/or vertical edges of a column.

4.1.1. *Dimensions of a SE*

In the initial, undeformed configuration a single SE is defined by four parameters:

1. total length, C, of two arms of a SE, $C = a + b$,
2. central angle, Φ
3. wall thickness t_a of the arm of the length a
4. wall thickness t_b of the arm of the length b

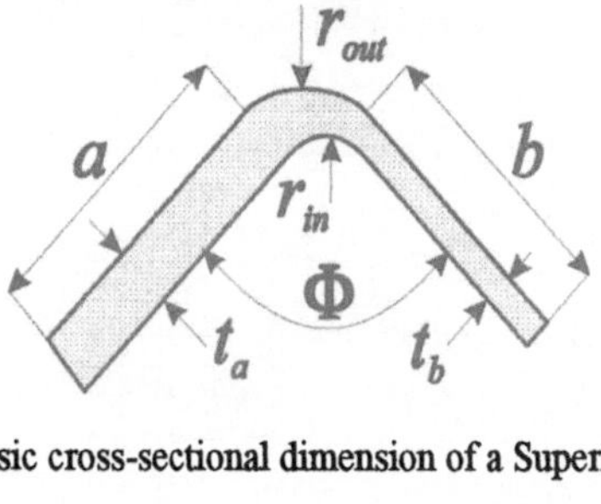

Figure 4. Basic cross-sectional dimension of a Superfolding Element.

In stamped sheet metal structures $t_a = t_b = t$ while in extruded aluminium structures thickness of the two adjacent walls might be different.

It should be noted, at this point, that the height of a SE which corresponds to the length of a plastic folding wave, *2H*, is not given a priori and must be calculated as a part of the solution. Accordingly, a computerized implementation of the SE method requires an adaptive meshing algorithm.

4.1.2. *An Active Layer of Folds*

A set of Superfolding Elements located between two horizontal planes defines a single layer of plastic folds (also referred to as a deformable cell, Fig.5).

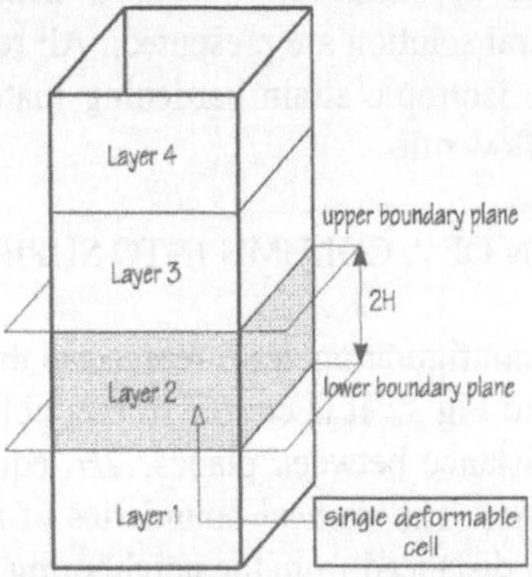

Figure 5 A deformable cell represents a single layer of active plastic folds in a progressively crushed prismatic column.

The number of SE in a given layer corresponds to the number of corners in a column. In progressive crushing of real columns plastic deformation are always spread over two neighboring layers. However, consideration of a single layer at a time is an useful approximation which leads to accurate results. An example of possible deformation patterns of a single active layer for various regimes of loading is shown in Fig.6.

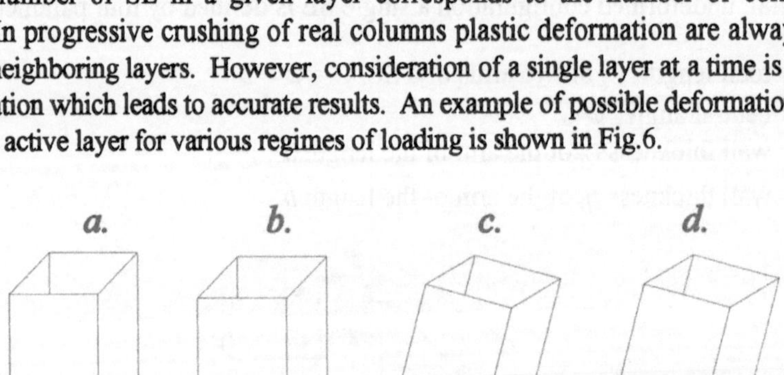

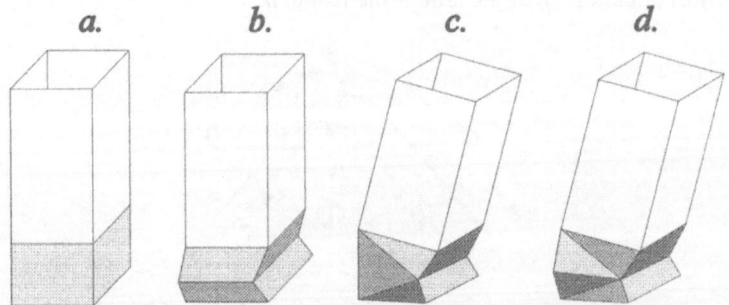

Figure 6 Various defomation modes of an active layer of folds in a square prismatic beam subjected to different loading histories. The symmetric mode *(b)* corresponds to the progressive axial crushing process.

The present section is concerned with axial loading only. In this case the deformation of all SE is symmetric and the boundary planes remain parallel throughout the deformation. process.

4.2. FOLDING MODES OF A SE

The most general deformation mode of a single SE is shown schematically in Fig.7. The plastic folding of the element involves activation of five different deformation mechanisms. These are:

- continuous deformation of a section of the floating toroidal surface, 1, at the so called corner point
- bending along horizontal stationary hinge lines, 2
- rolling deformations along moving inclined hinge lines, 3
- extensional deformations of the conical surface, 4, in the terminal phase of the deformation process and
- bending deformations along inclined hinge lines, 3, in the terminal phase of the folding process, when the moving hinge line is locked within the element.

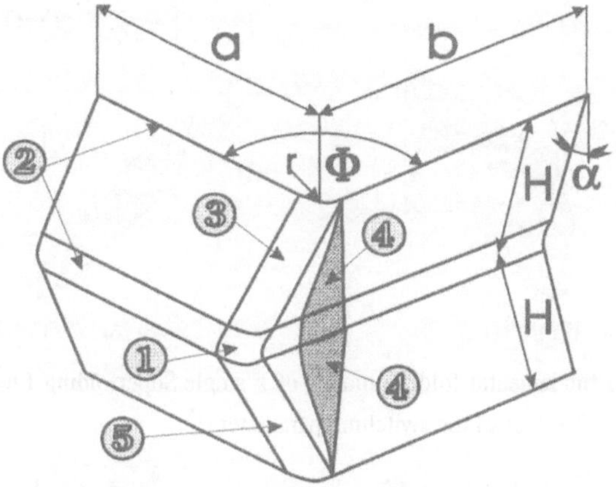

Figure 7 Basic folding mechanisms in a deformed Superfolding Element (SE):
① Deformation of a floating toroidal surface.
② Bending along stationary hinge lines.
③ Rolling deformations.
④ Opening of a conical surfaces.
⑤ Bending deformations along inclined, stationary, hinge lines following locking of the traveling hinge line 3

The general, folding mechanism in Fig.7 is constructed from two simpler folding modes, illustrated in Fig.8. These modes are referred to as an asymmetric and symmetric deformation mode, respectively. The mode shown in Fig.7 is called an asymmetric mixed mode. A progress of the deformation process in each mode is controlled by a single process parameter α, $0 \leq \alpha \leq \alpha_f$, which defines the rotation of a side face of an element from the initial upright position, Fig. 7. At the initiation of the folding process $\alpha = 0$. The process terminates when $\alpha = \alpha_f = \pi/2$. The asymmetric deformation mode is characterized by the absence of a conical surface ④ in Fig.7. Consequently, the entire folding process is controlled by the propagating hinge line ③. The symmetric deformation mode, on the other hand, lacks the propagating hinge line ③ in Fig.7. In this case local extensional plastic deformations are confined to the conical surface ④.

The development of a particular folding mode. in Fig's.7 and 8, is controlled by a single switching parameter α^*, $0 \leq \alpha^* \leq \alpha_f$. This parameter defines a configuration at which a symmetric mode takes over the control of the folding process. If $\alpha^* = \alpha_f$ then the folding of a SE is controlled by an asymmetric mode alone while the case $\alpha^* = 0$ corresponds to a purely symmetric mode, see Fig.8. For $0 < \alpha^* < \alpha_f$ both mechanisms are involved in the folding process: folding starts as an asymmetric mode and continues up to the point where the moving hinge line ③ is locked within an element. At this point the conical surface ④ starts to grow.

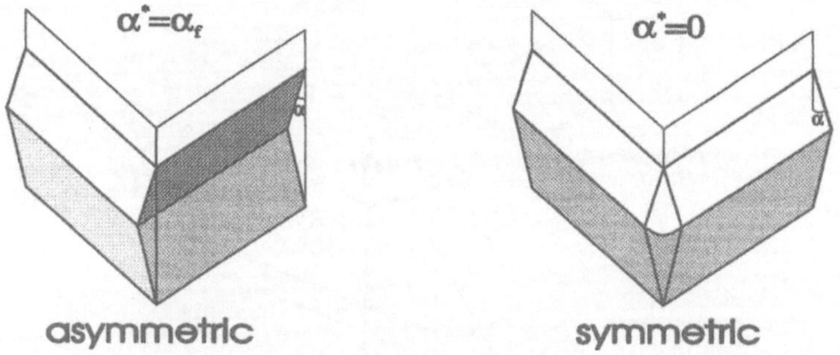

Figure 8 Two fundamental folding modes of a single Superfolding Element controlled by limit values of the switching parameter α^*.

An actual value of the switching parameter, α^*, depends on both the input parameters, $\{C, t, \Phi\}$, and constraints imposed onto deforming faces of a SE. In the case of an unconstrained or standing alone SE the asymmetric mode of deformation, $\alpha^* = 0$, is predominant for right angle and acute elements, $\Phi \leq \pi/2$, while the symmetric mode controls the folding process of obtuse elements with the central angle, Φ, larger then 120 degrees, $\Phi \geq 3\ \pi/2$, approximately. In the intermediate range of central angles both modes coexist while the fractional contribution of each mode to the total energy dissipation depends on the central angle, Φ, and the width to thickness aspect ratio, C/t. The folding modes of a standing alone SE are referred to as natural folding modes.

A SE which is a member of a deformable cell is constrained by neighboring elements. Kinematic constraints are introduced into the element either by imposing the deformation of one or two arms of the element in a pre-defined direction or by constraining the deformation of the element's corner line. The former case is typical for an assemblage of elements which model the deformation of a column with closed cross-section. In this case the requirement of the circumferential continuity of the deformation field may force an element to the deformation mode different then the natural folding mode. Similarly, constraints imposed onto the corner line can change the natural deformation mode of an element. For example, during the deformation of 'X' and 'Y' sections the continuity conditions imposed onto the common corner line of all contributing flanges prevent the development of asymmetric modes.

4.3. TRIAL DEFORMATIONS FOR A SINGLE SE

An example of a kinematically admissible configuration, $x^*=\chi^*(X, \beta, t)$; $t\in[t_0, t_f]$, of a single SE is shown schematically in Fig.7. In general vector β has four components.

$$\beta = \{r, H, \alpha^*, \delta_{eff}\} \tag{15}$$

These are (refer to Fig.7):

- average rolling radius, *r*,
- length of the plastic folding wave, *2H*,
- switching parameter, α^*
- effective crushing distance, δ_{eff}.

The effective crushing distance, δ_{eff}, defines an actual shortening of a SE from the initial configuration up to the point where internal contacts prevent further deformation of the element. Separate calculations show, Ref. [2-*Vol. II*], that the aspect ratio of the effective crushing distance, δ_{eff}, to the length of a plastic folding wave, *2H*, equals $\delta_{eff} / 2H = 0.73$, approximately, for all progressive crushing processes of practical importance. Consequently, the effective number of free parameters of the process is reduced to three, $\beta = \{r, H, \alpha^*\}$. The velocity field, v^*, and the corresponding rate of deformation tensor, d^*, in Eq.4 can now be calculated form the postulated deformation, $x^*=\chi^*(X, \beta, \tau)$; $\tau\in[t_0, t_f]$, by appropriate spatial and time differentiation. The calculated trial fields are then used to construct trial solutions E^*_{int} discussed in section 2. In the next section the trial dissipation E^*_{int} for a SE will be expressed in terms of the rate of curvature and rate of extensions defined at the middle surface of the element. Particulars of the transition from a three-dimensional description of continuum to a two-dimensional formulation of a shell element (in Eulerian description) are given in Ref.[11].

4.4 TRIAL SOLUTIONS FOR THE INTERNAL ENERGY DISSIPATION

The rate of internal energy dissipation in a deformed shell element results, in general, from the continuous and discontinuous velocity fields.

$$\dot{E}_{\text{int}} = \int_S \left(M_{\alpha\beta}\dot{\kappa}_{\alpha\beta} + N_{\alpha\beta}\dot{\varepsilon}_{\alpha\beta} \right) dS + \sum_{i=1}^{n} \int_{L^i} M_o^i \left[\dot{\theta}_i \right] dl^i \tag{16}$$

In Eq.16 S denotes the current shell mid surface, n is the total number of plastic hinge lines, L_i is the length of the i-ht hinge while $[\dot{\theta}_i]$ denotes a jump of the rate of rotation across a moving hinge line. Components of the rate of curvature and rate of extensions tensors are denoted, respectively, as $\dot{\kappa}_{\alpha\beta}$ and $\dot{\varepsilon}_{\alpha\beta}$ while $M_{\alpha\beta}$ and $N_{\alpha\beta}$ are the corresponding conjugate generalized stresses. For the sake of simplicity in the remaining part of this chapter the superscript '*' is omitted in all expressions for trial fields. The arguments τ and β will be specified whenever necessary.

Since the assumed deformation fields are axisymmetric and, in addition, inextensible in one of the principal directions, tangent to the shell's mid surface, the expression for (trial) rate of internal energy dissipation, Eq.16, reduces to

$$\dot{E}_{\text{int}} = \int_S N_o \dot{\varepsilon}_1 \, dS + \sum_{i=1}^{n} \int_{L^i} M_o^i \left[\dot{\theta}_i \right] dl^i \tag{17}$$

where $\dot{\varepsilon}_1$ is the rate of straining in a principal direction, tangent to the shell's mid surface and equals to a corresponding component of the rate of deformation tensor $\dot{\varepsilon}_1 = d_{11}$. The strain rate component perpendicular to the mid surface does not contribute to the internal energy dissipation due to specific form of the yield condition for plastic shells Ref. [11]. Integrating Eq.17 in the interval $0 \leq \alpha \leq \alpha_f$ renders the expression for (trial) energy dissipation per single plastic lobe (particulars of the calculation procedure are given in Ref.[2] and [3]).

$$E_{\text{int}}(\beta) = \int_0^{\alpha^*} {}_{(1)}\dot{E}_{\text{int}} \, d\alpha + \int_{\alpha^*}^{\alpha_f} {}_{(2)}\dot{E}_{\text{int}} \, d\alpha \tag{18}$$

Two integrals on the r.h.s. of Eq.18, defined through the switching parameter α^*, correspond to the asymmetric and symmetric modes of deformation, respectively.

In practical calculations it is convenient to define following expressions for the membrane and bending contributions to the internal energy dissipation

$$E_{\text{int}}^N \equiv \int_0^{\alpha} d\alpha \int_S N_0 \dot{\varepsilon} \, dS = \sigma_0^N(\bar{\varepsilon}) \int_0^{\alpha} d\alpha \int_S t \, \dot{\varepsilon} \, dS$$

$$E_{\text{int}}^M \equiv \sum_{i=1}^{n} \int_{L^i} M_o^i \left[\dot{\theta}_i \right] dl^i = \sum_{i=1}^{n} {}_{(i)}\sigma_0^M(\bar{\varepsilon}) \frac{t}{4} \int_0^{\alpha} d\alpha \int_{L^i} \left[\dot{\theta}_i \right] dl^i \tag{19}$$

where $\sigma_0{}^N$ and $\sigma_0{}^M$ denote, respectively, an average level of the flow stress in the entire crushing process. This stress is referred to as an energy equivalent stress. The functional dependence of the energy equivalent stresses on the constitutive relation and characteristic deformations of the SE are discussed in the next section.

4.5 CONSTITUTIVE RELATIONS AND THE ENERGY EQUIVALENT STRESS MEASURE

The material properties enter the energy relations, Eq.19, through the equivalent stresses σ_0^M and σ_0^N. These stresses are defined, respectively, as

$$\sigma_0{}^N(\bar{\varepsilon}) = \frac{1}{\bar{\varepsilon}}\int_0^{\bar{\varepsilon}} \sigma(\varepsilon)\, d\varepsilon$$
$$\sigma_0{}^M(\bar{\varepsilon}) = \frac{1}{\bar{\varepsilon}^2}\int_0^{\bar{\varepsilon}} \sigma_0{}^N(\varepsilon)\, \varepsilon\, d\varepsilon \qquad (20)$$

and correspond to an average level of the plastic flow stress in regions subjected to uniaxial tension/compression or bending, characterized by a representative strain $\bar{\varepsilon}$, Ref.[2].

The constitutive relation $\sigma(\varepsilon)$ is postulated here as a product of two functions

$$\sigma(\varepsilon,\dot{\varepsilon}) = \sigma_v(\varepsilon,\dot{\varepsilon}_v)\,\gamma(\varepsilon,\dot{\varepsilon}) \qquad (21)$$

where $\sigma_v(\cdot)$ corresponds to the standard quasi-static tensile characteristic of a given material, determined at the constant strain rate, $\dot{\varepsilon}_v$, *(10^{-3} [1/s] $\le \dot{\varepsilon}_v \le 10^{-4}$ [1/s])* while $\gamma(\cdot)$ describes the strain rate effects. The stress and strain measures, used in the above definitions, Eq.'s.21, correspond, respectively, to the Cauchy stress σ and logarithmic strain ε. These measures are also referred to as a natural stress and strain measure. The derivation procedure and examples of calculation of the energy equivalent stress for various materials are discussed, in full details, in Ref. [12].

4.5.1. *Approximate Constitutive Relations for Low Ductility Materials*

In all preceding considerations it has been tacitly assumed that the response of material, given in terms of natural measures, σ - ε, is symmetric with respect to ε regardless of the amount of straining. For large deformations, however, the material response is inherently asymmetric due to different mechanisms leading to the damage of material under compressive and tensile loading.

This section presents a simple method to calculate an approximate crushing response of elements made of low ductility materials. The underlying idea of this method is explained in fig.9.

The stress-strain relation, σ - ε, is symmetric, for both compressive and tensile strains, up to a critical strain $\varepsilon = \varepsilon_f$, where ε_f corresponds to a fracture of material under tensile loading. In the range $-\varepsilon_f \le \varepsilon \le \varepsilon_f$ the energy equivalent stresses are calculated using Eq.'s.20. Obviously, this part of the equivalent stress-strain relation is symmetric. It is further assumed that beyond the critical point, $\varepsilon = -\varepsilon_f$, the equivalent *compressive stress* is constant and equals the equivalent stress at the critical point, $\varepsilon = -\varepsilon_f$.

$$\sigma_0{}^i(\varepsilon) = \sigma_0{}^i(\varepsilon_f) \quad \text{for } \varepsilon > \varepsilon_f \quad \text{and } i = M, N. \qquad (22)$$

For *tensile deformations* the equivalent stress σ_o^i, $i = M, N$, is calculated using Eq.20. It is assumed, however, that there is no plastic resistance of material beyond $\varepsilon = \varepsilon_f$. The above approximation is used successfully to calculate crushing response of columns made of aluminium where the critical strain, ε_f, should be interpreted as a point of the initiation of internal damage of the material rather then as a point of fracture initiation. In columns made of mild steel ε_f provides for a reasonable estimate of the deformation of a SE which may result in a visible through - thickness cracking or fracture. The above method has been used successfully to calculate crushing response for strains slightly exceeding critical strain, ε_f. However, the present approach may result in erroneous predictions in the case of elements showing extensive fracture and/or tearing deformations.

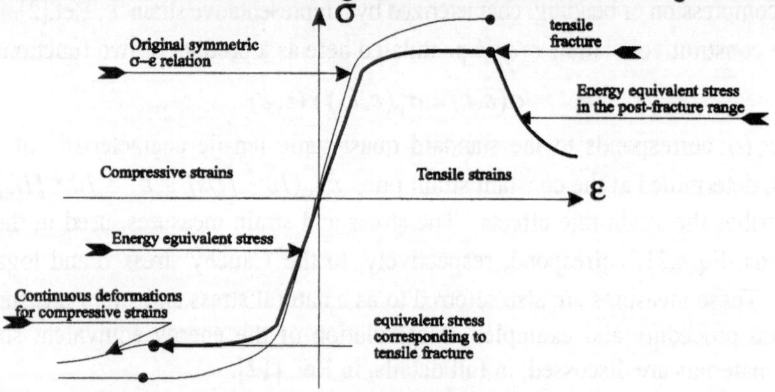

Figure 9 Modified, asymmetric σ–ε relation used to calculate crushing response for large plastic defomations beyond instability point.

4.6. THE FUNDAMENTAL SOLUTION FOR A SINGLE SE

The explicit expression for a mean crushing force, P_m, corresponding to a complete folding of a standing alone SE is derived by substituting expressions for the internal energy dissipation, Eq.'s.18 and 19, into the governing conditions, Eq.12. The complete derivation of the following result is given in Ref. [1] and [3]. Here only the final expression is presented.

$$P_m = \frac{t^2}{4}\{\sigma_o{}^N(\bar{\varepsilon}_1)\,A_1\frac{r}{t} + \sigma_o{}^M(\bar{\varepsilon}_2)\,A_2\frac{C}{H} + \sigma_o{}^M(\bar{\varepsilon}_3)\,A_3\frac{H}{r} + \sigma_o{}^M(\bar{\varepsilon}_4)\,A_4\frac{H}{t} + \sigma_o{}^N(\bar{\varepsilon}_5)\,A_5\}\,\frac{2H}{\delta_{eff}} \tag{23}$$

where $A_i = A_i(\Phi, \alpha^*)$, $i = 1, 2, \dots 5$

The five terms in parenthesis on the right hand side (r.h.s.) of Eq.23 describe, respectively, fractional contributions to the total energy dissipation resulting from five elementary deformation mechanisms, identified in Fig.8. The five factors, A_i $i = 1,2...5$ result from the surface-time integration. Some of the A_i factors, e.g. A_2 or A_4, are easily calculated as a closed-form functions of geometrical parameters, other, e.g. A_1 or A_3, are functions of elliptic integrals and must be calculated numerically. The meaning of other variables appearing in Eq.23 is explained in section 4.1.1.

Parameters corresponding to the equilibrium of a SE are determined from the set of three nonlinear algebraic equations, resulting from the minimum condition, Eq.12

$$\frac{\partial P_m}{\partial H} = 0\ ; \quad \frac{\partial P_m}{\partial r} = 0\ ; \quad \frac{\partial P_m}{\partial \alpha^*} = 0 \tag{24}$$

In general the set of governing equations, Eq.'s 24, has no closed-form solution. Such a solution can be found ,however, for two fundamental folding modes of a Superfolding Element illustrated in Fig.8. Such a solution will be discussed later in this section in conjunction with the crushing behavior of square sections.

4.6.1. *The SE with Different Thicknesses of Side Faces*

Eq.23, can be easily generalized to the case of a Superfolding Element with various thicknesses, t_a and t_b, of two arms, *a* and *b* respectively, refer to Fig.4.

$$\begin{aligned} P_m = \frac{1}{4}\{ & t_1^2\, \sigma_o{}^N(\bar{\varepsilon}_1)\, A_1 \frac{r}{t_1} + A_2(\sigma_o{}^M(\bar{\varepsilon}_2)\frac{a\,t_a^2}{H} + \sigma_o{}^M(\bar{\varepsilon}_2)\frac{b\,t_b^2}{H}) + \\ & t_3^2\, \sigma_o{}^M(\bar{\varepsilon}_3)\, A_3 \frac{H}{r} + t_4^2\, \sigma_o{}^M(\bar{\varepsilon}_4)\, A_4 \frac{H}{t_4} + t_5^2\, \sigma_o{}^N(\bar{\varepsilon}_5)\, A_5\}\ \frac{2H}{\delta_{eff}} \end{aligned} \tag{25}$$

Such a cross-section is typical for extruded aluminium elements where the difference in thicknesses of neighboring walls can be as large as one hundred percent or even more. The plastic energy dissipation in consecutive folding lobes depend now on the direction of folding of the SE. For example, in the case illustrated in Fig.7 the moving hinge line ③ sweeps the face *a* of the element and, thus, involves the deformation of a shell of the thickness t_a. At the some time the conical surface ④ is developed at the boundary between both faces and propagates into a thinner face. This feature of the folding process is reflected in the governing equation, Eq.25, by specifying an appropriate thickness for each contributing mechanism. For example, the deformation pattern in Fig.7 corresponds to the following set of thicknesses, t_i $i = 1, ...5$:

- $t_1 = t_3 = t_5 = t_a$ and
- $t_4 = t_a$ for $t_a \le t_b$ or $t_4 = t_b$ for $t_b < t_a$.

Similarly as in the case of an uniform-thickness element the equilibrium solution to the Eq.25 is obtained via the minimization procedure, Eq.12. It should be noted, however, that in this case a useful approximation of the folding mode by pure asymmetric or symmetric folding modes does not apply, even in the case of a right-angle element, and consequently, the solution must be found by minimizing the complete governing equation, Eq.25.

4.6.2 *The Governing Equation for an Assemblage of SE*
The crushing response of a single layer of folds, compare Fig's. 5 and 6, is calculated by summing up fractional contributions of all SE in a given layer. Accordingly, the governing equation of the problem is

$$P_m = \sum_{i=1}^{J} \frac{t_i^2}{4} \{ \sigma_0^N(\bar{\varepsilon}_1)\, A_1^i \frac{r}{t_i} + \sigma_o{}^M(\bar{\varepsilon}_2)\, A_2^i \frac{C_i}{H} + \sigma_o{}^M(\bar{\varepsilon}_3)\, A_3^i \frac{H}{r} + \sigma_o{}^M(\bar{\varepsilon}_4)\, A_4^i \frac{H}{t_i} + \sigma_o{}^N(\bar{\varepsilon}_5)\, A_5^i \} \frac{2H}{\delta_{eff}} \tag{26}$$

where the summation is extended over the *J* contributing SE. It is assumed here that the column is made of one material, so that, all average stresses in Eq.25 are calculated on the basis of a single constitutive relation. Each element, however, may have different geometrical dimensions: C_i, Φ_i and t_i, $i = 1, 2, \ldots J$.

Since, all elements in a given layer of folds deform with the same length of the folding wave, *2H*, there is only one *'H'* parameter in the governing equation. In order to simplify the calculation routines it is also assumed that values of the two other free parameters, i.e. the rolling radius *r* and the switching parameter α^*, are also the same in all contributing elements. Consequently, the solution procedure for an assemblage of SE parallels the corresponding procedure for a single SE.

5. Closed form solution for right angle SE made of a strain-hardening material

Theoretical procedures developed in preceding sections are applied here to the crushing problem of a rectangular column of uniform thickness, *t*, and cross-sectional dimensions *2a* and *2b*, Fig.10.

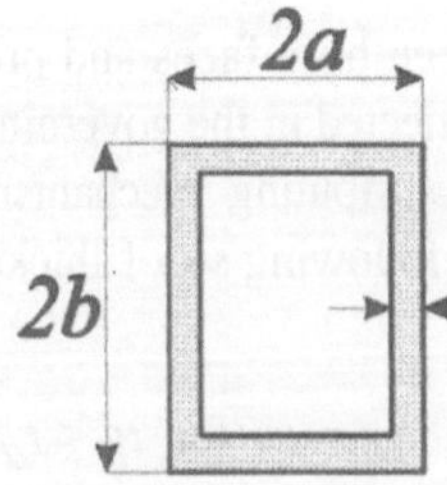

Figure 10 Cross-sectional dimensions of a rectangular column.

The column is made of a strain hardening material defined by the power-type constitutive law.

$$\frac{\sigma(\varepsilon)}{\sigma_y}=\left(\frac{\varepsilon}{\varepsilon_o}\right)^n \tag{27}$$

The governing equation of the crushing problem, Eq.25, can now be rewritten in a simplified form, applicable to a single, right-angle SE.

$$P_m=\frac{t^2}{4}(\sigma_o^1 A_1\frac{r}{t}+\sigma_o^2 A_2\frac{C}{H}+\sigma_o^3 A_3\frac{H}{r})\frac{2H}{\delta_{eff}} \tag{28}$$

where $\sigma_o{}^i$, $i = 1,2,3$, denote equivalent stresses in three main regions of plastic deformations, refer to Fig.7, while C is the length of a representative SE, $C=(a+b)$, refer to Fig.10. For a power-type constitutive relation, Eq.27, energy equivalent stresses, $\sigma_o{}^N$ and $\sigma_o{}^M$ are given as

$$\frac{\sigma_o{}^N(\varepsilon_o)}{\sigma_y}=\frac{1}{n+1}\left(\frac{\varepsilon_o}{\varepsilon_y}\right)^n$$

$$\frac{\sigma_o{}^M(\varepsilon_o)}{\sigma_y}=\frac{2}{(n+1)(n+2)}\left(\frac{\varepsilon_o}{\varepsilon_y}\right)^n \tag{29}$$

where ε_o denotes the characteristic final strain. In the case of an uniform tensile-compression deformations ε_o corresponds to the final strain measured at any point of the deformed region whereas ε_o equals to the final strain at an outer fiber in the case of bending deformations, see Ref.[12] for details.

The relevant strains for three main mechanisms of plastic deformation in Eq.28 are readily determined from the kinematics of the folding process. Consider first rolling deformations. The final strain at the outer fiber of a traveling hinge line ③ is inversely proportional to the rolling radius, r, and equals

$$\varepsilon_1=\ln(1+\frac{t}{2r}) \tag{30}$$

Similarly, the final strain at the outer surface of horizontal stationary hinge lines is

$$\varepsilon_2=\ln(1+\frac{t}{2R}) \tag{31}$$

where R is the large radius of the toroidal surface. It follows from the kinematics of the process and in particular from the formula for the effective crushing distance that radius R equals $R= 0.54H$ where $2H$ is the length of the plastic folding wave, Ref. [2]. Thus, the representative strain in the region ②, Fig.8, equals

$$\varepsilon_2=\ln(1+0.93\frac{t}{H}) \tag{32}$$

Determination of a compressive strain in the moving segment of a toroidal surface, ①, in Fig.7 involves complex calculation which exclude the possibility of getting a desired closed-form solution. However, separate numerical and experimental studies show that strains in this region are of the same order as those in moving hinge lines ③. Therefore, in the following analytical calculations Eq. 32 is used as an useful approximation of final strains at material points on the toroidal surface.

Further simplifications to the expression for a characteristic strain, ε_i, essential in analytical calculations, are achieved by expanding Eq.'s 30 and 32 into power series

$$\varepsilon_1 \cong \frac{t}{2r}; \quad \varepsilon_2 \cong 0.93\frac{t}{H}; \quad \varepsilon_3 \cong \frac{t}{2r} \tag{33}$$

where only the linear terms are retained. Substituting Eq.'s. 33 back into the expressions for energy equivalent flow stresses, Eq.29, finally yields

$$\sigma_1 \equiv \sigma_o{}^N(\varepsilon_1) = \sigma_y a_1 \left(\frac{t}{r}\right)^n \quad where \quad a_1 \equiv \frac{1}{n+1}\left(\frac{0.5}{\varepsilon_y}\right)^n$$

$$\sigma_2 \equiv \sigma_o{}^M(\varepsilon_2) = \sigma_y a_2 \left(\frac{t}{H}\right)^n \quad where \quad a_2 \equiv \frac{2}{(n+1)(n+2)}\left(\frac{0.93}{\varepsilon_y}\right)^n \tag{34}$$

$$\sigma_3 \equiv \sigma_o{}^M(\varepsilon_3) = \sigma_y a_3 \left(\frac{t}{r}\right)^n \quad where \quad a_3 \equiv \frac{2}{(n+1)(n+2)}\left(\frac{0.5}{\varepsilon_y}\right)^n$$

Having determined energy equivalent stresses, σ_i, as functions of kinematical parameters, r and H, the governing equation, Eq.28, can be rewritten in the form

$$P_m = \frac{t^2\sigma_y}{4}\left\{A_1\left(\frac{r}{t}\right)^{1-n} + A_2\frac{C}{t}\left(\frac{t}{H}\right)^{1+n} + A_3\frac{H}{t}\left(\frac{t}{r}\right)^{1+n}\right\}\frac{2H}{\delta_{eff}}; \quad A_i \equiv A_i\, a_i \tag{35}$$

where the constant coefficients A_i, $i=1, 2, 3$, are functions the central angle of a SE while coefficients a_i, $i=1, 2, 3$, defined in Eq.'s.34, are functions of material parameters only. For example, a set of coefficients A_i pertinent to the crushing response of a SE with the central angle Φ 90° is, Ref.[2-*Vol. II*]:

$$A_1 = 32\, I_1 \qquad I_1 = 0.555$$

$$A_2 = 4\,\pi$$

$$A_3 = 8\, I_3 \qquad I_3 = 1.148 \tag{36}$$

The equilibrium of the system, governed by Eq.35 is defined from the minimum conditions, Eq.12. The final results for optimal free parameters of the process, i.e., length of the folding wave $2H$ and rolling radius r, are:

$$\frac{r}{t} = A_1^{-\frac{2+n}{3+n}} A_2^{\frac{1}{3+n}} A_3^{\frac{1+n}{3+n}} K_1 \left(\frac{C}{t}\right)^{\frac{1}{3+n}} \quad \text{where} \quad K_1 = (1+n)(1-n)^{-\frac{2+n}{3+n}}$$

$$\frac{H}{t} = A_1^{-\frac{1+n}{3+n}} A_2^{\frac{2}{3+n}} A_3^{-\frac{1-n}{3+n}} K_2 \left(\frac{C}{t}\right)^{\frac{2}{3+n}} \quad \text{where} \quad K_2 = (1+n)(1-n)^{-\frac{1+n}{3+n}} \quad (37)$$

Substituting Eq.'s 37 back into Eq.35 gives the closed-form for the average crushing force P_m.

$$\frac{P_m}{M_y} = A_1^{\frac{(1+n)^2}{3+n}} A_2^{\frac{(1-n)}{3+n}} A_3^{-\frac{1-n^2}{3+n}} \left(\frac{C}{t}\right)^{\frac{1-n}{3+n}} \left[K_1^{1-n} + K_2^{-(1+n)} + K_2 K_1^{-(1+n)}\right] \frac{2H}{\delta_{eff}} \quad (38)$$

In Eq.38 the fully plastic bending moment, M_y, is calculated with respect to the yield stress of material, σ_y, and equals, $M_y = \sigma_y t^2/4$. For the perfectly-plastic material $n = 0$ and results given in Eq.'s.37 and 38 converge to the solution for a perfectly plastic material.

$$\frac{H}{t} = \sqrt[3]{\frac{A_2^2}{A_1 A_3}} \sqrt[3]{\left(\frac{C}{t}\right)^2} \qquad C = (a + b)$$

$$\frac{r}{t} = \sqrt[3]{\frac{A_2 A_3}{A_1^2}} \sqrt[3]{\frac{C}{t}} \qquad (39)$$

$$\frac{P_m}{M_y} = 3\sqrt[3]{A_1 A_2 A_3} \sqrt[3]{\frac{C}{t}} \frac{2H}{\delta_{eff}}$$

which for given values of constants A_i, Eq.36, and a given non-dimensional ratio of the length of the plastic folding wave to the effective crushing distance, $2H/\delta_{eff} \cong 1.37$, finally yields.

$$\frac{H}{t} \cong \sqrt[3]{\left(\frac{C}{t}\right)^2} \qquad C = (a + b)$$

$$\frac{r}{t} = 0.715 \sqrt[3]{\frac{C}{t}} \qquad (40)$$

$$\frac{P_m}{M_y} = 13.052 \sqrt[3]{\frac{C}{t}}$$

The later result is valid for a single right-angle element and differs form the result for the entire column in the level of non-dimensional mean force, P_m/M_y. For four SE contributing to the crushing strength of a rectangular column the mean crushing force is simply four times larger then the resistance of a single SE and equals

$$\frac{P_m}{M_y} = 52.2 \sqrt[3]{\frac{C}{t}} \qquad C = (a + b) \qquad (41)$$

while the length of the plastic folding wave, $2H$, and the magnitude of rolling radius, r, are the same as in the case of a standing alone element.

The effect of strain - hardening, discussed in this section, is summarized in Fig.11 for a material with the strain-hardening exponent, n, equal $n = 0.2$. Two graphs in this figure show variation of the mean crushing force, P_m, and plastic folding wave, $2H$, with the width to thickness aspect ratio, C/t. Both quantities, i.e. P_m and $2H$, are normalized with respect to the corresponding values calculated for a perfectly-plastic material. It is evident from diagrams in Fig.11 that the strain-hardening effect plays an important role in the crushing response of a SE. For example, the mean crushing force, P_m, is at least two times larger then the force calculated for the nominal yield stress of the material, σ_y. Also, the folding wave, $2H$, is at least 20% longer then the corresponding wave in perfectly plastic column. Both functional relations in Fig.11 are nonlinear functions of the C/t aspect ratio and are quite sensitive to the variation of the hardening exponent n. The above results illustrate the importance of the effect of strain-hardening in crash calculations of aluminium columns and emphasize the need of a precise determination of the constitutive material response in all practical crash calculations.

6. Dynamic Crushing Response of a SE

Theoretical results discussed in preceding sections apply to a static response of compressed members. In an actual calculations, however, it is important to retain the material strain-rate sensitivity, especially in the analysis of structures made of mild steel. For such structures the strain rate effects may increase the level of mean crushing force by the factor of 1.2 - 1.4 as compared to the quasi-static loading. For example, a prismatic column made of mild steel will typically absorb up to 30% more energy when compressed with a constant velocity of the order of 10 [m/s].

Aluminium alloys, on the other hand, are characterized by a weak strain rate sensitivity and the influence of this effects may be neglected in practical engineering calculations.

The strain rate effects are described here via the Cowper-Symonds uniaxial constitutive relation

$$\frac{\sigma^d}{\sigma^s} = 1 + \left(\frac{\dot{\varepsilon}}{D}\right)^{1/p} \tag{42}$$

where σ^d/σ^s denotes the ratio of instantaneous dynamic and quasi-static flow stress, respectively, while two material constants D and p are defined in uniaxial dynamic material tests.

The Cowper-Symonds formula, Eq.42, with the frequently used constants $D = 40.4[s^{-1}]$ and $p = 3.9$ is used to describe the strain rate sensitivity of mild steel in the neighborhood of the yield stress, σ_y. The meaning of the constant D, in Eq.42, is that an actual level of the quasi-static flow stress is doubled when the strain rate $\dot{\varepsilon}$ is equal to the constant D. For example, the yield stress σ_y in Eq.42 is doubled for a strain rate $\dot{\varepsilon} = 40.4\ [1/s]$.

As the plastic strain, ε, increases the sensitivity of material to the elevated strain rate decreases rapidly.

$I1 := 0.555 \quad \kappa1 := 0.5 \quad n := 0.2 \quad Ac1 := 32 \cdot I1 \quad Ct := 1,2..100$

$I3 := 1.148 \quad \kappa2 := 0.93 \quad \varepsilon u := 0.001 \quad Ac2 := 4 \cdot \pi$

$\kappa3 := 0.5 \quad Ac3 := 8 \cdot I3$

$$a1 := \frac{\kappa1^n}{(n+1)\cdot \varepsilon u^n} \qquad a2 := \frac{2\cdot \kappa2^n}{(n+1)\cdot(n+2)\cdot \varepsilon u^n} \qquad a3 := \frac{2\cdot \kappa3^n}{(n+1)\cdot(n+2)\cdot \varepsilon u^n}$$

$A1 := Ac1 \cdot a1$

$A2 := Ac2 \cdot a2$

$$A3 := Ac3 \cdot a3 \qquad K1 := (1+n)\cdot(1-n)^{-\frac{2+n}{3+n}} \qquad K2 := (1+n)\cdot(1-n)^{-\frac{1+n}{3+n}}$$

Mean crushing force, Pm, and plastic folding wave, 2H, for strain-hardening material.

$$Pn(Ct) := A1^{\frac{(1+n)^2}{3+n}} \cdot A2^{\frac{1-n}{3+n}} \cdot A3^{\frac{1-n^2}{3+n}} \cdot Ct^{\frac{1-n}{3+n}} \cdot \left[K1^{1-n} + K2^{-(1+n)} + K2 \cdot K1^{-(1+n)}\right]$$

$$Hn(Ct) := A1^{-\frac{1+n}{3+n}} \cdot A2^{\frac{2}{3+n}} \cdot A3^{-\frac{1-n}{3+n}} \cdot (1+n)\cdot(1-n)^{-\frac{1+n}{3+n}} \cdot Ct^{\frac{2}{3+n}}$$

Mean crushing force, Pm, and plastic folding wave, 2H, for perfectly plastic material.

$$Po(Ct) := 3\cdot(Ac1 \cdot Ac2 \cdot Ac3 \cdot Ct)^{\frac{1}{3}} \qquad Ho(Ct) := \left(\frac{Ac2^2}{Ac1 \cdot Ac2}\right)^{\frac{1}{3}} \cdot Ct^{\frac{2}{3}}$$

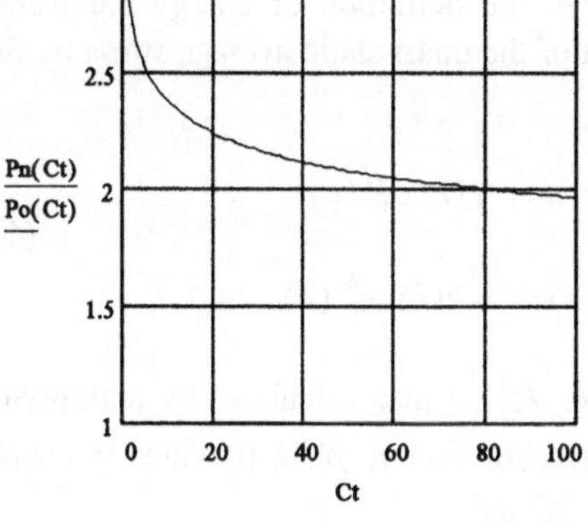

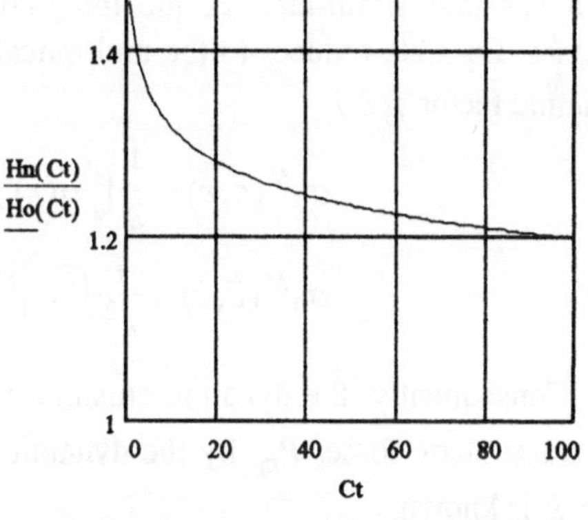

Figure 11 A MathCAD document illustrating the functional dependence of the mean crushing force, P_m, and the length of plastic folding wave, *2H*, versus the *C/t* aspect ratio. The subscripts 'n' and 'o' denote respectively the strain-hardening and perfectly-plastic response of the column's material. Calculations in this document are done for a rectangular column with $C = (a + b)$.

For example, the constants D and p in Eq.42 equal, respectively

$$D = 6844\ [1/s]$$

$$p = 3.91 \tag{43}$$

for the UTS (Ultimate Tensile Strength) of a typical mild steel, see Ref. [5]. In this case the strain rate $\dot{\varepsilon}$ necessary to double the quasi-static stress equals $\dot{\varepsilon} = 6844\ [1/s]$.

The above example illustrates a complex dependence of the constants D and p on an actual strain ε. However, it has been shown (see Ref. [5]) that the constitutive equation, Eq.42, together with material constants D and p relevant to the UTS of a given material, can be used as a practical estimate of strain rate effects in progressively crushed mild steel columns. This conclusion is also confirmed by hundreds of experiments performed by the present author.

In practical calculations it is convenient to rewrite the constitutive equation, Eq.42, in the form,

$$\sigma(\varepsilon, \dot{\varepsilon}) = \sigma_s(\varepsilon)\gamma(\dot{\varepsilon}) \quad where \quad \gamma(\dot{\varepsilon}) \equiv 1 + \left(\frac{\dot{\varepsilon}}{D}\right)^{1/p} \tag{44}$$

(compare the general form of constitutive equation, Eq.21). In Eq.44 the function $\sigma_S(\varepsilon)$ corresponds to the standard quasi-static tensile characteristic of material and, by the definition, does not depend on the strain rate $\dot{\varepsilon}$. On the other hand, the factor $\gamma(\dot{\varepsilon})$, referred to as a dynamic correction factor, does not depend on an actual strain state. Thus, in the present formulation the strain-hardening and strain-rate effects are effectively decoupled. This feature of the constitutive relation, Eq.44, greatly simplifies calculations for dynamic crush. Indeed, for a dynamic crushing process characterized by a constant strain-rate $\dot{\varepsilon}$, the integration in the definition of energy equivalent stresses, Eq.'s.20, reduces to the multiplication of the quasi-static average stress by the dynamic factor $\gamma(\dot{\varepsilon})$

$$\begin{aligned} \sigma_0{}^N(\bar{\varepsilon}, \dot{\varepsilon}) &= \frac{1}{\bar{\varepsilon}}\int_0^{\bar{\varepsilon}} \sigma(\varepsilon)\,\gamma(\dot{\varepsilon})\,d\varepsilon = \gamma(\dot{\varepsilon})\,\sigma_0^N(\bar{\varepsilon}) \\ \sigma_0{}^M(\bar{\varepsilon}, \dot{\varepsilon}) &= \frac{1}{\bar{\varepsilon}^2}\int_0^{\bar{\varepsilon}} \sigma_0{}^N(\varepsilon, \dot{\varepsilon})\,d\varepsilon = \gamma(\dot{\varepsilon})\,\sigma_0^M(\bar{\varepsilon}) \end{aligned} \tag{45}$$

Consequently, the dynamic crushing force, P_m^d, is also calculated by multiplying the quasi-static force, P_m, by the dynamic correction factor, $\gamma(\dot{\varepsilon})$, provided the strain rate $\dot{\varepsilon}$ is known.

The average rate of strain, $\dot{\varepsilon}$, is calculated from the known kinematics of the folding mechanism and assumed constant velocity, $\dot{\delta}$, of compression of a single lobe. The average strain rates depend on the deformation mode of a SE and equal

$$\dot{\varepsilon} \cong 0.33 \frac{\dot{\delta}}{C}$$

$$\dot{\varepsilon} \cong 0.43 \frac{\dot{\delta}}{C} \tag{46}$$

$$\dot{\varepsilon} \cong 0.25 \frac{\dot{\delta}}{C}$$

for the asymmetric, mixed and symmetric modes, respectively (details of the derivation procedure are given in Ref.[5]). In Eq.46 the velocity of axial compression $\dot{\delta}$ must be given in meters per second, *[m/s],* while *C* is the total cross-sectional length of a representative SE given in meters, *[m]*. For a cross-section composed of more then one SE the representative length, *C,* should be calculated by dividing the entire cross-sectional length, *L*, by the number of contributing elements, *J*.

Having determined a representative level of the strain-rate in a given cross-section the mean dynamic crushing force, P_m^d, is calculated from the formula

$$P_m^d = P_m \left\{ 1 + \left(\frac{\dot{\varepsilon}}{D} \right)^{1/p} \right\} \tag{47}$$

where P_m is the quasi-static mean crushing force defined in Eq.26.

7. Conclusions

In preceding sections a systematic introduction to the Macro Element method was presented. Then, the general method has been used to specify governing relations for the crushing response of a standing alone SE under axial loading. Subsequently, the application of the SE formulation was illustrated on an elementary example of a rectangular column subjected to axial crush loading.

The existing applications of the SE method, not discussed in the present chapter, go much further. The methodology outlined in section 4 can be used to predict crushing response of prismatic members of arbitrary complexity. Furthermore, the concept of a deformable cell, illustrated in Fig.5, lead to the formulation of a Superbeam Element, shown schematically in Fig.6. The Superbeam Element is capable of modeling crushing response of prismatic members under combined loading and in fact has already made its way to a commercial software. The concept of such a software is quite different then the concept of classical FE codes. Since the governing relations are relatively simple and do not require massive computing the main effort of designers of the software is devoted to the implementation of routines which not only calculate the crashing response but, first of all, help the user to optimize crash performances of an analyzed member.

Apart from the Superfolding and Superbeam elements there exists nowadays a whole family of Macro Elements which are dedicated to specific structural components used in many segments of transportation industry. The implementation of such elements in a new generation of computer codes poses a serious challenge to the software developers. Since an assemblage of arbitrary Macro Elements lacks the symmetry of interaction of contributing elements, the corresponding software must deal with object interaction rather then with an automated assemblage of similar elements.

Therefore, the development of such a software is linked, in a natural way, with object oriented programming and object interaction techniques.

8. References

1. J. M. Alexander, An approximate analysis of the collapse of thin cylindrical shells under axial loading, *Q. J. Mech. Appl. Math.*, **13**, 1, 10-15 (1960).
2. T. Wierzbicki, W. Abramowicz, *The Manual of Crashworthiness Engineering*,
   Volume I - *Theoretical Foundation.*
   Volume II - *Quasi-Static Progressive Crushing.*
   Volume III - *Stability of Progressive Collapse.*
   Volume VII - *Dynamic Crush - Methods and Results.*
   Center for Transportation Studies, Massachusetts Institute of Technology, (1987-1991).
3. W. Abramowicz and T. Wierzbicki, Axial crushing of multi-corner sheet metal columns, *J. App. Mech.*, **56**, 1, 113-120 (1989).
4. T. Wierzbicki and W. Abramowicz, Deep plastic collapse of thin-walled structures, in *Structural Failure* (T. Wierzbicki and N. Jones Ed.'s), John Wiley, New York (1989).
5. Abramowicz, W., Jones, N., Dynamic Progressive Buckling of Circular and Square Tubes., *Int. J. Impact Engng.*, 4, 4, 243 - 270, 1986.
6. T. Wierzbicki, *Concertina tearing of metal plates,* Joint MIT-Industry Program on Tanker Safety, Report 12, March 1993.
7. M. Kleiber and C. Wozniak, *Nonlinear Mechanics of Structures*, Kluwer Academic Publishers, Dordrecht (1991).
8. L. E. Malvern, *Introduction to the Mechanics of a Continuous Medium*, Prentice-Hall, New Jersey (1969).
9. R. Hill, Extremal paths of plastic work and deformation, *J. Mech. Phys. Solids*, **34**, 5, 511-523 (1986).
10. H. Petryk, A consistent energy approach to defining stability of plastic deformation processes, in *Stability in the Mechanics of Continua* (F. H. Schroeder Ed.), 262-280, Springer-Verlag, Berlin (1981).
11. W. Abramowicz, *Crushing Mechanics of Shell Structures*, Institute of Fundamental Technological Research, Polish Academy of Sciences, Report 35 (1981) - in polish.
12. W. Abramowicz, Crush Resistance of 'T' 'Y' and 'X' Sections, Joint MIT-Industry Program on Tanker Safety, Massachusetts Institute of Technology, Report 24 (1994).

CRASHWORTINESS OF BUS STRUCTURES AND ROLLOVER PROTECTION

MATYAS MATOLCSY
IKARUS Vehicle Manufacturing Co.
1165 Budapest, Hungary

Abstract

The paper gives a statistically based survey about the main categories of bus accidents and the major technical questions belonging to the design and construction of the body structure. General definition of the plastic hinges is given, the characteristics of which can be approached by mathematical functions having some basic parameters which depend on the geometrical and material properties of the tube or rod on which they take place. Experimental results are shown about plastic hinge characteristics. On the basis of the plastic hinge theory the front collision of buses are discussed. The rollover protection problem is discussed on the basis of the existing ECE Regulation 66. which is criticised either the test methods or the calculation method given in it. This criticism gives a starting point for the development of the regulation in the future.

1. Introduction

The passive safety problems of buses, has been arisen - in the early 70`s - on international level. The GRSA (expert group of bus safety, ECE/WP29 in Geneva) started to work on two main subjects: the rollover of buses (required strength of the superstructure) and the frontal impact (strength of seats and their anchorage, the retention of passengers) The goal of this work was to produce international regulations for bus type-approvals. In the same time there was an other international forum - the Meeting of Bus and Coach Experts, organised in every 3rd year in Budapest - which provided a good opportunity for the researches, bus manufacturers and other experts to discuss the whole subject (accident statistics, test methods, test results, computer simulation, etc.) on a technical-scientific level. This paper tries to give a survey about the most important subjects belonging to the passive safety of buses, mostly based on the Hungarian experiences, test results and research works.

J. A. C. Ambrósio et al. (eds.),
Crashworthiness of Transportation Systems: Structural Impact and Occupant Protection, 321–360.

2. Accident Analysis

When we started to work in the subject of bus crashworthiness, the first important step was to get information about real bus accidents. Organising some of the main operators and the police we collected accident statistics about buses and we have made a lot of detailed technical analysis of damaged bus structures. A general summary of the bus accident official statistics is given in Table 1. for a five year period. The table does not contain the passenger accidents when boarding and alighting, or staying on board. Table 2. gives some injury data belonging to the accidents of Table 1. The most dangerous bus accident is the collision with train, the next one is the rollover, after that the collision with rigid obstacle and with heavy vehicle. It is interesting to get some information about the impact directions acting on buses. Considering only the collisions with cars and vans, heavy vehicles and rigid obstacles as well as the rollovers from Table 1. (770 events) the following ratios were established:

frontal collision	57,2 %
rear impact	22,1 %
side impact	17,8 %
rollover	2,9 %

Co-operating with bus transport companies operating in and in the vicinity of Budapest, we had the opportunity to examine in details some serious accidents. Table 3. summarises the statistical data, focusing on the front impact. The first column gives the official statistics of an intercity bus company in the years 1975-76 while the second one the five year data of the accident examined in details. Fig.1. shows the result of a typical front impact.

Figure 1. Typical frontal impact on driver's cab

TABLE 1. General bus accident statistics

Type of accident	1978	1979	1980	1981	1982	Total	%
running over pedestrain	99	103	102	104	91	499	27,7
collision with bycicle,motorcicle	110	94	106	80	103	493	27,3
collision with car and van	131	92	98	89	88	498	27,7
collision with heavy vehicle	60	31	45	48	35	219	12,1
collision with train	2	3	1	1	2	9	0,5
collision with rigid obstacle	10	10	3	6	2	31	1,7
rollover	2	6	5	6	3	22	1,2
other	11	4	8	5	4	32	1,8
Total	425	343	368	339	328	1803	100,0

TABLE 2. Collisions with heavy vehicles and rollover

	1978	1979	1980	1981	1982	Total	
Collision with heavy vehicle							219
light injury	86	72	128	153	90	529	
serious injury	30	14	42	53	24	163	
fatality	6	7	3	3	2	21	
Rollover							22
light injury	44	29	17	38	4	132	
serious injury	9	12	5	12	1	39	
fatality	-	2	-	-	1	3	

TABLE 3. Bus frontal collisions data

	Intercity bus service 1975-1976		Analized bus accidents 1974-1978	
	No.of events	%	No. of events	%
Accident (all kind)	910	100	82	100
Front impact front wall damage	478	52	56	68
Front impact with front wall damage	478	100	56	100
Serious damage	305	64	48	82
Serious damage of driver`s cab	240	50	26	68
Serious damage of driver`s cab	240	100	26	100
Front impact	103	43	19	73
Left side angular impact	137	57	7	27

German statistical data from 1979 - given in Table 4. - call the attention to the endangered position of the driver. Considering that the average passenger capacity of a bus is 50, it means: the ratio of the drivers is 2%, while their ratio in injury and fatality is between 11-17%. These data show that the injury and fatality risk of the drivers is 5-8 times higher than that of the passengers, and mainly because of the front impact accidents. As it was mentioned. one of the most dangerous bus accident is the rollover. Table 2. shows that one rollover produces - in average - more than twice higher number of injuries, comparing it to the collision with heavy vehicle. Some further data are given about rollovers in Table 5.

TABLE 4. Injures data

	Fatality		Serious injury		Light injury	
	Events	%	Events	%	Events	%
Driver	4	14,3	99	16,9	435	11,0
Passenger	24	85,7	487	83,1	3521	89,0
Total	28	100,0	586	100,0	3956	100,0

TABLE 5. Rollover statistics

Source of data	Observed time period	Number of	
		rollover	fatalities
Cranfield Inst of Techn. U.K.	1975-78	21	77
HUK- Verhand Germany	1979-83	7	34
AUTÓKUT Hungary	1970-76	31	15

TABLE 6. Comparasion of bus and car accident rates

Relative accident parameters	Czeho-Slovakia (1970-82)		Japan (1970)		Germany (1979-83)	
	bus	car	bus	car	bus	car
accident/1000 vehicle,	63-97	24-37	62	34		
accident/10^7km			22	20		
fatality/1000 vehicle,	0,7-1,2		1,5		0,4-0,7	
fatality/10^7 km			0,5			
fatality/1000 accident	11,5	8,2			5	
serious injury/1000 accident	44,5	29			103	

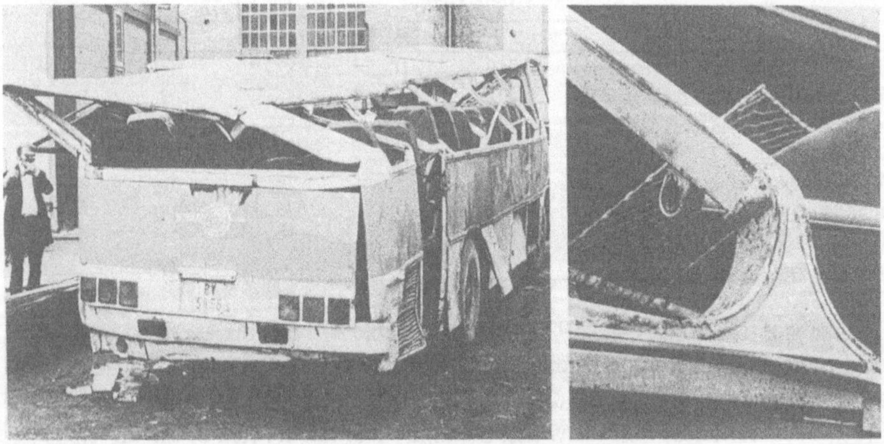

Figure 2. Bus superstructure after rollover

It should be emphasized that these data are not representative, but they are very useful to underline a very serious type of bus accident pointing out that we have to do more to protect the passengers in the case of rollover. Fig.2. shows the result of a typical rollover accident, the collapse of a bus superstructure. It could be interesting to mention that the Hungarian statistics showed in Table 5. gave the following ratio considering the three, internationally accepted bus categories:

Category I.	City bus	11
Category II	Intercity bus	11
Category III.	Long-distance coach	9

The categories in this case relate not to the service circumstances but to the construction, design of the bus. For example among the 11 city buses rolled over, 9 were in suburban and intercity service and only 2 in real city service. Table 4. shows that the bus transportation system is at last so dangerous as the cars, its responsibility is much higher. To collect and analyse the accident statistics has a great importance from many aspects, giving well based initiatives to the:

- governments and international organisations to work out new, or to develop existing bus safety regulations, requirements, standards, etc.
- leaders of the traffic-police and traffic experts how to organise the traffic, how to change the traffic rules to avoid certain type of accident
- bus manufacturers how to make buses safer, to increase their active and passive safety
- insurance companies to make more precise risk evaluation

The risk (risk of injury, risk of fatality) belongs to a certain type of accident (e.g. the injury risk in frontal collision). The more precise accident category the clearer risk definition. The reduction of injury /fatality risk by developing the construction of the bus can not be done generally. A design modification which is very effective in rollover, can be absolutely effectless in the case of a frontal impact or fire, or vice versa. The risk of injury may be expressed by injury probability. Fig.3. shows typical injury probability distribution functions, for example in case of frontal collisions. Fig.3/a gives the injury probability in the function of the bus speed. At v = 0 (the bus stands) the accident probability has a certain value P_o, which gives the probability of the accidents when an other vehicle runs into the bus. Theoretically these curves are limited at the maximum speed of the bus (v_{max}) Practically there is a certain limit speed ($v_{lim}<v_{max}$) where the accident probability reaches the value of 100% (The driver, realising the accident situation, applies the brake) Fig.3/b shows similar curves in the function of the relative impact speed. In this case the injury probability curves have meaning only if $v_r>0$, and there is no well defined maximum speed value.

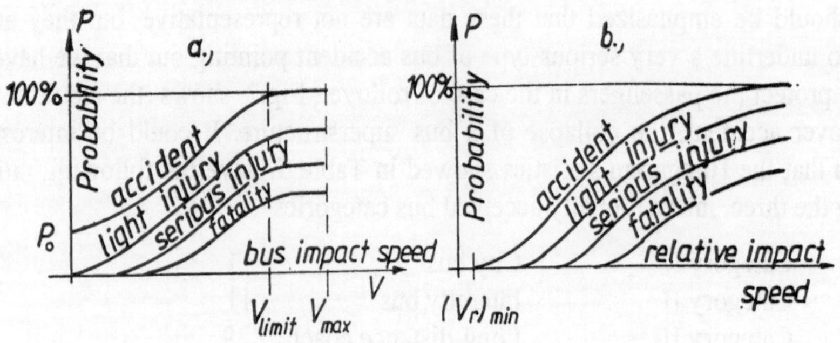

Figure 3. Accident and injury probability functions

3. Passive safety subjects of buses

3.1. FRONT IMPACT ACCIDENTS

Considering that the front impact accidents can be divided into different sub-groups (e.g. total front impact, partial impact, angular impact, impact with pole type object, etc.) a lot of safety subjects may be considered:

3.1.1. Safety Bumper

The goal of having safety front bumper can be more (see Fig.4.)

- if the relative weighted impact speed is smaller than a given value ($v_r \leq v_{r1}$) no permanent deformation is allowed on the front wall, or in other words: the bumper should have only elastic deformation. This requirement serves economical goal: to reduce the cost of repair
- if the relative impact speed is in a given range ($v_{r1} < vr \leq v_{r2}$) the permanent deformation of the safety bumper is acceptable, but no deformation on the front wall. In this case the bumper can be replaced. This requirement serves both economical (see above) and technical goal: to reduce and limit the deceleration.
- if the impact speed is high ($v_r > v_{r2}$) the permanent deformation of the front wall and the whole body structure should be designed and controlled. This requirement serves mainly technical goal: the limitation of the deceleration.

3.1.2. Strength of Passenger Seats and Their Anchorage

According to the deceleration coming from the front impact, the passenger seats get different forces and bending moments. These loads (e.g. mass forces, knee impact and chest impact from the passenger seating behind) have different time-history. The seats and seat anchorages must be strong enough when the loads belonging to certain deceleration are acting.

3.1.3. Retention of Passengers

This problem partly belongs to the appropriate strength of the seats and partly to other retaining systems (e.g. safety belt, partition, etc.) The biomechanical limits (knee force, chest and head deceleration) are connected to this safety subject.

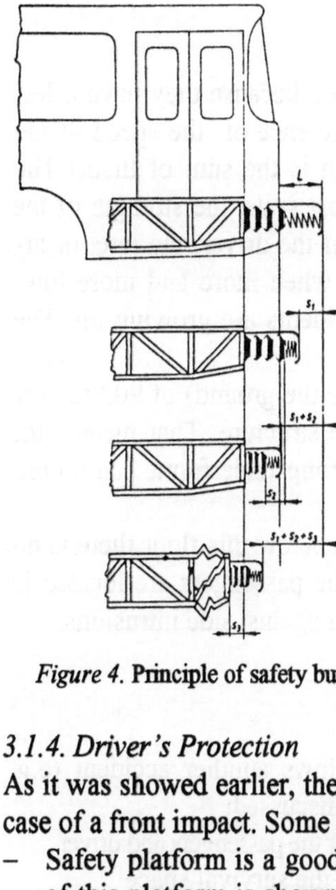

Figure 4. Principle of safety bumper

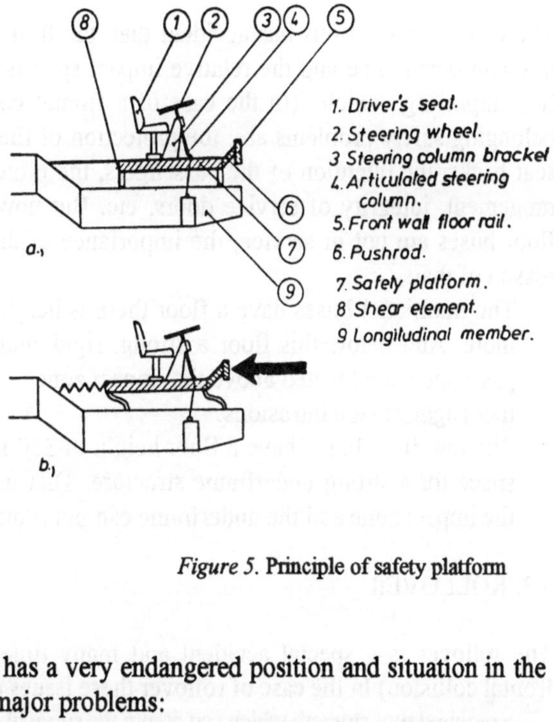

Figure 5. Principle of safety platform

3.1.4. Driver's Protection

As it was showed earlier, the driver has a very endangered position and situation in the case of a front impact. Some of the major problems:

- Safety platform is a good tool to reduce the injury risk of the driver. The principle of this platform is shown on Fig.5. The driver's seat and the supporting bracket of the dashboard and steering column are fixed to the rigid safety platform so they „move back" together with the platform when the impact force is acting. Shearing elements (plastic hinges) help to the platform to move back easily.
- Retention of the driver: strength of the driver's seat, use of safety belt or air bag, etc.
- The strength of the driver's cab against impact forces having different angular directions simulating different partial front impacts. The main goal is to avoid any structural intrusion into the driver's body

3.1.5. Integrity of Important Structural Elements

There are some very important structural elements - e.g. steering gear, break control device, front service door, front wheels - the working capability of which has an essential importance from the safety point of view of the bus during and after the frontal accident.

3.1.6. Interior Arrangement

The location and strength of the handles, the padding of rigid structural elements help to avoid the passenger's injury when moving uncontrolled in the case of an impact. The protection of the standing passengers has an essential importance.

3.2. REAR AND SIDE IMPACT

These are a less serious bus accident than the frontal impact, because they have a less frequent occurrence and the relative impact sped is the difference of the speed of the two impacting vehicle. (In the case of a frontal collision it is the sum of them) The belonging safety problems are: the protection of the rearmost seats, the strength of the seat backs, the retention of the passengers, the protection of the driver, the interior arrangement, integrity of service doors, etc. But nowadays, when more and more low-floor buses are put in service, the importance of these accidents are growing up. The reason of that:

- The traditional buses have a floor there is height (above the ground) of 900 mm or more. And below this floor a strong, rigid underframe structure. That means: the passengers are located above the impact zone and the strong underframe can protect them against side intrusions.
- The low-floor buses have a floor height of 350 mm and below this floor there is no space for a strong underframe structure. That means: the passengers are located in the impact zone and the underframe can not protect them against side intrusions.

3.3. ROLLOVER

The rollover is a special accident and many times it follows another accident (e.g. frontal collision) In the case of rollover these issues can be discussed:

- a required roof strength which can assure the survival space for the passengers and driver
- to avoid the ejection of the passengers, holding them in the survival space
- to avoid the uncontrolled motion of the passengers inside the bus, during rollover

3.4. FIRE

The fire is a very dangerous - relatively rear - accident, but it is not directly related to the problem of crash, therefor it will not be discussed in details. The following safety questions belong to the fire:

- constructional features (isolation of engine compartment, location of fuel tank, etc.)
- the use of fire-resistance materials
- the danger caused by the increasing heat, dense of smoke and poisoning gas mixture.
- quick evacuation of the bus, the number, location and operation of emergency exits.

4. International Safety Regulations

4.1. PRINCIPLE OF PASSIVE SAFETY

The goal of the passive safety developments and regulations is to reduce the injury risk of the passengers down to an acceptable limit in a measurable way, to ensure their survival possibility in a certain accident situation. To understand this general statement, some ideas have to be discussed and cleared.

4.1.1. Standard Accident

It is clear that every safety development (and regulation) is connected to a certain type of accident, but every type of accident has very large number of variations in respect to the severity and the way of process. For example: a bus overturn is when it turns down to its side with a speed closing to zero, and another overturn is when the bus rolls down into a precipice having a depth of 20 m. It is obvious that in the first case the passengers must be protected and in the second case this is almost impossible. Talking about injury risk and its reduction a standard accident has to be defined on the basis of the followings. It must be based on statistical evidences, technically well-defined, producible repeatable and agreed by the society in respect of its severity.

4.1.2. Risk of the Passengers

This means the injury probability in a given accident situation, including the degree of injury severity (e.g. light, - serious injury, fatality) The acceptable risk means the implicit agreement of the society about the tolerable injury probability in a certain standard accident. It is not the goal of the safety developments to reduce the risk down to zero in every case. Not because the risk depends not only on the standard accident but also on the individual health condition of the passengers and on a lot of accidental things.

4.1.3. Life Danger

The life danger is a special kind of risk, which can be caused by the following major effects:

a) loosing the required survival space (when structural elements, even sharp, broken edges penetrate into this space, into the passenger)
b) the deceleration creates undesirable motion of the passengers in passenger compartment impacting different structural parts and these inner impact forces can cause fatality.

The individual capability of the passengers, their tolerance limits against different kind of impacts appears in the accident statistics, in the number of fatalities, but it can not be involved into the international regulations. These figures are replaced by the average biomechanical parameters of the human being.

4.1.4. Survival Possibility

It means the main components of circumstances which give a chance to the passengers for survive. The survival possibility and the injury risk may be determined by detailed analysis of real accidents, standard accidents (tests) and the biomechanical behaviours of the human being.

4.1.5. Test Methods

It is very important to have reproducible test methods - mainly the standard accident or equivalent tests - which can prove that the vehicle is acceptable in respect to the injury risk or survival possibility.

4.2. ECE SAFETY REGULATIONS

One of the most important international safety regulations are belonging to the ECE Geneva Agreement. (Another very important group of the safety requirements are the EEC Directives, but from the technical side they are almost the same as the ECE Regulations) Among these regulations There are 7 regulations dealing with the active safety of buses and 4 regarding to the passive safety. These are:

Regulation 36	Some safety issues, like emergency exits, interior arrangement, etc. regarding to large buses
Regulation 52	The same for mini and midi buses.
Regulation 66	Roof-strength in case of rollover
Regulation 80	Strength of seats and seat anchorage

Comparing the subject of the existing regulations and the passive safety subjects mentioned in para.3., it can be seen that many of very important problems (like driver's protection, safety bumper, integrity of important structural element, side impact of low-floor buses, general use of safety belts, etc.) is not regulated yet.

5. Plastic Hinges

Studying the damaged bus frame structures after a collision, it can be established that the deformations are not evenly distributed; some structural elements are deformed and others are not and also in one individual element, there are locally strongly deformed sections and large rigid undeformed parts.

5.1. DEFINITION OF PLASTIC HINGES

The concept of „plastic hinge"(PH) is widely used in the technical publications, even when there is no exact, internationally accepted definition of it. In this paper the following definitions will be used for PH:

5.1.1. Elementary Hinge

On rod-like elements including the thin-walled tubes (where the length of the element is much bigger than the two other dimensions) that relative small section is called as PH where:

a) the local buckling results in the loss of stability
b) the plastic deformation is concentrated being larger than the other parts of the rod
c) the original cross section of the rod is essentially distorted
d) the length of which is bigger than the rod cross section, where it is formed

Fig.2. shows elementary PH formed on a bus superstructure. It should be emphasised that the local, high scale plastic deformations of PH results in large scale rotation, displacement, shortening, etc. of the rigid undeformed part(s) of the rod, on which the hinge is formed.

5.1.2. Combined Hinge

This is developed on special safety structural elements (e.g. safety rings for rollover, safety platform in the driver's cab, or safety bumper system, etc.). These special structural elements generally are not rod-like type; they contain more tubes and connecting plates. Therefore, the basic assumption that their length is much bigger than the two other dimensions is not valid anymore. Trying to generalise the PH definition given above, we have to comment the four conditions:

a) more than one local buckling take place and the loss of stability is the collective result of them. It means, for example that the first (and second..) local buckling does not cause the loss of stability yet.
b) the high scale plastic deformations are connected to the narrow surroundings of the individual local bucklings
c) while the special elements are not rod-like type, the distortion can not be common to all of the neighbouring local bucklings
d) generally it is meaningless to talk about the length of the plastic hinge

The reason, why we can call this sophisticated phenomenon as PH is the fact that the other parts of the special structural elements are undeformed (rigid) while they endure large scale rotation, displacement, shortening, etc. because of the intensive plastic deformations of the hinge region. Fig.6. shows an example for the combined PH formed on a bus safety ring, which is connected to the cross member of the underframe structure. The first local buckling formed on the inner column of the plated safety ring (No.1.) after that the plate was strongly deformed in the neighbourhood of the first buckling and influenced by the weldings. (No.2.) Local buckling can be observed on the side-wall column, too (No.3.) Finally the upper horizontal rail of the cross member was deformed (No.4.) The final result of these local buckling and deformations produced a rotation of the safety ring compared to the cross member. The other parts of the safety ring and the cross member remained underformed.

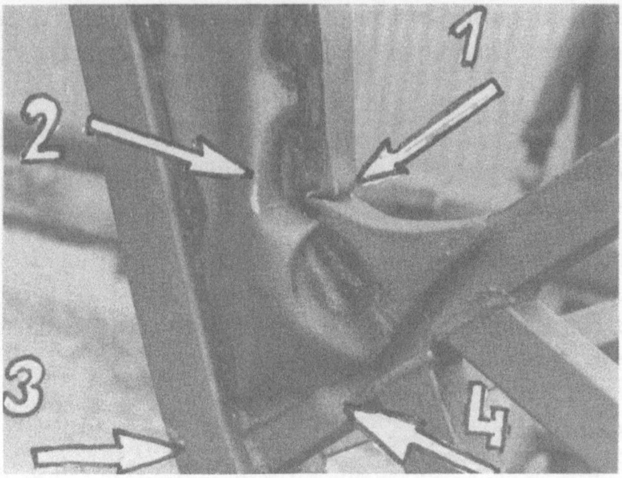

Figure 6. Combined plastic hinge

5.2 TYPES OF PLASTIC HINGES

There are many possibilities to categorise PH-s; the basis of our categorisation is their motion capability.

5.2.1. Linear Plastic Hinge

When a compressive force, acting parallel to the longitudinal axis of the rod creates the PH and its deformation results a compressive displacement, the original longitudinal axis of the rod does not change, there is no angular distortion. Fig.7. shows two examples from real bus structures. „a" is a folding type PH while „b" is a buckling type one. The buckling type linear PH, formed on „T" joint of thin-walled tubes, can be seen on Fig.8. where the tube dimensions are 40x40x1,5 mm. It is interesting to mention that depending on the tube thickness different kind of buckling mechanisms can be observed. Fig.9. shows the same „T" joints with different thickness (40x40x3 mm) and with different buckling mechanism. Fig.10. gives a new variation with „T" joint: if the joint is rounded (like in the case of bus window columns), the PH is formed in the transition range and its shape, depending on the technology differs from the two others discussed above. There are two kinds of linear PH (see Fig.11.):

a) linear hinge with unlimited displacement. In the needed working range of the hinge (during the considered deformation) there is no limitation
b) linear hinge with limited displacement. The working range of the hinge is clearly limited by geometrical reasons.

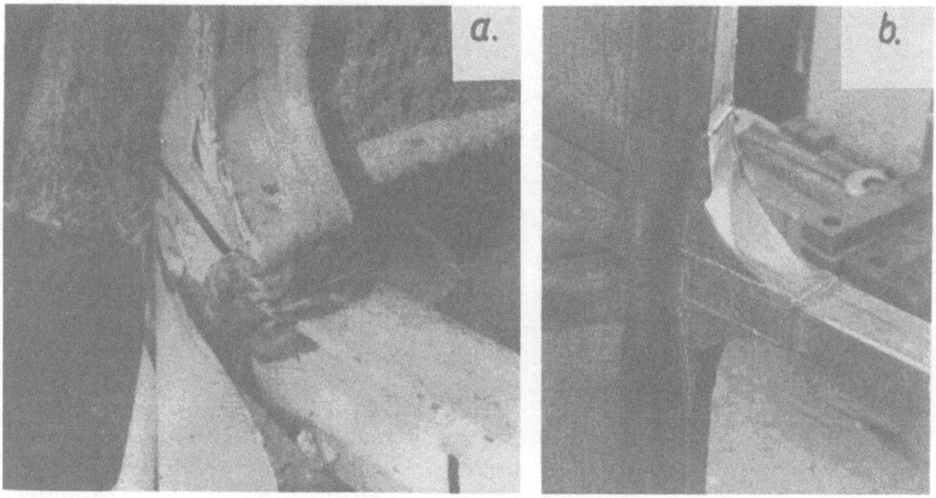

Figure 7. Linear PH-s on real bus structures

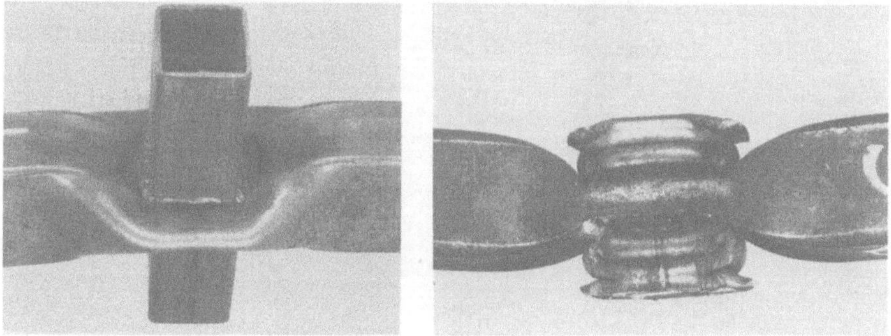

Figure 8. Plastic hinges on „T" joint

5.2.2. Rotational Plastic Hinge

When a bending moment creates the PH and it has an angular deformation or rotation, or in other words, the original longitudinal axis of the rod is broken and there is an angular distortion. This kind of hinges can be observed not only on window and door columns (See Fig.2) but for example on a lattice type underframe structure when tested against front impact forces (Fig.12.) There are two kinds of rotational PH-s: (See Fig.11.)

a) rotational hinge with unlimited angular displacement.
b) rotational hinge with limited angular displacement. The limitation is given by the geometry or by fracture.

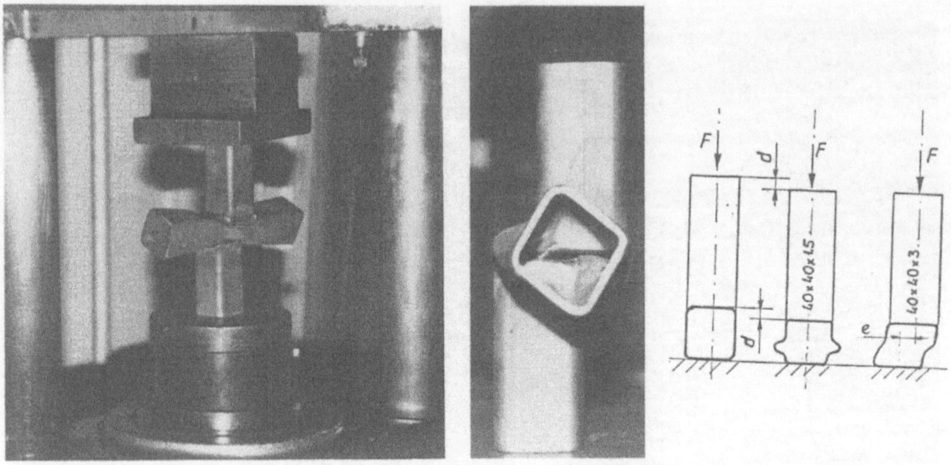

Figure 9. **Different types of „T" joint**

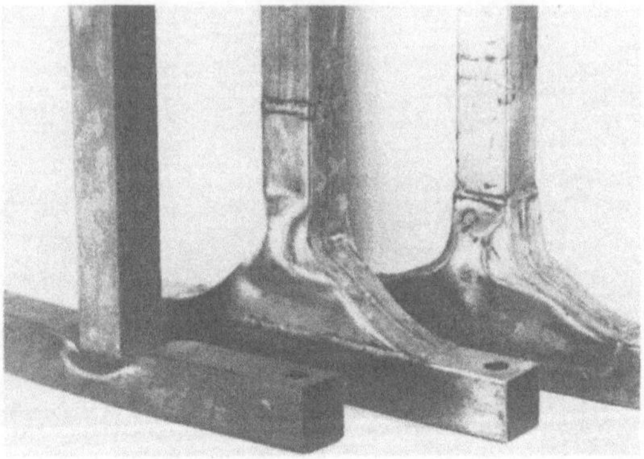

Figure 10. **Different window columns**

5.2.3. Combination of Elementary Plastic Hinges

There are hinges, which start to work as linear hinges and after that they get rotational displacement and change into a rotational hinge. (See Fig.11.) Examples are given on Fig.13. showing the front wall - side wall connection of a full scale specimen after a pendulum impact test, simulating a front impact accident. The white arrows show combinations of elementary PH -s. Fig.8. shows a combination of two linear - one folding and one buckling type - PH working after each other.

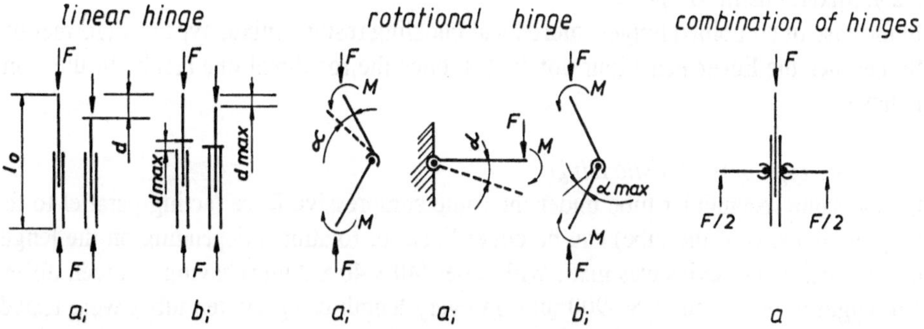

Figure 11. **Different types of plastic hinges**

Figure 12. **Compression test of underframe structure**

5.2.4. Mixed Plastic Hinge

In the case of combined hinges, more local buckling result a mixed type of deformation. Sometimes the linear behaviour dominates, once the rotational one results in the same manner.

5.2.5. The Type of the Plastic Hinge

On the same rectangular tube under the same compressive force (acting parallel to the longitudinal axis of the tube) can be either linear or rotational depending on the length of the tube. A test series was made with tubes (40 x 40 x 2 mm) having fourteen different length in the range of 50-900 mm. In every length category, ten tubes were tested. On the short tubes linear plastic hinge was formed and on the long ones rotational hinges occurred. In the middle length-range, a certain ratio of the tubes with a given length showed rotational and the others linear plastic hinges. Fig.14. (upper line) gives the probability of forming linear PH on the tube, when the two ends of the tube are free:

- if the length of the tube is less than the critical value of $l_{c1} = 200$ mm, the probability of forming linear hinge is 100%
- if the length is more than another critical value of $l_{c2} = 700$ mm, the probability of forming linear hinge is 0% or in other words: rotational hinge is formed
- between the two critical values the formation of linear hinge has a decreasing probabi lity.

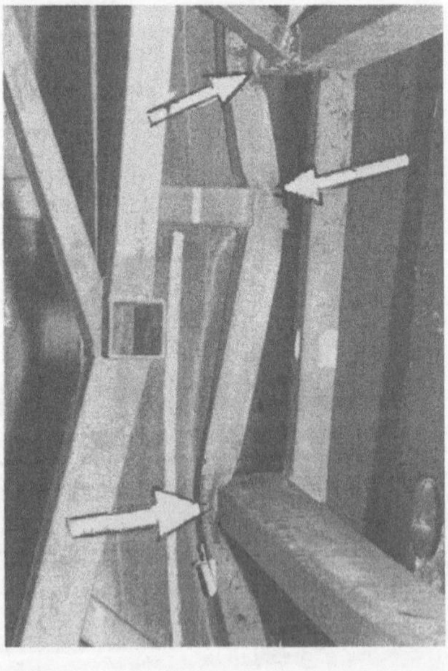

Figure 13. Combined PH on front wall frame

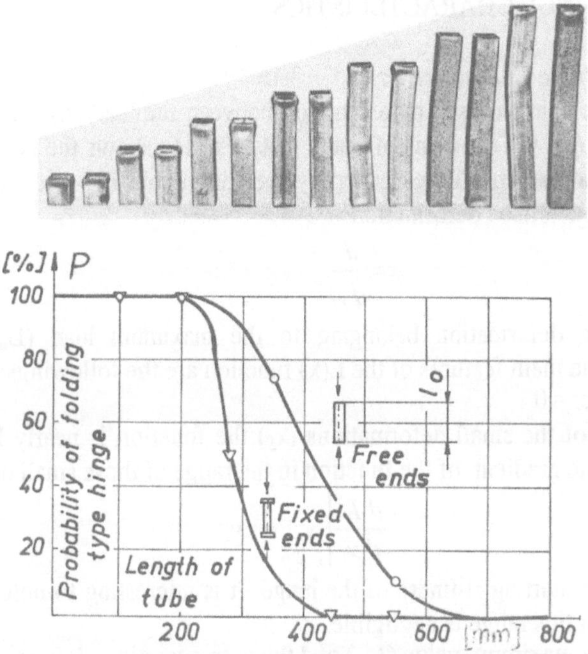

Figure 14. **Probability of forming folding type plastic hinge**

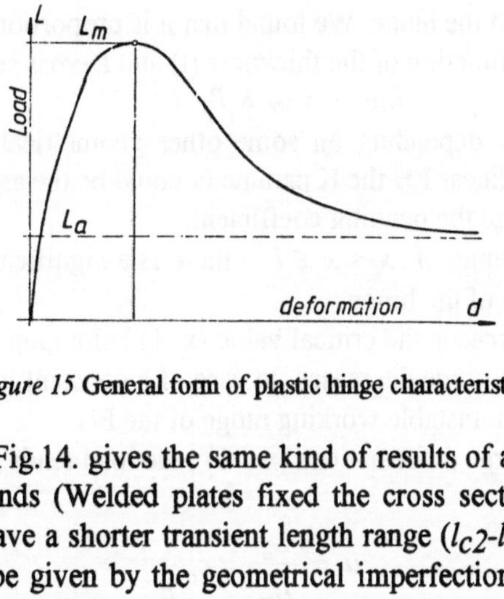

Figure 15 **General form of plastic hinge characteristic**

The lower line on Fig.14. gives the same kind of results of a parallel test series using tubes with fixed ends (Welded plates fixed the cross sections at the two ends). These kinds of tubes have a shorter transient length range (l_{c2}-l_{c1}) The explanation of this phenomenon can be given by the geometrical imperfections of the tubes coming from the manufacturing process.

The longer tube provides a higher probability of a significant bending moment when acting a parallel compressive force, and this moment creates a rotational PH

5.3. PLASTIC HINGE CHARACTERISTICS

5.3.1. General Hinge Characteristic

Every PH can be characterised by a function between the load (*L*) and the corresponding deformation (*d*). On the basis of many tests, Fig.15. shows the general form of the plastic hinge characteristic. When generally speaking about PH-s, it is useful to use the concept of the generalised deformation:

$$x=\frac{d}{d_m} \tag{1}$$

where d_m is the deformation belonging to the maximum load (L_m) of the hinge characteristic. The main features of the L(x) function are the followings:

- if $L=0$ then $x=0$
- in the range of the small deformations (x_S) the function is nearly linear, it can be linearized. The gradient of the function in the range of these small deformations:

$$\left.\frac{dL}{dx}\right|_{x=0} = v \tag{2}$$

  relates to the starting stiffness of the hinge. It is interesting to note that the energy absorption in this range is negligible.
- the load has a maximum value (L_m) and the corresponding deformation (x=1)

$$\left.\frac{dL}{dx}\right|_{x=1} = 0 \tag{3}$$

gives the starting point of loss of stability. The load maximum is an important value showing the loadability of the hinge. We found that it is proportional to the yield strength (R_y) of the material, the function of the thickness (t) and a cross section parameter (K)

$$L_m = C_m\, K R_y\, t \tag{4}$$

where C_m is a constant, depending on some other geometrical ratios ($C_m < 1$) For example in the case of a linear PH the K parameter could be the area of the cross section, while in a rotational hinge the bending coefficient.

- in the deformation range of $x_S < x \leq 1$, there is a significant plastic deformation, non-linear behaviour of the hinge.
- if the deformation exceeds the critical value (x>1) belonging to the loss of stability, the loadability of the hinge decreases. In case of a constant load reaching the maximum value, this is an unstable working range of the PH
- in the range of the large deformations (x>>1) , the load tends to an asymptotic value (L_a) The ratio:

$$\mu = \frac{L_a}{L_m} = c_a \frac{t}{b} \tag{5}$$

can be related to the distortion of the cross section at the plastic hinge and it is proportional to the relative thickness (t/b), while c_a is a constant

- the energy absorbed by the hinge, until a certain deformation (x) is

$$W(x) = \int_0^x L(x)dx \tag{6}$$

The average energy density of the hinge is the relative absorbed energy:

$$\overline{W} = \frac{W(x)}{x} = \frac{1}{x}\int_0^x L(x)dx \tag{7}$$

which is a good characteristic when comparing different PH-s from the point of view of energy absorption.

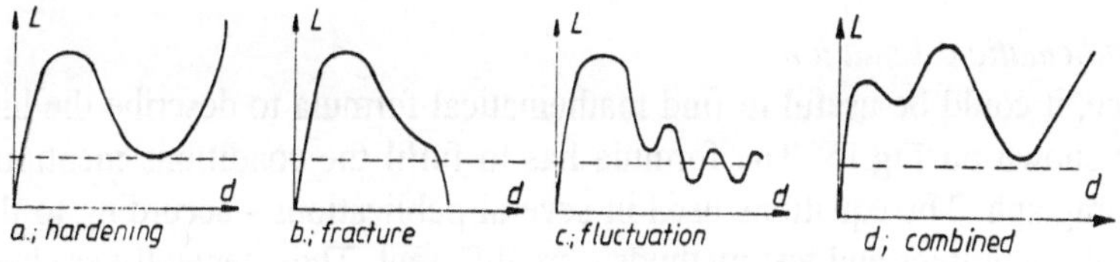

Figure 16. Deviations from the general characteristic

Figure 17. Fractures on safety rings of bus frame

5.3.2. Deviations from the General Form

There are at least four kinds of major deviations:

a) in the case of limited deformation, a hardening process takes place, when the deformation capability is exhausted, the deformation-load curve starts to increase very rapidly and the gradient can reach or exceed the starting stiffness ν. (see Fig.16/a.)

b) if a fracture takes place in a plastic hinge, the hinge loses its loading capability; the deformation - load curve quickly decreases and can reach the zero, see on Fig.16/b. The fracture starts on the tension part of the hinge, as Fig. 17. shows.
c) in the case of folding type hinges, waves are superimposed on the general deformation-load curve. This fluctuation is related to the individual folds following each other. It can be seen on Fig.16/c.
d) in the case of a combined hinge, when different local bucklings follow each other (see Fig.6) dominating in different way the PH characteristic, (See on Fig. 16/d.) Although the individual local buckling („sub-hinge") can be described by Eq.(8), the final formula generally can not be given in a simple, explicit form.

5.3.3. Mathematical Equation

In practice, it could be useful to find mathematical formula to describe the hinge characteristic shown on Fig.15. The formula has to fulfil the conditions mentioned in the earlier paragraph. The equations used in several publications - according to the several types of plastic hinges and test methods - are different. They generally are based on an exponential type function. The following five-parametric empirical equation can satisfy fairly well the conditions discussed above:

$$L(x) = L_a\left(1-e^{-\nu x}\right)\left(1+\beta x^{\alpha} e^{-\gamma x}\right) \quad (8)$$

where α, β, γ are constants, depending on the geometrical and material properties of the rod, on which the plastic hinge takes place, while ν is the starting stiffness defined in Eq.(2). Unfortunately α, β *and* γ are not independent from each other, there are very complicated relations among them. This are proved either by test results or by Eq.(8) from which it comes:

$$\beta = \frac{w(u-u\mu+1)}{\mu(u-1)} \qquad \gamma = \frac{\nu}{(u-1)\left(1-\mu+\dfrac{\mu}{u}\right)}+1 \quad (9)$$

where $u = e^{\nu}$ and $w = e^{\gamma}$. Eq.(8) involves the formula used by Tidbury [1]: if $\alpha = 0$, $\beta = 1$ and $x > 1$. This approach is based on the assumption that the initial part of hinge characteristic is linearly elastic and $x<1$ negligible small. On the basis of our experiences this assumption generally is not valid. In the range of small deformations, where the effect of the horizontal asymptote can be neglected, the unit in the second bracket is negligible compared to the product, and if $\nu \to \infty$, $\beta = 1$, we can get the equation suggested by Voith [2].

5.3.4. Probability Approach

The local buckling and the loss of stability are influenced by a lot of accidental effects, like the geometrical, technological and material defects, and imperfections. When loading more, nominally identical rods and tubes in the same way, the main parameters of the hinge characteristic have a significant scatter. The deformation process can be considered as a random process and the main parameters - like the maximum load, asymptotic load, etc. - should be described by probability functions. On the basis of test results the Gaussian normal distribution can be proposed for this purpose. From this fact, it follows that the absorbed energy, and also the other technical features of the vehicles depending on these essential PH parameters (impact forces, decelerations, deformations around the survival space, injuries, etc.) have a certain occurrence probability. It is very important to know when designing and calculating a required safety level that it can be done only with a well-defined probability value, so the absolute safety does not exists theoretically; practically it is very expensive.

5.3.5. Dynamic Characteristics

The plastic hinge characteristic (see Eq.8) is based on the results of static or quasi-static tests; but in the real accidents the deformation process is quick and the impact is a dynamic process. The comparison of dynamic and static test results [4]. has shown that strain rate has an influence on the PH characteristic, especially on the maximum load (L_m) and on the average energy density (see Eq.7) When the strain rate is discussed, it is important to emphasise, there is only a very weak correlation between the impact speed of the vehicle and the strain rate in the PH deformation. The real impact speed can be modified by two ways:

- the structural stiffness between the impact surface and the PH can have a strong influence; the softer structure leads to higher reduction in the strain rate.
- the buckling and folding mechanism is the other essential effect on the strain rate.

The shape and the general form of the dynamic hinge characteristic is similar to the static one; the ratio between the two parameters mentioned above, can be expressed as fo llows:

$$\frac{L_m}{L_{md}} = a_1 \mathrm{E}^{c_1} \qquad \text{and} \qquad \frac{\overline{W}}{\overline{W_d}} = a_2 \mathrm{E}^{c_2} \tag{10}$$

where L_{md} is the maximum dynamic load, $\overline{W_d}$ is the dynamic energy density and E is the strain rate, while a_1, a_2 c_1 and c_2 are constants, depending on the structural geometry and material properties as well as on the mass ratio between the impacting and impacted parts. Fig.18. shows static and dynamic (pendulum) bay tests: the location and the shape of the PH are similar.

5.3.6. Repeated Loading of a Hinge

When a complete bus rolls over, when it makes a rotation at least a degree of 270°, or in other words the roof of structure hits the ground on both cant rails, following each other, the PH formed for example on a window pillar first gets a bending moment from right side causing a certain deformation (rotation) and after that a bending moment from left side when the earlier deformed PH has to deform „backwards". It is interesting to emphasize that both hinge characteristics can be described by Eq.(8) but the „backword" curve has lower load values (L_m and L_α) and different parameters. It has an „alongated" shape on the same deformation scale.

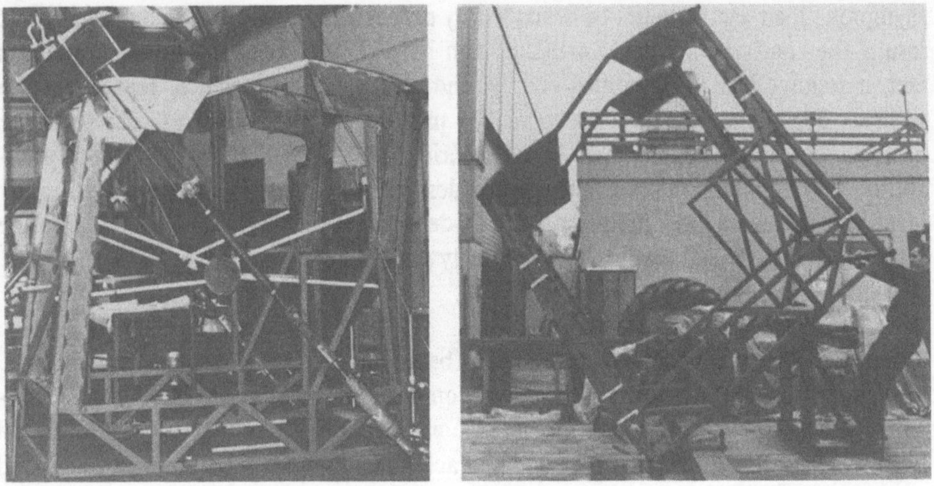

Figure 18. Static and dynamic pendulum bay tests

5.4. SOME CONSTRUCTIONAL VIEW-POINTS OF FORMING PLASTIC HINGES

Having the same outside contour and same loading direction but different construction of bus safety rings, the location and the form of the PH can be very different. Fig.19. compares three different kind of safety rings from bus superstructures:

a) simple ring containing an underframe cross member, two side walls with a window pillar and a cross rod of the roof. Acting a cant rail load, four PH -s formed similarly to Fig.2. Two rotational hinges at the upper cant rail and two on the window column, just above the lower rail of the window.
b) having a strongly reinforced safety ring, the location and the form of the hinges drastically changed: two combined hinges at the floor level, one on the roof cross member and a broken one at the right cant rail.
c) in the case of a smaller renforcement and simulating the supporting effect of seats and partitions connected to the side walls, the PH formed on the roof cross member and at the lower rails. Fig.18. also gives very good evidences about these, because the tested bays had asymmetric safety ring.

If the safety ring is too wide (deep), the fracture in the tension part has a high probability. Fig.17. gives good examples: having a 320 mm wide safety ring the combined PH on the compression side worked well but on the other side the tension caused fracture. It should not be forgot that the broken PH is exhausted, no further energy absorption and no more loading capability. This phenomenon calls the attention to a very important fact: in this case „the stronger the better" is not valid. Fig.20. shows a lot of interesting details. The position of the PH formed on the rear wall column is different from that one being on the window column in front of the rear wall. The rear wall hinge is doubled: one is at the lower window rail and the other at the upper rail of the rear wall cross member. Two interesting PH -s can be seen on the figure and a torsional hinge was formed on the diagonal of the rear wall cross member.

Figure 19. Different safety rings after pendulum impact

Figure 20. Deformed safety ring in the rear wall

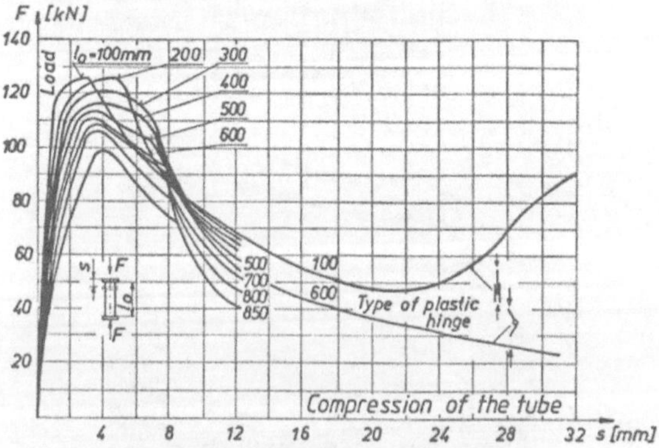

Figure 21. Effect of tube length on the hinge characteristic

6. Test Results

6.1. TESTING ELEMENTARY HINGES

Fig.21. gives the results of a test series with compressed rectangular tubes of different lengths Two types of PH can be observed on this figure: the linear (folding type) hinge and the rotational one. The short tubes produced a hardening type characteristic, when the folding capabilities of the tube was exhausted. These test results came from the test series mentioned in para 5.2.5. The probability distribution function of the maximum load as well as that of the following local maximum values, measured on a sample of 10 rectangular tubes (40x40x2 mm with a length of 150 mm) forming a folding type linear PH is shown on Fig.22/a. The measured values are plotted on a special Gaussian normal probability paper, which helps to linearise the test results. The same phenomenon can be observed in the case of dynamic impact test. Fig.22/b. shows the distribution functions of the maximum load measured on linear PH under dynamic impact load. (The pendulum had an impact velocity of 14,3 m/s and impact energy of 8 mkN.) Fig.23. compares the energy absorption in the function of the maximum deformation for static and dynamic load . Fig.24. shows the relations between the impact speed and the maximum compression of the linear PH. The last two figures relates inplicite to Eq.(9) Fig.25. gives the characteristic of rotational PH. The bending moment is plotted against of the angle of rotation. On figure „a"the specimens have modelled two kinds of bus window column: a rectangular joint and a rounded one. The.figure „b"shows the effect of different tube geometry. Fig. 26. demonstrates the phenomenon of repeated bending of a rotational hinge (see para 5.3.6.) The results of another test series with „T" joints (linear PH, see para 5.2.1) are given on Fig.27. with different geometrical configur ations.

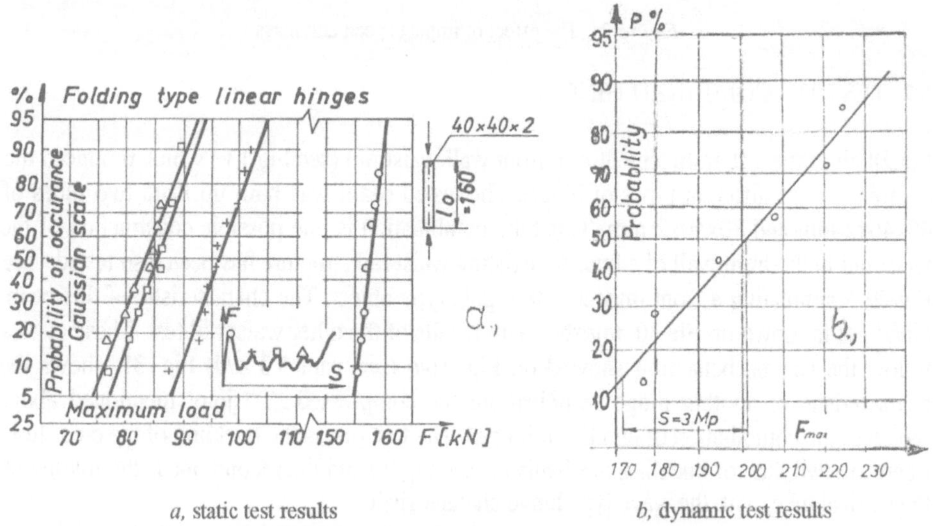

a, static test results *b*, dynamic test results

Figure 22. **Distribution functions of load maximums**

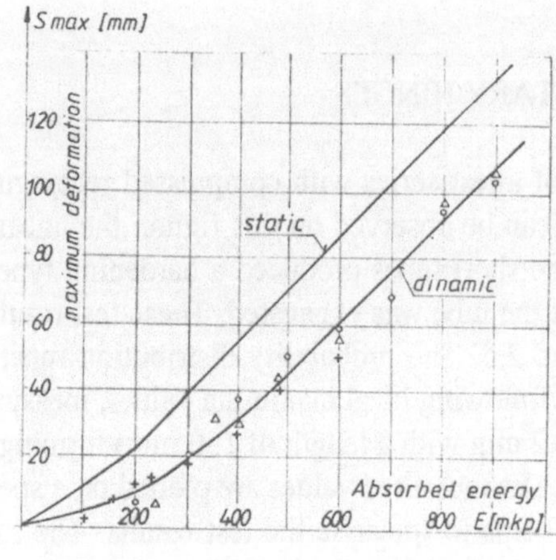

Figure 23 **Static and dynamic enery absorption**

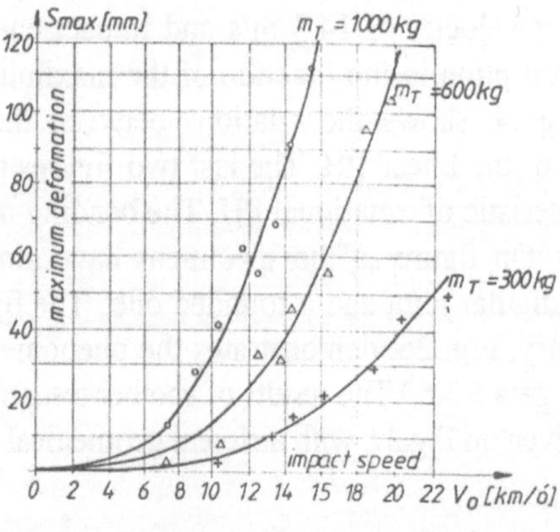

Figure 24. **The effect of impact speed and mass**

6.2. TESTING COMBINED HINGES

Fig.28 gives the test arrangement of a front wall waist rail (see Fig.13.) which produced the combination of different types of hinges The tested beam was built up from two tubes of 40x40x2 mm and 80x40x2 mm. This tube combination is one possible construction of the waist rail in the front wall of a bus. An existing waist rail structure has been also tested - see Fig 29. - simulating a front impact with a pole type object. The characteristic of this co m-bined PH is shown on Fig.30. together with results of the other waist rail test. These curves follow the general behaviour showed on Fig. 16/d (see para 5.3.2./d) Fig. 31. shows the characteristic of another hinge combination: the compressed „T" joint mentioned above produces the combination of two linear hinges. The first one is the buckling of the cross tube (there are two kind of buckling mechanisms, see Fig.9.) and the second one is the folding of the vertical tube, with the wave type hinge characteristic.

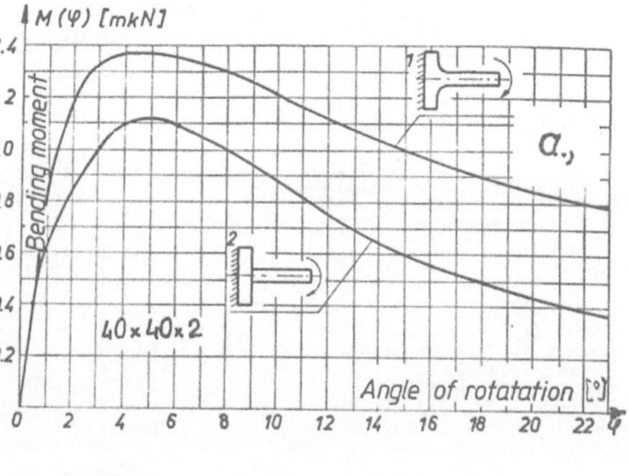

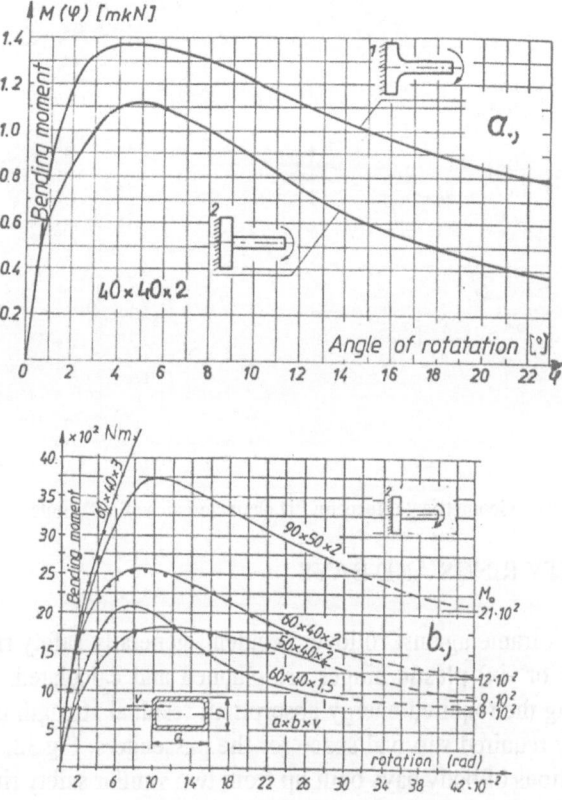

Figure 25. **Rotatinal PH characteristics**

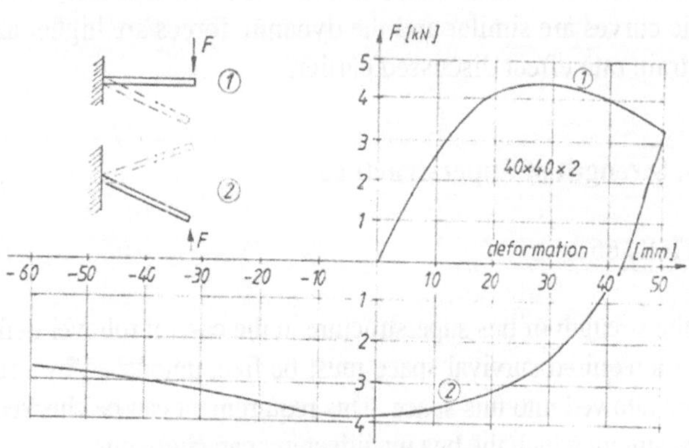

Figure 26. **Repeated bending of PH**

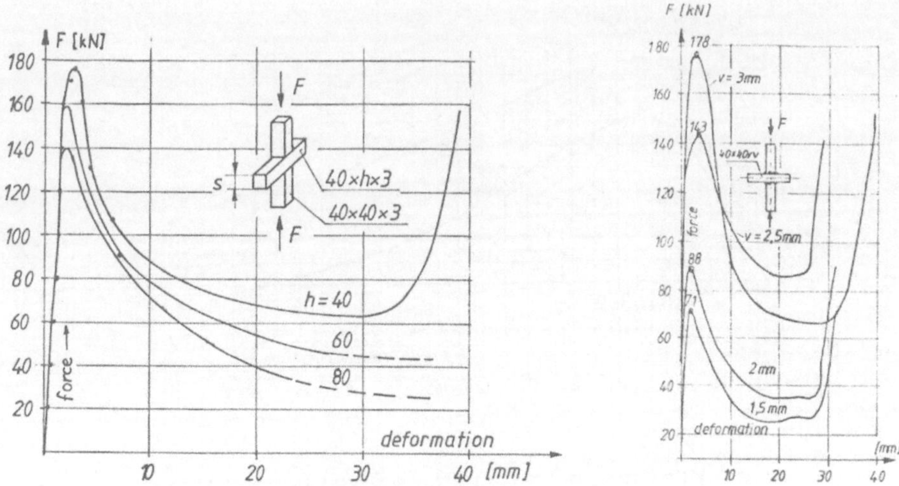

Figure 27 Geometrical effects on PH characteristics of „T" joints

6.3. TESTING SAFETY RINGS AND BAYS

When designing a bus frame against rollover accident, generally safety rings are used, on which more (four or six) plastic hinges are planned and calculated. These hinges work together assuring the required energy absorption, residual strength and controlled deformations, and the required survival space for the passengers. Fig.32. compares the summarised deformations of body bays built up from two similar safety rings loaded by static and dynamic forces. The test of these bays, the deformations and the plastic hinges can be seen on Fig.18. The two curves on each figures belong to two different load directions. The safety rings are not symmetrical; that is the reason for the different curves. The static load curves are plotted as the function of the deformation while the dynamic curves as the function of time. These test results show two things: the shape of the static and dynamic curves are similar and the dynamic forces are higher and this can be correlated to the strain rate effect discussed earlier.

7. Rollover of Buses, Strength of Superstructure

7.1. ECE REGULATION 66

Reg. 66 dealing with the strength of bus superstructure in the case of rollover defines a sta n-dard accident in which a required survival space must be free, untouched by structural el e-ments, no intrusions are allowed into this space. This requirement can be checked, tested by four different methods, among which the bus manufacturer can chose one:

Figure 28. Laboratory test of a front wall waist rail

- full scale roll over test with complete bus
- full scale over test with body bays, body segments
- full scale pendulum test with body bays, body segments
- calculation method

The theoretical background of these four methods is the „principle of equivalence". Although it is not written in the Reg. 66, the „principle of equivalence" means the following: the basic test method is the full scale rollover test with complete bus and it is assumed that the other three methods give the same results, produce equivalent deformations of the superstructure. (Bigger deformation is acceptable from the point of view of safety).

Figure 29 **Static pole type collision simulation with front wall waist rail**

In other words: the three substitutional test methods can replace the basic test method but they can not, must not substitute each other. If a bus has been legislated - for example - on the basis of a bay test, an other authority - in the case of any doubt - has to repeat the bay test to check the test results. The pendulum test is „out of the row" when trying to enforce the „principle of equivalence". There are essential differences comparing it to the full scale rollover test with complete vehicle, to whom it should be equivalent:

- the angle of impact is artificial and not related to the bus being tested
- the change of the impact angle during the pendulum test differs from that in a rollover test.
- the ground effect (secondary impact in the waist rail) is absolutely neglected
- the mass effect of the body section, the height of its centre of gravity does not play any role during the pendulum test
- the anchorage of the body bay for pendulum test can completely change the whole deformation and energy absorption process

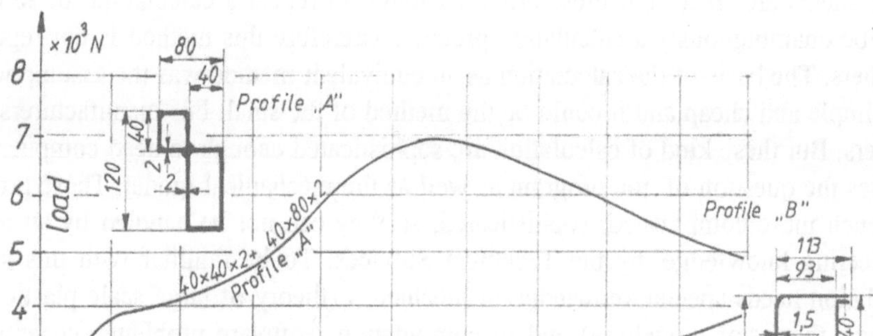

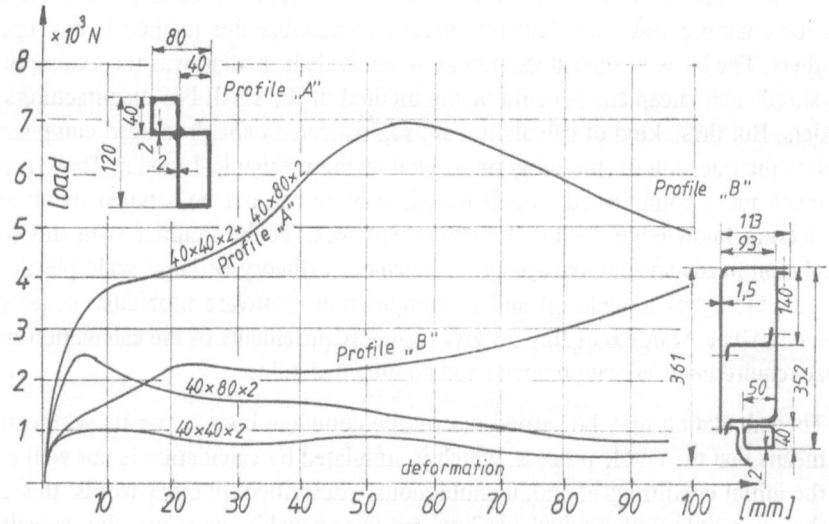

Figure 30. **Test results of front wall waist rails**

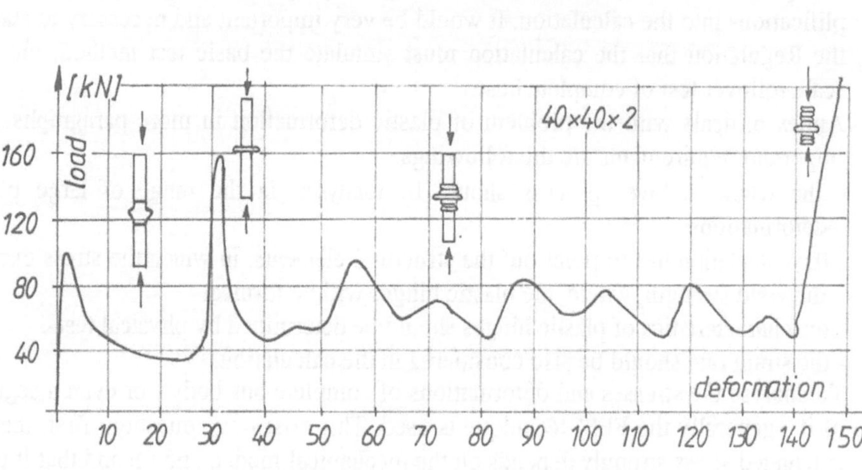

Figure 31. **PH characteristic of combined linear hinges**

The reason, why the pendulum test has been accepted in Reg.66. is historical. It was the first test in mind when GRSA (Working Group of Experts) started to work in that subject. The main argument for it was that it is a simple and cheap test comparing to the full scale rollover test, at that time which was rather complicated and not agreed (Hungarian method, U.K: methods, etc.) When agreeing in the full scale rollover test method described in Reg.66. another simple and cheap method came into the picture: the rollover test of full scale body sections, bays. But the pendulum test remained in force.

The fourth equivalent method: the calculation is not well defined in Reg.66. Some requirements are fixed, but these are not enough to repeat a calculation, these do not describe unambiguously a calculation process. Therefore this method is not repeatable by others. The basis of the calculation as an equivalent method was the assumption that it is simple and cheap and it could be the method of the small bus manufacturers, body builders. But these kind of calculation are sophisticated enough to need computer , and it raises the question of the program as well as the mechanical model. These problems are much more complicated, sophisticated, so they can not be handled by an average engineering knowledge, by the Technical Services. To be familiar with this kind of calculation needs special knowledge in mechanics (theory of large scale plastic deformations, theory of modelling) and in computation (software problems, generation of input data) The Annex 6 of Reg.66 gives some requirements of the calculation method. These requirements are summarized and commented below:

a) The calculation may be carried out on the complete body or on its segments. This means that the whole process, which is simulated by calculation is not well defined, the initial conditions are not unambiguously described. In other words: this extends the „principle of equivalence" to an unacceptable measure, the substitutional method is (can be) replaced by another substitutional method. But this replacement -trough the mechanical model and the computer program - brings unacceptable simplifications into the calculation. It would be very important and necessary to state in the Regulation that the calculation must simulate the basic test method, the full-scale rollover test of complete buses.
b) Annex 6. deals with the problem of plastic deformation in more paragraphs. The important requirements are the followings:
   - the whole rollover process should be analysed in the range of large plastic deformations
   - the calculation has to point out the structural elements, in which the stress exceeds the yield strength, where the plastic hinges will be formed.
   - the characteristics of plastic hinges should be determined by physical tests.
   - the strain rate should be also considered in the calculation.

   To analyse the stresses and deformations of complete bus body - or even a segment of it - generally the FEM technique is used. This arises two question: first, that the calculated stress strongly depends on the mechanical model, and second that it is not sure that plastic hinges are forming on that - and only on that - structural elements where the stress exceeds the yield strength in the very early stage of the calculation.

A more realistic solution is - Annex 6. refers to that possibility but it contradicts the automatic indication of plastic hinges by FEM analysis - to determine the location of the plastic hinges by engineering analysis of the structure. To be sure that this kind of engineering modelling is not disputable either by the Technical Service or on international level, the Regulation should contain more detailed rules and instructions about the possible location of the plastic hinges.

c) If the calculation does not involve the fracture process of the structural elements, it must be proved by tests that the fracture will not occur in the range of the plastic deformation. It is a very important requirement and - while the mechanical models and calculation methods are not, and will not be able to prove that in the near future - these physical test should be generally required in any case.

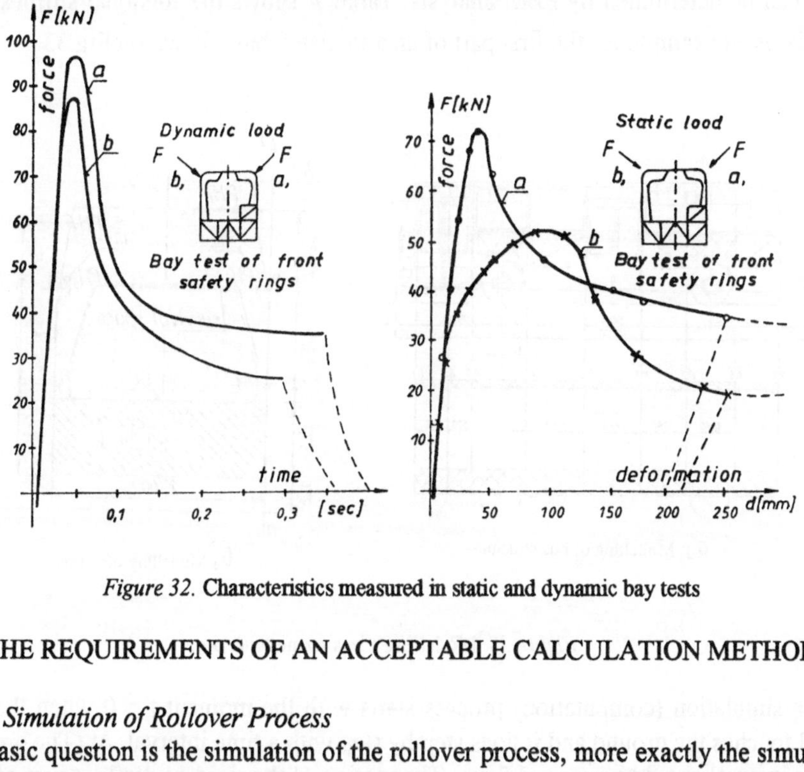

Figure 32. Characteristics measured in static and dynamic bay tests

7.2. THE REQUIREMENTS OF AN ACCEPTABLE CALCULATION METHOD [5]

7.2.1. Simulation of Rollover Process

The basic question is the simulation of the rollover process, more exactly the simulation of the full scale rollover test of the complete bus. The structure and the capability of the computer program determines the mechanical model of the superstructure. The process simulation should start when the cant rail of the roof just touches the ground, this is the starting time t = 0. In this moment the longitudinal symmetry plane of the bus has an angle φ_0 related to the horizontal ground, and the bus has a kinetic energy E_0 angle of velocity ω_0. Along the contrail a supporting force F is built up, which causes a certain angular deceleration ε and all characteristics of the program should be the following:

– The position of the plastic hinges should be determined by engineering analysis, engineering decision. They can be placed on the so called rings, which are part of the body
– The rings, representing the whole superstructure of the bus are not independent from each other, their motion and deformation are linked through the rigid chassis and the in-plane rigid roof and side wall structures which have only elastic deformations in their planes. The consequence of the above said is that the rings produce a certain amount of torque to each other promoting or hindering the neighbours deformation. These torque $\underline{T}$ can be got from the next linear equation system:

$$\underline{T} = \underline{\underline{s}}^{-1}\underline{\Phi} \tag{11}$$

where $\underline{\underline{s}}$ is the torsional stiffness matrix of the body, while $\underline{\Phi}$ is the relative torsional deformation of the rings related to each other. The elements of the matrix $\underline{\underline{s}}$ can be determined by FEM analysis. Table 7. shows the torsional stiffness matrix as an example for the first part of an articulated bus, shown on Fig.33.

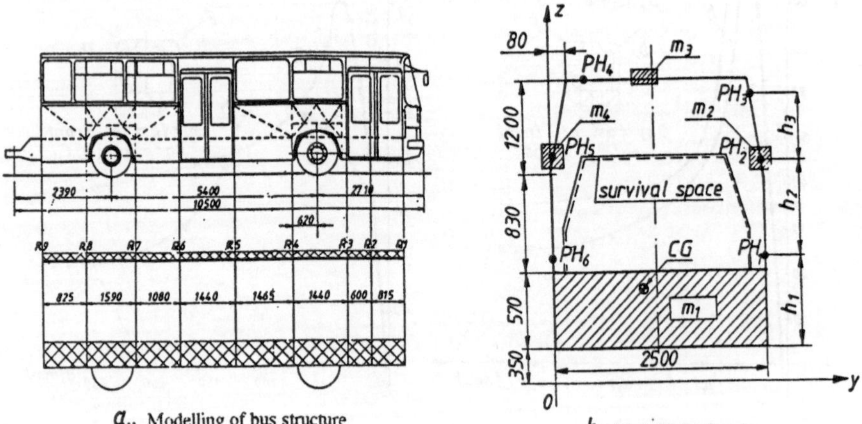

a., Modelling of bus structure b., Modelling of a ring

Figure 33. Modelling a bus frame with rings

– The simulation (computation) process starts with the moment t = 0 when the cant rail touches the ground and it goes step by step with a time interval Δt (The order of this interval should be around 5 ms) The motion of the rigid underfloor part of each ring which represents a mass m_i can be derived from the following equations:

$$\begin{aligned} \varepsilon_i &= (M_{iF} - M_{im} + T_{i-1} - T_i) I_i^{-1} \\ \Delta\omega_i &= \varepsilon_i \Delta t \\ \Delta\varphi_i &= (\omega_i + \Delta\omega_i/2)\, \Delta t \end{aligned} \tag{12}$$

where M_{iF} is the moment of the reaction force F_i while M_{im} is the moment coming from the free masses m_i and T_i and$T_{(i-1)}$ are the torque coming from the body in front of and behind of the ring examined, and I_i is inertia of the masses m_i being located on the ring.

TABLE 7. Torsional stiffness matrix

	ϑ_{ij} elements of elasticity matrix (x 10^{-10} rad/Nm)							
j \ i	1	2	3	4	5	6	7	8
1	690	183	130	103	4	6	22	16
2		350	95	84	18	9	21	14
3			369	320	7	5	13	14
4				504	30	21	2	10
5					145	48	38	3
6						268	265	112
7							516	225
8								184

The bending moments of each plastic hinge can be calculated from the coordinates of the moving, rotating rings - using the principle of the energy minimum. Knowing the plastic hinge characteristics the ring deformations can be determined.

7.2.2. Modelling the Body Structure

All the elements contributing to the strength of superstructure should be consired. Very useful way is to determine complete rings being in planes. Two remarks on that:

- ring is a structural part which contains a cross member in the underframe structure, side wall columns, window (or door) pillars, cross rib in the roof, or any other closed ring being similar to that.
- sometimes, in the practice these rings are not in a very plane, e.g. because of the unification of the windows, the window pillars are „shifted out" from the plane. It is acceptable to reduce these pillars into the planes, („in plane" reduction) but these planes should be linked to the cross members of the underframe structures.

Fig.33/a shows the model of the first part of an articulated bus with nine rings: R_1 .. R_i...R_9 The underframe structure is considered as a rigid part between two rings, the direct effect of the wall and roof structure on the strength of superstructure can be neglected (but indirectly, trough the torsional stiffness they play a very important role) The mass distribution should be cleared, the following concept may be proposed:

m_r	=	mass of the complete roof and everything which is connected to the roof, this mass is evenly distributed
m_{lwu}	=	mass of the left upper wall, above the window rail, evenly distributed
m_{rwu}	=	the same for the right side wall
m_{lwl}	=	mass of the left lower wall, under the window rail, evenly distributed
m_{rwl}	=	the same for the right side wall
m_u	=	mass of the underfloor structure containing the piping, electrical wiring and smaller components, being evenly distributed

m_{um} = mass of the main units having an individual mass more than 100 kg. The location of these masses according to the CG of the units, their input to the structure through their supporting brackets, and these masses should be proportionally reduced to the neighbour rings.

m_w = mass of the rear wall, evenly distributed in its plane

m_{cw} = mass of the front wall, evenly distributed in its plane

The sum of all of these masses should result the complete empty mass of the bus. For the reproducibility of the calculation, the relation and connection between the real superstructure and the model must be shown clearly: e.g. the number and location of the rings, the „in plane" reduction of rings if it happened, the neglections (which must increase the safety) and the mass distribution.

7. 2.3. The Mechanical Model

It should be built up according to the followings:

- the geometry of the rings should be determined (e.g. the length of the rigid rods, etc.)
- the place of the PH-s should be pointed out . The possibility of at least six PH-s should be on every ring: two on the floor level, two at the window rail and two near to the cant rail. The hinges on the two sides could be not on the same level (same height) as well as they can have different height on every ring. The hinges near to the cant rail should not be the point touching the ground (the acting point of the supporting force F)
- the proportional masses around the rings must be determined and positioned.

Fig.33/b shows the mechanical model of a ring. The rings, which need only four PH-s - like the front wall ring in our example, because the front panel below the wind screen is rigid enough and there is no need for plastic hinges on the floor level - can be handled with six hinges, among which two hinges have very rigid characteristic (with one order more rigid than the others) The deformation, rotation of these rigid, fictitious hinges is neglectible to the real other hinges. Special rings can be used as well, like the partition behind the driver's cab. Two principles should be considered:

- as much as possible (related to the capacity and capability of the computer program) the model should be „close" to the real structure
- simplifications are acceptable if and only if they serve the safety in the calculation, that means: the model, the rings must be weaker, less energy absorbing capacity than the real structure.

For the reproducibility - as input to the calculation - the main data should be fixed and given to the presentation of the calculation, so the distribution of the masses, too. The following idea, following equations could be used for the ring R_i having four mass components shoved on Fig.33/b:

$$\begin{aligned} m_{i1} &= m_{iu} + m_{ium} + 2/3(m_{ilwl} + m_{irwl}) \\ m_{i2} &= 2/3\, m_{ilwu} + 1/3\, m_{ilwl} \\ m_{i3} &= m_{ir} + 1/3\,(m_{ilwu} + m_{irwu}) \\ m_{i4} &= 2/3\, m_{irwu} + 1/3\, m_{irwl} \end{aligned} \tag{13}$$

where the mass elements relate to the main body parts described in para 7.2.2. It is important to emphasize that the sum of the four mass components has to result the total mass of ring R_i, and the sum of the ring masses has to give the total empty mass of the bus (m_o)

$$\sum_{j}^{4} m_{ij} = m_{Ri} \qquad \text{and} \qquad \sum_{i} m_{Ri} = m_o \tag{14}$$

TABLE 8. Mass elements of the rings

Ring	masses (kg)				
	m_1	m_2	m_3	m_4	Σ
R1	332	93	73	93	591
R2	358	189	46	20	443
R3	735	27	65	29	857
R4	109	39	93	41	1271
R5	1274	39	94	41	1448
R6	918	34	81	36	1069
R7	1458	36	86	38	1618
R8	1223	32	76	34	1367
R9	116	11	27	12	166
					8830

TABLE 9. Types of PH-s

Rings	Types of plastic hinges					
	PH1	PH2	PH3	PH4	PH5	PH6
R1	∞	A	A1	A	A1	∞
R2	B	C	C1	C	C1	B1
R3	D	E	E1	F	F	G
R4	H	F	F	F	F	H1
R5	J	E	E1	F	F	K
R6	L	E	E1	F	F	M
R7	H	F	F	F	F	H1
R8	H	F	F	F	F	H1
R9	N	O	P	P1	O1	N1

Table 8. summarizes the mass components for our example. This bus is a simple one having flat and horizontal floor without seat platforms, horizontal straight line window rail, etc. There are many other constructions which are more complicated for modelling point of view.

It must be noted that the effect of the windscreen and the rear wall window should not be considered. Many roll over tests (as well as real accidents) proved that these glasses jump out from their housing or break very often and it is almost impossible to judge theoretically whether they will fail or not.

7.2.4. Plastic Hinge Characteristics

One of the major point of the calculation is the way of dealing with the PH-s. All of the PH-s characteristics should be well documented data of the technical report about the calculation. Table 9. summarizes the different kind of plastic hinges in our example, where the same capitals represents the same hinges and the different capitals different hinges. The number fixed to a capital (e.g. A1) shows that the two hinges geometrically are the same, but they are bent in different directions. In many cases the characteristics could be different when changing the direction of the bending moment. Using six plastic hinges on one ring, having 9 rings as in our example, the possible number of the characteristics could be 54. But generally there are more similar structural element, similar plastic hinges, therefore this maximum number can be reduced, e.g. in our example to 23. Regulation 66. requires physical tests for every PH so in our case in means 50-70 laboratory tests (2-3 tests with one type of hinge) This example shows that the calculation method which was supposed to be a simple cheap method, in the practice is not so simple and not so cheap. There are two possibilities to reduce the number of real physical tests:

a) „substitution" of an unknown hinge characteristic by a known one, if it is unambiguously clear and proved (e.g. from the geometry, thickness, material, etc.) that the unknown hinge has higher resistance and energy absorbing capacity. One example: it known (tested) a hinge characteristic on a tube of 40x40x2 mm. It can be used for a PH to be formed on a tube 40x40x2,5 mm or 40x50x2 mm.
b) „Similarity concept" can be used when the range of the similarity is earlier proved by tests. For example: it is known (tested) the hinge characteristics of two tubes 40x40x1,5 mm and 40x40x3 mm. Using the „similarity concept" the characteristic of the tube 40x40x2 mm can be derived. On the basis of Fig.18. and Eq (8) the „similarity concept" means to determine the five parameters of the hinge characteristic. Some ideas:

- the starting stiffness (see Eq.2) is proportional to the elastic stiffness of the plastic hinge, or in other words to moment of inertia
- the maximum load can be estimated on the basis of Eq.(4) while the asymptotic load value using the approach of Eq.(5)
- knowing the geometry and material of the tubes, the parameters α, β, γ can be estimated.

7.2.5. The Output Data of Calculation

The output of the calculation has to prove and show at least the followings:

- the intact survival space at all of the rings like it is shown on Fig.34. If it is needed, between two rings linear interpolation may be done.
- the energy absorption of the individual rings and balance the energy of the whole bus
- representative parameters of the rollover process as the function of time Δt_0 Fig.34 shows - as an example - the supporting forces F_1 F_2 and the angular velocity ω for the ring R_4 are given on Fig.4.

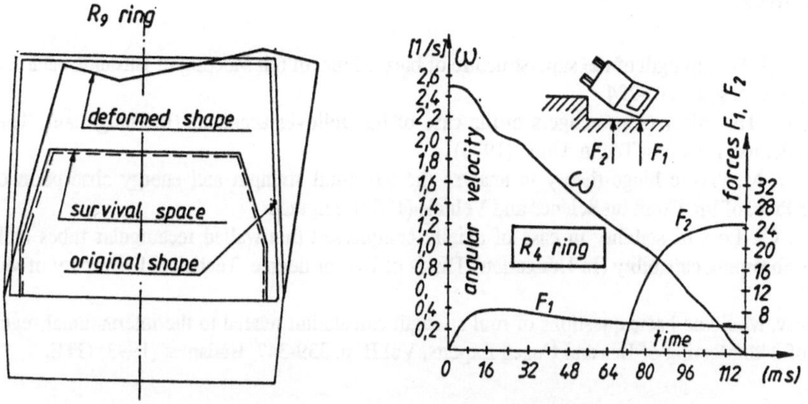

a., Deformation of a ring

b.) Parameters of roll over process

Figure 34. Output results of calculation

7.2.6. *Questions to be Cleared*

As it was mentioned earlier, the reproducability requirement of the calculation method does not exist in Reg. 66. therefor the mutual controllability of the calculations does not work, or in other words the principle of the mutual legislation is deaply hurt. To improve Reg. 66. from the point of view of the calculation method the further international efforts should be made in Geneva:

a) It should be stated and fixed that the calculation must simulate full-scale roll-over of complete buses, Only this kind of computer program is acceptable.
b) Computer programs should be approved and accepted by an international expert group. The main requirements of the approval should be earlier established and published.
c) Every program, which is approved and used by one of the contracting parties must be available for every other countries (Technical Services) for legislation. The copyright of these programs should be cleared.
d) the documentation of the legisletive calculation should be cleared and fixed (see model, input data and output evaluation)

Until these problems are not internationally cleared and solved the calculation method is not acceptable for international legislation

9. References

1 Tidbury, G. The strength of the superstructure of buses. Proc. of 6th Meeting of BusandCoach Experts (1975) Budapest, p. 216-224
2 Voith, A. The safety of passengers in the case of bus rollover accidents (in Hungarian). Thesis of Doctor degree, Budapest Techn. Univ. (1976)
3 Matolcsy, M. Plastic hinge theory to analyse the structural strength and energy absorption of bus frames Proc. of Int. Conf. on Science and Vehicle (1987) Belgrade
4 Molnár, Cs. Loss of stability in case of axially compressed thin-walled rectangular tubes and their energy absorping capability (In Hungarian) Thesis of Doctor degree, Technical University ofMiskolc (1977)
5 Matolcsy, M. Some basic questions of roof strength calculation related to the international regulation Proc. of 24th Meeting of Bus and Coach Experts, Vol.II. p. 339-347, Budapest (1993) GTE.

PART V
Finite Element Modelling in Crashworthiness

VEHICLE CRASHWORTHINESS AND OCCUPANT PROTECTION IN FRONTAL IMPACT BY FE ANALYSIS – AN INTEGRATED APPROACH

T. B. KHALIL[1] and M.Y. SHEH[2]
[1]*NAO Engineering Center, General Motors Corporation*
Warren, Michigan, USA
[2]*Cray Research Corporation,*
Southfield, Michigan, USA

Abstract

This paper provides an overview of Finite Element (FE) analysis of motor vehicle crashworthiness and occupant protection technology for frontal crash simulation. Particular attention is devoted to the development of an integrated FE model combining the vehicle structure, interior components, dummy and air bag in one model. The model was developed to simulate head-on crash into a rigid barrier from an initial velocity of 13.4 m/s (30 m.p.h.). In this model, similar to the test, a normally seated dummy is restrained only by the deploying air bag and knee bolster. The predicted vehicle velocity-time response agreed with available experimental data from one prototype test. The model was subsequently exercised to simulate the National Highway Traffic Safety Administration's (NHTSA) New Car Assessment Program (NCAP) test. In this simulation, the vehicle speed was increased to 15.6 m/s (35 m.p.h.) and the dummy was restrained by a three-point manual lap/shoulder belt, in addition to the air bag and knee bolster. Two additional simulations were conducted with the base model: impact into a 50 % offset rigid barrier; impact into a rigid pole. Vehicle deformations and dummy kinematics are presented and discussed. The information presented in this paper should be viewed as preliminary simulation results of a very complex nonlinear mechanics problem by FE analysis where several approximations and assumptions were introduced. These generally influence the accuracy of the analysis predictions, especially for the occupant response. Also, no attempt was made to validate/correlate the model response with test data.

1. Introduction

Crashworthiness measures the vehicle structure's capability to absorb the crash kinetic energy and the restraint systems (belts, air bags, knee bolsters , etc.) ability to provide adequate protection to the vehicle occupants in a survivable crash. Car collisions occur with many objects, over a wide range of speeds and along numerous impact directions.

J. A. C. Ambrósio et al. (eds.),
Crashworthiness of Transportation Systems: Structural Impact and Occupant Protection, 363–399.

In 1994, the National Safety Council estimated that there were 20 million vehicle crashes on the US roads, resulting in 43,000 fatalities and 2.1 million injuries requiring hospitalization [1]. The estimated annual financial loss from traffic accidents is equivalent to about 2 % of the US GNP [2]. The ultimate goal of vehicle safety engineers has been and still remains - to reduce occupant harm. This is typically accomplished by a combination of crash avoidance and crashworthiness measures. The crashworthiness engineer aims at a structural design that provides an acceptable vehicle deceleration pulse and minimizes intrusion of the vehicle into the occupant space, in the event of crash. In addition the passenger compartment is designed with numerous energy management features including restraint systems to reduce contact loads on the occupant due to sudden deceleration and to minimize the potential of ejection. Since the dominant direction in vehicle collisions in traffic accidents is frontal, this paper is limited to analysis of such collisions. In addition, crash avoidance is beyond the scope of this paper, and therefore will not be considered here.

Accident reconstruction and/or analysis of motor vehicle crashes, while providing important information regarding the safety performance of vehicle in the traffic environment, unfortunately does not provide sufficient quantitative information - deceleration pulse, occupant kinematics, occupant loads, etc., necessary for vehicle design. So, design engineers rely on a combination of laboratory tests and analysis to achieve their safety objectives.

The tests can be classified into component tests, sled tests, and full-scale barrier collisions. The component tests determine the dynamic or quasistatic response of an isolated component to loading. In a sled test, a vehicle buck representing the passenger compartment with all or some of its interior components and restraint systems is used. Mechanical surrogates of humans (anthropomorphic test devices - "dummies") or cadaver subjects are seated in the buck to simulate a driver and/or passenger and subjected to dynamic loads, similar to a vehicle deceleration-time pulse, to evaluate the occupant response in a frontal impact. The dummy kinematics - deformation, velocity and acceleration and associated forces and moments are measured to help determine the impact severity and the effectiveness of the restraint system in reducing loads transferred to the occupant.

The typical full-scale barrier test involves collision of a guided vehicle, propelled into a rigid flat barrier at a predetermined initial velocity and angle. This test is typically run to insure vehicle structural integrity and compliance with government mandated regulation, for example, US Federal Motors Vehicle Safety Standard (FMVSS) 208. A fully instrumented vehicle with numerous load cells, accelerometers and instrumented dummy (or dummies) in the driver (and passenger) seat(s) impacts a rigid barrier at zero degrees, at a velocity of 13.4 m/s (30 m.p.h.). The barrier face is instrumented with several load cells to monitor the impact force-time history. For compliance with FMVSS 208, the dummies in the driver and right front passenger positions must meet the injury criteria in both restrained (with lap/shoulder belt) and unrestrained (without lap/shoulder belt) conditions. A similar, non mandatory test is typically run at 15.6 m/s (35 m.p.h.) in the NHTSA's New Car Assessment Program (NCAP). In this latter test, the dummy is restrained by the manual three-point lap/shoulder belt. Other frontal tests that also have been conducted by some manufacturers include impacts at plus and minus 30 degrees from 30 m.p.h., central pole impact at zero degrees, and recently offset impact with 40-50 % overlap onto rigid or deformable barriers.

Such tests are not only time consuming, but also expensive, particularly at the early stages in vehicle development, where only prototypes are available. To ensure the crashworthiness and compliance with US and international safety regulations of one vehicle program, the manufacturer may test over 100 prototype vehicles, with each early prototype cost ranging from $400,000 - $750,000 [3]. The need to simulate the crash event by an analytical procedure has been expressed by design engineers for many decades. New, robust analytical tools using state-of-the-art in computational mechanics and computer hardware are indispensable for accurate crash simulations.

Analytical simulation of vehicle crashworthiness has been evolving over the past 25 years. Three types of models are used to simulate vehicle structures - Lumped Parameter (LP) models [4], developed in the early seventies , hybrid models [5], appeared in the early eighties and Finite Element (FE) models [6] also started in the early eighties and gained considerable impetus by the end of that decade. The FE models can be divided into two groups, heuristic beam models [7] and analytically based models which use beam, solid and shell elements. The progression of these models followed a pattern of increasing geometric details since it was realized that a simple analytical model of the crash event, developed and tuned to fit one or more parameters from a specific test, does not ensure accurate prediction for other impact conditions. In fact, the most detailed models (LP and FE) developed to date should be considered approximations of a highly complex non linear system. Obviously advances in understanding complex system performance such ascrashworthiness can be achieved by increasingly including more details that capture realistic vehicle kinematics and loads encountered in general crash conditions.

Similar to vehicle structures, occupant models using lumped parameters were developed to simulate occupant response to a crash pulse, using computer codes such as CAL 3D [8], developed in the early seventies and later MADYMO [9]. This occupant dynamics technology proved to be quite appealing because of its accuracy and simplicity. However, it requires physical understanding of the system and experimental measurements of force-deformation characteristics representing contact and interactions between the occupant and vehicle interior. Later, in the early nineties, hybrid models combining FE models for vehicle interior components with rigid body models for the occupant were introduced [10]. More recently FE models of the air bag [11] and of the Hybrid III dummy [12] were developed.

Currently, several FE models of vehicle structures (both component and full vehicle models), restraint systems (knee bolsters, belts and air bags) and Hybrid III dummies are available in the open literature. However, few models have attempted to simulate an integrated system where the occupant, occupant restraints and the vehicle structure are included in one model to simulate frontal impact into a barrier.

This paper provides an overview of structural component models, restraint models, a Hybrid III dummy model, and full vehicle models, using FE analysis. Next the development of an integrated model which includes a full vehicle structure with interior passenger compartment components (seat, collapsible steering column, instrument panel, knee bolster, driver air bag) and a detailed representation of the Hybrid III dummy is briefly presented. More details of the integrated model development are documented in reference [13]. The integrated model used a single FE code to simulate a 13.4 m/s frontal barrier test, similar to the FMVSS 208 test procedure. Developing such an integrated model allows for assessing how changes in the vehicle structure may influence occupant response in real time; evaluating the efficacy of the restraint systems on occupant protection; understanding the differences between sled and full scale barrier tests and quantifying the difference in occupant response resulting from vehicle collision with offset versus full barriers. However, it should be pointed out that modeling the influence of the restraint subsystems (air bag, belts, knee bolster, etc) on occupant response, in the event of crash, requires several approximations and assumptions.

2. FE Technology in Crashworthiness Analysis

Crashworthiness analysis of transportation vehicles in general and of ground vehicles in particular is among the most challenging nonlinear problems in structural mechanics. Vehicle structures are typically manufactured from many stamped thin shell parts and subsequently assembled together by various welding and fastening techniques. The structure may contain steel, aluminum and/or plastic materials. During the crash incident, the structure is subjected to high impact loads which produce localized plastic hinges and buckling which ultimately may lead into large deformations and rotations with contact and stacking among the various components. The deformations initially involve wave effects, associated with high stresses. Once these stresses exceed the yield strength of the material and/or its critical buckling load, localized structural deformations occur during a few wave transits in the structure. This is followed by inertial effects, which dominate the subsequent transient response. Of particular interest here are structural integrity and associated kinematics and stacking of components, forces transmitted through the various members, stresses, strains, and energy absorption. In addition the crash event may be considered as a low to medium speed dynamic event (5-100 m.p.h.), persisting for a short duration of 100-200 ms. Closed form analytical solutions for this class of problems in structural mechanics present insurmountable challenge to the analyst. Numerical techniques is the only practical option, at this time.

The FE method of structural dynamics solves numerically a set of nonlinear partial differential equations of motion in the space-time domain, coupled with material stress-strain relations along with definition of appropriate initial and boundary conditions. The solution, first, discretizes the equations in space by formulating the problem in a weak variational form and assuming an admissible displacement field. This yields a set of second order differential equations in time. Next, the system of equations are solved by discretization in the time domain. The discretization is accomplished by the classical Newmark-Beta method [14]. The technique is labeled implicit if the selected integration parameters render the equations coupled; and in this case the solution is unconditionally stable. If the integration parameters are selected to decouple the equations, then the solution is labeled explicit; and it is conditionally stable. Earlier developments in nonlinear FE technology used primarily implicit solutions [15]. FE simulation for structural crashworthiness by explicit solvers appear to be first introduced by Belytschko [16]. Later, Hughes et. al. [17] discussed the developments of mixed explicit-implicit solutions.

The explicit FE technique solves a set of hyperbolic wave equations in the zone of influence of the wave front, and accordingly does not require coupling of large numbers of equations. On the other hand, the unconditionally stable implicit solvers provide a solution for all coupled equations of motion, which require assembly of a global stiffness matrix. The time step for implicit solvers is about 2-3 orders of magnitude of the explicit time step. For crash simulations involving extensive use of contact, multiple material models and a combination of non traditional elements, it turned out that explicit solvers are more robust and computationally more efficient than implicit solvers. The discretized equations of motion for explicit FE formulation can be written as:

$$\underline{M}\,\ddot{\underline{x}} = \underline{f}^{(ext)} - \underline{f}^{(\text{int})} \tag{1}$$

where: $\underline{M}$ is a diagonal inertia matrix of the structure, $\ddot{\underline{x}}$ is the nodal acceleration vector, $f^{(\text{ext})}$ is the external force vector and $f^{(\text{int})}$ is the internal nodal force vector.

Time integration of equations (1) is obtained by a central difference technique:

$$\left.\begin{aligned} \dot{\underline{x}}^{(n)} &= \underline{M}^{-1}\left(\underline{f}^{(\text{ext})^{(n)}} - \underline{f}^{(\text{int})^{(n)}}\right) \\ \underline{x}^{(n+1)} &= \underline{x}^{(n)} + \dot{\underline{x}}^{(n+1)/2}\,\Delta t^{(n+1)/2} \\ \underline{x}^{(n+1)} &= \underline{x}^{(n)} + \dot{\underline{x}}^{(n+1)/2}\,\Delta t^{(n+1)/2} \\ \Delta t^{(n+1)/2} &= \left(\Delta t^{(n)} + \Delta t^{(n+1)}\right)/2 \end{aligned}\right\} \tag{2}$$

Where n is the integration step, Δt is the time step, $\dot{\underline{x}}$ and x are nodal velocity and displacement vectors , respectively.

Using the initial conditions the nodal kinematics can be computed. Next, compute the strain rate, spin and Jaumann stress rate for each element:

$$\left.\begin{aligned} &\text{Strain Rate:} && \dot{\underline{\varepsilon}}^{(n)} = 0.5\left(\underline{\nabla \dot{x}}^{(n)} + \underline{\nabla \dot{x}}^{(n)^T}\right) \\ &\text{Spin:} && \underline{\omega}^{(n)} = 0.5\left(\underline{\nabla \dot{x}}^{(n)} - \underline{\nabla \dot{x}}^{(n)^T}\right) \\ &\text{Jauman Stress Rate:} && \overset{\nabla}{\underline{t}} = \underline{C\dot{\varepsilon}} \end{aligned}\right\} \tag{3}$$

then, update the Cauchy stress for each element:

$$t_{ij}(t+\Delta t) = t_{ij}(t) + \dot{t}_{ij}(t) \tag{4}$$

where

$$\dot{t}_{ij} = \overset{\nabla}{t}_{ij} + t_{ik} + \omega_{kj} + t_{jk}\omega_{ki} \tag{5}$$

then the solution proceeds to the next time increment and so on until the desired solution time is reached.

3. Evolution of Crashworthiness Models

The first application of FE technology to crashworthiness analysis was in 1973 for aircraft structures. A computer program (KRASH) [18] was developed to analyze the impact response of a helicopter aircraft by a combination of beams, springs and discrete masses using implicit integration. In 1981, another FE code (DYCAST) [19] was also developed for crashworthiness analysis of aircraft structures. However, in this case the code included membrane and plate elements in addition to one-dimensional elements. The equations were integrated by one of three options of implicit integration techniques. This section briefly overviews the history of crashworthiness analysis by FE modeling for vehicle structures, occupants and restraint systems. In addition, integrated vehicle structures and occupant models which were developed to run in a single FE code are outlined.

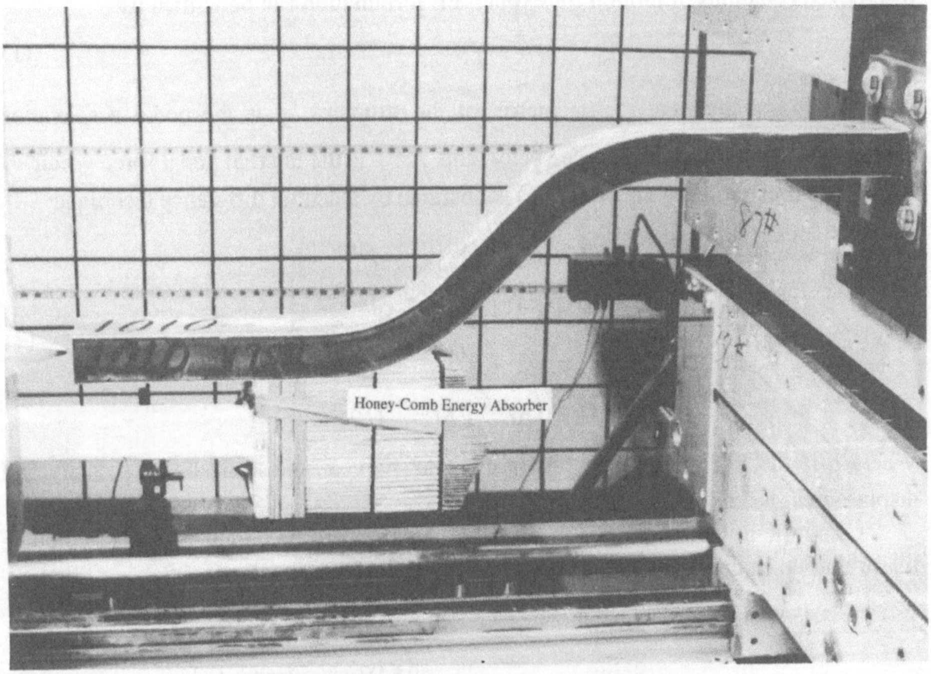

Figure 1. Dynamic test set-up for S-Rail using a sled Fixture

3.1. MOTOR-VEHICLE STRUCTURES

3.1.1. *Component Models*

Motor-vehicle structural models can be grouped into: generic components such as S-rails and rectangular tubes, actual isolated components such as upper rails, lower rails, hoods, etc. and subassemblies representing parts of the front structure. These components are typically tested in both quasistatic and dynamic modes to identify their crush performance. In dynamic testing, a "drop-silo" or a sled is used. In drop-şilo testing the component is fixed to the ground on a load cell and loading is typically applied from the gravitational fall of a rigid mass onto the free end of the component. In sled testing of components, the component is mounted horizontally onto the sled (Figure 1)which is subsequently launched to impact a rigid or deformable surface with the component making first contact. Figure 2 shows the S-rail final deformations corresponding to initial impact speeds of 2, 4.5 and 8.2 m/s into a rigid wall. The rail deformations exhibited two plastic hinges at the rail curvatures, rotation at the free end and plastic hinge at the fixed end.

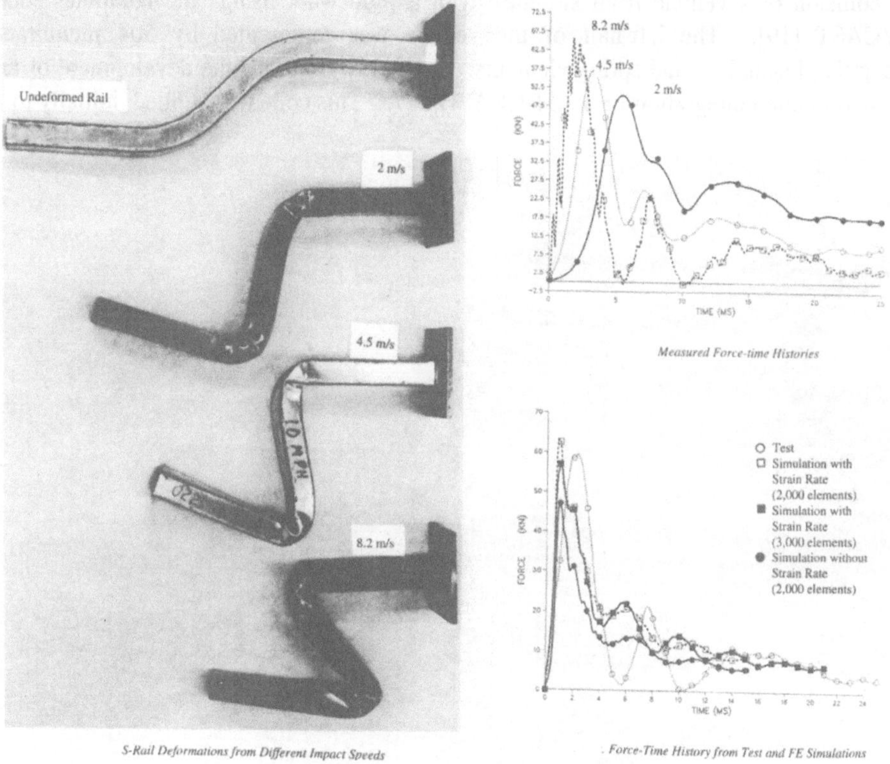

S-Rail Deformations from Different Impact Speeds

Force-Time History from Test and FE Simulations

Figure 2. S-Rail response

The measured peak force increased with impact speed from about 50 to 60kPa, due to strain rate effects of the material used. The FE simulation corresponding to 8.2 m/s impact agreed quite well with the test result when the strain rate effects were included in the simulation. Increasing the number of shell elements from 2,000 to 3,000 showed minor influence on the overall response.

The deformed mid-rail shapes of a passenger car are shown in Figure 3, corresponding to test speeds of 2.2, 4.4 and 6.7 m/s, respectively [20]. The FE predictions of the force-time pulse show a reasonable agreement with the test data. Like the S-rail response, the peak force increased with increasing the impact speed due to strain rate effects. Figure 4 shows a comparison between the predicted and measured force pulse from impact of a front structure subassembly onto a rigid wall [21]. In this case the impact speed was 13.4 m/s. In this simulation no strain rate effects were included in the analysis since the mid-rails were manufactured from high strength steel which typically exhibits little strain rate effects.

3.1.2. *Full-scale vehicle models*

FE models attempting to simulate motor vehicle structural crashworthiness have been evolving since the early 1980's. Evidently, the first crash model [22], simulated a head-on collision of a vehicle front structure with a rigid wall, using the computer code DYCAST [19]. The left-half of the vehicle was represented by 504 membrane triangular, beam, bar, and spring elements. Haug [23] discussed the development of an implicit-explicit integration FE PAM-CRASH code. This code was applied initially to

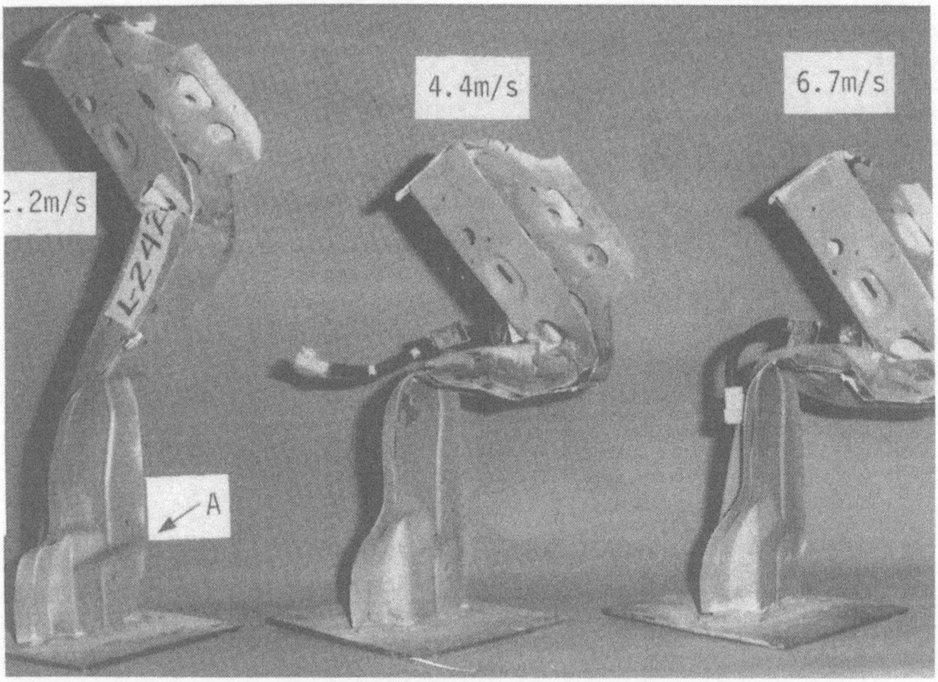

Figure 3a. Vehicle component deformations

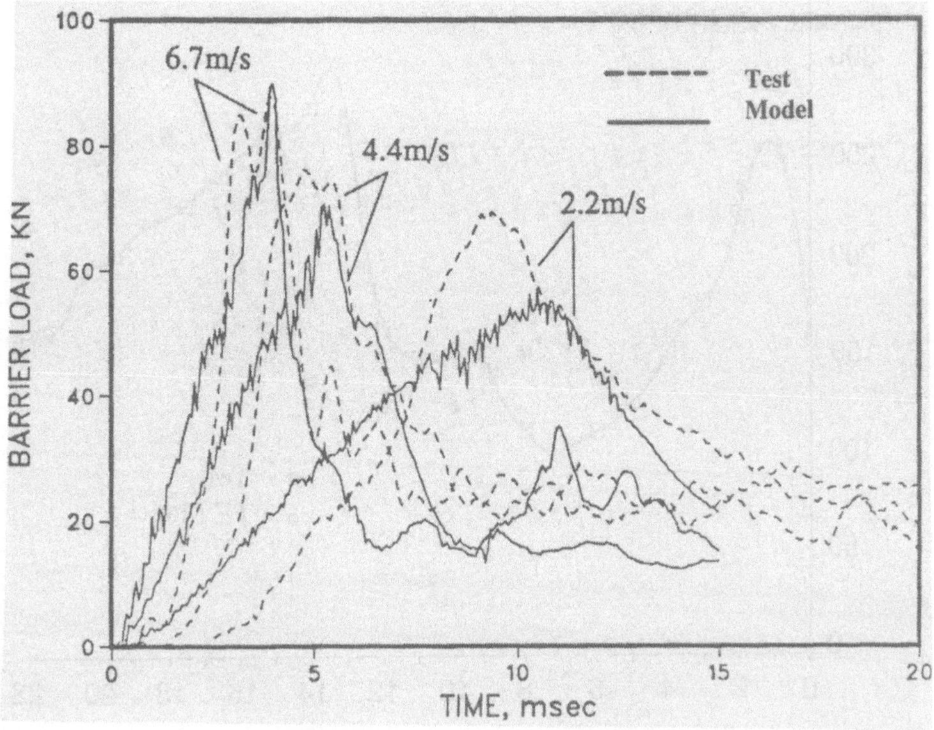

Figure 3b. Measured and FE calculated forces

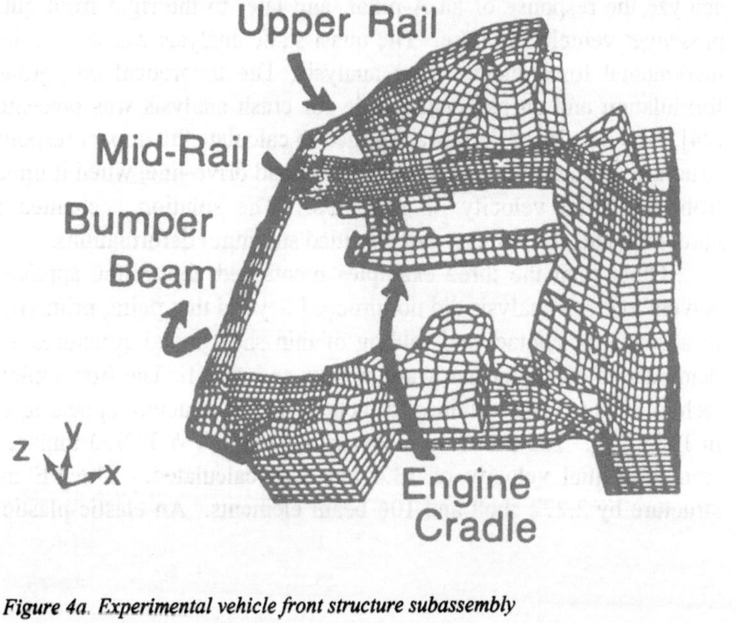

Figure 4a. Experimental vehicle front structure subassembly

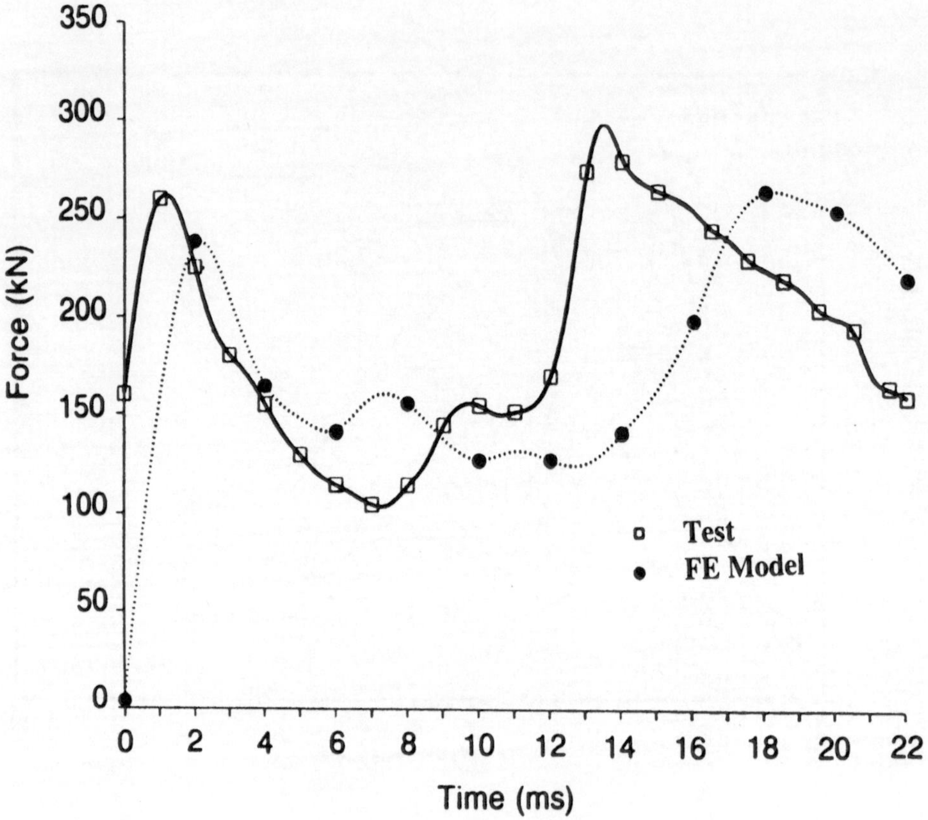

Figure 4b. Force-time history

analyze the response of an A-pillar, and later to the right front quarter of a unit-body passenger vehicle structure. The quasi static analysis was accomplished by an iterative incremental force/displacement analysis. The theoretical background for implicit FE formulation and an associated code for crash analysis was presented by Argyris et al [24]. The developed code was applied to calculate the impact response of a vehicle front structure, devoid of engine, transmission and drive-line, when it impacted a rigid barrier from an initial velocity of 13.4 m/s. The solution accounted for material strain hardening and rate effects, and provided structural deformations.

Other than the three examples mentioned above, the application of implicit FE solvers to crash analysis did not proceed beyond that point, primarily due to its inability to account for contact and folding of thin sheet metal structures and due to excessive demands on computer hardware storage and speed. The first application of explicit FE technology to crashworthiness of actual vehicle structure appear to be attempted by ESI in 1985 [25]. The frontal vehicle structure of a VW POLO impact into a rigid barrier, from an initial velocity of 13.4 m/s was calculated. The FE model simulated the structure by 2,272 shell and 106 beam elements. An elastic-plastic constitutive model with strain hardening was used to describe the sheet metal behavior. The analysis provided vehicle kinematics and barrier force-time pulse. Subsequent to this simulation, many crashworthiness calculations have been attempted by automotive manufacturers. In addition to ESI's PAM-CRASH code, two other commercial codes - LS DYNA3D [26] and RADIOSS [27] - are commonly used in crashworthiness simulations.

These codes include an extensive library of one, two and three-dimensional elements, in addition to discrete elements which allows for modeling of just about all structures in modern transportation vehicles. Also, many constitutive models for linear and nonlinear metallic and nonmetallic materials with failure criteria is a common feature in these codes. Perhaps the most attractive option in the codes is their ability to simulate contact in thin shell structures with minimum geometric definition. Since the codes were expected to be used in analysis of structures with hundreds of thousands of degrees of freedom over 100-200 thousand time steps, optimized soft ware that required minimum storage and CPU was a high priority. To this end, reduced integration elements with hour-glass control are available and commonly used in all crash models.

Table 1 provides a summary of representative, front structure models, available in the open literature. These models simulated vehicle crash with a rigid wall. They clearly indicate the evolution of the models in size and complexity. It may also be observed that detail modeling of vehicle structures has become an integral step in the vehicle design process. Other models have been developed to simulate side impact with deformable barriers [40], vehicle-to-vehicle offset and angle frontal and rear impacts, respectively [41,42]. These models will not be discussed here as they fall beyond the scope of this paper.

TABLE 1. Summary of crashworthiness models for frontal impact

Year/Ref.	Vehicle Manuf.	No. of Elements	Sim Time (ms)	Analysis Code	CPU (hours)/Computer
1986/[28]	BMW	2,800	90	CRASHMAS	89/ --
1986/[29]	SUZUKI	3,439	60	DYNA3D	24/CRAY-XMP
1986/[30]	VW (POLO)	5,661	60	PAM-CRASH	4/ CRAY 1
1986/[31]	CITROEN (BX)	7,900	80	PAM-CRASH	12/CRAY-XMP
1986/[32]	OPEL (VECTRA)	7,991	80	PAM-CRASH	16/CRAY-1S
1987/[33]	GM (GM 80)	9,550	80	PAM-CRASH	12/CRAY-XMP
1987/[34]	ISUZU (760 COUP)	10,000	70	PAM-CRASH	14/CRAY-XMP
1989/[35]	SAAB (900)	7,500	80	DYNA3D	22/CRAY-XMP/28
1989/[36]	MAZDA	18,000	80	PAM-CRASH	18/CRAY-XMP
1989/[37]	NISSAN	12,000	120	PAM-CRASH	20/CRAY-XMP
1989/[38]	GM (W COUP)	17,000	80	DYNA3D	20/CRAY-XMP
1992/[39]	GM-Exp. Vehicle	17,000	80	DYNA3D	9/CRAY-YMP/4
1994/[53]	OPEL (Omega)	16,000	100	RADIOSS	?/CRAY-YMP/4
1994/[54]	FORD (WINDSTAR)	60,000	100	RADIOSS	12/CRAY-C90/16

3.2. OCCUPANT MODELS

Anthropomorphic (human like) Hybrid III dummies are routinely used in experimental crashworthiness assessment of motor vehicles in frontal impact for both sled and barrier tests. These dummies are instrumented with accelerometers and load cells, to measure the dummy response to the crash deceleration pulse, and allow for a comparison with established biomechanical human tolerance data. Analytical simulation of the occupant response is traditionally calculated by lumped-parameter models, introduced since the early seventies [8]. These models simulated the occupant by a group of rigid bodies interconnected by appropriate joints. The model is subjected to the crash pulse and contact constraints. The corresponding kinematics and associated loads are calculated and compared with test data.

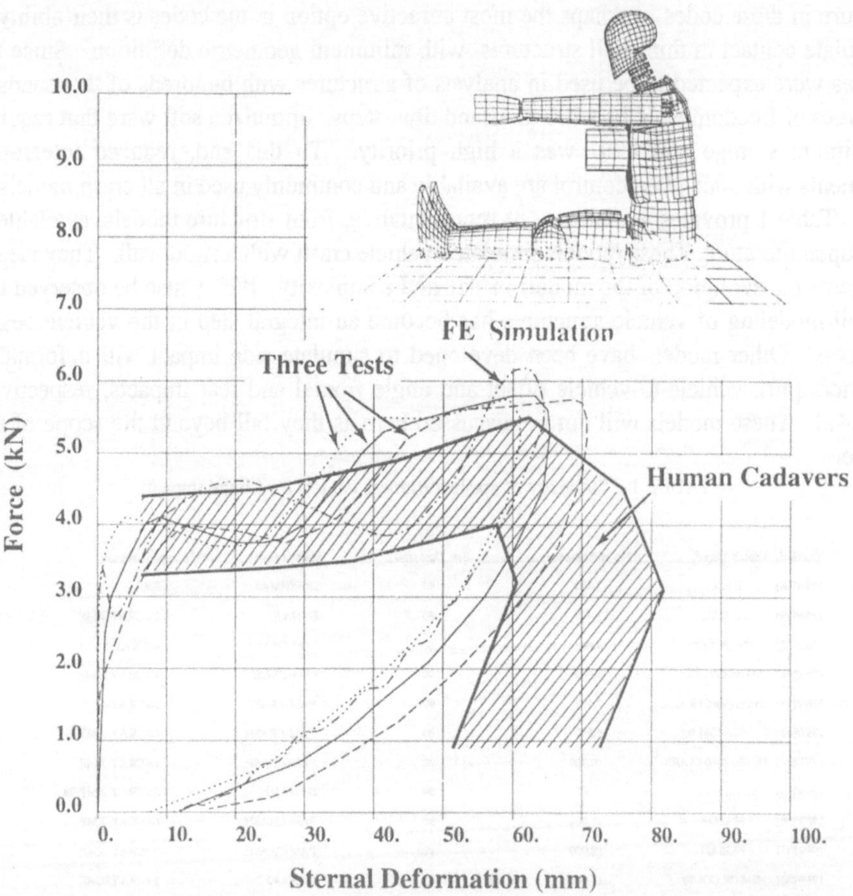

Figure 5. Force-deformation response of Hybrid III dummy chest

The limitations of the lumped-parameter models are documented [11]. This motivated the development of an FE model of the Hybrid III dummy. The development and validation of the dummy's individual components and of the whole assembly's response to thoracic impact are discussed in detail in reference [11]. Figure 5 shows an FE model of the Hybrid III dummy and associated Force-Deformation resulting from pendulum impact on the thorax. The predicted response agreed well with data from dummy experiments and was generally within the response corridors of cadaver data. Further validation of the dummy was carried out by comparing its response with sled and barrier tests [43]. In these tests the dummy was restrained by a three-point belt, knee-bolster and air bag .

3.3. RESTRAINT MODELS

Restraint systems, belts and air bags, are introduced to reduce the forces on the occupant and distribute these forces to body regions that can sustain relatively higher loads. Lap belts were introduced first in the mid-fifties as optional equipment to help in occupant "ride down" and to minimize the potential for ejection. These belts were the forerunners of current three-point belt systems. The early development of belt systems were achieved by experimental methods using sled and barrier tests. Analytical techniques for belts were introduced into the occupant simulation LP models as early as the mid-sixties [44]. All current LP occupant models include mathematical formulations to accommodate belt stiffness, belt slip, belt slack, etc. This technology has been basically duplicated in the explicit FE codes.

Air bag development for the driver and passenger has been evolving since the late sixties. They were introduced as passive restraint systems which can isolate the human from concentrated impact loads resulting from potential contact with the vehicle interior subsequent to a crash. Current air bags, used in frontal impact protection, are designed to fully inflate in about 20-40 ms from the onset of sensing a crash. The exact time depends on the vehicle velocity prior to crash, the crash pulse, the design of theinflater system, etc. At the onset of a crash, a normally seated occupant would move forward and contact the deploying bag which attenuates the impact. The bag fabric porosity and vents allow for gas leakage which in turn reduces the gas pressure and occupant rebound. Similar to belt development, early design of air bags used sled and barrier tests. Analytical techniques for air bag simulations were developed first as added features to existing LP occupant codes. In this case the analysis focused on two issues: gas dynamics [45] and interaction loads [46]. Ideal gas equations were used to calculate the pressure and temperature for an expanding control volume. Inflater mass flow rate and gas temperature in time are the required inputs to calculate the evolution of bag volume, pressure and temperature. Uniform gas pressure and temperature were assumed throughout the bag volume. Both bag material stretch and gas leakage were included in the analysis by introducing empirical formulas. In the early developments [45], the air bag shape was fixed as an ellipsoid that can expand during inflation. Occupant contact forces are calculated from the bag pressure acting on the contact zone between the bag and the occupant. Some of the limitations of this development included: neglecting the bag inertia, bag unfolding, membrane loads, changes in bag shape, bag slap force etc. These drawbacks limited the applications to modeling of the in-position seated occupant. The need for accurate simulation of air bag deployment which allows for changes in bag shape as a result of occupant interactions particularly for the out-of-position occupant led into the development of air bag modeling by FE technology.

The early FE models of air bag deployment simulated the bag by wrinkle-free membrane elements, which would inflate due to internal uniform pressure [11,47,48]. The gas dynamics equations used were identical to those introduced for LP models. Single surface contact algorithms were used to allow for unfolding of folded bags. Later developments included gas simulation by Euler discretization, coupled with Lagrangian discretization of the bag fabric. Although the FE technology alleviated some of the disadvantages of air bag technology in the LP codes, it increased the analysis complexity and required significantly more computer resources. Figure 6 shows the initial (0 ms) and final (130 ms) configurations of Hybrid III dummy interactions with a deploying driver air bag. The air bag initially was folded with 16 folds and located at the center of the steering wheel, with the vinyl bag cover removed. The dummy was in a normal seating position, unrestrained, and exposed to a sled pulse simulating a small car crash. The predicted chest compression agreed reasonably well with the test result, showing a peak deformation of 45 mm. A similar study was conducted with the dummy leaning forward to simulate an out-of-position driver. In this case (Figure 7) the predicted chest compression was 40 mm which was slightly higher than the measured peak of 37 mm.

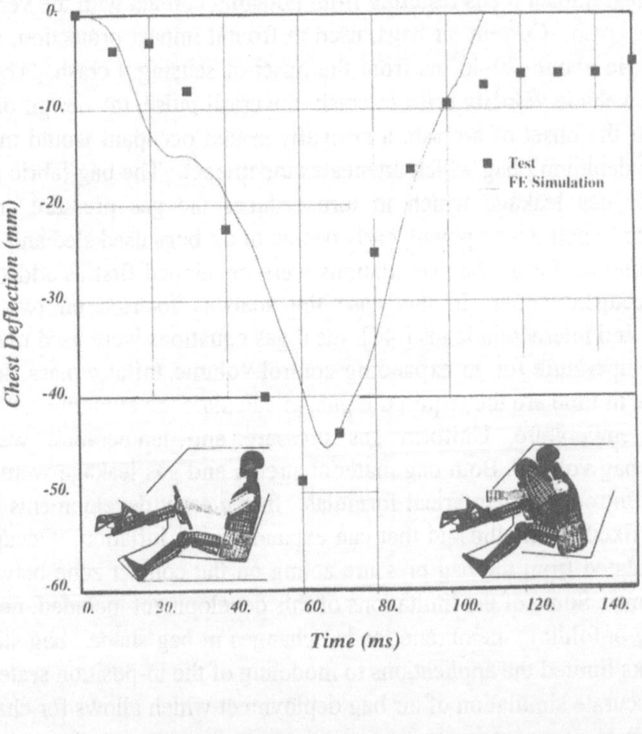

Figure 6. Chest deformation from in-position dummy interactions with air bag

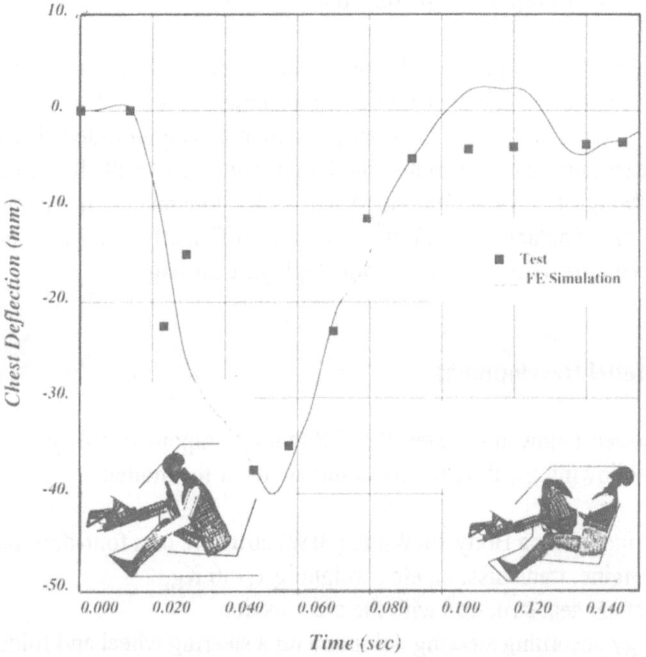

Figure 7. Chest deformation from out-of--position dummy interactions with air bag

3.4. INTEGRATED MODELS

Traditionally crashworthiness analysis has focused on sequential simulations of vehicle structures and occupants, with the occupant simulation being driven by the vehicle (calculated or measured) deceleration pulse. Consequently the analysis techniques and associated tools have been developed to meet the needs of the two communities. Obviously this process has many shortcomings. Among which are: real time interactions between the occupant and vehicle structure during the crash event are not simulated, the analyst must learn and use two analysis techniques which is an inefficient use of human and hardware resources, etc. To remedy this situation, as indicated earlier, attempts have been made to couple occupant and structural codes [10]. This process allowed for real time interactions between the vehicle and occupant, but left the analysis community with two analytical tools to deal with.

It has always been and still remains the desire of safety analysts to simulate the crash event in one model which can predict occupant response associated with the crash event in real time. Conceptually all FE codes can accomplish this task, since rigid body equations of motion are a special case of the FE formulation. The only missing components were dummy and restraint models. Once these are developed, then crashworthiness simulations which include structures, occupants and restraint systems in one model become a reality. The development of a subsystem model of the Hybrid III thoracic interactions with a steering wheel using a single code has been demonstrated

[49]. The Ove-Arup group discussed the advantages of using a single code in safety analysis, and presented graphics depicting dummy models and an integrated vehicle-dummy model [50].

Schelkle and Remensperger [51] discussed the development of an integrated model of a passenger compartment and occupant using a single FE code. The model simulated a sled test by 8,700 elements, in frontal crash; and included an unfolded air bag, steering system, knee bolster and seat. A 100 ms simulation required 13 CPU hours on a CRAY-2 computer. Although the model exhibited reasonable kinematics, it was indicated by the authors that the contact algorithms were not sufficiently robust to account for interactions between the dummy arms and the deploying air bag.

4. Integrated Model Development

The model discussed below integrated the following components/subsystems into one FE model. The following are the primary components of the model:

- Vehicle including a Body-In-White (BIW) structure of a four-door passenger sedan, engine, transmission, etc., weighing 1,750 Kg,
- A bucket car seat structure with the seat cushion,
- An energy absorbing steering column with a steering wheel and folded air bag,
- An instrument panel, including a driver side knee bolster,
- A door structure, and
- A Hybrid III dummy.

The model development and analysis were accomplished in 4 phases:

Phase 1: Dealt with the passenger compartment interior components consisting of the Hybrid III dummy, seat, energy absorbing steering column, knee bolster, steering wheel, and folded driver air bag. This subsystem model will be referred to as the Sled Model (SM), since it approximates a sled test environment.

Phase 2: Involved the BIW structure, engine, drive shafts, transmission, tires, instrument panel, knee bolster, and two front doors. This subsystem model is referred to as the Vehicle Model (VM).

Phase 3: Focused on integrating the two subsystem models and providing the appropriate interfaces and boundary conditions. This model is referred to as the Integrated Vehicle Occupant Model (IVOM). Upon integrating the model a frontal impact at zero degrees into a rigid barrier, corresponding to FMVSS 208 test, was performed.

Phase 4: Dealt with exercising the IVOM in a limited parameter study to investigate the model robustness. This consisted of three frontal impact simulations: a "New Car Assessment Program" (NCAP) calculation at 15.6 m/s after introducing a three-point belt harness, a 13.4 m/s impact into a 50 % offset rigid barrier and a 13.4 m/s impact into a centrally located rigid pole of 355 mm diameter.

4.1. PHASE 1 : DESCRIPTION OF SLED MODEL (SM)

The SM consisted of the following components:

4.1.1. *Hybrid III Dummy*

A component level validation was previously conducted for the head, neck, thorax, lumbar spine, and knee. The calculated component responses were in good agreement with experimental data [12]. Joint definitions similar to LP occupant models were introduced to assure proper articulation among the dummy segments. The model has approximately 6,700 brick elements, 260 beam elements and 450 shell elements. In addition, it was necessary to add a neck collar containing 220 solid elements with viscoelastic behavior, similar to the dummy jacket, to allow for smooth contact between the air bag and the neck. Otherwise the air bag fabric gets tangled with the segmented neck and causes numerical difficulties for the contact algorithm.

4.1.2. *Collapsible Steering Column and Self-Aligning Steering Wheel*

The steering wheel model was previously used in a study to validate the development of the Hybrid III thorax [11]. The steering column model represented the essential geometry of the steering system by 10 beam elements, 140 solid elements and 1,704 shell elements. The self-aligning wheel rim has 108 solid elements, with a rigid material specified. The energy absorbing ball and sleeve joint of the steering column was represented by a special beam element, which allows for axial stroking and binding, if lateral bending moments are generated at the ball and sleeve joint.

4.1.3. *Driver Air Bag*

The driver air bag model was represented by 2,220 membrane elements. The geometry of this model simulated a folded bag with 16 folds. The bag was located in its compartment in the steering wheel. The bag cover was not included in this simulation.

4.1.4. *Bucket seat*

The bucket seat model represented the steel frame and a foam cushion. The seat frame model contained approximately 4,000 shell elements, and the cushion was modeled by 960 solid elements.

4.1.5. *Knee Bolster*

The knee bolster was simulated by a 45deg. rigid wall, located at about 0.1m from the knee.

4.1.6. *Model Statistic*

The approximate number of nodes and elements are:

- 8,400 shell elements
- 8,100 solid elements
- 300 beam elements
- 25,000 nodes
- 30 joints for dummy articulation
- 100 discrete spring and damper elements
- 100 parts

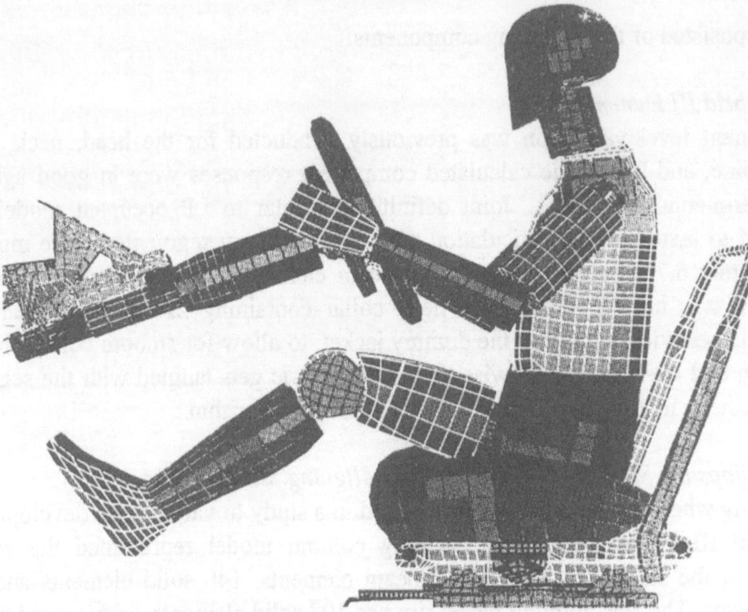

Figure 8. Initial position of sled subsystem model

Figure 8 shows the initial configuration of the SM. The time step was two microseconds, which could only be achieved by considering certain parts of the seat and steering column as rigid bodies.

4.1.7. *Contact Definitions*

The following contact surfaces and constraints have been introduced:

- Rigid walls were defined to simulate the floor panel, instrument panel and knee bolster.
- Contact surfaces were defined between the dummy and seat cushion, between the seat cushion and the seat structure and between the outer and inner steering column jackets.
- A contact entity of toroidal shape was defined to account for contact between the steering wheel and dummy and between the steering wheel and air bag.
- One-sided contact with "a soft constraint" penalty used to define contact between the air bag and dummy deformable parts (head, neck collar and jacket).
- Automatic contact with time dependent gap is used to simulate the contact within the folded air bag, to account for no penetration during bag deployment.

4.1.8. *Boundary and Initial Conditions*

The boundary and initial conditions imposed on the model are: (a) The entire model without the dummy is subjected to a trapezoidal deceleration pulse of 20 G amplitude, 10 ms rise time, 5 ms decay time and persisting for a duration of 35 ms, in the horizontal direction; (b) The dummy is subjected to 1 G acceleration in the vertical direction; and (c) An initial velocity of 13.4 m/s was also assigned to all nodal points. The lower part of the steering column was connected to a linear spring joint. The energy absorbing ball and sleeve joint was represented by a special spring element which accounts for sliding (and binding) motion of the telescopic inner and outer shafts of the steering assembly.

4.1.9. *Computing Requirements*

Approximately 20 runs were made during the development of this subsystem model. Most of the effort dealt with:

- proper air bag deployment
- contact between the air bag and the head/neck region
- contact between the air bag and the steering wheel
- contact between the air bag and the dummy thorax

Approximately six CPU hours on Cray Y-MP8E were needed to complete a 100 ms simulation, and 8 million words of memory were required to accommodate the model.

4.1.10. *Results*

Figure 9 shows a mid sagittal section cut through the air bag and dummy's upper torso at 0 and 100 ms. This type of display is routinely obtained during calculations involving complex contact, such as those encountered by the dummy's soft skin and a deploying air bag, to insure that no (very small) penetration has occurred between the air bag fabric and the dummy's outer skin. These results clearly demonstrate proper air bag deployment and forward motion of the dummy and subsequent interactions with the air bag. Figure 10 shows that the force on the thorax reached a peak of 1 kN at about 60 ms. It must be pointed out that this simulation is highly idealized and does not represent what an occupant experiences in a real crash.

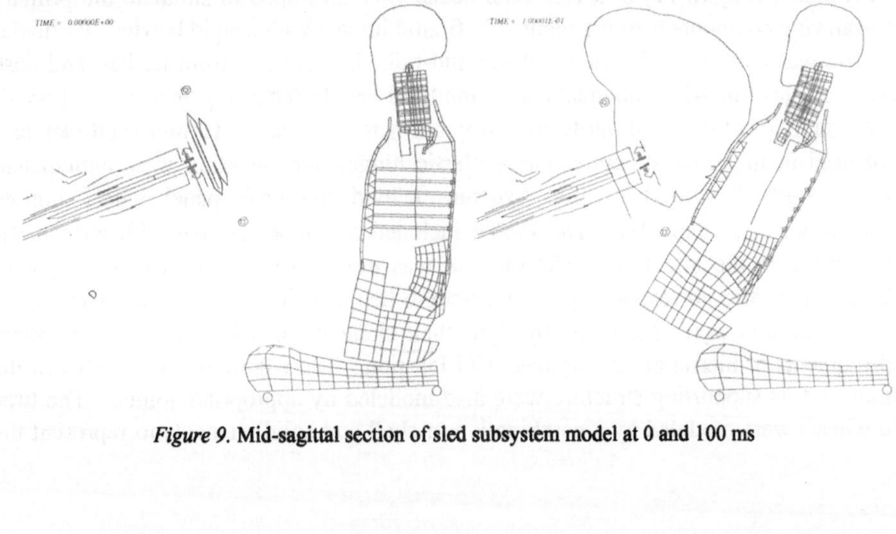

Figure 9. Mid-sagittal section of sled subsystem model at 0 and 100 ms

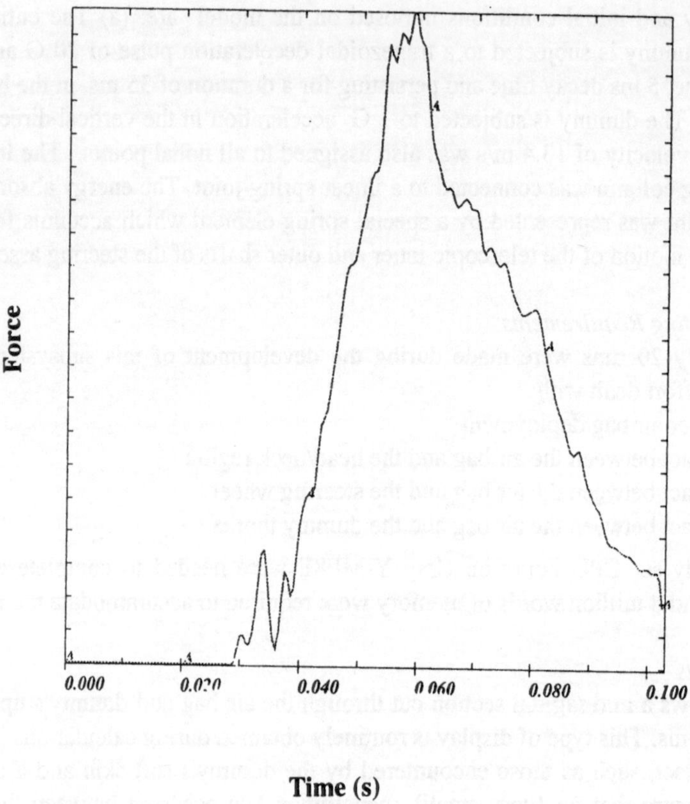

Figure 10. Calculated contact force on the chest from air bag deployment

4.2. PHASE 2: VEHICLE STRUCTURE MODEL

An FE model (Figure 11) of a four-door sedan was developed to simulate the primary load carrying components of the vehicle in frontal impact with a rigid barrier. Particular attention was paid to the FE mesh of the mid-rails, lateral rails, front tie bar, and dash panel. All sheet metal components were modeled and fastened together by spot welds corresponding to the actual hardware. Approximately 10 mm x 10 mm shell elements were used in the front structure where plastic hinges and buckling were anticipated. Coarser mesh was used for the structure behind the dash panel where limited deformations were expected. The model included a bumper system with appropriate connectivity to the mid rails. The radiator was modeled using solid elements, with material properties corresponding to honeycomb behavior in compression. The engine and transmission were simulated by rigid shell elements, which represented the mass and moments of inertia at the engine's CG location. The engine mounts between the engine and its supporting structure were also modeled by appropriate joints. The tires and wheels were modeled by a combination of shell and solid elements to represent the compliance and load transmission characteristics of these components. Two front door models were included with appropriate hinge properties. The rear doors were excluded, as their influence on frontal crash performance was assumed negligible. An instrument panel was also included along with appropriate structures for knee restraint. Finally the inertia characteristics of the whole vehicle model was checked against the actual vehicle and concentrated masses were added to ensure agreement between the model's inertia and the corresponding hardware.

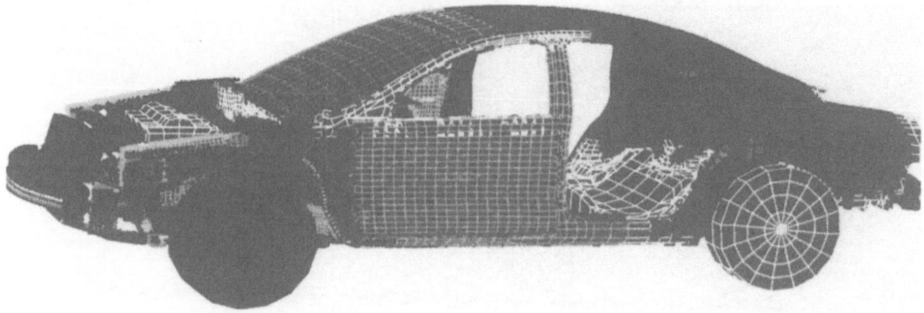

Figure 11. FE model of undeformed vehicle

4.2.1. *Model Statistics:*

- 61,500 shell elements
- 500 solid elements
- 25 beam elements
- 1,200 spot welds
- 15 joints
- 40 concentrated nodal masses
- 10 springs
- 200 parts
- 66,000 nodes

4.2.2 *Contact Definitions*

A rigid wall was defined in front of the vehicle with stick condition. Automatic contact surfaces were defined in six zones as follows:

- Front-left corner (up to front body hinge pillar)
- Front-right corner (up to front body hinge pillar)
- Front-center (include up to the middle of the engine)
- Rear-center (from middle of the engine to the fire wall)
- Driver side center-pillar to door
- Passenger side center-pillar to door

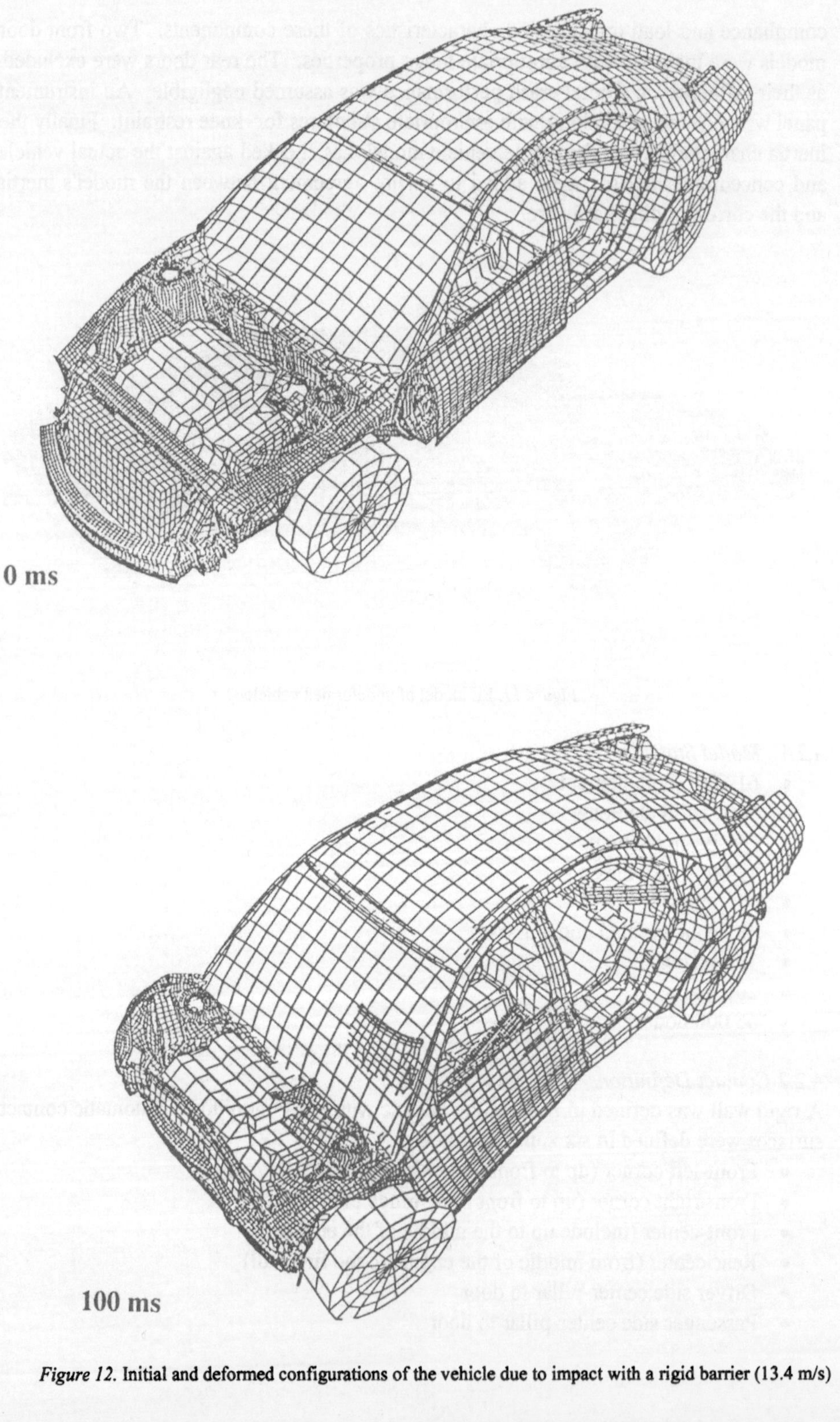

Figure 12. **Initial and deformed configurations of the vehicle due to impact with a rigid barrier (13.4 m/s)**

4.2.3. *Initial Condition*
An initial velocity of 13.4 m/s was defined for the entire vehicle structure which impacted a rigid wall.

4.2.4. *Results*
The model response was calculated on a CRAY Y-MP8E system. The time step was approximately 0.7 microsecond. A 100 ms simulation was completed in about 45 hours on one processor. In this run, it was necessary to refine the radiator model since severe "hourglassing" was observed. The initial (at time 0) and final (at 100 ms) vehicle deformed shapes are shown in Figure 12. Intermediate vehicle configurations (not shown due to space limitation) exhibited realistic sequential deformations as seen in high speed film analysis of barrier crashes. The time histories of the global energy balance, velocity at the front rocker, and barrier force provided very reasonable results, comparable to test data. These results will be discussed in more detail in phase 3.

4.3. PHASE 3: INTEGRATED MODEL

This phase dealt with integrating the sled model with the vehicle structure to simulate a 13.4 m/s frontal barrier impact, similar to FMVSS 208 test. The integration process was accomplished with PAM CRASH software [52]. The following steps were necessary to integrate the sled and vehicle models:

- Nodes on the seat rails were merged with the floor of the vehicle structure;
- The steering column was connected to the rear engine cradle cross member with joint definitions;
- The steering column support bracket was allowed to deform; and was connected to the instrument panel;
- Boundary and initial conditions were redefined for the sled components;
- Air bag to windshield contact definition was added;
- A foam knee bolster was added to the instrument panel on the driver side.

4.3.1 *Model Statistics*:
The integrated model has approximately:

- 70,000 shell elements
- 9,000 solid elements
- 300 beam elements
- 1,300 spot welds
- 300 parts
- 91,000 nodes
- 20 contact segments

The whole model was given an initial velocity of 13.4 m/s; and launched into the rigid barrier. Air bag deployment was triggered at 20 ms. The CPU requirements for a 100 ms simulation were- 65 CRAY-YMP hours and 40 million words (280 Megabytes) of memory.

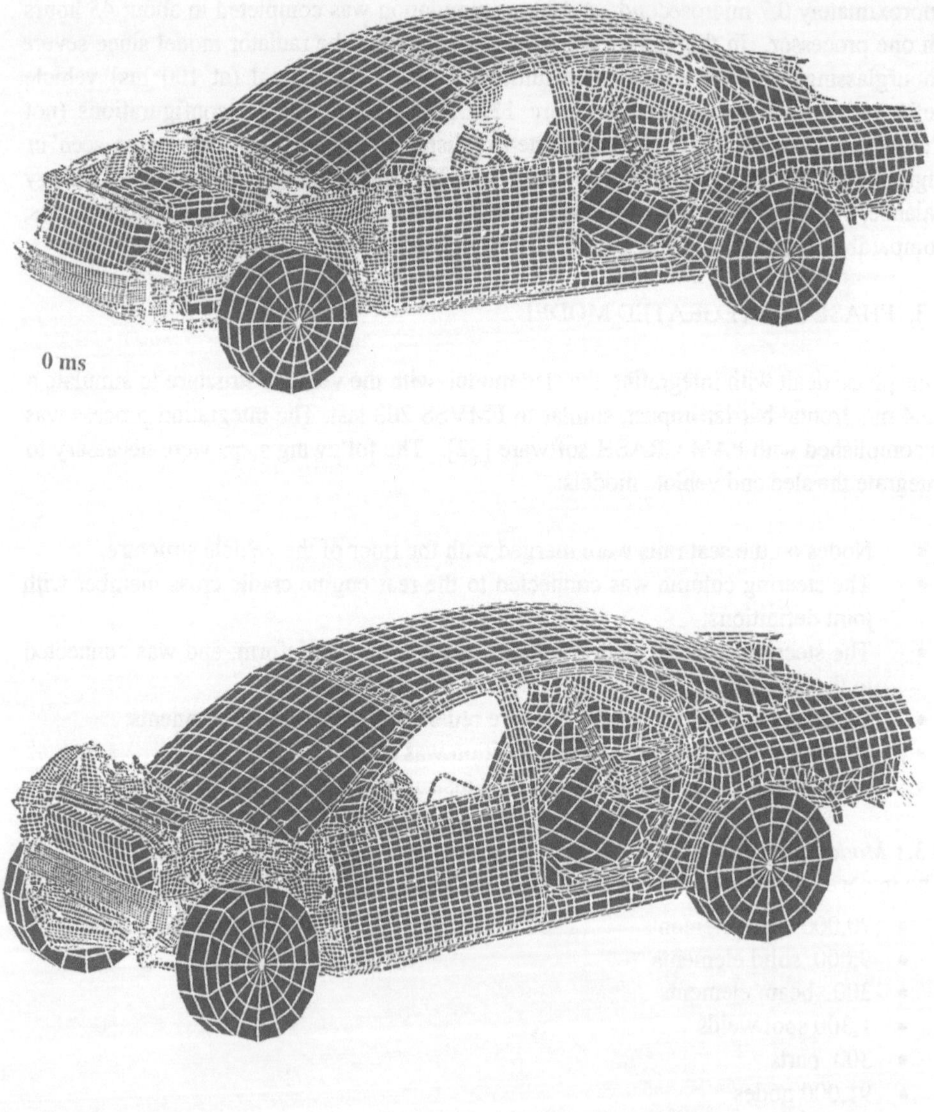

Figure 13. Initial and deformed integrated model configurations due to impact with a rigid wall from 13.4 m/s

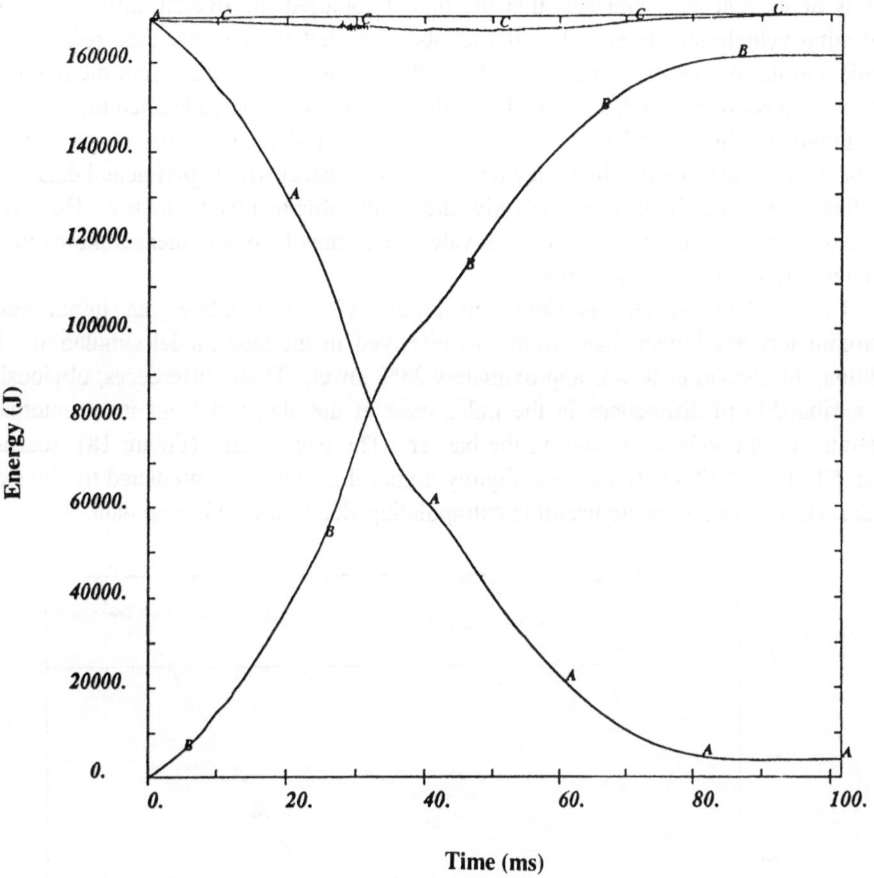

Figure 14. Energy balance of integrated model simulation

4.3.2. *Results*

The following results should be viewed as preliminary simulation results, yet to be verified against test data, of a very complex nonlinear mechanics problem by FE analysis where several approximations and assumptions were introduced. These may and do influence the accuracy of the analysis predictions, especially for the occupant kinematics and associated loads.

Figure 13 shows the integrated model at time 0 and a deformed configuration at 100 ms. The model kinematics, showing vehicle deformations, air bag deployment and forward motion of the dummy and subsequent interactions with the air bag and knee bolster are qualitatively similar to a barrier test. Figure 14 shows the energy balance for a typical run. It can be observed that the total energy remained approximately constant throughout the 100 ms duration, the rise in internal energy and the decay in kinetic energy are smooth; attesting to very little interpenetrating among the contacted segments. The vehicle velocity response in time, at the rear rocker, is shown in Figure 15, along with experimental data from one test. The velocity trace agreed quite well with test data, particularly at the point where the vehicle velocity crosses the zero line.

This is of interest as it indicates that the model captured the overall stiffness of the deforming vehicle structure. Also, it was observed that the velocity-time response is identical to its counter part obtained in Phase 2, since the added mass from the dummy-seat-steering column complex was added to the model as distributed lumped masses.

Figure 16 shows the barrier unfiltered force-time pulse. The pulse shape with its two peaks and the times at which they occurred is consistent with experimental data. The first peak force almost coincided with the results obtained from one test. However, the second peak was only 60% of the test value. This may be due to inexact modeling of the engine-to-dash panel interactions.

The air bag pressure is shown in Figure 17. It exhibited an initial peak approximately 5% higher than what was observed in the sled model simulation . In addition, the second peak was approximately 33% lower. These differences, obviously, are attributable to differences in the pulse used in the sled test from its counterpart generated by the vehicle impact on the barrier. The bag volume (Figure 18) reached about 53 liters at 70 ms. It also was slightly higher than what was predicted by the sled mode. These differences are useful in extrapolating sled results to barrier data.

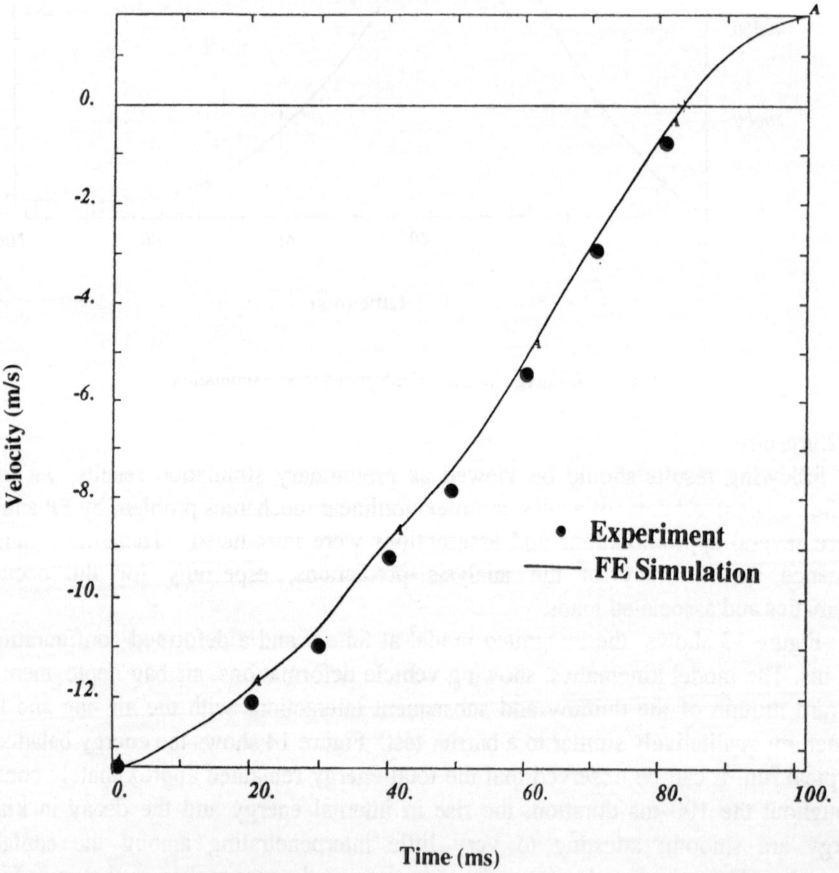

Figure 15. Comparison between measured and calculated rear rocker velocity of integrated model

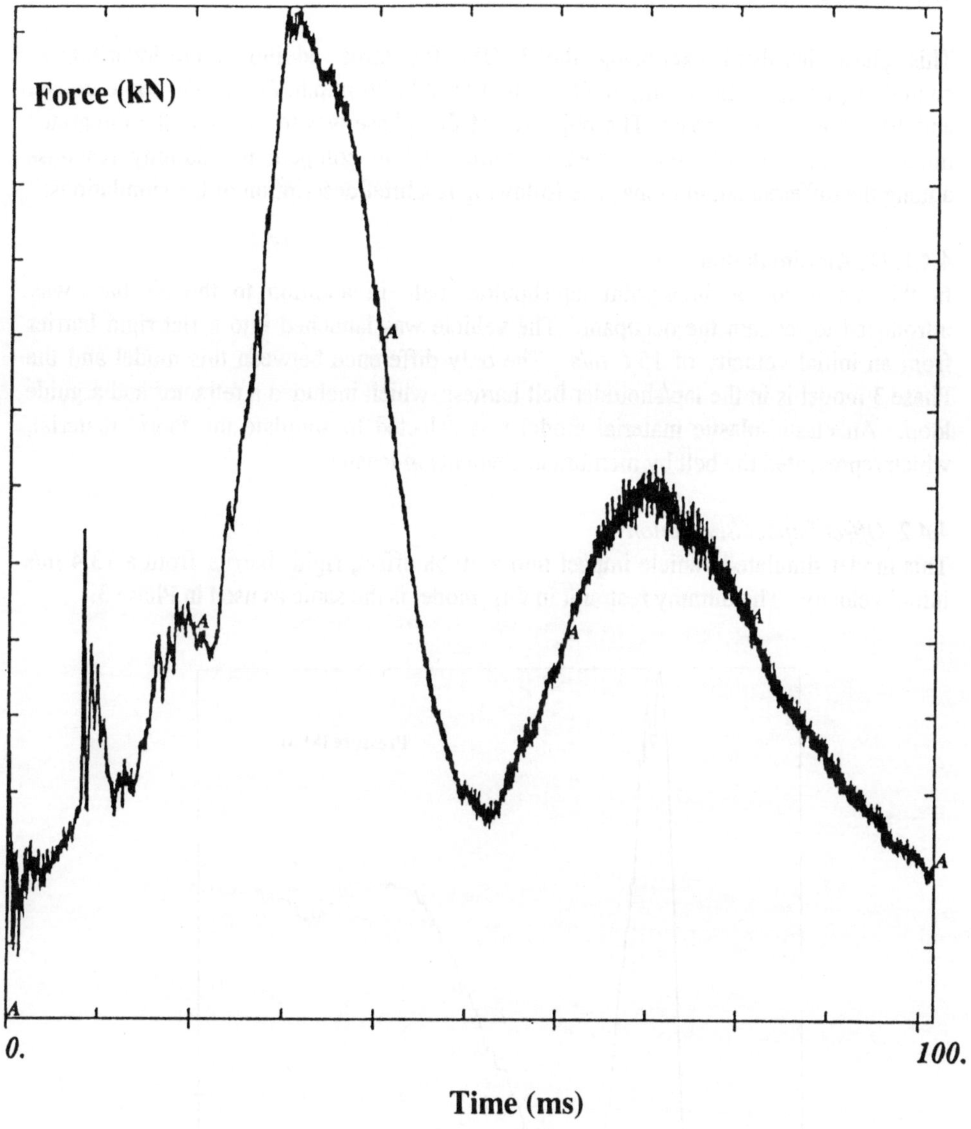

Figure 16. Calculated force pulse on rigid barrier for integrated model

Figure 19 depicts the calculated dummy kinematics (head acceleration, chest acceleration, and femur loads) in time. Observe that the head acceleration has not attained its peak at 100 ms; this is likely to occur at 110-120 ms in this class of simulations. No attempt was made to compare this dummy response with experimental data, since it is felt that this is beyond the scope of this project. Also, due to using some interior components that were not intended for this body structure.

4.4. PHASE 4: PARAMETRIC STUDY

This phase involved exercising the IVOM for three additional crashworthiness conditions, namely: the 35 m.p.h. (15.6 m/s) NCAP, 30 m.p.h. (13.4 m/s) pole impact and 30 m.p.h. offset impact. The objective of this phase was to examine the integrated model response to different impact scenarios and to compare the dummy response among the different simulations. The following is a brief description of the simulations.

4.4.1. NCAP Simulation

In this simulation a three-point lap/shoulder belt, in addition to the air bag, was introduced to restrain the occupant. The vehicle was launched into a flat rigid barrier from an initial velocity of 15.6 m/s. The only difference between this model and the Phase 3 model is in the lap/shoulder belt harness, which included a retractor and a guide loop. An elastic-plastic material model was selected to simulate the fabric material, which represented the belt by membrane elements in tension.

4.4.2. *Offset Impact Simulation*

This model simulated vehicle impact into a 50 % offset, rigid barrier from a 13.4 m/s initial velocity. The dummy restraint in this model is the same as used in Phase 3.

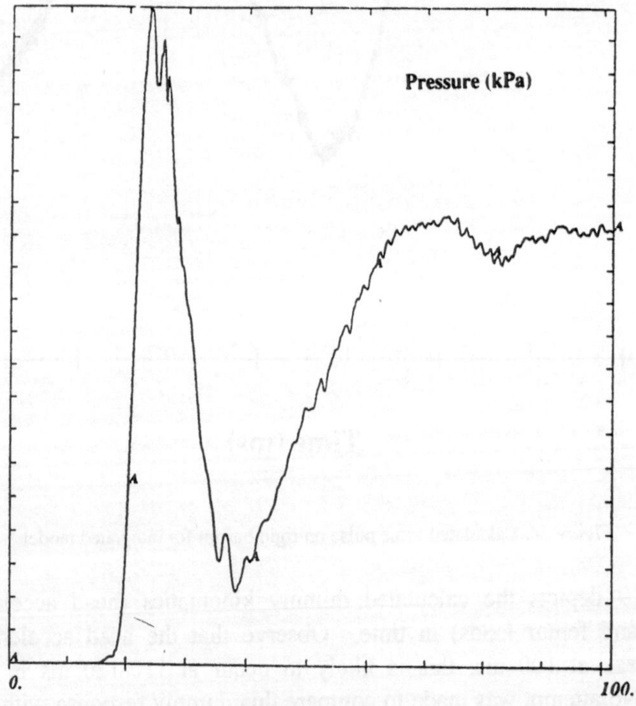

Figure 17. Air bag pressure-time history of integrated model

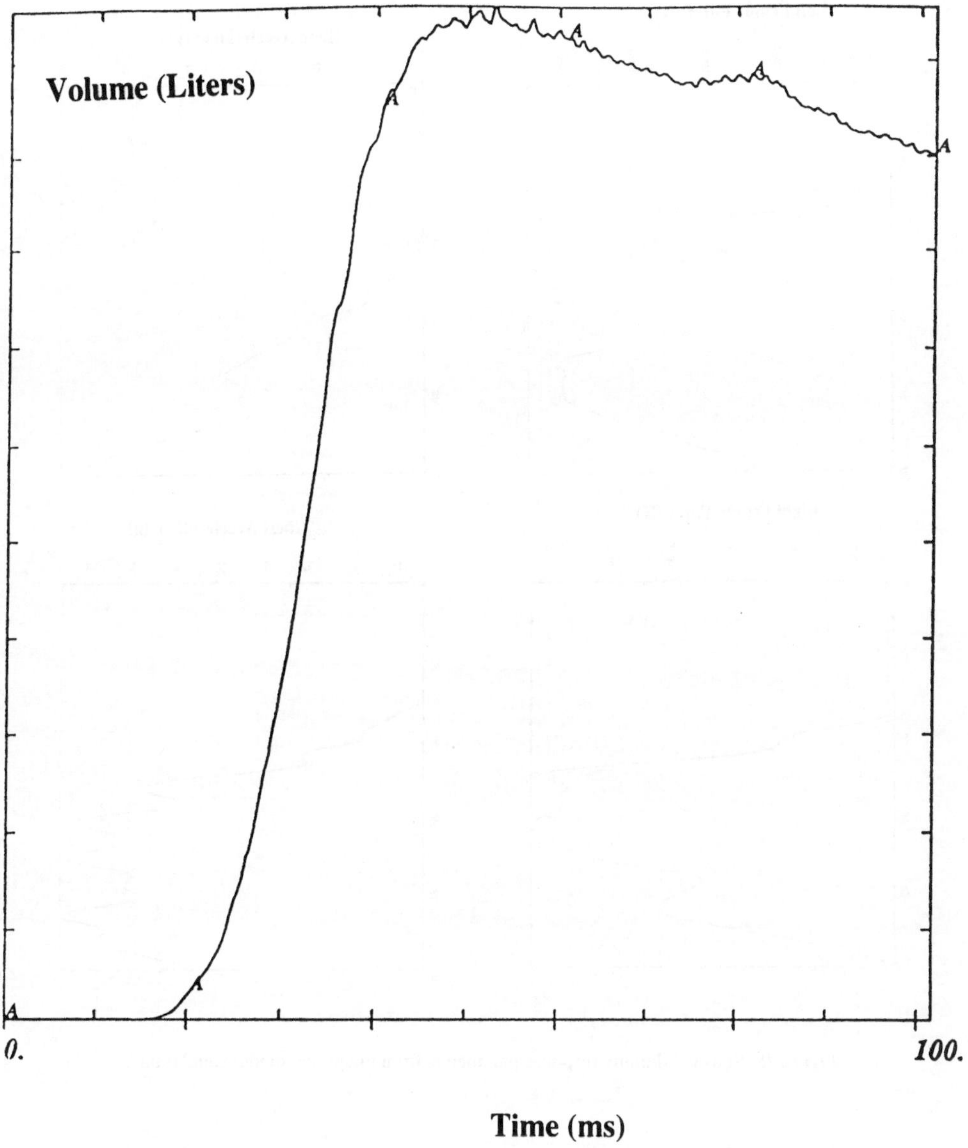

Figure 18. Air bag volume evolution of integrated model

4.4.3 *Pole Impact*

This model represented vehicle impact into a rigid cylindrical pole of 355 mm diameter. Again, this model is the same as used in Phase 3. In the last two simulations it was necessary to redefine the contact surfaces in the radiator area. The air bag firing was initiated at 20 ms for all runs. All three simulations ran successfully to 100 ms.

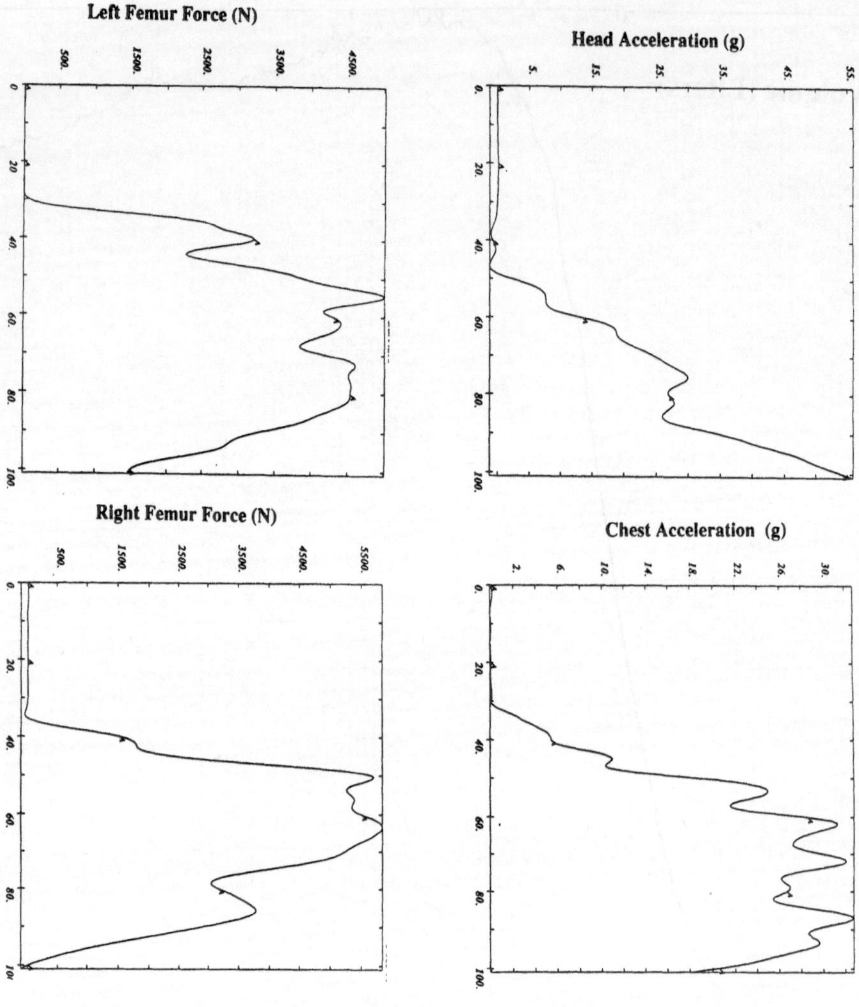

Figure 19. Selected dummy response parameters from integrated model simulation

4.4.4. *Results*

Similar results to what was provided for phase 3 were obtained here. The vehicle deformations exhibited qualitatively acceptable results. Figure 20 provides a comparison of the initial and final (100 ms) deformations. It was observed that in the NCAP run excessive deformations occurred in the passenger compartment, particularly in the floor area. This is conceivably due to a coarser mesh used in the model in this section of the vehicle. Verification of the model accuracy remains to be checked against experimental data. Accordingly, this result should be considered preliminary.

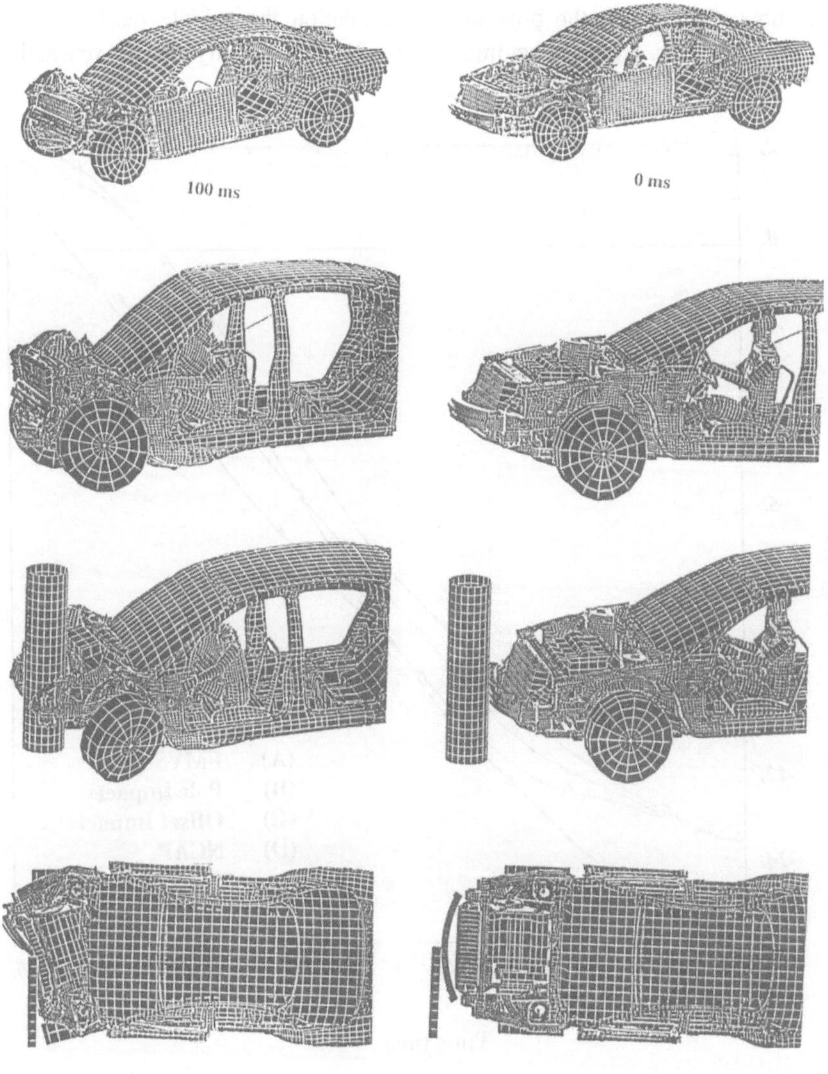

Figure 20. **Final and initial vehicle configurations from four simulations - impact with rigid barrier from 13.4 m/s, impact with rigid barrier from 15.7 m/s, impact with rigid pole from 13.4 m/s, and impact with offset rigid barrier from 13.4 m/s**

Figure 21 shows the velocity-time responses from the four simulations. It can be observed that the FMVSS 208 and NCAP simulations reached the zero velocity point at the same time, 85 ms. In the pole impact simulation the vehicle reached the zero velocity point at 97 ms. In the offset impact case, the zero velocity was not reached until 102 ms.

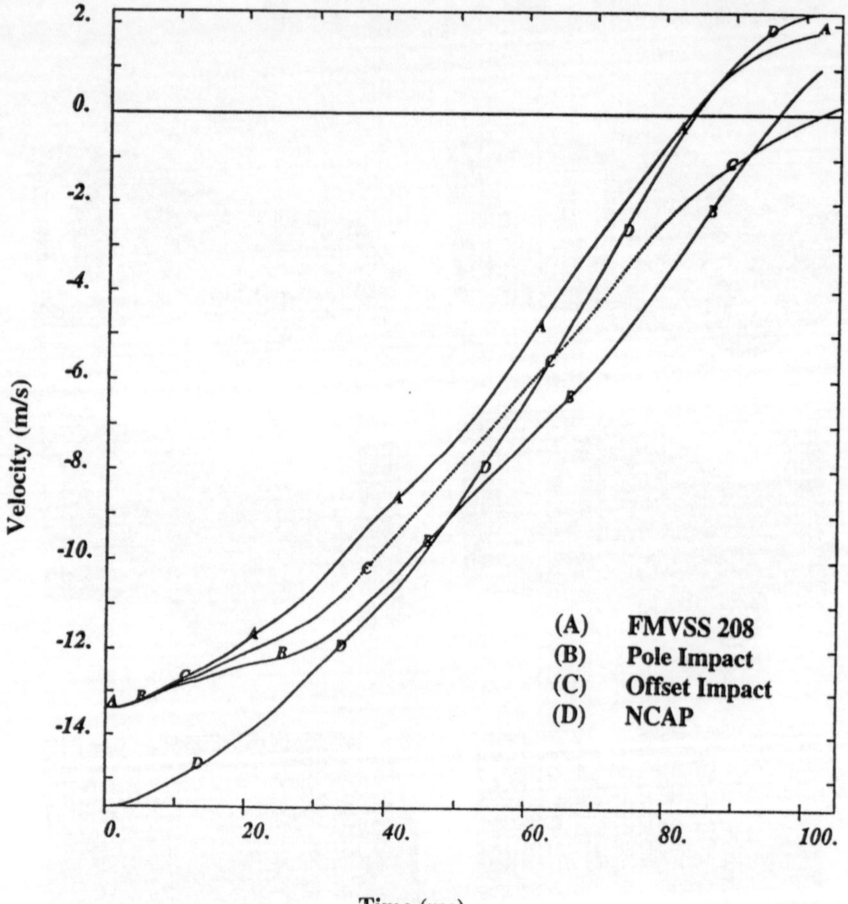

Figure 21. Rear rocker velocity from four simulations

Also, difference in vehicle deceleration among the four runs were observed. These zero time crossings are consistent with experimental observations.

Figure 22 provides a summary of the peak values for dummy chest acceleration, pelvis acceleration, left femur, and right femur loads. The head acceleration was not included since the peak value was not reached at 100 ms, as indicated earlier. No experimental data was available to verify the level of accuracy of the predicted dummy responses.

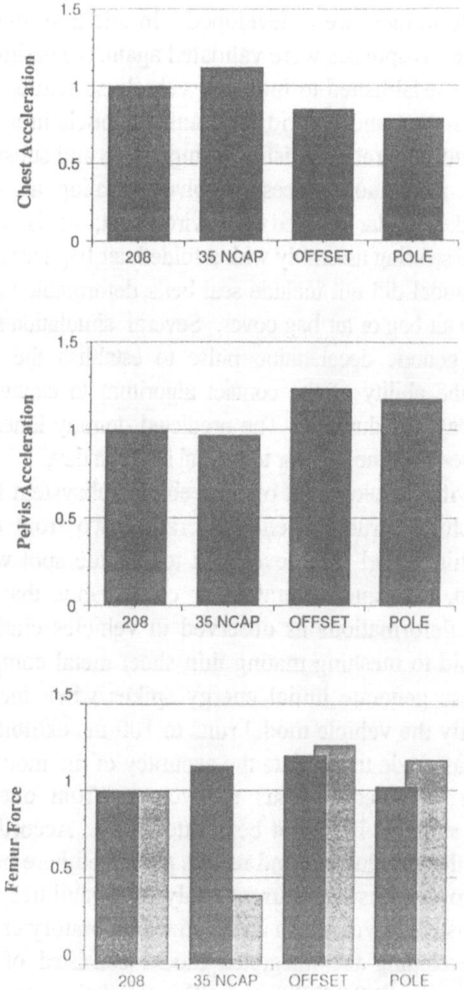

Figure 22. Normalized dummy response parameters from four simulations

5. Summary

Although the FE technology for structural mechanics was introduced in the early sixties, it took about 25 years of additional development to apply it successfully to crashworthiness simulation of automobile structures. The developments were mainly in nonlinear problem formulation, reduced integration elements, explicit time integration and contact-impact treatments.

The mid-eighties to the mid-nineties time span can be characterized as the renaissance period of crashworthiness models. Generic and actual components of vehicle structures as well as full scale vehicle models were developed. In addition, dummy and air bag models were created and their responses were validated against experimental data.

In 1995, a process was established to integrate vehicle structure, instrument panel, steering assembly, driver air bag and Hybrid III dummy models in a single FE model. This process centered around integrating existing components and subsystem models.

The first step in the integration process involved creating an interior passenger compartment subsystem model, similar to a sled test environment; consisting of the Hybrid III dummy, seat, and collapsible steering assembly with a folded air bag located in the middle of the steering wheel. This model did not include seat belts, deformable knee bolster, or the compartment that houses the air bag or air bag cover. Several simulation runs were obtained with this model using a generic deceleration pulse to establish the robustness of the simulation, particularly of the ability of the contact algorithm to capture the interactions between the deploying air bag and dummy. The predicted dummy kinematics and contact forces exhibited realistic pattern in time, similar to typical sled test data.

The next step dealt with development of the vehicle subsystem model; consisting of a four-door (sedan) vehicle structure, engine, cradle, two front doors, instrument panel and knee bolster. This model was developed to include spot welds, suspension, engine mounts, drive train, tires and several other components that are necessary to obtain realistic structural deformations as observed in vehicles crashed into barriers. Particular attention was paid to meshing mating thin sheet metal components to avoid initial penetrations, as these generate initial energy spikes when included among the contact segments. Currently the vehicle model runs to 100 ms exhibiting good overall kinematics. No attempt was made to validate the accuracy of the model response except for one macro parameter - velocity versus time curve, from one test. Complete validation of the vehicle submodel has not been attempted. Accordingly, no attempt should be made to extend the conclusions and results presented here beyond the primary objective of the paper - to demonstrate current analysis capabilities in integrating and predicting the gross response of a vehicle in an idealized laboratory crash environment.

The final step in developing the integrated model consisted of merging the two subsystem models to simulate an FMVSS 208 test. The simulation ran successfully to 100 ms; and provided qualitatively comparable response, not identical, to a typical barrier test.

Three simulations were next performed to investigate the model robustness: 1. A 15.6 m/s (to represent an NCAP test) vehicle to flat rigid barrier impact; 2. A 13.4 m/s vehicle impact into a 50 % offset rigid barrier; and 3. A 13.4 m/s impact into a rigid pole located in the mid-plane of the vehicle. A three-point belt system, in addition to the air bag, was used in the NCAP calculation. All simulations ran successfully and provided realistic system crash performance. In particular, the predicted dummy kinematics and associated loads and accelerations were quite realistic.

In conclusion, explicit integration FE modeling techniques can simulate both structural and occupant response resulting from a vehicle crash in a single integrated model. This is clearly demonstrated in this study, albeit the results are preliminary

6. References

1. National Safety Council (1995) *Accident Facts*, Itasca, IL: Author.~
2. Martinez, R. (1995) *Hamonazied Motor Vehicle Safety Agenda*, Geneva.
3. Kippola, W.J. and Stando, M.J. (1994) Emerging Trends in Safety Design and Technology, *14Th.International Technical Conference on Enhanced Safety of Vehicles*, Munich, Germany, Paper No. 94-S9-O- 08.
4. Kamal, M.M. (1970) Analysis and Simulation of vehicle to Barrier Impact, *Society of Automotive Engineers*, SAE Paper No. 700414
5. Ni, C.M. (1981) A General Purpose Technique for Nonlinear Dynamic Response of Integrated Structures, *Fourth International Conference on Vehicle Structural Mechanics*, SAE Publisher.
6. Khalil , T.B. and Bennett, J.A. (1989) Nonlinear Finite Element Analysis in Crashworthiness Automotive Shell Structures - An Overview", in*Analytical and Computational Models of Shells*, A. K. Noor, T. Belytschko and J.C. Simo, (eds.), ASME Publisher, CED Vol. 3. PP 569-582.
7. Mahmood H.F. and Paluzeny A. (1986) Analytical Technique for Simulating Crash Response of Vehicle Structures Composed of Beam Elements, *Sixth International Conference on Vehicle Structural Mechanics*, SAE Publisher.
8. Fleck, J.T., Butler F. E. and Vogel, S.L. (1974) Three Dimensional Computer Simulation of Motor Vehicle Crash Victims, Vol I-IV, Report Nos DOT-HS-810507, 508, 509 and 510.
9. Wismans, J. and Hermans, J.H.A. (1988) MADYMO 3D Simulations of the Hybrid III dummy Sled Tests, *SAE International Congress and Exposition*, Paper No. 880645.
10. PAM-CVS User's Manual, Version 1.0 (1990), Engineering Systems International, Paris, France.
11. Khalil, T.B. et. al. (1991) Development of a Three-Dimensional Finite Element Model of Air Bag Deployment and Interactions with an Occupant using DYNA3D, in*Proceedings of 35 Th. Stapp Car Crash Conference*, SAE Publisher, pp. 251-269.
12. Khalil, T.B. and Lin, T.C. (1994) Simulation of the Hybrid III Dummy Response to Impact by Nonlinear Finite Element Analysis, in *Proceedings of 38 Th. Stapp Car Crash Conference*, SAE Publisher, pp. 325-345.
13. Khalil, T.B., Sheh, M.Y., Chalons, P. and Du Bois, P.A. (1995) Integrated Vehicle-Dummy-Airbag Model for Frontal Crash Simulation by FE Analysis, in*Symposium on Crashworthiness and Occupant Protection in Transportation Systems*, J. Ried, et. al. (eds.), ASME Publisher, pp. 355-383.~
14. Newmark, N.M. (1959) A Method of Computation for Structural Dynamics, *J. Engineering Mechanics Division*, ASCE, 67, 67-94.
15. Farhoomand, I. (1972) Nonlinear Dynamic Stress Analysis of Two-dimensional Solids,*Ph.D. Thesis*, University of California, Berkeley, Graduate Division.
16. Belytschko, T.B. and Hsieh, B.J. (1974) Nonlinear Transient Analysis of Shells and Solids of Revolution by Convected Elements, *AIAA Journal*, 12, pp. 1031-1039.
17. Hughes, T.J.R., Pister, K.S. and Taylor, R.L. (1979) Implicit-Explicit Finite Elements in Nonlinear Transient Analysis, *Computer Methods in Applied Mechanics and Engineering*, 17/18, 159.
18. Wittlin, G. (1983) Aircraft crash dynamics: modeling, verification and application, in N. Jones and T. Wierzbicki (eds.), *Structural Crashworthiness*, Butterworths and Co. Publisher, London, pp. 258-281.
19. Pifko, A.B. and Winter, R. (1981) Theory and applications of finite element analysis to structural crash, in *Structural and Nonlinear Solid Mechanics*, A.K. Noor and H.G. McComb (eds.), Pergamon Press, Oxford.
20. Sheh, M.Y. and Khalil, T.B. (1991) The Impact Response of a Vehicle Structural Rail by Experiments and Finite Element Analysis in *Symposium on Crashworthiness and Occupant Protection in Transportation Systems,* T.B. Khalil, H.F. Mahmood, C.M. Ni andA.I. King (eds.*),* ASME Publication, AMD-Vol. 126/BED-Vol. 19, pp. 195-207.
21. Khalil, T.B. and Vander Lugt, D.A. (1989) Identification of Vehicle Front StructureCrashworthiness by Experiments and Finite Element in *Symposium on Crashworthiness and Occupant Protection in Transportation Systems,* T.B. Khalil, and A.I. King, (eds.*),* ASME Publisher, AMD-Vol. 106 and BED-Vol. 13, pp. 41-53.
22. Winter, R., Mantus, M. and Pifko A.B. (1981) Finite Element Crash Analysis of a Rear Engine Automobile, *4tth International conference on Vehicle Structural Mechanics*, SAE Publisher, pp. 55- 61.

23. Haug, E., Arnadeau, F., Dubois, J. and De Rouvray A. (1983) Static and Dynamic Finite Element Analysis of Structural Crashworthiness in the Automotive and Aerospace industries", in *Crashworthiness*, N. Jones and T. Wierzbicki (eds.), Butterworths Publisher, pp. 175-218.
24. Argyris, J. , Balmer, H.A., ST. Doltsinis, J. and Kruz, A. (1986) Computer Simulation of Crash Phenomena", *Int. J. for Numerical Methods in Engineering*, 22, pp. 497-519.
25. Haug, E., Scharnhorst T. and Du Bois P. (1986) FEM- Crash Simulation of a Frontal Impact (VW-POLO), Translated From German, VDI Report 613, pp. 479-505.
26. LS-DYNA3D User's Manual (1994), Version 920, Livermore Software Technology, Livermore California.
27. RADIOSS User's Manual (1989), MECALOG SAR, Paris, France.
28. Bertz, G., Jarzab W. and Raasch I. (1986) Computation of A Frontal Impact of A Rear Wheel Drive Car, Translated From German, *VDI Report* ,613, pp. 507-525.
29. Benson, D.J. and Hallquist J.O. (1986) The Application of DYNA3D in Large ScaleCrashworthiness Calculations, Lawrence Livermore National Laboratory, *Report No. UCRL*-94028.
30. Scharnhorst, T. I. Rassch, I. and E. Schelke, E. (1986) Conclusivness of Math Methods for Crash Behavior Simulation of Cars, Translated from German, *VDI Report*, 613.
31. Chedmail, J.F. et. al. (1986) Numerical Techniques, Experimental Validation and Industrial Applications of Structural Impact and Crashworthiness Analysis with Supercomputers for the Automotive Industry, *International Conference on Supercomputing*, Zurich, Switzerland.
32. Nalepa, E. (1989) Crashworthiness Simulation of the Opel Vectra using the Explicit FE Method, *Int Journal of Vehicle Design, IAVD Congress on Vehicle Design and Components*, pp. 191-201.
33. Vander Lugt, D.A. et. al. (1987) Passenger Car Frontal Barrier Simulation Using NonlinearElemen Method, *SAE Passenger Car Meeting and Exposition*, Dearborn, Michigan, Paper No. 871958, 1987
34. Isuzu Motors Limited,Crash Analysis of 760 Coupe , Body Engineering Department, 1987, Personal Communication.
35. Nilsson, L. (1989) Computational Crash Analysis at the Saab Car Division, *12 Th. International Technical Conference on Experimental Safety Vehicles*, Goteborg, Sweden.
36. Kurimoto, K. et.al. (1989) Simulation of Crashworthiness and its Application, *12 Th. International Technical Conference on Experimental Safety Vehicles*, Goteborg, Sweden.
37. T. Futamata, T. et. al. 1989) Crash Simulation Methods for Vehicle Development at NISSAN ",*12 Th International Technical Conference on Experimental Safety Vehicles*, Goteborg, Sweden.
38. Johnson, J.P. and Skynar, M.J. (1989) Automotive Crash Analysis using the Explicit Integration Finite Element Method", in *Proc. of Crashworthiness and Occupant Protection in Transportation Systems*, T.B. Khalil, and A.I. King, (eds.), ASME Publisher, AMD-Vol. 106 and BED-Vol. 13, pp. 27-33
39. Sheh, M.Y. et. al. (1992) Vehicle Crashworthiness Analysis Using Numerical Methods and Experiments, *Eighth International Conference on Vehicle Structural Mechanics and CAE*, Traverse City, Michigan, SAE Publisher, pp. 119-128
40. teyer, C., Diet, S. and Du Bois, P. (1989) Numerical Simulation of the Proposed Side Impact Procedure, in *Proceedings of Crashworthiness and Occupant Protection in Transportation Systems*, AMD-Vol. 106 and BED-Vol. 13, T.B. Khalil, and A.I. King, (eds.), ASME Publisher, pp. 53-95.
41. A. Kaiser, A. (1992) Some Examples on Numerical Simulation in Vehicle Safety Development,*Eight, International Conference on Vehicle Structural Mechanics and CAE*, Traverse City, Michigan, SAE Publisher, pp. 119-128.
42. Calso, S.M. et. al. (1993) Simulation of Offset and In-line car to Car Rear Impact" , in*Proceedings of Crashworthiness and Occupant Protection in Transportation Systems*, AMD-Vol. 169 and BED-Vol. 25, J. Reid, and K. Yang (eds.), ASME Publication, pp. 149-161.
43. Lin, T-C, Wawa, C. and Khalil, T.B. (1995) Application of a FE Model of the Hybrid III Dummy for Frontal Crash with Air Bag, in *Stapp Car Crash Conference Proceedings*, SAE Publisher, pp. 37-51.
44. McHenry, R.R. (1965) Analysis of the Dynamics of Automobile Passenger Restraint Systems, in *Seventh Stapp Car Crash Conference Proceedings*, SAE Publisher, pp. 207-249.
45. Wang, J.T. and Nefske, D.J. (1988) A New CAL 3D Airbag Inflation Model, SAE Technical Paper Series, paper Number 880654.
46. Nieboer, J.J., Wisman, J. and Rraterman,E. B. (1988) Status of the MADYMO 2D Airbag, in 32nd *Stapp Car Crash Conference Proceedings*, SAE Publisher, pp. 223-235
47. Hoffman, R. et. al. .(1990) Finite Element Analysis of Occupant Restraint System Interaction with PAM-CRASH, in 34 th *Stapp Car Crash Conference Proceedings*, SAE Publisher, pp. 289-300

48. Nieboer, J.J., Wisman, J. and De Coo, P.J.A. . (1990) Airbag Modeling Techniques, in Proceedings of 34 Th. *Stapp Car Crash Conference Proceedings,* SAE Publisher, , pp. 243-259.
49. Khalil, T.B. and Lin, K.H. (1991) Hybrid III Thoracic Impact on Self-Aligning Steering Wheel by Finite Element Analysis and Mini-Sled Experiments", *In Proceedings of 35 Th. Stapp Car Crash Conference,* SAE Publisher, pp. 73-85.
50. Ove Arup & Partners International, (1992) An integrated Approach To VehicleCrashworthiness and Occupant Protection, Report, Issue 1, London, England.
51. E. Schelkle, E. and R. Remensperger, R. (1991) Integrated Occupant-Car Crash Simulation with the Finite Element Method: The Porsche Hybrid III-Dummy and Airbag Model, in*Frontal Crash Safety Technologies for the 90's,* SAE Paper No. 910654.
52. PAM-CRASH User's Manual (1994) Version 12.1, Engineering Systems International, Paris, France.
53. Kohlhoff, St et.al. (1994) A Computational Approach to an Investigation of Frontal Car-To-Car Collision for the Development of Barrier Test, Numerical Analysis in Automobile Engineering,*VDI Conf.* Wurzburg, Germany, pp. 467-484.
54. Cheng, J.C. and Doong, J. (1994) Numerical Analysis of Vehicle Crashworthiness in Various Configurations, *Cray Channels*, Vol. 16, No. 2 , pp. 8-11

RECENT TRENDS AND ADVANCES IN CRASH SIMULATION AND DESIGN OF VEHICLES

E. HAUG[1], J. CLINCKEMAILLIE[1], X. NI[1], A. K. PICKETT[2], T. QUECKBORNER[2]

[1]*ESI Group, 20 rue Saarinen, Silic 270, 94578 Rungis-Cedex, France*

[2]*ESI GmbH, Frankfurter Str. 13-15, D-65760 Eschborn, Germany*

Abstract

An overview on some recent trends and advances in the crash simulation and design of transport vehicles is given. This overview highlights selected algorithmic solver code advances in the used simulation tools, the use and the modelling of new materials for crash energy absorption, concept car design techniques, massive parallel programming and performance gains, side impact barrier modelling, mechanical occupant surrogate modelling (dummies), biomechanical models of human parts, as well as extensions of crash simulation techniques to the simulation of drop tests for appliances, shock absorption of a Mars lander, etc. The shown examples and descriptions testify the extreme progress and diversification crash simulation techniques have undergone in the past ten years.

1. Introduction

A modern crash simulation software package consists of pre- and post-processing modules and of a numerical solver module. The beneficial industrial use of a crash simulation package depends not only on the effectiveness of the numerical solver code, but to a large extent also on the quality, effectiveness and user-friendliness of the associated pre- and post-processing modules. Two such modules, PAM-GENERIS™ (pre-processing) and PAM-VIEW™ (post-processing) are presently offered with the PAM-SOLID™ solver codes. At the time of writing, both modules undergo extensive restructuring and enhancements that are necessary to assure the full industrial applicability and success of the simulation package. This paper concentrates on recent advances and trends of the numerical solver code, PAM-CRASH™, a member of the PAM-SOLID™ family of codes. The PAM-SOLID™ family of codes consists in PAM-CRASH™, the PAM-SAFE™ occupant safety analysis code, the PAM-STAMP™ sheet stamping code and the PAM-SHOCK™ impact and high frequency response analysis code.

The industrial simulation of vehicle crash events started 10 years ago with the first successful VW Polo frontal crash overnight simulation on a CRAY 1 computer, using PAM-CRASH™. Meanwhile all major passenger car makers employ crash simulation to assist them in the design of crashworthy vehicles within competitive delays, and the

J. A. C. Ambrósio et al. (eds.),
Crashworthiness of Transportation Systems: Structural Impact and Occupant Protection, 401–417.

simulation of impacts of different natures has spread to other branches of industry (truck crash, railway car crash, container drop, packaging design, airplane crash, etc.).

The next logical step has been to extend the methodology to the simulation of passenger safety via the modelling of passive safety devices such as airbags, seat belts, cushions, etc., that restrain models of vehicle occupant surrogates (dummies) during crash events. Progress in modelling techniques for dummies has lead to four identifyable "generations" of dummies, perhaps soon followed by a fifth generation of biological dummies.

Moreover, the used explicit finite element technique is now being applied to related processes, such as thin sheet metal forming, thermoplastic composite shell forming, pressure and hydroforming, etc., that benefit from the numerical techniques elaborated for large strain/large displacement nonlinear material/ dynamic analyses of structures, modelled by large numbers of thin shell finite elements (up to 100 000 and more).

The fulgurant success of the explicit thin shell modelling technique is now reinforced by porting the codes to massively parallel computers that will bring super-computer power into the design departments, and by increased porting to inexpensive PC-level workstations that will make crash simulation accessible to the smallest suppliers.

Numerous extensions in meshing from CAD data, in macro beam elements for concept design, in new material models, strain rate and damage sensitive models, for foams, composites, plastics, etc. render the codes more effective and efficient for simulation of crash and impact of a wider class of industrial problems.

The major thrust areas for continued developments comprise the linking of the simulation packages into CAD systems, more material models, more accurate element formulations and contact algorithms, made possible through increased computing power, larger models, the use of multi-physics (e.g., thermo-coupling, solid-fluid interaction) and continued porting to massively parallel computers and clustered workstations.

2. Algorithmic Advances

Despite a considerable level of industrial efficiency and robustness reached by the modern explicit finite element crash simulation solver codes, there is still much room for algorithmic improvements and advances. Some of such efforts are outlined next.

2.1. CONTACT SIMULATION

It should be mentioned that the correct and robust simulation of physical contacts in crash simulation is a prerequisite of its success. Notoriously, however, contact simulation still belongs to the most challenging and complex tasks of programming, not only in mono-processor (or shared memory parallel) code versions, but, in particular, in massively parallel programs using distributed memory. In the first case "rigid wall" constraint or "sliding interface" treatments of contact must be streamlined as much as possible to be computationally efficient, while not loosing accuracy and robustness. For sliding interface type contact algorithms the proven penalty method is the most widely used, and todays development efforts aim at rendering this method as transparent as possible to the end user, including the difficult to achieve success criteria of both, computational efficiency and robustness.

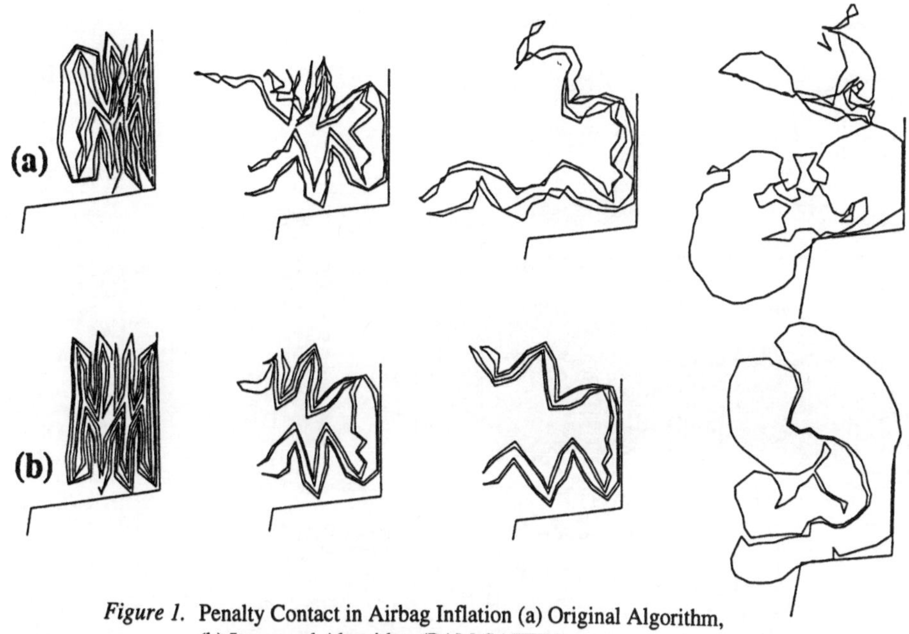

Figure 1. Penalty Contact in Airbag Inflation (a) Original Algorithm, (b) Improved Algorithm (PAM-SAFE™)

Figure 1 shows a comparison of contact treatment with an older and an improved penalty contact algorithm for airbag inflation.

For massively parallel programming, the explicit finite element programs are readily parallelized using effective techniques of automatic domain decomposition, where each of the parallel processors is assigned to work on similar chunks of the model with a minimal amount of communication required with other processors. Since contact domains are not static by definition (initially remote parts of the structure do become connected as the structure deforms), initial domain decompositions can quickly become obsolete and the calculation of contacts can lead to serious communications overheads and load-imbalances between processors. For these reasons, the major efforts in massively parallel ports of crash simulation codes goes into the effective parallelization of contact algorithms and is pursued actively.

2.2. ADAPTIVE MESHING

In crash simulation of thin-walled components and structures, adaptive meshing techniques aim at automatic local refinements of thin shell finite element meshes near areas of large deformations. In the simulation of industrial sheet metal stamping this problem has been solved, and the PAM-STAMP™ member of the PAM-SOLID™ family of simulation codes is equipped with industrial options of uniform and automatic adaptive thin shell mesh refinement and derefinement.

The first reason why adaptive meshing algorithms are more difficult to implement for crash simulation is that mesh adaptivity must be made compatible with a wider variety of contact treatments (again !) and rigid wall contact/impact options. Although contact is the principal driving mechanism in sheet stamping, its treatment is simpler due to the imposed tool geometries and the virtual absence of sheet self-contact. In crash simulation, self-contact is often preponderant but less amenable for implementation with automatic adaptive meshing.

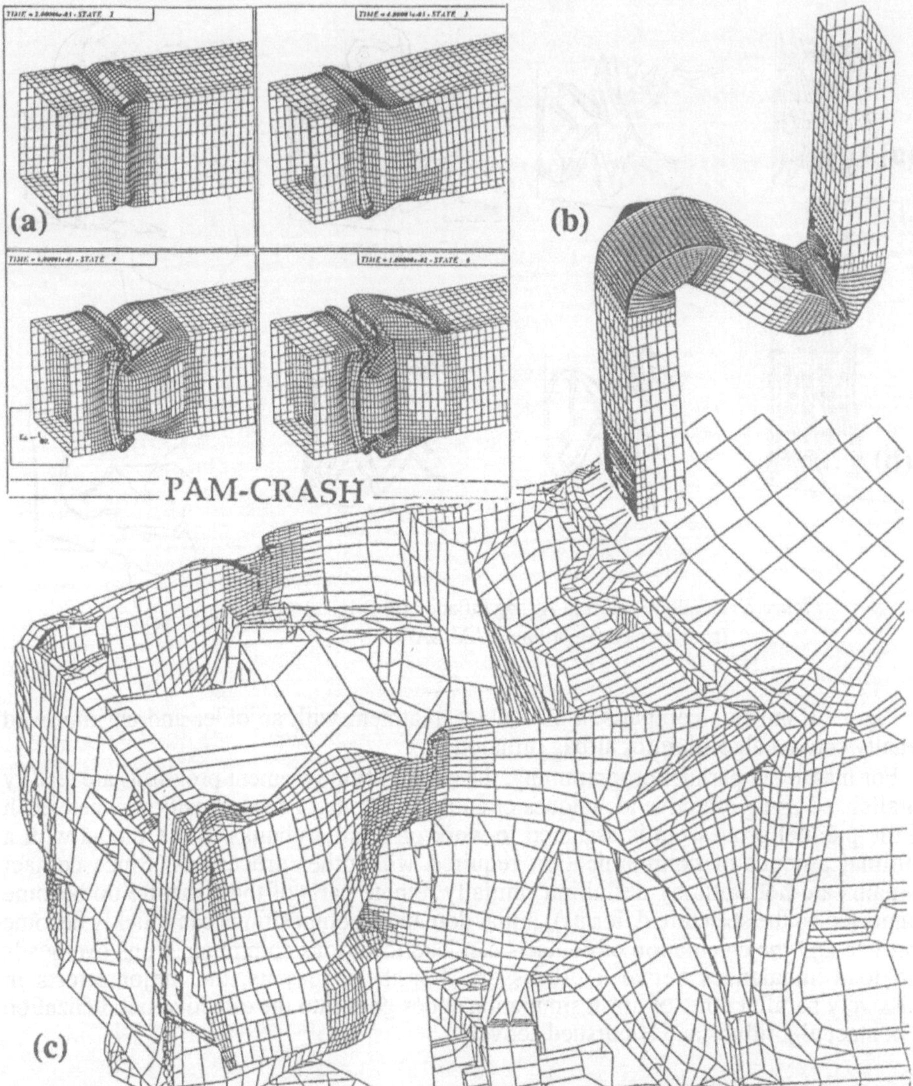

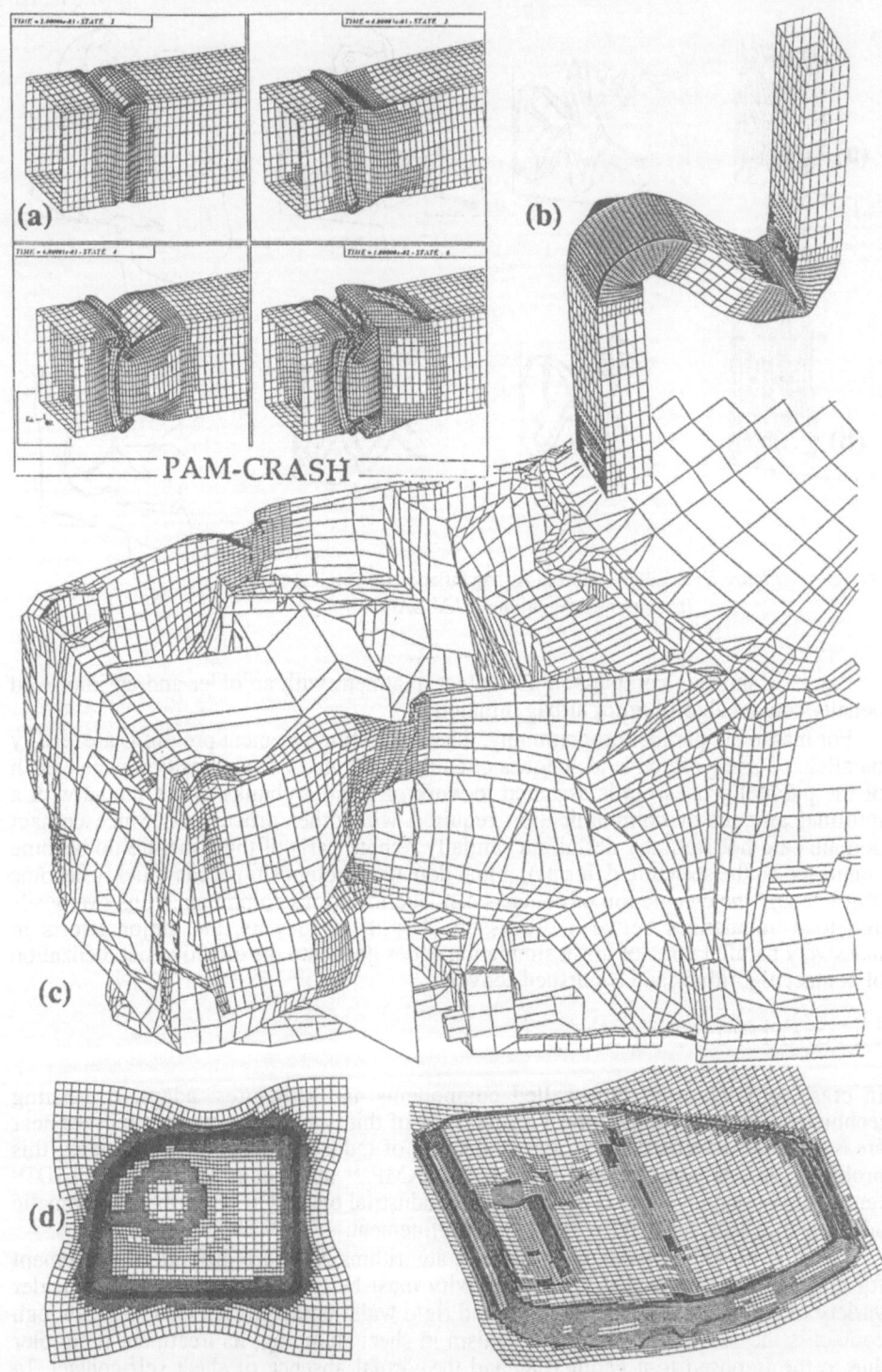

Figure 2. Adaptive Meshing, (a) Box-column, (b) S-frame, (c) Frontal Crash (PAM-CRASH™), (d) Stamped Parts (PAM-STAMP™)

The second reason why the utilization of adaptive meshing may be less obvious, even dangerous, in crash simulation is the possibility of localization of plastic hinges or crush zones. The reason for such localization may reside in the fact that refined mesh areas become numerically less stiff than the original coarser grid, which is exactly what is expected, but that subsequent plastic hinges and crush zones in the not yet refined mesh may be masked by the added flexibility or reduced numerical resistance of the refined areas, which is an undesirable effect. The refinement criteria must therefore be chosen carefully, and the original mesh must not be too coarse in order to avoid such localization.

The method is demonstrated in Figure 2 on the examples of the axial crushing of a thin-walled box column (a), in deep-collapse plastic hinge development in an S-frame (b), in a full passenger car frontal crash (c), and in examples from sheet metal stamping (d).

3. Material Models

More and more emphasis is given to the design, calibration and validation of material models for the description of the behaviour of aluminum alloys, plastics, foams, rubbers and composites. Material models that describe the behaviour of all these types of materials are available in the PAM-SOLID™ code family, and some models are highlighted below.

3.1. GENERAL ELASTO-PLASTIC/STRAIN RATE/DAMAGE MODEL

Several isotropic and anisotropic elasto-plastic with strain-rate dependent plasticity and damaging material models are implemented for the finite elements of the PAM-SOLID™ family. Their response is summarized in Figure 3(a), where in a material stress-strain diagram strain-rate and damage dependent stress-strain curves are plotted. This approach can be applied in principle to any type of material. The basic undamaged stress-strain law, $\sigma_0(\varepsilon, \dot{\varepsilon}_{ref})$ being a small or zero reference strain rate, is post-multiplied by a strain rate dependency function, $f(\varepsilon, \dot{\varepsilon})$ and by a damage function, $g(d(\varepsilon, \dot{\varepsilon})$, where the damage parameter $0 \leq d(\varepsilon, \dot{\varepsilon}) \leq 1.0$ and $g = 1-d$. The strain-rate function tends to increase the yield stress and the damage function reduces the material resistance. The indicated general damage description is due to Lemaitre-Chaboche and applies to arbitrary materials (metals, plastics, composites, foams, etc.). For ductile metals, the Gurson damage law is presently implemented in collaboration with the University of Valenciennes. It acts on plastic strain and translates the effects of nucleation, growth and finally coalescence of micro-voids on the material behaviour. The effect of voids or damage on stress at different scales is depicted in Figure 3(b).

3.2. ALUMINUM ALLOYS

The plastic behaviour of aluminum alloys can be described by the Hill 1990 non-quadratic yield function for anisotropic materials, Figure 4. This law has been implemented in the PAM-STAMP™ member of the PAM-SOLID™ family and it has been applied to the simulation of deep drawing of aluminum sheets.

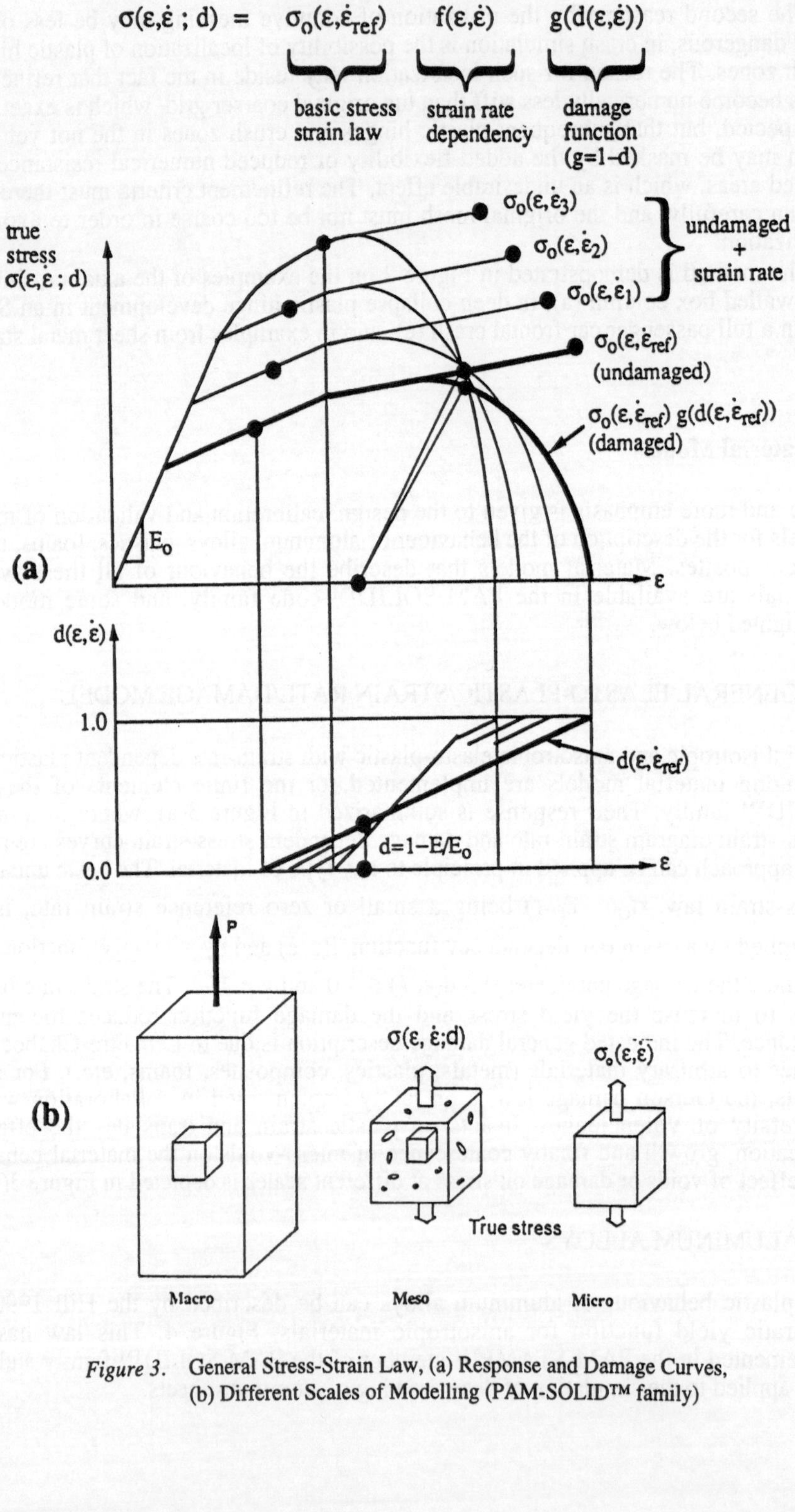

Figure 3. General Stress-Strain Law, (a) Response and Damage Curves, (b) Different Scales of Modelling (PAM-SOLID™ family)

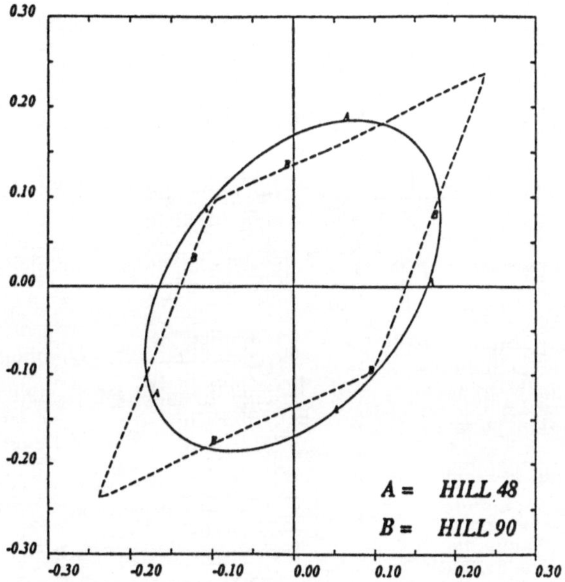

Figure 4. Yield Function for Aluminum Alloys (PAM-STAMP™ ; Andrew Heath ESI GmbH)

3.3. PLASTICS

The consequence of a softening and then stiffening elastic-plastic behaviour, common in plastics, is demonstrated in Figure 5 where a plane-stress tensile test is simulated. First, a purely softening true stress-strain law is applied in the finite element simulation of a tensile test (a), where plastic instability ("necking") occurs as soon as the numerical value of the plastic stress, σ, has fallen to the value of the tangent modulus, E_T. This instability soon leads to localization of plastic strain and specimen rupture after the plastic strain has reached a specified rupture limit. Next, the original plastic hardening curve is modified (b) to stiffen for higher plastic strains, which limits the plane-stress condition of necking, $\sigma = E_T$, to the range $\varepsilon_1 < \varepsilon < \varepsilon_2$. The original neck develops as before, but it is now arrested by the stiffening of the material after the finite elements of the necked area have been stretched beyond a plastic strain of value ε_2. The neck then spreads to the neighbour elements and it is seen to gradually invade the entire test piece.

3.4. MORE MATERIALS

For the description of side impact barriers, paddings, flesh foam of mechanical dummies, etc., a variety of foam materials is available in the PAM-SOLID™ package that can adequately model the collapse, compaction, viscous, hysteresis and rate-dependent behaviour of these materials. Figure 6 summarizes the application of barrier foam models to the simulation of side-impact barriers.

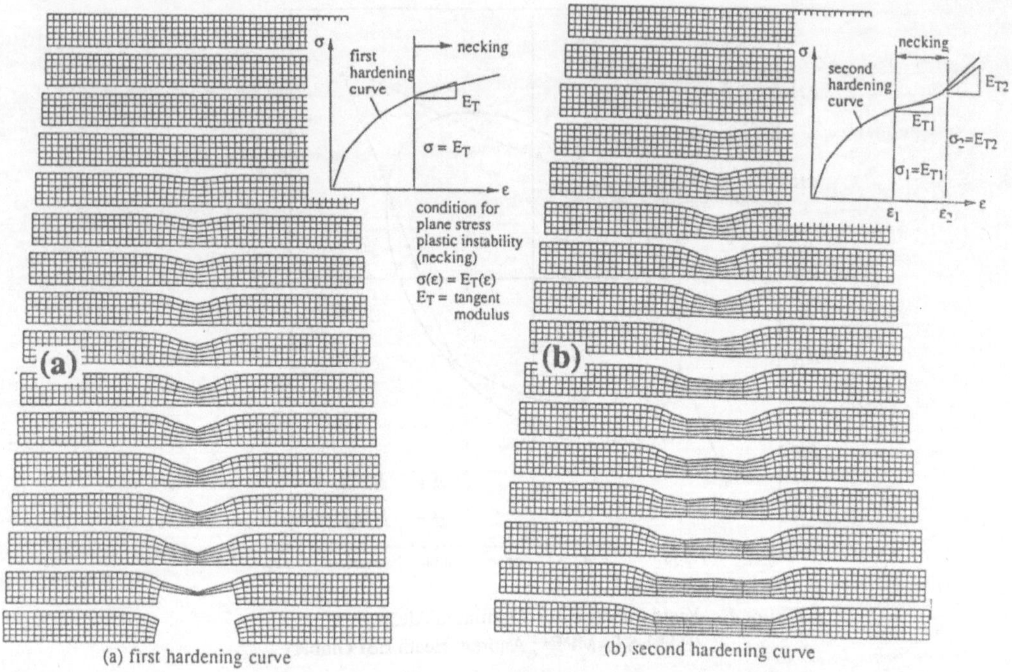

Figure 5. Plastic Strain Hardening in a Plane-Stress Specimen, (a) Softening Material, (b) Softening/Stiffening Material (PAM-SOLID™ family)

For the description of rubbers and rubber-like materials in the modelling of tires, engine supports, parts of dummies, etc., a variety of hyper-elastic quasi-incompressible material models has been implemented. Figure 7 summarizes the behaviour of these models.

For the simulation of crash events that involve fiber and fabric reinforced composites, bi-phase brittle damaging material models have been implemented, that are now being extended to incorporate plastic (metal-composite hybrid) and viscous (thermoforming of composites) matrix materials. Figure 8 gives an overview on the crash behaviour of composites.

All material models will certainly undergo further refinement, and new material models will be added to answer the growing demand of more and more industry to cope with their specific products.

4. Concept Crash Design

An interesting trend in crash simulation is the growing demand of industry to employ simplified crash simulation tools, at least during the early design phases of transport vehicles. Figure 9 outlines a successive crashworthiness design and verification approach, ranging from pre-crash concept design (Stage 0) to full crash models (Stage 5). Current crash simulation practice addresses mostly the last stage, i.e., the fabrication

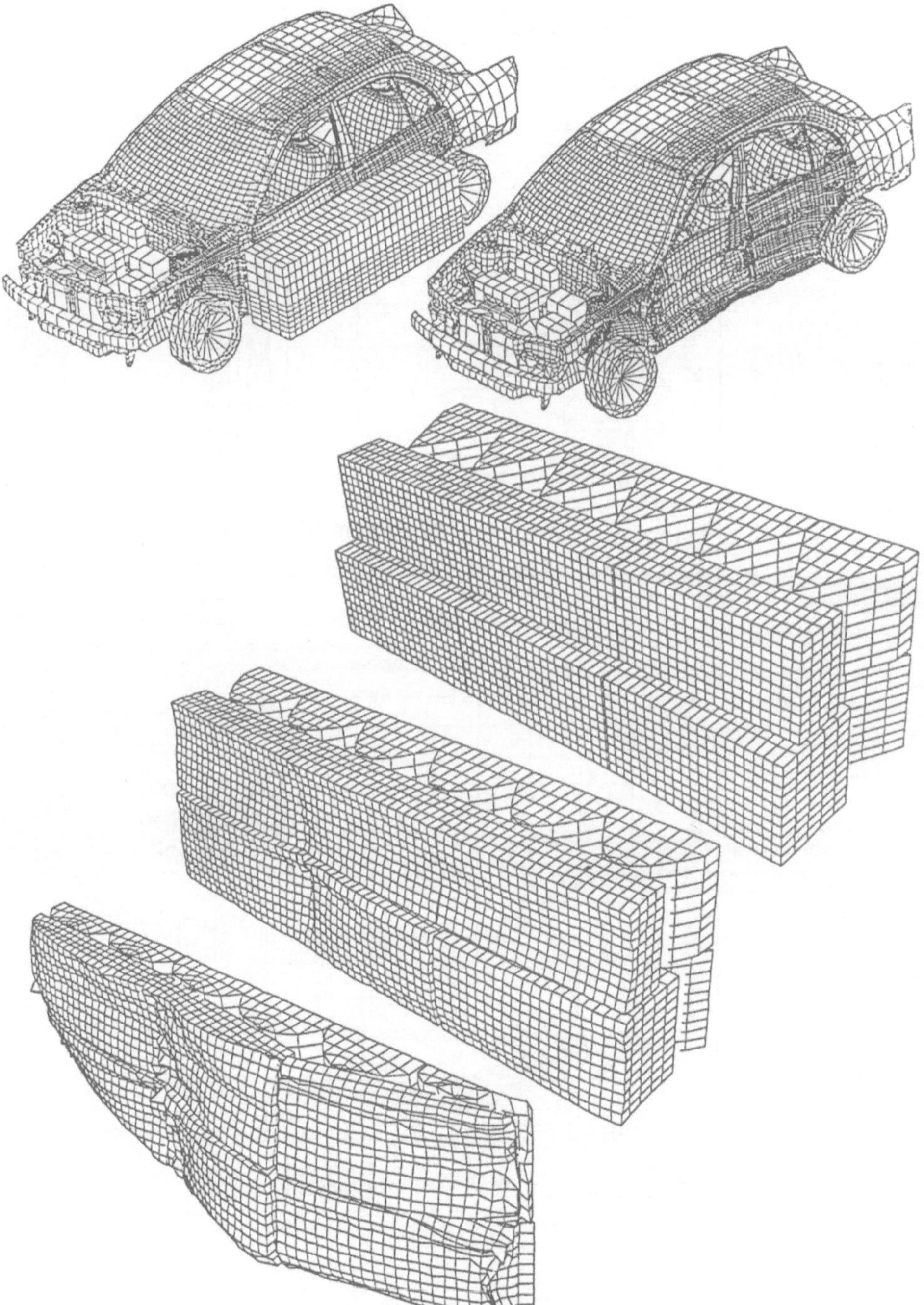

Figure 6. Side Impact Barrier Simulation using Foams (PAM-SAFE™, ESI GmbH)

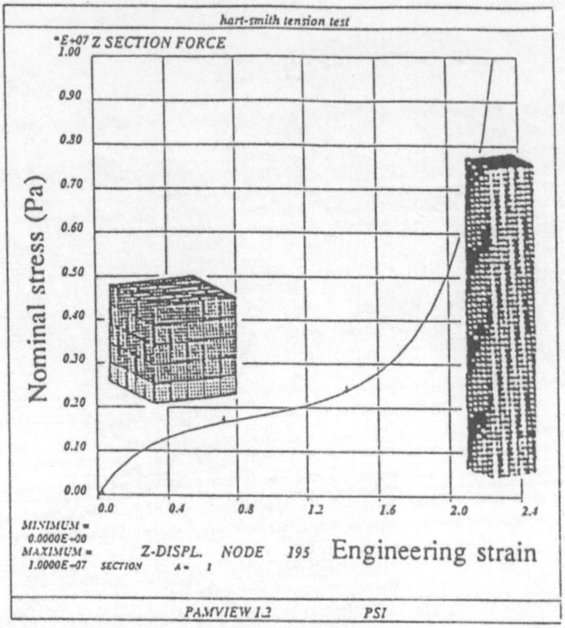

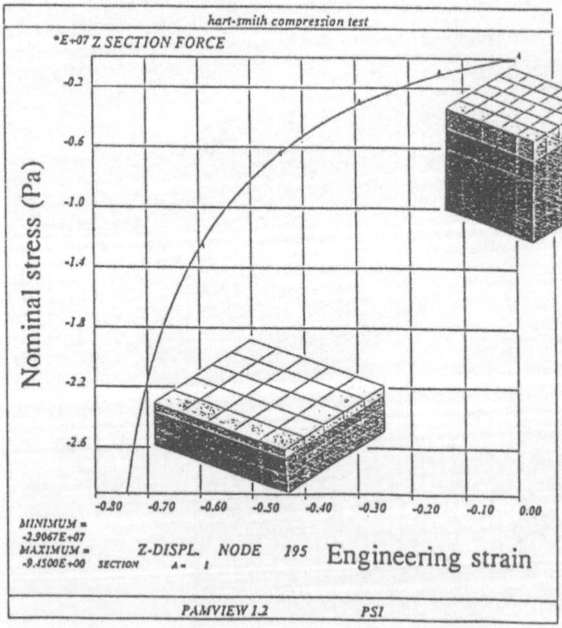

Figure 7. Hyper-elastic Material Models (PAM-SOLID™ family)

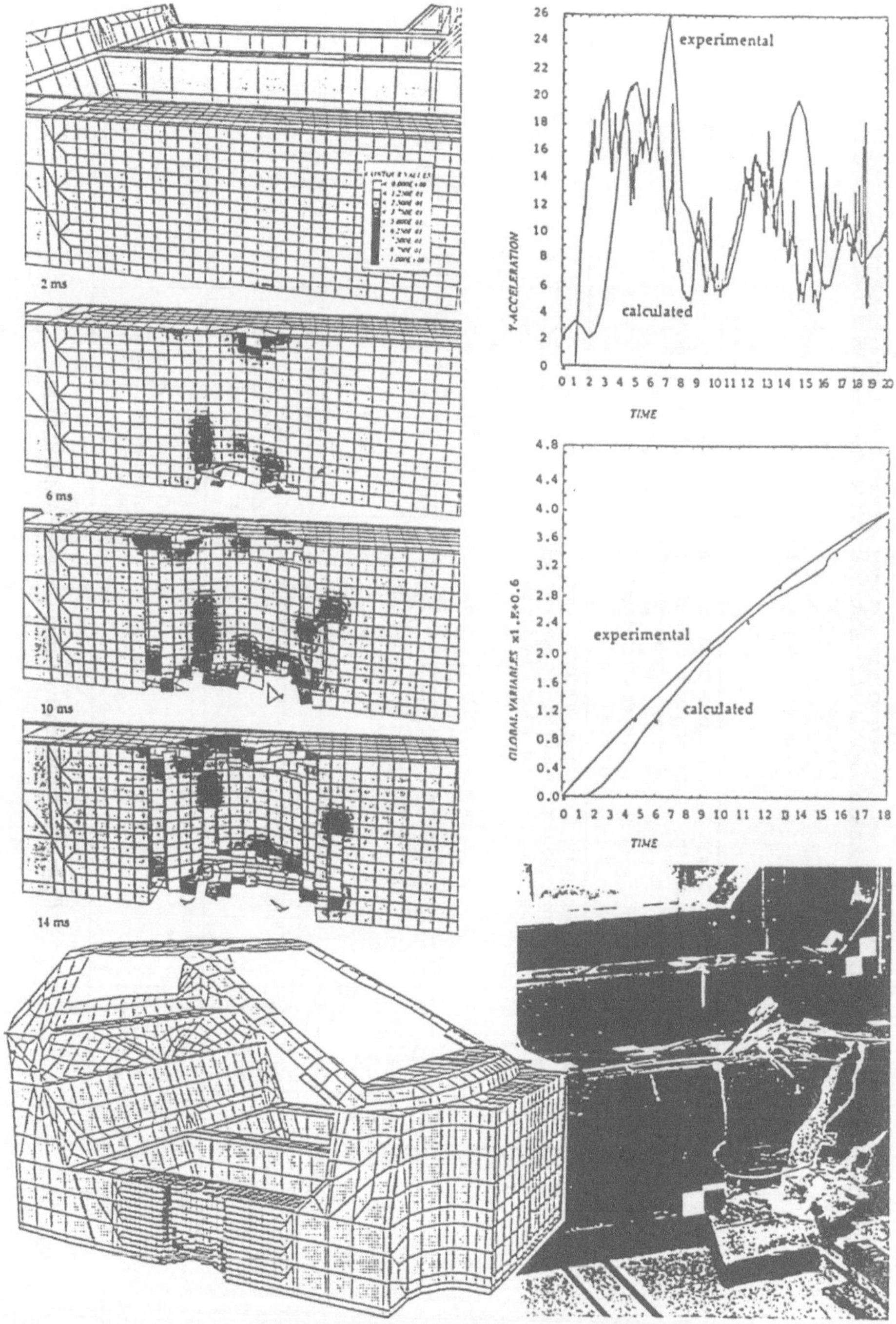

Figure 8. Composite Crash Modelling (PAM-CRASH™)

	Stage	Objectives	Type of Analysis	Tools
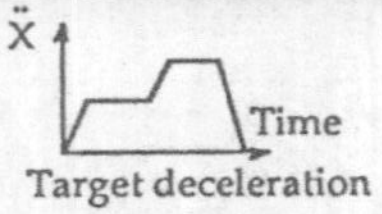	0. Pre-Crash Concept Design	Elastic design of skeleton (force paths and fluxes) under operating conditions.	Simple : hand calculations, etc.	Handbooks, ...
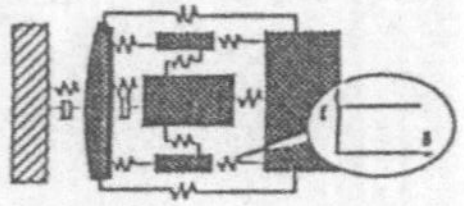	1. Early Crash Management	Determine crash resistance of major members needed to meet target decelerations (for front, side, rear, etc. impacts)	1-D Lumped Parameter (LP) models (Spring - mass)	• dedicated Runge Kutta solvers • PAM-CRASH with a library of standard LP models
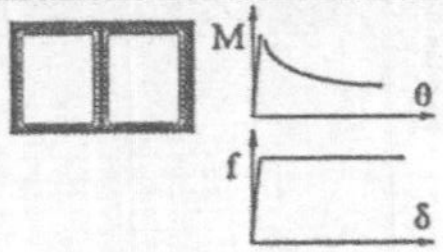	2. Preliminary Member Section Design	Design sections of crash members to meet required resistances found in stage 1.	2-D (sections) • Analytical section collapse models, or • Detailed FE shell models	• CRASH-SFE or CRASHCAD • PAM-CRASH (Exceptionally : physical tests)
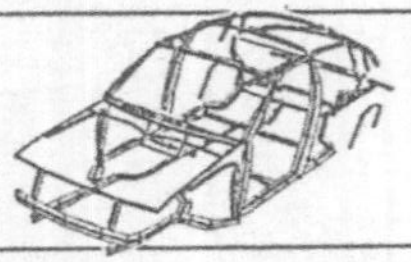	3. Simplified 3D Analysis	Validate member section design of stage 2 on the assembled car	3-D beam model of skeleton	PAM-CRASH
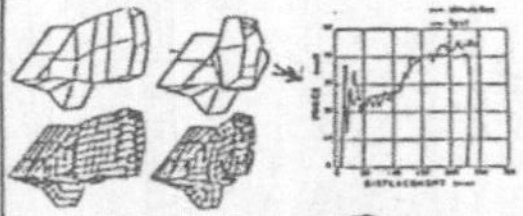	4. Detailed Member Design	Improve 3D crash response by iterations on member and joint design, using detailed geometry	3-D Hybrid (beam / shell) models, detailed FE submodels, macro-models	PAM-CRASH
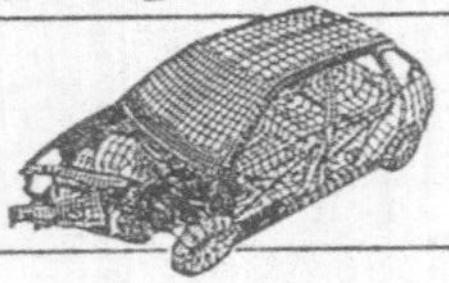	5. Verification	Verify crashworthiness design	3-D Detailed full crash simulations using FE models	PAM-CRASH

Figure 9. Comprehensive Crash Design (PAM-SYSTEM)

of full crash models with large numbers of finite elements (20000–100000 and more), useful for the ultimate verification of the crashworthiness of a designed structure. In order to design a structure for crashworthiness, however, information and time needed to fabricate detailed crash verification models may be absent and simplified methods are welcome.

One of these methods permits to quickly design the cross sections of energy absorbing members, using analytical superfolding or kinematic collapse theories for the axial crushing and deep collapse bending failure of arbitrary thin-walled components (Stage 2). Figure 10 outlines a crash concept station graphic user interface that is currently under development at ESI, based on the kinematic superfolding software, PAM-SUPERFOLD™, developed in collaboration with the University of Valenciennes. In this interface, the user is allowed to specify the cross section of a box column and the material properties. The software can then access kinematic folding software (CRASH-CAD of IMPACT DESIGN ; PAM-SUPERFOLD™) to obtain an estimate of the crash response in axial crushing or deep collapse bending. If desired, another module under development, PAM-OPT, can be employed to obtain optional responses through automatic variation of an assigned set of optimization parameters and constraints, Figure 11. The resulting section design of the crush member can then be verified by an automatically generated thin shell finite element model using the PAM-CRASH™ solver code. Stage 3 of the crash design process can then employ the crash member response curves of Stage 2 in three-dimensional beam/nonlinear spring or global beam models, designed for that purpose in the crash solver code. Stage 4 forms the transition from early crash design to the design of structural nodal points, members with true geometries, cutouts and connections, etc. During this hybrid phase zooms on critical areas may be made where a fine mesh is applied while the remaining parts of the structure may still be modelled coarsely. Stage 5, finally, will verify the overall design using the finest models.

5. Safety Aspects

Besides the development of deformable impact barriers (side and offset front) and of models of mechanical dummies (Hybrid III, EUROSID, DOT-SID, BIOSID, Child Dummies, etc.), biomechanical models of human parts are now being developed. This activity will probably grow considerably in the near future, because simulation with calibrated biomechanical models appears to be the only way to evaluate the direct response of the human body to crash events. At ESI several such models are under development (head, lower leg, etc.) and the preliminary model of the lower leg, including bones, ligaments, tendons and muscles is shown in Figure 12. Further details concerning these models will be published in the near future.

6. Related Topics

Crash simulation techniques and codes can be applied successfully not only to vehicle crash, but also to related topics such as refrigerator drop to design energy absorbing packaging materials, electronic appliance drop tests (e.g. laptop computers) to assess damage to sensitive interiors, planetary mission hard landers, tire responses when rolling over bumps, impact response of sports articles, impact response on bridges,

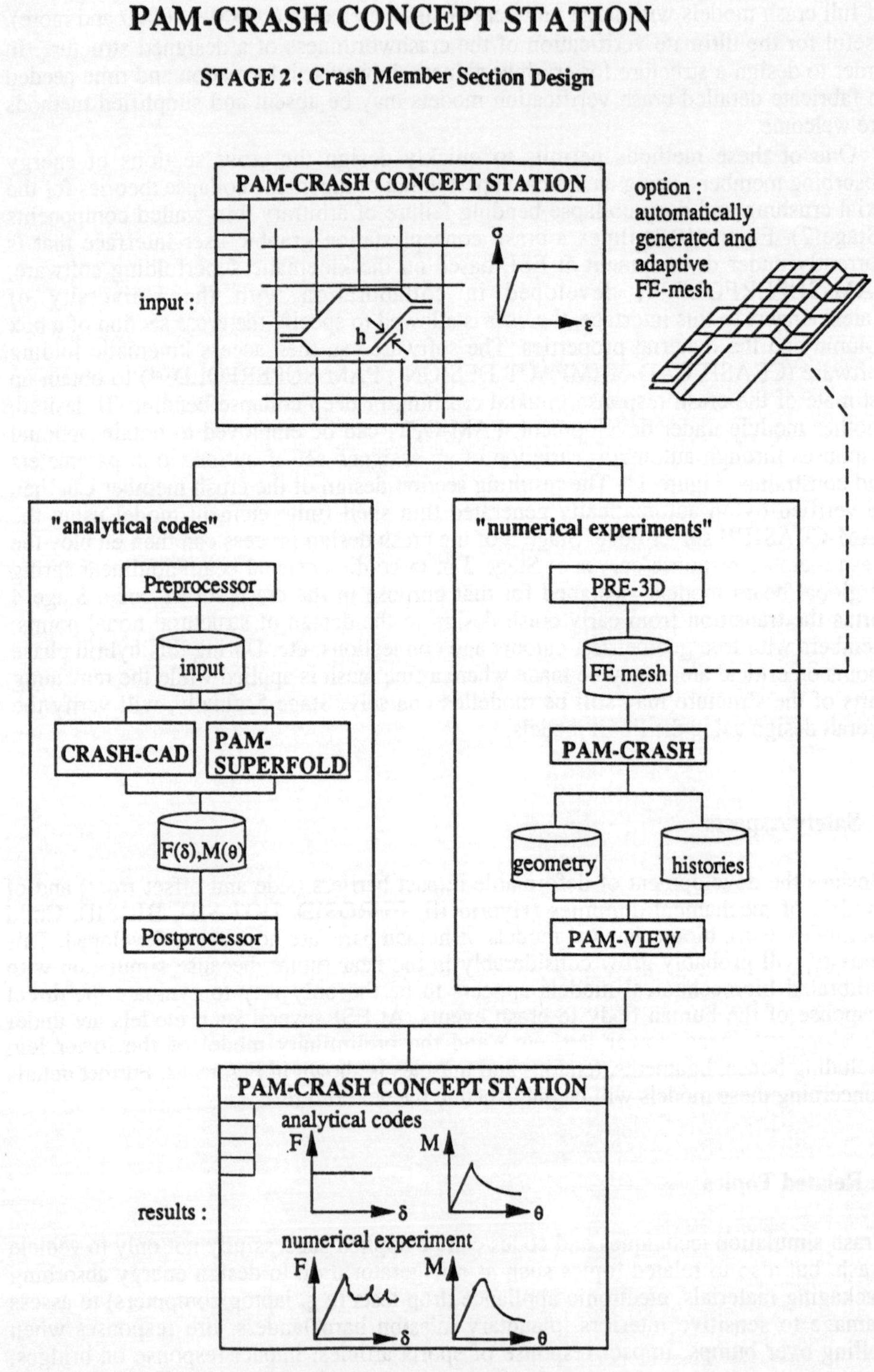

Figure 10. Crash Concept Station

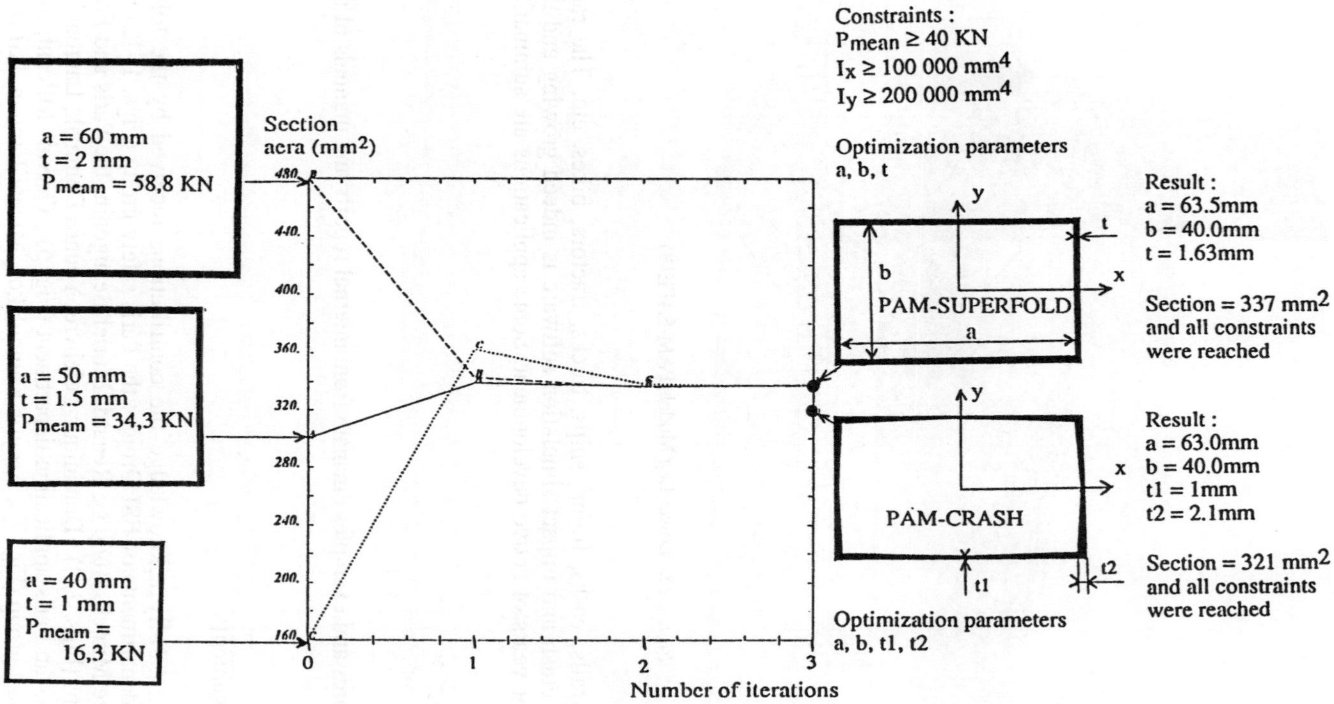

Figure 11. Section Optimization (PAM-OPT™)

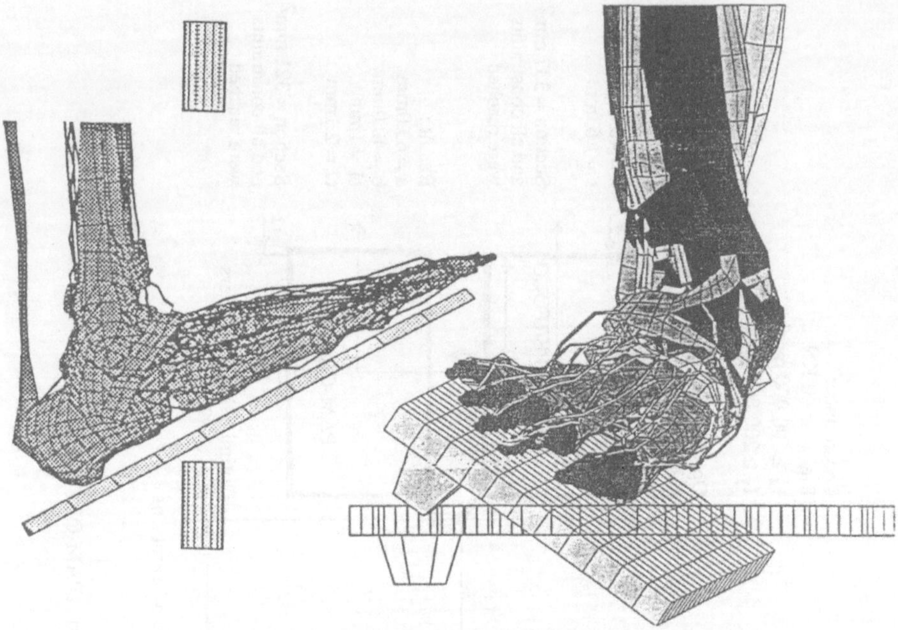

Figure 12. Lower Leg Model (PAM-SAFE™)

cranes, guard rails, casks, trains, ships, trucks, tractors, bikes, etc. The range of application of crash and impact simulation software is indeed growing and there is ample room for years of active development. Some applications are summarized on Figure 13.

7. References

All shown figures and examples emanate from internal reports and manuals of the ESI Group.

8. Acknowledgements

The authors gratefully acknowledge the contributions received by the following members and departments of ESI Group : the Paris safety group (Figs. 1, 12) : David Lasry, Grégoire Munck, Aude Le Renard, Muriel Beaugonin, the Paris and Frankfurt stamping groups (Figs. 2, 4) : Dominique Lefebvre, Younes Dammak, Laurent Taupin, Andrew Heath, the Paris application department (Fig. 5) : Gervais Milcent, the Paris shock and impact group (Fig. 7) : Argyris Kamoulakos, Michel Jamjian, PSI (Fig. 11), Pierre Culière, and Nihon ESI (Fig. 13), Thomasz Kisielewicz, Kohei Andoh.

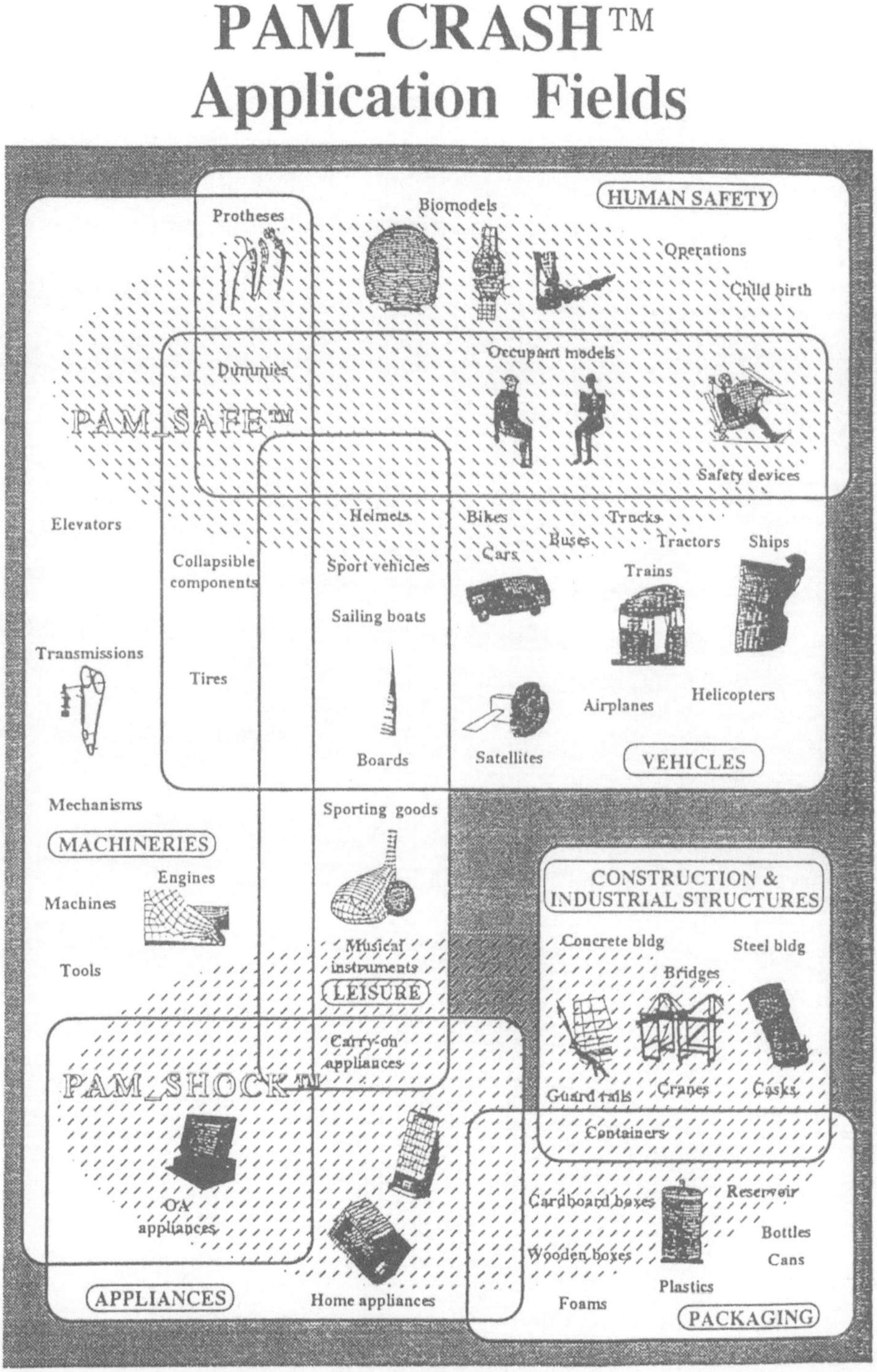

Figure 13. PAM-CRASH™ Application Fields

PART VI
Multibody Dynamics Approaches

RIGID-FLEXIBLE MULTIBODY EQUATIONS OF MOTION SUITABLE FOR VEHICLE DYNAMICS AND CRASH ANALYSIS

PARVIZ E. NIKRAVESH
Department of Aerospace and Mechanical Engineering
University of Arizona
Tucson, AZ 85721 USA

In this paper several formulations for automatic generation of the equations of motion for rigid and flexible multibody systems are reviewed. These formulations are of special interest for vehicle dynamics and crash analysis. We first discuss a so-called body-coordinate formulation to construct the Newton-Euler equations of motion for constrained rigid multibody systems. Then we review a nonconventional point-coordinate formulation. An easy-to-use method for deriving the equations of motion for flexible bodies in a multibody environment is also presented. Several ideas on how a body-fixed frame can be attached to a flexible body are discussed. It is shown how these formulations for rigid and flexible bodies could be mixed in order to construct the complete set of equations of motion. The constructed equations are normally a large set of mixed differential-algebraic equations. The method of joint-coordinates for transforming these equations to a smaller set is briefly discussed. Two application examples for ride/handling and crash analyses are also presented.

1. Introduction

The derivation of equations of motion for computational multibody dynamics has been the topic of many research activities. The scope of these activities has been quite broad. Some techniques allow us to generate the equations of motion in terms of a large set of differential-algebraic equations. Other techniques yield the equations of motion as a minimal set of ordinary differential equations. Many other "in between" approaches provide us with various alternatives. Each formulation has its own advantages and disadvantages depending on the application and our priorities.

One of the most interesting applications of multibody dynamics is in vehicle dynamics. Multibody modeling of vehicles allows us to analyze a vehicle for ride comfort, handling and maneuverability, and most importantly for crash analysis. The ideal form of the equations of motion for models used in ride and handling simulations may be quite different from those used in crash simulations. These models may share some of the objectives such as accuracy of results, but some other objectives may be different. For example, in a model for ride and handling simulations one of the objectives may be computational efficiency, since the model may be used for simulating various driving scenarios. Where as in a crash analysis, computational efficiency may not be as important as the model capability to handle structural deformation in various crash scenarios. For these reasons, in this paper we discuss more than one multibody modeling method which could be mixed and matched to describe a variety of multibody systems.

J. A. C. Ambrósio et al. (eds.),
Crashworthiness of Transportation Systems: Structural Impact and Occupant Protection, 421–443.

Although computational efficiency has not been emphasized in these formulations, other techniques have been referenced for transforming these equations to more efficient forms.

The equations of motion can be constructed in the form of a large set of differential-algebraic equations. The configuration of a rigid body is normally described by a set of translational and rotational coordinates. Algebraic constraints are introduced to represent kinematic joints connecting bodies, and then the Lagrange multiplier technique is used to describe joint reaction forces. Although these formulations are easy to construct, one of their main drawback is their computational inefficiency. One such formulation is called the body-coordinate formulation which is discussed in this paper. This formulation has also been referred to as the absolute or Cartesian coordinate formulation [1].

Another method discussed in this paper is called the point-coordinate (or natural coordinate) formulation [2, 3]. The method exhibits many interesting and useful features. This method takes advantage of a rudimentary idea of describing a body as a collection of points (and vectors [2]). The coordinates of these points are dependent on each other through constant-length constraints due to the assumption of rigid body. Some of the kinematic joints between two bodies may not require any algebraic constraints by allowing bodies to share some of the points. However, some kinematic joints may require a few simple algebraic constraints. Similar to the body-coordinate formulation, this formulation yields a large set of loosely coupled differential-algebraic equations of motion.

The equations of motion for a deformable body in a multibody environment are also presented in this paper. Finite element technique is used to construct the nodal mass and stiffness matrices of the deformable body. This formulation does not enforce any restrictions on the stiffness matrix--it may exhibit linear or nonlinear characteristics of the deformable body. The equations of motion for a flexible body can be linked to the equations of motion for other flexible or rigid bodies quite easily. The form of these equations has proven to be highly effective in ride, handling, and crash analysis of vehicles.

In this paper we first consider rigid body dynamics where both body- and point-coordinate formulations are discussed. Following that, we present the deformable body formulation. The process of linking the equations of motion from different formulations is discussed. Some important features regarding the application of these formulations in vehicle dynamics and crash analysis are reviewed. Two examples for ride-handling and crash analyses are also provided.

2. Notation

One-dimensional vectors and arrays are denoted by lower-case bold-face characters ($\mathbf{q}$, $\dot{\mathbf{r}}$, $\boldsymbol{\omega}$). Matrices are denoted by upper-case bold-face characters ($\mathbf{A}$, $\mathbf{D}$). Scalars are denoted by light-face characters (m, m, α). A right-subscript denotes body index ($\mathbf{s}_i$). A right-superscript denotes a point, a point index, or a node index ($\mathbf{s}^j$). A left-superscript denotes the index of a reference frame (${}^k\mathbf{s}$); if the reference frame is a nonmoving system, then the left-superscript is omitted. An over-score "tilde" indicates the conversion of a 3-vector to a 3 x 3 skew-symmetric matrix ($\tilde{\mathbf{s}}$).

3. Rigid Bodies

In this paper, two formulations for modeling rigid bodies for dynamic analysis are reviewed. These formulations are of interest due to their simplicity and ease of use. The form of equations can be modified and adjusted for different applications. One of these formulations is called body-coordinate formulation due to the fact that we define coordinates describing position and orientation of each body with respect to a nonmoving reference frame. The second formulation that is discussed here is referred to as the point-coordinates. A body is defined as a rigid collection of two or more points, and the absolute coordinates of these points are used to position a body in space.

3.1. BODY COORDINATE FORMULATION

For specifying the position of a rigid body in a global nonmoving xyz coordinate system, it is sufficient to define the spatial location of the origin and the angular orientation of a body-fixed $\xi\eta\zeta_i$ coordinate system. The origin of the body-fixed coordinate system can be positioned at the body mass center (centroidal frame) or at any other point on the body (non-centroidal frame) as shown in Fig. 1. For a typical body i, vector $\mathbf{c}_i$ denotes a vector of coordinates containing the Cartesian translational coordinates of the origin, $\mathbf{r}_i$, and a set of rotational coordinates such as Euler angles, Euler parameters, etc. A 3 x 3 rotational transformation matrix, $\mathbf{A}_i$, denotes the angular orientation of the $\xi\eta\zeta_i$ relative to the xyz system, which is a function of the defined rotational coordinates. With this transformation matrix, the components of any vector described in the body-fixed coordinate system can be transformed to the xyz coordinate system as $\mathbf{s}_i = \mathbf{A}_i\, {}^i\mathbf{s}_i$. A vector of velocities for body i is defined as $\mathbf{v}_i$, which contains the translational velocities, $\dot{\mathbf{r}}_i$, and the angular velocities, ω_i. A vector of acceleration for this body is denoted as $\dot{\mathbf{v}}_i$, containing $\ddot{\mathbf{r}}_i$ and $\dot{\omega}_i$. In some formulations, it may be more convenient to describe the angular velocity and acceleration in terms of the body-fixed components; i.e., ${}^i\omega_i$ and ${}^i\dot{\omega}_i$. A point P on body i, which is positioned with respect to the origin of $\xi\eta\zeta_i$ coordinate system by a vector $\mathbf{s}_i^P$ as shown in Fig. 1, has the global coordinates

$$\mathbf{r}_i^P = \mathbf{r}_i + \mathbf{s}_i^P = \mathbf{r}_i + \mathbf{A}_i\, {}^i\mathbf{s}_i^P \tag{1}$$

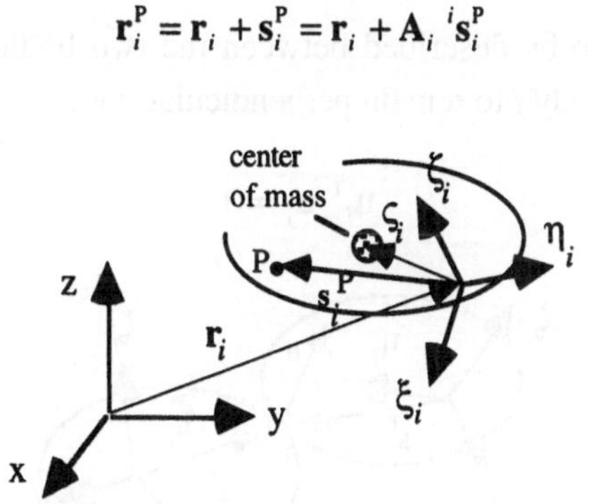

Figure 1. Locating a body with respect to a nonmoving reference system.

3.1.1. *Equations of Motion For A Body*

The Newton-Euler equations of motion for body *i* are written as

$$\mathbf{M}_i \, \dot{\mathbf{v}}_i = \mathbf{g}_i \tag{2}$$

where,

$$\mathbf{M}_i \equiv \begin{bmatrix} \mathrm{m}_i \, \mathbf{I} & -\mathrm{m}_i \, \tilde{\varsigma}_i \\ \mathrm{m}_i \, \tilde{\varsigma}_i & \boldsymbol{J}_i \end{bmatrix}, \quad \mathbf{g}_i \equiv \begin{bmatrix} \mathbf{f}_i - \mathrm{m}_i \, \tilde{\omega}_i \, \tilde{\omega}_i \, \varsigma_i \\ \mathbf{n}_i - \tilde{\omega}_i \, \boldsymbol{J}_i \, \omega_i \end{bmatrix} \tag{3}$$

In these equations, m_i and $\boldsymbol{J}_i$ are the mass and inertia tensor. Vectors $\mathbf{f}_i$ and $\mathbf{n}_i$ are the sum of forces and moments acting on the body. The inertia tensor $\boldsymbol{J}_i$ is defined with respect to a noncentroidal reference frame parallel to the xyz axes. This matrix is related to a centroidal inertia tensor, $\mathbf{J}_i$, as $\boldsymbol{J}_i = \mathbf{J}_i - \mathrm{m}_i \tilde{\varsigma}_i \tilde{\varsigma}_i$. Furthermore, $\mathbf{J}_i$ is related to the commonly used constant inertia tensor, ${}^{i}\mathbf{J}_i$, as $\mathbf{J}_i = \mathbf{A}_i \, {}^{i}\mathbf{J}_i \, \mathbf{A}_i^{\mathrm{T}}$. It is obvious that if a centroidal frame is used, since $\varsigma_i = \mathbf{0}$, some of the terms in Eq. 3 will vanish. Furthermore, if we use a centroidal reference frame, the Euler equations are commonly expressed in terms of the $\xi\eta\zeta_i$ components and not the xyz components.

The forces and moments acting on a body can be the result of springs, dampers, or other force elements in the system. Knowing the coordinates and the velocity of the attachment points of these elements, and also knowing the characteristics of the elements, the corresponding forces and moments can be computed. Specific formulations for standard force elements are referred to [1].

3.1.2. *Kinematic Joints*

A kinematic joint can be described as a set of holonomic constraints between the coordinates of two bodies. These constraints are normally nonlinear algebraic equations. For example, a spherical joint between two bodies *i* and *j*, as shown in Fig. 2, can be described by allowing the two bodies to share a common point at the center of the spherical joint:

$$\mathbf{r}_i + \mathbf{A}_i \, {}^{i}\mathbf{s}_i^{\mathrm{P}} - \mathbf{r}_j - \mathbf{A}_j \, {}^{j}\mathbf{s}_j^{\mathrm{P}} = \mathbf{0} \tag{4}$$

Other conditions can also be described between the two bodies, such as vector $\mathbf{u}_i$ on body *i* and vector $\mathbf{u}_j$ on body *j* to remain perpendicular; i.e.,

$$\mathbf{u}_i^{\mathrm{T}} \, \mathbf{u}_j = 0 \tag{5}$$

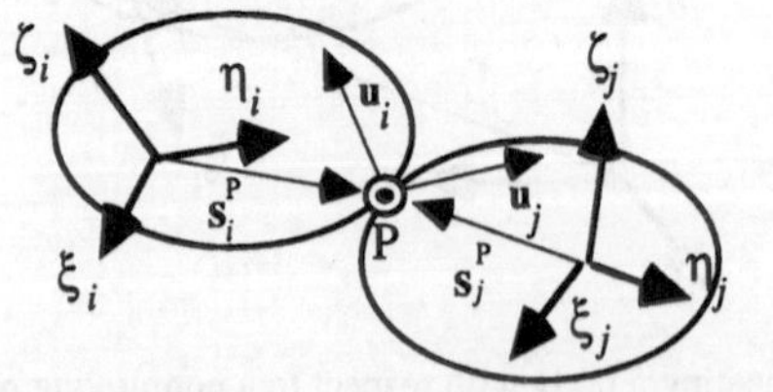

Figure 2. A kinematic joint between two rigid bodies.

If these two vectors, for example, define the axes of the cross of a universal joint and point P defines the intersect of these two axes, then Eqs. 4 and 5 together describe four constrain equations for a universal joint. Similar conditions can be written to define other types of kinematic joints, such revolute, cylindrical, and prismatic [1].

3.1.3. *Multibody Equations of Motion*

The complete set of kinematic constraints between bodies in a multibody system is denoted as a set of m position constraints:

$$\mathbf{\Phi}(\mathbf{c}) = \mathbf{0} \tag{6}$$

The first time derivative of these constraints yields m velocity constraints as:

$$\dot{\mathbf{\Phi}} \equiv \mathbf{\Phi}_{\mathbf{c}}\,\dot{\mathbf{c}} = \mathbf{D}\mathbf{v} = \mathbf{0} \tag{7}$$

where $\mathbf{\Phi}_{\mathbf{c}} \equiv \partial\mathbf{\Phi}/\partial\mathbf{c}$ is the Jacobian matrix of the constraints, and if we use angular velocity vectors instead of the time derivative of the rotational coordinates, the Jacobian matrix will find a slightly different form which is denoted as matrix $\mathbf{D}$. The time derivative of the velocity equations yields m acceleration equations as:

$$\ddot{\mathbf{\Phi}} \equiv \mathbf{D}\dot{\mathbf{v}} + \dot{\mathbf{D}}\mathbf{v} = \mathbf{0} \tag{8}$$

If the kinematic joints are represented as a set of algebraic equations, as in Eq. 6, then Eq. 2 can be written for all the bodies in the system, with the help of Lagrange multiplier technique, as

$$\mathbf{M}\,\dot{\mathbf{v}} - \mathbf{D}^{\mathrm{T}}\boldsymbol{\lambda} = \mathbf{g} \tag{9}$$

where $\mathbf{M}$ is a block-diagonal matrix containing the 6 x 6 inertia matrices, $\mathbf{g}$ contains the 6-vectors of forces for the bodies, and λ is a vector of m Lagrange multipliers. Note that the equations for each body can be described either in the centroidal or non-centroidal form. Equations 6-9 represent a set of differential-algebraic equations of motion.

3.2. POINT COORDINATE FORMULATION

The position and orientation of a rigid body in a nonmoving xyz coordinate system can be described by the position of several points on the body. These points will be referred to as *primary points*. Other points on the body will be called *secondary* or *nonprimary* points, where their coordinates can be described in terms of the coordinates of the primary points. In this paper we consider only bodies described by four primary points. The formulations for bodies described by two or three primary points can be found in [3]. Furthermore, due to space limitation, in this paper we only provide the minimum number of formulas and equations without any proof or derivation. Interested reader may refer to [3] for more detail.

Figure 3(a) shows a rigid body represented by four primary points. Each point, such as i, has a Cartesian position vector, $\mathbf{r}^i$, representing the coordinates of that point in the nonmoving xyz reference frame. These Cartesian coordinates are referred to as the *primary coordinates*. It is clear that between every two points on a body there exists a constraint equation as:

$$(\mathbf{r}^i - \mathbf{r}^j)^{\mathrm{T}} (\mathbf{r}^i - \mathbf{r}^j) - \ell^{i,j2} = 0 \tag{10}$$

We will refer to such constraints as *primary constraints* in order to distinguish them from the *kinematic constraints* associated with kinematic joints. One major advantage of using primary coordinates for a body, instead of body-coordinates, is the elimination of rotational coordinates and the corresponding rotational transformation matrix.

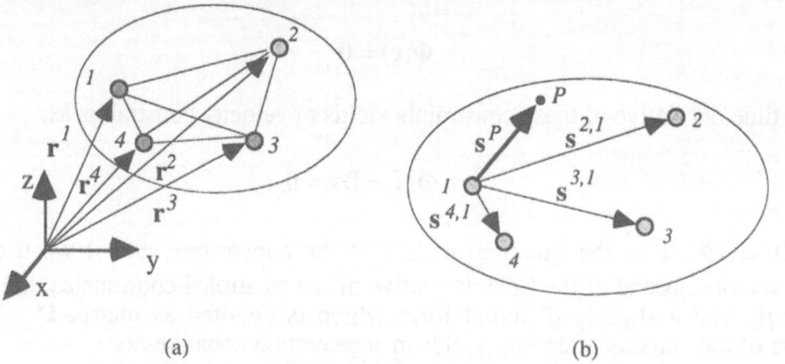

Figure 3. (a) Locating a body by its primary points and (b) locating a nonprimary point *P* in terms of the primary coordinates.

In order to locate a secondary point, such as point *P* shown in Fig. 3(b), as a function of the primary coordinates of the same body, assume that at a given time; e.g., the initial time, $\mathbf{r}^1$, $\mathbf{r}^2$, $\mathbf{r}^3$, $\mathbf{r}^4$, and $\mathbf{r}^P$ are known. The components of vectors $\mathbf{s}^{2,1}$, $\mathbf{s}^{3,1}$, $\mathbf{s}^{4,1}$, and $\mathbf{s}^P$ are computed first as $\mathbf{s}^{i,1} = \mathbf{r}^i - \mathbf{r}^1$, $i = 2, 3, 4$, and $\mathbf{s}^P = \mathbf{r}^P - \mathbf{r}^1$. Then, vector $\mathbf{s}^P$ is described in terms of the components of $\mathbf{s}^{i,1}$ vectors as $\mathbf{s}^P = a^1\,\mathbf{s}^{2,1} + a^2\,\mathbf{s}^{3,1} + a^3\,\mathbf{s}^{4,1}$, or

$$\mathbf{a} = \begin{bmatrix} \mathbf{s}^{2,1} & \mathbf{s}^{3,1} & \mathbf{s}^{4,1} \end{bmatrix}^{-1} \mathbf{s}^P$$

where $\mathbf{a} \equiv \begin{bmatrix} a^1 & a^2 & a^3 \end{bmatrix}^{\mathrm{T}}$. These coefficients are computed only once at the initial time. Then, at any later time, we can determine $\mathbf{r}^P$ known the instantaneous values of $\mathbf{r}^1$, $\mathbf{r}^2$, $\mathbf{r}^3$, and $\mathbf{r}^4$ as:

$$\mathbf{r}^P = \mathbf{r}^1 + a^1(\mathbf{r}^2 - \mathbf{r}^1) + a^2(\mathbf{r}^3 - \mathbf{r}^1) + a^3(\mathbf{r}^4 - \mathbf{r}^1)$$

3.2.1. *Equations of Motion For A Body*

Since a body is represented by several primary points, the mass and any force or moment acting on the body must be distributed to the primary points. For a force or a moment distribution, the resultant forces must be equivalent to the original force and/or moment.

Consider a single force acting at point *P* as shown in Fig. 4. Point *P* is positioned from the mass center by vector $\mathbf{s}^P$. Assume that an equivalent set of four forces are found that act on the four points. We may assume that the four forces are all parallel to the original force $\mathbf{f}^P$, i.e., $\mathbf{f}^j = \alpha^j\,\mathbf{f}^P$; $j = 1, ..., 4$, where α's are four unknown coefficients. If we enforce the equivalency conditions for the sum of forces and moments, the four coefficients can be found from the following set of equations:

$$\begin{bmatrix} \xi_i^1 & \xi_i^2 & \xi_i^3 & \xi_i^4 \\ \eta_i^1 & \eta_i^2 & \eta_i^3 & \eta_i^4 \\ \zeta_i^1 & \zeta_i^2 & \zeta_i^3 & \zeta_i^4 \\ 1 & 1 & 1 & 1 \end{bmatrix} \begin{bmatrix} \alpha^1 \\ \alpha^2 \\ \alpha^3 \\ \alpha^4 \end{bmatrix} = \begin{bmatrix} \xi_i^P \\ \eta_i^P \\ \zeta_i^P \\ 1 \end{bmatrix}$$

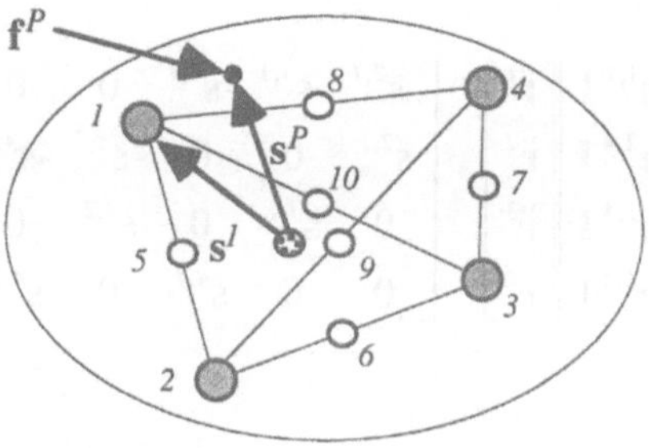

Figure 4. Force and mass distribution to primary and secondary points.

The unknown coefficients are a function of the position of point P and not a function of the magnitude or the direction of the applied force. Therefore, as long as the point of application of the force does not change, the coefficients are found only once.

For a pure moment distribution, there is more than one way to find a set of equivalent forces acting on the primary points. The simplest method, however, would be to consider a couple instead of a pure moment, then to apply the preceding formulation for each force of the couple. Other methods can be computed in [3].

The mass distribution to primary points must satisfy the total mass, the first moment, and the second moment conditions. These conditions provide ten algebraic equations; therefore, we can have up to ten unknowns . Four of the unknowns are the mass of the four primary points, and the other six unknowns could be the mass of six secondary points. There are infinite possibilities for positioning the six secondary points, where one is shown in Fig. 4. Each secondary point is located exactly halfway between two primary points. If we number the primary points *1, ..., 4* and the secondary points *5, ..., 10*, then the position of the secondary points can be described as $\mathbf{s}^5 = \frac{1}{2}(\mathbf{s}^1 + \mathbf{s}^2)$, etc.

If the mass of each point is denoted as m^j, Then the ten equations are written as:

$$\sum_{j=1}^{10} \mathrm{m}^j = \mathrm{m}_i \tag{11}$$

$$\sum_{j=1}^{10} \mathbf{s}^j \, \mathrm{m}^j = \mathbf{0} \tag{12}$$

$$-\sum_{j=1}^{10} \tilde{\mathbf{s}}^j \, \tilde{\mathbf{s}}^j \, \mathrm{m}^j = \mathbf{J}_i \tag{13}$$

The first expression yields one, the second expression yields three, and the third expression yields six equations. These ten equations are solved only once to find the unknown masses.

In order to derive the point coordinate equations of motion for a rigid body, we first write the equations of motion for the system of ten particles (four primary and six secondary points) using Lagrange multiplier technique for the constraints between the ten particles. We assume that the mass of the primary and secondary points have already been determined. After eliminating the Lagrange multipliers, constraints, and accelerations associated with the secondary points, the equations of motion are written as:

$$\begin{bmatrix} m^{1,1}\mathbf{I} & m^{1,2}\mathbf{I} & m^{1,3}\mathbf{I} & m^{1,4}\mathbf{I} \\ m^{2,1}\mathbf{I} & m^{2,2}\mathbf{I} & m^{2,3}\mathbf{I} & m^{2,4}\mathbf{I} \\ m^{3,1}\mathbf{I} & m^{3,2}\mathbf{I} & m^{3,3}\mathbf{I} & m^{3,4}\mathbf{I} \\ m^{4,1}\mathbf{I} & m^{4,2}\mathbf{I} & m^{4,3}\mathbf{I} & m^{4,4}\mathbf{I} \end{bmatrix} \begin{bmatrix} \ddot{\mathbf{r}}^1 \\ \ddot{\mathbf{r}}^2 \\ \ddot{\mathbf{r}}^3 \\ \ddot{\mathbf{r}}^4 \end{bmatrix} - \begin{bmatrix} -\mathbf{s}^{2,1} & -\mathbf{s}^{3,1} & -\mathbf{s}^{4,1} & 0 & 0 & 0 \\ \mathbf{s}^{2,1} & 0 & 0 & -\mathbf{s}^{3,2} & -\mathbf{s}^{4,2} & 0 \\ 0 & \mathbf{s}^{3,1} & 0 & \mathbf{s}^{3,2} & 0 & -\mathbf{s}^{4,3} \\ 0 & 0 & \mathbf{s}^{4,1} & 0 & \mathbf{s}^{4,2} & \mathbf{s}^{4,3} \end{bmatrix} \begin{bmatrix} \lambda^1 \\ \lambda^2 \\ \lambda^3 \\ \lambda^4 \\ \lambda^5 \\ \lambda^6 \end{bmatrix} = \begin{bmatrix} \mathbf{f}^1 \\ \mathbf{f}^2 \\ \mathbf{f}^3 \\ \mathbf{f}^4 \end{bmatrix} \tag{14}$$

where,

$$m^{1,1} = \mathrm{m}^1 + \frac{\mathrm{m}^5 + \mathrm{m}^8 + \mathrm{m}^{10}}{4}, \quad m^{2,2} = \mathrm{m}^2 + \frac{\mathrm{m}^5 + \mathrm{m}^6 + \mathrm{m}^9}{4},$$

$$m^{3,3} = \mathrm{m}^3 + \frac{\mathrm{m}^6 + \mathrm{m}^7 + \mathrm{m}^{10}}{4}, \quad m^{4,4} = \mathrm{m}^4 + \frac{\mathrm{m}^7 + \mathrm{m}^8 + \mathrm{m}^9}{4},$$

$$m^{1,2} = m^{2,1} = \frac{\mathrm{m}^5}{4}, \quad m^{1,3} = m^{3,1} = \frac{\mathrm{m}^{10}}{4}, \quad m^{2,3} = m^{3,2} = \frac{\mathrm{m}^6}{4}$$

$$m^{1,4} = m^{4,1} = \frac{\mathrm{m}^8}{4}, \quad m^{2,4} = m^{4,2} = \frac{\mathrm{m}^9}{4}, \quad m^{3,4} = m^{4,3} = \frac{\mathrm{m}^7}{4}$$

Note that the remaining Lagrange multipliers and the associated Jacobian matrix correspond to the six primary constraints. This means that the complete set of equations of motion for a body needs the six primary constraints and their first and second derivatives.

3.2.2. *Kinematic Joints*

Kinematic joints between rigid bodies can be described as algebraic constraints between the primary coordinates. For example if the primary point k on body i coincides with the primary point l on body j; e.g., in the case of a spherical joint, then the associated constraint is:

$$\mathbf{r}_i^k - \mathbf{r}_j^l = \mathbf{0}$$

However, the idea behind the use of primary coordinates is to eliminate the need for defining some, if not all, of such constraints. This is achieved by allowing bodies to share primary points and hence, reducing the total number of primary coordinates. If two bodies are connected by a spherical joint, for example, then one primary point is shared between the two bodies at the center of the joint, as shown in Fig. 5(a). Similarly, for a revolute joint, two primary points on the joint axis can be shared by the two bodies, as shown in Fig. 5(b).

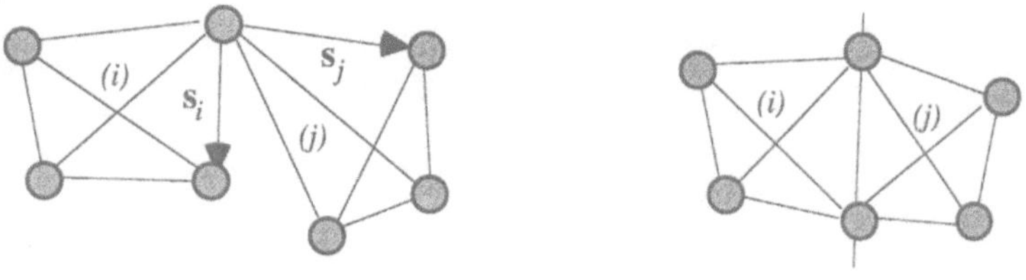

(a) (b)

Figure 5. Shared primary points for bodies connected by (a) spherical (or universal) or (b) revolute joints.

In the case of a universal joint between two bodies, we need to properly place some of the primary points on the intersect and on the axes of the cross, as shown in Fig. 5(a). This allows us to use the primary coordinates to define vectors $\mathbf{s}_i$ and $\mathbf{s}_j$ on the joint axes. Since these two vectors must remain perpendicular, we can write:

$$\mathbf{s}_i^{\mathrm{T}}\,\mathbf{s}_j = 0$$

This is the only constraint that we need for the universal joint. For some other joints, such as cylindrical or prismatic, since the bodies translate relative to each other, no primary points are shared. For such joints we need to describe several kinematic constraints between vectors defined on the bodies [3].

3.2.3. *Dynamics of Multibody Systems*

The primary coordinate representation allows us to construct the equations of motion of a system of multibodies interconnected by kinematic joints. Here we demonstrate the process with a simple example. Assume two bodies connected by a spherical joint are represented by seven primary points, as shown in Fig. 6. Since the primary point *4* is shared by both bodies, its mass and force vector receive contributions from both bodies. It can be shown that the mass matrix and the force vector are written as:

$$\mathbf{M}^{1-7} = \begin{bmatrix} m^{1,1}\mathbf{I} & m^{1,2}\mathbf{I} & m^{1,3}\mathbf{I} & m^{1,4}\mathbf{I} & & & \\ m^{2,1}\mathbf{I} & m^{2,2}\mathbf{I} & m^{2,3}\mathbf{I} & m^{2,4}\mathbf{I} & & & \\ m^{3,1}\mathbf{I} & m^{3,2}\mathbf{I} & m^{3,3}\mathbf{I} & m^{3,4}\mathbf{I} & & & \\ m^{4,1}\mathbf{I} & m^{4,2}\mathbf{I} & m^{4,3}\mathbf{I} & m^{4,4}\mathbf{I} & m^{4,5}\mathbf{I} & m^{4,6}\mathbf{I} & m^{4,7}\mathbf{I} \\ & & & m^{5,4}\mathbf{I} & m^{5,5}\mathbf{I} & m^{5,6}\mathbf{I} & m^{5,7}\mathbf{I} \\ & & & m^{6,4}\mathbf{I} & m^{6,5}\mathbf{I} & m^{6,6}\mathbf{I} & m^{6,6}\mathbf{I} \\ & & & m^{7,4}\mathbf{I} & m^{7,5}\mathbf{I} & m^{7,6}\mathbf{I} & m^{7,7}\mathbf{I} \end{bmatrix}, \quad \mathbf{f}^{1-7} = \begin{bmatrix} \mathbf{f}^1 \\ \mathbf{f}^2 \\ \mathbf{f}^3 \\ \mathbf{f}^4 \\ \mathbf{f}^5 \\ \mathbf{f}^6 \\ \mathbf{f}^7 \end{bmatrix}$$

where $m^{4,4} = m_i^{4,4} + m_j^{4,4}$ and $\mathbf{f}^4 = \mathbf{f}_i^4 + \mathbf{f}_j^4$.

This example shows how the mass and the force of shared primary points are constructed. For a multibody system with *b* rigid bodies interconnected by kinematic joints, assume that we have defined *p* primary points. Therefore, we need three 3*p* vectors of primary coordinates, velocities, and accelerations as:

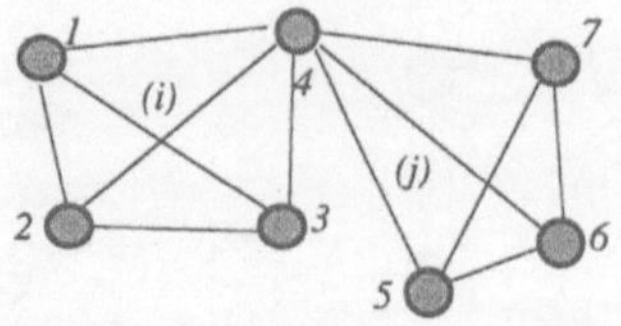

Figure 6. Two bodies connected by a spherical joint.

$$\mathbf{r} = \begin{bmatrix} \mathbf{r}^1 \\ \mathbf{r}^2 \\ \vdots \\ \mathbf{r}^p \end{bmatrix}, \quad \dot{\mathbf{r}} = \begin{bmatrix} \dot{\mathbf{r}}^1 \\ \dot{\mathbf{r}}^2 \\ \vdots \\ \dot{\mathbf{r}}^p \end{bmatrix}, \quad \ddot{\mathbf{r}} = \begin{bmatrix} \ddot{\mathbf{r}}^1 \\ \ddot{\mathbf{r}}^2 \\ \vdots \\ \ddot{\mathbf{r}}^p \end{bmatrix}$$

The primary coordinates are dependent through the primary and the kinematic joint constraints. Assume that there are m independent constraints of the form,

$$\boldsymbol{\Phi}(\mathbf{r}) = \mathbf{0} \tag{15}$$

These include the kinematic as well as the primary constraints. The first and second time derivatives of these constraints yield the velocity and acceleration constraints:

$$\dot{\boldsymbol{\Phi}} \equiv \mathbf{D}\dot{\mathbf{r}} = \mathbf{0} \tag{16}$$
$$\ddot{\boldsymbol{\Phi}} \equiv \mathbf{D}\ddot{\mathbf{r}} + \dot{\mathbf{D}}\dot{\mathbf{r}} = \mathbf{0} \tag{17}$$

where $\mathbf{D} \equiv \boldsymbol{\Phi}_{\mathbf{r}} \equiv \partial\boldsymbol{\Phi}/\partial\mathbf{r}$ is the Jacobian of the constraints and due to the linear or quadratic nature of the constraints, it exhibits a very simple form. In general, the equations of motion for a multibody system are written as,

$$\mathbf{M}\ddot{\mathbf{r}} - \mathbf{D}^{\mathrm{T}}\boldsymbol{\lambda} = \mathbf{f} \tag{18}$$

where vector λ contains the Lagrange multipliers associated with all of the constraints.

4. Deformable Bodies

For analyzing flexible bodies or structures, it is common to use finite element method. However, most standard finite element formulations do not allow for gross rigid body motion of the structure. In multibody systems, the finite element formulation of a flexible body requires the inclusion of terms representing the gross motion, distributed flexibility, and the coupling between them. In this section, we discuss a simple representation of the equations of motion for a flexible body in a multibody environment.

4.1. NODAL KINEMATICS

For specifying the position of a flexible body in a global nonmoving xyz coordinate system, it is sufficient to specify the spatial location of the origin and the angular orientation of a body-fixed $\xi\eta\zeta_i$ coordinate system, as shown in Fig. 7. We will clarify later how the frame is attached to the body. For notation simplification, we will avoid showing the body index in the following formulations. The position and orientation, velocity, and acceleration of the body-fixed frame with respect to the xyz frame are defined the same way as for a rigid body (refer to Sec. 3.1).

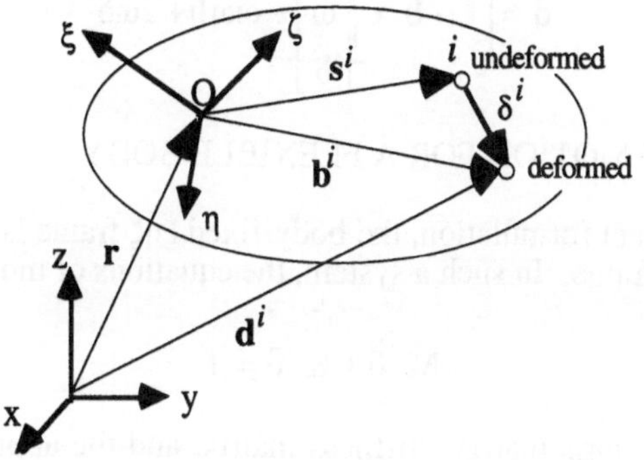

Figure 7. Vectors defining a node on a flexible body.

Assume that the flexible body is modeled by n nodes using finite element method. A typical node i at its undeformed state is located from the origin of the body-fixed frame by vector $\mathbf{s}^i$, with a nodal deformation vector δ^i. Without any loss of generality, in our discussion we consider only the translational degrees of freedom for a node. If we denote the deformed position of a node from the body origin and from the xyz frame as $\mathbf{b}^i$ and $\mathbf{d}^i$ respectively, the following formulas can be written for the position, velocity, and acceleration of a typical node i:

$$\mathbf{d}^i = \mathbf{r} + \mathbf{b}^i = \mathbf{r} + \mathbf{s}^i + \delta^i \tag{19}$$

$$\dot{\mathbf{d}}^i = \dot{\mathbf{r}} - \tilde{\mathbf{b}}^i \omega + \dot{\delta}^i \tag{20}$$

$$\ddot{\mathbf{d}}^i = \ddot{\mathbf{r}} - \tilde{\mathbf{b}}^i \dot{\omega} + \ddot{\delta}^i + \tilde{\omega}\tilde{\omega}\mathbf{b}^i + 2\tilde{\omega}\dot{\delta}^i \tag{21}$$

We can write Eqs. 19-21 for all the nodes of the flexible body as:

$$\mathbf{d} = \hat{\mathbf{I}}\mathbf{r} + \mathbf{b} \tag{22}$$

$$\dot{\mathbf{d}} = \hat{\mathbf{I}}\dot{\mathbf{r}} - \tilde{\mathbf{b}}\omega + \dot{\delta} \tag{23}$$

$$\ddot{\mathbf{d}} = \hat{\mathbf{I}}\ddot{\mathbf{r}} - \tilde{\mathbf{b}}\dot{\omega} + \ddot{\delta} + \tilde{\omega}\tilde{\omega}\mathbf{b} + 2\tilde{\omega}\dot{\delta} \tag{24}$$

where $\mathbf{d}$, $\mathbf{b}$ and δ are $3n$ arrays, and $\tilde{\mathbf{b}}$ and $\hat{\mathbf{I}}$ are $3n$ x 3 matrices defined as:

$$\mathbf{d} \equiv \begin{bmatrix} \vdots \\ \mathbf{d}^i \\ \vdots \end{bmatrix}, \quad \mathbf{b} \equiv \begin{bmatrix} \vdots \\ \mathbf{b}^i \\ \vdots \end{bmatrix}, \quad \delta \equiv \begin{bmatrix} \vdots \\ \delta^i \\ \vdots \end{bmatrix}, \quad \tilde{\mathbf{b}} \equiv \begin{bmatrix} \vdots \\ \tilde{\mathbf{b}}^i \\ \vdots \end{bmatrix}, \quad \hat{\mathbf{I}} \equiv \begin{bmatrix} \vdots \\ \mathbf{I} \\ \vdots \end{bmatrix}, \quad \bar{\mathbf{I}} \equiv \begin{bmatrix} \ddots & & \\ & \mathbf{I} & \\ & & \ddots \end{bmatrix}$$

The $3n$ x $3n$ identity matrix **I** will be used in the next equation. We understand that a 3 x 3 matrix such as $\tilde{\omega}\tilde{\omega}$ in front of **b** is treated as a scalar which gets multiplied by every 3-vector in the array. With this compact notation, the global acceleration vector for all the nodes is described as:

$$\ddot{\mathbf{d}} = \begin{bmatrix} \hat{\mathbf{I}} & -\tilde{\mathbf{b}} & \mathbf{I} \end{bmatrix} \begin{bmatrix} \ddot{\mathbf{r}} \\ \dot{\omega} \\ \ddot{\delta} \end{bmatrix} + \tilde{\omega}\tilde{\omega}\mathbf{b} + 2\tilde{\omega}\dot{\delta} \tag{25}$$

4.2. EQUATIONS OF MOTION FOR A FLEXIBLE BODY

In standard finite element formulation, the body-fixed $\xi\eta\zeta$ frame is also considered as the nonmoving reference frame. In such a system, the equations of motion are written as:

$$\mathbf{M}\,{}'\ddot{\delta} + \mathbf{K}\,{}'\delta = {}'\mathbf{f}$$

where **M**, **B**, **f** are the mass matrix, stiffness matrix, and the applied force vector. The left-superscript "prime" indicates that the vectors are referenced in the body-fixed frame. For a typical node i, the equations of motion are written as:

$$m^i\,{}'\ddot{\delta}^i + [\mathbf{k}^i]\,{}'\delta = {}'\mathbf{f}^i$$

where m^i is the mass and $[\mathbf{k}^i]$ represents the three rows of the stiffness matrix associated with node i. Note that in order for this equation to be valid, the reference frame must be stationary. Now if we allow the body-fixed frame to move relative to the xyz frame, the equations of motion for the node must be expressed in terms of the absolute acceleration of the node as:

$$m^i\,\ddot{\mathbf{d}}^i = \mathbf{f}^i - \mathbf{A}\,[\mathbf{k}^i]\,{}'\delta \tag{26}$$

where all the vectors are defined in terms of their absolute components. Note that the structural force vector, $[\mathbf{k}^i]\,{}'\delta$, is transformed from the body-fixed frame to xyz frame using the rotational transformation matrix **A**. For notation simplification, we will denote the structural force vector as:

$$\mathbf{f}^{i(s)} = -\mathbf{A}\,[\mathbf{k}^i]\,{}'\delta \tag{27}$$

Then, Eq. 26 is written as

$$m^i\,\ddot{\mathbf{d}}^i = \mathbf{f}^i + \mathbf{f}^{i(s)} \tag{28}$$

Equation 28 is written for all the nodes as (expression in parentheses is in compact form):

$$\begin{bmatrix} \ddots & & \\ & m^i\mathbf{I} & \\ & & \ddots \end{bmatrix} \begin{bmatrix} \vdots \\ \ddot{\mathbf{d}}^i \\ \vdots \end{bmatrix} = \begin{bmatrix} \vdots \\ \mathbf{f}^i + \mathbf{f}^{i(s)} \\ \vdots \end{bmatrix} \qquad (\mathbf{M}\ddot{\mathbf{d}} = \mathbf{f} + \mathbf{f}^{(s)}) \tag{29}$$

representing $3n$ equations in $3n$ accelerations. Substitution of Eq. 24 into Eq. 29 yields:

$$\mathbf{M}\begin{bmatrix}\hat{\mathbf{I}} & -\tilde{\mathbf{b}} & \ddot{\mathbf{I}}\end{bmatrix}\begin{bmatrix}\ddot{\mathbf{r}}\\ \dot{\omega}\\ \ddot{\delta}\end{bmatrix} = \mathbf{f}+\mathbf{f}^{(s)}-\tilde{\omega}\tilde{\omega}\mathbf{M}\mathbf{b}-w \tag{30}$$

where,

$$w = \begin{bmatrix}\vdots\\ w^i\\ \vdots\end{bmatrix} = \begin{bmatrix}\vdots\\ 2\mathrm{m}^i\tilde{\omega}\dot{\delta}^i\\ \vdots\end{bmatrix}$$

Equation 30 represents *3n* equations in *3n + 6* accelerations, therefore it cannot be solved in its present form for the unknown accelerations. We can premultiply Eq. 30 by the transpose of the transformation matrix of Eq. 25; i.e., $[\hat{\mathbf{I}} \quad -\tilde{\mathbf{b}} \quad \ddot{\mathbf{I}}]^{\mathrm{T}}$, to get:

$$\begin{bmatrix}\hat{\mathbf{I}}^{\mathrm{T}}\mathbf{M}\hat{\mathbf{I}} & -\hat{\mathbf{I}}^{\mathrm{T}}\mathbf{M}\tilde{\mathbf{b}} & \hat{\mathbf{I}}^{\mathrm{T}}\mathbf{M}\\ -\tilde{\mathbf{b}}^{\mathrm{T}}\mathbf{M}\hat{\mathbf{I}} & \tilde{\mathbf{b}}^{\mathrm{T}}\mathbf{M}\tilde{\mathbf{b}} & -\tilde{\mathbf{b}}^{\mathrm{T}}\mathbf{M}\\ \mathbf{M}\hat{\mathbf{I}} & -\mathbf{M}\tilde{\mathbf{b}} & \mathbf{M}\end{bmatrix}\begin{bmatrix}\ddot{\mathbf{r}}\\ \dot{\omega}\\ \ddot{\delta}\end{bmatrix} = \begin{bmatrix}\hat{\mathbf{I}}^{\mathrm{T}}(\mathbf{f}+\mathbf{f}^{(s)}-\tilde{\omega}\tilde{\omega}\mathbf{M}\mathbf{b}-w)\\ -\tilde{\mathbf{b}}^{\mathrm{T}}(\mathbf{f}+\mathbf{f}^{(s)}-\tilde{\omega}\tilde{\omega}\mathbf{M}\mathbf{b}-w)\\ \mathbf{f}+\mathbf{f}^{(s)}-\tilde{\omega}\tilde{\omega}\mathbf{M}\mathbf{b}-w\end{bmatrix} \tag{31}$$

Although this equation represents *3n + 6* equations in *3n + 6* unknown accelerations, we cannot solve the equations for the unknowns since six of the equations are redundant. The main reason that neither Eq. 30 nor Eq. 31 is solvable is that we have not yet defined how the body-fixed frame is attached to the flexible body.

4.2.1. *Nodal-Fixed Reference Frame*

We can always follow the procedure used in standard finite element modeling by defining boundary conditions on several nodes. We can attach the origin of the body-fixed frame to a node, such as node *o*, as shown in Fig. 8. Then, we can enforce another node, such as node *i*, to remain on the ξ-axis. Next we specify another node, such as node *j*, to remain on the ξ-η plane. These conditions are stated as:

$$\delta^o_{(\xi)} = \delta^o_{(\eta)} = \delta^o_{(\zeta)} = 0; \quad \delta^i_{(\eta)} = \delta^i_{(\zeta)} = 0; \quad \delta^j_{(\zeta)} = 0 \tag{32}$$

These six conditions uniquely define the position and the orientation of the body-fixed frame on the flexible body. Note that these conditions do not apply any kinematic constraints on the flexible body.

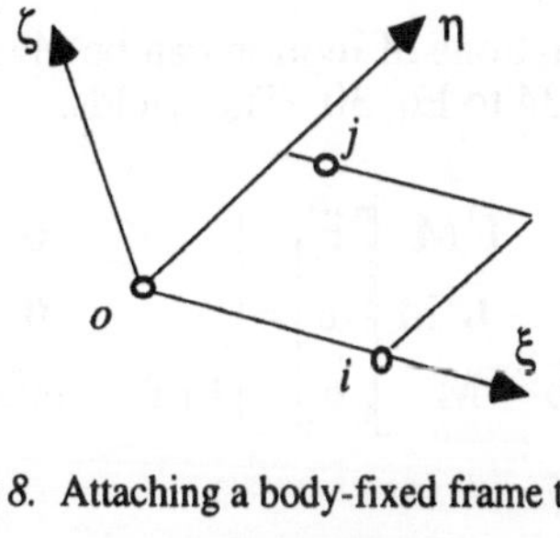

Figure 8. Attaching a body-fixed frame to nodes.

4.2.2. *Mean-Axes Reference Frame*

A very convenient technique to attach a body-fixed frame to a flexible body is to use the so-called mean-axes conditions [4]. These conditions enforce the sum of the linear momentum and the sum of the angular momentum of all the nodes, due to structural deformation (not rigid body motion), to be zero. These conditions are expressed as (expressions in parentheses are in compact form):

$$\sum m^i \dot{\delta}^i = \mathbf{0} \qquad (\hat{\mathbf{I}}^T \mathbf{M}\dot{\delta} = \mathbf{0}) \tag{33}$$

$$\sum m^i \tilde{\mathbf{b}}^i \dot{\delta}^i = \mathbf{0} \qquad (\tilde{\mathbf{b}}^T \mathbf{M}\dot{\delta} = \mathbf{0}) \tag{34}$$

These conditions can be derived by minimizing the kinetic energy of the flexible body with respect to the translational and angular velocities of the body-fixed frame; i.e., with respect to $\dot{\mathbf{r}}$ and ω.

Equation 33, which is the condition on the linear momentum, is also the condition that keeps the origin of the body-fixed frame to the instantaneous mass center of the flexible body. Furthermore, Eqs. 33 and 34 yield the following useful identities:

$$\sum m^i \mathbf{b}^i = \mathbf{0} \qquad (\hat{\mathbf{I}}^T \mathbf{M}\mathbf{b} = \mathbf{0}) \tag{35}$$

$$\sum m^i \delta^i = \mathbf{0} \qquad (\hat{\mathbf{I}}^T \mathbf{M}\delta = \mathbf{0}) \tag{36}$$

$$\sum m^i \dot{\mathbf{b}}^i = \mathbf{0} \qquad (\hat{\mathbf{I}}^T \mathbf{M}\dot{\mathbf{b}} = \mathbf{0}) \tag{37}$$

$$\sum m^i \dot{\delta}^i = \mathbf{0} \qquad (\hat{\mathbf{I}}^T \mathbf{M}\dot{\delta} = \mathbf{0}) \tag{38}$$

$$\sum m^i \ddot{\mathbf{b}}^i = \mathbf{0} \qquad (\hat{\mathbf{I}}^T \mathbf{M}\ddot{\mathbf{b}} = \mathbf{0}) \tag{39}$$

$$\sum m^i \ddot{\delta}^i = \mathbf{0} \qquad (\hat{\mathbf{I}}^T \mathbf{M}\ddot{\delta} = \mathbf{0}) \tag{40}$$

$$\sum m^i \tilde{\mathbf{b}}^i \ddot{\delta}^i = \mathbf{0} \qquad (\tilde{\mathbf{b}}^T \mathbf{M}\ddot{\delta} = \mathbf{0}) \tag{41}$$

4.2.3. *Solvable Equations of Motion*

The first set of solvable equations of motion is Eq. 28 which represents *3n* equations in *3n* unknown absolute accelerations. Although this is a solvable set of equations, it is not a very useful set. The difficulty comes from the fact that the structural forces are computed from nodal deflections. However, the nodal deflections cannot be extracted easily from the second integral of vector $\ddot{\mathbf{d}}$.

A second form of solvable equations of motion can be obtained if we consider the fixed-axes conditions. The six conditions given in Eq. 32 (actually their second time derivatives) can be used in Eq. 30 to obtain the same number of equations as the number of unknown accelerations.

A third form of solvable equations of motion can be obtained by appending the mean-axes conditions of Eqs. 33 and 34 to Eq. 30. This yields:

$$\begin{bmatrix} \mathbf{0} & \mathbf{0} & \hat{\mathbf{I}}^T\mathbf{M} \\ \mathbf{0} & \mathbf{0} & -\tilde{\mathbf{b}}^T\mathbf{M} \\ \mathbf{M}\hat{\mathbf{I}} & -\mathbf{M}\tilde{\mathbf{b}} & \mathbf{M} \end{bmatrix} \begin{bmatrix} \ddot{\mathbf{r}} \\ \dot{\omega} \\ \ddot{\delta} \end{bmatrix} = \begin{bmatrix} \mathbf{0} \\ \mathbf{0} \\ \mathbf{f} + \mathbf{f}^{(s)} - \tilde{\omega}\tilde{\omega}\mathbf{M}\mathbf{b} - \boldsymbol{w} \end{bmatrix} \tag{42}$$

Or in expanded form this equation is written as:

$$\left[\begin{array}{cc|c} & & \cdots \;\; \mathbf{m}^i\mathbf{I} \;\; \cdots \\ & & \cdots \; -\mathbf{m}^i\tilde{\mathbf{b}}^{iT} \; \cdots \\ \hline \vdots & \vdots & \ddots \\ \mathbf{m}^i\mathbf{I} & -\mathbf{m}^i\tilde{\mathbf{b}}^i & \mathbf{m}^i\mathbf{I} \\ \vdots & \vdots & \ddots \end{array}\right]\left[\begin{array}{c} \ddot{\mathbf{r}} \\ \dot{\omega} \\ \hline \vdots \\ \ddot{\delta}^i \\ \vdots \end{array}\right] = \left[\begin{array}{c} \mathbf{0} \\ \mathbf{0} \\ \hline \vdots \\ \mathbf{f}^i + \mathbf{f}^{i(s)} - \mathbf{m}^i\tilde{\omega}\tilde{\omega}\mathbf{b}^i - \boldsymbol{w}^i \\ \vdots \end{array}\right]$$

These are *3n* + 6 equations in *3n* + 6 unknown accelerations.

The fourth form of solvable equations of motion is obtained by observing the following definitions and identities:

-- The sum of nodal masses is equal to the mass of the flexible body:

$$\mathrm{m} = \sum \mathrm{m}^i \qquad (\mathrm{m}\mathbf{I} = \hat{\mathbf{I}}^{\mathrm{T}}\mathbf{M}\hat{\mathbf{I}}) \tag{43}$$

-- The instantaneous (variant) inertia tensor of the body is defined as:

$$\mathbf{J} = -\sum \mathrm{m}^i\tilde{\mathbf{b}}^i\tilde{\mathbf{b}}^i \qquad (\mathbf{J} = \tilde{\mathbf{b}}^{\mathrm{T}}\mathbf{M}\tilde{\mathbf{b}}) \tag{44}$$

-- The sum of internal structural forces is equal to zero:

$$\sum \mathbf{f}^{i(s)} = \mathbf{0} \qquad (\hat{\mathbf{I}}^{\mathrm{T}}\mathbf{f}^{(s)} = \mathbf{0}) \tag{45}$$

-- The sum of internal structural moments is equal to zero:

$$\sum \tilde{\mathbf{b}}^i\mathbf{f}^{i(s)} = \mathbf{0} \qquad (-\tilde{\mathbf{b}}^{\mathrm{T}}\mathbf{f}^{(s)} = \mathbf{0}) \tag{46}$$

-- Other identities:

$$\tilde{\mathbf{b}}^i\,\tilde{\omega}\tilde{\omega}\,\mathbf{b}^i = -\tilde{\omega}\,\tilde{\mathbf{b}}^i\tilde{\mathbf{b}}^i\,\omega \tag{47}$$

$$\hat{\mathbf{I}}^{\mathrm{T}}\mathbf{M}\tilde{\mathbf{b}} = \tilde{\mathbf{b}}^{\mathrm{T}}\mathbf{M}\hat{\mathbf{I}} = \mathbf{0} \tag{48}$$

$$\tilde{\omega}\tilde{\omega}\,\hat{\mathbf{I}}^{\mathrm{T}}\mathbf{M}\mathbf{b} = \mathbf{0} \tag{49}$$

$$\hat{\mathbf{I}}^{\mathrm{T}}\boldsymbol{w} = 2\tilde{\omega}\,\hat{\mathbf{I}}^{\mathrm{T}}\mathbf{M}\dot{\delta} = \mathbf{0} \tag{50}$$

$$\tilde{\mathbf{b}}^{\mathrm{T}}\,\tilde{\omega}\tilde{\omega}\,\mathbf{M}\mathbf{b} = -\tilde{\omega}\,\mathbf{J}\,\omega \tag{51}$$

Using these identities and the identities of Eqs. 33-41, Eq. 31 is simplified to:

$$\begin{bmatrix} \mathrm{m}\mathbf{I} & \mathbf{0} & \mathbf{0} \\ \mathbf{0} & \mathbf{J} & \mathbf{0} \\ \mathbf{M}\hat{\mathbf{I}} & -\mathbf{M}\tilde{\mathbf{b}} & \mathbf{M} \end{bmatrix}\begin{bmatrix} \ddot{\mathbf{r}} \\ \dot{\omega} \\ \ddot{\delta} \end{bmatrix} = \begin{bmatrix} \hat{\mathbf{I}}^{\mathrm{T}}\mathbf{f} \\ -\tilde{\mathbf{b}}^{\mathrm{T}}\mathbf{f} - \tilde{\omega}\,\mathbf{J}\,\omega + \tilde{\mathbf{b}}^{\mathrm{T}}\boldsymbol{w} \\ \mathbf{f} + \mathbf{f}^{(s)} - \tilde{\omega}\tilde{\omega}\,\mathbf{M}\mathbf{b} - \boldsymbol{w} \end{bmatrix} \tag{52}$$

Or, in expanded form this equation is written as:

$$\left[\begin{array}{c|c|c} m\mathbf{I} & & \\ & \mathbf{J} & \\ \hline \vdots & \vdots & \ddots \\ m^i\mathbf{I} & -m^i\tilde{\mathbf{b}}^i & m^i\mathbf{I} \\ \vdots & \vdots & \ddots \end{array}\right]\left[\begin{array}{c}\ddot{\mathbf{r}} \\ \dot{\omega} \\ \hline \vdots \\ \ddot{\delta}^i \\ \vdots\end{array}\right]=\left[\begin{array}{c}\sum \mathbf{f}^i \\ \sum \tilde{\mathbf{b}}^i\mathbf{f}^i-\tilde{\omega}\mathbf{J}\omega-\sum \tilde{\mathbf{b}}^i \boldsymbol{w}^i \\ \hline \vdots \\ \mathbf{f}^i+\mathbf{f}^{i(s)}-m^i\tilde{\omega}\tilde{\omega}\mathbf{b}^i-\boldsymbol{w}^i \\ \vdots\end{array}\right]$$

Equation 52 represents *3n + 6* equations in *3n + 6* unknown accelerations. The terms $\sum \mathbf{f}^i$ and $\sum \tilde{\mathbf{b}}^i\mathbf{f}^i$ (or $\hat{\mathbf{I}}^{\mathrm{T}}\mathbf{f}$ and $-\tilde{\mathbf{b}}^{\mathrm{T}}\mathbf{f}$) represent respectively the sum of external forces and the sum of external moments acting on the body. The first row of Eq. 52 is Newton's equation of motion for the mass center of the flexible body. The second row of this equation is Euler's equation for the rotation of the mean-axes frame, where the inertia tensor **J** is a function of the instantaneous shape of the body. Furthermore, a Coriolis term, $\tilde{\mathbf{b}}^{\mathrm{T}}\boldsymbol{w}$, appears in this equation which is not present in Euler's rigid body equation.

5. Rigid-Flexible Multibody Systems

In a rigid-flexible multibody system, rigid bodies may be modeled by the body- or the point-coordinate formulation. We may mix the two formulations in a single multibody model. The flexible bodies may be modeled by any set of equations of motion discussed in Sec. 4. As an example of how we may mix different formulations, consider the schematic presentation of a multibody system shown in Fig. 9(a) which requires four rigid and one flexible bodies. For simplifying the discussion, assume that all of the kinematic joints are of spherical type. The bodies are numbered *1*, ..., *5*. In order to demonstrate how different formulations can be mixed, we use body-coordinate formulation for bodies *1* and *2*, point-coordinate formulation for bodies *3* and *4*, and body *5* is the only flexible body in the system.

The body coordinate representation of bodies *1* and *2* requires the spherical joint S_1 to be described with the constraint equations of Eq. 4:

$$\mathbf{r}_1+\mathbf{A}_1\,{}^1\mathbf{s}_1^C-\mathbf{r}_2-\mathbf{A}_2\,{}^2\mathbf{s}_2^B=\mathbf{0}$$

Bodies *3* and *4* are modeled by seven points where point *4* is shared, therefore the spherical joint S_3 does not require any constraint equations. For the spherical joint S_2 between point *D* on body *1* and point *2* on body *3*, we can write the following constraints:

$$\mathbf{r}_1+\mathbf{A}_1\,{}^1\mathbf{s}_1^D-\mathbf{r}_3^2=\mathbf{0}$$

The spherical joint S_4 is between point *3* on body *3* and node *j* on the flexible body. For this joint we can write the following constraints:

$$\mathbf{r}_3^3-\mathbf{d}_5^j=\mathbf{0}$$

Similarly, for joint S_5 we can write:

$$\mathbf{r}_2+\mathbf{A}_2\,{}^2\mathbf{s}_2^A-\mathbf{d}_5^i=\mathbf{0}$$

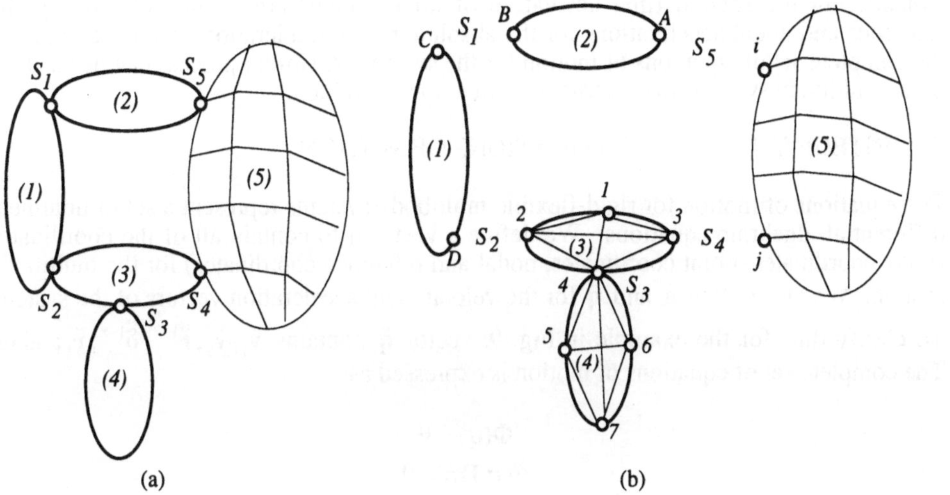

Figure 9. Schematic representation of a multibody system.

In addition to these kinematic constraints, we must include twelve primary constraints, as given by Eq. 10, between the seven point coordinates. The time derivative (first or second) of these twenty four constraints (twelve primary and twelve kinematic constraints) provides the Jacobian matrix which will be denoted here as matrix **D**.

The Equations of motion for this multibody system can be written as:

$$\begin{bmatrix} \mathbf{M}_1 & \mathbf{0} & \mathbf{0} & \mathbf{0} \\ \mathbf{0} & \mathbf{M}_2 & \mathbf{0} & \mathbf{0} \\ \mathbf{0} & \mathbf{0} & \mathbf{M}^{1-7} & \mathbf{0} \\ \mathbf{0} & \mathbf{0} & \mathbf{0} & \mathbf{M}^{1-n} \end{bmatrix} \begin{bmatrix} \dot{\mathbf{v}}_1 \\ \dot{\mathbf{v}}_2 \\ \ddot{\mathbf{r}}^{1-7} \\ \ddot{\mathbf{d}}^{1-n} \end{bmatrix} - \mathbf{D}^{\mathrm{T}}\boldsymbol{\lambda} = \begin{bmatrix} \mathbf{g}_1 \\ \mathbf{g}_2 \\ \mathbf{f}^{1-7} \\ \mathbf{f}^{1-7} + \mathbf{f}^{(s)} \end{bmatrix}$$

where $\mathbf{M}_1$ and $\mathbf{M}_2$ are the 6 x 6 matrices for bodies *1* and *2* as given by Eq. 3, $\mathbf{M}^{1-7}$ is the mass matrix associated with seven primary points as stated in Sec. 3.2.3, and $\mathbf{M}^{1-n}$ is the nodal mass matrix associated with the n nodes of the flexible body as given in Eq. 29. Similarly, the force vectors on the right-hand-side are determined associated with the bodies, points, and nodes. At this point we can define a body-fixed frame for the flexible body, using the mean-axes conditions. Substituting Eq. 25 in the above equation for the absolute nodal accelerations, then appending the mean-axes conditions (refer to Eq. 42), and after a slight rearrangement of terms, we obtain:

$$\begin{bmatrix} \mathbf{M}_1 & \mathbf{0} & \mathbf{0} & \mathbf{0} & \mathbf{0} & \mathbf{0} \\ \mathbf{0} & \mathbf{M}_2 & \mathbf{0} & \mathbf{0} & \mathbf{0} & \mathbf{0} \\ \mathbf{0} & \mathbf{0} & \mathbf{M}^{1-7} & \mathbf{0} & \mathbf{0} & \mathbf{0} \\ \mathbf{0} & \mathbf{0} & \mathbf{0} & \mathbf{M}^{1-n} & \mathbf{M}^{1-n}\hat{\mathbf{I}} & -\mathbf{M}^{1-n}\tilde{\mathbf{b}} \\ \mathbf{0} & \mathbf{0} & \mathbf{0} & \hat{\mathbf{I}}^{\mathrm{T}}\mathbf{M}^{1-n} & \mathbf{0} & \mathbf{0} \\ \mathbf{0} & \mathbf{0} & \mathbf{0} & -\tilde{\mathbf{b}}^{\mathrm{T}}\mathbf{M}^{1-n} & \mathbf{0} & \mathbf{0} \end{bmatrix} \begin{bmatrix} \dot{\mathbf{v}}_1 \\ \dot{\mathbf{v}}_2 \\ \ddot{\mathbf{r}}^{1-7} \\ \ddot{\boldsymbol{\delta}}^{1-n} \\ \ddot{\mathbf{r}} \\ \dot{\boldsymbol{\omega}} \end{bmatrix} - \mathbf{D}^{\mathrm{T}}\boldsymbol{\lambda} = \begin{bmatrix} \mathbf{g}_1 \\ \mathbf{g}_2 \\ \mathbf{f}^{1-7} \\ (\mathbf{f} + \mathbf{f}^{(s)} - \tilde{\boldsymbol{\omega}}\tilde{\boldsymbol{\omega}}\mathbf{M}\mathbf{b} - w)^{1-7} \\ \mathbf{0} \\ \mathbf{0} \end{bmatrix}$$

Considering the second time derivative of all the constraints, and substituting the reference and nodal accelerations for the absolute nodal accelerations, we have obtained the complete set of equations of motion for the system. A similar process can be used for other multibody systems to construct the equations of motion.

5.1. INTEGRATION OF THE EQUATIONS OF MOTION

The equations of motion for rigid-flexible multibody systems represent a set of nonlinear differential-algebraic equations. We define a vector $\mathbf{q}$ to contain all of the coordinates (body coordinates, point coordinates, nodal and reference coordinates) for the multibody system. We also define $\dot{\mathbf{q}}$ and $\ddot{\mathbf{q}}$ for the velocity and acceleration vectors of the system. To clarify this, for the example in Fig. 9, vector $\ddot{\mathbf{q}}$ contains $\dot{\mathbf{v}}_1, \dot{\mathbf{v}}_2, \ddot{\mathbf{r}}^{1-7}, \ddot{\boldsymbol{\delta}}^{1-n}, \ddot{\mathbf{r}}$, and $\dot{\boldsymbol{\omega}}$. The complete set of equations of motion is expressed as:

$$\boldsymbol{\Phi}(\mathbf{q}) = \mathbf{0}$$
$$\dot{\boldsymbol{\Phi}} \equiv \mathbf{D}\dot{\mathbf{q}} = \mathbf{0}$$
$$\begin{bmatrix} \mathbf{M} & -\mathbf{D}^{\mathrm{T}} \\ \mathbf{D} & \mathbf{0} \end{bmatrix} \begin{bmatrix} \ddot{\mathbf{q}} \\ \boldsymbol{\lambda} \end{bmatrix} = \begin{bmatrix} \mathbf{g} \\ -\dot{\mathbf{D}}\dot{\mathbf{q}} \end{bmatrix}$$

where the second derivative of the constraints is appended to the equations of motion. Here, the kinematic constraints and the primary constraints are all appended into one set of constraints. Now consider that the equations of motion can be put in the standard form:

$$\dot{\mathbf{y}} = \mathbf{f}(\mathbf{y}, \mathrm{t})$$

where $\mathbf{y}$ and $\dot{\mathbf{y}}$ arrays contain the coordinates, velocities, and accelerations as:

$$\dot{\mathbf{y}} \equiv \begin{bmatrix} \dot{\mathbf{q}} \\ \ddot{\mathbf{q}} \end{bmatrix}, \quad \mathbf{y} \equiv \begin{bmatrix} \mathbf{q} \\ \dot{\mathbf{q}} \end{bmatrix}$$

The solution of the equations of motion requires a numerical integration process that predicts the elements of $\mathbf{y}$ at any time-step t. The solution of the equations of motion determines the accelerations that are needed as the elements of $\dot{\mathbf{y}}$ to be returned to the integration algorithm. Since here we deal with a set of differential-algebraic equations, the issue of *constraint violation* must be considered carefully. Many methods have been presented in the past two decades for the solution of such equations [1].

5.2. TRANSFORMATION TO JOINT COORDINATES

A powerful method that can transform the differential-algebraic equations of motion from a large set to a smaller set of either differential-algebraic or ordinary-differential equations is called the *joint-coordinate* method [5-7]. In rigid multibody systems, a set of joint coordinates is defined describing the relative degrees of freedom between the interconnected bodies. For example, revolute or prismatic joints require one joint coordinate, universal or cylindrical joints require two joint coordinates, and a spherical joint needs three joint coordinates. The time derivative of the joint coordinates is referred to as the *joint velocities*. For open-loop systems (systems with no closed kinematic loops), this vector has a dimension equal to the number of degrees-of-freedom of the system. A velocity transformation matrix can be constructed relating the body- and/or point-velocities to the joint velocities. This matrix plays a fundamental role in transforming the equations of motion from body- and/or point-coordinate form to joint coordinate form. For open-loop systems, all of the kinematic constraints and their associated Lagrange multipliers in the equations of motion vanish as the result of this transformation, and the result is a system of ordinary-differential equations. For closed-loop systems, the transformation process yields another set of differential-algebraic equations, but much smaller in number than the original set. In the new set, only a few of the constraints and Lagrange multipliers remain present. However, it is possible to follow through with a second transformation and convert the resultant differential-algebraic equations to a highly nonlinear set of ordinary-differential equations without any constraints or Lagrange multipliers.

The transformation of the equations of motion to joint coordinates can also be applied to systems containing flexible bodies. The steps involved in the process are such that we could transform either the entire set or only a subset of the equations of motion to the joint coordinates. For example, the rigid body portion of the equations may be transformed to the joint coordinates but the flexible body portion may be kept unchanged. Due to the space limitation in this paper, experimentation with this issue is left to the interested reader. One point to emphasize here is that substantial computational efficiency can be gained by the use of joint coordinate formulation.

6. Application Examples

This paper has presented a computational procedure for systematic construction of the equations of motion for rigid-flexible multibody systems. The main application of this formulation is in the area of vehicle dynamics and crash analysis. In this section two examples are provided to assist the reader in better understanding the wide range of possibilities with multibody dynamic formulations.

6.1. VEHICLE DYNAMICS

Multibody formulations presented in this paper have been used extensively in simulating ride and handling maneuvers of a variety of vehicles. Although both the body-coordinate and the point-coordinate formulations can be used effectively in vehicle dynamic analysis, transformation of the equations of motion to the joint-coordinate space is highly recommended. Such a transformation not only reduces the computational time, but it also reduces the difficulties associated with a phenomenon known as constraint violation.

Another important issue to be considered in ride and handling simulations is the *tire model*. In many simulations, even a sophisticated and accurate multibody model may yield poor results if we do not incorporate a good tire model. In the past decade, several analytical and empirical based tire models have been introduced for ride and handling simulations.

6.1.1. *An Example for Ride and Handling Analysis*

This example is a three dimensional rigid/flexible multibody vehicle model that closely represents a typical mid-size sedan. A schematic of the model, with some body numbers and joint types, is shown in Fig. 10. The vehicle has super-strut front suspensions and quadra-link rear suspensions, both of which are variants of the McPherson suspension.

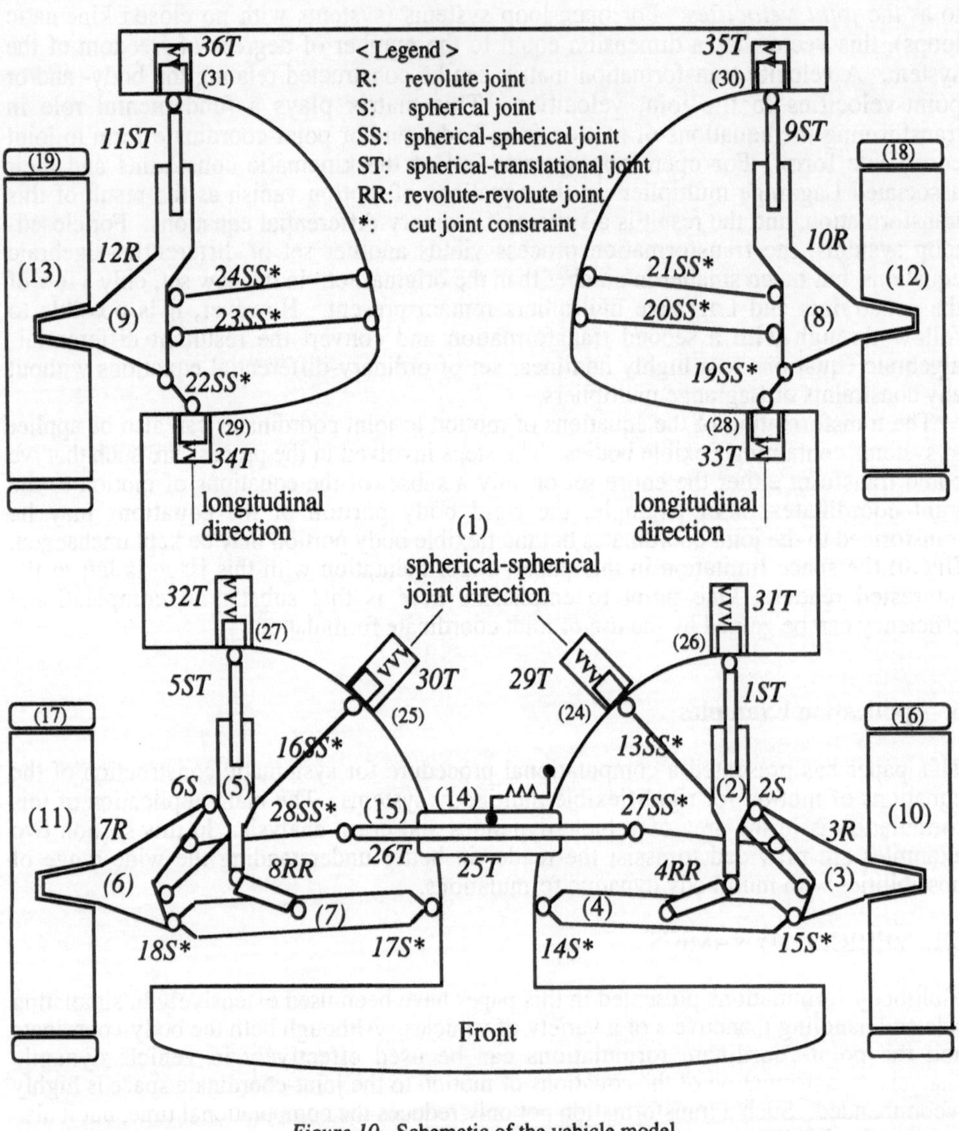

Figure 10. Schematic of the vehicle model.

The front wheels are steered by a rack-and-pinion steering mechanism, where the shift of the rack-bar relative to the rack-housing determines the steering angles of the front wheels through the tie-rods.

The multibody modeling of this system starts with the body-coordinate formulation and then the equations are transformed to the joint-coordinate formulation. The car body (body 1) can be modeled either as a rigid or as a flexible body. All the links of the suspension and steering systems are assumed to be rigid. Suspension rubber bushings are modeled by supplementary bodies (24-31), kinematic joints (29T-36T), and translational spring-damper elements. The rack-housing is kinematically joined to the chassis by a translational joint and a rubber-bushing element representing the compliance between the housing and the chassis. A kinematic steering constraint is defined between the rack-bar and rack-housing to govern the relative shift of the rack-bar.

For this study, the Magic Formula tire model is used [8]. This particular tire model was selected since a complete set of experimentally measured data for the particular tires used on this vehicle was available. The tire model is defined on bodies 16-19. In the tire model, the interacting forces and moments between the tire and the road surface are determined based on empirical formulas as a function of tire orientation, velocity, and radial stiffness. The tire lateral flexibility is modeled based on a *ring-body* concept [9]. Spring-damper elements that represents the tire lateral stiffness and damping are defined between the ring bodies (16-19) and the cylinder bodies (10-13). The ring-body concept allows the contact patch of a tire to deform longitudinally and/or laterally with respect to the main body of the tire. The vehicle model represents a front wheel drive car. Therefore, the torque of the two front wheels are feedback controlled to maintain the desired vehicle speed. Speed-dependent rolling resistance is applied to all the wheels.

In some ride and handling simulations, small deformations of the main body may not be of importance therefore, the chassis (body 1) could be modeled as a rigid body. However, in simulating some specific driving scenarios, deformation of the chassis may be a contributing factor to the handling response of the vehicle. In such situations the main body is modeled as a flexible body. As described in the previous sections, a flexible body can be described as a finite element model and hence a multi-rigid/flexible body formulation can be constructed for the complete system.

Typical scenarios that were used for dynamic analysis of this particular model are: driving around a circular path at a constant speed; lane-change maneuvers; and pulse steering maneuvers. Results from such simulations have been validated against experimentally measured data. It has been shown that the dynamic response of such complex multibody systems can be predicted with high accuracy.

6.2. CRASH ANALYSIS

Vehicle models for crash analysis range from very simple and crude to highly complex and possibly accurate. If a simple vehicle structure is considered for crash analysis, it may be possible to apply rigid multibody formulation to construct its corresponding computational model. In such models the concept of plastic hinge may be employed. However in more accurate models, a detailed description of the vehicle structure must be considered. In the formulation of Sec. 4 for flexible bodies, no restrictions were imposed on the system stiffness matrix. The stiffness matrix can be constructed by any finite element program exhibiting linear or nonlinear characteristics [10-12]. Furthermore, both the plastic hinge approach and the finite element approach may be combined in the same analytical model.

6.2.1. *An Example for crash Analysis*

Application of multibody and plastic hinge modeling techniques to an actual vehicle crash event was presented in [13]. The modeled event was a frontal barrier crash test of a particular automobile. For the test, the frontal sheet metal, radiator, grill, and engine were removed as depicted schematically in Fig. 11. The impact velocity was 13.4 m/sec and the total mass of the tested vehicle was approximately 1300 kg.

Body-coordinate formulation was considered in the simulation model for constructing the equations of motion for the main body, suspension systems, and the wheels. The stub frame was modeled as a collection of rigid bodies, connected by revolute or universal joints with torsional springs. Revolute joints were imbedded in channel and z-section frames along the weak axis of the cross section, since a frame tends to bend easily along that axis. For square box-sections, universal joints were imbedded along the two principal axes. Elasto-plastic characteristics of the torsional springs were obtained from experimental tests. The bumper was modeled as a rigid body translating relative to the stub frame. A nonlinear spring model was placed along the translational axis of the bumper to represent its plastic deformation.

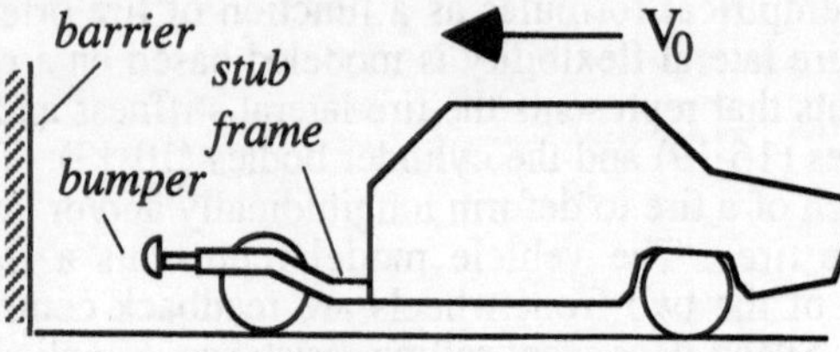

Figure 11. Schematics of a dynamic crash test.

Several simulations for frontal and oblique impacts were performed. The results from the frontal impact simulations were compared against test results. As reported in [13], reasonable agreement between the experimental and simulated results was observed. The simulated collapsed shape of the vehicle, after 40 ms, is shown in Fig. 12. This and other similar models have shown that multibody modeling techniques can effectively be used to predict the dynamic response of complex structures under crash conditions.

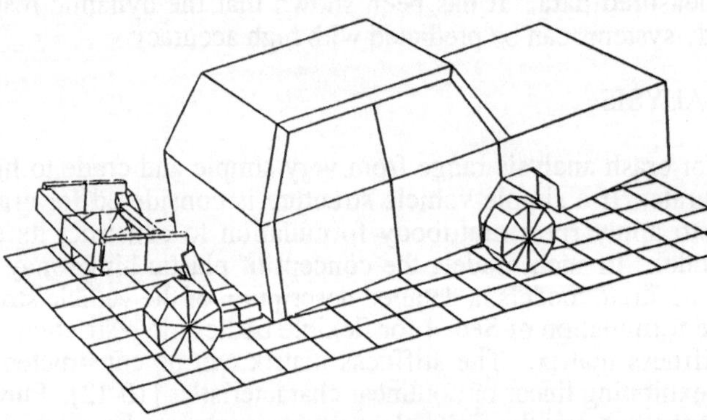

Figure 12. The simulated crash after 40 ms.

7. Summary

In summary, the methodologies presented in this paper have many interesting characteristics that may be found useful in a variety of applications, specially in vehicle dynamics and crash simulations. These methodologies can be combined with other methods to develop even more efficient, accurate, and flexible procedures. It should be pointed out that there is no single multibody formulation to be considered as the *best formulation* for general dynamic analysis. Each formulation has its own unique or common features and, therefore, selected features should be adopted to our advantage.

8. References

1. Nikravesh, P.E. (1988) *Computer-Aided Analysis of Mechanical Systems*, Prentice-Hall.
2. Garcia de Jalon, J. And Bayo, E. (1994) *Kinematic and Dynamic Simulations of Multibody Systems*, Springer-Verlag.
3. Nikravesh, P.E. and Affifi, H.A. (1994) Construction of the Equations of Motion for Multibody Dynamics Using Point and Joint Coordinates, *Computer-Aided Analysis of Rigid and Flexible Mechanical Systems*, Kluwer academic publishers, NATO ASI Series E: Applied Sciences - Vol. 268, pp.31-60.
4. Agrawal, O.P. and Shabana, A.A. (1986) Application of Deformable-Body Mean Axis to Flexible Multibody System Dynamics, *Computer Methods in Applied Mechanics and Engineering* 56, 217-245.
5. Jerkovsky, W. (1978) The Structure of Multibody Dynamics Equations, *J. Guidance and Control*, Vol. 1, No. 3, pp. 173-182.
6. Kim, S.S. and Vanderploeg, M.J. (1986) A General and Efficient Method for Dynamic Analysis of Mechanical Systems Using Velocity Transformation, *ASME J. Mech., Trans., and Auto. In Design*, Vol. 108, 176-182.
7. Nikravesh, P.E. and Gim, G. (1993) Systematic Construction of the Equations of Motion for Multibody Systems Containing Closed Kinematic Loops, *ASME J. Of Mechanical Design*, Vol. 115, No. 1, 143-149.
8. Bakker, E., Nyborg, L. And Pacjeks, H.B. (1987) Tyre Modeling for Use in Vehicle Dynamics Studies, *SAE paper*, No. 870421.
9. Park, J. And Nikravesh, P.E. (1997) A Multibody Approach to Modeling Tire Longitudinal/Lateral Flexibility, February 1997 SAE International Congress and Exposition.
10. Ambrósio, J.A.C. (1991) *Elastic-Plastic Large Deformation of Flexible Multibody Systems in Crash Analysis*, Ph.D. Dissertation, University of Arizona, Tucson, AZ.
11. Nikravesh, P.E. and Ambrósio, J.A.C. (1991) Systematic Construction of the Equations of Motion for Rigid-Flexible Multibody Systems Containing Open and Closed Kinematic Loops, Int. J. Of Num. Meth. In Eng., Vol. 32, 1749-1766.
12. Ambrósio, J.A.C. and Nikravesh, P.E. (1992) Elasto-Plastic Deformation in Multibody Dynamics, *Nonlinear Dynamics*, Vol. 3, 85-104.
13. Nikravesh, P.E. and Chung, I.S. (1984) Structural Collapse and Vehicular Crash Simulation Using a Plastic Hinge Technique, *J. Struct. Mech.*, 12(3), 371-400.

CONTACT/IMPACT DYNAMICS APPLIED TO CRASH ANALYSIS

H. M. LANKARANI
Associate Professor and Bombardier/Learjet Fellow
Mechanical Engineering Department
and National Institute for Aviation Research
Wichita State University
Wichita, KS 67260-0035, USA

Abstract

The methods for predicting the impact responses of mechanical systems, such as a vehicle's response in a crash, can primarily be classified into two groups. In one, the impact is treated as a discontinuous event. A simple form of this type of analysis is usually conducted for accident reconstruction purposes for which each vehicle is modeled as a single body or object. For impact analysis of constrained or jointed systems such as a truck-and-trailer, and also for determination of the structural crash responses of the individual components of a mechanical system, such as the impact responses of the vehicle chassis, suspension systems, steering mechanism, etc., the extension of these traditional procedures are presented here, which involve both normal and frictional impulses, for impact analysis of any general multibody mechanical system. This method might sometimes be of limited use, since no significant change must occur in the system's configuration before and after impact. In the second group of methods used for impact analysis of mechanical systems, the local deformations and the contact forces are treated as continuous. The application of these methods requires the knowledge of the variations of the contact forces. Several models, including nonlinear visco-elastic, Hertzian-based, and visco-plastic models are presented. The suitability of each model depends on the material properties of the surfaces in contact and also the range of impact velocities. The unknown parameters are determined utilizing the energy and momentum considerations. These models are then generalized to the impact between two bodies of a multibody mechanical system. Experimental testing procedures, utilizing a quasi-static approach to measure impact responses of fixed plates and simply supported beams, are conducted for verification of the developed contact force models. Examples, including a two-car side-swipe collision and a complex vehicle rollover are presented as applications of the presented methodologies.

1. Introduction

Knowledge of contact/impact mechanics is a crucial step in the prediction or assessment of the crash response of a mechanical system or a structure. In an impact, nonlinear

J. A. C. Ambrósio et al. (eds.),
Crashworthiness of Transportation Systems: Structural Impact and Occupant Protection, 445–473.

contact forces of unknown nature are created, which act and disappear over a short period of time. The methods for predicting the impact responses of mechanical systems, such as a complete model of an automobile or a multibody model of a vehicle occupant, can primarily be classified into two groups. In one, the impact is treated as a discontinuous event [1-3]. A simple form of this type of analysis is usually conducted for accident reconstruction purposes for which each vehicle is modeled as a single body or object. While Brach pioneered such analysis [4,5], this classical approach has been followed by many for direct-central and oblique impacts of two free objects not connected to other objects [6-8].

For impact analysis of constrained or jointed systems such as a truck-and-trailer, this conventional method needs to be expanded. Furthermore, if the interest is in determination of the structural crash response of the individual components of a mechanical system, such as the responses of the chassis, suspension systems, steering mechanism, etc. as a result of a crash, these traditional procedures need to be expanded to multiple object (or multibody impact) problems. Impact within a multibody system was first studied by Wittenburg [9], based on which a set of momentum balance-impulse equations was later developed by Wehage [10]. Khulief and Shabana extended these equations to flexible multibody systems [11]. In [12], it was shown that in order to gain numerical efficiency and stability, the equations of motion for a multibody system may be assembled in a canonical form. All these methods were however limited to impact situations with only a direct or normal component of impact velocity of the surfaces in contact. In the real-world, most impact situations have both relative normal and tangential impact velocity components. In the presence of any tangential component of impact velocity of the surfaces, the effect of friction on the impact response must be included. The inclusion of friction makes the analysis of impact much more complicated. Studies that have included fiction in multibody impact situations, such as Whitaker [13], Keller [14], and Pereira and Nikravesh [15], have resulted in inconsistencies in the energy balance, and at times have shown energy gains. This phenomenon was first pointed out by Kane while analyzing the impact of a compound pendulum on the ground [16]. The inconsistencies in the energy balance is due to the inherent problems associated with the use of the Newton's hypotheses in the definition of the coefficient of restitution.

In this paper, a momentum-balance/impulse set of equations is presented for impact analysis of any general multibody mechanical system. The equations are formulated by integrating the canonical form of the governing differential equations of motion for the system using the system's joint coordinates and joint momenta. The solution to these equations gives the jump in the system velocities or momenta, with which the resulting motion of the mechanical system after impact can be determined. The use of joint coordinates eliminates the necessity for treatment of the constraint equations associated with the system kinematic joints. The canonical formulation is a natural way of balancing the momenta of the system under impact. The Poisson's hypothesis, instead of the Newton's hypothesis, is used for the definition of the coefficient of restitution. Han and Gilmore used this method for frictional impact analysis of multiple free objects but not jointed system of bodies [17]. An algorithm for the multibody frictional impact, based on the Routh's graphical technique, is presented here so that the friction model be capable of detecting all modes of impact including: sliding, sticking, and reverse sliding. Example of the use of frictional impact analysis, including a two-car side swipe collision and also a vehicle rollover, are presented and the analysis results are verified compared to experimental tests results. It is shown that depending on the nature of analysis, this method might sometimes be of limited use, since the duration of the contact period is usually unknown before-hand, and that during the period of contact no significant change must occur in the system configuration before and after impact.

In the second group of methods used for impact analysis of mechanical systems, the local deformations and the contact forces are treated as continuous. The application of these methods requires the knowledge of the variations of the contact forces. Different models have been postulated to represent this variation. In the most simple one, the contact force is modeled by a parallel linear spring-damper element [18]. This model, known as the Kelvin-Voigt viscoelastic model, has been used for the impact between two bodies within a multibody system [19]. However, this linear model may not be very accurate since it does not represent the overall nonlinear nature of an impact. Furthermore, the half-sine shape solution that it provides for the local deformation of the two bodies in the direction of impact suggests that the two bodies exert tension on each other right before the rebounding stage. A more suitable model of the contact force is the nonlinear Hertzian force-deformation law [20]. Although the Hertzian theory is a purely elastic one, some studies have been performed to extend the theory to include energy dissipation [18, 21-23]. In this paper, several models, including linear and nonlinear visco-elastic models, Hertzian-based models, and visco-plastic models are presented for the direct-central impact of two solid particles. The suitability of each model will be discussed as related to the material properties of the surfaces in contact and also the range of impact velocities. The unknown parameters in the contact force models are determined utilizing the energy and momentum considerations. Experimental testing procedures and corresponding results for verification of these contact force models are presented for the impact of a steel sphere on aluminum or steel bases. The boundary conditions of the base are varied to represent the impact of the sphere on a rigid base and also on the simply supported beam. For impact situations in a multibody system, the variations of the contact forces are predicted accordingly. The resulting forces are directly included in the mechanical system's equation of motion, or using an effective mass concept in a two-particle model of system during the period of impact. The system's response is evaluated by integrating the system's equations over the period of contact. The suitability of the prescribed methods is discussed by application of these procedures to some examples, including the joint clearance internal impact response of an aircraft landing gear mechanism and the structural crash response of a complex vehicle model in a rollover situation.

2. Discontinuous Event Methods

The most commonly used method of predicting the impact responses in a mechanical system is the so-called, "discontinuous" or "piecewise" analysis method. At the time of impact, momentum-balance/impulse equations are formulated and solved for the departing velocities or momenta of the system after impact. One basic assumption underlying such analysis is that the impact occurs in such a short duration that the system configuration does not change before and after impact. In this section, the problem and the solution for a two-object impact, such as a two-car collision, are presented first. For determination of the impact response of the system's individual components, a set of momentum-balance/impulse equations are then described for a multibody impact situation. Examples, including a real-world two-car collision accident reconstruction problem (impact of two free objects), Kane's compound pendulum impacting the ground (a simple multibody system impact), and a complete vehicle model undergoing a rollover (a complex multibody system impact) are presented. The results from the analyses are compared to the experimental results from actual tests.

2.1. TWO-BODY IMPACT (TWO-VEHICLE ACCIDENT RECONSTRUCTION)

The basic tools necessary for impact analysis of two free objects are the laws of conservation of momentum and the definitions of the coefficients of restitution and friction. Figure 1 depicts such scenario for a two-car collision. Assumptions underlying the piecewise analysis for such problems include: the tire and external forces are much smaller compared to the contact forces and do not produce any impulses (conservation of momentum holds); the vehicles' masses, moments of inertia, and CG's (centers of gravity or mass) do not change significantly before and after impact; and that the resultant impulse acts at one common point of the two vehicles, known as the "impact center." As Figure 1 shows, a normal-tangential coordinate system, **n-t**, is erected at the point of impact or the impact center C, normal and tangential to the impact surfaces. Right before impact, each car 'i' (i=1,2) has a normal velocity V_{in}^-, a tangential velocity V_{it}^-, and an angular velocity ω_i^-. Each car 'i' has a mass m_i, a radius of gyration k_i, a moment of inertia I_i, and the car's CG is located from the impact center by coordinates (a_i, b_i) on the **n-t** coordinate system. The objective is to determine the departing translational velocities V_{in}^+, V_{it}^+, and the angular velocity ω_i^+ for the two vehicles. During the contact period, contact forces will generate impulses Π_n and Π_t in the normal and tangential directions. The six impulse-momentum equations are:

$$m_1 V_{1n}^- + \Pi_n = m_1 V_{1n}^+ \qquad \text{(a)}$$
$$m_1 V_{1t}^- + \Pi_t = m_1 V_{1t}^+ \qquad \text{(b)}$$
$$I_1 \omega_1^- - \Pi_n a_1 + \Pi_t b_1 = I_1 \omega_1^+ \qquad \text{(c)}$$
$$m_2 V_{2n}^- - \Pi_n = m_2 V_{2n}^+ \qquad \text{(d)}$$
$$m_2 V_{2t}^- - \Pi_t = m_2 V_{2t}^+ \qquad \text{(e)}$$
$$I_2 \omega_2^- + \Pi_n a_2 - \Pi_t b_2 = I_2 \omega_2^+ \qquad \text{(f)}$$

Eliminating the impulses, the resulting momentum-balance equations will of the form:

$$m_1(V_{1n}^+ - V_{1n}^-) + m_2(V_{2n}^+ - V_{2n}^-) = 0 \qquad (1)$$
$$m_1(V_{1t}^+ - V_{1t}^-) + m_2(V_{2t}^+ - V_{2t}^-) = 0 \qquad (2)$$
$$I_1(\omega_1^+ - \omega_1^-) + a_1 m_1(V_{1n}^+ - V_{1n}^-) - b_1 m_1(V_{1t}^+ - V_{1t}^-) = 0 \qquad (3)$$
$$I_2(\omega_2^+ - \omega_2^-) + a_2 m_2(V_{2n}^+ - V_{2n}^-) - b_2 m_2(V_{2t}^+ - V_{2t}^-) = 0 \qquad (4)$$

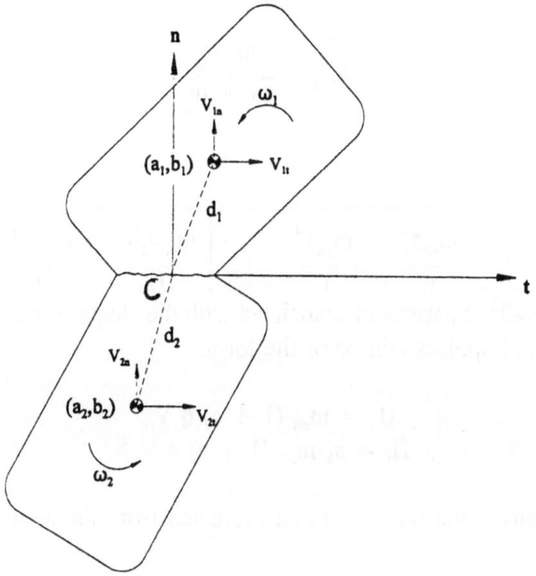

Figure 1. Schematics of a two-car collision.

A coefficient of restitution "e" is defined as the ratio of the departing relative velocity to the approach relative normal velocity as

$$e = -\frac{V_{rn}^{+}}{V_{rn}^{-}} = -\frac{(V_{2n}^{+}-a_2\omega_2^{+})-(V_{1n}^{+}-a_1\omega_1^{+})}{(V_{2n}^{-}-a_2\omega_2^{-})-(V_{1n}^{-}-a_1\omega_1^{-})} \tag{5}$$

This description of the coefficient of restitution is known as the "Newton's hypothesis," which is defined as the ratio of the normal relative departing velocity V_{rn}^{+} to the normal relative approach velocity V_{rn}^{-} of the impact center. The final relationship necessary to solve for the 6 unknown departing velocity is formed by relating the normal and tangential impulses based on the law of limiting friction as

$$\Pi_t = \mu_k \Pi_n \tag{6}$$

where μ_k is the kinetic coefficient of friction. The solution of the six linear equations, (1 through 6), gives the velocities of the two cars after impact as [5]:

$$V_{1n}^{+} = V_{1n}^{-} + m_{eff}(1+e)\, q\, V_{rn}^{-}/m_1 \tag{7}$$
$$V_{1t}^{+} = V_{1t}^{-} + \mu_k m_{eff}(1+e)\, q\, V_{rn}^{-}/m_1 \tag{8}$$
$$\omega_1^{+} = \omega_1^{-} + m_{eff}(1+e)(a_1 - \mu_k b_1)\, q\, V_{rn}^{-}/I_1 \tag{9}$$
$$V_{2n}^{+} = V_{2n}^{-} - m_{eff}(1+e)\, q\, V_{rn}^{-}/m_2 \tag{10}$$
$$V_{2t}^{+} = V_{2t}^{-} - \mu_k m_{eff}(1+e)\, q\, V_{rn}^{-}/m_2 \tag{11}$$
$$\omega_2^{+} = \omega_2^{-} - m_{eff}(1+e)(a_2 - \mu_k b_2)\, q\, V_{rn}^{-}/I_2 \tag{12}$$

where

$$V_{rn}^{-} = (V_{2n}^{-} - a_2\omega_2^{-}) - (V_{1n}^{-} - a_1\omega_1^{-}) \tag{13}$$

is again the relative normal approach velocity of the two cars at the impact center, m_{eff} is the effective or equivalent mass of the system,

$$m_{eff} = \frac{m_i\, m_j}{m_i + m_j} \tag{14}$$

and

$$q = \frac{1}{1 + \frac{m_{eff} a_1^2}{I_1} + \frac{m_{eff} a_2^2}{I_2} - \mu_k \left[\frac{m_{eff} a_1 b_1}{I_1} + \frac{m_{eff} a_2 b_2}{I_2} \right]} \tag{15}$$

is a non-dimensionalized parameter associated with the shape or eccentricity of the two cars. The resulting impulses will be of the form:

$$\Pi_n = m_{eff}\,(1 + e)\, q\, V_{rn}^- \tag{16}$$
$$\Pi_t = \mu_k\, m_{eff}\,(1 + e)\, q\, V_{rn}^- \tag{17}$$

The energy loss during the impact can be evaluated from the approach and departing velocities as

$$\Delta T = \sum_{i=1}^{2} \left[\frac{1}{2} m_i \left(V_{in}^{-2} + V_{it}^{-2} - V_{in}^{+2} - V_{it}^{+2} \right) + \frac{1}{2} I_i \left(\omega_i^{-2} - \omega_i^{+2} \right) \right] \tag{18}$$

The relationships presented here are the basic tools necessary to perform an accident reconstruction of a two-dimensional two-car collision, which constitute majority of the real-world crashes. Similar analysis of a one- or two-object impact can be conducted in three dimensions using Euler angles, Bryant angles, or quaternions [24], for situations that involve a vehicle in flight [25].

As an example of the use of the preceding relationships, the impact configuration of a two-car "side swipe" from an actual test condition is shown in Figure 2 [26]. The data for the striking car, Car 1, include: a mass of 1057 kg, a radius of gyration of 1.25 m, a length of 4.36 m, a width of 1.69 m, a CG position of 1.69 m from the front of the car, an angle of +60 degrees (counterclockwise) from the **n**-axis to the vehicle's heading axis, a distance of 2.00 m from the car CG to the impact center, and an impact velocity of 43 km/hr = 11.9 m/s. Car 2 has a mass of 1031 kg, a radius of gyration of 1.23 m, a length of 4.36 m, a width of 1.69 m, a CG location of 1.69 m from the front end, an angle of -90 degrees from the **n**-axis to the vehicle's heading axis, a distance of 1.55 m from the impact center to the front of the vehicle (maximum crush), a distance of 0.88 m from the CG to the impact center, and an impact velocity of 43 km/h = 11.9 m/s. The coefficient of restitution for this "side swipe" impact was approximate to be 0.32 [6,7]. A friction coefficient for metal-to-metal contact of 0.8 was used for our analysis [6,7]. Based on the analysis method presented, the resulting velocities of Car 1 are: 2.10 m/s, -7.26 m/s, and 0.36 rad/s for the normal, tangential, and angular velocities respectively. For Car 2, the corresponding departing velocities are: 3.90 m/s, 8.78 m/s, and -2.21 rad/s. All the velocities obtained are very close to the velocities obtained in the test as measured from high-speed cameras [26]. The kinetic energy loss calculated from the departing velocities is 65.6 kJ, very close to the measured test value of 66 kJ [26] obtained from the departing measured velocities. Of course in accident reconstruction scenarios, one or more of the *velocities before impact* is usually the unknown, and the velocities after impact are known quantities evaluated form the work-energy principles. In these situations, the presented momentum balance impulse equations, Eqs. (7-12), in these situations are solved for the unknown velocities of the vehicle(s) before impact.

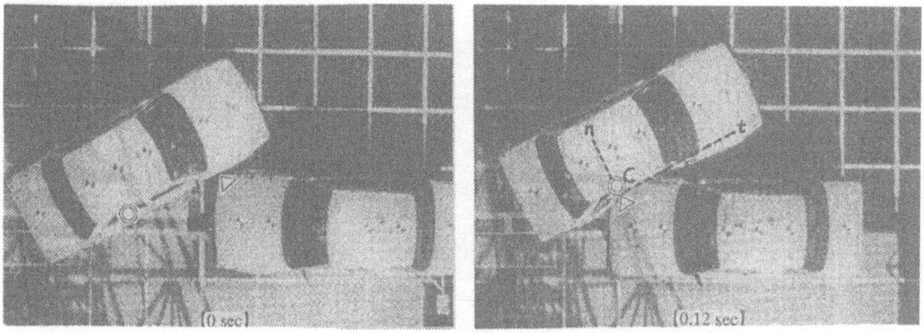

Figure 2. A two-car side swipe test configuration [26].

2.2. MULTIBODY IMPACT

Previous "discontinuous-event" analysis methods of the frictional impact in jointed multibody mechanical systems have shown energy gains in the results partly due to the failure of recognition of the correct impact mode; i.e., sliding, sticking, and reverse sliding, and partly due to the inherent problem in the use of Newton's hypothesis for the definition of the coefficient of restitution. The use of the Newton's hypothesis for the coefficient of restitution might result in energy gains for frictional impact of jointed systems, as pointed out by Kane [16]. In this section, a methodology is presented for the analysis of impact problems with friction in jointed multibody systems using the Poisson's hypothesis. The methods presented here are suitable for impact analysis of constrained or jointed systems such as a truck-and-trailer system, or determination of the structural crash response of the individual components of a mechanical systems, such as the responses of the chassis, suspension systems, steering mechanism of a vehicle in a crash. The formulation shrinks the large system impact problem to a more manageable form that resembles a two-body frictional impact problem. The method also recognizes the correct mode of impact based on the initial slip velocity and the configuration of the impacting bodies at the time of impact.

The formulation is developed using a canonical form of the equation of motion with joint coordinates and joint momenta, which results in a natural way of balancing the momenta. The use of joint coordinates eliminates the complications arisen from the kinematic constraint equations. The configuration of the impact of a system is shown schematically in Figure 3. The canonical-joint form of the equation of motion for a mechanical system is [12,15,27]:

$$\mathbf{p} = \mathbf{M}\dot{\boldsymbol{\theta}} \tag{19}$$

$$\dot{\mathbf{p}} = \mathbf{f} + \dot{\mathbf{M}}\dot{\boldsymbol{\theta}} \tag{20}$$

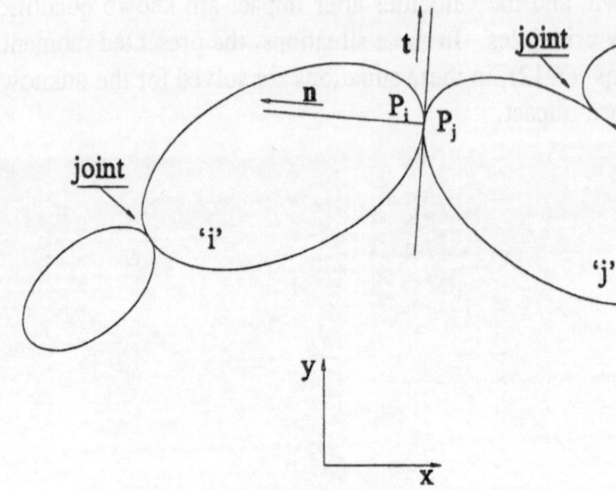

Figure 3. Schematic configuration of a multibody impact.

where **M** is the coordinate-dependent system generalized mass matrix, **f** is the generalized force vector, $\boldsymbol{\theta}$ is the vector of joint coordinates, **p** is the joint momenta, and the symbol '.' performs time differentiation. A normal tangential **n-t** coordinate system is established at the point of impact, perpendicular and parallel to the impacting surfaces. The system relative departing velocities of the contact point (impact center) after impact, V_n^+ and V_t^+, are obtained in terms of the relative velocities before impact V_n^- and V_t^- as [28]:

$$V_n^+ = V_n^- + m_{nn}\Pi_n + m_{nt}\Pi_t \tag{21}$$
$$V_t^+ = V_t^- + m_{nt}\Pi_n + m_{tt}\Pi_t \tag{22}$$

where

$$m_{nn} = c_n^T M^{-1} c_n \tag{23}$$
$$m_{nt} = c_n^T M^{-1} c_t \tag{24}$$
$$m_{tt} = c_t^T M^{-1} c_t \tag{25}$$

are three generalized impact parameters, which are functions of the inertia properties and the configuration of the system at the time of impact; c_n and c_t are the composite normal and tangential vectors for all the coordinates; and $\Pi_n = \Pi_n(t_f)$ and $\Pi_t = \Pi_t(t_f)$ are the accumulated total normal and tangential impulses at the end of the period of contact t_f (final time). Variables m_{nn} and m_{tt} are always positive while m_{nt} could have either sign. In order to update the relative departing velocities V_n^+ and V_t^+, the impact parameters m_{nn}, m_{nt}, m_{tt}, and also the composite normal-tangential vectors c_n and c_t are evaluated in terms of the system state at the time of impact. The impulses Π_n and Π_t are determined based on the Routh's graphical method [29], which is described next.

Even though the impact is assumed to occur instantaneously, for the evaluation of the normal and tangential impulses, the time axis is temporarily expanded. In general, the contact may be considered to occur in two phases: the compression phase and the restitution phase. During the compression phase, the relative normal velocity of the two bodies in contact is reduced to zero. The end of the compression phase is referred to as the instant of maximum compression, or maximum approach. The restitution phase starts at this point and lasts until the two bodies in contact separate. Let us denote the instant of initial contact t^-, the instant of maximum compression t^m, and the instant of departure or restitution t^+ or t_f. The Poisson's hypothesis is used for the definition of coefficient of restitution "e" as the ratio of the accumulated normal impulse at the end of the restitution phase $\Pi_n(t_r)$, from t^m to t^+, to the accumulated normal impulse at the end of the compression phase $\Pi_n(t_c)$, from t^- to t^m; i.e.,

$$e = \frac{\Pi_n(t_r)}{\Pi_n(t_c)} \tag{26}$$

The accumulated impulse during the contact period is plotted on the Π_n-Π_t plane for different modes of impact, as shown in Figure 4 [28,17].

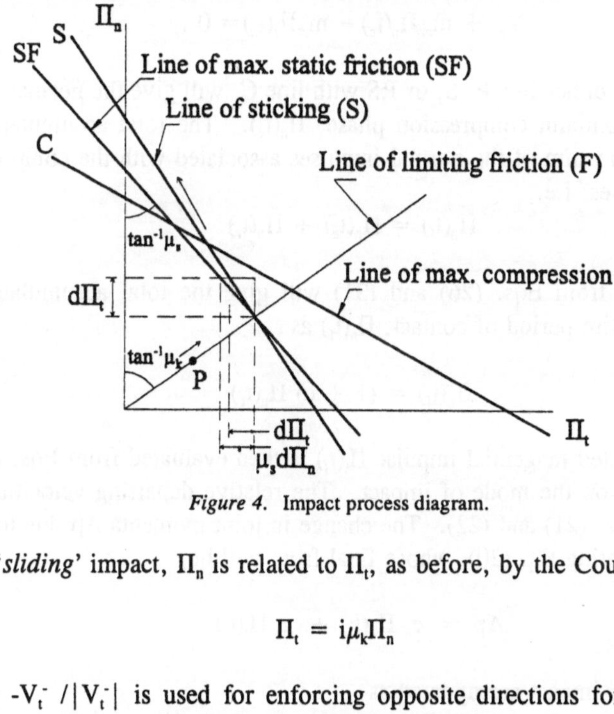

Figure 4. Impact process diagram.

In case of '*sliding*' impact, Π_n is related to Π_t, as before, by the Coulomb's law as

$$\Pi_t = i\mu_k\Pi_n \tag{27}$$

where $i = -V_t^- / |V_t^-|$ is used for enforcing opposite directions for Π_t and the slip velocity V_t^-, and μ_k is the kinetic coefficient of friction. A plot of Eq. (27) is represented by the line of limiting friction (line F) in Figure 4. In case of a '*sticking*' impact, the departing relative tangential velocity is zero. Hence by setting $V_t^+ = 0$, Eq. (22) reduces to:

$$V_t^- + m_{nt}\Pi_n(t_f) + m_{tt}\Pi_t(t_f) = 0\ , \tag{28}$$

which is represented by the line of sticking (line S) in Figure 4. Another mode of impact is the '*reverse sliding*' mode, when the slip velocity changes its direction during the impact. This could happen if after following line F, the impact reaches line S and the incremental impulse $d\Pi_t$ on line S, is greater than $\mu_s d\Pi_n$ (μ_s is the static coefficient of friction), as shown in Figure 4. The reversal of the slip velocity also causes the reversal in the direction of $d\Pi_t$. Thus in case of reverse sliding, Coulomb's law governs as (line RS in Figure 4)

$$\Pi_t - \Pi_{ts} = - i\mu_k(\Pi_n - \Pi_{ns}) \qquad (29)$$

where Π_{ns} and Π_{ts} are the values of Π_n and Π_t at the intersection of lines F and S (for sticking).

Thus the impact will follow one of the lines F, S, or RS depending on the mode of impact corresponding to 'sliding', 'sticking', or 'reverse sliding' respectively. To determine the accumulated impulse, a line C, which represents the end of the compression phase in Figure 4, is drawn. At the end of the compression phase, the normal relative velocity is zero. Hence, line C, representing the line of maximum compression, is obtained by taking $V_n^+=0$ in Eq. (21), which gives:

$$V_n^- + m_{nn}\Pi_n(t_c) + m_{nt}\Pi_t(t_c) = 0 \, . \qquad (30)$$

The intersection of either line F, S, or RS with line C, will give the normal impulse at the end of the maximum compression phase, $\Pi_n(t_c)$. The total accumulated normal impulse $\Pi_n(t_f)$ is the sum of the normal impulses associated with the compression and the restitution phases; i.e.,

$$\Pi_n(t_f) = \Pi_n(t_c) + \Pi_n(t_r) \qquad (31)$$

Eliminating $\Pi_n(t_r)$ from Eqs. (26) and (31) will give the total accumulated normal impulse for the entire period of contact, $\Pi_n(t_f)$ as

$$\Pi_n(t_f) = (1 + e)\, \Pi_n(t_c) \qquad (32)$$

The total accumulated tangential impulse $\Pi_t(t_f)$ is then evaluated from Eqs. (27), (28), or (29) depending on the mode of impact. The relative departing velocities are then evaluated from Eqs. (21) and (22). The change in joint momenta $\Delta\mathbf{p}$ due to impact is obtained by integrating Eq. (20), whose final from will be,

$$\Delta\mathbf{p} = \mathbf{c}_n\, \Pi_n(t_f) + \mathbf{c}_t\, \Pi_t(t_f) \, . \qquad (33)$$

The momenta after impact are updated as

$$\mathbf{p}^+ = \mathbf{p}^- + \Delta\mathbf{p} \, . \qquad (34)$$

The joint velocities $\dot{\theta}$ are also updated from Eq. (19), and the analysis of the system's state is resumed with the new momenta (or velocities).

Based on the path that an impact follows on the "impact process diagram", Figure 4, the impact can be classified into seven different cases as indicated in Table 1. This classification is for both direct and oblique impacts and is based on the mode of impact: sliding, sticking, and reverse sliding; and also whether the mode of impact occurs in the compression or the restitution phase. Note that for any impact situation, the case of impact can be identified from the conditions listed in the table, which can be evaluated from the impact geometry, speed, and configuration.

TABLE 1. Seven possible cases of impact.

Impact Type		Oblique Impact $V_t^- \neq 0$		Direct Impact $V_t^- = 0$
Conditions	$0 < \Pi_{ns} < \Pi_{nc}$ or: $\Pi_{nc} < 0$	$0 < \Pi_{nc} < \Pi_{ns}$ $= < (1+e)\Pi_{nc}$	$0 < (1+e)\Pi_{nc} < \Pi_{ns}$ or: $\Pi_{ns} < 0$	$\Pi_{ns} = 0$
$\lvert m_{nt} \rvert / m_{tt} < \mu_s$	**Case 1**	**Case 2**	**Case 5**	**Case 6**
$\lvert m_{nt} \rvert / m_{tt} >= \mu_s$	**Case 3**	**Case 4**	**Case 5**	**Case 7**

Case 1: sliding and sticking in the compression phase
Case 2: sliding and sticking in the restitution phase
Case 3: sliding and reverse sliding in the compression phase
Case 4: sliding and reverse sliding in the restitution phase
Case 5: forward sliding
Case 6: sticking with no relative approach tangential velocity
Case 7: sliding with no relative approach tangential velocity

For illustration of the methodology, two impact problems are solved using the presented methodology. In the first, the impact of a compound or double pendulum striking the ground is considered, as shown in Figure 5, (a simple jointed multibody system impact). Kane [16] used this problem to point out the energy gain obtained by using the formulation based on the Newton's hypothesis. A sample of the results for the cases followed during the impact and the energy loss in terms of the coefficients of restitution and friction are shown in Figure 6. The results from this study was also verified using a high-speed Ektapro camera capable of analyzing 1000 frames per second. The results obtained from the present study have indicated that by considering the correct mode of impact and using the Poisson's hypothesis (current study) instead of the Newton's hypothesis [15], energy gains can be avoided. It has also been observed that when the slip velocity does not change its direction during the impact (such as the two-car collision example presented earlier), the approaches based on Newton's or Poisson's hypotheses give identical results.

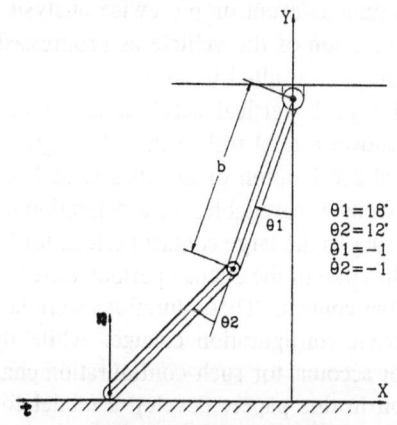

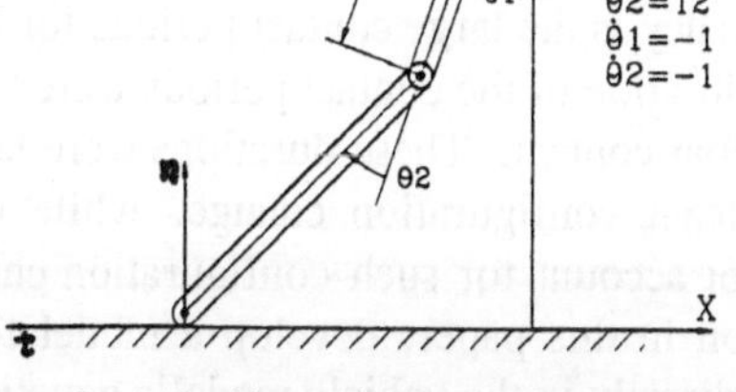

Figure 5. A double pendulum striking the ground.

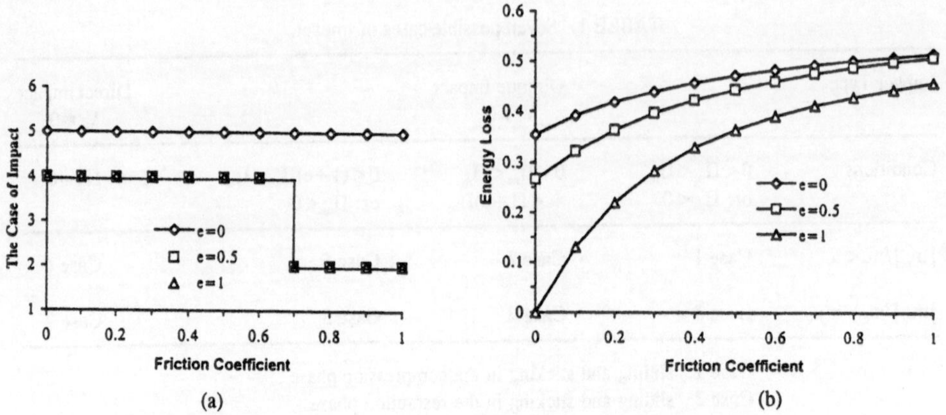

Figure 6. Sample results: (a) Cases of impact, (b) percentage energy loss.

As a second example, a study of a vehicle rollover was conducted. A simulation was performed to duplicate the experimental rollover test of a vehicle at 30 mph on a FMVSS 208 rollover cart impacting a water-filled decelerator system, thereby throwing the vehicle off the cart [30]. The initial roll angle was 23 degrees. The vehicle, at the time of departure, had a velocity of approximately 25 mph in the translational direction and 1.5 rad/s in the roll direction, as shown in Figure 7. A number of different analyses with rigid and flexible bodies have been performed on this vehicle [30,31]. An additional simulation related to the topic of discontinuous-event analysis of frictional impact in multibody mechanical systems, presented here, was also performed and the results were compared to those from the experiment. The model of the vehicle included 13 bodies including: the chassis combined with the rollbar cage, 4 wheels combined with the hubs, two double A-arm suspension systems in front, two single A-arms for the rear suspension. Shock absorbers with nonlinear behavior were included in the model in front and rear, and the tire model was a simple unilateral nonlinear spring model. For the tire contact on the ground, a coefficient of restitution of 0.6 and a friction coefficient of 1.0 were estimated from the results of the actual tests. Similarly for the rollbar cage contact of the ground, a coefficient of restitution of 0.8 and a friction coefficient of 1.2 were estimated. A discontinuous-event or piecewise analysis was conducted on the vehicle model. The configuration of the vehicle as progressed during the rollover is shown in Figure 8. The analysis resulted in a total roll of the 247 degrees, a settling time of 2.4 seconds, and a peak vertical acceleration of the chassis of 9G. The experimental results have shown a total roll of the 427 degrees, a settling time of 3.0 seconds, and a peak vertical acceleration of the chassis of 15G. Although the results for the positions and paths seemed reasonable, the acceleration did not have close match. The reason for this discrepancy is the large contact periods for both the tires contact and the rollbars contact. The duration of the contact periods were 0.5 s for first tire contact, and 0.25 s for the first rollbar contact. These durations were large enough such that the vehicle underwent significant configuration change, while the presented piecewise analysis presented does not account for such configuration change. We will consider this example again later on in this paper, develop a model for the generated contact forces, include the model directly in the vehicle model's equations of motion, and make further comparisons of the analyses with the experimental results.

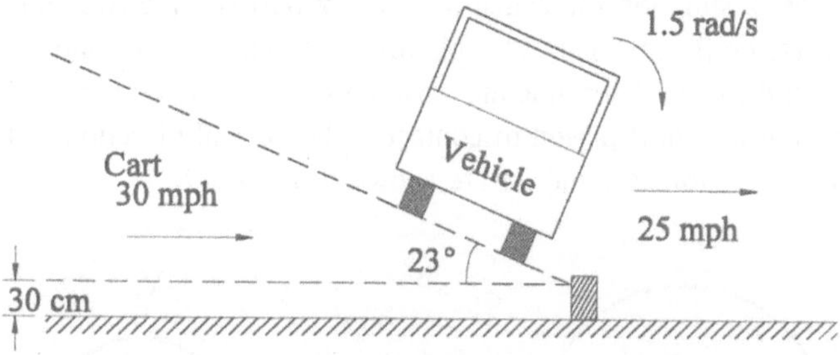

Figure 7. **Initial configuration of the vehicle for the rollover test.**

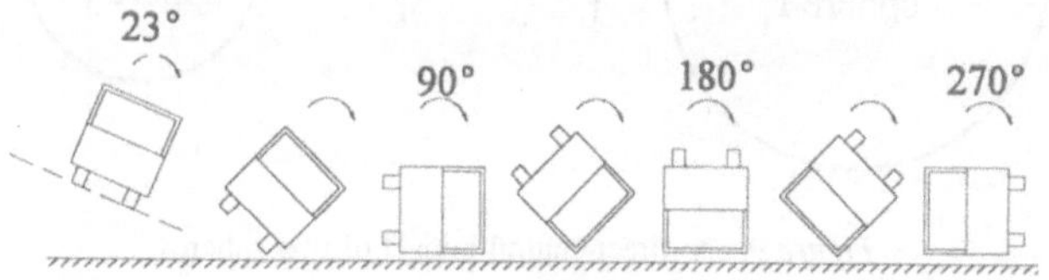

Figure 8. **Rollover configuration of the vehicle predicted from the "piecewise" analysis.**

3. Continuous Event Methods

3.1. MODELING CONTACT FORCES

When two solids are in contact, deformation takes place in the local contact zone resulting in a contact force. This suggests that the contact force is directly related to the amount of local deformation or indentation of the two solids. The best-known force model for the contact between two spheres of isotropic material was developed by Hertz based on the theory of elasticity [20]. With radii R_i and R_j of the two spheres 'i' and 'j', and masses m_i and m_j, the contact force f follows the relation

$$f = K\,\delta^n\,, \tag{35}$$

where δ is amount of the relative penetration or indentation between the surfaces of the two spheres and n = 1.5. The generalized impact parameter K depends on the material properties and the radii of the spheres:

$$K = \frac{4}{3\pi(h_i + h_j)}\sqrt{\frac{R_i R_j}{R_i + R_j}}\,, \tag{36}$$

where the material parameters h_i and h_j are

$$h_l = \frac{1 - \nu_l^2}{\pi E_l}\,; \quad l = i, j\,. \tag{37}$$

Variables ν_l and E_l are, respectively, the Poisson's ratio and the Young's modulus associated with each sphere.

Consider now a situation for which the contact between the two spheres is caused by a direct-central collision. The two spheres have velocities V_i^- and V_j^- right before impact. We would like to determine the variations of the interaction force between the two spheres during the short period of contact. The normal direction **n** to the contact surfaces and a pair of forces f and -f are shown in Figure 9.

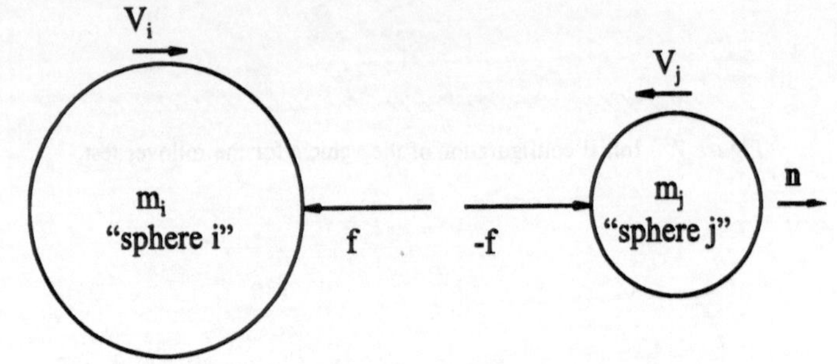

Figure 9. A direct-central impact of two spheres.

Generally, the two spheres will not rebound with the same initial velocities, because part of the initial kinetic energy is dissipated in the form of permanent deformation, heat, etc. It is apparent that the contact force model of equation (35) cannot be used during both phases of contact, compression and restitution, since this would suggest that no energy is dissipated in the process of impact. One popular treatment is based on the idea that dissipation of energy occurs in the form of internal damping of colliding solids. This assumption is valid for low impact velocities; i.e., those impact situations for which impact velocities are negligible compared to the propagation speed of deformation waves across the solids. The contact force model will then be in terms of a damping coefficient D,

$$f = K\,\delta^n + D\,\dot{\delta} \tag{38}$$

where $\dot{\delta}$ is the relative (or indentation) velocity of the solids. A hysteresis form for the damping coefficient was proposed by Hunt and Grossley [21] as

$$D = \eta\,\delta^n\,, \tag{39}$$

where the parameter η is called the "hysteresis damping factor." The contact force model of equation (38) may be used for the entire period of contact. With this model, the energy loss is assumed to be due to the material damping of the bodies, which would dissipate energy in the form of heat. For known parameters K and D (or η), the shape of the hysteresis loop corresponding to this force model and the solution corresponding to the variations of the indentation with time are shown in Figure 10 [32]. In this figure, variables δ_m and f_m refer to the values of indentation and the contact force at the instant of maximum compression t^m. In the contact force model of Eq. (38), the damping coefficient D or the hysteresis damping factor η must be determined. An estimate of the parameter η based on the classical impulse-momentum equation and the work-energy principle can be determined.

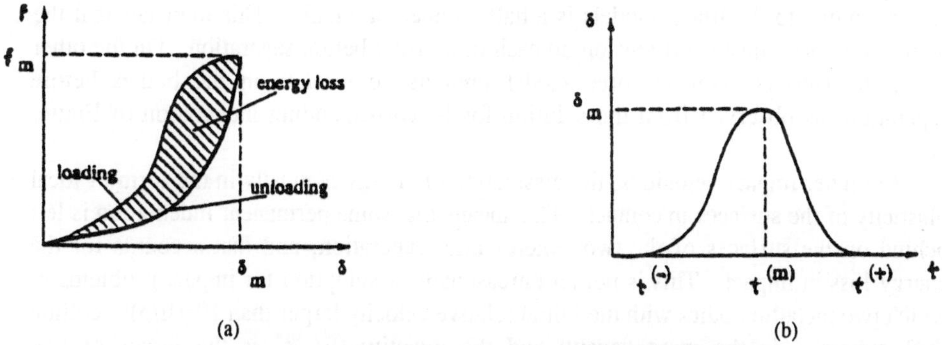

Figure 10. Hertz contact force model with hysteresis damping:
(a) contact force versus indentation, and (b) indentation versus time.

From consideration of the kinetic energies before and after impact, the energy loss ΔT may be expressed in terms of the coefficient of restitution 'e' and the relative approach velocity $\dot{\delta}^- = V_i^- - V_j^-$ as

$$\Delta T = \frac{1}{2} m_{eff}\, \dot{\delta}^{-2} (1 - e^2)\,, \tag{40}$$

where again m_{eff}, given by Eq. (14), is the system's equivalent or effective mass. The energy loss may also be expressed by integration of the contact force around the hysteresis loop as [33]

$$\Delta T = \oint D\dot{\delta}\, d\delta = \oint \eta\, \delta^n\, \dot{\delta}\, d\delta \cong 2 \int_0^{\delta_m} \eta\, \delta^n\, \dot{\delta}\, d\delta = \frac{2}{3} \frac{\eta}{K} m_{eff}\, \dot{\delta}^{-3}\,, \tag{41}$$

where $\oint$ refers to the integration around a hysteresis loop for a contact force of the form shown in Figure 10(a). The hysteresis damping factor η may be evaluated by comparing the right sides of equations (40) and (41),

$$\eta = \frac{3K(1 - e^2)}{4\dot{\delta}^-}\,, \tag{42}$$

which shows a direct relationship between the coefficient of restitution and an equivalent damping factor. The contact force in conjunction with the damping representation may be written in an alternative form as

$$f = K\, \delta^n \left[1 + \frac{3(1 - e^2)}{4} \frac{\dot{\delta}}{\dot{\delta}^-} \right], \tag{43}$$

which shows the effect of impact speed and the coefficient of restitution on the variations of the contact force.

The clear advantage of the Hertz contact force model, $f = K\delta^n$, with its damping representation in equation (43) over the Kelvin-Voigt viscoelastic model is its nonlinearity. The overall pattern of impact is far from linear, while the Kelvin-Voigt model and its damping representation are linear models. The solution for indentation corresponding to the linear models is a half-damped harmonic. This indicates that the bodies in impact must exert tension on each other right before separation. On the other hand, the Hertzian contact force model predicts no tension on the bodies before separation, as observed from the solution for its corresponding indentation of Figure 10(b).

At higher impact velocities, the dissipation of energy is mostly in the form of local plasticity of the surfaces in contact. This means that some permanent indentation is left behind on the surfaces of the two spheres after separation, and that accounts for the energy loss in impact. This is not an unreasonable assumption for impact problems in which two metallic bodies with the initial relative velocity larger than $10^{-5}(E/\rho)^{0.5}$ collide [34], where ρ is the mass density and the quantity $(E/\rho)^{0.5}$ is the larger of two propagation speeds of the elastic deformation waves in the colliding solids. With this condition, the contact force loads according to equation (35) during the compression period, and unloads according to

$$f = f_m \left[\frac{\delta - \delta_p}{\delta_m - \delta_p}\right]^n \quad \text{during restitution,} \tag{44}$$

where variable δ_p is the permanent indentation of the two spheres after separation. The shape of the hysteresis loop corresponding to this contact force model and the solution corresponding to the variation of the indentation with time are shown in Figure 11.

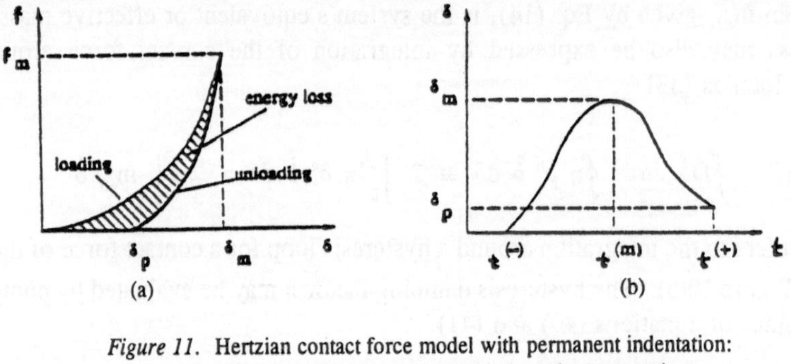

Figure 11. Hertzian contact force model with permanent indentation: (a) contact force versus time, and (b) indentation versus time.

The proposed contact force model can be used for the impact between two spheres, if the parameters in the model are known. The generalized parameter K may be evaluated from the radii and the material properties of the two spheres using equation (36). The remaining parameters are δ_m, f_m, δ_p, which can be determined by integrating the relative indentation equations of motion twice, substitution in the contact force expression, and integrating the contact force around the hysteresis loop and equating it to the kinetic energy loss, as [32]

$$\delta_m = \left[\frac{n+1}{2K}\, m_{eff}\, \dot{\delta}^{-2}\right]^{\frac{1}{(n+1)}} \tag{45}$$

$$f_m = K\, \delta_m^n \tag{46}$$

$$\delta_p = \frac{(n+1)\, m_{eff}\, \dot{\delta}^{-2}}{2\, f_m}\,(1 - e^2)\ . \tag{47}$$

Hence maximum indentation of the two spheres and maximum contact force depend on the material properties, masses, radii, and velocities of the two spheres right before impact. The permanent indentation is evaluated from the initial approach velocities of the spheres and a known coefficient of restitution between the spheres.

A continuous analysis may now be performed by numerically integrating the equations of motion of the two spheres forward in time in conjunction with the developed contact force model. A solution is thus obtained in the form of positions, velocities, and accelerations of the spheres at any instant of time during the contact period. As a by-product of the preceding parameter identification process, one can approximate the duration of the contact period between the two spheres as [35]

$$\Delta t \cong 2\frac{\delta_m}{\dot{\delta}^-}\int_0^1 \frac{dz}{\sqrt{1 - z^{n+1}}} = 2.94\,\frac{\delta_m}{\dot{\delta}^-}\,. \tag{48}$$

which illustrates that the duration of the contact can be predicted from the initial speed of the impact and the geometry and material properties of the surfaces in contact.

The preceding discussion on the evaluation of the contact forces can easily be generalized to impact situations within a multibody system, as illustrated by the impact of two bodies 'i' and 'j' in the system of Figure 3. The points of contact on the two bodies are P_i and P_j, $\mathbf{n}$ is a unit vector in the normal direction, and $\mathbf{t}$ is a unit vector in the tangential direction to the contact surfaces of the two bodies. No matter which type of coordinates are used to assemble the equations of motion for the multibody system, the coordinates and the velocities of the bodies can be calculated, at any instant of time, from the solution of the equations of motion. For a known system configuration at the initial time of contact, the location of the contact points $\mathbf{r}_i^{P-}$ and $\mathbf{r}_j^{P-}$ and the components of the algebraic unit vectors $\mathbf{n}$ and $\mathbf{t}$, with respect to a non-moving xyz coordinate system, may be calculated. From the known body velocities, velocities of the contact points in the xyz coordinate system, $\dot{\mathbf{r}}_i^{P-}$ and $\dot{\mathbf{r}}_j^{P-}$, may also be calculated at that time. Hence, the indentation and the indentation velocity at the initial time of contact are

$$\delta^- = 0 \tag{49}$$

$$\dot{\delta}^- = \mathbf{n}^T\,(\dot{\mathbf{r}}_i^{P-} - \dot{\mathbf{r}}_j^{P-}) \tag{50}$$

in which the symbol T performs the transpose operation and velocities of the contact points are projected in the normal direction to the contact surface. The expression in equation (36) for the parameter K can be used for the contact between any two bodies if the local surfaces of contact are both spherically shaped. Similar expressions have been obtained by Hertz [20] and others [18,36] for other shapes of the local contact surfaces such as sphere on plane, parallel cylinders, and plane on plane. Once the generalized parameter K is calculated, with a given coefficient of restitution "e" and the initial approach velocity $\dot{\delta}^-$, all other parameters in the contact force model can be determined. The friction force is then evaluated from the normal contact forces and the law of limiting friction. With known variations of the contact and friction forces during the contact period, a continuous analysis of the system can be performed simply by adding these forces to the multibody system equations of motion. This analysis method provides accurate results, since all of the equations of motion are integrated over the period of contact. It thus accounts for the changes in the configuration and the velocities of the system during that period. To avoid computational inaccuracies and inefficiencies associated with the integration of the system equations of motion over the period of contact, an effective mass concept for a two-particle model of the system during the contact period as well as scaling of the time axis are sometimes performed [32].

3.2. EXPERIMENTAL VALIDATION

For verification of the developed contact force models, impact tests were conducted on aluminum and steel plates with a steel indentor [37]. The configuration of the impact tests setup is shown in Figure 12. The plates were kept flat on a rigid base and clamped thus avoiding all bending and shearing effects. Since the radius of curvature of the plate is infinite, the generalized contact parameter K in Eq. (36) reduces to

$$K = \frac{0.424\sqrt{R}}{h_i + h_j} \tag{51}$$

where R is the radius of the steel impacting mass. Also, since the plate is stationary and fixed, its mass tends to infinity, and hence the system effective mass is the mass of the impactor; i.e.,

$$m_{eff} = m \tag{52}$$

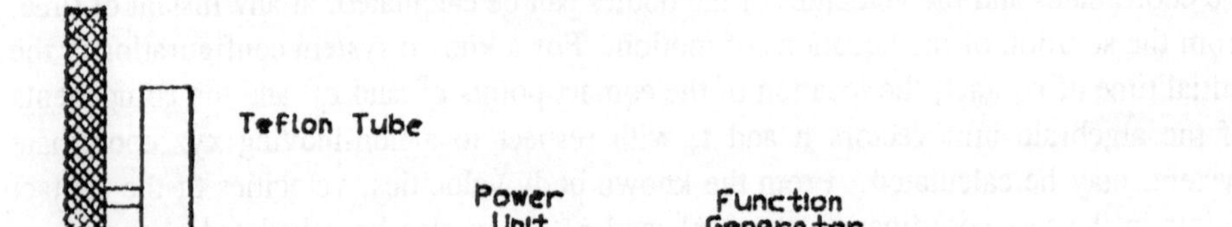

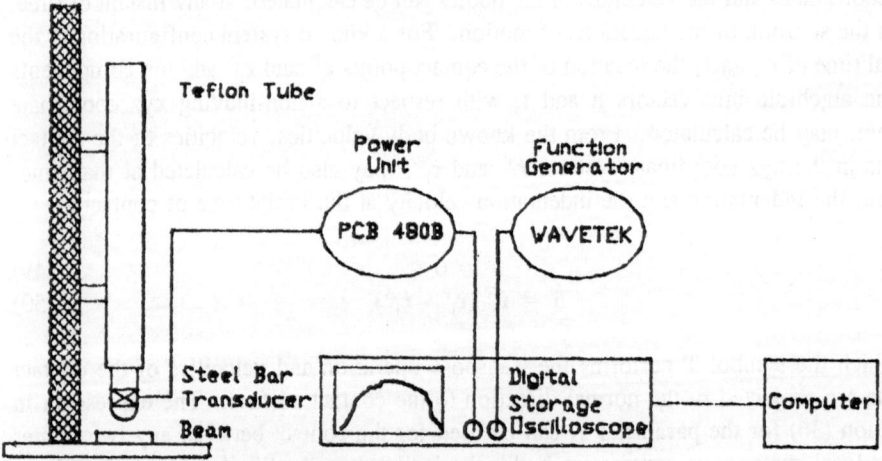

Figure 12. Schematics of the drop tower instrumentation.

The steel impactor was dropped from different heights normally onto the plate. The impact velocity of the impactor was determined from the dropped height h as

$$V = \dot{\delta} = (2gh)^{0.5} \tag{53}$$

For the small range of the drop heights, the effect of aerodynamic drag is minimal; hence, Eq. (53) provides a good measure of the required height for a desired impact velocity. Nonetheless, a high-speed camera (1000 frames per second) was used in the experiments to evaluate the velocity of impact even more accurately. The impact force pulse was sensed through a transducer bonded to the impactor. The impact force was digitized and stored in a computer. The data was then processed through a computer program to yield the acceleration, velocity and displacement by numerical integration, and other impact parameters δ, δ_p, δ_m, and e were determined. The indentation diameter 'd' on the target plate was measured using a caliper. As shown in Figure 13, the permanent indentation δ_p was calculated from the radius of indentation r and the radius of the impactor R as

$$\delta_p = 0.5r^2/R \tag{54}$$

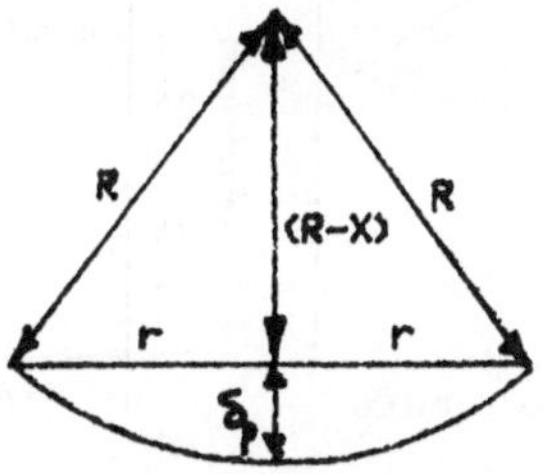

Figure 13. Geometry of the indentation diameter and radius of the indentor.

The coefficient of restitution "e" was evaluated from the ratio of velocity of the impactor after to before impact, or as the square of the ratio of the rebound height to the drop height, as Eq. (53) suggests. Note that the use of Newton's hypothesis for the definition of the coefficient of restitution provided consistent results for this two free object impact situation. The contact forces generated were then modelled as described earlier, and the results form the theory and experiment were then compared.

Figure 14 shows the results for an impact velocity of 2.0 m/s on an aluminum plate. As both the contact force versus time and contact force versus deformation plots show, Figures 14(a) and (b), the theoretical Hertzian contact force model with permanent indentation matches closely the experimental results. The power index 'n' in Eq. (35) was found to be 1.70. The displacement and velocity response of the impactor are shown in Figures 14(c) and (d). From the velocity plot, it can be observed that the coefficient of restitution is 0.54. A permanent indentation of $0.51(10)^{-4}$ m was measured, as observed from the displacement plot. Experiments were repeated with plates made of several other materials instead of aluminum. In each case, the models showed reasonable correlation with the experiments. A sample of the results for an impact velocity of 2.0 m/s on a steel plate for example includes a coefficient of restitution e, a permanent indentation δ_p, and a power index n were 0.47, $0.54(10)^{-4}$ m, and 1.78 respectively. Similar tests were conducted at other impact velocities. Figure 15 shows a summary of the tests results including the coefficient of restitution e, contact duration Δt, power index n, and the permanent indentation δ_p at different drop heights h or velocities $\dot{\delta}^{-}$ on an aluminum plate. As observed, with an increase in the impact velocity, the coefficient of restitution slightly reduces, the contact period reduces, the permanent deformation increases, and the power index nearly remains the same.

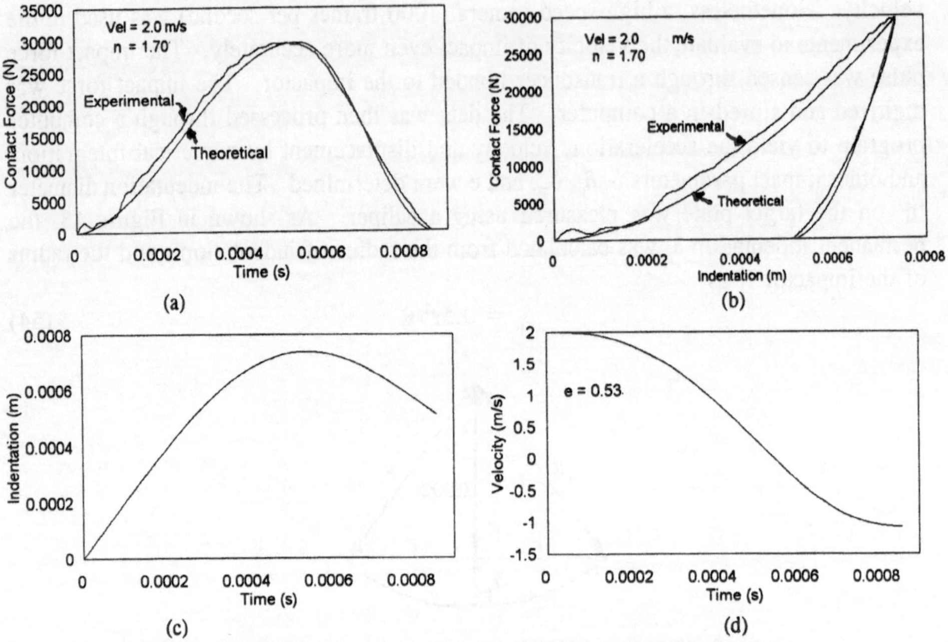

Figure 14. **Impact on an aluminum plate with an impact velocity of 2.0 m/s: (a) contact force versus time, (b) contact force versus indentation, (c) indentation response, (d) impact velocity response.**

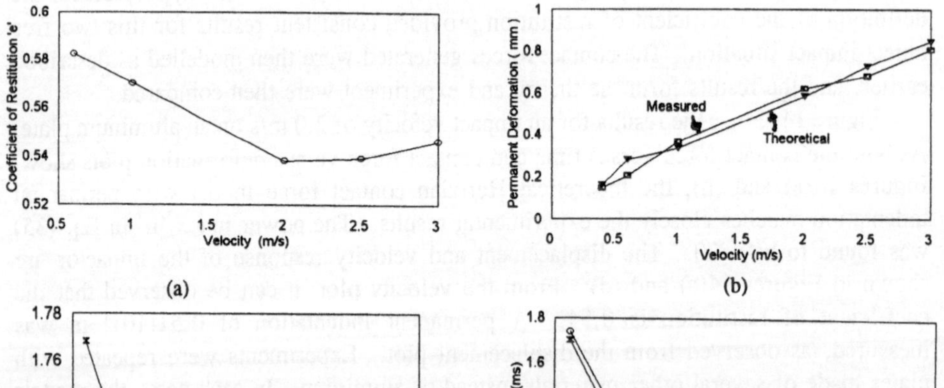

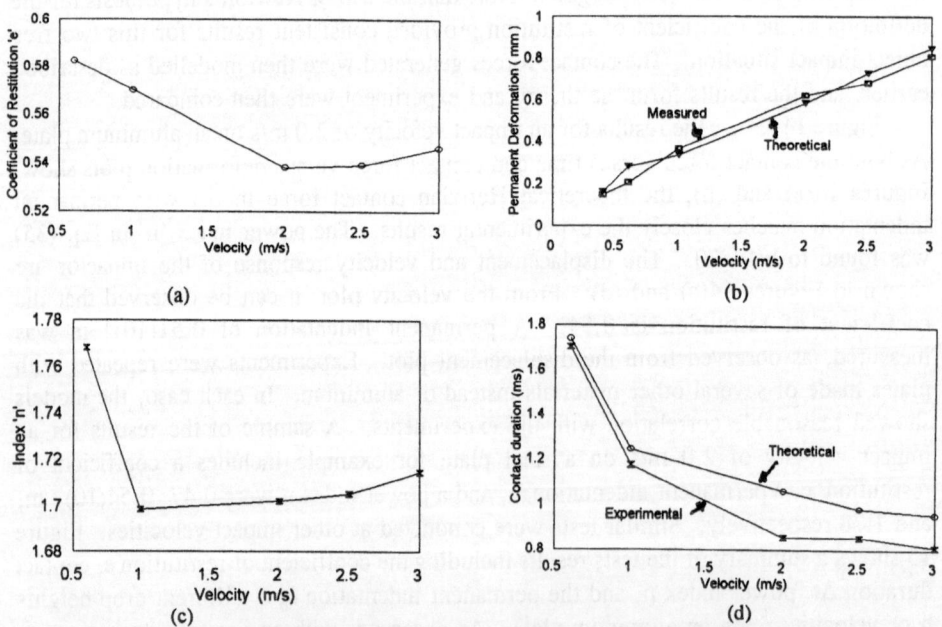

Figure 15. **Impact on an aluminum plate with different impact velocities: (a) coefficient of restitution versus impact velocity, (b) permanent indentation versus impact velocity, (c) power index versus impact velocity, (d) contact duration versus impact velocity.**

To investigate the coupling between the local or small deformation (indentation) and large or global deformation of a beam or plate, the fixed plate was replaced with a beam simply supported at the two ends. The contact force due to the transverse impact of a spherical mass on the center of the beam was modelled. As shown in Figure 16, the local indentation $\delta = (f/K)^{1/n}$ due to the contact force has the contribution from three components. The first component is due to the free-fall displacement of the sphere. The second component comes from the retardation or contact force, which must be subtracted from the first term. The third term is the contribution due to the displacement or vibration of the beam, which again needs to be subtracted from the first term. Hence the total indentation δ in this case is [38]

$$\delta = \left[\frac{f(t)}{K}\right]^{1/n} = V_0 t - \frac{1}{m}\int_0^t f(\tau)(t-\tau)d\tau - \sum_{i=1,3,...}^{\infty} \frac{1}{i^2\omega_1 M}\int_0^t F(\tau)\sin(i^2\omega_1(t-\tau))d\tau \quad (55)$$

where V_0 is the impact velocity, t is the time, m is the mass of the sphere, f(t) is the contact force generated, M is the mass of the beam, K is the generalized impact parameter given by Eq. (51), and ω_1 is the fundamental frequency of vibration of the beam,

$$\omega_1 = \frac{\pi^2}{l^2}\sqrt{\frac{E_2 I_2}{A_2 \rho_2}} \quad (56)$$

where E_2, I_2, A_2, ρ_2, and l_2 respectively refer to the beam's Young's modulus, cross-sectional second moment of area, cross-sectional area, mass density, and length. Using the drop tests tower of Figure 12, tests were conducted on a simply supported aluminum beam. The contact force variation was measured from the setup. The displacement of the beam was numerically evaluated from the last term in Eq.(55).

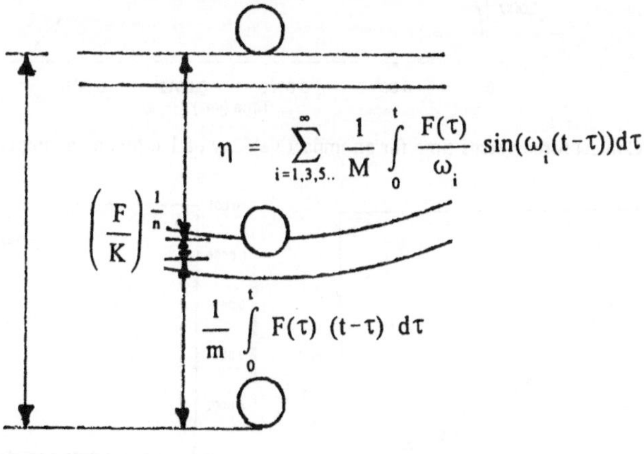

Figure 16. Description of the different displacements for impact of a sphere on a simply supported beam.

Figure 17 shows the variation of the contact force versus time at an impact velocity of 1.6 m/s (a drop height of 13 cm). The ratio of the contact force to the mass of the impactor gives the acceleration of the impacting mass with which successive integration is conducted to obtain velocity and displacement of the impacting mass, as displayed in Figure 18. The coefficient of restitution was evaluated as before to be 0.85 using the square of the ratio of the rebound height to the drop height as measured from the high speed photography data. The displacement of the beam was evaluated numerically from the last term in Eq. (55), which is shown in Figure 19. The local deformation or indentation is obtained as the difference between the indentor displacement (Figure 18(b)) and the beam displacement (Figure 19). The impactor local deformation or indentation response is shown in Figure 20. The power index "n" was evaluated from the maximum contact force and the maximum indentation to be equal to 2.12. Finally, the generated contact force was modelled using the Hertzian contact force model both with permanent indentation and with internal damping. A comparison between the experimental results on the contact force and the two models are shown in Figure 21. The permanent indentation model does not predict the contact force variation as well as the damping model since the contact is occurring at a relatively slow speed. The damping model has a better representation of the energy-dissipation phenomenon in this situation, as most of the energy of the impact is dissipated by the system's internal damping. The energy dissipated in the process of impact was evaluated numerically integrating the force-indentation curve to be equal to 1.7 J. The strain energy going into the vibration of the beam was evaluated to be very minimal, 10^{-5} J.

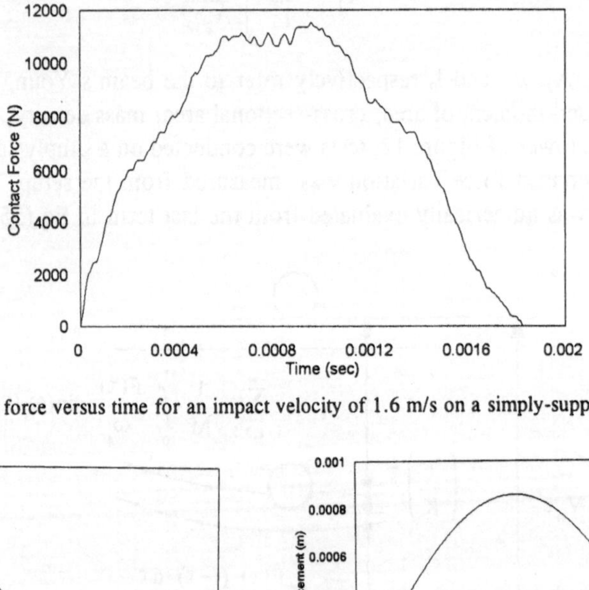

Figure 17. Contact force versus time for an impact velocity of 1.6 m/s on a simply-supported beam.

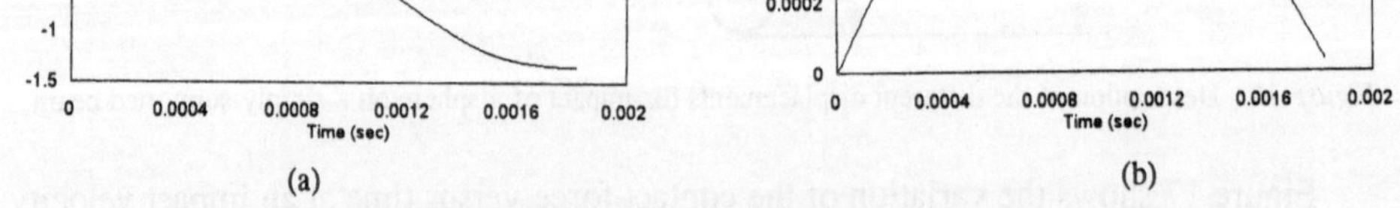

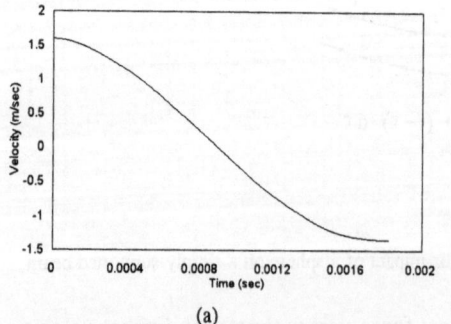

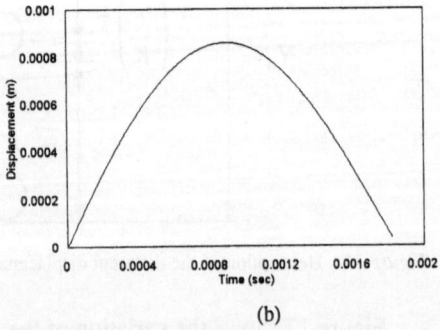

Figure 18. Impactor response for an impact velocity of 1.6 m/s on a simply-supported beam: (a) impactor velocity versus time, (b) impactor displacement versus time.

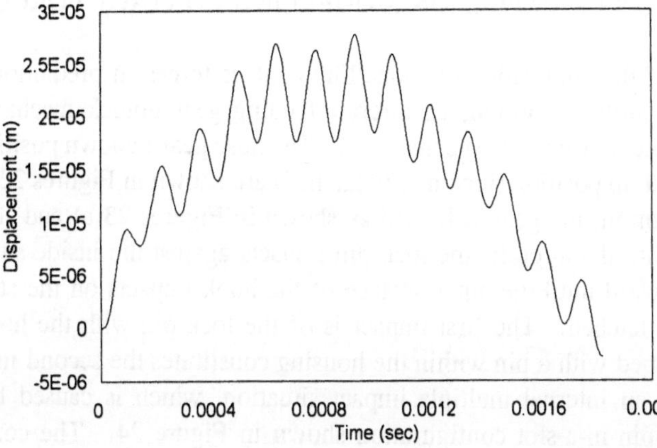

Figure 19. **Beam displacement for an impact velocity of 1.6 m/s on a simply-supported beam.**

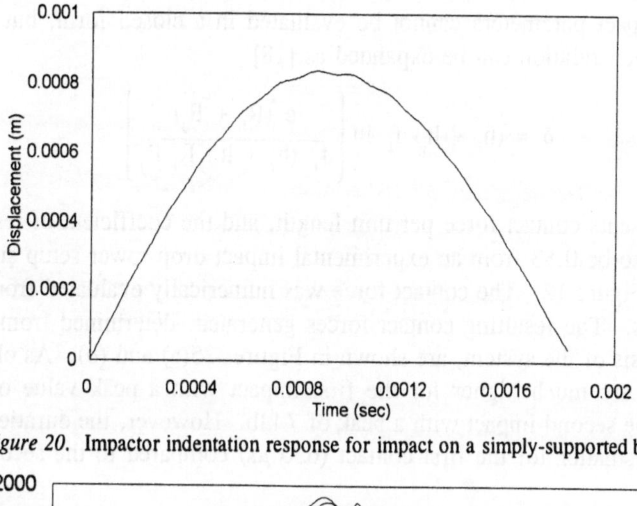

Figure 20. **Impactor indentation response for impact on a simply-supported beam.**

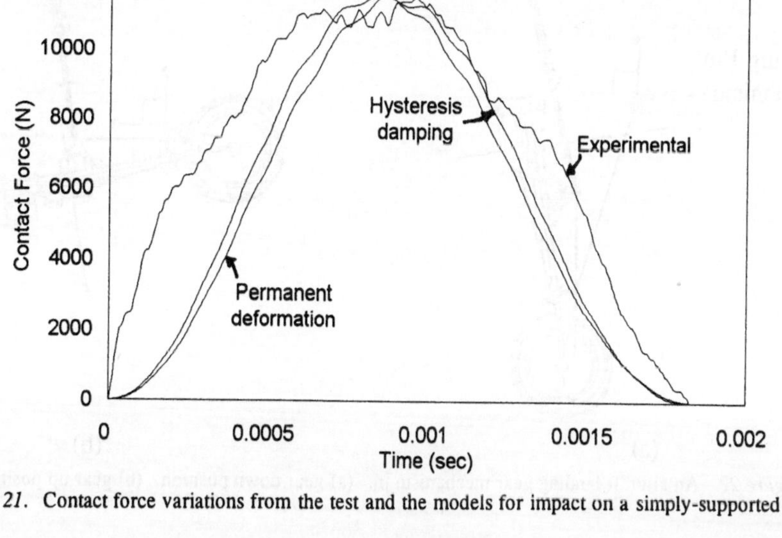

Figure 21. **Contact force variations from the test and the models for impact on a simply-supported beam.**

3.3. APPLICATION EXAMPLES ON MODELLING CONTACT FORCES

As an example of the application of modelling contact forces in prediction of impact responses of a multibody system, an aircraft landing gear uplock mechanism which encounters multiple impacts is first considered. The nose gear in down position (locking pin unlatched) and up position (locking pin latched) are shown in Figures 22(a) and (b). During the up motion, the gear is locked as shown in Figures 23(a) and (b). As the wheel is pulled into the aircraft, the lock pin impacts against the inside surface of the hook, pushed upward until the outer surface of the hook impacts on the stop pin, and the lock pin gets latched. The first impact is of the lock pin with the hook, and the hook getting stopped with a pin within the housing constitutes the second impact of the system. This is an internal multiple impact situation, which is caused by the joint clearance in the pin-in-a-slot configuration shown in Figure 24. The contact forces generated in the pin-in-a-slot impact was modelled using a nonlinear viscoelastic type Hertzian contact force model. For two cylinders (pin and the slot) in contact, the generalized impact parameters cannot be evaluated in a closed form, but the contact force-deformation relation can be expanded as [18]

$$\delta = (h_i + h_j)\, f_L \ln\left[\frac{e\,(R_i + R_j)}{f_L\,(h_i + h_j)\, R_i\, R_j}\right] \qquad (57)$$

where f_L represents contact force per unit length, and the coefficient of restitution "e" was estimated to be 0.83 from an experimental impact drop tower setup similar to the one shown in Figure 12. The contact force was numerically evaluated from a table of δ and f_L values. The resulting contact forces generated, determined from a complete dynamic analysis of the system, are shown in Figures 25(a) and (b). As observed, the contact forces are much higher for the first impact with a peak value of 37 klb as compared to the second impact with a peak of 7 klb. However, the duration of contact force is much smaller for the first contact (650 μs) compared to the second one (8.5 ms).

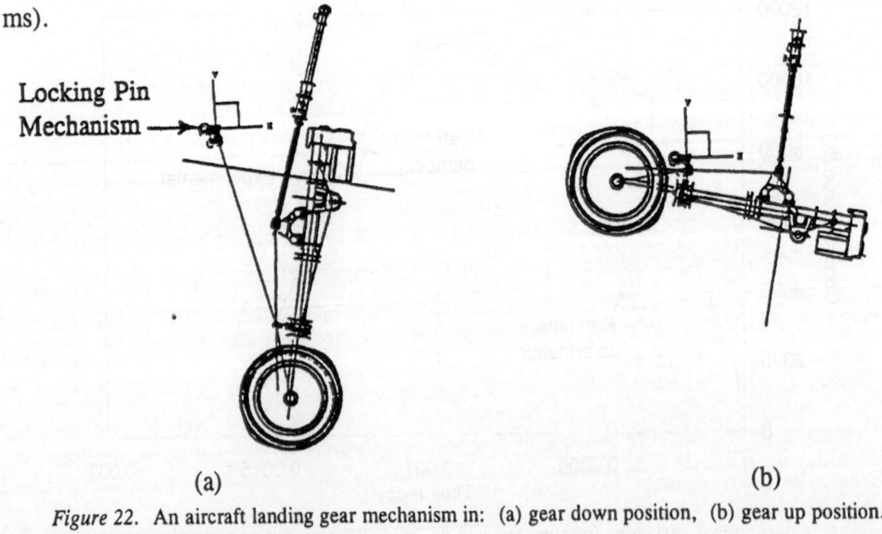

(a) (b)

Figure 22. An aircraft landing gear mechanism in: (a) gear down position, (b) gear up position.

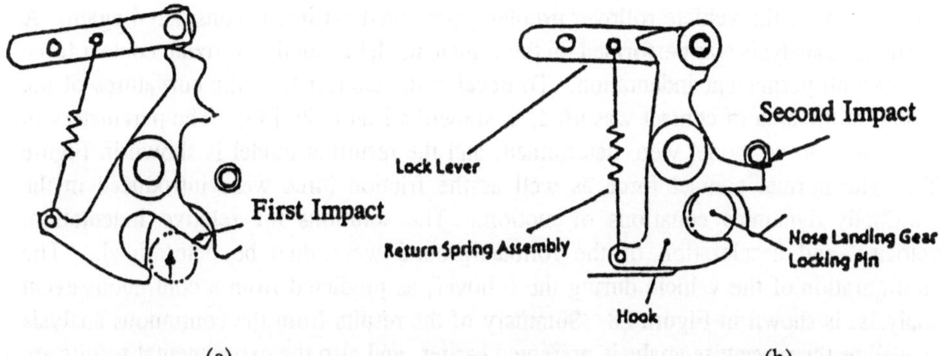

(a) (b)

Figure 23. Locking pin mechanism in: (a) upgoing position, (b) locked position.

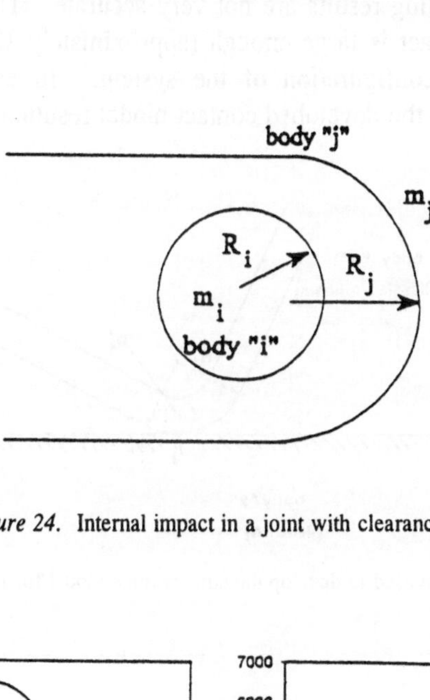

Figure 24. Internal impact in a joint with clearance.

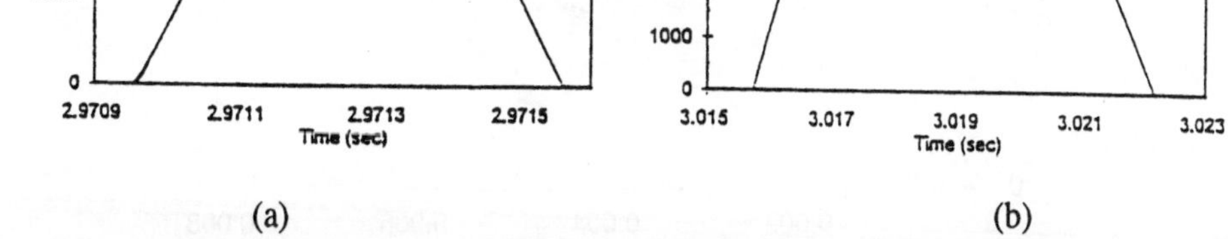

(a) (b)

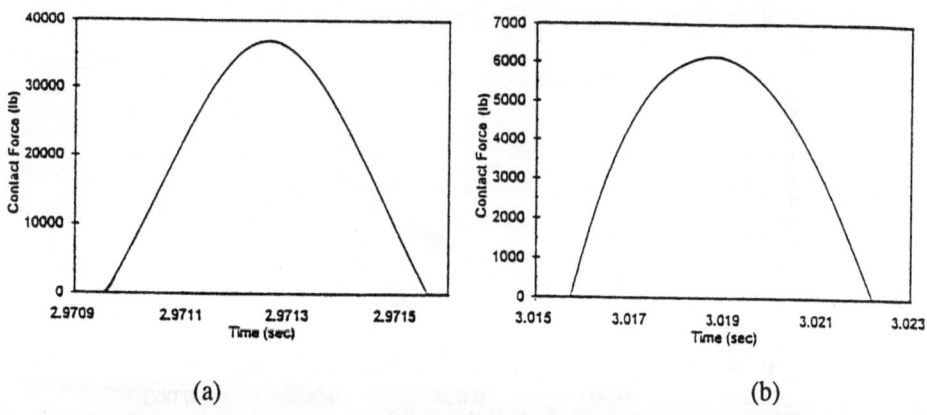

Figure 25. Contact forces generated during: (a) first impact, (b) second impact.

As a second example, for illustration of impact analysis of a multibody system using contact forces, the vehicle rollover problem, presented earlier, is considered again. A continuous analysis was performed on the vehicle model using the Hertzian contact force model with permanent indentation. To develop the contact law, the curvatures of the rollbar at the area of contact was used, as shown in Figure 26 [36]. The parameters in the contact force model were determined, and the resulting model is shown in Figure 27. The normal contact force as well as the friction force were introduced in the multibody dynamics equations of motion. The solutions for relative indentation, velocity, and acceleration of the rollbars/ground were then be determined. The configuration of the vehicle during the rollover, as predicted from a continuous event analysis, is shown in Figure 28. Summary of the results from the continuous analysis as well as the piecewise analysis presented earlier, and also the experimental results are shown in Table 2. As observed, although piecewise analysis is computationally more efficient, the corresponding results are not very accurate. This is due to the fact that the duration of the contact is large enough (approximately ¼ second) that significant changes occur in the configuration of the system. In comparison however, the continuous analysis with the developed contact model resulted in much closer results to the experimental values.

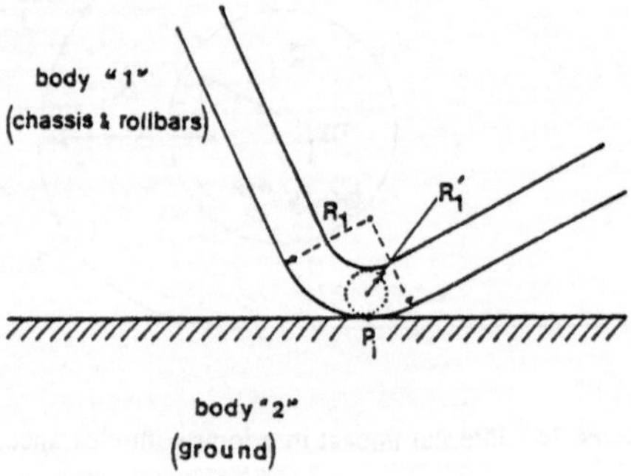

Figure 26. Curvature used to develop the contact force model for rollbar/ground contact.

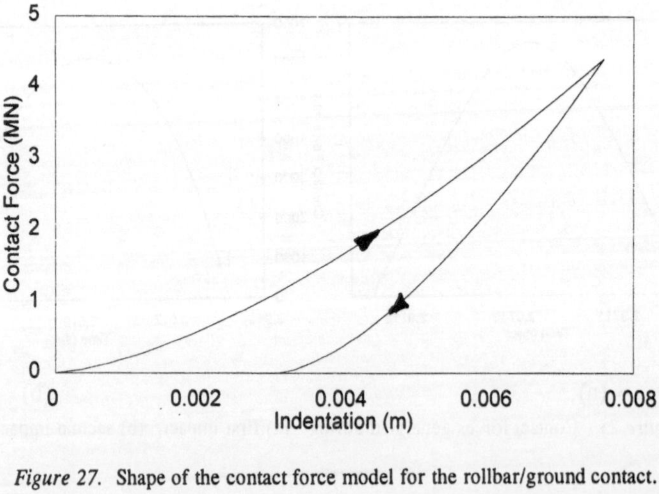

Figure 27. Shape of the contact force model for the rollbar/ground contact.

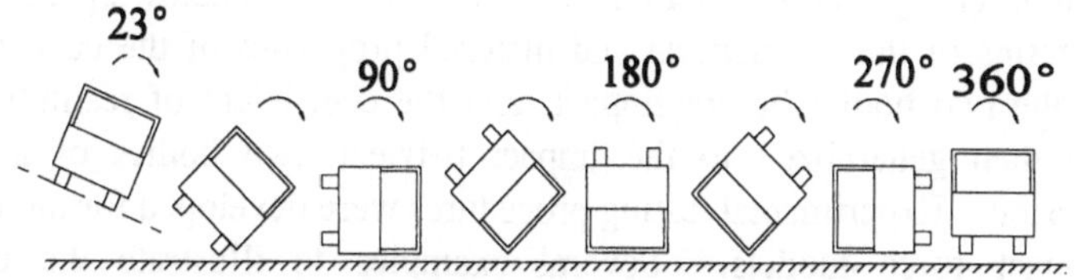

Figure 28. Rollover configuration of the vehicle as predicted from a continuous analysis.

TABLE 2. Summary of the results from experiment and analyses of the vehicle rollover.

Variables	Piecewise	Continuous	Experiment
Total angle of roll (deg)	247	337	427
Settling time (s)	2.4	3.0	3.0
Rollbar contact period (s)	0.0	0.15	0.25
Max. rollbar contact force (10^6 N)	---	4.4	---
Max. rollbar indentation (m)	---	0.007	---
Rollbar velocity before impact (m/s)	4.0	3.3	---
Rollbar velocity after impact (m/s)	-3.2	-3.4	
Coefficient of restitution	0.79	0.79	0.79
Coefficient of friction	1.0	1.0	---
Peak vertical acceleration (G)	9	17	15
CPU ratio	1.0	2.1	---

4. Conclusion

Two different methodologies, namely the discontinuous event method and the continuous analysis method, were presented in this paper for predicting the impact responses of mechanical systems. This treatment of the contact/impact mechanics was shown to be applicable to crash analysis of vehicles. An accident reconstruction technique was first presented for a two-vehicle collision. For impact analysis of constrained or jointed systems and also for the determination of the structural crash response of the individual components of a vehicle in a crash, such as the suspension systems, steering mechanism, etc., the conventional system momentum-balance/impulse equations were then extended to multiple object (or multibody) impact problems. Both normal and frictional impulses were treated in this methodology. Depending on the nature of the problem, the piecewise analysis method was shown to be of limited use at times, since the duration of the contact period is usually unknown before-hand and that no significant change must occur in the system's configuration during that small period.

A continuous analysis method was presented for the impact of two solids. To represent the variation of the contact force during the contact period, two Hertzian models may be used, one with hysteresis damping and the other with local plasticity effects. At low impact velocities, energy is dissipated in the form of internal damping or heat. If the initial indentation velocity is not negligible compared with the propagation speed of deformation waves across the two impacting solids, then permanent indentation is the dominant factor accounting for the energy dissipation. For both models, based on energy and momentum considerations, the unknown parameters were evaluated in terms of the geometrical and material properties of the contact surfaces, velocities of the two bodies before impact, and the coefficient of restitution. These models were then generalized to the impact between two bodies of a multibody mechanical system. Experimental testing procedures were developed for the verification of these contact force models. Several examples to illustrate the use of the methodologies were presented, including the compound pendulum impacting the ground and modelling of contact forces for the pin-in-hole mechanism of an aircraft landing gear mechanism. For the application of the developed methodologies to crash analysis, a two-car side-swipe collision and a complex vehicle model undergoing a rollover were studied for which the results from analyses showed good correlation with the experimental results.

5. References

1. Greenwood, D.T. (1965) *Principles of Dynamics*, Prentice-Hall, Englewood Cliffs, New Jersey.
2. Meirovitch, L. (1970) *Methods of Analytical Dynamics*, McGraw-Hill, New York.
3. Kane, T.R. (1968) *Dynamics*, Holt, Rinehart & Winston, New York.
4. Brach, R.M. (1984) Friction, restitution, and energy loss in planar collision, *ASME J. of Applied Mechanics* **51**, 164-170.
5. Brach, R.M. (1989) Rigid body collision, *ASME J. of Applied Mechanics* **56**, 133-138.
6. Bakaitis, S.H. (ed.) (1989) Reconstruction of motor vehicle accidents: a technical compendium - I, SAE PT-34, Society of Automotive Engineers.
7. Bakaitis, S.H. (ed.) (1991) Reconstruction of motor vehicle accidents: a technical compendium - II, SAE PT-35, Society of Automotive Engineers.
8. SAE. (1991, 1992, 1993, 1994, 1995) Accident reconstruction: technology and animation, I, II, III, IV, and V, SAE SP853, SP907, SP946, SP1150, and SP1083, Society of Automotive Engineers.
9. Wittenburg, J. (1977) *Dynamics of a systems of Rigid Bodies*, Teubner, Stuttgart, Germany.
10. Wehage, R.A. (1980) Generalized coordinate partitioning in dynamic analysis of mechanical systems, PhD Dissertation, University of Iowa, Iowa City, IA.
11. Khulief, Y.A. and Shabana, A.A. (1986) Dynamic analysis of constrained systems of rigid and flexible bodies with intermittent motion, *ASME J. of Mechanisms, Transmissions, and Automation in Design* **108**, 38-44.
12. Lankarani, H.M. and Nikravesh, P.E. (1992) Canonical impulse-momentum equations for the impact analysis of multibody systems, *ASME J. Mechanical Design* **114**, 180-186.
13. Whitaker, E.T. (1937) *Analytical Dynamics*, 4th ed., Cambridge University Press, London.
14. Keller, J.B. (1986) Impact with friction, *ASME J. of Applied Mechanics* **53**, 1-4.
15. Pereira, M.S., and Nikravesh, P.E. (1994) Impact dynamics of multibody systems with frictional contact and joint coordinates, NATO Advanced Science Institute on Computer-aided Analysis of Rigid and Flexible Mechanical Systems, Contributed Papers 2, Troia, Portugal.
16. Kane, T.R. (1984) A dynamic puzzle, Stanford Mechanics Alumni Club Newsletter, 6.
17. Han, I., and Gilmore, B.J. (1990) Multibody impact motion with friction - analysis, simulation and experimental validation, Proc. ASME Design Automation Conference, Chicago, IL.
18. Goldsmith, W. (1960) *Impact, the Theory and Physical Behavior of Colliding Solids*, E. Arnold Ltd., London.
19. Khulief, Y.A. and Shabana, A.A. (1987) A continuous force model for the impact analysis of flexible multibody systems, *ASME J. of Mechanism and Machine Theory* **22**, 213-224.
20. Hertz, H. (1895) *Gesammelte Werke* **1**, Leipzig, Germany.
21. Hunt, K.H. and Grossley, F.R.E. (1975) Coefficient of restitution interpreted as damping in vibroimpact, *ASME J. of Applied Mechanics* **7**, 440-445.

22. Crook, A.W. (1952) A study of some impacts between metallic bodies by a piezoelectric method, *Royal Society London* **A212**, 377-390.
23. Bernhart, K.E. (1955) Transverse impact on elastically supported beams, PhD Dissertation, University of California, Berkeley, CA.
24. Nikravesh, P.E. (1988) *Computer-aided Analysis of Mechanical Systems*, Prentice Hall, Englewood Cliffs, New Jersey.
25. Paulsen, R.E., and Fay, R.J. (1993) Three-dimensional analysis of a vehicle in flight, Accident Reconstruction: Technology and Animation III, SAE SP-946, Society of Automotive Engineers.
26. Ishikawa, H. (1993) Impact model for accident reconstruction - normal and tangential restitution coefficients, Accident Reconstruction: Technology and Animation III, SAE SP-946, Society of Automotive Engineers.
27. Pars, L.A. (1965) *A Treaties of Analytical Dynamics*, William Heinemann, Ltd., London.
28. Shakil, A., Lankarani, H.M., and Pereira, M.S. (1996) Frictional impact analysis in constrained multibody systems, to appear in the Proc. of ASME Design Automation Conference, University of California, Irvine, CA.
29. Routh, E.J. (1891) *Dynamics of Systems of Rigid Bodies*, fifth ed., McMillan and Co., London.
30. Gim, G., Pereira, M.S., Lankarani, H.M. and Nikravesh, P.E. (1987) Rollover analysis of vehicles with safety rollbars, Technical Report No. CAEL-87-5, University of Arizona, Tuscon, AZ.
31. Gim, G., Pereira, M.S., Lankarani, H.M. and Nikravesh, P.E. (1987) Technical data and plastic hinge model for M151-A2 rollbar model, Technical Report No. CAEL-87-4, Univ. of Arizona, Tuscon, AZ.
32. Lankarani, H.M. and Nikravesh, P.E. (1993) Continuous contact force models for impact analysis in multibody systems, *International J. of Nonlinear Dynamics* **5**, 193-207, Kluwer Academic Publishers, Dordrecht.
33. Lankarani, H.M. and Nikravesh, P.E. (1990) A contact force model with hysteresis damping for impact analysis of multibody systems, *ASME J. of Mechanical Design* **112**, 369-376.
34. Love, A.E.H. (1944) *A Treatise on the Mathematical Theory of Elasticity*, 4th ed., Dover Publications, New York.
35. Lankarani, H.M. and Nikravesh, P.E. (1992) A Hertzian contact force model with permanent indentation in impact analysis of solids, ASME Advances in Design Automation, Design Technical Conferences, Scottsdale, AZ.
36. Lankarani, H.M. (1988) Canonical equations of motion and estimation of parameters in the analysis of impact problems, PhD Dissertation, University of Arizona, Tucson, AZ.
37. Shivaswamy, S., and Lankarani, H.M. (1994) Impact analysis of plates using a quasi-static approach, Proc. ASME Design Automation Conf., Boston, MA.
38. Timoshenko, S., and Goodier, J.N. (1951) *Theory of Elasticity*, 2nd ed., McGraw Hill, New York.

MULTIBODY DYNAMIC TOOLS FOR CRASHWORTHINESS AND IMPACT

J. A. C. AMBRÓSIO AND M. S. PEREIRA
IDMEC Polo Instituto Superior Técnico,
Avenida Rovisco Pais,
1096 Lisboa CODEX, Portugal

Abstract

Formulations on multibody dynamics are shown to be suitable for the development of methodologies for the impact simulation and crashworthiness design of vehicles. The proposed design methods comprise a range of computer aided tools of increasing complexity and accuracy which can be used with greater advantage and efficiency in the different crashworthiness design stages. The key issue in the use of such rigid and flexible multibody formulations is their capability to model and simulate efficiently the behavior of complex systems experiencing material and geometric nonlinear deformations while undergoing gross motion. The proposed multibody based crashworthiness design methods and associated multibody dynamics tools require information about the structural nonlinear behavior of specific parts of the vehicle structures which can be obtained from numerical or experimental tests of specific structural components and subsequently used in the formulations. Alternatively, the behavior of such components can be directly and efficiently incorporated through the use of appropriate nonlinear finite element procedures. This hybrid feature lends to the present design tools flexibility, ease of use and efficiency gains, as a result of a better understanding of the crash phenomena with particular emphasis in the interaction of the gross motion with different collapse mechanisms. Formulations for the sensitivity analysis and optimization of mechanical systems are also presented allowing for the design of optimum crash characteristics of energy absorption devices. The capabilities of the design tools presented herein are demonstrated with several applications to different vehicles types in different crash-impact scenarios. It is shown that the same formulations are applicable to enhanced occupant multibody models allowing extended injury assessment capabilities.

J. A. C. Ambrósio et al. (eds.),
Crashworthiness of Transportation Systems: Structural Impact and Occupant Protection, 475–521.

1. Introduction

Vehicle systems are generally a complex arrangement of structural and mechanical sub-systems with very different design purposes and different mechanical behavior. Modern vehicle design include crashworthiness features which involve complex arrangements of structural and mechanical parts with the aim to absorb in a controlled manner a maximum amount of kinetic energy with low acceleration levels and acceptable deformations of the vehicle structural components. For crash-impact simulation purposes some parts are so small or sufficiently rigid that flexibility effects can be neglected. Other vehicle sub-systems, such as the main structural body of a road car or the passenger compartment of a railway vehicle, must be considered as flexible, and, because of their complex geometry, finite element methodologies are required to be described so that small elastic deformations are taken into account and superimposed to large rigid body rotations and translations.

For most of the vehicle impact problems involving structural crash, it is crucial not only to capture the coupling between the overall gross motion of the system and its structural failure mechanisms, which include elastic-plastic deformations but it is also important to access the mechanisms of occupant protection for realistic scenarios. During the last twenty five years, computer aided analysis of crashworthiness and structural impact has received a large attention and it is now emerging as a powerful methodology. These formulations have been successfully applied in practical and industrial situations.

Different numerical formulations with varying degree of complexity and accuracy have been proposed using spring-mass models [1-5] and finite element methodologies [6-9]. Hybrid approaches [10,11] utilizing data obtained from quasi-static crushing of different segments of the colliding structure have also been developed. In these methods the generalized nonlinear load-displacement characteristics are kinematically coupled to the global structural system to obtain the overall dynamic response of the structure. For local plastic deformations the concept of plastic hinges has been widely used. Here lumped deformable elements (nonlinear spring-damper elements) having a force deflection curve similar to that of the real structural system are used to describe the system deformations [12-14]. The drawback of these formulations is that they require an early knowledge of the pattern of deformation of the flexible component.

For more complex plastic deformation mechanisms, an integrated methodology has been developed which is able to capture the full nonlinear structural behavior of the system components [15]. This formalism is based on an updated Lagrangean formulation of the equations of motion for flexible bodies undergoing large rotations and displacements. In standard finite element formulations the large displacements and deformations of the gross motion are not generally taken into consideration. However recent efforts in the field of nonlinear structural dynamics have contributed for the development of well known commercially available codes such as DYCAST [16], PAM-CRASH [17], DYNA-3D [18] and WHAMS-3D [19]. These programs are now able to simulate with improved accuracy several different structural impact phenomena such as large localized deformations, structural instabilities, transient vibrations, stress wave propagations and eventually structural collapse due to material damage and loads causing stresses above the ultimate strength. However these codes, , require large computer resources and normally involve time consuming modeling data preparation which make them rather unsuitable as a design tool during the initial design stages. The integrated methodology adds to the multibody description of the system the generality of the nonlinear finite element modeling, preserving the coupling between structural deformations and the rigid body motion, rather than superimposing them.

This paper summarizes some developments on multibody dynamics based methodologies, applicable to crashworthiness analysis and design, which have been developed by the team in the Instituto Superior Técnico and IDMEC (Lisbon, Portugal).

2. Formulations for multibody dynamics [25]

In the last two decades a great deal of research has been done in computer simulation of complex multibody systems, most of them summarized in recent books by Nikravesh [20], Roberson and Schwertassek [21], Haug [22], Shabana [23], García de Jalon and Bayo [24] and Pereira and Ambrósio [25]. As a result of this research several general-purpose computer programs have been developed [26]. However these codes show important limitations, in what is concerned with crashworthiness modeling capabilities.

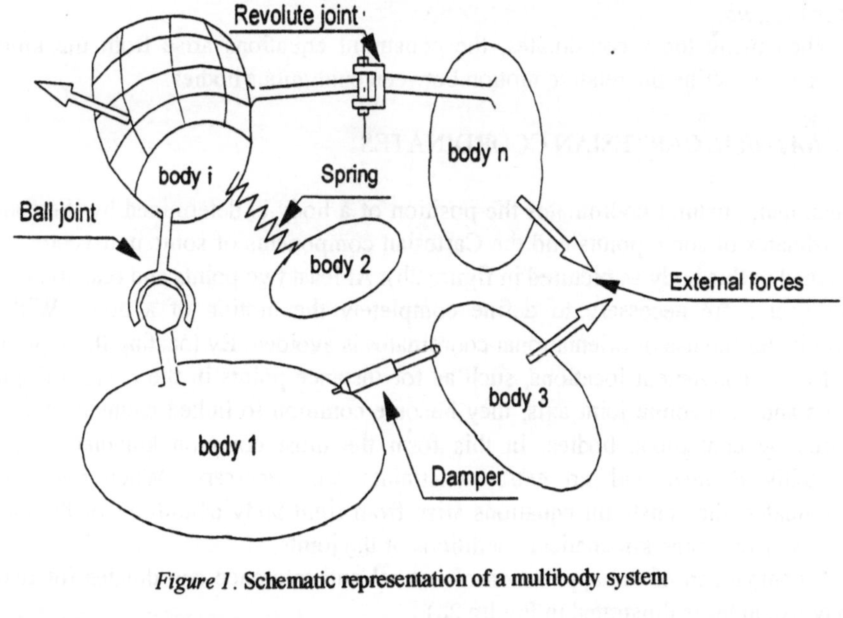

Figure 1. Schematic representation of a multibody system

A multibody system is a collection of rigid and flexible bodies joined together by kinematic joints and force elements as depicted in Figure 1. The presence of the kinematic joints is the main reason why it is normally necessary to use an extended set of coordinates to describe the system. These are called the dependent coordinates and define uniquely the position and orientation of the body in the system. Two different kinds of dependent coordinates that have been used in the formulations presented in this paper introduced here: the reference Cartesian coordinates and the natural Cartesian coordinates.

2.1. *REFERENCE* CARTESIAN COORDINATES

Reference Cartesian coordinates identify the position of each body through the Cartesian coordinates of a point and a set of parameters to define its angular orientation as depicted in Figure 2a). For the ith body in the system $\mathbf{q}_i$ denotes a vector of coordinates which contains the Cartesian translational coordinates $\mathbf{r}_i$, a set of rotational coordinates $\mathbf{p}_i$, and a set of nodal coordinates $\mathbf{q}_f'$, $\mathbf{u}'$ or δ', if body i is flexible. A vector of velocities for a rigid body i is defined as $\mathbf{v}_i$ contains a 3-vector of translational velocities $\dot{\mathbf{r}}_i$ and a 3-vector of angular velocities $\boldsymbol{\omega}_i$ (defined in the XYZ coordinate system). If body i is flexible then the vector of velocities $\mathbf{v}_i$ contains $\dot{\mathbf{r}}_i$, $\boldsymbol{\omega}_i$ (defined in the $\xi\eta\zeta_i$ coordinate system) and a vector of nodal velocities $\dot{\mathbf{q}}_f'$ or $\dot{\delta}'$. The vector of accelerations for the body is denoted by $\dot{\mathbf{v}}_i$ and it is simply the time derivative of $\mathbf{v}_i$. For a multibody system containing *nb* bodies, the vectors of coordinates, velocities, and accelerations are $\mathbf{q}$, $\mathbf{v}$ and $\dot{\mathbf{v}}$ which contain the elements of $\mathbf{q}_i$, $\mathbf{v}_i$ and $\dot{\mathbf{v}}_i$, respectively, for i=1, ..., *nb*.

When using these coordinates, the constraint equations arise from the kinematic joints and describe the relative motion between contiguous bodies.

2.2. *NATURAL* CARTESIAN COORDINATES

When using natural coordinates the position of a body is determined by the Cartesian coordinates of some points and the Cartesian components of some unit vectors rigidly attached to this body as pictured in figure 2b). At least two points and one non co-linear unit vector are necessary to define completely the motion of a body. With these coordinates the use of orientational coordinates is avoided. By locating these points and vectors in convenient locations, such as for instance points in the center of spherical joints and in revolute joint axis, they become common to linked elements and can be shared by contiguous bodies. In this form the most common kinematic joints are implicitly defined and no extra constraints are necessary. When using natural coordinates, the constraint equations arise from rigid body conditions of the elements and also from some kinematical conditions of the joints.

A comparison of the application of natural and reference coordinates for two rigid body examples is illustrated in Figure 2b).

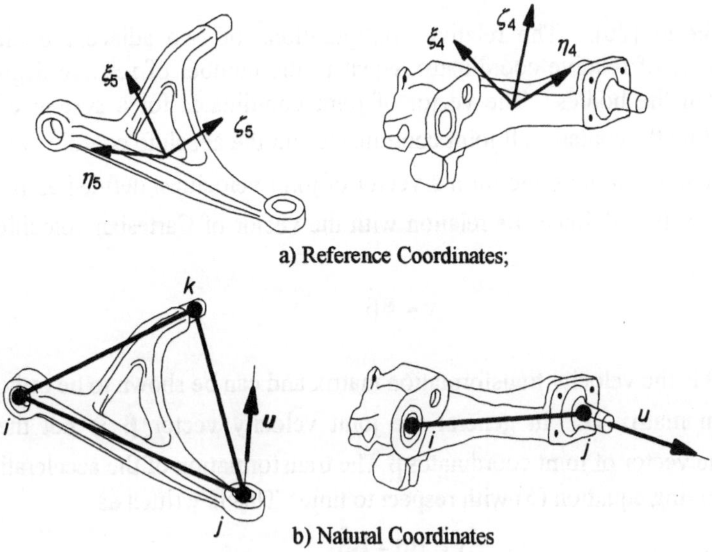

a) Reference Coordinates;

b) Natural Coordinates

Figure 2. Coordinates describing a single body:

2.3. CONSTRAINED DYNAMIC EQUATIONS

Regardless of the coordinates used let the kinematic constraints representing joints between rigid bodies or rigid body conditions be described by *mr* independent equations as

$$\mathbf{\Phi}(\mathbf{q}) = \mathbf{0} \tag{1}$$

The first and second derivatives of the constraints yield the kinematic velocity and acceleration equations

$$\dot{\mathbf{\Phi}} \equiv \mathbf{D}\mathbf{v} = \mathbf{0} \tag{2}$$

$$\ddot{\mathbf{\Phi}} \equiv \dot{\mathbf{D}}\mathbf{v} + \mathbf{D}\dot{\mathbf{v}} = \mathbf{0} \tag{3}$$

where **D** is the Jacobian matrix of the constraints. The equation of motion for the system of rigid bodies are written [20]

$$\mathbf{M}\dot{\mathbf{v}} - \mathbf{D}^{\mathrm{T}}\boldsymbol{\lambda} = \mathbf{g} \tag{4}$$

where **M** is the inertia matrix, $\boldsymbol{\lambda}$ is a vector of Lagrange multipliers, and $\mathbf{g} = \mathbf{g}(\mathbf{q}, \mathbf{v})$ contains the forces and moments that act on the bodies, and the gyroscopic terms.

The constrained equations of motion expressed by equations (1) to (4) can be converted to a smaller set of equations in terms of a set of coordinates known as joint coordinates. For open loop systems the set of equations obtained is minimal. Such transformation is briefly discussed here but for a more detailed discussion the interested reader can refer to [26]. The relative configurations of two adjacent bodies are described by a set of relative coordinates, equal to the number of relative degrees of freedom between the bodies. The vector of joint coordinates for a system of rigid bodies, denoted by $\boldsymbol{\beta}$, contains all joint coordinates and the absolute coordinates of the floating base bodies. In the same form a vector of joint velocities, defined as $\dot{\boldsymbol{\beta}}$, is the time derivative of $\boldsymbol{\beta}$ is defined. Its relation with the vector of Cartesian velocities $\mathbf{v}$ is given by

$$\mathbf{v} = \mathbf{B}\dot{\boldsymbol{\beta}} \tag{5}$$

where matrix $\mathbf{B}$ is the velocity transformation matrix and can be shown to be orthogonal to the Jacobian matrix $\mathbf{D}$. In general the joint velocity vector $\dot{\boldsymbol{\beta}}$ is not the time derivative of the vector of joint coordinates $\boldsymbol{\beta}$ The transformation of the accelerations is obtained by deriving equation (5) with respect to time. This is written as

$$\dot{\mathbf{v}} = \dot{\mathbf{B}}\dot{\boldsymbol{\beta}} + \mathbf{B}\ddot{\boldsymbol{\beta}} \tag{6}$$

Substituting equation (6) into equation (4), premultiplying by $\mathbf{B}^T$, and using the orthogonality condition between $\mathbf{B}$ and $\mathbf{D}$ yield

$$\boldsymbol{M}\ddot{\boldsymbol{\beta}} = \boldsymbol{f} \tag{7}$$

where

$$\boldsymbol{M} = \mathbf{B}^T\mathbf{MB} \tag{8}$$

$$\boldsymbol{f} = \mathbf{B}^T(\mathbf{g} - \mathbf{M}\dot{\mathbf{B}}\dot{\boldsymbol{\beta}}) \tag{9}$$

Equation (7) represents the generalized equation of motion for an open-loop system of rigid bodies. This equation, containing the minimum number of second-order differential equations, can be used instead of the mixed set of differential-algebraic equations given by equations (1) through (4). Closed-loops can be analyzed as open-loops provided that the loop is cut at a joint and the corresponding kinematic constraint is not eliminated from the equations of motion. In a second step, the methodology described here can be used to eliminate the explicit use of the kinematic constraint that represents the cut joint [27]. However, it is computationally more advantageous not to perform this second step of velocity transformations in most of the problems.

2.4. NUMERICAL INTEGRATION PROCEDURES

Constrained dynamic equations can be seen as formed by two coupled sub-systems: the first describes the dynamics of the rigid or flexible structural system, while the second sub-system describes the kinematic constraints that are present in the mechanical system. The set of second order differential equations describing the dynamics of flexible mechanical systems are normally "stiff", i.e.., the system eigenfrequencies are distributed over a broad frequency range. This is mainly due to the high frequency content of the structural flexibility behavior and also to the physical properties of the impact situations under study.

For the numerical solution of the system equations of motion a predictor-corrector, variable order algorithm is used [28] together with a constraint violation stabilization scheme [29] or with an Augmented Lagrangean methodology [24]. This algorithm also includes an error estimation and a step size control feature which is particularly suitable for impact simulations where sudden forces may appear as a result of contact between different parts of the impacting structures.

2.5. BIOMECHANICAL BEHAVIOR IN A CRASH SITUATION [30]

Multibody systems with rigid components provide the most of the numerical tools required to develop biomechanical models for crashworthinaess applications. With the exception of some models used in aerospace applications were the spine of the human body is modeled is described as a flexible body most of the other applications only use systems of rigid bodies.

2.5.1. *Biomechanical model*

A three-dimensional, whole body response, biomechanical model of the human body is presented as an application of the multibody formulations discussed. This model is developed using a general purpose multibody code that uses the methodology based on natural coordinates. All the information required to assemble the equations of motion of the model is hold within a database created for the purpose. The necessary information is basically the mass, the dimension, the principal moments of inertia and the center of mass location of each rigid body. Using this database, models with different sizes and masses can be easily obtained. In this work, two biomechanical models have been considered: the *50% Anthropomorphic Dummy* and the *50% Human Male*.

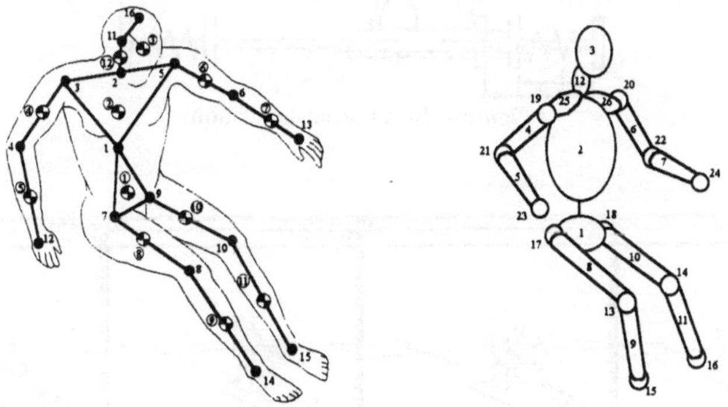

a) Bodies and joints b) Contact surfaces

Figure 3. Three-dimensional biomechanical model.

The model is described using twelve rigid bodies interconnected by eleven kinematic joints, as shown in figure 3a). A set of contact surfaces is defined for the calculation of the external forces exerted on the model by the seat cushions and the floor or any other object coming in contact with a particular biomechanical components, including other members of the model. These surfaces are ellipsoids and cylinders and are also depicted in figure 3b). When contact is detected a dissipative continuous contact force model is used [31]. The contact force is a nonlinear function of the apparent penetration of the contacting bodies.
The kinematic joints used in the model are of two types: spherical and revolute joints. The hand and foot segments were not included in the model because their importance in crash prediction injuries is not very relevant.

Joint resistance torques that characterize the passive muscle action are modeled by viscous torsional dampers located in each kinematic joint. When the two contiguous bodies interconnected by one of these joints reach physically unacceptable positions, a penalty torque is applied to each body. This torque is represented by a nonlinear torsional spring and also includes a dissipative term represented by a torsional damper.

2.5.2. *Sled impact simulation*
A vehicle impact situation, where the biomechanical model is seated on a rigid car seat is represented in figure 4. The car with an energy absorbing structure mounted on the front impacts a wall with 25km/h . The car occupant has a lap belt and a shoulder belt. The graphical results of the simulation are displayed in figure 5.

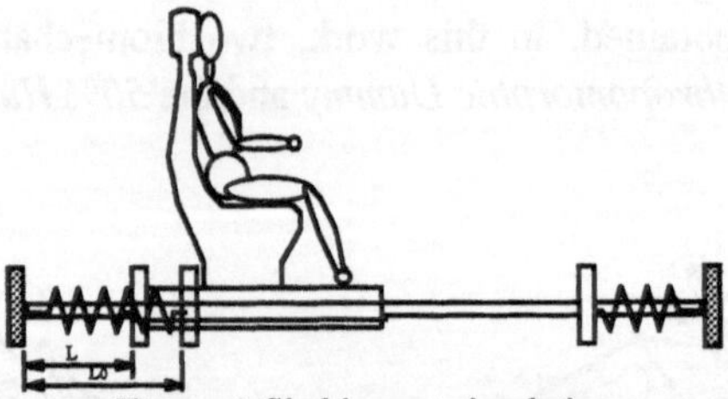

Figure 4. Sled impact simulation.

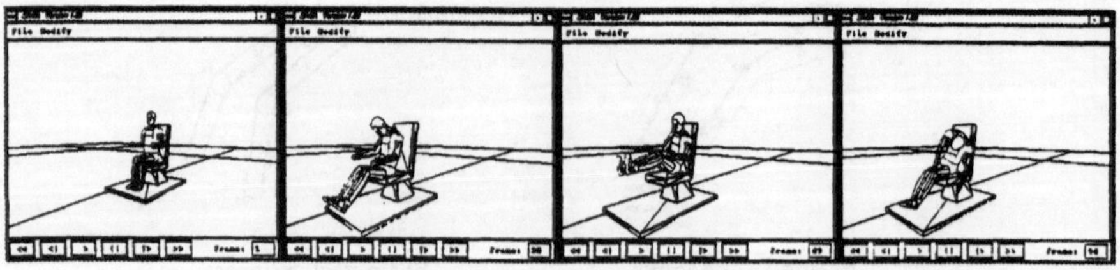

Figure 5. Graphical results.

3. Flexible multibody dynamics [14]

For the crashworthiness and impact analysis, using a multibody formalism, the description of the flexibility of its components is necessary. The behavior of systems subjected to impact is characterized by zones of large deformations and by zones where only elastic deformations take place. For the purpose of describing this behavior, linear and nonlinear formulations of multibody systems are reviewed here.

3.1. LINEAR DEFORMATIONS

It has been shown [32, 33] that the configuration of a deformable body in a multibody system can be described by a set of global reference coordinates $\mathbf{q}_{ri}$ and local elastic coordinates $\mathbf{u}_i$ which are defined using the finite element methodology. As shown in figure 6, the position of a flexible body in the non-moving reference frame XYZ is specified by the spatial location $\mathbf{r}_i$ of a body fixed frame xhz and a set of angular orientation coordinates $\mathbf{f}^i$, thus the coordinates describing the gross motion of the body are $\mathbf{q}_{r_i}^T = [\mathbf{r}_{r_i}^T, \boldsymbol{\phi}_{r_i}^T]$

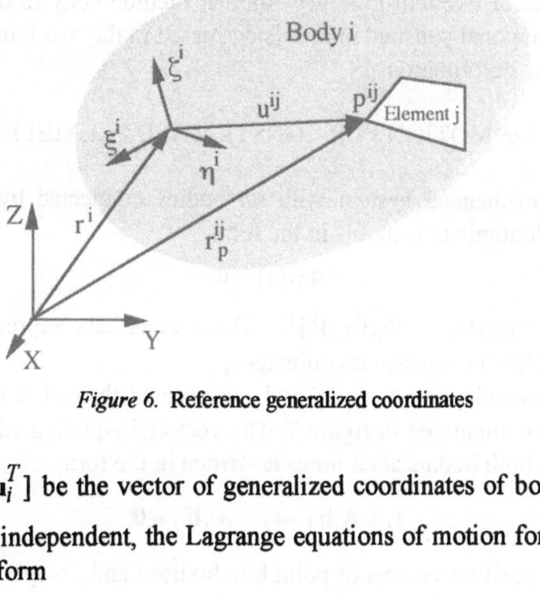

Figure 6. Reference generalized coordinates

Let $\mathbf{q}_i^T = [\mathbf{q}_{r_i}^T, \mathbf{u}_i^T]$ be the vector of generalized coordinates of body i. Assuming all coordinates to be independent, the Lagrange equations of motion for this flexible body are written in the form

$$\frac{d}{dt}\left(\frac{\partial T_i}{\partial \mathbf{v}_i}\right) - \left(\frac{\partial T_i}{\partial \mathbf{q}_i}\right) + \left(\frac{\partial U_i}{\partial \mathbf{q}_i}\right) - \mathbf{g}_i = \mathbf{0} \tag{10}$$

Using the finite element method to describe the flexibility of body i , the kinetic energy T^i is a function of the velocities $\mathbf{v}^i$ and positions $\mathbf{q}^i$, while the elastic energy U^i is function of $\mathbf{q}^i$. The equations of motion (10) for body i take the form [32-34]

$$\mathbf{M}_i(\mathbf{q}_i)\dot{\mathbf{v}}_i + \mathbf{K}_i\mathbf{q}_i = \mathbf{g}_i(\mathbf{v}_i, \mathbf{q}_i, t) + \mathbf{s}_i(\mathbf{v}_i, \mathbf{q}_i) \tag{11}$$

where $\mathbf{M}_i$, $\mathbf{K}_i$ are the mass and stiffness matrices of body i, respectively, $\mathbf{g}_i$ is the vector of generalized forces of body i, $\mathbf{s}_i$ is a vector containing velocity quadratic terms and other acceleration independent terms. In a less compact form, equation (11) is written as:

$$\begin{bmatrix} \mathbf{M}_{rr} & \mathbf{M}_{r\phi} & \mathbf{M}_{rf} \\ \mathbf{M}_{\phi r} & \mathbf{M}_{\phi\phi} & \mathbf{M}_{\phi f} \\ \mathbf{M}_{fr} & \mathbf{M}_{f\phi} & \mathbf{M}_{ff} \end{bmatrix} \begin{bmatrix} \ddot{\mathbf{r}} \\ \omega' \\ \ddot{\mathbf{u}}' \end{bmatrix} + \begin{bmatrix} \mathbf{0} & \mathbf{0} & \mathbf{0} \\ \mathbf{0} & \mathbf{0} & \mathbf{0} \\ \mathbf{0} & \mathbf{0} & \mathbf{K} \end{bmatrix} \begin{bmatrix} \mathbf{0} \\ \mathbf{0} \\ \mathbf{u}' \end{bmatrix} = \begin{bmatrix} \mathbf{g}_r \\ \mathbf{g}'_\phi \\ \mathbf{g}'_f \end{bmatrix} - \begin{bmatrix} \mathbf{s}_r \\ \mathbf{s}'_\phi \\ \mathbf{s}'_f \end{bmatrix} \tag{12}$$

In this equation the invariant submatrices $\mathbf{M}_{rr}$, $\mathbf{M}_{\phi r}$, $\mathbf{M}_{r\phi}$ and $\mathbf{M}_{\phi\phi}$, are associated with the gross motion of the body-fixed coordinate frame, and $\mathbf{M}_{ff}$, the standard finite element mass matrix. Assuming small linear elastic deformations for the flexible body, the stiffness matrix $\mathbf{K}_i$ is also constant and it is the same as obtained in a standard finite element procedure. The remaining terms of the mass matrix are time variant and must be calculated every time step which can be a computational burden if the appropriate procedures are not implemented. The mean axis conditions can be applied to equation (12) resulting in a constant mass matrix weakening the inertia coupling between rigid and flexible degrees of freedom [32, 34]. Another methodology to transform the mass matrix $\mathbf{M}_i$ into a diagonal constant matrix is discussed in this work in the framework of nonlinear multibody deformations.

3.2. EQUATIONS OF MOTION FOR CONSTRAINED FLEXIBLE BODIES

Consider now a mechanical system with *nb* bodies connected by joints which are described by *m* holonomic constraints in the form

$$\Phi(\mathbf{q},t) = \mathbf{0} \tag{13}$$

where $\Phi(\mathbf{q},t) = [\Phi_1(\mathbf{q},t)^T,\ldots,\Phi_m(\mathbf{q},t)^T]^T$. These equations express the dependency between the generalized Cartesian coordinates $\mathbf{q}$.

Consider, for example, two bodies i and j connected through a spherical joint in a common point k, as illustrated in figure 7. The vectorial equation which forces point k to be coincident in both bodies at all times is written in the form

$$\mathbf{r}_i + \mathbf{A}_i\mathbf{b}'_i - \mathbf{r}_j - \mathbf{A}_j\mathbf{b}'_j = \mathbf{0} \tag{14}$$

where $\mathbf{b}'_i$, $\mathbf{b}'_j$ are position vectors of point k in bodies i and j respectively; $\mathbf{A}_i$, $\mathbf{A}_j$ are transformation matrices from the body coordinate systems to the global inertia frame.

This joint has two algebraic constraint equations. If both bodies are flexible $\mathbf{b}'_i$, $\mathbf{b}'_j$ depend on the generalized elastic coordinates, implying that these vectors have to be calculated at each time for the current deformation state. Then

$$\mathbf{r}_i + \mathbf{A}_i(\mathbf{b}'_{0_i} + \delta_{k_i}) - \mathbf{r}_j - \mathbf{A}_j(\mathbf{b}'_{0_j} + \delta_{k_j}) = \mathbf{0} \tag{15}$$

where $\mathbf{b}'_{0_i}$, $\mathbf{b}'_{0_j}$ correspond to the position vectors of point k in the undeformed state, δ_{k_i}, δ_{k_j} are the flexible displacements of the connection node (point k) of bodies i and j, respectively.

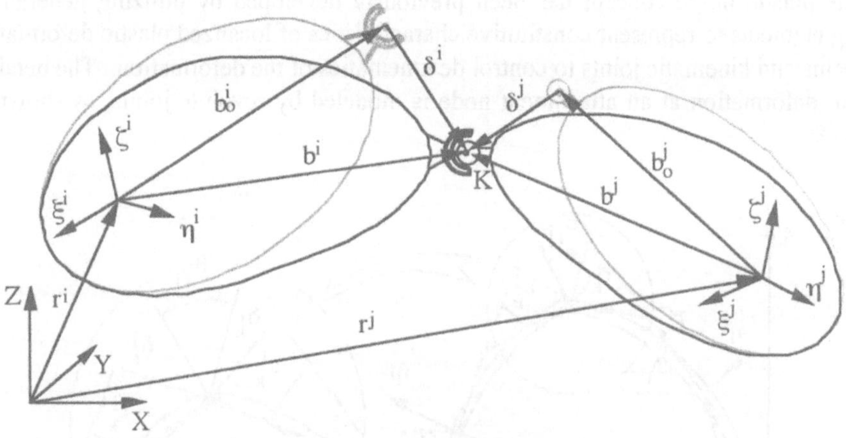

Figure 7. Spherical Joint

The holonomic kinematic constraints are introduced in the variational form of the equations of motion of body i using Lagrange multipliers. After substituting all energy expressions in these equations, the dynamic equations of motion for a flexible body are written in a compact form as:

$$\mathbf{M}_i(\mathbf{q}_i)\dot{\mathbf{v}}_i + \mathbf{D}^T\boldsymbol{\lambda} = \mathbf{g}_i(\mathbf{v}_i, \mathbf{q}_i, t) + \mathbf{s}_i(\mathbf{v}_i, \mathbf{q}_i) - \mathbf{K}_f\mathbf{q}_i \tag{16}$$

where $\mathbf{D} = (\partial\boldsymbol{\phi}/\partial\mathbf{q}_i)$ is the Jacobian matrix for the constraints.

Once these equations have been obtained for each body, it is necessary to assemble them for all bodies of the mechanical system. For the equations obtained, the angular acceleration of the body fixed coordinate frame must be transformed to global components such that the accelerations in equation (11) are consistent with the rigid body accelerations used in the transformations of the joint coordinate method, expressed by equations (5). The constraint equations and their corresponding Lagrange multipliers can be eliminated from the equations of motion by using the velocity transformations referred before. The interested reader is directed to references [32, 33] for a more detailed discussion on the use of the joint coordinate method with flexible bodies.

4. Concentrated plastic deformations [35]

4.1. THE PLASTIC HINGE CONCEPT

In many impact situations, the individual structural members are overloaded, principally in bending giving rise to plastic deformations in highly localized regions, called plastic hinges. These deformations, called hinges develop at points where maximum bending moments occur, or at load application points and at joints and locally weak areas. Therefore, for most practical situations, their location can be predicted well in advance.

The plastic hinge concept has been previously developed by utilizing generalized spring elements to represent constitutive characteristics of localized plastic deformation of beams and kinematic joints to control de kinematics of the deformation. The bending plastic deformation at an attachment node is modeled by revolute joints, as shown in figure 8.

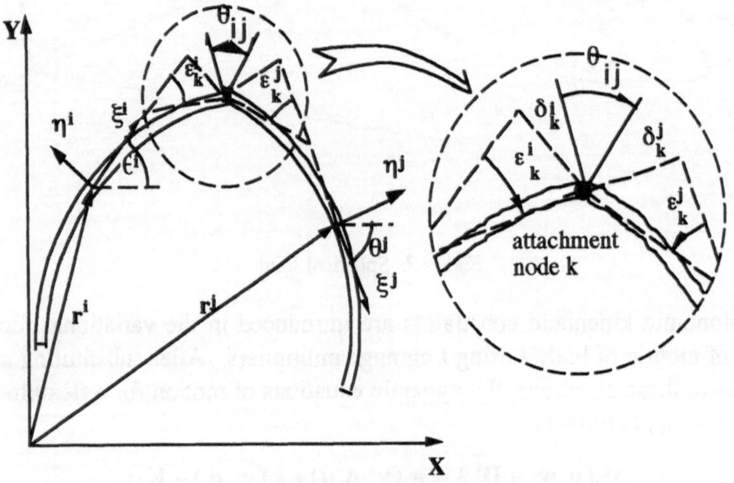

Figure 8. Plastic Hinge concept

The revolute joint must be simultaneously perpendicular to the neutral axis of the beam and to the plastic hinge bending plane. From figure 8 the following relationship can be written

$$\theta^{ij} = \theta^{i} + \varepsilon_k^{\,i} - \theta^{j} - \varepsilon_k^{\,j} \qquad (17)$$

which shows the dependency of the plastic hinge angle on the rigid bodies relative rotation and on the elastic rotations of body *i* and body *j* at the attachment node k. The angle values are directly obtained as relative coordinates from the integration process and correspond to the relative degree of freedom, θ^{ij}, of the revolute joint under consideration.

A typical torque-angle constitutive relationship illustrated in Figure 9, was obtained in an earlier work [36] for the case of a steel tubular cross section based on a kinematic folding model [37]. This model was modified to take into account elastic plastic material properties including strain hardening.

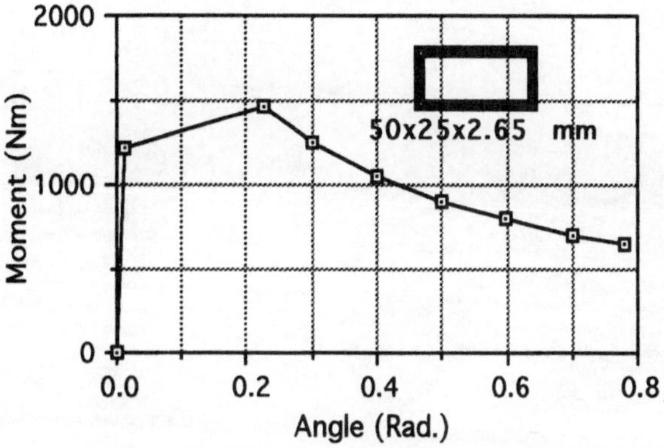

Figure 9. Plastic Hinge constitutive relationship

The plastic hinge modeling technique generally requires that the spring data must be multiplied by a dynamic correction factor in cases where the material is strain rate sensitive. Currently, the formula suggested by Winmer has been used [38]

$$P_d / P_s = 1 + 0.07\, V_0^{0.82} \tag{18}$$

Here P_d and P_s are the dynamic and static forces, respectively, and V_0 is the impact velocity. The coefficients indicated above are dependent on the type of cross section. The advantage of this procedure resides in its generality for accounting strain rate effects. The user, however, may be able to incorporate other correction factors.

4.2. VALIDATION WITH AN IMPACTING BEAM

The formulations for rigid and flexible body dynamics including the plastic hinge model have been implemented in a computer program. In order to verify the proposed analytical technique a comparison with an experimental test is carried out [40, 41]. The experimental set up for this test case and the test specimen are shown in figure 10 being the procedure similar to the pendulum ram impact test. The bar is accelerated until an angular velocity of 11.85 rad/s is reached before colliding with the edge of a rigid block located at a distance of 0.5 m from the axis of the revolute joint. During the impact, accelerations are measured using accelerometers mounted on the ballasting mass. A post impact observation clearly indicates the existence of a localized plastic bending zone. A final value of the permanent bending angle was measured and was found to be 22°. The result supports the assumption of localized plastic deformations described by plastic hinges in numerical models.

Figure 10. Experimental configuration

A series of computer simulations are performed for the analysis of the impacting beam referred to above. Different models made of rigid bodies and rigid-flexible bodies are considered. These are shown schematically in figure 11. The models have two bodies and two revolute joints. Each flexible body, is discretized with 5 nodes and 4 beam elements. The contact between the impacting beam and the rigid block was modeled using a Hertz contact law [31]. Figure 12 illustrates a sequence of computer generated positions for the two rigid bodies model during the simulation .

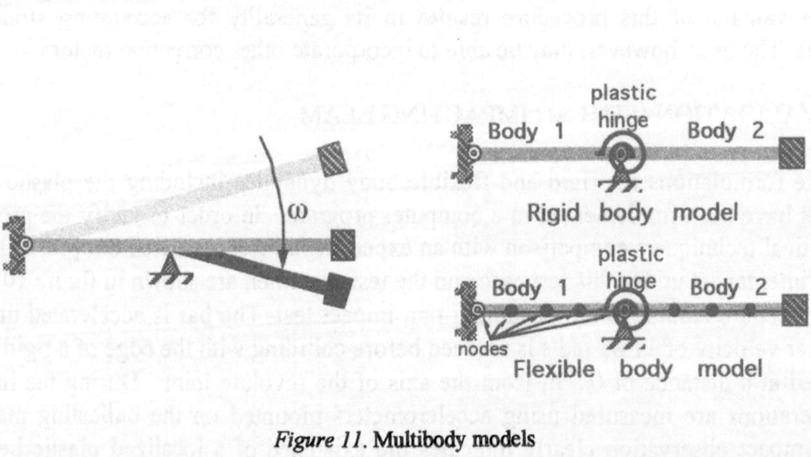

Figure 11. Multibody models

The predicted gross motion shows a similar trend when compared with high speed photographs taken during the experimental test. An assessment of the accuracy of the theoretical models is made on the basis of the results obtained for the permanent bending angles. This is justified as the final configuration of the beam is strongly dependent on the dynamics of the problem and also on the mechanisms of energy absorption.

An excellent agreement was observed when comparing the experimental value of the deformed angle θ between bodies 1 and 2 (21.5°) with the results obtained in the different numerical simulations (21.25°). The flexible body model predicts a similar permanent deformation.

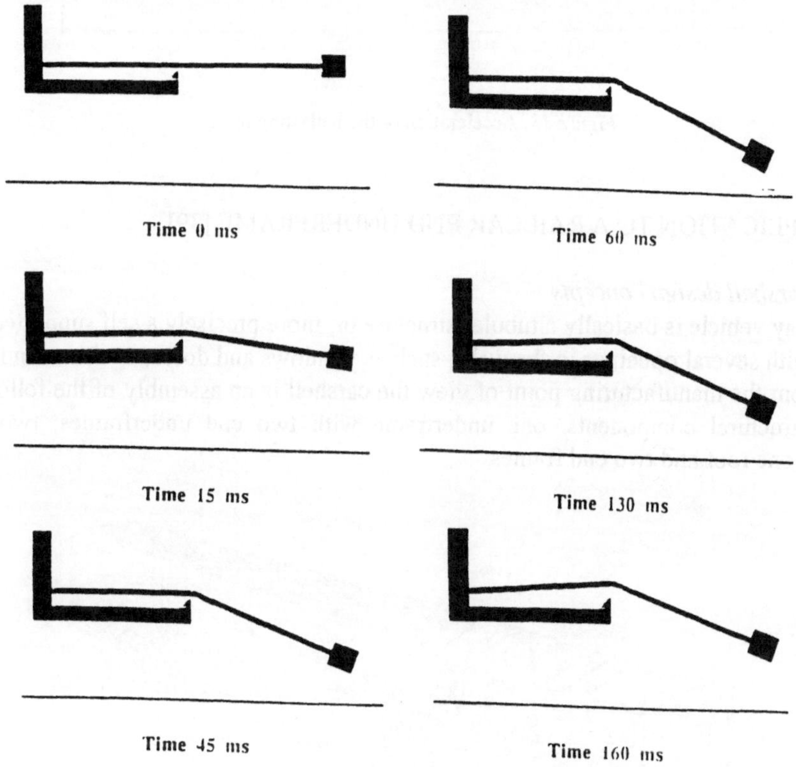

Figure 12. Evolution of the bending angle. Rigid body model

The acceleration of the ballasting mass, obtained for the rigid and flexible body simulations and for the experimental results are shown in Figure 13. It can be observed that, in what the acceleration levels are concerned, the rigid body model averages the experimental curve. The results obtained with the flexible body model clearly depict the acceleration peaks observed experimentally during the initial contact stages (t<0.005s).

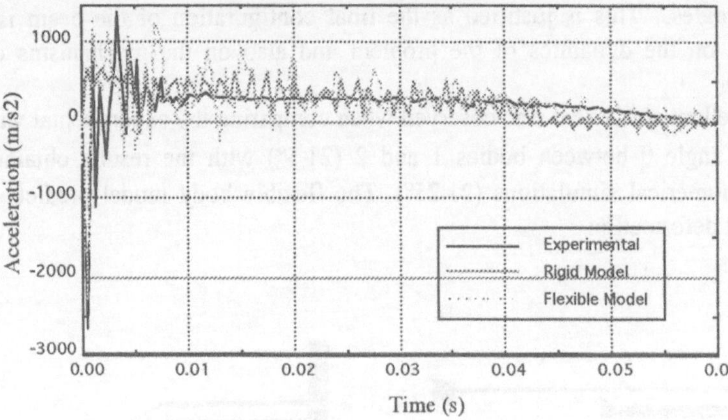

Figure 13. Acceleration of the ballasting mass

4.3. APPLICATION TO A RAILCAR END UNDERFRAME [39]

4.3.1 *Carshell design concepts*

A railway vehicle is basically a tubular structure or, more precisely a self supported box beam with several openings in the webs, such as windows and doors, as shown in figure 14. From the manufacturing point of view the carshell is an assembly of the following main structural components: one underframe with two end underframes; two side panels; one roof and two end frames

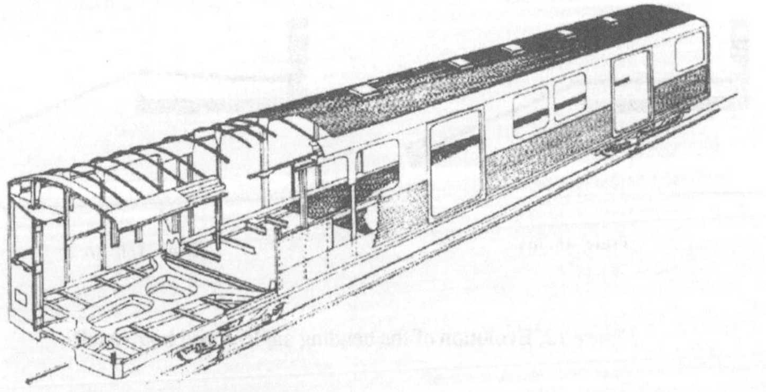

Figure 14. Main structural components of a stainless steel railway vehicle

The design guidelines towards a crashworthy vehicle behavior must consider the following requirements [42]:

- The crushed zones must be concentrated at the extremities. The predicted collapse mechanisms must be progressive and cannot develop further than the bolster beam (interface with the bogie). These collapse mechanisms shall be designed in order to maximize the absorption of the kinetic energy.
- The passengers compartment must be designed considering that no permanent deformations occur in frontal collisions up to 30 km/h.
- The longitudinal carshell strength between the bolster beams must be, at least, 15% higher than the strength at the extremities.

This design rule aims to ensure that the passenger compartment remains undeformed in cases of a crash situation. The main design objective is to minimize the vehicle and passengers accelerations during the impact period, subject to constraints on the maximum allowable deformations. To achieve this some structural collapse mechanisms have been explored to be used at the carbody extremities.

The two end underframes located at the vehicle extremities account for about 20% of the total mass of the vehicle structure. They have a crucial influence in the carshell crashworthiness behavior. Consequently the main design efforts focused on these substructures and end underframes are developed, regarding the crush plastic mechanisms prediction and the energy absorption levels. Figure 15 depicts the layout of possible structural configurations: one for central automatic coupling (a) and another for frontal side buffers (b).

In these initial designs, plastic hinges will develop initiating the first collapse mechanisms. A maximum displacement of 0.4 m, is defined before the second collapse mechanisms is initiated by crushing the "Y" shaped substructure in front of the bolster beam. The total crush displacement at the extremity is a function of the impacting speed, ranging from 0.35 m to 1.00 m.

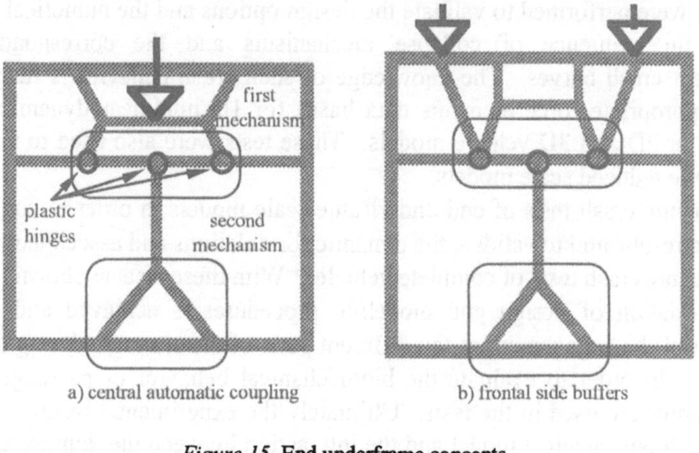

Figure 15. End underframe concepts

Note that overriding occurs in train accidents. To avoid the telescoping effects between carbodies, some design principles are established regarding the failure strength of the connections of the front and rear end collision and corner posts with both, the end underframe top beam and the roof. Though these crash mechanisms influence the ultimate crashworthy behavior of the train, they are no discussed here.

4.3.2 *Experimental tests*

To support a new carshell crashworthy design, and also with the objective of validating modeling and simulation multibody dynamic codes, a range of quasi-static and dynamic tests have been carried out:

i) Quasi-static and dynamic bending tests of typical superstructure members such as sideframe, roof and underframe, provide a database of their plastic moments curves. This database is then used to greater advantage in different design tools. Typical plastic bending moment results of commonly used stainless steel sections are shown on figure 16

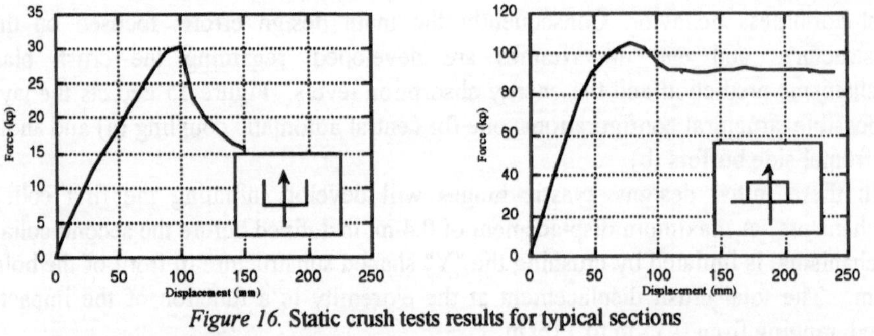

Figure 16. Static crush tests results for typical sections

ii) Quasi-static crush tests of end underframes scale models are shown in figure 17. These tests were performed to validate the design options and the numerical approaches regarding the sequence of collapse mechanisms and the corresponding force-displacement crush curves. The knowledge of such crush behavior is fundamental to generate appropriate force elements data bases for 1D nonlinear dynamic models of train rakes or 2D and 3D vehicle models. These tests were also used to validate such cost effective reduced scale models.

iii) Dynamic crash tests of end underframe scale models in order to consolidate the quasi-static results and to validate the dynamic formulations and associated models.

iv) Dynamic crash tests of complete vehicles. With these tests as shown in figure 18, a final validation of design and modeling procedures is achieved and a complete assessment of the engagement of the different parts of the structure during the collision is possible. In order to evaluate the biomechanical behavior of passengers during a crash dummies are used in the tests. Ultimately the experimental results contribute to validate the biomechanical model and the interaction between the dummy and the front seat.

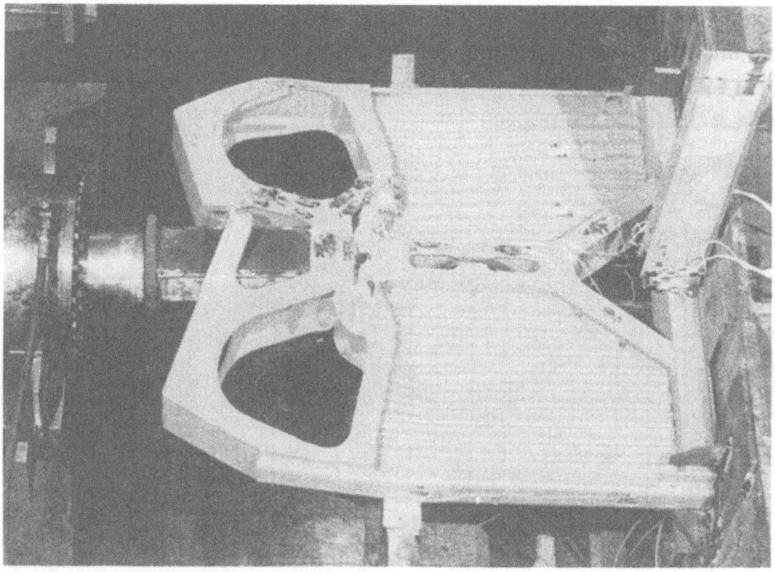

Figure 17. Experimental test. 1:3 scale model

Figure 18. Dynamic crash test (impacting speed of 30 km/h)
(crushed extremity)

4.3.3. *2D Rigid and Flexible Models with Plastic Hinges*
2D multibody models are used to access and explore the different collapse mechanisms of the end underframe structures that deform essentially in a plane. Figure 19 illustrates a rigid body model of one end underframe, designed by SOREFAME, used in a train tested for a 30 km/h frontal impact.

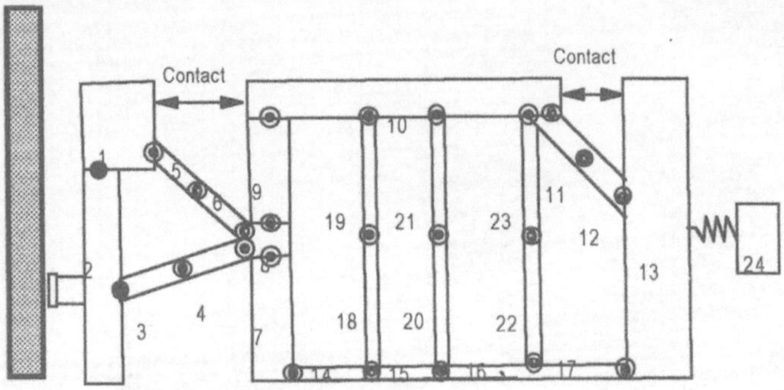

Figure 19. Discrete model of half of the end underframe

The model is composed of 24 bodies, connected by 24 plastic hinges. The plastic hinge characteristics are obtained from the elasto-plastic characteristics of the cross sections and corrected with a dynamic correction factor in order to take into account the strain rate sensitivity of the stainless steel. In the model body 13 accounts for the remaining mass of the entire carbody shell while body 24 models sand bags placed in the floor during the experimental test to take into account the mass of the equipments and interior trimming. The attachment of this mass to the end underframe is modeled by a stiff spring. In the extremity of the structure, near body 2, a buffer is placed. The contact between the buffer and the wall is modeled by a Hertz contact model [31]. In a second of the endunderframe , bodies 3-6, 7 and 9 are assumed to be flexible.

The simulation of the train crash is carried out from an instant before the carbody touches the wall until the train stops as a result of the impact. During the contact period, that lasts about 120 ms, the end underframe of the carbody dissipates all the kinetic energy of the train as a result of its deformation. The deformation of the structure obtained with the discrete deformation model composed of rigid bodies is described by Figure 20. It is patent that the mechanism of deformation is a result of successive plastic hinges that are formed as the crash progresses. The assumption for the plastic hinges locations used in the discrete deformation models are consistent with the observed deformations shown in figure 20. The progress of the deformation obtained with the second model, i.e., discrete deformations using linear flexible bodies, is similar to the deformation exhibited by the model composed of rigid models.

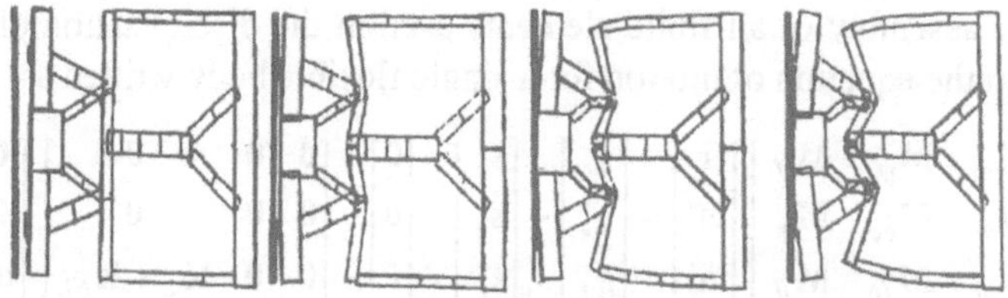

Figure 20. Deformation of the end underframe with the plastic hinge model

Accelerometers mounted in different locations of the carshell provided the accelerations of the center of the carbody. It is observed in Figure 21 that the numerical results obtained with the two models are similar to those measured.

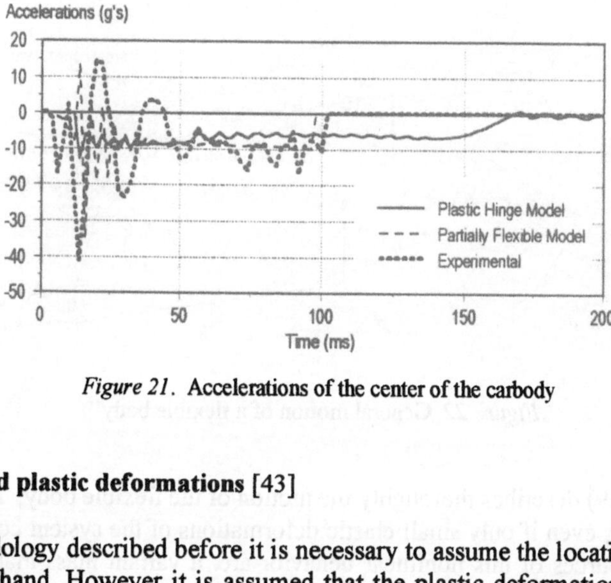

Figure 21. Accelerations of the center of the carbody

5. Distributed plastic deformations [43]

In the methodology described before it is necessary to assume the location of the plastic hinges beforehand. However it is assumed that the plastic deformations are localized. In many practical situations this is not possible and another description of the multibody flexibility is necessary. With this purpose it is proposed here a nonlinear formulation using the finite element method to describe the nonlinear geometric and material deformations of the flexible bodies of the system.

5.1. NONLINEAR DEFORMATIONS OF A FLEXIBLE BODY

The most general motion of a flexible body involves a continuous change of its shape while undergoing large rotations and displacements, as depicted by Figure 21. The equations of motion of the flexible body are obtained using the principle of virtual works followed by an updated Lagrangean formulation [43]. Let the finite element method be used to represent the equations of motion of the flexible body. Referring to Figure 22, the assembly of all finite elements used in the discretization of the flexible body results in the equation of motion for a single flexible body written as:

$$\begin{bmatrix} \mathbf{M}_{rr} & \mathbf{M}_{rf} & \mathbf{M}_{rf} \\ \mathbf{M}_{\phi r} & \mathbf{M}_{\phi\phi} & \mathbf{M}_{\phi f} \\ \mathbf{M}_{fr} & \mathbf{M}_{f\phi} & \mathbf{M}_{ff} \end{bmatrix} \begin{bmatrix} \ddot{\mathbf{r}} \\ \dot{\omega}' \\ \ddot{\mathbf{u}}' \end{bmatrix} = \begin{bmatrix} \mathbf{g}_r \\ \mathbf{g}'_\phi \\ \mathbf{g}'_f \end{bmatrix} - \begin{bmatrix} \mathbf{s}_r \\ \mathbf{s}'_\phi \\ \mathbf{s}'_f \end{bmatrix} - \begin{bmatrix} \mathbf{0} \\ \mathbf{0} \\ \mathbf{f} \end{bmatrix} - \begin{bmatrix} \mathbf{0} & \mathbf{0} & \mathbf{0} \\ \mathbf{0} & \mathbf{0} & \mathbf{0} \\ \mathbf{0} & \mathbf{0} & \mathbf{K}_L + \mathbf{K}_{NL} \end{bmatrix} \begin{bmatrix} \mathbf{0} \\ \mathbf{0} \\ \mathbf{u}' \end{bmatrix} \qquad (19)$$

where $\ddot{\mathbf{r}}$ and $\dot{\omega}'$ are respectively the translational and angular accelerations of the origin of the body fixed reference frame and $\ddot{\mathbf{u}}'$ denotes the nodal accelerations measured in the body fixed coordinate system. Note that the local coordinate frame $\xi\eta\zeta$ attached to a point of the flexible body is used to represent the body's gross motion and it is also used to describe the deformation of the body. Vector $\mathbf{u}'$ denotes the increments of displacements from the previous configuration to the current configuration measured in the body fixed coordinate system.

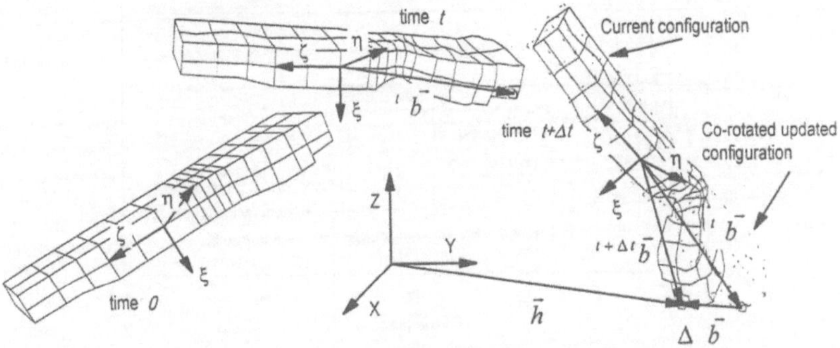

Figure 22 General motion of a flexible body

Equation (19) describes thoroughly the motion of the flexible body, However, it is highly nonlinear, even if only small elastic deformations of the system components are present. The sources of this nonlinear behavior are: a variant mass matrix; changing external applied forces; gyroscopic and centrifugal forces; non-constant stiffness matrices. The variant mass matrix for the flexible body results from the assembly of the individual contributions of each finite element. The contribution of a single finite element for the mass matrix is described by the sub-matrices

$$\begin{aligned} \mathbf{M}_{rr_j} &= \mathbf{I}\int_{V_j} \rho \, dv & \mathbf{M}_{r\phi_j} &= -\mathbf{A}\int_{V_j} \rho \, \tilde{\mathbf{b}}' \, dv \\ \mathbf{M}_{rf_j} &= \mathbf{A}\int_{V_j} \rho \, \mathbf{N} \, dv & \mathbf{M}_{\phi\phi_j} &= -\int_{V_j} \rho \, \tilde{\mathbf{b}}'\tilde{\mathbf{b}}' \, dv \\ \mathbf{M}_{\phi f_j} &= \int_{V_j} \rho \, \tilde{\mathbf{b}}'\mathbf{N} \, dv & \mathbf{M}_{ff_j} &= \int_{V_j} \rho \, \mathbf{N}^T\mathbf{N} \, dv \end{aligned} \qquad (20)$$

Here, **A** is the transformation matrix from the body fixed coordinate system to the inertial frame, **N** is the matrix of shape functions of element *j* and ρ is the mass density. In these equations the submatrices $\mathbf{M}_{rr}$ and $\mathbf{M}_{ff}$ are constant and represent respectively the mass of the entire element and the standard finite element mass matrix. $\mathbf{M}_{rf}$ and $\mathbf{M}_{\phi f}$ are the time variant matrices responsible for the inertia coupling between the gross motion of the body and its deformations. In the numerical implementation of the mass matrix special attention must be paid to the evaluation of $\mathbf{M}_{\phi\phi}$ as large deformation develop. This sub-matrix, representing the inertia tensor of the flexible body is approximately constant if the body deformations are small, otherwise its time variance cannot be neglected. All other submatrices in equation (19) are either null or constant, provided that a proper choice is made for the location and orientation of the body fixed coordinate frame.

The right-hand side of equation (19) contains, a time variant vector of gyroscopic and centrifugal forces which are evaluated for each element as

$$\begin{bmatrix} \mathbf{s}_r \\ \mathbf{s}'_\phi \\ \mathbf{s}'_f \end{bmatrix}_j = \begin{bmatrix} \mathbf{A}\tilde{\boldsymbol{\omega}}'\tilde{\boldsymbol{\omega}}' \int_{V_j} \rho\ \mathbf{b}'\ dv \\ \int_{V_j} \rho\ \tilde{\mathbf{b}}'\tilde{\boldsymbol{\omega}}'\tilde{\boldsymbol{\omega}}'\mathbf{b}'\ dv \\ \int_{V_j} \rho\ \mathbf{N}^T\tilde{\boldsymbol{\omega}}'\tilde{\boldsymbol{\omega}}'\mathbf{b}'\ dv \end{bmatrix} + 2 \begin{bmatrix} \mathbf{A}\tilde{\boldsymbol{\omega}}' \int_{V_j} \rho\ \mathbf{N}\ dv \\ \int_{V_j} \rho\ \tilde{\mathbf{b}}'\tilde{\boldsymbol{\omega}}\mathbf{N}\ dv \\ \int_{V_j} \rho\ \mathbf{N}^T\tilde{\boldsymbol{\omega}}'\mathbf{N}\ dv \end{bmatrix} \dot{\mathbf{u}}'_j \tag{21}$$

The vector of the external generalized applied forces $\mathbf{g}_j$ for each element is:

$$\begin{bmatrix} \mathbf{g}_r \\ \mathbf{g}'_\phi \\ \mathbf{g}'_f \end{bmatrix}_j = \begin{bmatrix} \int_{A_j} \mathbf{f}_s\ da \\ \int_{A_j} \tilde{\mathbf{b}}'\mathbf{A}^T\ \mathbf{f}_s\ da \\ \int_{A_j} \mathbf{N}^T\mathbf{A}^T\ \mathbf{f}_s\ da \end{bmatrix} + \begin{bmatrix} \int_{V_j} \rho\ \mathbf{f}_b\ dv \\ \int_{V_j} \rho\ \tilde{\mathbf{b}}'\mathbf{A}^T\ \mathbf{f}_b\ dv \\ \int_{V_j} \rho\ \mathbf{N}^T\mathbf{A}^T\ \mathbf{f}_b\ dv \end{bmatrix} \tag{22}$$

In equation (22), $\mathbf{f}_b$ and $\mathbf{f}_s$ are respectively the body and the surface forces. Matrices $\mathbf{K}_L$ and $\mathbf{K}_{NL}$ are the linear and nonlinear stiffness matrices respectively, and **f** denotes the vector of equivalent nodal forces due to the actual state of stress, respectively written as:

$$\mathbf{K}_L = \int_{V_j} \mathbf{B}_L^T\ \mathbf{C}\ \mathbf{B}_L\ dv \tag{23}$$

$$\mathbf{K}_{NL} = \int_{V_j} \mathbf{B}_{NL}^T\ \boldsymbol{\tau}'\ \mathbf{B}_{NL}\ dv \tag{24}$$

$$\mathbf{f} = \int_{V_j} \mathbf{B}_L^T\ \hat{\boldsymbol{\tau}}'\ dv \tag{25}$$

In these equations $\mathbf{B}_L$ and $\mathbf{B}_{NL}$ denote the linear and nonlinear strain matrices respectively and $\boldsymbol{\tau}'$ is the Cauchy stress tensor for the updated configuration. Note that the reference to the linearity of the stiffness matrices $\mathbf{K}_L$ and $\mathbf{K}_{NL}$ is concerned to their relation with the displacements. For a constitutive tensor $\mathbf{C}$ not constant both $\mathbf{K}_L$ and $\mathbf{K}_{NL}$ are not linear. This is the case when a multibody system experiences elasto-plastic deformations of one or more of its components. For these problems, an elasto-plastic constitutive tensor $\mathbf{C}$ must be used in the equation (23).

5.2. GENERALIZED FLEXIBLE COORDINATES

Equation (19) is not efficient for numerical implementation due to the variance of the mass and stiffness matrices and the need to invert the mass matrix every time step during the integration process to solve the equations explicitly for the accelerations. A simpler form of the equations of motion for a flexible body is obtained using a lumped mass formulation and substituting the local nodal accelerations $\ddot{\mathbf{u}}'$ are substituted by the nodal accelerations relative to the inertial frame $\ddot{\mathbf{q}}_f'$.

Let the vectors of nodal accelerations be partitioned into translational and angular accelerations as:

$$\mathbf{u}' = \left[\boldsymbol{\delta}'^T, \boldsymbol{\theta}'^T\right] \tag{26}$$

The relation between the relative and absolute nodal accelerations for a node k is described by:

$$\ddot{\mathbf{q}}'_{f\,k} = \begin{bmatrix} \ddot{\mathbf{d}}' \\ \ddot{\boldsymbol{\alpha}}' \end{bmatrix}_k = \ddot{\mathbf{u}}'_k + \begin{bmatrix} \mathbf{A}^T & -\left(\tilde{\mathbf{x}}_k + \tilde{\boldsymbol{\delta}}_k\right)' \\ \mathbf{0} & \mathbf{I} \end{bmatrix} \begin{bmatrix} \ddot{\mathbf{r}} \\ \dot{\boldsymbol{\omega}}' \end{bmatrix} + \begin{bmatrix} \tilde{\boldsymbol{\omega}}'\tilde{\boldsymbol{\omega}}'(\mathbf{x}_k + \boldsymbol{\delta}_k)' + 2\tilde{\boldsymbol{\omega}}'\dot{\boldsymbol{\delta}}'_k \\ \tilde{\boldsymbol{\omega}}'\dot{\boldsymbol{\theta}}'_k \end{bmatrix} \tag{27}$$

where $\mathbf{x}_k$ is the vector containing the node position in the reference configuration. Equation (27) is evaluated for all n nodes and substituted into equation (19) yielding

$$\sum_{k=1}^{n} m_k \ddot{\mathbf{d}}'_k = \mathbf{g}_r \tag{28a}$$

$$\sum_{k=1}^{n} \left[m \left(\tilde{\mathbf{x}} + \tilde{\boldsymbol{\delta}}\right)' \ddot{\mathbf{d}}' \right]_k = \mathbf{g}'_\theta \tag{28b}$$

$$\mathbf{M}_{ff}\, \ddot{\mathbf{q}}'_f = \mathbf{g}'_f - {}^t_{t'}\mathbf{F} - \left({}^t_{t'}\mathbf{K}_L + {}^t_{t'}\mathbf{K}_{NL}\right)\mathbf{u}' \tag{28c}$$

Equations (28a) and (28b) describe the motion for the center of mass of a system of particles [44]. If the origin of the body fixed coordinate system is coincident with the center of mass of the flexible body, equations (28a) and (29b) describe the motion of the origin of the ξηζ referential. Equation (28c) is the equation of motion for the nodes of the flexible body, expressed in the body fixed coordinate system. Due to the use of the lumped mass formulation the mass matrix $\mathbf{M}_{ff}$ is diagonal

$$\mathbf{M}_{ff} = \mathrm{Diag}\,(m_1\,\mathbf{I}\,, \mathbf{0}\,, \ldots, m_k\,\mathbf{I}\,, \mathbf{0}\,, \ldots, m_n\,\mathbf{I}\,, \mathbf{0}) \tag{29}$$

where m_k is the lumped mass of node k, and **I** and **0** are 3×3 identity and null matrices associated with the translational and rotational degrees of freedom, respectively.

Very often it is useful to locate the origin of the body fixed coordinate system in a point of the flexible body different from its center of mass. For this purpose assume that the flexible body has a rigid part and a flexible part being the body fixed coordinate frame attached to the center of mass of the rigid part, as shown in Figure 23. The flexible part is attached to the rigid part by the nodes that belong to boundary ψ and shares the body fixed coordinate frame with the rigid part. For a rigid body, the Newton-Euler equations of motion are written as

$$m\ddot{\mathbf{r}} = \mathbf{f}_r \tag{30a}$$

$$\mathbf{J}'\dot{\boldsymbol{\omega}}' = \mathbf{n}' - \tilde{\boldsymbol{\omega}}'\mathbf{J}'\boldsymbol{\omega}' \tag{30b}$$

where $\mathbf{f}_r$ and $\mathbf{n}'$ are the external forces applied over the center of mass of the rigid part and moments applied over the body. Equations (30a) and (30b) can be used instead of (28a) and (28b) provided that proper kinematic constraints are introduced between the flexible and rigid parts of the body. These kinematic constraints enforce the nodes in the boundary ψ to have null displacements, velocities and accelerations with respect to the body fixed coordinate frame

$$\mathbf{u}' = \dot{\mathbf{u}}' = \ddot{\mathbf{u}}' = \mathbf{0} \tag{31}$$

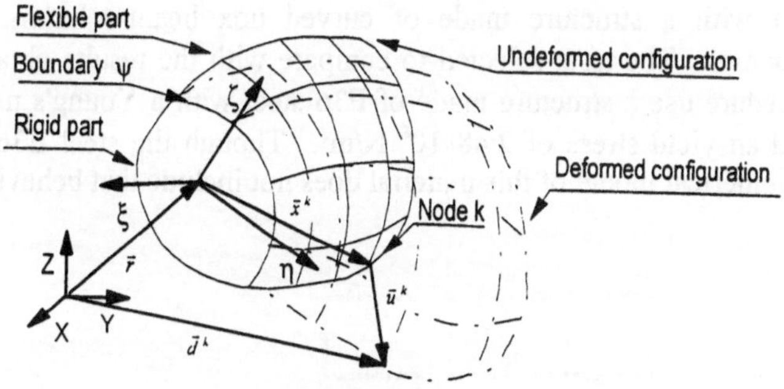

Figure 23 Flexible body with a rigid part

The constraint equations can be applied to the equation (28c) for each one of the boundary nodes using the Lagrange multiplier technique. At a latter step these multipliers are eliminated from the equations of motion resulting in the dynamic equations [45]

$$\begin{bmatrix} m\mathbf{I}+\overline{\mathbf{A}}\underline{\mathbf{M}}^*\overline{\mathbf{A}}^T & -\overline{\mathbf{A}}\underline{\mathbf{M}}^*\mathbf{S} & \mathbf{0} \\ -\left(\overline{\mathbf{A}}\underline{\mathbf{M}}^*\mathbf{S}\right)^T & \mathbf{J}'+\mathbf{S}^T\underline{\mathbf{M}}^*\mathbf{S} & \mathbf{0} \\ \mathbf{0} & \mathbf{0} & \mathbf{M}_{ff} \end{bmatrix} \begin{bmatrix} \ddot{\mathbf{r}} \\ \dot{\boldsymbol{\omega}}' \\ \ddot{\mathbf{q}}'_f \end{bmatrix} = \begin{bmatrix} \mathbf{f}_r + \overline{\mathbf{A}}\mathbf{C}'_\delta \\ \mathbf{n}' - \tilde{\boldsymbol{\omega}}'\mathbf{J}'\boldsymbol{\omega}' - \mathbf{S}^T\mathbf{C}'_\delta - \overline{\mathbf{I}}^T\mathbf{C}'_\theta \\ \mathbf{g}'_f - \mathbf{f} - \left(\mathbf{K}_L + \mathbf{K}_{NL}\right)\mathbf{u}' \end{bmatrix} \tag{32}$$

where $\underline{\mathbf{M}}^*$ is a diagonal mass matrix containing the mass of the $\underline{n}$ boundary nodes. Matrices $\overline{\mathbf{A}}^T$, $\mathbf{S}$ and $\overline{\mathbf{I}}$ are made from (3x3) matrices as:

$$\overline{\mathbf{A}}^T = [\mathbf{A} \;\; \mathbf{A} \;\; \cdots \;\; \mathbf{A}]^T \; ; \; \mathbf{S} = \left[(\bar{\mathbf{x}}'_1 + \tilde{\delta}'_1)^T \;\; (\bar{\mathbf{x}}'_2 + \tilde{\delta}'_2)^T \;\; \cdots \;\; (\bar{\mathbf{x}}'_{\underline{n}} + \tilde{\delta}'_{\underline{n}})^T \right]^T \; ; \; \overline{\mathbf{I}} = [\mathbf{I} \;\; \mathbf{I} \;\; \cdots \;\; \mathbf{I}]^T$$

Vectors $\mathbf{C}'_\delta$ and $\mathbf{C}'_\theta$ represent respectively the reaction force and moment of the flexible part of the body over the rigid part. These reaction force/moments are

$$\mathbf{C}'_\delta = \mathbf{g}'_{\underline{\delta}} - \mathbf{F}_{\underline{\delta}} - (\mathbf{K}_L + \mathbf{K}_{NL})_{\underline{\delta}\delta} \delta' - (\mathbf{K}_L + \mathbf{K}_{NL})_{\underline{\delta}\theta} \theta' \quad (33\text{ a})$$

$$\mathbf{C}'_\delta = \mathbf{g}'_{\underline{\theta}} - \mathbf{F}_{\underline{\theta}} - (\mathbf{K}_L + \mathbf{K}_{NL})_{\underline{\theta}\delta} \delta' - (\mathbf{K}_L + \mathbf{K}_{NL})_{\underline{\theta}\theta} \theta' \quad (33\text{ b})$$

In these equations subscripts δ' and θ' refer to the partition of the vectors and matrices with respect to the translational and rotational nodal degrees of freedom. The underlined subscripts are referred to the boundary nodal displacements of the nodes fixed to the rigid part.

5.3. APPLICATION TO THE SIMULATION OF A TORQUE-BOX IMPACT

A simple torque box shown in figure 24 is considered here to demonstrate the application of the distributed deformation model to a structural impact of a multibody system. In reference [46] Ni presents the experimental setup and results of a moving barrier impact with a structure made of curved box beam-columns for different incoming velocities. The tests selected to compare with the results obtained with the proposed procedure use a structure made of E36 steel with a Young's modulus of 2.1 10^{11} N/m^2 and an yield stress of 2.48 10^8 N/m^2. Though the steel E36 is strain rate sensitive the numerical model of this material does not include that behavior.

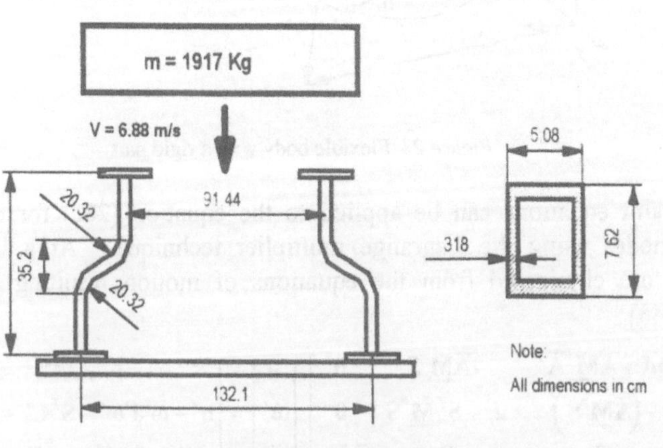

Figure 24 **Experimental setup for the torque-box**

In the simulations carried out in this work the moving barrier is moving towards the torque box with a velocity of 6.88 m/s. Different models were considered for the torque box modeled with nonlinear thin wall beam elements. The results for the accelerations displacement and velocity for the models using 7 and 11 nodes to describe half of the torque box structure are presented in Figures 25 and 26, respectively.. Though simulations with finer meshes were carried out, their results do not differ considerably from those obtained with the 11 node mesh for each beam-column. Comparing the predicted behavior of the structure with the experimental results it is noticed that a reasonable agreement is obtained for the displacements with the 11 node mesh. The predicted velocities show a slightly stiffer behavior of the numerical model of the torque box than the tested structure. The accelerations obtained present a peak similar to that of the experimental test and during the initial part of the deformation the average values are similar. After a while, the computational model accelerations start to drift away from the experimental accelerations remaining in average higher than these. This behavior is expected because of all the local nonlinear effects, such as local instabilities in the region of the plastic hinges, that exist in the experimental structure cannot be accurately modeled with beam elements.

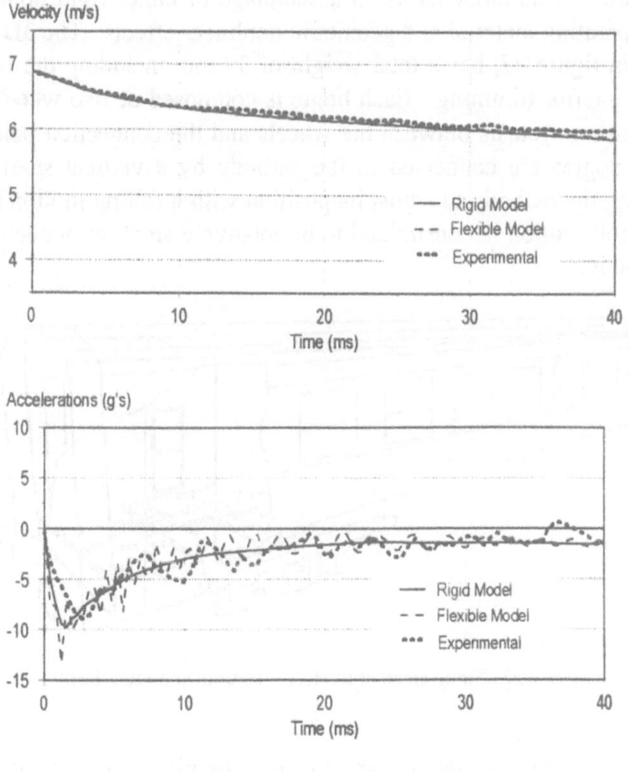

Figure 25. **Velocity and acceleration of the top nodes of the torque box**

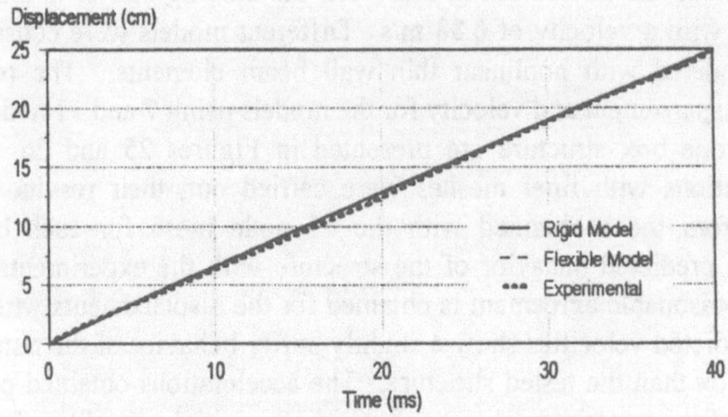

Figure 26 Displacement of the top nodes of the torque-box

5.4. STRUCTURAL IMPACT OF A TRAIN CARBODY

The spatial flexible multibody models are used to access the overall behavior of a railcar extremities. This modeled as an assemblage of either rigid bodies or flexible members incorporating material and geometric nonlinear effects. The 3D train carbody model, shown in figure 27, has a total weight of 39 ton. including the two bogies, all equipment and interior trimming. Each bogie is composed of two wheel sets and two levels of suspension systems between the wheels and the connection point to the train carbody. The bogies are connected to the carbody by a vertical shaft in an elastic bushing allowing the carbody to adjust its position with a rolling motion relative to the railroad. This roll motion is constrained to be relatively small by jounce stops in each side of the carbody.

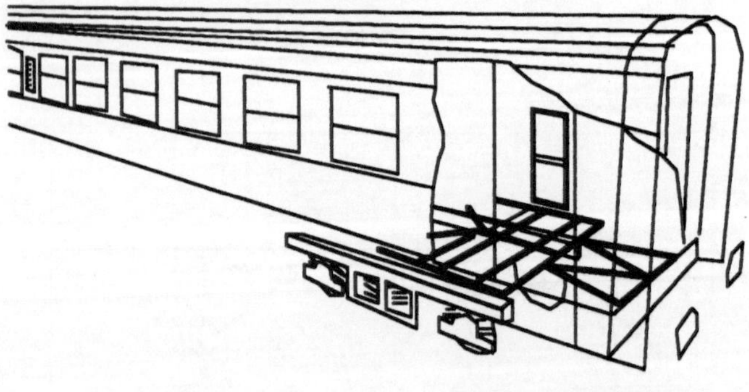

Figure 27. Finite element model of the train end-underframe

Considering that crushing takes place in the extremity, it is assumed in the finite element model of the carbody that the end underframe is deformable. The rest of the carbody between the bolster beams is the rigid part of a partially flexible body. The computer model of the train carbody used in the simulation is composed of 13 moving bodies and 20 sets of spring-dampers corresponding to the primary and secondary suspension systems. The interaction between wheels and rails is modeled by nonlinear spring-damper elements, high lateral sliding friction coefficients and very low rolling friction coefficients. The initial conditions of the simulation correspond to a train carbody moving with a speed of 30 Km/h towards a rigid wall.

The finite element model of the end underframe, shown in figure 27, is made of 54 nodal points, corresponding to 150 degrees of freedom for the flexible part and 62 nonlinear beam elements. The material used in this model is a stainless steel with a plastic modulus that is one tenth of its elastic modulus.

During the contact period, that lasts about 175 ms, the end underframe of the carbody dissipates all the kinetic energy of the train by deformation, as described in figure 28. The deformation patterns clearly indicate that the end underframe collapse mechanism is a result of the development of successive plastic hinges as the crash progresses. The kinetics of the end-underframe deformation is confirmed using a detailed nonlinear finite element model and the nonlinear finite element code [47]. A nonlinear static analysis of this model is carried on with the front nodes of the end-underframe loaded until the structure exhausts its capability to deform any further. Though the type of finite elements and the mesh used in this analysis are completely different from those used with the methodology described previously, the progress of the deformation, shown in figure 29, is similar to the one predicted by the model using the partially flexible body.

The limit load obtained with both analysis is about 120 ton., for half of the end underframe, when the first plastic hinges are formed, as plotted in figure 30. However, the prediction by the detailed finite element model for the limit load corresponding to the second plastic hinge, appearing close to the bolster beam, is about 30% higher then what is predicted in the current application.

In the experimental test which was carried out with this train carbody [48] the maximum load carried by each buffer of the vehicle was measured to be 120 tons (refer to figure 30). Here the pattern of deformation observed in the end underframe is similar to the one predicted by numerical model of the train. However neither a second peak of load was measured nor a second plastic hinge closer to the bolster beam was formed. That is because in the experimental test a collision of the top of the carbody with the rigid wall occurred forcing the kinetic energy to be absorbed by a different region of the carbody. Consequently the deformation of the end underframe did not progress any further. The maximum displacement obtained for the deformation of the end-underframe in the experimental test was 0.309 m, which corresponds to the deformation described by figure 28(d). In the numerical model the total displacement of the front of the carbody with respect to its center of mass is 0.875 m. After the first 90 ms of the impact the numerical models behavior diverges from that of the experimental test. These results are consistent with the observation that the second part of the impact observed is not modeled.

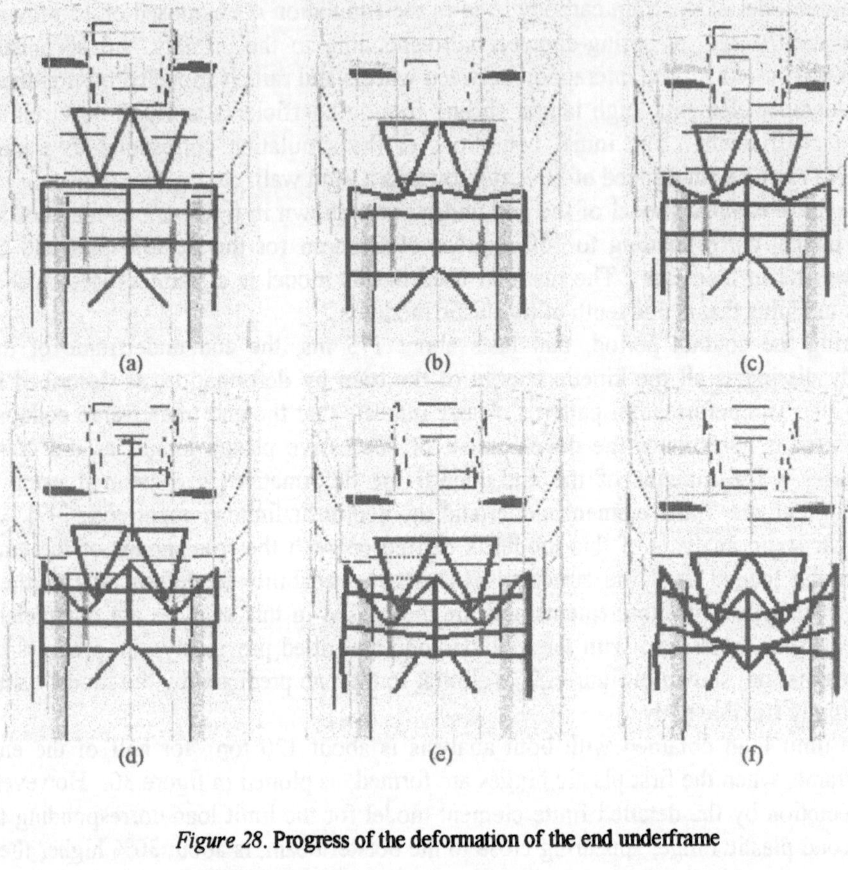

Figure 28. **Progress of the deformation of the end underframe**

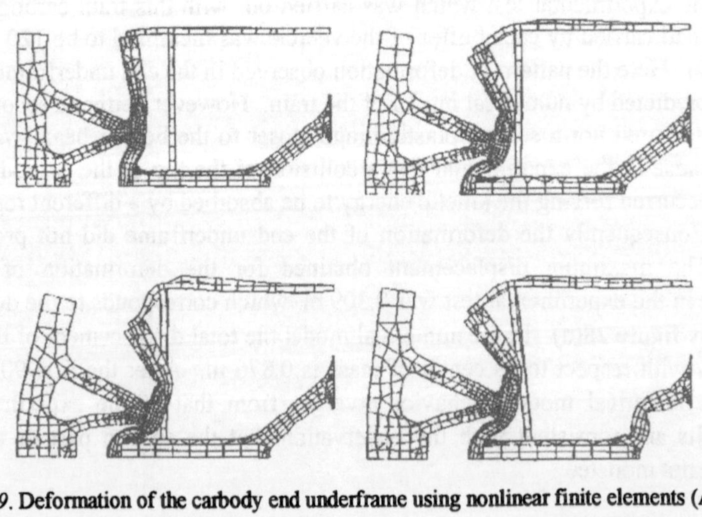

Figure 29. **Deformation of the carbody end underframe using nonlinear finite elements (ABAQUS)**

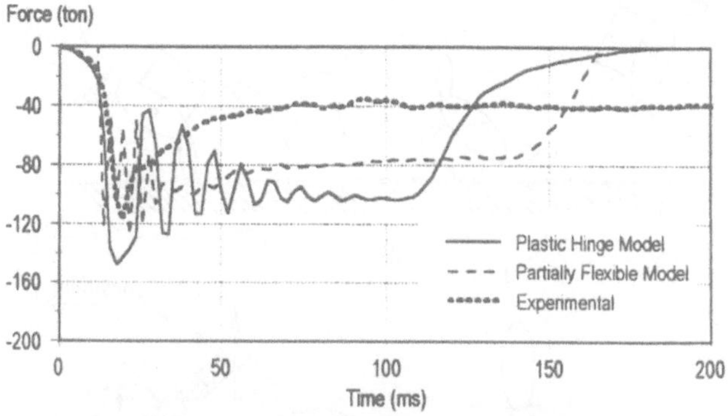

Figure 30. Reaction force over the wall for each front buffer

6. The kinetostatic method [49]

In some crashworthiness applications the inertia of the part of the flexible body that actually deforms during impact is small when compared with the inertia of the undeformed part. This assumption is the basis of the kinetostatic method. The deformable part of the body is included in the model as a massless structure attached to a rigid body as depicted by Figure 31. Consequently any coupling between the rigid body motion and the body deformation is neglected.

6.1. KINETOSTATIC EQUATIONS AND CONTACT CONDITIONS

Using the assumption of a massless flexible part, equation (32) is re-written as:

$$\begin{bmatrix} m\mathbf{I} & \mathbf{0} & \mathbf{0} \\ \mathbf{0} & \mathbf{J}' & \mathbf{0} \\ \mathbf{0} & \mathbf{0} & \mathbf{0} \end{bmatrix} \begin{bmatrix} \ddot{\mathbf{r}} \\ \dot{\boldsymbol{\omega}}' \\ \ddot{\mathbf{q}}'_f \end{bmatrix} = \begin{bmatrix} \mathbf{f}_r + \bar{\mathbf{A}}\mathbf{C}'_\delta \\ \mathbf{n}' - \tilde{\boldsymbol{\omega}}'\mathbf{J}'\boldsymbol{\omega}' - \mathbf{S}^T\mathbf{C}'_\delta - \bar{\mathbf{I}}^T\mathbf{C}'_\theta \\ \mathbf{g}'_f - \mathbf{f} - (\mathbf{K}_L + \mathbf{K}_{NL})\mathbf{u}' \end{bmatrix} \tag{34}$$

where m and J' are the mass and the inertia tensor of the complete body respectively. Clearly the equations of motion for the rigid part are uncoupled from those of the flexible part and are re-written as

$$\begin{bmatrix} m\mathbf{I} & \mathbf{0} \\ \mathbf{0} & \mathbf{J}' \end{bmatrix} \begin{bmatrix} \ddot{\mathbf{r}} \\ \dot{\boldsymbol{\omega}}' \end{bmatrix} = \begin{bmatrix} \mathbf{f}_r + \mathbf{f}^* \\ \mathbf{n}' - \tilde{\boldsymbol{\omega}}'\mathbf{J}'\boldsymbol{\omega}' + \mathbf{n}^* \end{bmatrix} \tag{35}$$

The equilibrium equations for the flexible part are nonlinear static equations expressed in the body fixed coordinate system

$$(\mathbf{K}_L + \mathbf{K}_{NL})\mathbf{u}' = \mathbf{g}'_f - \mathbf{f} \tag{36}$$

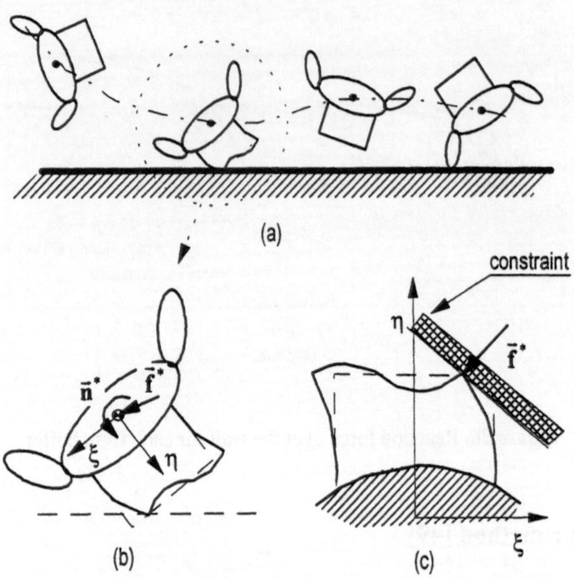

Figure 31. **Deformation of a flexible body using a kinetostatic method.**

When the flexible part impacts another object, for example a rigid nonmoving obstacle, it undergoes some deformations, as depicted by Figure 31(b). In turn the effect of the deformation of the flexible part over the attached rigid part is described by a force/moment. The reaction force and moment of the flexible part over the rigid part, denoted by f^* and n^* respectively, are given by

$$\mathbf{f}^* = \overline{\mathbf{A}}\mathbf{C}'_\delta \tag{37}$$

$$\mathbf{n}^* = -\mathbf{S}^T\mathbf{C}'_\delta - \overline{\mathbf{I}}^T\mathbf{C}'_\theta \tag{38}$$

The vector of applied nodal forces $\mathbf{g}'_f$ still contains the external forces applied over the nodes. For the purpose of deriving an expression for the nodal forces due to the impact with an obstacle assume for the moment that all applied forces over the flexible part are due to the impact, i.e., they are the reaction forces of the obstacle over the structure plus the contact friction forces.

Let the jth node of the finite element representation of the flexible part come in contact with the surface of an obstacle. This surface is defined by the global coordinates of a point Q_j, denoted by d_j^Q and a normal unit vector n_j as shown in Figure 32. At a given instant, the global coordinates of the position of node j for the undeformed flexible body, denoted by d_j is calculated from the global configuration of the rigid part. The pseudo penetration α_j of node j into the contacting surface is calculated as

$$\alpha_j = \mathbf{n}_j^T\left(\mathbf{d}_j^Q - \mathbf{d}_j\right) \tag{39}$$

In reality, the structure deforms such a way that node j remains on the surface of the obstacle. Let δ_j denote the vector of nodal displacements of node j with respect to its undeformed position. Its projection onto the normal to the surface must be equal to the pseudo penetration α_j

$$\alpha_j = \mathbf{n}_j^T \boldsymbol{\delta}_j \tag{40}$$

Equation (40) is substituted into equation (31) to yield a constraint equation for node j

$$\mathbf{n}_j^T \left(\mathbf{d}_j^Q - \mathbf{d}_j - \boldsymbol{\delta}_j \right) = 0 \tag{41}$$

In order to apply this constraint to the static equilibrium equations of the flexible part given by equation (36), the nodal constraint equation (40) is first written in terms of the nodal displacements with respect to the body fixed coordinate system

$$\mathbf{n}_j^T \left[\mathbf{d}_j^Q - \mathbf{A}\left(\boldsymbol{\delta}'_j + \mathbf{b}'_j \right) - \mathbf{r} \right] = 0 \tag{42}$$

where, according to Figure 32, the result $\mathbf{d}_j = \mathbf{A}\mathbf{b}'_j + \mathbf{r}$ is used. Equation is rearranged as

$$\mathbf{n}_j^T \mathbf{A} \boldsymbol{\delta}'_j = -\mathbf{n}_j^T \left(\mathbf{A}\,\mathbf{b}'_j + \mathbf{r} - \mathbf{d}_j^Q \right) \tag{43}$$

Equation (43) represents a kinematic constraint on the displacement of node j. If more than one node simultaneously contact one or more obstacles, equation (43) is written for each node and the resulting set of constraints is denoted in a compact form as

$$\mathbf{G}\,\mathbf{u}' = \boldsymbol{\alpha} \tag{44}$$

where the rows of matrix G contain the components of vectors normal to the obstacle, u' is the vector of the nodal displacements for all of the nodes, and α contains α's for all contacting nodes that are described by the right-hand side of equation (43). Note that equation (44) is a set of linear equations with the displacement increments u' as unknowns.

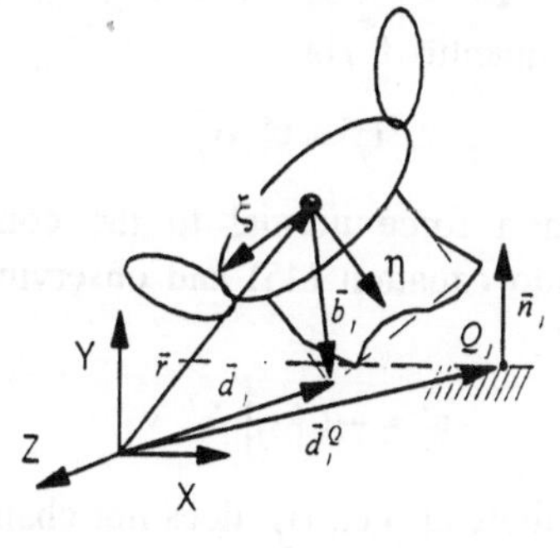

Figure 32. Definition of the position constraint posed by the rigid surface

The reaction force at node j due to the contact is denoted by f^*_j. The friction force acting on the flexible part in the contacting node is expressed as

$$\mathbf{g}_j = -\mu \left|\mathbf{f}^*_j\right| \mathbf{A}\, \mathbf{v}_j \tag{45}$$

where μ is the friction coefficient and v_j is a unit vector in the direction of the velocity of node j projected on the constraint surface. This force is only valid if there is sliding of node j. If the friction force given by equation (45) is smaller than the product of dynamic friction coefficient by the normal reaction force due to the ground normal reaction, then stiction occurs. This case is not considered here.

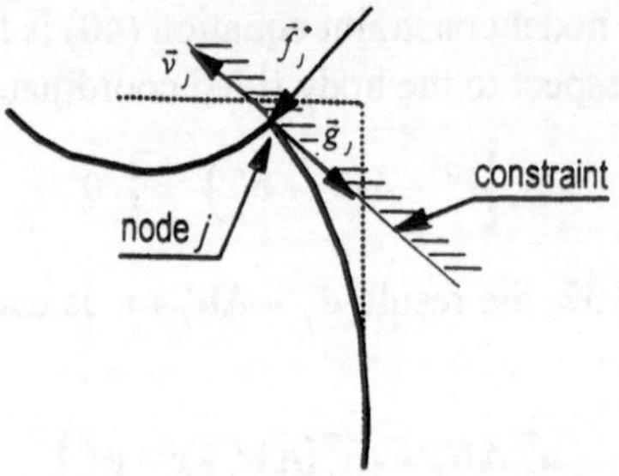

Figure 33. Reaction and friction forces at node *j*

The constraint equation represented by equation (44) and the friction forces are introduced in the static equilibrium equations of the flexible part using the Lagrange multipliers technique. Denoting by σ the Lagrange multipliers associated with the constraints of the contacting nodes, the system of constrained equilibrium equations becomes

$$(\mathbf{K}_L + \mathbf{K}_{NL})\mathbf{u}' = \mathbf{g}'_f - \mathbf{f} - \mathbf{G}^T\sigma \tag{46}$$

Note that the term $\mathbf{G}^T\sigma$ in equation (44) represents the constraint forces due to the impact. For the jth node, the quantity f^*_j is

$$\mathbf{f}'^*_j = \mathbf{G}_j \sigma_j \tag{47}$$

which is exactly the reaction force normal to the constraint surface at node j. Substituting this equation into equation (45) and observing that n_j is an unit vector yields the friction force as

$$\mathbf{g}'_j = -\mu \left|\sigma_j\right| \mathbf{A}^T\, \mathbf{v}_j \tag{48}$$

The nodal constraints are unilateral; i.e., σ_j does not change sign and it is a positive quantity as long as there is contact. Therefore, equation (48) is written for each contacting nodes as

$$\mathbf{g}'_j = \mathbf{H}_j^T \, \sigma_j \tag{49}$$

where $\mathbf{H}_j^T = -\mu \; \mathbf{A}^T \; \mathbf{v}_j$. If more than one node is in contact with the obstacle and friction forces are the only external forces on the flexible part, besides the reaction forces, equation (49) is evaluated for all those nodes yielding:

$$\mathbf{g}'_{fr} = \mathbf{H}^T \, \sigma \tag{50}$$

where matrix H^T contains all H_j^T's. The friction forces are substituted in equation (38) results in

$$(\mathbf{K}_L + \mathbf{K}_{NL})\mathbf{u}' = \mathbf{g}'_f - \mathbf{f} - (\mathbf{G} - \mathbf{H})^T \sigma \tag{51}$$

In order to calculate the displacement increments u' and the Lagrange multipliers σ equation (51) and equation (44) must be solved simultaneously. After evaluating the Lagrange multipliers, equations (46) and (47) are used to calculate the reaction and friction forces at every contacting node. Since the structure is in static equilibrium, the set of reaction forces f^* given by equation (37) is equivalent to the set of forces directly applied to the rigid part of the body. For a typical node *j*, in contact, the forces act at a point considered as a natural extension of the rigid part with its local coordinates coincident with those of the node. These forces cause a moment on the rigid part of the body equal to the cross product of the vector locating point *j* with respect to the local coordinate frame by the reaction force vector.

During a simulation, as long as none of the nodes in the structure is in contact with any obstacles, f^* is a null vector. This means that the dynamic analysis proceeds as a multi-rigid body system. In order to detect if a particular node *j* contacts or penetrates a surface at a certain time step, the term α_j is calculated from equation (38). A positive α_j means no contact while a null or negative α_j indicates a penetration. When a penetration is detected, the corresponding reaction force/moment is calculated and included in the vector of forces. This reaction force/moment is updated as long as the node is in contact with the obstacle. When more than one node is detected to be in penetration, the sign of all Lagrange multipliers in vector σ must be verified. If any of these multipliers turns out to be negative, its corresponding constraint must be removed and equation (51) must be solved again.

The methodology presented is equivalent to a finite element analysis of the flexible part with imposed displacements. For a general case the kinematics of the imposed displacements of the contact nodes is obtained by the dynamics of the multibody system. Provided that the pattern of the deformation of the structural component does not differ from that of the flexible body during impact, it is possible to use the results of a finite element analysis, previously done for the flexible component, instead of solving directly equation (50). In this case it is assumed that the loading path of the structural component is similar in both cases. This is the procedure used in the application discussed next.

6.2. APPLICATION TO THE DESIGN OF ENERGY ABSORBING COMPONENTS FOR LOW SPEED IMPACT

During the frontal or oblique impact of a car moving with a low speed against an obstacle the structural components most likely to be damaged are those in the vicinity of the impact region. The criteria for an acceptable performance of the energy absorbing device is that all the kinetic energy of the vehicle is dissipated when it impacts a non-moving obstacle up to 20 Km/h. Moreover, all the plastic deformations of the vehicle components must be contained within the device and the vehicle occupants must be subjected to low accelerations. It is assumed here that the car design is such that in a normal crash situation the deformation progresses from the exterior to the interior of the vehicle. Consequently any energy absorbing device must be placed close to the impact region of the car.

6.2.1. *Component shape and location*

The energy absorbing device to mount in the front of the car must imply a minimum number of structural modifications on the base model of the vehicle. For this vehicle, with a mass of 1348 Kg moving with a velocity of 20 Km/h, the new structural component must dissipate an energy superior to 20835 J. After an analysis of the structural components of the front of the car represented by Figure 35, the volume contained between the bumper and the resistant structure of the front, 0.00291 m^3, is chosen to locate the energy absorbing device. Considering that the material selected for this equipment can dissipate 15270 KJ/m^3 [49] it is predicted that the component can absorb 4.3 times the required kinetic energy. For the kinetostatic model of the vehicle the chassis is considered to be a rigid body while the energy absorbing device is treated as the massless flexible part.

6.2.2. *Structural analysis procedure*

For the purpose of describing the nonlinear behavior of the energy absorbing device the finite element method is used. The model of the device is composed of 463 continuum elements and beam elements that share 540 nodes in order to describe the energy absorbing device, the bumper and the mounting beams, as depicted by Figure 35. Contact elements are used in order to describe the contact between a moving plane and the bumper in the different simulations. The finite element model adds up to a total of 210 degrees of freedom. The material used in the finite element model is a foam which presents an elastic-perfectly plastic behavior, null resistance under shear loading and a null Poisson's coefficient.

The structure of the device is analyzed for five different quasi-static loading cases corresponding to different orientations of a moving plane. This allows for the evaluation of the X and Y components of the force necessary to crush completely the energy absorbing device. In Figure 35 the evolution of the deformation is described for three orientations of the moving plane.

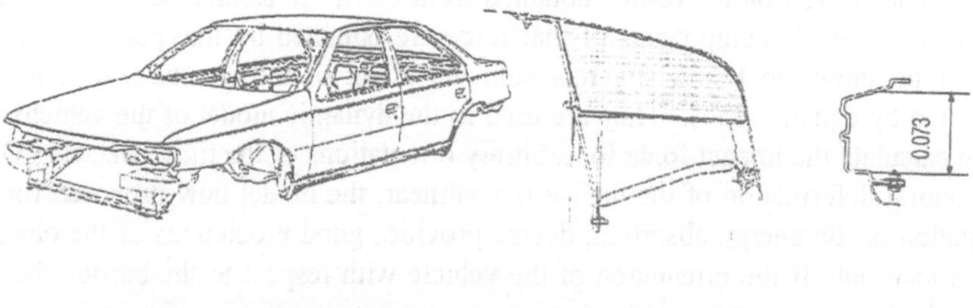

Figure 34. **Shape and location of the energy absorbing device**

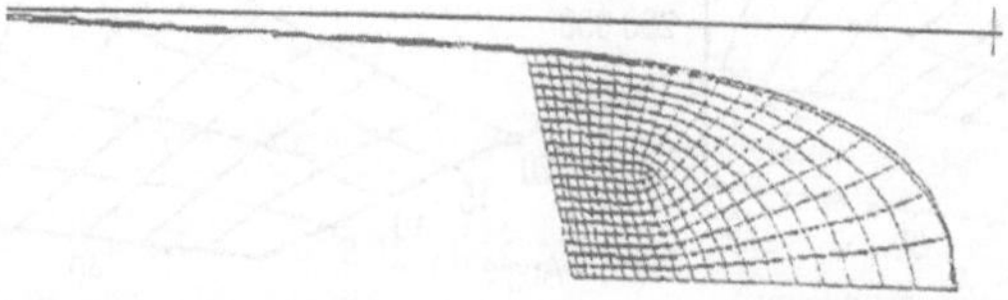

Figure 35. **Finite element model for the energy absorbing device and bumper**

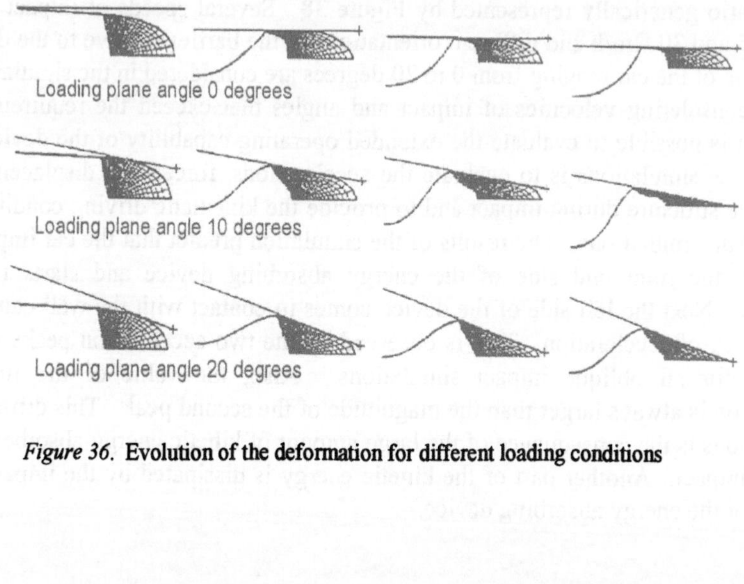

Figure 36. **Evolution of the deformation for different loading conditions**

The analysis of the device in different loading conditions allow the prediction of the evolution of the load exerted by the plane over the device as a function of its displacement. Based on the results obtained from the finite element analysis a carpet plot for the X and Y components of that force are obtained by interpolation. These carpet plots, shown in Figure 36, represent the effect of the structural deformation represented by equation (51). They are used in the dynamic model of the vehicle as a form to calculate the impact force for arbitrary orientations of the rigid obstacle. Once the structural deformation of the device is nonlinear, the model now proposed for the deformation of the energy absorbing device provides good predictions of the obstacle reaction load only if the orientation of the vehicle with respect to the barrier changes moderately during impact.

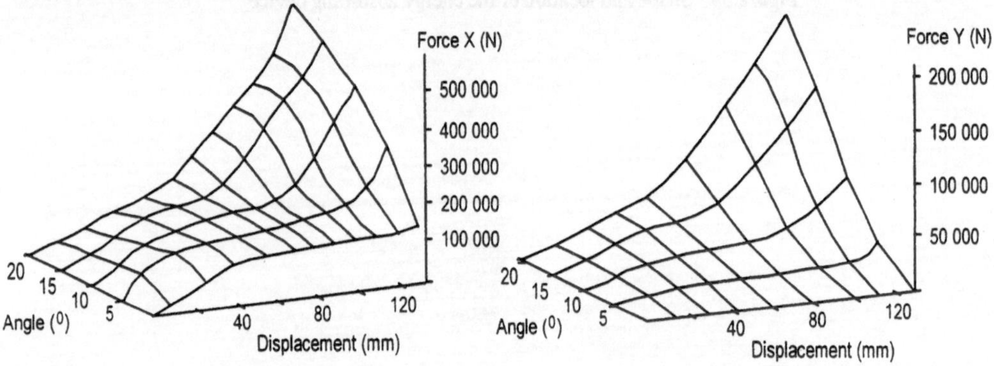

Figure 37. **Components of the impact force: (a) along X; (b) along Y**

6.2.3. *Vehicle simulations*

Various simulations were performed for the frontal impact of the car with a rigid wall in a scenario generically represented by Figure 38. Several speeds of impact ranging between 5 and 30 Km/h and different orientations of the barrier relative to the direction of traveling of the car ranging from 0 to 20 degrees are considered in the simulations.

By considering velocities of impact and angles that exceed the requirements of analysis it is possible to evaluate the extended operating capability of the device. The aim of these simulations is to evaluate the accelerations, forces and displacements of the vehicle structure during impact and to provide the kinematic driving conditions for the occupant simulations. The results of the simulation predict that the car impacts the wall with the right end side of the energy absorbing device and starts to rotate clockwise. Next the left side of the device comes in contact with the wall causing the second peak of acceleration. This is observed by the two acceleration peaks in the X direction for all oblique impact simulations. Also, the value of the first peak acceleration is always larger than the magnitude of the second peak. This difference in accelerations is the consequence of the large amount of kinetic energy absorbed during the first impact. Another part of the kinetic energy is dissipated by the impact of the left side of the energy absorbing device.

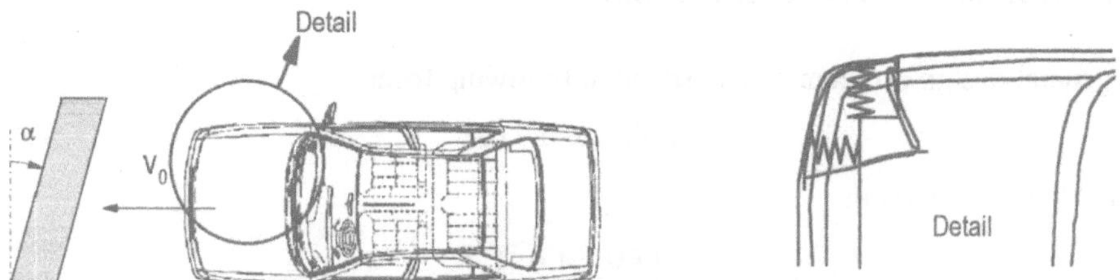

Figure 38. **Initial configuration for the impact simulations of the vehicle**

The remaining kinetic energy of the vehicle is dissipated due to the friction of the tires with the ground during the rotation and sliding of the car. Typical results from these analysis for the car acceleration with an angle of 10° for the barrier are shown in Figure 39. It is observed there that for a speed of 20 Km/h a maximum acceleration of 25 g in the X direction is predicted. Moreover the second peak of acceleration occurs about 0.1 second later.

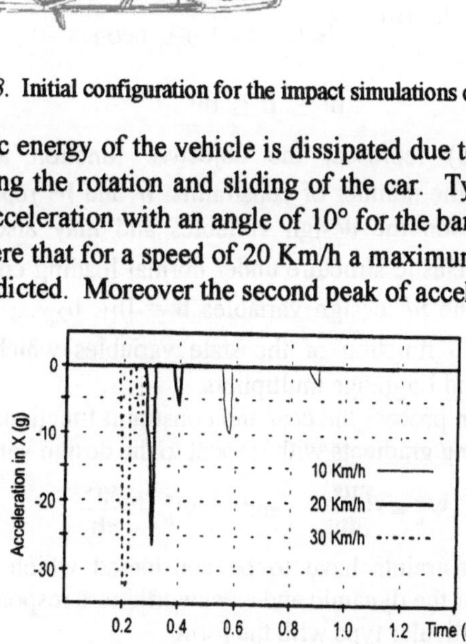

Figure 39. **Car acceleration for the center of mass of the chassis along X**

The different simulations performed show the same qualitative results and the same sequence of motion. However, impact speeds over 25 Km/h for impact angles over 10° lead to deformations of the device larger than the limit deformations predicted by the finite element analysis. Consequently, for these initial conditions, the results predicted by the model are meaningless.

7. Crashworthiness optimization [50]

Although some studies are reported in the literature where optimization techniques are used in crashworthiness design [51-54] it is not surprising that sensitivity analysis and optimization techniques have not been extensively applied directly to the vehicle crashworthiness design. Here a crashworthiness design methodology is reviewed. The structure is approximated by a 2D multibody constrained mechanical system with revolute joints and nonlinear revolute springs representing plastic hinges. For selected structural components, design variables corresponding to the size of the different cross sections are defined.

7.1. DESIGN SENSITIVITY ANALYSIS

The general design problem can be stated in following form

$$\text{Minimize } \Psi^o(\mathbf{b}) \tag{52}$$

subject to the constraints

$$\Psi_i^c(\mathbf{b}) \Rightarrow \begin{cases} = 0 & i = 1, k \\ \leq 0 & i = k+1, \text{ncon} \end{cases} \tag{53}$$

$$\mathbf{b}^l \leq \mathbf{b} \leq \mathbf{b}^u \tag{54}$$

where $\Psi^o(\mathbf{b})$ and $\Psi^c(\mathbf{b})$ represent the objective function and the constraints respectively, and *ncon* is the number of constraints. $\mathbf{b}^l$ and $\mathbf{b}^u$ represent technological lower and upper bounds on the design variables and may also reflect acceptable structural designs for the elastic structure under normal loading conditions. $\Psi^o(\mathbf{b})$ and $\Psi_i^c(\mathbf{b})$ are functions of the *nb* design variables $\mathbf{b} = [b_1, b_2, ..., b_{nb}]^T$. The design variables vector **b** is also function of the state variables which include positions, velocities, accelerations and Lagrange multipliers.

During the optimization process the cost and constraint functional must be evaluated as well as the corresponding gradients with respect to the design variables

$$\Psi_b^o = \frac{\partial \Psi^o}{\partial \mathbf{b}} \quad \text{and} \quad \Psi_b^c = \frac{\partial \Psi^c}{\partial \mathbf{b}} \tag{55}$$

Different types of constraints have to be considered which may correspond to different design criteria for the dynamic and crashworthiness response problem [54]:

- Functionals of the particular type with the form

$$\Psi = \max f(\mathbf{q}, \mathbf{v}, \dot{\mathbf{v}}, \lambda, t, \mathbf{b}) \quad \text{for } t^0 \leq t^j \leq t^1 \tag{56}$$

represent, for example, the maximum acceleration that occurs during the impact, in a particular point of interest in the structure at distinct time t^j, and t^0 and t^1 are the initial and final time, respectively.

- Functionals of the integral type with the form

$$\Psi_i = \int_{t^0}^{t^1} f_i(\mathbf{q}, \mathbf{v}, \dot{\mathbf{v}}, \lambda, t, \mathbf{b})\, dt \tag{57}$$

represent either the total energy absorbed or a standard measure of vehicle crash severity index which is given by [52]

$$VCSI = \frac{1}{t^1 - t^0} \int_{t^0}^{t^1} a(t)^2\, dt \tag{58}$$

where a(t) is the deceleration of the passenger compartment or the passenger itself when striking an obstacle such as an interior division, a table or the back of the front seat. Design sensitivities can be efficiently obtained using the direct differentiation method [50], or standard finite difference techniques. The latter are easier to implement but normally more expensive in terms of computer resources. Several optimization algorithms, such as the feasible direction method, the gradient projection method, and the linearization method are coded in a multi-purpose program ADS (Advanced Design Synthesis) [55], which is used in the present work. A modified feasible directions method [56], is applied in the present numerical example.

7.2. OPTIMIZATION OF AN END UNDERFRAME

The end underframe model of figure 19 indicates the members that are to be optimized in their cross section dimensions which in turn are used as design variables. Due to symmetry, only the left side of this model is considered in the simulation of the frontal impact against a rigid wall. Two colliding velocities have been considered in the simulation: V=30 Km/h (8.33 m/s) and V=50 Km/h (13.833 m/s). Plastic hinges are modeled by nonlinear torsional springs mounted over revolute joints. The nonlinear response of the torsional springs are described in terms of a quasi-static moment-angle relationship based on the curves for elastic-plastic behavior, where the bending stiffness of the elastic part is obtained from the elementary beam theory as

$$k = \frac{EI}{\ell} \tag{59}$$

where E is the modulus of elasticity, I is the cross sectional moment of inertia and l is assumed to be the flange width. The limit plastic moment is obtained for each section. For sections 1 to 4, shown in figure 40, IPE profiles (European Profile) are considered in the optimization process. In this case the plastic moment is calculated as

$$M_p = \sigma_y \left[\frac{1}{2} e\ell^2 + \frac{1}{4} a^2 (H - 2e) \right] \tag{60}$$

where σ_y is the yielding stress, and the other constants are the cross sections dimensions. The material is a cold-rolled steel with σy = 500 MPa.

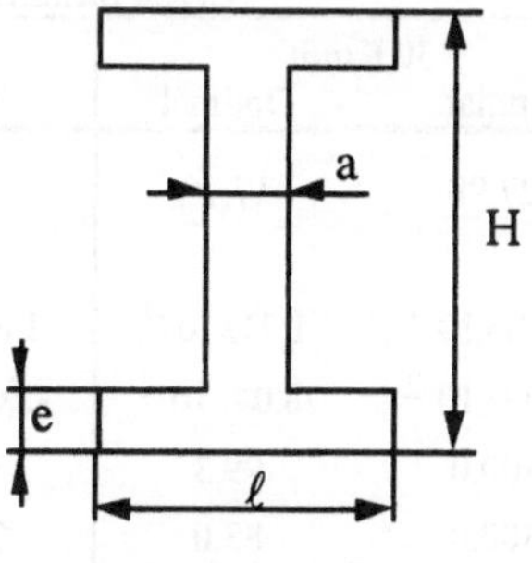

Figure 40. Beam cross sections

In the present case, the design criterion for the crashworthiness is based only on the structural response predicted by the crashworthiness analysis for each design problem.

Consequently, an objective function is defined as the acceleration peak experienced at the end underframe in the compartment zone, i.e. body 13 of figure 19 Because the acceleration levels are related with human injury tolerance levels, its minimization is directly related to the reduction of the human injuries.

A set of design variables is considered corresponding to the different beam heights, H, of the cross sections 1 to 4. Other geometric parameters e, a, and ℓ, are obtained as functions of H, considering the range of cross sections commercially available.

Longitudinal constraints are also defined to impose limits on the maximum structural crash deformation in order to avoid failure of the cross sections and overlapping of the structural components. The objective function and constraints are summarized as:

Find **b** such that $(|\ddot{y}_{13}(t,\mathbf{b})|)$ is minimized

Subject to: $\max. |(y_{13}-y_1)-(y_{13}^0-y_1^0)|-X_1 \leq 0$

$\max. |(y_{13}-y_{10})-(y_{13}^0-y_{10}^0)|-X_2 \leq 0$

The subscripts indicate the body number, y is the displacement of the body in the Oy direction and the superscript indicates the initial position. The maximum deformations allowed are respectively, $X_1 = 0.937$ m and $X_2 = 0.254$ m. The purpose of the second constraint is to avoid the overlapping of the structure members in the zone of sections 1, 2 and 3. Four design variables are considered corresponding to the sections represented in Figure 40, as $\mathbf{b} = \{H_1, H_2, H_3, H_4\}$. The initial design vector is $\mathbf{b} = \{300, 300, 400, 500\}$ (mm) and the lower and upper bounds are, $\mathbf{b_l} = \{60, 60, 60, 60\}$ (mm) and $\mathbf{b_u} = \{600, 600, 600, 600\}$ (mm). The lower and upper bound vectors correspond respectively, to the minimum and maximum depth of the sections available commercially. The results obtained for the two colliding velocities are summarized in Table 1.

TABLE 1. Crashworthiness Design of The End Underframe of a Train

	Crash Situation			
Velocity	30 Km/h		50 Km/h	
	Initial	Optimal	Initial	Optimal
Objective Function (G)	27.98	17.79	130.8	31.18
Violations(m)				
Constraint 1	$-3.77\text{x}10^{-1}$	$1.71\text{x}10^{-4}$	$1.64\text{x}10^{-1}$	$-5.7\text{x}10^{-3}$
Constraint 2	$-3.11\text{x}10^{-2}$	$-8.04\text{x}10^{-4}$	$-2.02\text{x}10^{-2}$	$-2.2\text{x}10^{-2}$
$b_1 = H_1$ (mm)	300.0	99.3	300.0	352.6
$b_2 = H_2$ (mm)	300.0	82.0	300.0	319.0
$b_3 = H_3$ (mm)	400.0	120.9	400.0	469.0
$b_4 = H_4$ (mm)	500.0	459.0	500.0	600.0
CPU Time (s) Workstation HP 735	6 h 46 m		10 h 4 m	

For both velocities in the acceleration levels decrease as a result of the optimization process

For the impact at 30 Km/h, the acceleration results and are presented in figure 41. These results show that a decrease in the maximum acceleration is obtained at the cost of a larger deformation. Globally, with a reduction of the sections size is obtained with the optimized structure.

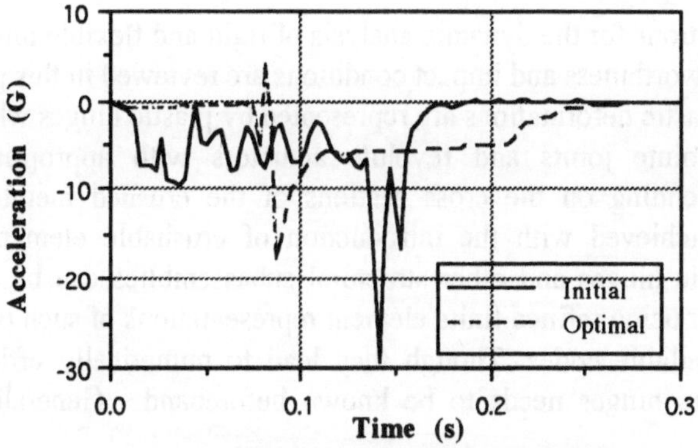

Figure 41. Acceleration for the impact at V=30 Km/h

The acceleration results for the impact at V=50 Km/h, are shown in Figure 42. In this case the global deformation constraint (constraint 1) is violated in the initial design.

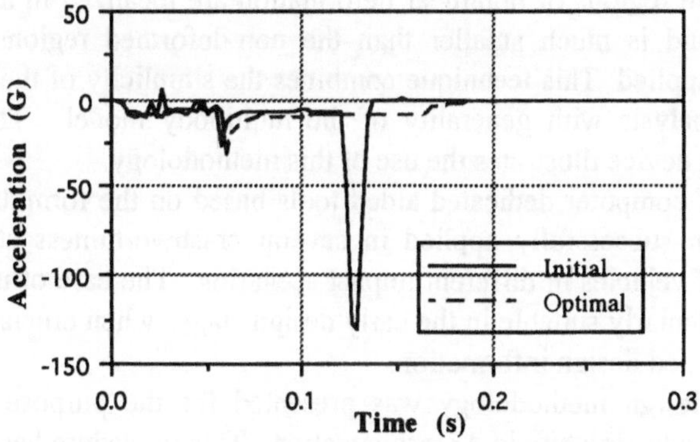

Figure 42. Acceleration for the impact at V=50 Km/h

For this simulation the initial design yields a very large deformation of the structure as a result of the strong collision of the lower extremity of central beam (body

10) with the lower beam (body 13). Another result of this contact ins a large acceleration peak. The optimal design is associated with an increase of the dimensions of all sections, corresponding to higher levels of energy absorption in sections 1, 2 and 3. A subsequent decrease in the acceleration peak that occurs when bodies 10 and 13 collide is observed in the simulation.

8. Conclusions

Different formulations for the dynamic analysis of rigid and flexible multibody systems suitable for crashworthiness and impact conditions are reviewed in this paper.

Localized plastic deformations are represented by plastic hinges which, in turn, are modeled by revolute joints and revolute actuators with appropriate constitutive relationships depending on the cross sections of the crushed members. Extended capabilities are achieved with the introduction of crushable elements. The crush behavior of plastic hinges and other structural subassemblies can be obtained, either experimentally, or using refined finite element representations of such components with commercially available codes. Though they lead to numerically efficient codes the location of plastic hinges needs to be known beforehand. Generally this is not a problem.

Formulations allowing for the combined multibody and nonlinear finite element description of the flexible parts have also been developed and presented. This approach is particularly useful in situations where only localized parts of the flexible bodies collapse as a result of an impact. The advantage of this formulation is that no a priori knowledge of the flexible behavior is necessary. However the computational costs of this procedure may be high.

In cases were regions of nonlinear deformation are localized in a limited volume of the system, and is much smaller than the non-deformed regions, a kinetostatic methodology is applied. This technique combines the simplicity of the nonlinear static finite element analysis with generality of the multibody model. The design of an energy absorbing device illustrates the use of this methodology.

The suite of computer dedicated aided tools based on the formulations presented herein have been successfully applied in several crashworthiness studies involving different types of vehicles in different impact scenarios. The ease of use of such tools makes them particularly suitable in the early design stages when crucial decisions must be made with limited design information.

At last, a design methodology was presented for the purpose of studying an optimal vehicle body structure in a crash situation. This procedure has been developed within the framework of mathematical programming methods involving an efficient and inexpensive dynamic analysis module. The capability of this methodology is demonstrated for an end underframe structure of a train in a frontal impact. Such complex design efforts using conventional structural optimization techniques are still far from being feasible with the power available in current computer platforms.

Further research efforts are still required in order to improve the efficiency of multibody dynamics as applied to crashworthiness and structural impact analysis and design problems. These are: more adequate modeling of nonlinear flexible bodies in multibody systems with formulations using the structural stiffness and mass terms as obtained in standard finite element formulations, thus allowing a greater interaction with commercially available codes with the possibility to introduce more complex crushable elements; introduction of piecewise analysis techniques based on momentum balance methods for an efficient evaluation of velocity jumps during impact; development of new methodologies for constraint addition and deletion capabilities to deal in a more general manner with the contact problem and finally, the exploitation of more efficient integration algorithms.

ACKNOWLEDGMENTS

The authors acknowledge the support of: AGARD - Advisory Group for Aerospace and Research Development, project P77; JNICT - Junta Nacional de Investigação Científica e Tecnológica, Project STRD/C/TPR/569/92. Some of the work reported herein, namely the experimental tests and design tools development and enhancement, was carried out within the framework of project BRITE/EURAM BE-3385-89 - TRAINCOL.

References

1. Emori, R. I. (1968) Analytical Approach to Automobile Collision, *Automobile Engineering Conference*, SAE Paper No. 680016, .
2. Tani, M. and Emori, R. I. (1970) A Study of Automobile Crashworthiness, *SAE Trans.*, SAE Paper No. 700175, .
3. Kamal, M. M. (1970) Analysis and Simulation of Vehicle to Barrier Impact, *International Automobile Safety Conference*, SAE Paper No. 700414.
4. Lin, K. H. (1973) A Rear-end Barrier Impact simulation Model for Unibody Passenger Cars, *SAE Trans.* **82**: Paper 730156.
5. Dressler, C. J. and Schorry, R. E. (1979) High Speed Impact and Aggressivity Analysis of the CALSPAN/Chrysler Research Safety Vehicles (RSV), *3rd International Conference on Vehicle Structural Mechanics*, SAE Paper No. 790993.
6. Pifko, A. B. and Winter, R. (1981) Theory and Applications of Finite Element Analysis to Structural Crash Simulation" *J. Computers and Structures*, **13**, pp277-285.
7. Hayduk, R. J. , Winter, R. , Pifko, A. B. , and Fesanella, E. L. (1983) Application of the Non-linear Finite Element Computer Program DYCAST to Aircraft Crash Analysis, *Structural Crashworthiness*, Eds. N. Jones and T. Wierzbicky, Butterworths, London, England, pp.283-307.
8. Haug, E. , Arnaudea, F. , Du Bois, J. and Rouvay, A. (1983) Static and Dynamic Finite Element Analysis of Structural Crashworthiness in the Automotive and Aerospace Industries, *Structural Crashworthiness*, Eds. N. Jones and T. Wierzbicky, Butterworths, London, England, pp.175-217.
9. Du Bois, P. and Chedmall, J. F. (1987) Automotive Crashworthiness Performance on a Supercomputer, *SAE Trans.*, SAE paper No. 870565.
10. Pereira, M. S. , Nikravesh, P., Gim, G. and Ambrósio, J. (1987) Dynamic Analysis of Roll-over and Impact of Vehicles, *XVIII Bus and Coach Experts Meeting*, Budapest, Hungary.

11. Sato, T. B. (1966) Dynamical Considerations in Automobile Collisions, *J. of the Society of Automotive Engineers of Japan*, **20**, No. 5.
12. Nikravesh, P.E. ,Chung, I.S. and Benedict, R.L. (1983) Plastic hinge approach to vehicle simulation using a plastic hinge technique, *J. Comp. Struct.* **16**, 385-400.
13. Wittlin, G. (1983) Aircraft crash dynamics: modeling, verification and application, *Structural Crashworthiness* (N. Jones, T. Wierzbicki Eds.), Butterworths, London, England, 259-282.
14. Ambrósio, J.A.C. and Pereira, M.S. (1994) Flexibility in multibody dynamics with applications to crashworthiness, *Computer-Aided Analysis of Rigid and Flexible Mechanical Systems* (M.S. Pereira and J.A.C. Ambrósio Eds.), Series E: Applied Sciences - Vol.268, Kluwer, Dordrecht, Netherlands, 199-232 .
15. Ambrósio, J.A.C. , Nikravesh, P.E. (1992) Elastic-plastic deformations in multibody dynamics, *Nonlinear Dynamics* **3**, 85-104 .
16. Pifko, A. B. and Winter, R. (1981)Theory and Applications of Finite Element Analysis to Structural Crash Simulation, *J. Computers and Structures*, **13**, pp277-285.
17. Haug, E. (1989) The PAM-CRASH Code as an Efficient tool for Crashworthiness Simulation and Design, *Second European Cars/trucks Simulation Symposium*, Schliersee, Germany, May 22-24.
18. Halquist, J. O. (1982) *Theoretical Manual for DYNA-3D*, Lawrence Livermore Laboratory.
19. Belytschko, T. and Kenedy, J. M. (1986) *WHAMS-3D, An Explicit 3D Finite Element Program*, KBS2 Inc. P. O. Box 453, Willow Springs, IL 60480.
20. Nikravesh, P. E. (1998) Computer Aided Analysis of Mechanical Systems, Prentice-Hall.
21. Roberson , R. E. and Schwertassek, R. (1988) *Dynamics of Multibody Sytems*, Springer-Verlag.
22. Haug, E. J. (1989) *Computer Aided Kinematics and Dynamics of Mechanical Systems*, Volume 1: Basic Methods, Allyn and Bacon.
23. Shabana, A. (1989) *Dynamics of Multibody Sytems*, Wiley.
24. García de Jalon, J. and Bayo, E. (1993) Kinematic and Dynamic Simulation of Multibody Systems - The Real Time Challenge-, Springer-Verlag, N.Y..
25. Pereira M. S. and Ambrósio J. A. C., Eds., (1994) *Computer-Aided Analysis of Rigid and Flexible Mechanical Systems* , Series E: Applied Sciences - Vol.268, Kluwer, Dordrecht, Netherlands.
26. Schielen, W. (1990) *Multibody Systems hanbook*, Springer-Verlag, Heidelgerg, Germany.
27. Nikravesh, P. E. (1990) Systematic Reduction of Multibody Equations of Motion To a Minimal Set, *Int. J. Non-Linear Mechanics*, **25**, 143-151.
28. Shampine, L. F. and Gordon, M. K. (1975) Computer Solution of Ordinary Differential Equations: The initial Value Problem, Freeman, San Francisco, USA.
29. Baumgarte, J.(1972) Stabilization of Constraints and Integrals of Motion, *Computer Methods in Applied Mechanics Engineering*, **1**.
30. Ambrósio, J. A. C. Silva, M. P. T., and Gonçalves, J. P. (1995) Development of Energy Absorbing Devices Using a Kinetostatic Multibody Dynmics Methodology, *Int. J. Crashworthiness*, **1**, No. 2.
31. Lankarani, H.M. and Nikravesh, P.E.(1994) Continuous contact force models for impact analysis in multibody systems, *Nonlinear Dynamics*, **5**(2), , 193-207.
32. Pereira, M. S. and Proença, P. L. (1991) Dynamic Analysis of Spatial Flexible Multibody Systems Using Joint Coordinates", *Int. J. Num. Meth. in Engng.*, **32**, 1799-1812.
33. Ambrósio, J. A. C. and Nikravesh, P. E. , (1992) Elastic-Plastic Deformation In Multibody Dynamics, *Nonlinear Dynamics*, **3**, 85-104.
34. Shabana, A. , and Wehage, R. (1983) Variable Degree of Freedom Component Mode Analysis of Inertia Variant Flexible Mechanical Systems, *ASME J. of Mech. Trans. and Auto. Design*, **105**, pp371-378.
35. Dias, J.M.and Pereira, M.S. (1995) Dynamics of Flexible Mechanical Systems With Contct-Impct and Plastic Deformations, *J. of Nonlinear Dynamics*, **8**, 491-512.
36. Anceau, J. H. , Drazetic, P. and Ravalard, I. (1992) Plastic Hinges Behaviour in the Multibody Systems, *Mécanique Matériaux Électricité*, nº 444, France.
37. Kecman, D. (1983) Bending Collapse of Rectangular and Square section Tubes, *Int. J. of Mech. Sci.*, **25**(9-10), 623-636.
38 Winmer, A. (1977) Einfluss der Belastungsgeshwindigheit auf das Festigkeits und Verformungsverhaten am Beispiel von Kraftfarhzengen", *ATZ* **77**(10),281-286.

39. Pereira, M.S., Ambrósio, J..A.. and Dias, J.M., Crashworthiness Analysis and Design Using Rigid-Flexible Multibody Dynamics With Application to Train Vehicles, *Int. J. Numerical Methods in Engng*, (accepted).
40. Ambrósio, J.A., Pereira, M.S. and Dias, J.M., Distributed and Discrete Nonlinear Deformations on Multibody Dynamics, *Nonlinear Dynamics*, (accepted).
41. Pereira M. S. , Drazetic P. and Ravalard Y. , "An Hybrid Method For Impact Analysis of Rigid-Flexible Structural Systems Undergoing Gross Motion", submited J. *Impact Engng.*
42. BRITISH RAIL (1992) Structural Design Load Cases for Passenger Carrying Vehicle Bodies, BRB-GM/TT0079, Issue 1, Rev.A.
43. Ambrósio J A and Nikravesh P E .(1992) Elastic-plastic deformations in multibody dynamics, *Nonlinear Dynamics*, , **3**, 85-104.
44. Greewood D T (1965) *Principles of Dynamics*, Englewood Cliffs, Prentice-Hall.
45. Ambrósio J A (1991) *Elastic-Plastic Large Deformation of Flexible Multibody Systems in Crash Analysis*, University of Arizona.
46. Ni, C.-M. (1973) Impact Response of curved box beam columns with large global and local deformations, *AIAA/ASME/SAE 14th Structures Structural Dynamics. and Material Conference*, Williamsburg, Virginia, U.S.A. March 20-22.
47. *ABAQUS, Version 5.3* (1993) Hibitt, Karlson & Sorensen, Inc., .
48. *Dynamic Crash Tests: Dynamic Test of a Vehicle Against an Obstacle* (1993) Technical Report n. TRAINCOL/T7.2-F, SOREFAME, Portugal.
49 Silva M. T., Gonçalves J. P. , Ambrósio J A and Faria, L. (1994) *New shock absorbing device for automobiles*, Technical Report CEMUL/ECIA, Lisbon, Instituto Superior Técnico.
50. Dias, J.M. and Pereira, M.S. (1995) Application of Multibody Dynamics to the Crashworthiness Optimization of Vehicle Structures, *Computational Dynamics in Multibody Systems*, Eds. M.S. Pereira and J.A.C. Ambrósio, KLUWER.
51. Bennett, J. A. , Lin, K. H. and Nelson, M. F. (1976) The Application of Optimization Techniques to Problems of Automobile Crashworthiness", *SAE Trans.*, **86**, 2255-2262.
52. Song, J. O. (1986) An Optimization Method for Crashworthiness Design", *Proc. Sixth Int. Conf. Vehicle Structural Mechanics*, Detroit, MI, 39-46.
53. Lust, R. (1992) Structural Optimization with Crashworthiness Constraints", *Structural Optimization*, **4**, 85-89.
54. Dias, J. and Pereira, M. S. (1994) Design for Vehicle Crashworthiness Using Multibody Dynamics, *I. J. Vehicle Design*, **15**, 3/4, 563-577.
55. Vanderplaats, G. (1987) *ADS- A Fortran Program for Automated Design Synthesis*, Version 2.01, Eng. Design Optimization Inc., Ca, USA.
56. Vanderplaats, G. , and Moses, F. (1973) Structural Optimization by Methods of Feasible Directions", *J. Comp. & Struct.*, **3**, 739-755.

PART VII
Aircraft Crash Protection

AIRCRAFT AND HELICOPTER CRASHWORTHINESS: DESIGN AND SIMULATION

C.M. KINDERVATER
German Aerospace Research Establishment (DLR)
Institute of Structures and Design
Pfaffenwaldring 38-40
70569 Stuttgart, Germany

Abstract

In aeronautics, first structural design requirements for better crash protection were established for military helicopters and light fixed-wing aircraft. Also, for all other aircraft categories further progress has been initiated to improve crash survivability in potentially survivable impact scenarios. Crash resistance covers the energy absorbing (EA) capability of crushing structural parts to limit decelerative forces to the occupants, the demand to provide a protective shell around the occupants, and to minimize the riscs in the occupant environment and in the post crash regime. Designing for aircraft crash resistance: System design aspects including the airframe, the landing gear, seats and overhead mass retention are outlined. However, design aspects will be focused in more detail on the airframe and airframe substructures. The use of high performance composite materials needs specific design philosophy and concepts due to the brittle nature but high EA capability of composites. Verified design concepts of highly efficient EA and load bearing composite airframe components will be demonstrated. Analytical methods and simulation tools: The demonstrated crash analysis tools will include a hybrid simulation technique (KRASH) and a crash code based on Finite Element Analysis (FEA), i.e. PAM-CRASH. Hybrid modelling and simulation case studies will include full scale composite airframes (BK 117 composite airframe and full-scale carbon composite Lear Fan 2100). FEA crash code case studies will be focused on metal and composite airframe sections (transport aircraft and Lear Fan 2100 section). The simulation studies are correlated with crash test data. Specific emphasis will be addressed to the used and modified material models and calibration techniques of the models.

J. A. C. Ambrósio et al. (eds.),
Crashworthiness of Transportation Systems: Structural Impact and Occupant Protection, 525–577.

1. Introduction

Crash safety for automotive vehicles has been a well established subject since the 1950's and occupant protection within a crashworthy structure is integral to the vehicle design process. Especially composite materials offer a high potential for tailored designs by a wide variety of matrices and fibres, various preforms, and laminate architecture, i.e. fibre orientation and stacking sequence of the single laminae. Composite materials also have a considerable potential for absorbing kinetic energy during a crash. The composite energy absorption capability offers a unique combination of reduced structural weight and improved vehicle safety by higher or at least equivalent crash resistance compared to metal structures. Crash resistance covers the energy absorbing capability of crushing structural parts as well as the demand to provide a protective shell around the occupants (structural integrity). This basic principle of occupant crash protection is used in the automotive field, where crash safety is meanwhile a well established car design requirement. In aeronautics, first structural design requirements for better crash protection were established for military helicopters and light fixed-wing aircraft in the form of the *Aircraft Crash Survival Design Guide* (ACSDG) [1] and the *MIL-STD-1290 A* [2]. For all other aircraft categories further progress can be expected in the future to improve the requirements for structural crash resistance.

In aeronautics, especially composite helicopter and light fixed-wing airframe structures crashworthiness is important due to the existing and increasing application of primary composite structures in these aircraft. The considered material systems are higher performance types such as epoxy resins reinforced with glass-, and especially with carbon- and aramid-fibres or hybrid laminates thereof. Besides the classical thermoset polymers thermoplastic systems with continous fibre reinforcements come into use, owing to faster and cheaper fabrication suited for high degree of automation and better recycling possibilities.

The arrival of faster computers and more efficient explicit FE codes has now led to complete vehicle crash simulations by FE methods, which can even include details such as airbag and occupant or dummy models. Rather surprisingly much less attention has been paid to the development of FE simulation of aircraft structures under dynamic and crash conditions, despite the extremely high cost of aircraft crash tests. The automotive industry was driven by consumer pressure for safety and by safety legislation. More attention is now being given to aircraft structural crashworthiness since the establishment of structural design requirements, thus the development and validation of reliable simulation tools for crashworthiness studies is an important means of reducing development costs to meet future aircraft safety requirements.

Current crashworthiness studies on metallic and composite airframe structures are based extensively on hybrid codes such as the Lockheed developed KRASH code, which models a complete aircraft or substructure with mass, beam and nonlinear spring elements. Such techniques have been successfully applied to study the crashworthiness of helicopters, transport aircraft, commuter aircraft , and even sailplanes. In this chapter KRASH analysis of various aircraft is presented and , where applicable, is compared with results from a full scale drop tests. A hybrid code such as KRASH provides a global dynamic analysis for a highly idealised aircraft structure, and is a fast, low-cost method of assessing global structural crashworthiness. However, the code requires extensive calibration tests on the structure or critical structural elements and is thus not a fully predictive tool.

FE methods allow the detailed modelling of structural geometry and are being used with considerable success as a fully predictive crash simulation tool by the automotive industry. Analyses to validate FE methods for aircraft substructures in this chapter are based on the explicit dynamic FE code PAM-CRASH. One analysis studies the crash behaviour of the riveted aluminium rear fuselage bay structure of a transport aircraft below the passenger floor between two frames. Models for the aluminium dynamic materials properties are included in the code, however a critical aspect of the analysis was found to be the rivet modelling since rivet failure controlled the structural collapse. PAM-CRASH is further applied to analyse the crash response of a carbon fibre reinforced plastic (CFRP) fuselage section of the Lear Fan 2100. Computed responses are compared with a NASA structural test in which a composite fuselage section was drop tested from a height of 4.6m. Composites models are available for limited classes of materials in FE codes, but further research is needed to extend the models and obtain suitable failure data for materials of specific interest to the aircraft industry.

Modern explicit FE codes such as PAM-CRASH are valuable tools for aircraft crashworthiness studies. However, they are expensive to use and for structural failure studies very detailed FE geometry models are required. Although they could be used to simulate the complete crash response of an aircraft, it is not practical to do this without investing in super-computers and large teams of FE modellers as in the automotive industry. One way forward for the aircraft industry in the medium term is to consider coupled global-local simulations, based on both hybrid codes for a global analysis of the aircraft and FE codes for detailed local analyses of critical substructures. This is a practical method of reducing the cost of crashworthiness studies. This coupled simulation approach is discussed in the last part of the chapter by an example of a detailed FE analysis to calibrate the response of the floor beam elements for a KRASH simulation of an airframe section.

2. Structural Crushing and Energy Absorption Behaviour

2.1 VALUATION CRITERIA FOR CRUSHING BEHAVIOUR

A terminology has been developed to describe and compare the EA performance of collapsing or crushing material specimens or structural parts. The valuation criteria are derived basically from the force-deflection curve and the absorbed energy which is the area under this curve. The most common terms are:

Specific Energy (E_{spm}): The specific energy relates the absorbed energy to the mass of the absorber or structure and is therefore an important criterion for lightweight designs.

Mean Crushing Stress: The crushing stress is calculated by the average crush force (F_{avg}) divided by the original cross sectional area of the absorber.

Crush Force Efficiency (A.E.): This value relates the average crush force (F_{avg}) to the maximum force (F_{peak}) of the crush characteristic.

Stroke Efficiency (S.E.): The stroke efficiency is the ratio of the stroke at 'bottoming out' (l_s) to the initial length (l_0)of the absorber. High ratios indicate efficient use of the material.

Energy Efficiency (E.E.): A combination of the crush force efficiency and the stroke efficiency leads to the hypothetical E.E.-value. The E.E.-value correlates the actual absorbed energy (area under the force-deflection curve) to the product calculated from the peak force (F_{peak}) and the initial length (l_0) of the absorber structure.

2.2 COMPOSITE FAILURE AND ENERGY ABSORBING MECHANISMS

Structural crashresistance is dominated by the material behaviour, the structural design, and geometrical parameters. Prismatic metal structures typically fail by very regular folding mechanisms (diamond shape buckling or ring buckling). Along the folding lines plastic hinges are generated and the energy is absorbed by material yielding, plastification, and friction between the contact surfaces.

Due to the brittleness, i.e. low strain to failure of composites reinforced by glass- or carbon fibres, material plastification is almost not apparent. Fractures occur in fibres, in the matrix and at the fibre/matrix interface in tension, compression, and shear parallel and normal to the fibre direction. On the laminate level, also delaminations caused by shear or buckling of single layers or groups thereof occur. Crushing on the micromechanical level is therefore very complex, and combinations of various mechanisms are apparent. The key mechanism of high composite EA under crash loading is a strength controlled formation of a microcrack pattern which spreads out locally in the laminate. The formation of such a propagating crushfront must be activated by a stress concentration - a so called „trigger" - where cracks first occur. The morphology of the crushfront which is generated by a crush initiator determines the level of the crushforce. The crush process at a chamfered tip of a carbon/aramid-fibre epoxy laminate and the stabilisation of the crushfront is shown in Figure 1. This complex sequence of local cracking and fragmentation is controlled by the material properties, the laminate's architecture, and by the geometrical specimen- or structural configuration.

Instability dominated failures caused by Euler- and shell buckling generally lead to low EA. Failure in 'local buckling' occurs in material compositions with high failure strain matrices and/or fibres (aramid, high-performance polyethylene, hybrid laminates). Also, friction plays an important role as an EA process. Frictional forces act between the broken and sliding parts within the crushfront and also between the contact surface and the sliding fronds. Microfragmentation and friction in the crushfront

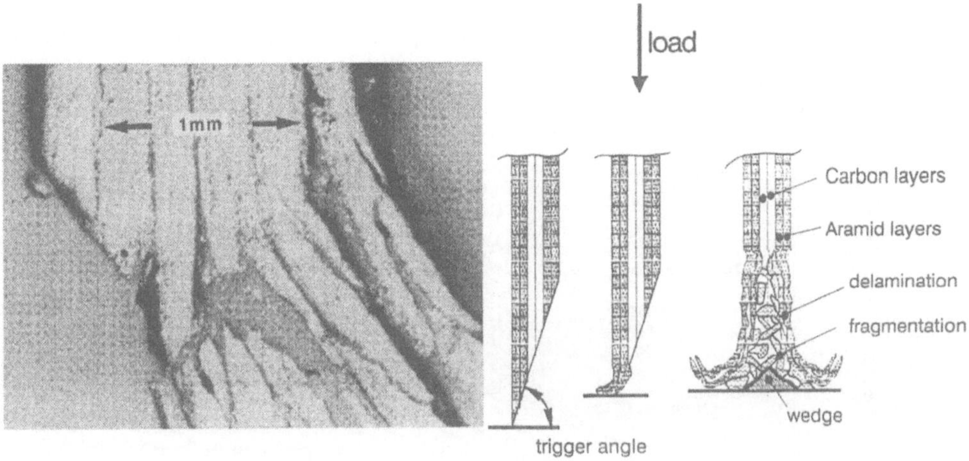

Figure 1 **Crushfront formation initiated by a bevel trigger; microscopic view of a crushfront**

generate a 'quasi-plastification' energy dissipation process in composite laminates. A microscopic view into a crushfront of a carbon/epoxy tube is shown also in Figure 1.

2.3 ENERGY ABSORPTION AND CRUSHING OF GENERIC ELEMENTS

To develop a data base on EA behaviour of composite generic elements many quasi-static and dynamic crush tests were performed in a drop tower to understand how composites absorb energy and how the numerous lamination parameters influence the crushing performance. Most of the test work was focused on tubular structural elements. The tube with a circular cross section (composite and metal) is obviously the most favorable shape for EA. Composite tubes can achieve specific energies above 100kJ/kg in combination with 95% crush force efficiency. Additionally, the force-deflection characteristics are almost 'ideal elastic-plastic'. Comprehensive contributions to tube crushing and major influencing parameters of the crushing process are reported in the literature [3-7].

The large potential of EA performance of composites is more evident in Figure 2 where composite and aluminium tubes are compared in terms of specific energy and crush force efficiency to other structural elements such as honeycombs, foams, stringer- and integrally-stiffened elements, sandwich designs, and cruciforms. It is evident that

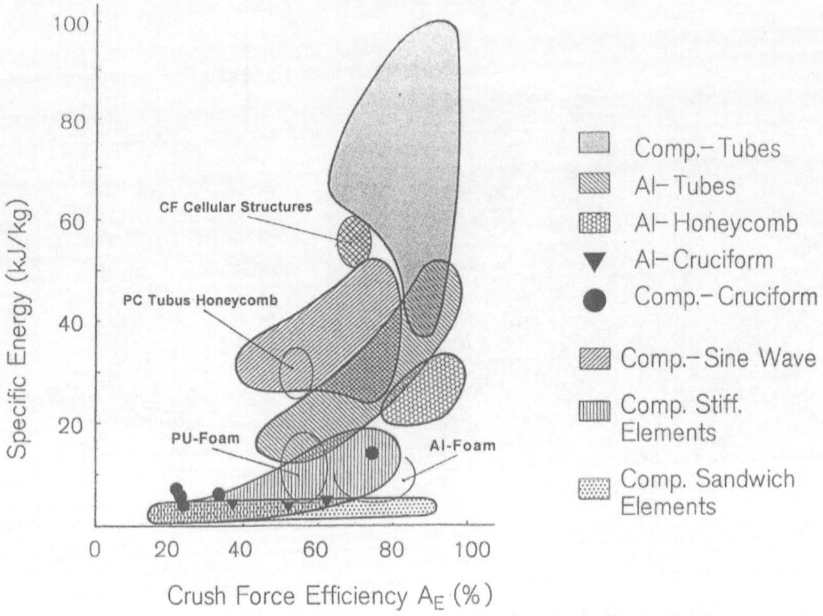

Figure 2 **Crushing performance of generic structural elements**

more complex structural assemblies have decreasing crushing performance compared to tubes. Composite elements are in general superior compared to metal configurations.

3. Designing for Aircraft Crash Resistance

3.1 DESIGN PHILOSOPHY AND REQUIREMENTS FOR AIRFRAME CRASH RESISTANCE

In aircraft crashes the kinetic energy in the horizontal direction is mainly absorbed by friction between the sliding structure and the ground, and - if possible - by a certain amount of soil deformation. A typical horizontal crash pulse shows the maximum deceleration level at fuselage major impact during the first third of the pulse duration. During the slide out phase the deceleration level is much lower. In crash accidents with a high vertical component of the impact velocity the crash loads have to be absorbed mainly by controlled structural deformation. A typical vertical crash pulse where the landing gear is involved shows a low g-level during gear stroking which increases

rapidly at the onset of subfloor crushing and seat stroking. The pulse shape generated by subfloor crushing is dependent on the structural design. The „classical" pulse shape is assumed to be triangular, however, a more crash resistant design with EA capability shows for a certain time duration (stroke) a deceleration (force) plateau. The EA-device in the seat controls the g-level felt by the occupant. It should not exceed 14-15g as recommended in the ACSDG [1].

For the control of decelerative loads on occupants, the type of aircraft will affect the crashworthy design approach which is an optimization process for a certain aircraft category, Figure 3. A systems approach should be applied whenever possible which comprises the landing gear, the subfloor and the high mass retention structure, seat/restraint systems with tuned EA characteristics, and other cabin furniture (bins, galleys). However, light fixed-wing general aviation aircraft, small passenger airplanes and helicopters, especially with retracted landing gear, have little crushable airframe structure. Such designs typically consist of a framework of longitudinal beams and lateral bulkheads covered by the outer skin and cabin floor. The total structural height is often only about 200mm. The design of intersections of beams and bulkheads (cruciforms), the beam webs and floor sections (boxes) and the bottom parts of the frame structures contribute essentially to the overall crash response of an airframe subfloor assemblage, Figure 4. Under vertical crash loads cruciforms are 'hard point' stiff columns which create high decelerative peak loads at the cabin floor level and cause dangerous inputs to the seat/occupant system. Frame and shell structures above the cabin floor are crucial elements for high mass retention (transmissions, engines, rotor hubs etc.) and for providing a livable volume in a crash sequence. Plastically deformable frames or side shell structures of the airframe offer the possibility of load limiting concepts for large overhead masses.

In the case of an impact on water a number of these structural elements may be less successful in absorbing energy. For instance the landing gear will meet little resistance when impacting a water surface. During water impacts the subfloor can only absorb crash energy when the skin panels transfer the water pressure loads to the EA subfloor elements (beams and bulkheads). This load transfer can only take place when the skin deflects substantially without rupture. Details of a „tensor skin" concept are reported in [9, 10], a tensor skin design concept is discussed further in chapter 3.2.3.

Crashworthy subfloor crushing characteristics: The design principle of the „controlled load concept" is shown in Figure 5 where the bottom part of the airframe comprises a high strength structural cabin floor and a crushing zone for EA. The force equillibrium in the bottom skin of the airframe under water impact is also indicated.

When the aircraft is crashing on more or less rigid surfaces the subfloor crush characteristics should have a moderate initial stiffness and then a slightly increasing or constant crush force level to achieve high EA. In the acceleration- (force-) time domain such crash pulses - called x-T- and xT-ya-pulses - are added to the crash pulse design chart of the ACSDG [1], Figure 6. This chart correlates the deceleration levels , the velocity changes and the stopping distances of various typical crash pulses such as

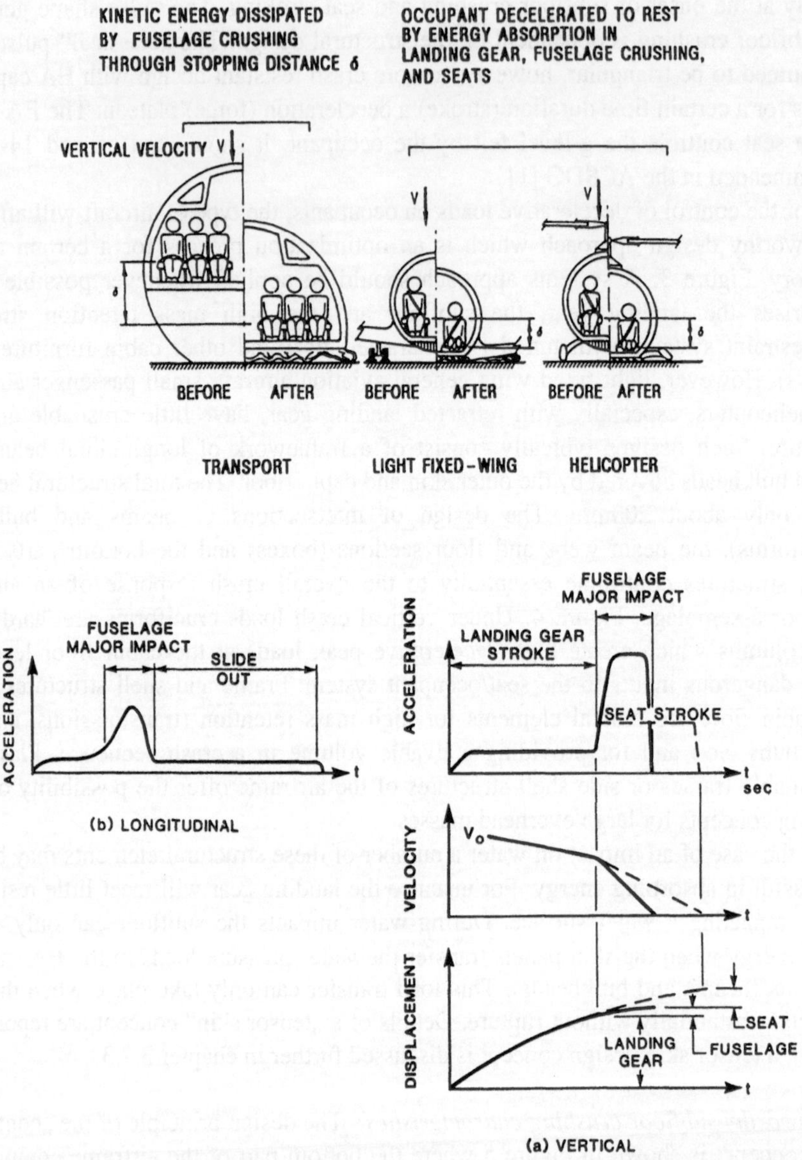

Figure 3 **Energy absorption concept for various aircraft;representative horizontal and vertical crash pulses**

rectangular, symmetric triangular, asymmetric triangular and sinusoidal pulse shapes. For a certain maximum deceleration- (a-) level, for all regarded pulse shapes the shortest stopping distance can be achieved by an rectangular pulse. However, this pulse has an infinite deceleration onset rate which is not tolerable to any occupant. The xT-

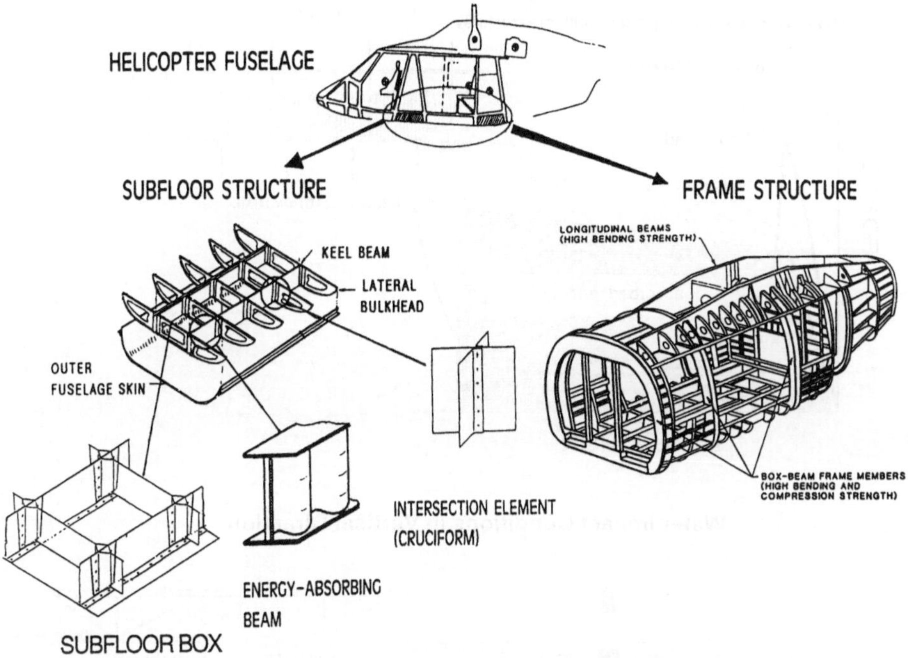

Figure 4 **Airframe substructures for crashresistant design improvements**

and xT-ya-pulses are parametric variants of the rectangular pulse. The selected pulses have both x-values of 0.1T and the xT-ya-pulse has an y-value of 0.75a. Both shapes are 'design golas' but are assumed, and it will be demonstrated later, that they could be realized by the design of a subfloor crushable structure. For the xT-pulse the stopping distance compared to the rectangular (same maximum a-level) is higher by the factor of 1.1 (time duration factor: 1.053) and for the xT-ya-pulse by the factor of 1.23 (time duration factor: 1.212), i.e. both are still very effective pulses compared to other exemplary shapes.

The basic requirements for crash resistant subfloors can be summarized as follows:

- Uniform distribution of ground reaction and seat loads.
- Limitation of the decelerative forces by structural deformation with a „controlled load" concept.
- The crushing characteristics of the subfloor components should have a moderate initial stiffness and then a constant or slightly increasing force level (xT- or xT-ya-pulses).
- Maintain cabin floor structural integrity.

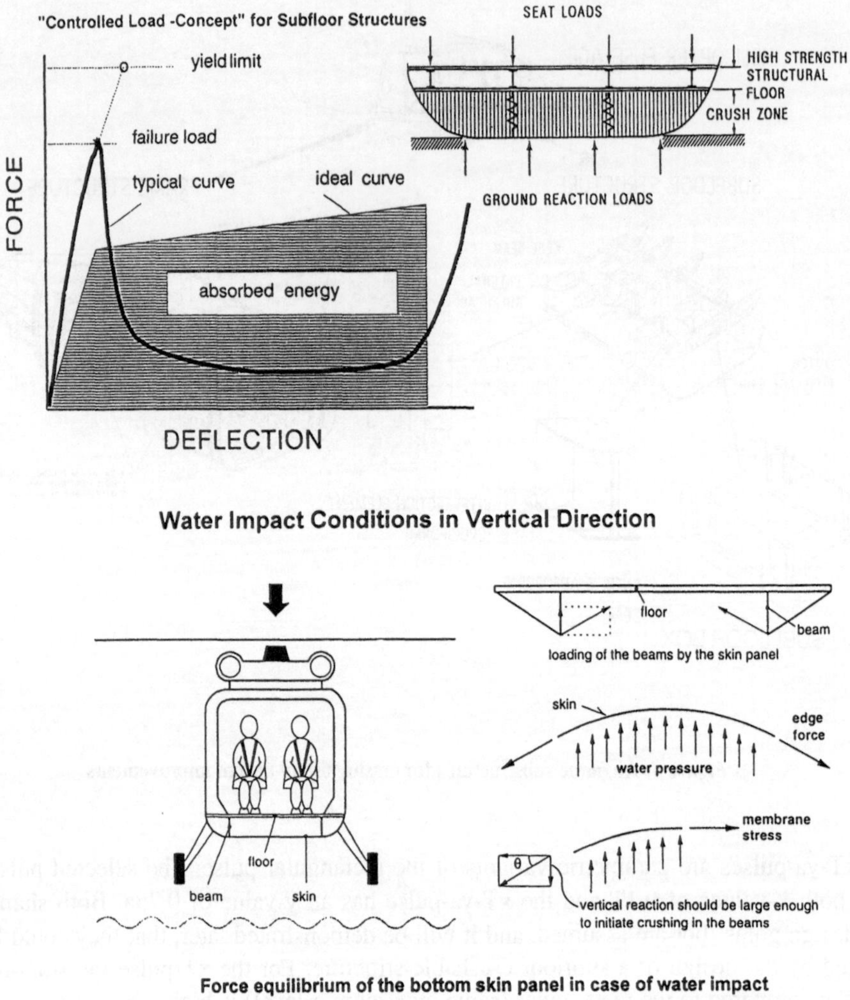

Figure 5 „Controlled load" concept for airframe subfloors

- To minimize cost and weight penalties a dual function structural concept should be realized: load carrying capability for normal operation and EA for crash cases.
- Water impact: outer subfloor skins between beams and bulkheads should be able to transfer water pressure loads via membrane stresses to EA elements without failure.

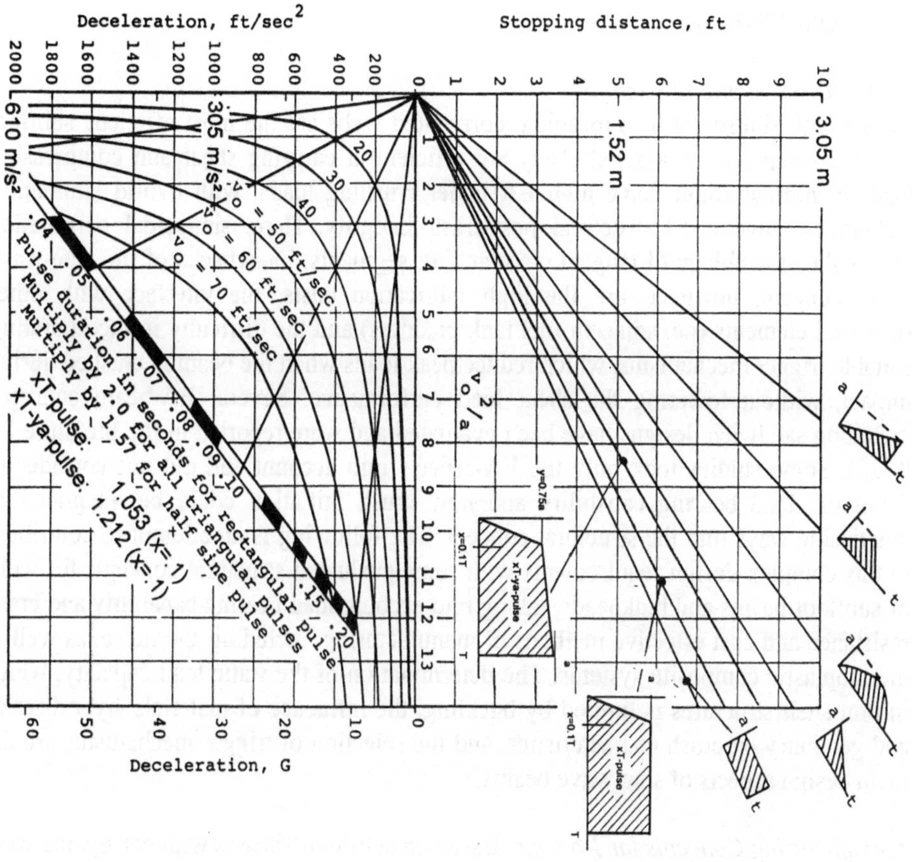

Figure 6 **Extended crash pulse design chart from ACSDG [1]**

Composite aircraft structures must be designed carefully to assure crash resistant features. Compared to metals, totally different design concepts have to be developed and verified. High EA with composites has been obtained for compressive loadings where brittle fracturing of the composite into sublaminates occurs. Under tensile or bending loads structural integrity may be lost at initial fracture and EA can be low. Structural crashworthiness comprises EA and structural integrity. Both are dependent upon material properties and design concepts. To guarantee post crash structural integrity, i.e. a protective shell around the occupants, composite structures, especially comprising carbon fibres, have to be hybridized with tougher fibres such as aramid fibres (Kevlar or Twaron) or high performance polyethylene (Dyneema/DSM).

3.2 DESIGN AND MANUFACTURING OF CRASHRESISTANT AIRFRAME COMPONENTS

3.2.1 *I-Beam Subfloor Sections*

Beams with sinusoidal or trapezoidal corrugated webs are the most efficient subfloor design concepts yet evaluated: They are efficient at carrying shear and compression loads in normal flight, have high EA under crushing loads, with hybrid lamination techniques, have good structural post-crash integrity. Here, sinusoidal corrugation means the assemblage of tangent circular ring segments. Drawbacks of the sine-wave beam concept, however, are the high fabrication costs, the interface with other structural elements (i.e. adjacent fuel tank bladders) and the difficulty in incorporating suitable trigger mechanisms, which reduce peak loads when the beam is crushed during impact, without lowering the shear load performance. Several EA-beam concepts including sandwich designs have been evaluated and were reported in the literature [4, 8, 23]. Some studies took only the EA-aspects into account and did not consider in detail the load bearing capability and how crush initiation could be integrated in practicable ways into the structural concept. The following is an attempt to contribute to this complex design requirements with some evaluated structural concepts for webs of subfloor beams and bulkheads, taking into account load bearing capability and crash resistance and cost effective methods of manufacturing, including thermoset as well as thermoplastic composite systems. The determination of the static load capacity, which in thin shell structures is limited by buckling, the influence of materials hybridization and geometry on crush characteristics, and the selection of trigger mechanisms are the main design aspects of sine-wave beams.

Manufacturing Concepts for I-beams. Based on subfloor I-beams with corrugated webs manufactured in thermoset composite prepreg autoclave technology a design and manufacturing concept for a geometrically similar thermoplastic variant was developed using the same carbon- and aramid-fibre fabrics, hybrid laminate layup in the stacking sequence and a polyetherimid (PEI) matrix system. The corrugated web geometry consists of an assemblage of 12 circular tangent ring segments (sine waves) which merge at both ends into flat parts in the midplane of the web. The opening angle of the single segments is 132,5°. The considered I-beam section has 45mm wide symmetric flanges on top and bottom, an overall length of 500mm and a height of 200mm. The web laminates comprise 4 carbon (c) and 4 aramid (a) fabric layers with the following stacking sequence: $[+/-45°_a;+/-45°_a;+/-45°_c;0/90°_c]_s$. The placement of the aramid fabric layers at the outside of the laminate is chosen due to the better post crash structural integrity of this stacking sequence compared to others.

Thermoset beam variant: For the autoclave manufacturing the Ciba Vicotex M10 prepreg system is used in combination with a carbon 5H satin and an aramid 1/3 crowfoot fabric. The Vicotex system is very flexible with respect to combinations of curing pressure and temperature. The lowest possible curing temperture is about 80°C, the highest about 180°C, and the pressures range between 1 and 3 bar. For manufacturing of the beam section a curing temperature of 120°C and 3 bar pressure are selected, acting 1 hour on the laminate. The 4 piece aluminium tool is designed for one shot fabrication of the I-beam, i.e. web and flanges are fabricated in-situ. However, this manufacturing technique needs a lot of manpower to perform the hand layup.

Thermoplastic beam variant: The design of the thermoplastic beam shown in Figure 7 is suited for automated manufacturing: the two caps, the web and four connecting angles are manufactured separately and the single parts are bonded or welded together in a second manufacturing step. The Tencate CETEX CD 0282 with carbon 5H satin and CETEX VD 0175 with aramid 1/3 crowfoot fabric, both with PEI-matrix are selected. For the prototype beams a thermoforming process with CNC-milled male and female aluminium molds is used for the corrugated web and the connecting angles. The strips for the flanges are cut from preconsolidated laminate plates. The contour of web and angle parts are also cut out with water jet from a prefabricated PEI laminate plate having the above mentioned laminate architecture for the web. The laminate layups for the connecting angles are: $[+/-45°_c; +/-45°_c]_s$. The shaped plates are infrared-heated 100s outside the press to forming temperature of 320°C. Then they are pulled automatically between the upper and lower part of the mold by means of a support frame system and the press is closed within 2-3s. The surface temperature of the molds is set to 200°C before the thermoforming process starts. A pressure of 60 bar is held constant for 60s in the closed mold. The parts can be taken out of the molds when they have been cooled down to 100°C. For series production for the caps a pultrusion and for the webs a roll-forming process would be more suitable and more economic compared to thermoforming.

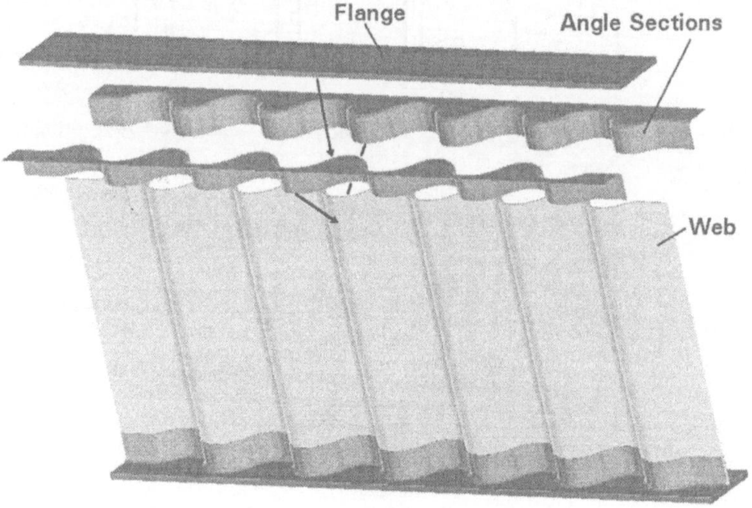

Figure 7 Design concept of thermoplastic sine wave beam

Trigger concepts. With respect to structural efficiency the trigger concept is the key feature in the design process. From the crash resistance point of view, the trigger has two functions: (1) to initiate compression strength dominated stable propagating crushing and (2) to reduce initial peak loads. For a pure absorber type structure such as a crushing tube, the trigger concept can be comparatively simple: very often just a bevel trigger at one tube end is used. From the load bearing aspect, the trigger should not affect seriously the beam's structural performance under operational loads, i.e. avoid major influence on shear strength/buckling behaviour. The trigger acts as a stress concentration where a strength controlled failure should start locally. A real design challenge is to maintain at least a certain post crash integrity in the trigger zone of a structural element. Selected designs of triggers are summarized in Figure 8. They include J-shaped connections at the bottom of beams, notch- and bead-triggers, embedded stress concentrators and bond shear triggers. Some of the trigger versions are taken into account in the following buckling/strength analysis. J-triggers, notches, and embedded stress concentrators are also evaluated experimentally, the last two under static shear as well as crushing loads.

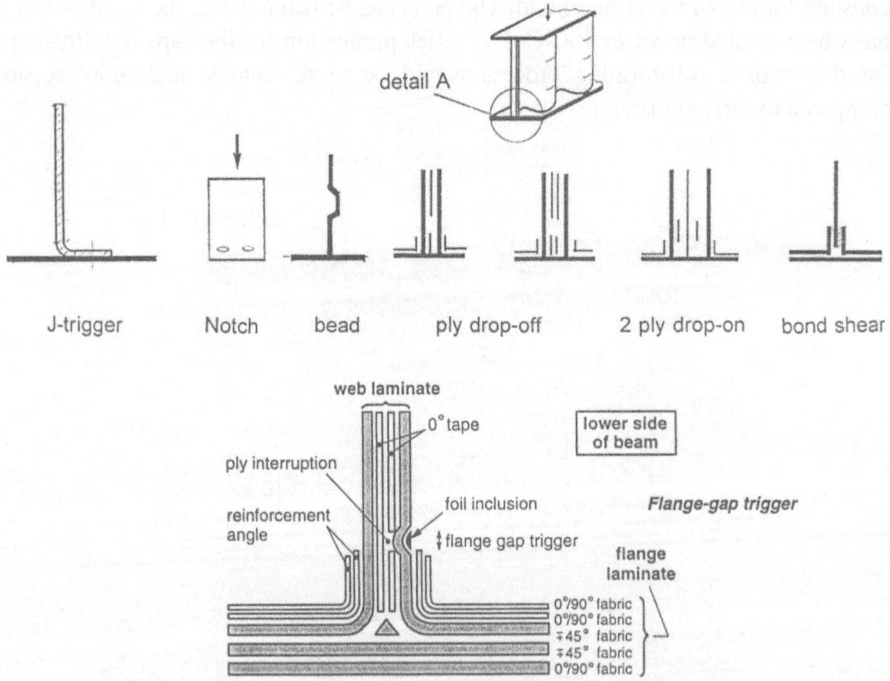

Figure 8 Trigger concepts

Shear-/compression buckling and strength-behaviour of subfloor I-beams. This has been studied by Finite Element analysis (FEA) of various corrugated web configurations in order to optimise the sine-wave geometry . Investigated were the effects of geometrical parameters such as free buckling length (spar height 150, 200, 300 and 400mm), laminate thickness (1.4, 1.75mm), the opening angle of the wave elements (120°, 150° and 180°), and the effect of hybridisation. The basic web laminate consisted of a 50/50 intraply woven carbon/aramid fabric with epoxy resin. The radius of the wave element was held constant at R=20mm. Buckling critical combinations of geometrical parameters are small laminate thickness, large spar height and opening angle of the tangent circular ring section (wave element) between 90° and 120°. The results are summarised in Figure 9. An optimum for all the analysed laminate thicknesses and spar heights is an opening angle of 150°. An increase of the angle up to 180° does not result in most cases to higher buckling load but leads to higher weight due to increasing cross section. Compression buckling loads are higher than strength failure loads, which is a crucial fact for crash resistance, because compression strength dominated failures result in more efficient crush EA.

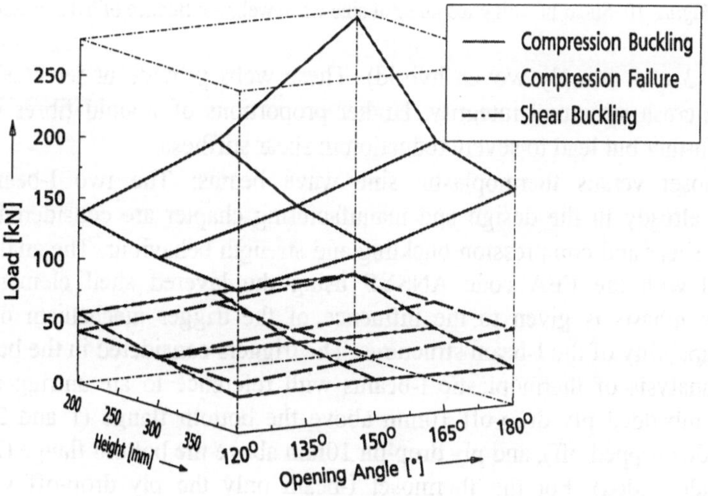

Figure 9 **Hybrid sine wave webs: shear and compression buckling and strength**

Influence of Hybridisation. Figure 10 shows further analytical results on hybridisation of carbon fibre and aramid fibre plies in a sinusoidal corrugated web with opening angle of 150° and constant laminate thickness 1.4mm based on 7 prepreg plies. Two spar heights (h = 200 and h = 400mm) are considered. The sine wave web with the lowest aramid share (laminate 6)shows the highest shear buckling loads. However, the post-crash integrity of this laminate is not satisfactory. Acceptable shear buckling resistance can be achieved when the proportion of aramid plies is between 30-50%

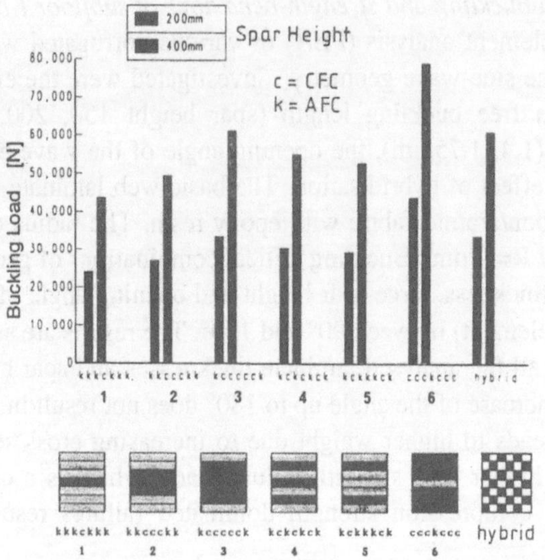

Figure 10 **Shear buckling response of sine wave webs - influence of hybridization**

(laminate 3 and intraply woven hybrid). These webs provide under crash loading a good post crash structural integrity. Higher proportions of aramid fibres improve the integrity further but lead to severe reduction in shear stiffness.

Thermoset versus thermoplastic sine wave beams: The two I-beam variants discussed already in the design and manufacturing chapter are considered here with respect to shear and compression buckling and strength behaviour. The analysis is also performed with the FEA code ANSYS using the layered shell element STIF91. Specific emphasis is given to the influence of the trigger mechanism on the load bearing capability of the I-beam structures. The triggers considered in the buckling and strength analysis of thermoplastic I-beams with reference to an untriggered version include: embedded ply drop-off 10mm above the bottom flange (1 and 2 ply-strips 10mm wide dropped off), and ply drop-on 10mm above the bottom flange (2 ply-strips 10mm wide added). For the thermoset I-beam only the ply drop-off versions are compared to the untriggered version. Figure 11 summarizes the strength analyses under compression and shear loading. The results are normalized with respect to the untriggered thermoset version. The buckling results are not considered here in detail, because they are all above the strength values and are much less affected with respect to trigger versions.

Compression: For both matrix systems with 1 ply dropped-off the triggered strength reduces about 17%, and for 2 plies dropped-off the reduction is about 70% compared to the untriggered version. A ply drop-on with the thermoplastic version does not reduce the strength behaviour and is therefore skipped for further trigger evaluation with the thermoset system and experimental crushing verification which is reported later.

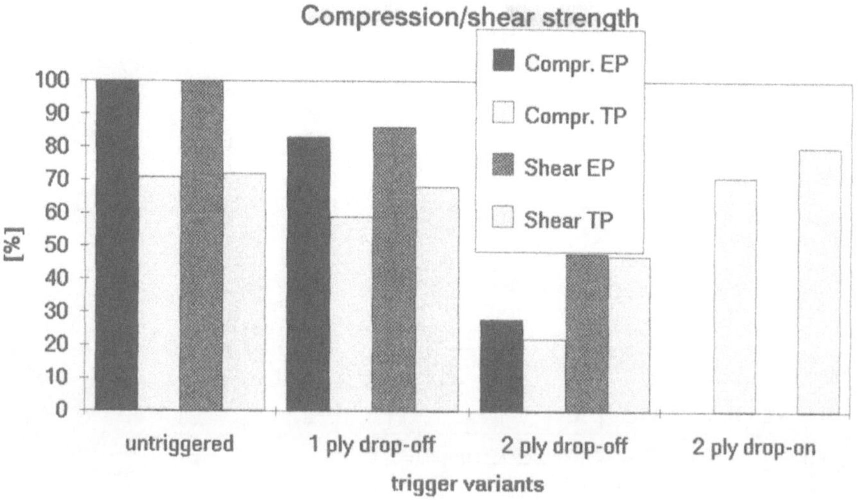

Figure 11 **Shear and compression response of hybrid sine wave webs - influence of triggers**

Shear: With the epoxi system the 1-ply drop-off reduces the shear strength about 14%, and the 2-ply-drop-off about 52%. With the thermoplastic system the reduction with triggers is less. 1-ply drop-off causes only 5% less strength, and 2-ply drop-off 46%. The 2-ply drop-on results in even slightly higher shear strength of the thermoplastic web.

Trapezoidal corrugation: Sine wave beam webs require high tooling and manufacturing efforts. Therefore, alternative geometries such as trapezoidal corrugation and sandwich with trapezoidal core have been studied analytically with repect to shear and compression buckling. Selected configurations have been investigated under crushing behaviour. The study was based on the standard carbon/aramid fibre hybrid laminate with 8 layers with the same web length (same weight) for all variants . A ratio l_1/l_2 of straight and inclined laminate portions was defined and was varied between values of 'zero' (triangular corrugation) and 5.5 (rectangular corrugation). A l_1/l_2-ratio of 2.0 resulted in the same compression buckling behaviour as the comparable sine wave web, but had an even slightly higher shear buckling performance, Figure 12. The l_1/l_2-ratio of 2.0 is an „quasi" optimum with regard to shear and compression buckling, because lower and higher ratios lead to lower loads. All trapezoidal core sandwich versions show low buckling loads, owing to the low laminate thickness of the face sheets in the unsupported areas.

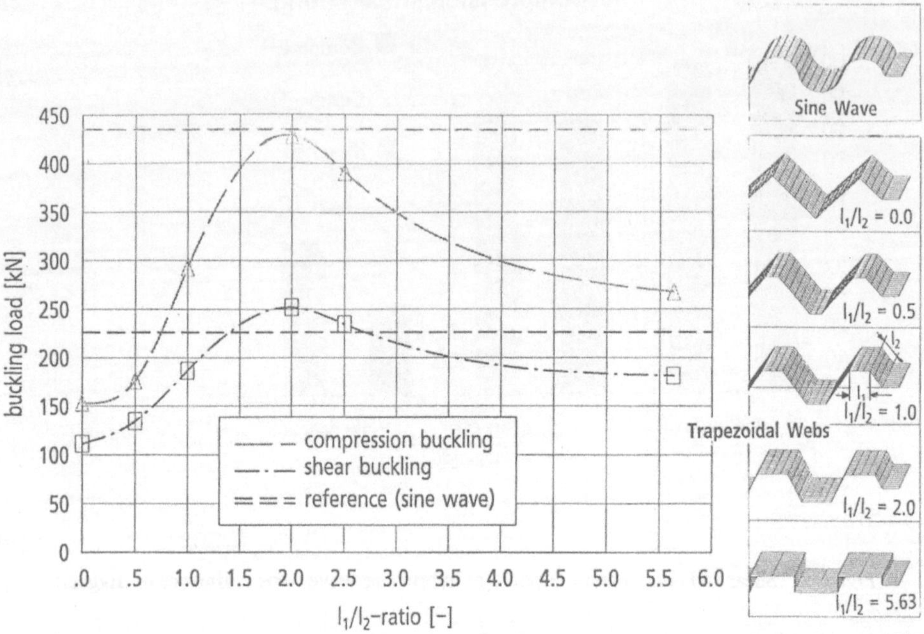

Figure 12 **Buckling of trapeziodal corrugated webs**

3.2.2 *Intersection elements (cruciforms) and box structures*

Aircraft Subfloor Intersections (Cruciforms). Cruciforms are formed by intersections of beams and bulkheads and represent typical floor structure subelements. Various cruciform designs were investigated under the aspect to retain the basic subfloor design concept and to limit the additional manufacturing efforts to achieve crash resistance.

Peak load triggering: Maximum load is determined by the stiffness of the cruciforms and may be intolerable for crew and passengers in the case of a crash. A stiffness reduction is possible if thereby the capability of governing the operational loads is not affected. Several triggering concepts were tested and are discussed below.

Starting from an aluminium baseline cruciform taken out from a commuter type aircraft (flat subfloor!) other aluminium and composite cruciforms having single and multiple notched edge joints, corrugated edge joints, and increasing aramid fibre share of the laminate at the mid-section are studied, Figure 13. A moderate initial stiffness and then a constant or slightly increasing crush force level in combination with post crash structural integrity are the major cruciform design goals for vertical energy absorption.

Further design improvements lead to a hybrid cruciform variant, the HTP-element. This element has a column-like mid-section formed by a Y-shaped split of the shear web laminate, an integrated bevel trigger at the bottom of the shear webs, and tapered egdejoints at the keel beam attachment. The keel beam and shear web laminates have a

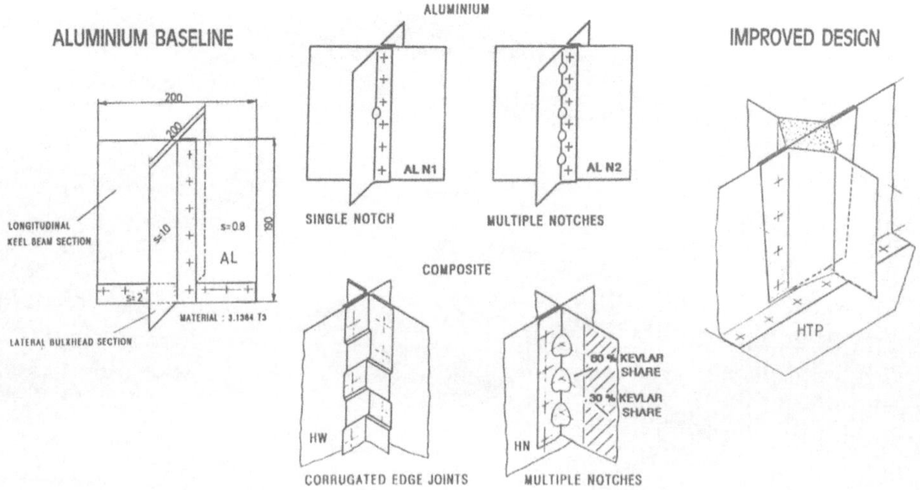

Figure 13 **Design variants of structural cruciforms**

J-shaped connection to the outer skin. In several static and dynamic crush tests 3-4 times higher absolute energy absorption compared to the other elements could be achieved. The specific absorbed energy of an aluminium cruciform with a single notchreaches only 28% of the value of the HTP-element. From the cruciform test series, the following conclusions for cruciform crash resistant designs can be drawn:

- Multiple notching for peak failure load reduction results in low EA.
- Post-crush structural integrity can be achieved by hybridization, e.g. a mixture of carbon and aramid fibre laminates.
- Pure carbon cruciforms have high EA and high weight savings of 30% compared to aluminium but disintegrate completely during crushing. Hybrid elements have weight savings between 15-20% compared to aluminium.
- Composite cruciforms show the same or even much higher (HTP element) absolute EA compared to metal elements.

Crash Resistant Composite Subfloor Boxes. Based on the HTP-cruciform test results a crash resistant carbon/aramid fibre hybrid floor box was designed in accordance to an aluminium baseline structure of a commuter airframe. The box represents a part of the midsection of the subfloor structure and each of the two keel beams carry the inner seat rails of the left and right seat row. The composite test articles were fabricated in autoclave technique, also using the Ciba Vicotex M10 prepreg system The two keel

Figure 14 Subfloor box design

beams together with the lateral bulkheads form 4 HTP- cruciform type elements. The keel beams and the bulkheads are connected to the outer skin with J-triggers at the bottom, Figure 14. All parts are bonded and riveted together. For testing, the cabin floor structure is represented by an aluminium clamping plate.

H-Box structures: H-Box teststructures were designed and statically and dynamically crushed to investigate EA improvements of the box concpepts caused by integrally stiffened webs on the level of a reduced test article size. At the intersections of the beam and shortened bulkheads HTP-elements as in the boxes were formed. Two different web designs were investigated: 1) An asymmetric design (HBOXAS1) with 5 vertical beads integrated into the laminate and a J-trigger connection to the outer skin panel. 2) A symmetric beam (HBOXSS1) with symmetrically integrated 5 vertical beads, also with a J-trigger of the web. Both designs can be considerd as "minor modifications" with the aim to improve lateral buckling stiffness and crushing behaviour of the web area.

3.2.3 *Tensor skins*

At the National Aerospace Laboratory (NLR) in Holland a composite sandwich skin concept ("tensor"-skin) has been developed, Figure 15. The tensor skin provides a mechanism which allows a composite skin to unfold and deflect before membrane stresses build up [9,10]. Hence, reaction forces to the membrane stress have a large

Figure 15 NLR „Tensor skin" concept for water impact [9,10]

vertical component from the start, limiting the membrane stresses in the skin needed to react to the water pressure and loading the beams with the compression forces needed to crush them. The tensor-skin concept makes use of high performance polyethylene (PE/DSM) fibres which can survive the large deformation during the unfolding process while maintaining their strength in tension. The faces of the tensor-skin can be carbon/epoxy, aramid/epoxy or hybrid carbon/aramid epoxy materials. The PE/epoxy core is configured as a corrugated plate, thereby containing the necessary folds. The design loads for the sandwich tensor skin panels are mainly in-plane shear with respect to the operational loads, and membrane tension of the stretched core with respect to crash loads.

3.3 CRUSHING AND ENERGY ABSORPTION PERFORMANCE

3.3.1 *Crushing and EA performance of corrugated webs (I-beams)*

Laminate architecture and element geometry. With hybrid fabric laminate layup crush tests are performed with tubes, tube segements and sine wave beam sections comprising three different stacking sequences respectively. In all symmetric laminates two layers carbon/aramid intrply fabrics are placed at the outside with a fibre orientation of +/-45°. In the midplane one (HSIN1), two (HSIN2), and three (HSIN3) additional carbon layers are positioned having 90% of the fibres in the loading and 10% in the transverse direction. All elements have a curvature radius of R=20mm. The sine wave segments and the tube segments have an opening angle of 145°. The sine wave beam sections comprise the bottom and top flanges; no triggers are used. The tubes and tube segments are triggered with a bevel trigger on top. Figure 16 compares the mean crushing stresses tested under static loading of the various configurations.

HSIN-tubes reach the highest mean crushing stresses, however, the values of the HSIN-segments are just 3-5% below this level. Therefore, crushing stress levels can be determined by means of much simpler test specimens compared to the tube. With the tube and tube segments the HSIN3-laminate does not lead to higher crushing stress levels. The crushing behaviour of untriggered sine wave sections can obviously not be determined by means of simple elements having the same radius and segment opening angle: the reached crushing stresses are much lower (55-78N/mm²). However, additional carbon layers in the web midplane increase the crushing performance. The difference in the EA performance has two reasons: 1) tube and tube segments show local crushing on the entire cross section. The laminates split up in the middle and the two halves bend in- and outwards with a large amount of fibre cracking. HSIN1-sine wave beams do fold showing high structural integrity. HSIN2 and HSIN3 beams crush only in the curved areas. In the areas of curvature changes the laminates are ruptured into strips which delaminate somehow and just bend sidewards without essential fibre cracking. 2) The sine wave section webs are not triggered. After failure initiation the force levels drop and built up again to an almost constant level after 15-20mm crushing. The average force level is therefore lower compared to tubes and tube segments which reach constant crushing levels within 5mm axial deformation.

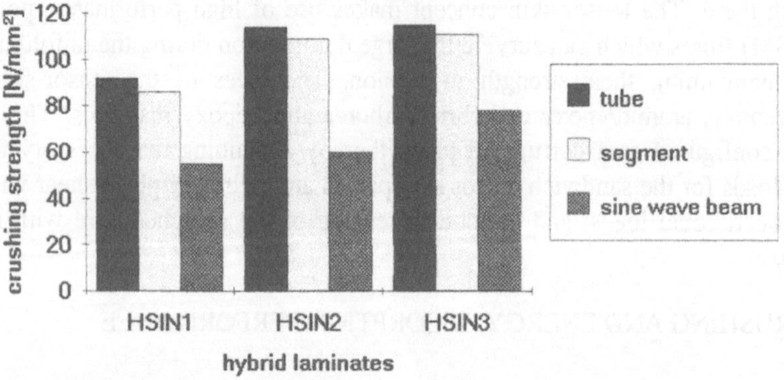

Figure 16 **Crushing behaviour of HSIN-elements**

Influence of Trigger Mechanisms and web geometry. Shear loading with embedded ply drop-off: Crush initiators should preferably be embedded smoothly and uniformly in the laminate architecture with the aim to reduce the compression strength of the laminate in the trigger zone. However, the shear strength and stiffness should not be affected very much. Such a concept was tested in a shear frame with thermoset sine wave beam sections having the standard 8-ply carbon/aramid fibre fabric laminate. Figure 17 shows the embedded trigger design and the comparison of a shear frame diagonal tension loads. The untriggered beam section failed at 149kN and the triggered beam at 142kN. This means only a shear load reduction of less than 5% which is a very acceptable compromise for crash resistance achievement. For comparison, notch triggers in a carbon sine wave beam under shear loading reduce the load bearing capability about 60% compared to the untriggered version. Quasi-static shear tests have been carried out on the first prototype thermoplastic sine wave beam which was assembled using a bonding technique with a paste type adhesive. Shear failure loads comparable, but a little below those on the thermoset reference beam, were obtained (115kN). Reinforcing the angle section laminate with the originally designed 4 plies of fabric will probably lead to a very similar shear load level as was found in the thermoset reference structure.

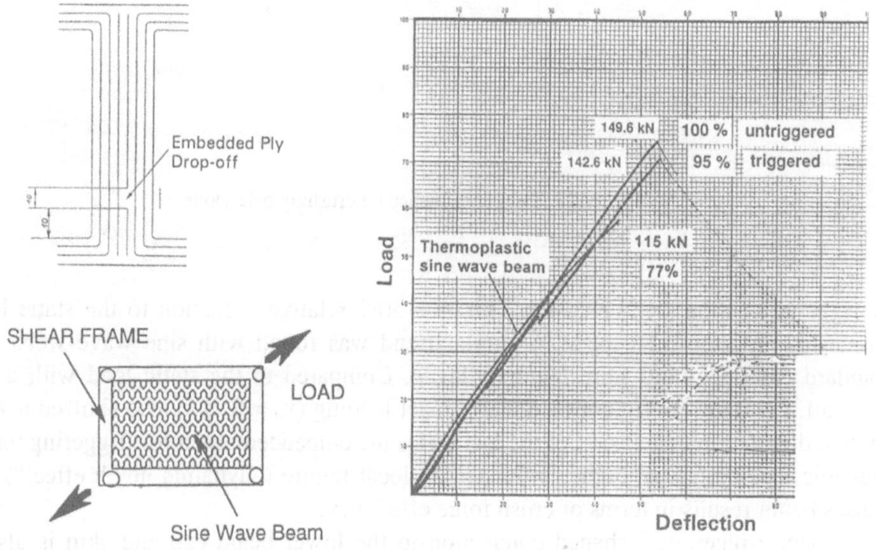

Figure 17 Static shear loading of sine wave beam sections

Notch (Slot) Trigger and embedded ply drop-off: Notch type trigger in a thermoset trapezoidally shaped web with standard laminate layup caused about 50% peak load reduction compared to untriggered webs under static crushing, Figure 18. An embedded ply drop-off under static load caused only 20% peak load reduction. However, under dynamic crushing (v_0 = 8 m/s) a peak load reduction of about 60% compared to the

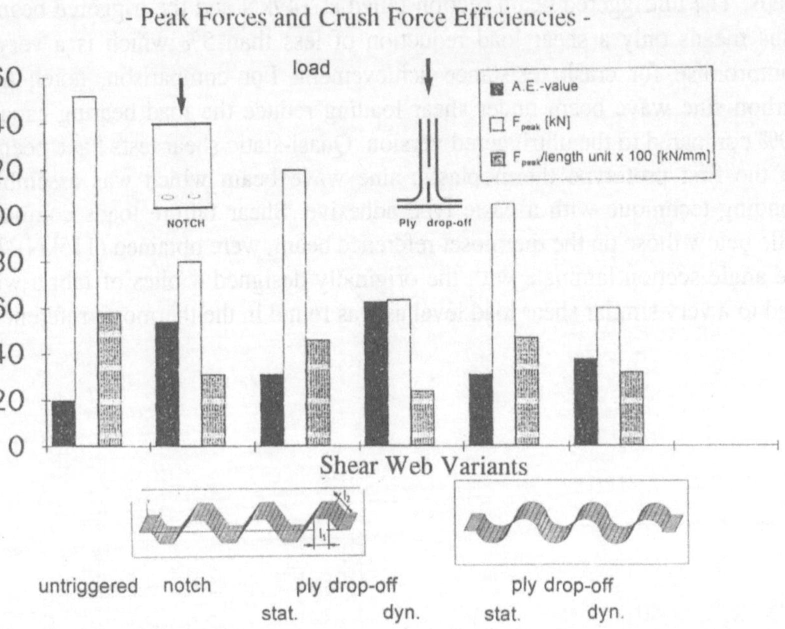

Figure 18 **Trigger influence on crushing behaviour**

untriggered version was observed. This is a 50% relative reduction to the static load case with ply drop-off trigger. A similar trend was found with sine wave webs and standard carbon/aramid fibre laminate layup. Compared to the static load with a ply drop-off, the same configuration under impact loading ($v_0 = 8$ m/s) also showed a 40% force reduction. In conclusion of the test series the embedded notch for triggering under dynamic loading has an higher influence on local failure („dynamic notch effect") but causes better results in terms of crush force efficiency.

J-shape trigger: A J-shaped connection to the lower beam cap and skin is also a promising solution for initial peak-load reduction and stable crush initiation. A systematic variation of the radii is evaluated (4, 6, 8, 10, 12mm) with a 75mm wide and 120mm long crush specimen having a 8 ply carbon/aramid fibre laminate, Figure 19.,Taking into account the failure loads and the EA at 10 and 25mm respectively the specimen having a radius at the J-connection of 8mm can be considered as the best geometry in combination with this laminate configuration.

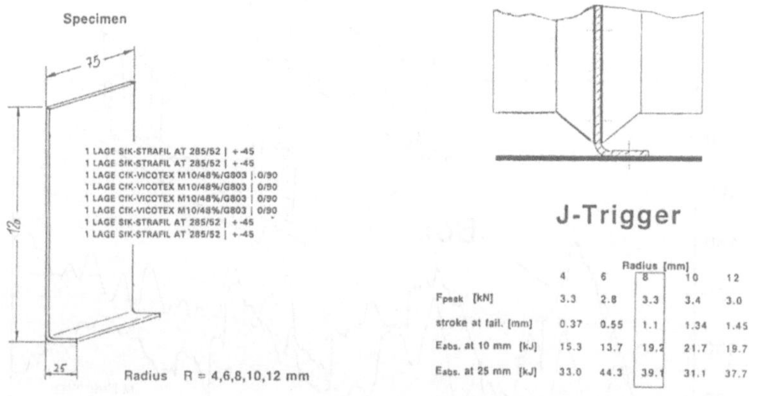

Radius [mm]	4	6	8	10	12
F_{peak} [kN]	3.3	2.8	3.3	3.4	3.0
stroke at fail. [mm]	0.37	0.55	1.1	1.34	1.45
$E_{abs.}$ at 10 mm [kJ]	15.3	13.7	19.2	21.7	19.7
$E_{abs.}$ at 25 mm [kJ]	33.0	44.3	39.1	31.1	37.7

Figure 19 **Crushing behaviour of J-triggers**

3.3.2 *Crushing performance of cruciform box structures*

Boxes. The static test result for the floor box compared with the fourfold test values of a single HTP-element are shown in Figure 20. The box fails first at 80kN, a second peak occurs at 140kN, and then a pretty constant average crushing force of about 100kN develops up to a stroke of 135mm. The superposition of the crush characteristics of four single HTP-cruciforms up to 80mm stroke shows that 80% of the energy is absorbed by folding and fracturing of the cruciforms, and about 20% by the beams and bulkheads between. The overall structural integrity of the box was poor which was especially observed in a drop test at 10m/s vertical impact speed.

A retrofit design floor box was reinforced with symmetric integrally stiffened panels in the beam and bulkhead areas. Also, for good ventilation of trapped air inside the box during a crash, additional holes have been foreseen in the lateral bulkhead laminates. The third box was tested at 8.8m/s impact velocity and an initial kinetic energy of 16,9kJ (drop weight mass: 436kg). The box started to fail at the foreseen crush initiators (J-triggers) and crushed from bottom to top to a maximum stroke of 76mm. An additional stroke of about 60 mm would have been available for more EA.

H-boxes. Two H-box structures were statically crushed with a maximum stroke of 135mm, Figure 21. The HBOXAS1 structure showed crushing initiation without severe peak forces; it reached at about 40mm stroke an average load level of 105kN which was held constant up to maximum deflection. The symmetric component HBOXSS1 showed a sharp load peak of 120kN at crush initiation followed by a drop of crushing force. This was caused by splitting of the web laminate into two halves along the midplane. The major EA was contributed in both structures by the intersections (HTP-

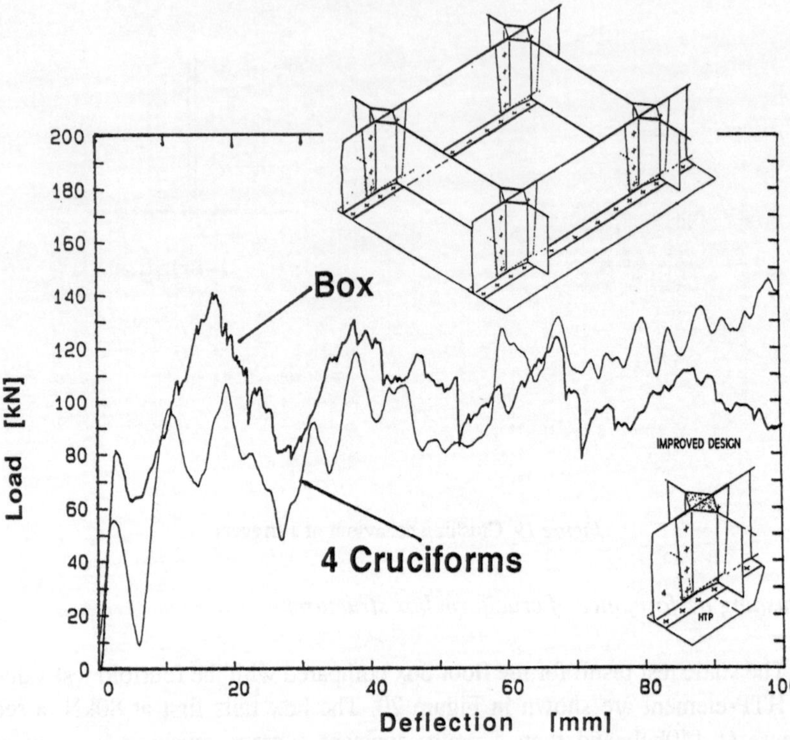

Figure 20 **Crushing performance of cruciforms and subfloor box**

elements). Design improvements could be achieved by a horizontal row of rivets above the J-trigger zone to avoid the peeling effects in the symmetric web.

The subfloor test plan was continued with two H-box drop tests (HBOXSD1 and HBOXAD1) at 9m/s impact speed. Compared to static tests, the results are reversed: the symmetric stiffened box showed higher EA under impact loading, where under static loading the asymmetric box was better. The improvements in the symmetric design is directed to additional rows of rivets in the web to avoid splitting up of the laminate which was observed in the static test. The table included in Figure 21 compares the results of the H-box test series.

3.3.3 Tensor skins

A sine-wave beam configuration with two layers of unidirectional carbon tape in the web was used for the design of the tensor skin panels. A design case was considered where the tensor skin panel is able to initiate crushing in two sine-wave beams in a box of the subfloor section. This means that the tensor skin panels should survive transverse loads of at least 2 × 57kN = 114kN. The inner face of the tensor skin panel consisted of

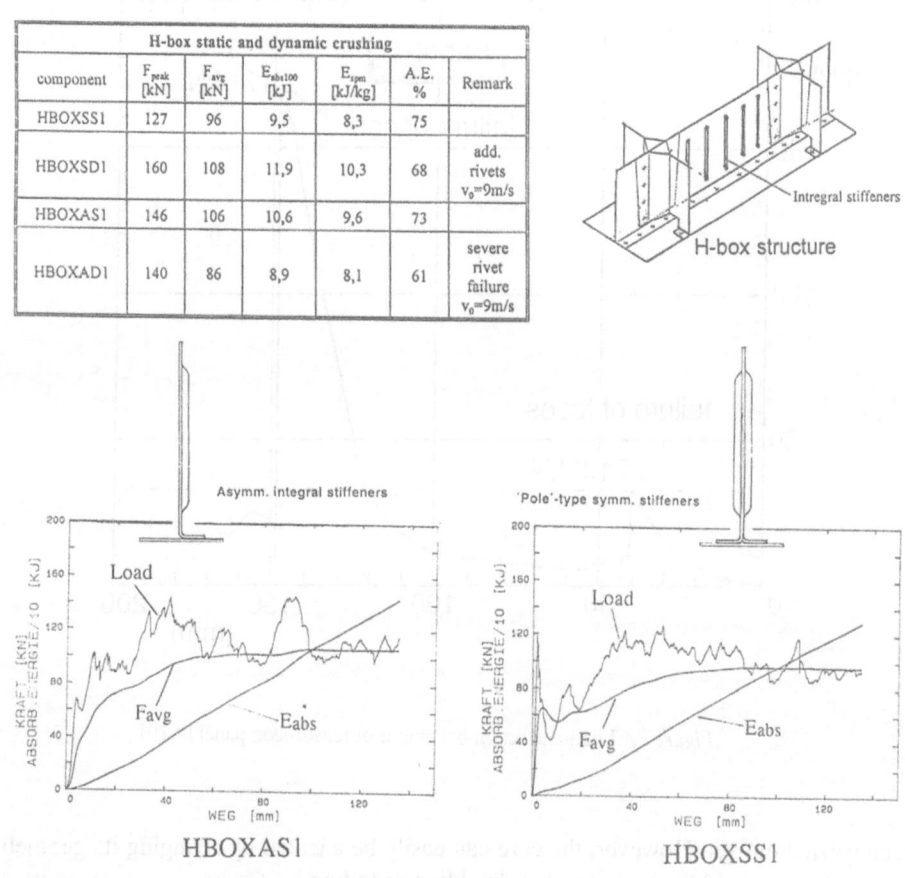

H-box static and dynamic crushing						
component	F_{peak} [kN]	F_{avg} [kN]	E_{abs100} [kJ]	E_{spm} [kJ/kg]	A.E. %	Remark
HBOXSS1	127	96	9,5	8,3	75	
HBOXSD1	160	108	11,9	10,3	68	add. rivets v_0=9m/s
HBOXAS1	146	106	10,6	9,6	73	
HBOXAD1	140	86	8,9	8,1	61	severe rivet failure v_0=9m/s

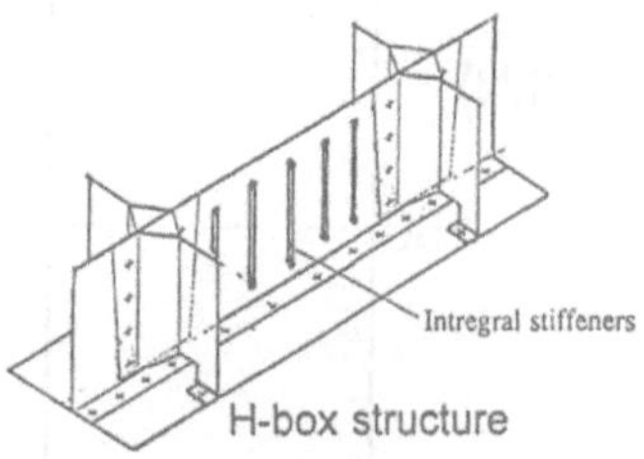

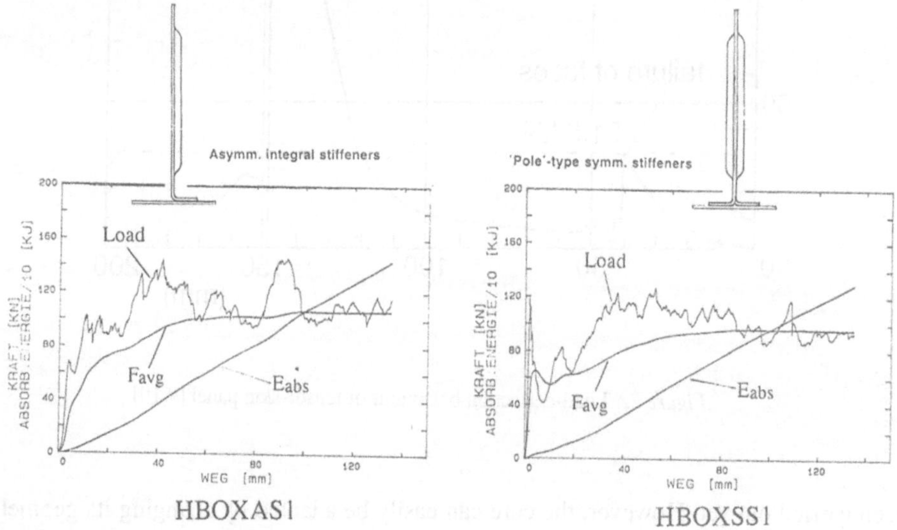

Figure 21 **H-box structure crushing behaviour**

carbon/aramid crowfoot fabric, the outer face consisted of aramid crowfoot fabric embedded in epoxy. Both faces had a three ply [45°,0°,45°] lay-up. The core of the tensor panel had a three ply $[45°]_3$ PE/epoxy lay-up.

The panel was clamped around all four sides, and indented by a blunt indenter simulating the water pressure in a quasi-static test. As expected the two faces failed early, the corrugated core unfolded and stretched as required to fail at a high load level. Figure 22 shows the load-deflection curve measured during the test of the tensor panel. First the outer face failed at 32kN, simultaneously with a more gradual tearing of the inner face, after which the load dropped, and the core unfolded. At the end of the unfolding process, the core stretched and the load increased again until final failure occurred at 171kN with a deflection of 150mm. Based on this test it was concluded that the sandwich tensor skin meets the design requirement of 114kN. A shear test has not

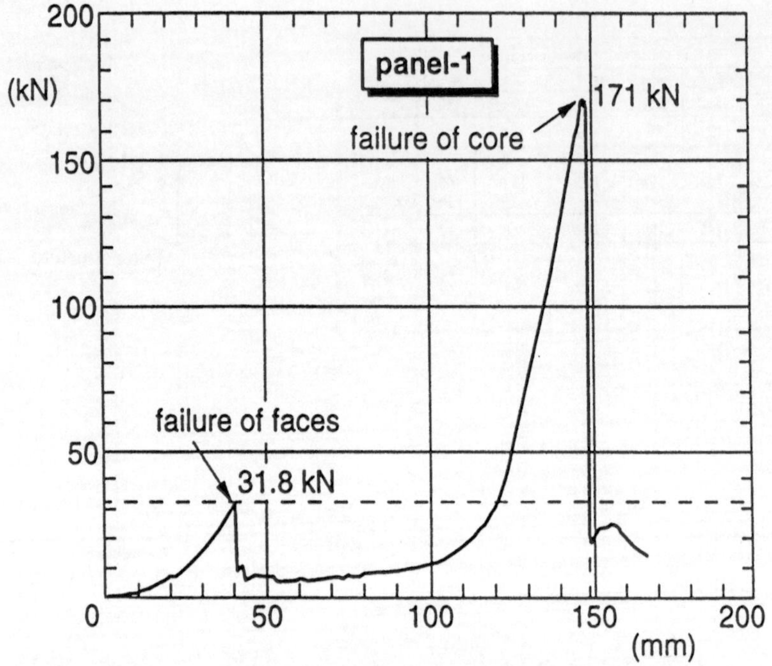

Figure 22 Load-deflection behaviour of tensor-skin panel [9,10]

been carried out yet. However, the core can easily be adapted by changing its geometry if shear loads would lead to premature buckling or failure.

4. Analytical Methods and Simulation Tools

4.1 HYBRID CRASH SIMULATION METHODS

4.1.1 *KRASH hybrid code*

The crash simulation program KRASH predicts the response of vehicles to multi-directional crash environments. KRASH provides the interaction between rigid bodies through interconnecting structural elements (beams), which are appropriately attached (pinned, clamped). These elements represent the stiffness characteristics of the structure between the masses. The equations of motion are explicitly integrated to obtain the velocities, displacements and rotations of the lumped masses under the influence of external and internal forces. In the hybrid modelling technique, large regions of structure are approximated in a simplified manner. Non-linear behaviour (e.g. force-deflection curves) of substructures, that is already known from tests or other analyses can be introduced into the model by use of macro elements like springs, non-linear beams or plastic hinges.

The program KRASH has a history of 25 years. The original version of KRASH was developed and experimentally verified under U.S. Army sponsorship between 1971 - 1974. Whereas the initial program was for application to rotorcraft, further developments sponsored by the FAA extended the capabilities of KRASH for application to general aviation and transport airplanes and culminated in the release of KRASH85 [12]. During the last 5 years KRASH was significantly improved by Dynamic Response Inc., California (DRI). A lot of new features have been added to the code, important especially for aircraft crash simulation. DRI-KRASH [13] now includes additional injury criteria, e.g. HIC and SI calculations, an expanded oleo-pneumatic landing gear module, a soft soil module as well as a water impact module [14].

KRASH has been validated by more full-scale aircraft crash tests than any other crash impact structures program, and although originally developed for aircraft applications, it has in the meantime also been used for the crash simulation of cars, trains and other vehicles. The FAA and the British Air Accident Investigation Board (AAIB) are sponsoring efforts to develop air accident reconstruction / investigation tools, with KRASH as its core program. Today all American and European helicopter manufacturers as well as some leading airplane manufacturers use the program KRASH. Representative KRASH models of various aircraft categories are shown in Figure 23.

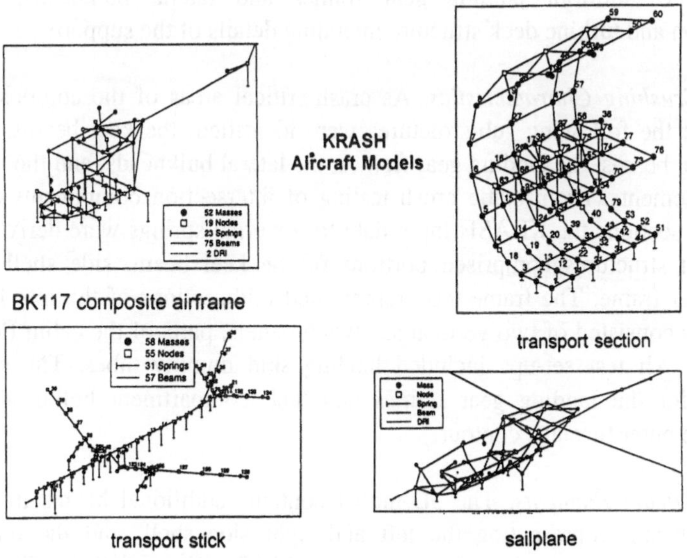

Figure 23 KRASH aircraft models

4.1.2 *BK 117 Airframe KRASH-Simulation Studies*

2D- and 3D-KRASH-Model Development. The full-scale simulation studies are directed to a composite aircraft based on the MBB/Kawasaki BK 117. Composite parts replace most of the basic helicopter structure; the experimental aircraft retains the engine deck, transmission deck, dynamic system, and tail section of the standard basic helicopter. The BK 117 is certificated following the FAR Part 29. Additionally, influenced by other design constraints, only limited efforts towards improved crash survivability could be performed within a joint Eurocopter/DLR activity [15].

2D-KRASH Model: The 2D-KRASH model represents the helicopter in the X-Z-symmetry plane. Only symmetric crash cases can be simulated with this model; it consists of 22 concentrated masses, 25 linear beams, 11 massless node points, 15 nonlinear springs, and 1 DRI-element (mass 22 and beam 25). The model is adapted to a maximum gross weight of 3200 kg and a forward C.G. location at FS 4380 to represent the highest loads and deformations in a crash case. All the structure above the cabin floor including tailboom behaves linear-elastically under the simulated crash conditions. The concentrated mass points are distributed along the cabin floor area, and above and along the turbine deck.The subfloor structure is considered to crush. The skid landing gear (spring 1 and 2) and tailboom absorber (spring 15) crush data together with all other relevant design details to generate the model were provided by Eurocopter Deutschland.

3D-KRASH Model: The 3D-model included in Figure 23, is suited for arbitrary crash impact conditions comprising roll and lateral velocity components. It consists of 52 masses, 75 beams, 20 massless nodes, 23 nonlinear springs, and two DRI elements placed in the cockpit area. The model accounts for structural details in the buttline direction, i.e. subfloor landing gear frames and lateral bulkheads, and lateral transmission and turbine deck structure including details of the supports.

Subfloor Crushing Characteristics. As crash-critical areas of the composite subfloor assemblage the following sub-structures were identified: the keel beams, the cockpit longitudinal beams, the landing gear frames, the lateral bulkheads, and the intersections of those elements. From static crush testing of intersection components and landing gear frame sections the KRASH input data for external springs were derived [15]. The intersection structures comprised portions of the keel beam, side shell frame, and landing gear frame. The frame tests represented mid-sections of the rear landing gear frame; they consisted of two vertical sandwich panels, parts of the cabin floor, and the outer skin. All test set-ups included landing skid dummy tubes. The spring curve modeling for the landing gear frames and tank compartment bulkheads took into account the outer fuselage contour.

Structural Beam Elements. The 3D-model contains additional beams in the buttline direction, beams representing the left and right side shell, and the rear bulkhead structure. The lowest position of beams is the cabin floor level (WL 1500). All structure below is represented by external springs. The tailboom and empennage are modeled by one single beam, respectively . In one KRASH-beam element the elastic or nonlinear behaviour of various structural parts is concentrated. The total cross section is summed up by the the cross sections of single structural elements. The elastic- and shear moduli are calculated taking into account a weighing factor of the particular cross section area related to the whole section of the beam. The moments of inertia are summed up including the Steiner moment of inertia of the single elements.

Mass Distribution. The mass distribution of the 2D-model is shown in Figure 24. From the data sources 189 single mass point dimensions and locations could be retrieved. The following-on steps to determine the final mass distribution are the definition of KRASH-mass locations, concentration (reduction) of masses, and the redistribution of masses following certain prediscribed criteria. The last two steps are supported at DLR by the computer programs CONMASS and MODMASS.

KRASH-Mass Locations: Certain large masses such as gear boxes, engines, and the rotor system, are fixed in their individual C.G.-locations. The other masses of the upper fuselage are fixed for the 2D-model in the z-direction on the level of the turbine deck, and the x-positions coincide with the intersections of the vertical structural members (side shell frames/bulkheads) and the turbine deck. More mass details on the transmission and turbine deck can be modeled in the 3D-version. The tailboom structure is concentrated at two mass locations (2D and 3D). In the crush zone area (subfloor) no KRASH-masses are locatéd. The z-position of the lower masses is the cabin floor level. The x-positions are fixed for the 2D-model in the middle at the intersection of vertical structural elements, keel beams, and the landing gear frames, and tank compartment bulkheads. In the 3D-model masses are also distributed along the right and left keel beams at relevant intersections. Those masses have massless node points where the vertical beams are connected to.

Concentration of Masses: The mass concentration procedure uses the desired KRASH-mass location input together with the originally defined mass distribution. The CONMASS-program finds all masses which are closest to a certain KRASH-mass location, and concentrates all this masses into one mass point and location.

Redistribution of Masses: With the concentration of masses and positioning the masses to the desired KRASH-mass locations the aircraft model is unbalanced in terms of overall C.G.-position, total mass, and moments of inertia. With the program MODMASS an iterative procedure can be performed to balance the model to a predescribed configuration by redistribution and changing the concentrated masses to KRASH-mass locations. Also, the difference between the identified mass distribution and Design Gross Weight (DGW) or changing of masses such as fuel can be corrected. The crash response of the final KRASH-model can be expected to be very accurate from the point of view of mass distribution, especially when the overall moments of inertia come into play in simulated roll- and pitch-conditions.

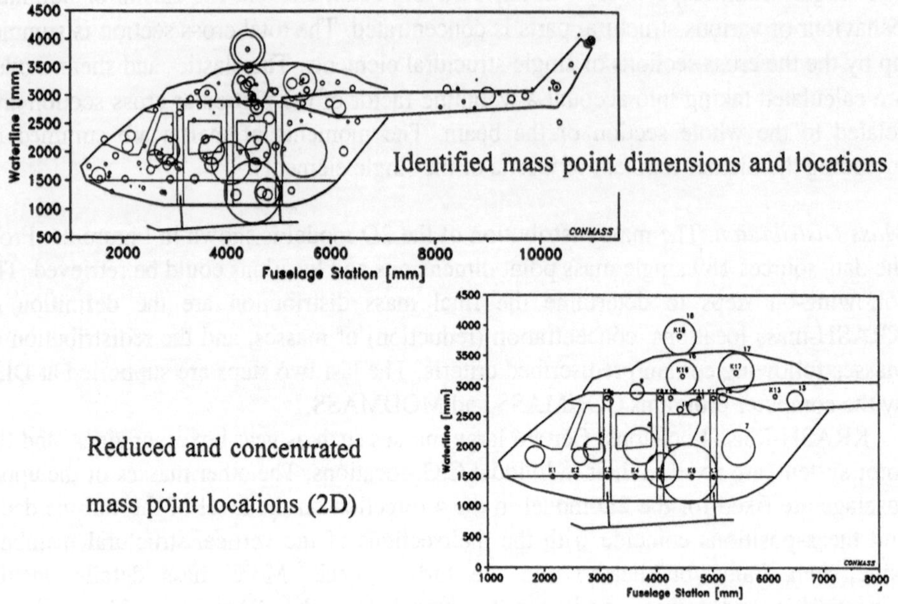

Figure 24 BK117 - mass distribution

Crashworthy Seat Study. The simulation studies were performed with the 2D-model under vertical impact (v_o=7.9m/s). In the standard configuration the normally installed seat structure was modeled. The EA-seat version had a load-limiting device installed, acting at 14.5g with a possible stroke of 120mm. Figure 25 summarizes the simulation results. Without an EA-seat the DRI-value is in the range of 24-25 which indicates a spinal injury potential of about 70%. The EA-seat reduces the DRI to 15.5 which is about 0.5% of injury potential. The stroke of the EA-seat is 110mm which is close to bottoming out. Plotting the results in the Eiband diagram (whole body +z-direction) shows that the acceleration response of the seat without an absorber hits the region with minor injuries, and the EA-seat curve remains completely in the area without injuries.

Simulation of the BK 117 Crash Test (2D-Model). A full-scale crash test with the BK117 metal airframe was performed at Kawasaki/Japan in 1985. No crash test with the composite airframe is planned. In fact, the test was considered to be helpful for verification, because the composite airframe is an one to one scale replacement of the metal structure, and the EA capability was demonstrated by component crush test. For the simulation the model was adapted to the test condition (C.G.-location and total

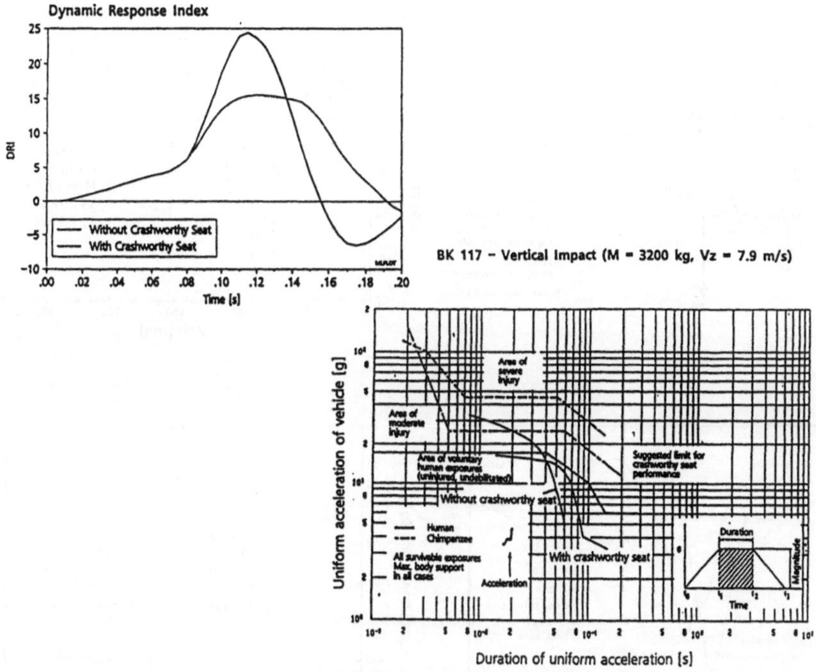

Figure 25 Crashworthy seat study: DRI-response and Eiband-curve injury potential

mass of 2800kg). Some representative correlations between test and simulation are summarized in Figure 26. The airframe showed good structural integrity and overhead mass retention in the test and simulation. Most deformation was observed in the skid landing gear, the subfloor, the rear inclined airframe bulkhead, and at the mounting area of the tailboom to the airframe. At selected mass locations very good response correlation between test and simulation could be achieved in terms of peak acceleration, pulse shape, and duration. Also the overall crash sequence could be simulated realistically. The comparison of simulation versus test concerning peak accelerations is at the floor level (mass 3) 42 to 45g, and at the turbine deck (mass 16) 27 to 19g respectively. At the turbine deck the difference might be influenced by the 2D-modeling where the deformation and damping of the lateral structural members is not taken into account. The pilots´ upper torso response of the simulation was about 10g higher compared to test.

4.1.3 KRASH Simulation of Complete LEAR FAN Aircraft

NASA Langley Research Center acquired two full-scale composite Lear Fan aircraft for crash test evaluation in the Impact Dynamics Research Facility (IDRF). One airplane has already been tested in essentially an "as is" condition to provide baseline data. First test results are reported in [16]. The second airplane will be provided with a modified subfloor structure that improves energy absorption and will also be tested at the IDRF.

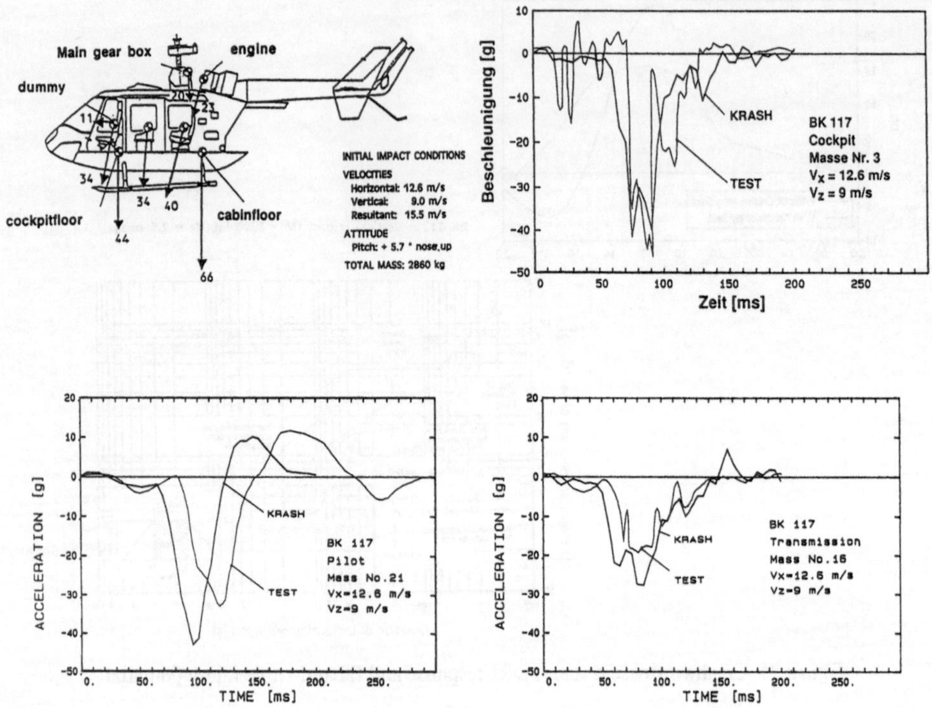

Figure 26 **BK117 - correlation of full-scale crash test (metal) and simulation (composite airframe)**

A full-scale KRASH 79 model of the Lear Fan aircraft was retrieved from the literature and was converted to the more recent KRASH 85 version. Details of this model and pretest simulation results are reported in [11]. Taking into account more recent full-scale test data from the full-scale aircraft and fuselage sections, and the latest PAM-CRASH simulations of one section, the hybrid model was updated again using the DRI-KRASH93 version. The latest model shown in Figure 27 comprises 40 masses, 52 nodes, 27 nonlinear springs and 39 beam elements.

The actual impact conditions reported from the crash test are:
$V_x = 80$ ft/s = 24.38 m/s; $V_z = 30$ ft/s = 9.14 m/s; 0° roll; yaw < 0.5°; -20.6° flight path; + 20.6° pitch relative to the flight path (⇒ 0° pitch in the KRASH simulation)

Figure 28 shows the simulated crash sequence with the above listed initial conditions. The fuselage hits the ground first in the middle of the cabin area with a tendency for the aircraft to rotate forward. Later in the crash sequence the aircraft rotates backwards during sliding out and the added springs at the rear fuselage come into contact with the ground.

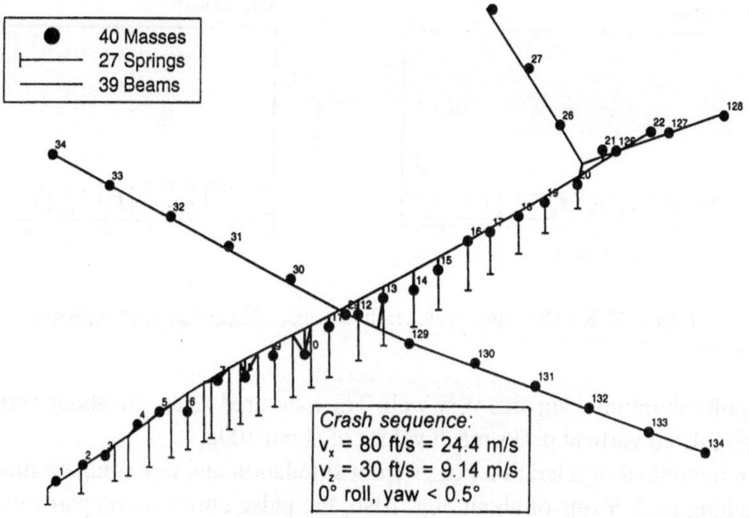

Figure 27 **KRASH model of the Lear Fan 2100**

Figure 29 shows a comparison of KRASH simulations with selected test data at test station 1 (cockpit area: seat rails of the left seat) and simulation results at mass 5, which comes closest to the test station 1. It must be mentioned for this comparison that the measured acceleration at the seat rails are local dynamic responses, whereas the acceleration behaviour of mass 5 of the stick model is a highly damped global structural response. However, the correlation of the longitudinal velocities is very good. The longitudinal velocity for the test was integrated from the unfiltered acceleration data. At 100ms the test and simulation show that the aircraft is still sliding forward with a velocity of about 19m/s.

The vertical and longitudinal acceleration of the test was filtered with a SAE class 60 filter. The measured vertical acceleration at test station 1 is much higher within the first 10ms of impact compared to the simulation. The measured response shows local rebound, whereas the simulated deceleration level stays high up to 25ms, and then reduces to levels which are very comparable to the measured signal, i.e. the overall

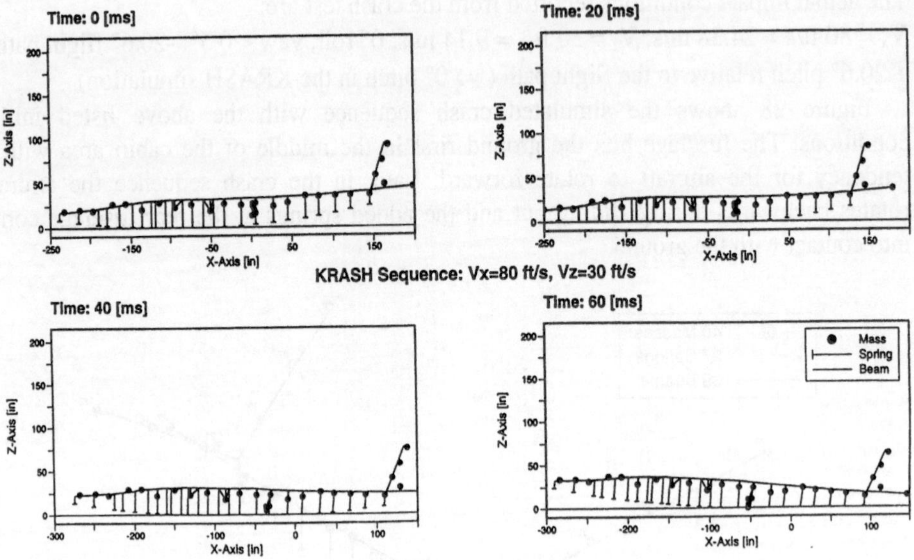

Figure 28 **KRASH simulation: crash sequence of Lear Fan 2100 crash test**

vertical pulse duration compares very well. The measured peaks are about 160g and the highest simulated vertical peaks reach values of about 100g.

The longitudinal deceleration pulses from simulation and test compare much better, both reaching peak levels of about 30g. Also, the pulse duration compares well. In the longitudinal acceleration during the test at 50ms is a clear acceleration peak which leads to a short increase in velocity at 50ms after performing the integration. This response is attributed to a local event at the seat rail during that time frame. These local events are not seen in the simulation runs with the global KRASH model.

4.2 FE TOOLS FOR CRASH SIMULATION

4.2.1 *Dynamic FE simulation*

FE simulation techniques allow the engineer to predict the crash response of a structure directly without having to make use of structural test data to calibrate elements in the analysis, as is the case with hybrid simulation methods. However, the crash behaviour of an aircraft structure is extremely complex, involving both nonlinear dynamic materials response and large structural deformations. Such analyses are at the limit of validity of current FE analysis codes, especially for nonmetallic structures. The established FE codes in the aircraft industry are the implicit FE codes, such as NASTRAN, which are used widely for structural analyses, buckling calculations, and for prediction of dynamic vibration responses. In the last 15 years several commercial FE codes such as LS-DYNA3D [17] and PAM-CRASH [18] have been developed

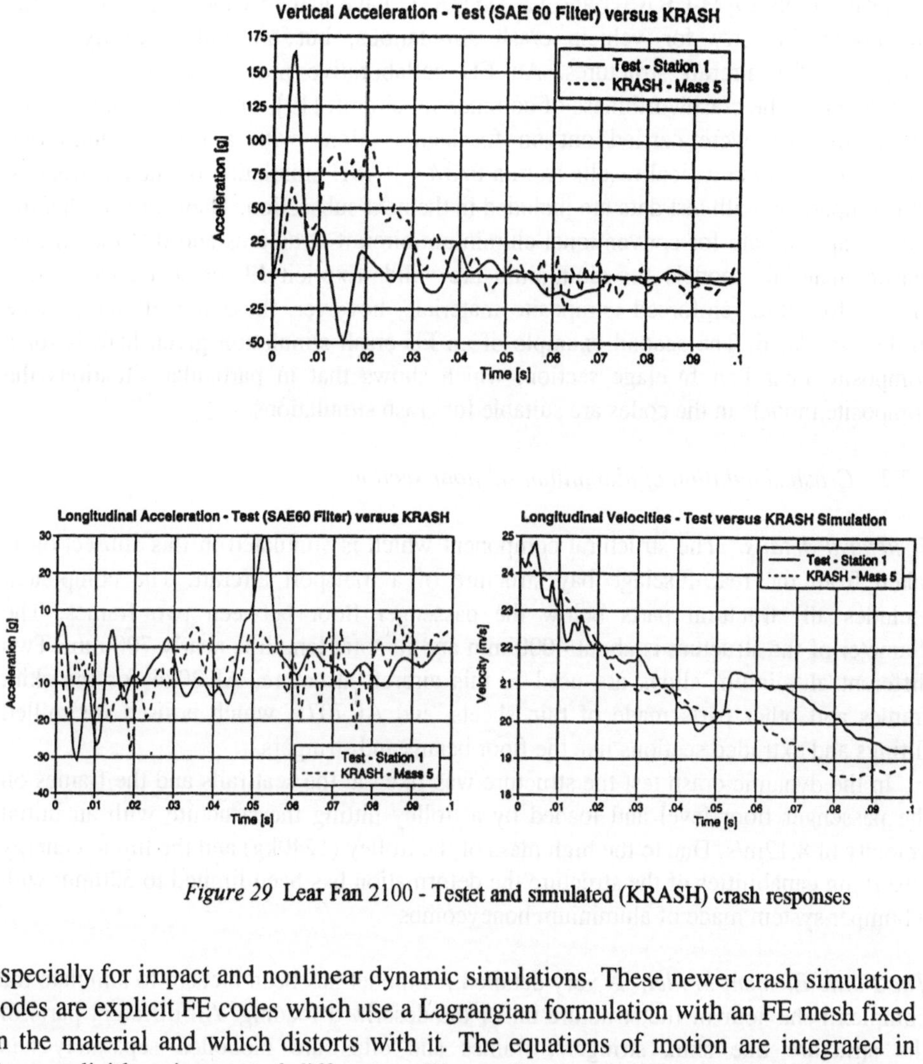

Figure 29 Lear Fan 2100 - Testet and simulated (KRASH) crash responses

especially for impact and nonlinear dynamic simulations. These newer crash simulation codes are explicit FE codes which use a Lagrangian formulation with an FE mesh fixed in the material and which distorts with it. The equations of motion are integrated in time explicitly using central differences. The method requires very small time steps for a stable solution, thus it is particularly suitable for impact and crash simulations and less appropriate for equilibrium structural analyses. The main advantages of the explicit method is that the governing equations are uncoupled allowing an 'element-by-element' solution, with no global stiffness matrix assembly or inversion required. The method is generally recognised to be very robust for highly nonlinear problems. The codes contain materials models for metals and composites, and most important for crash analysis, contact in the structure is easily and efficiently handled by introducing temporary 'penalty forces' as additional external forces to resist penetration and control sliding. In response to the needs of the automotive industry there are also models of safety features such as airbags and occupant dummies which can be incorporated into the structural analysis.

As design tools PAM-CRASH and LS-DYNA3D have gained wide acceptance in the automotive industry for vehicle crash simulations, but are only recently being considered for aircraft structures. An EU collaborative project has recently been completed on the crashworthiness of commercial aircraft [19], in which crash tests and FE simulations were carried out on fuselage sections with a riveted aluminium construction. Some typical results from a PAM-CRASH simulation of such a structure and comparison with test data are included in the next subsection. They confirm that the code is appropriate for conventional aluminium aircraft structures and that the elastic-plastic materials models for aluminium are valid. Explicit FE crash codes contain models for fibre reinforced composite materials, however, these models have to be further validated. The second example of an FE crash simulation given here is for a composite Lear Fan fuselage section, which shows that in particular situations the composite models in the codes are suitable for crash simulations.

4.2.2 *Crash simulation of aluminium subfloor section*

Structural details. The structural component which is simulated in this subsection is taken from the rear fuselage bay structure of a transport aircraft. The component includes all structural parts below the passenger floor between two frames. The diameter of the structure is about 4000 mm and its total length is nearly 700 mm. Two different aluminium alloys are used in this aircraft structure, Al 2024 for the skin, frames and other parts made of thin sheets, and Al 7075, which is used for milled fittings and extruded sections like the floor beams and stringers.

In the dynamic crash test the structure was fixed at the seat rails and the frames on the passenger floor level and loaded by a trolley hitting the structure with an initial velocity of 8.12m/s. Due to the high mass of the trolley (1240kg) and the limited energy absorbing capabilities of the structure the deformation has been limited to 320mm with a bumper system made of aluminium honeycombs.

The Finite Element Model. A very detailed FE model has been created to simulate the complex behaviour of the structure under the described loading. All different parts of the structure like skin, stringers, frames and clips are modelled separately and connected with rivet elements that can fail during the simulation. In Figure 30 the final mesh of the FE-model is displayed. This final mesh consists of 66440 nodes, 58884 shell elements and additional 3485 rivets.

The clamping and loading conditions in the simulations are similar to those realised in the crash test. The displacements of the nodes along all four seat rails and at the frames on the passenger floor level were suppressed. The trolley is represented by a moving rigid wall with a mass of 1240kg and an initial velocity as measured in the test just before the impact (8.12m/s).

The material properties used in the simulations were generated from the results of a comprehensive material test program performed within [19]. Quasi-static and dynamic material tests have been performed to investigate the influence of the material direction of roll and the sensitivity of the properties on the strain rate. Because only a small

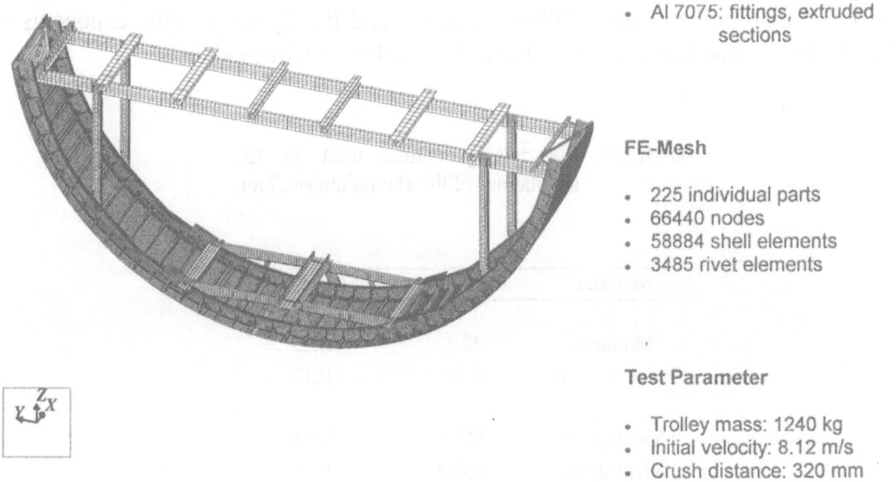

Figure 30 FE model of aluminium subfloor section

influence of both effects could be found, an elastic-plastic material model with isotropic hardening has been selected for the simulations. Strain rate effects were not taken into account.

The results of the material tests in engineering stresses and strains have been converted into true stresses and strains for further use in the simulations. The hardening of the aluminium alloys beyond the proportional limit is assumed to be governed by a power law of the following form:

$$\sigma = K(\varepsilon_0 + \varepsilon_p)^n \tag{1}$$

In this formulation σ and ε are true stress and true strain components, and K and n are material parameters. The total strain is interpreted as a constant initial yield strain ε_0 calculated from the Young's modulus and the initial yield stress of the material and the plastic strain ε_p which is calculated by PAM-CRASH for every element. Material rupture in the material model can be achieved by eliminating elements whose effective plastic strain reaches a value higher than a given limit strain. In Table 1 the material properties used for the simulations are summarised.

As already mentioned all parts of the fuselage structure were modelled separately and connected with rivet elements. In the FE code PAM-CRASH rigid bodies are used to represent the rivets. Two nodes are linked with such a rigid body and can move (translate and rotate) as a pair. The links between the nodes are allowed to separate upon violation of the following failure criterion:

$$(P_N/P_{N*})^{a_1} + (P_S/P_{S*})^{a_2} \leq 1 \tag{2}$$

In this formulation the calculated normal and shear loads in the rivet P_N and P_S are related to the normal and shear failure loads P_{N*} and P_{S*}. By varying the exponents a_1 and a_2 the interaction between normal and shear failure can be modelled.

TABLE 1. Materials properties used in the simulations of the aluminium subfloor

Material	Al 2024	Al 7075
Modulus	72.1	69.5
Poisson's ratio	0.33	0.33
Yield stress	326.8	505.0
Yield strain	0.453	0.73
Max. pl. strain	16.0	9.2
Parameter K	713.2	747.4
Parameter n	0.145	0.0796

Results of Simulations. A first simulation was performed without rivet failure and element elimination. This simulation was terminated after 45ms, which corresponds with a crushing distance of about 325mm. With a time increment of 10^{-3}ms, 45045 cycles were calculated. The total CPU-time on a HP9000/735 workstation was 37.6 h.

In Figure 31 the deformed structure is shown 45ms after the impact. Plastic deformations start on the left hand side of the structure between the second and third stringer position, very close to the location where the skin failed in the test. The plastic deformations in this region enlarge during the simulation and additional plastifications occur just beside reinforcements around the intersection of the struts with the frames. In general, the correlation between the simulated deformations and those found in the dynamic crash test was very good.

Parallel to the described simulation another calculation including rivet and material failure with element elimination was started. The rivet failure loads were estimated as P_{N*} = 8 kN and P_{S*} = 5 kN with a quadratic interaction of normal and shear loads (a_1=a_2=2). During this simulation the first rivet failure occurred after 17ms.Then many more rivets failed but only a few elements were eliminated because their plastic strain exceeded the limit. Although some clips and supports were separated from the frames and skin, there are only small differences in the global deformation compared with simulation without rivet failure.

A comparison of both simulations with measured test data is given in Figures 32 and 33. In Figure 32 the force between the impacting trolley and the structure is plotted against the time. Up to the first rivet failure (17ms) both simulations are absolutely identical. Then the load starts to decrease for the simulation including failure (Curve B, dotted line) compared with the other analysis (Curve A, solid line). Although a 300Hz filtered acceleration signal has been used to calculate the test curve, large oscillationscaused by lower frequency noise from the test rig or the trolley can be observed (Curve C, dashed line). However, the mean load from the test curve correlates well with the simulations.

In Figure 33 the time-integrated acceleration signal is plotted together with the velocity of the rigid wall which represents the trolley in the simulations. In the simulation which includes failure (Curve B) the velocity remains slightly higher after 17ms when failure of rivets and elements started. While the results from the first simulation (Curve A; solid line) start to separate from the test curve the simulation with rivet failure correlates very well with the test up to the point where the aluminium absorbers suddenly stopped the trolley (Curve C, 42ms). This comparison shows that a correct modelling of the failure of rivets is necessary to calculate the structural behaviour correctly. In related analyses with greater structural damage the influence of rivet failures in the model was found to be even more significant.

The presented simulations of the aluminium fuselage structure and their comparison with test data show that the explicit FE code PAM-CRASH is a valuable tool for studying the crash behaviour of aircraft structures. The strain hardening as well as the failure of the aluminium could be simulated correctly. Problems in the analysis arise in

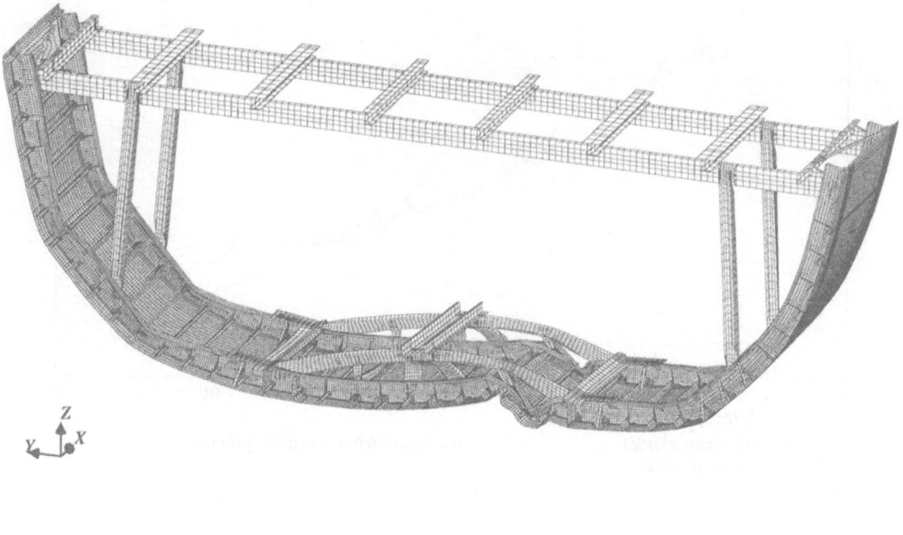

Figure 31 **Aluminium subfloor section - deformed structure**

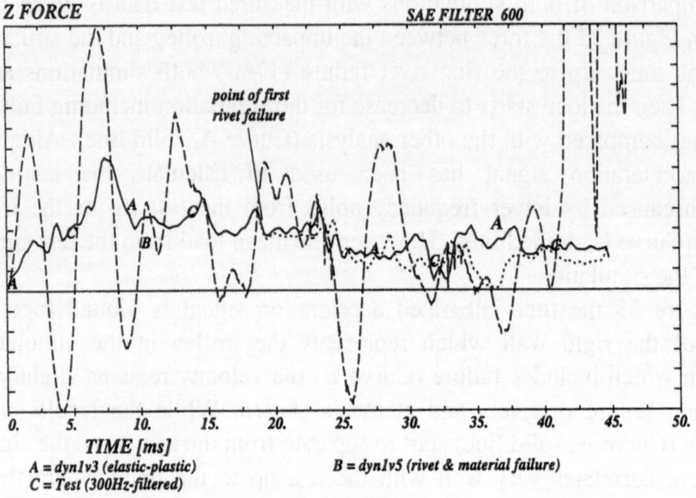

Figure 32 Aluminium subfloor section - test versus simulation: force-time response of trolley

regions where structural parts are connected with rivets. It was found that a model of the rivet failure had to be included in the simulations to obtain the realistic behaviour of the structure.

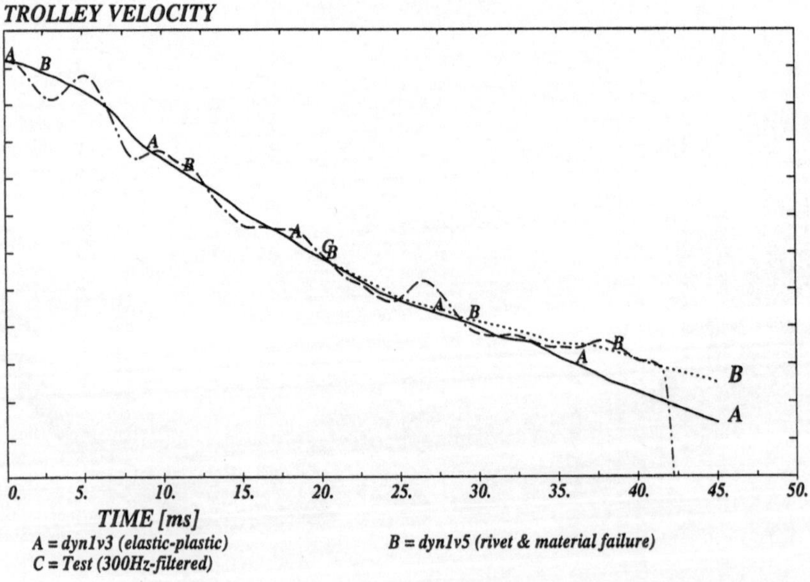

Figure 33 Aluminium subfloor section - test versus simulation: trolley velocity

4.2.3 *Crash simulation of composite fuselage section*

Structural details. The composite section modelled in the analysis consists of a short length of the Lear Fan 2100 fuselage containing two CFRP frames with CFRP skins, along with the longitudinal aluminium floor beams which support the two pairs of seat rails. This section was chosen because of its inclusion in a NASA Langley Impact Dynamics Branch test programme in which full scale crash tests are being carried out on composite airframes, as reported more fully in [11] and [16]. It consists of an oval shell with maximum diameter 1.54m, 0.3m in length, containing two frames 0.14m apart, as is seen schematically in Figure 34. The frames are fabricated as 'Z' profiles 50mm deep with a 25mm flange which are bonded to the shell thus giving an 'L'stiffener. At the base of the section the frame doublers are about 100mm in height where they are attached to four longitudinal floor beams which support the seat rails in pairs. The seat rails are attached to the frames and stiffened laterally and vertically by 'L' section beams or bulkheads. Investigation of materials samples taken from a crashed section enabled the CFRP materials and ply lay-up to be established. The floor beams are 1.25 mm thick aluminium plates and the remainder of the structure consists of

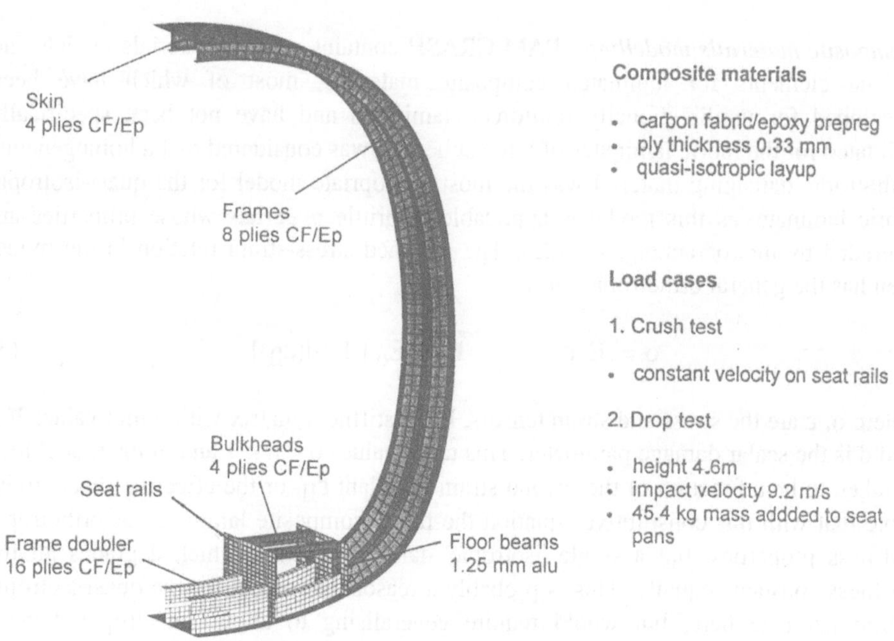

Figure 34 FE model of Lear Fan 2100 section

CFRP laminates based on 5 harness satin weave carbon fabric in an epoxy resin with a nominal ply thickness of 0.33mm. The fuselage skin and bulkheads consist of 4 plies carbon fabric/epoxy laminate with a quasi-isotropic lay-up [45°,0°,0°,45°], the frame doublers at the base of the section contain 16 plies with the same repeating quasi-isotropic lay-up, and the main frames are 8 fabric plies at 45° to the frame axis.

The test programme reported in [11] included a static crush test on the section subfloor applied at the seat rails, and a complete section drop test onto a rigid floor from a height of 4.6m giving an impact velocity of 9.2m/s. Some initial FE simulations of both tests were performed. These simulations were based on a limited knowledge of the composite materials in the structure and gave good qualitative agreement, but measured and predicted load and acceleration levels were quantitatively not very accurate [21]. In this paper new results from the FE simulation of the drop test are reported, based on plylay-ups and thicknesses determined from the actual structure. The corresponding results from the quasi-static crush test were found to show very good agreement with the crushing forces measured in the test. In the drop test 45.4kg masses to simulate the mass of an occupant were added to the seat pans, which were attached by very stiff aluminium plates to the seat rails. The section was instrumented to measure floor and occupant decelerations during impact. For the analysis of the drop test the seat rails are rigidly connected to a point 300mm above the rails at which the 45.4kg mass is added. The acceleration response at this point is then computed for comparison with test data.

Composite materials modelling. PAM-CRASH contains several materials models and special elements for laminated composite materials, most of which have been developed for unidirectionally reinforced laminates and have not been specifically validated for the fabric laminates of interest here. It was considered that a homogeneous orthotropic damaging material was the most appropriate model for the quasi-isotropic fabric laminate, as this model is applicable to brittle materials whose properties are degraded by microcracking, see [22]. The assumed stress-strain relation in the model then has the general orthotropic form

$$\sigma = \mathbf{E}\ \varepsilon \qquad \mathbf{E} = \mathbf{E_0}\,[\,1 - d(\varepsilon_{II})] \tag{3}$$

where σ, ε are the stress and strain tensors, E the stiffness matrix with initial values $\mathbf{E_0}$, and d is the scalar damage parameter. This takes values $0 < d < 1$ and in the model here is taken to be a function of the second strain invariant ε_{II}, or the effective shear strain. Note that with this constitutive equation the fabric composite laminate has orthotropic stiffness properties, but a single 'isotropic' damage function which degrades all the stiffness constants equally. This is probably a reasonable model for the quasi-isotropic fabric laminate here, but would require generalising to allow anisotropic damage functions for unidirectional and more general orthotropic fabric materials.

Thus in this materials model all the material moduli are reduced linearly from their initial value as the damage parameter increases from zero. A uniaxial stress-strain curve with the general form shown in Figure 35 can be modelled by a bilinear damage function, in which there are two damage constants d_1 and d_u to be determined. These damage parameters may differ in tension and compression. In the FE analysis the tensile values are used when the volume strain is positive and the compressive values when it is negative. The parameter d_1 measures the departure from linearity at the first knee in the curves, whilst the parameter d_u determines how quickly the stress falls from its peak value to the assumed small residual value σ_u.

The quasi-isotropic CFRP laminate was modelled here as a single orthotropic material, rather than as a multi-ply laminate, and materials tests were carried out to measure the basic mechanical properties of an equivalent carbon fabric/epoxy laminated plate consisting of 8 fabric plies with a $[45°, 0°, 0°, 45°]_s$ lay-up. Figure 36 shows the stress-strain curves obtained in tension and compression coupon tests in a symmetry direction of the quasi-isotropic fabric laminate. The tests show that the quasi-isotropic fabric laminates are approximately linearly elastic in tension, with a more nonlinear 'yielding' behaviour in compression. The material may be characterised as an elastic 'damaging' material with the generalised form of stress-strain curve shown in Figure 35. Table 2 summarises the measured properties taken from the curves. The material has a higher E-modulus in tension than compression, with tensile strengths

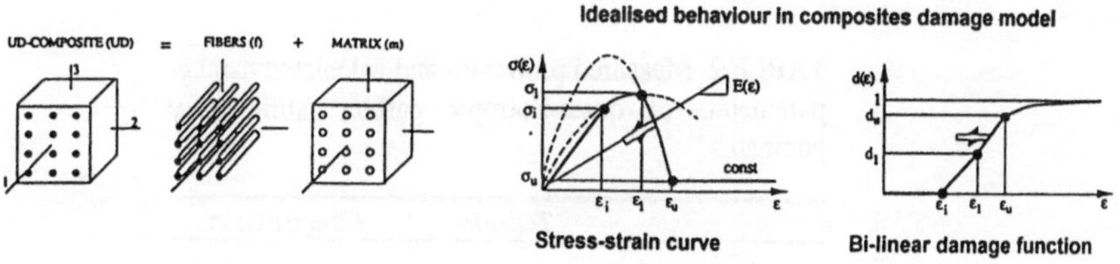

- PAM-CRASH contains a bi-phase composites model for UD plies
- fibre and matrix properties are modelled with different elastic moduli and damage properties in tension, compression and shear
- degenerate bi-phase model (without fibres) used to model quasi-isotropic CF/Ep fabric laminates
- material is initially orthotropic elastic but degraded by micro-cracks before brittle failure, modelled by a damage parameter d $(0 < d < 1)$
- orthotropic stress-strain relation has general form :

$$\sigma = E\varepsilon \qquad E = E_0[1 - d]$$

E is orthotropic stiffness matrix, E_0 its initial values

- d assumed to be bi-linear function of 2nd strain invariant II_ε
  - d_1 gives initial modulus reduction
  - d_u controls reduction from peak stress to residual σ_u

Figure 35 Composite model in Lear Fan simulations

significantly higher than compressive strengths. In the table, σ_1, ε_1 refer to the peak failure stresses and failure strains indicated in Figure 37, whilst the subscript 'i' refers to the values at the end of the linearly elastic region where some initial microdamage takes place in the laminate causing the initial elastic modulus E_0 to be reduced. The remaining composite data required for the analysis, such as shear moduli, shear strengths and Poisson's ratios were calculated from laminate analysis, based on typical CFRP fabric ply properties. For the aluminium floor beams, elastic-plastic properties of a typical airframe aluminium alloy 2024 T351 were assumed.

FE model and results. The structure is modelled by 4-node Mindlin shell elements, and contains about 2500 elements with 3035 nodes, each having 6 degrees of freedom. The fuselage shell and bulkheads were modelled as a single ply of the 4-ply quasi-isotropic sub-laminate elastic damaging material with the properties given in Table 2. The frames as 2 plies and the frame doublers as 4 plies of this sublaminate. The aluminium floor beams were assumed to be isotropic elastic-plastic plates with strain hardening. The FE modelling was kept relatively simple in this analysis in comparison with the detail required to model plastic collapse and folding of aluminium aircraft structures described above, since the primary interest here was to investigate the influence of the FE composites model on the structural response.

TABLE 2 Measured properties and calculated damage parameters of quasi-isotropic carbon fabric/epoxy laminates

	Tension	*Compression*
Modulus GPa		
E_0	54.5	45.0
stress MPa		
σ_i	327	112.5
σ_1	502	209
σ_u	30	57.4
Strain %		
ε_i	0.6	0.25
ε_1	0.94	0.67
ε_u	1.1	0.85
Damage		
d_1	- 0.02	0.31
u	0.95	0.85

In the simulation the half-structure was dropped vertically onto a rigid floor with initial impact velocity of 9.2m/s, with symmetry boundary conditions along the central symmetry axis, and with nodal constraints applied to the seat rails and the seat added mass. Impact of the section on the rigid floor causes the 45.4kg seat mass with its high inertia to impose crushing loads on the floor beams, where most of the materials failure

Measured stress-strain curves of quasi-isotropic CF/Ep laminate

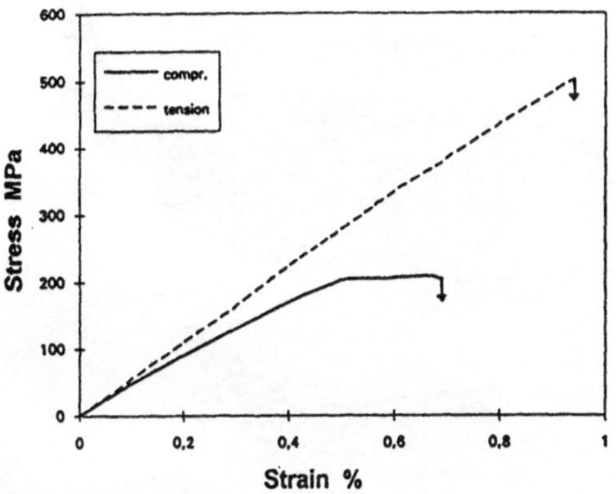

Moduli and damage parameters determined from test data

Modulus GPa	Stress MPa			Strain %			Damage	
E_o	σ_l	σ_1	σ_u	ε_l	ε_1	ε_u	d_1	d_u
Tension 54.5	327	502	30	0.6	0.94	1.1	0.02	0.95
Compression 45.0	112.5	209	57.4	0.25	0.67	0.85	0.31	0.85

Figure 36 Lear Fan simulations - determination of composite properties

takes place. Figure 37 shows typical computed deformations at 10ms, with contours superimposed of the total damage parameter d in the composite elements. The composite structure is severely damaged, as would be expected from its brittle material response and low failure strains. However most of the damage and fracture is in the bulkheads and the CFRP fuselage shell is not significantly damaged. Study of the deformed geometry and damage parameter values throughout the simulation indicates that first the frame doubler is damaged at the interface to the fuselage shell near the impact point. Then the inner bulkheads fracture as the folding of the inner aluminium floor beams starts. There is then fracture of the horizontal beams between the seat rails, and as load is transferred to the outer floor beams the outer bulkheads fracture as the aluminium beam begins to fold. This simulated damage sequence corresponds closely with the observed failure sequence in the NASA floor crush test and section drop test.

Figure 38 summarises the results for the seat pan accelerations from the drop test analysis and compares them with a filtered signal measured at the inner seat rail in the NASA tests as reported in [11]. The composite section simulation shows a strong, narrow 150g peak at about 2ms, which is absent in the test results, thereafter above 4ms

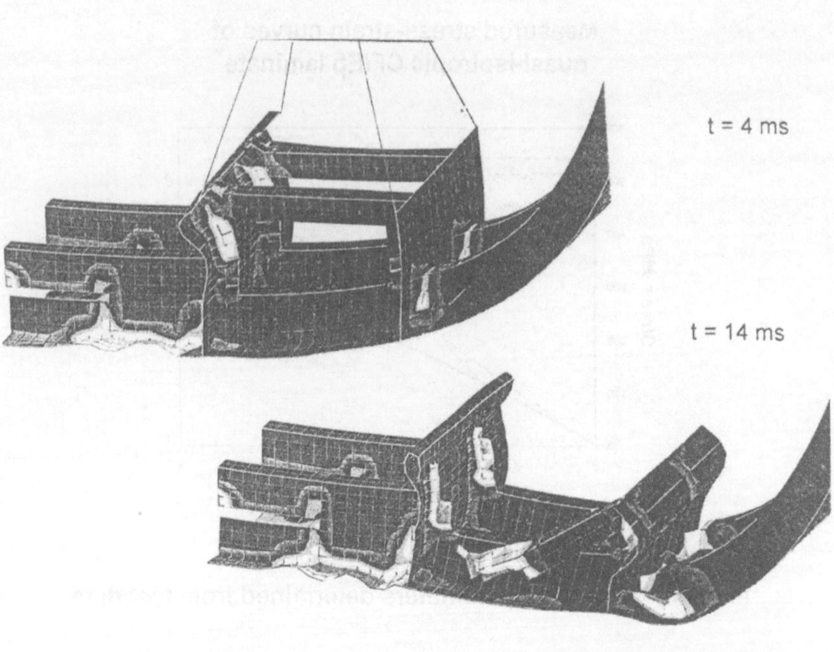

Figure 37 FE simulation of Lear Fan section drop test - composite damage parameter

there is excellent correlation with the test data, both in the predicted g-levels of about 40g and the pulse duration of about 35ms. Note that the unfiltered test data does indicate a 65g pulse at about 2ms which was filtered out of the test data plotted in Figure 38. An interesting feature of the simulated pulse is a secondary 90g peak at 10ms. Study of the structural deformations shows that this is the time when the horizontal beams fracture, as seen in Figure 37 which leads to a sudden transfer of load into the inner aluminium floor beam. Both test and simulation show values of about 40g at the stool lead mass for a duration of about 30ms.

We conclude from these results that the explicit FE code PAM-CRASH can successfully model the impact response of composite structures, at least for the quasi-isotropic fabric reinforced CFRP structure considered here. The software appears to be robust enough to handle brittle materials and the damaging materials model described in [22] used appropriately seems to be suitable for modelling these materials. The results confirm that a composite fuselage section is severely damaged, with high peak accelerations and a broad acceleration pulse so that more attention needs to be given to occupant safety in composite aircraft structures. In the case of the airframe section studied here, this may be achieved by better design of the floor beams which support the seat rails, for example by using specially designed composite sine wave beam structures [10, 23,]. Such energy absorbing devices to replace the aluminium plate floor beams in the Lear Fan fuselage are under consideration in ongoing work at NASA as reported in [20].

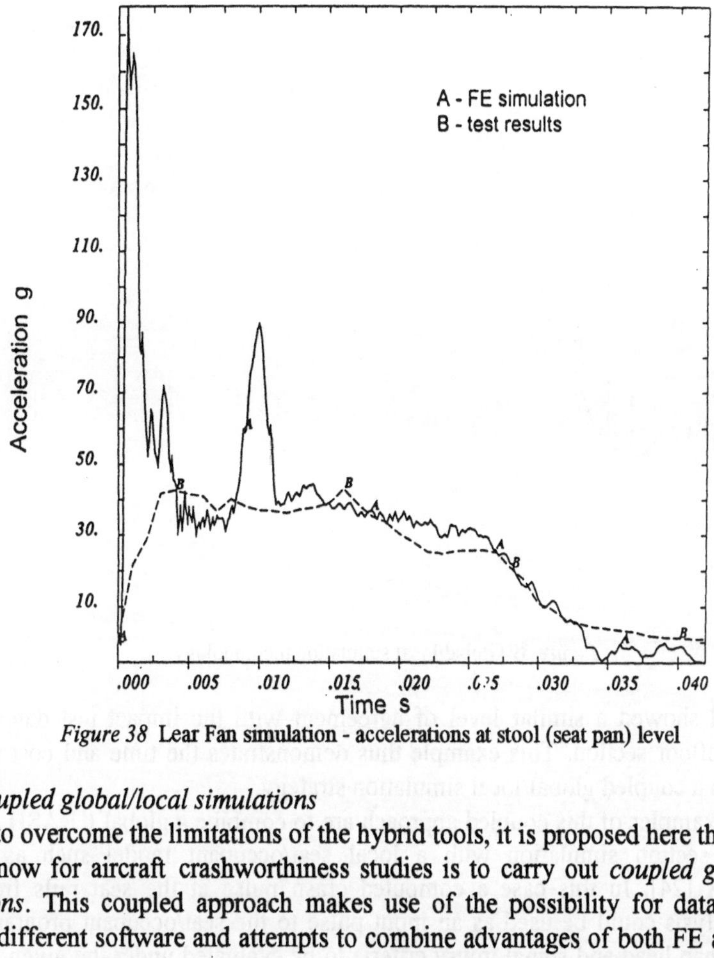

Figure 38 Lear Fan simulation - accelerations at stool (seat pan) level

4.2.4 *Coupled global/local simulations*

In order to overcome the limitations of the hybrid tools, it is proposed here that the way forward now for aircraft crashworthiness studies is to carry out *coupled global/local simulations*. This coupled approach makes use of the possibility for data exchange between different software and attempts to combine advantages of both FE and hybrid methods. In this approach a complete *global* KRASH aircraft model is developed, with *local* FE models used to calculate the nonlinear response of crush springs or plastic hinges required in the KRASH analysis. The FE models of the spring or hinge replaces an expensive substructure test programme to measure these characteristics, and provides a complete predictive tool. Such a coupled approach has been successfully carried out in [19] to simulate the crash response of the large subfloor section depicted in Figure 30. Instead of using the complete FE model with nearly 60000 elements, an idealised KRASH model was generated as shown in Figure 39 containing 32 masses, 52 beams and 10 nonlinear springs. The crush force-deflection information required for the springs was obtained by calibrating the spring response from the results of a detailed FE model of the lower subfloor region, also shown in Figure 39. This FE subfloor model contained about 20000 elements. A further FE simulation of the moment-rotation response of the plastic hinges formed at the lower end of the vertical struts was also used in the KRASH model. The predictions of the force-time and velocity-time responses from the KRASH analysis were found to be close to those of the full FE

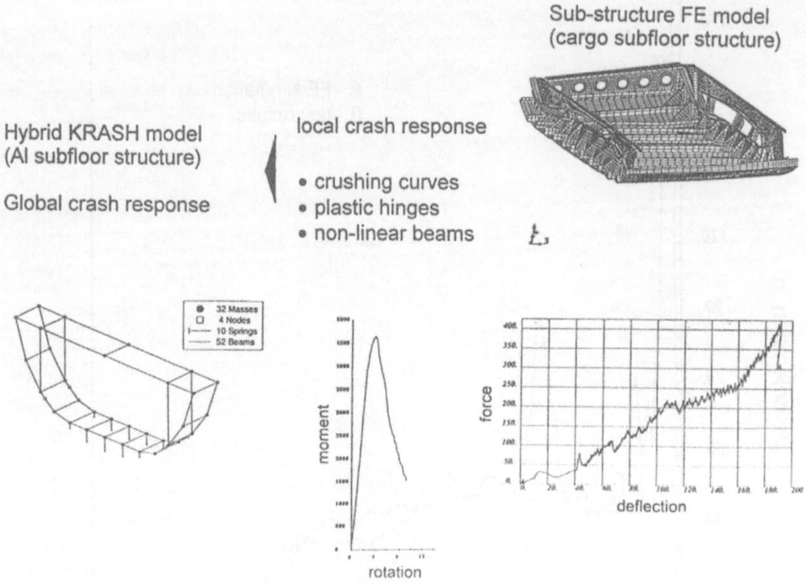

Figure 39 Global/local simulation methodology

analysis, and showed a similar level of agreement with the impact test data on the complete subfloor section. This example thus demonstrates the time and cost savings possible with a coupled global/local simulation strategy.

Further examples of this coupled approach are to combine a global KRASH aircraft or fuselage section simulation with a local seat/occupant model such as SOM-LA/SOM-TA [24]. In this case a computed crash pulse at the seat rails from the KRASH analysis could be used as an input pulse to the seat/occupant program, thus enabling human head and spinal injury criteria to be evaluated under the given aircraft crash conditions. Other variations of the method could include the incorporation of FE air-bag models, or FE occupant and dummy models, within a global hybrid simulation, all of which contribute to further efficiencies in crashworthiness studies.

5. Conclusions

Key features of composite crushing and EA are addressed to microfragmentation processes initiated by so called trigger mechanisms; smooth deceleration-time histories are generated.

Structural post crash integrity is achieved by hybridization techniques, i.e. a mixture of brittle and tougher material systems.

Typical airframe substructures are subfloor beams, bulkheads, cruciforms, and frames. Such structures are focus candidates for first design optimizations and the generation of a crash behaviour data base. Very promising composite concepts have already been investigated (corrugated webs, HTP-elements, boxes)

Thermoplastic versions of composite structures have economic manufacturing potential compared to thermoset counterparts. However, further studies will follow to enhance the data on EA performance and load bearing capabilty.

A hybrid code such as KRASH is well established in the aircraft industry and has been developed specifically for crashworthiness studies. Geometrical models are relatively simple as is demonstrated by the complete Lear Fan KRASH model The designer can carry out in short time parameter variation studies with the program. On the other hand considerable engineering experience is required to obtain good results with hybrid codes. Skill is required in the idealisation of the geometry model, in determining mass distribution and spring stiffness characteristics so that essential structural features are included. Where spring properties are highly nonlinear, as in crush elements, it is necessary to carry out crush tests on critical elements in order to characterise spring properties. Thus the hybrid method becomes semi-empirical and not a complete predictive tool.

By contrast FE codes such as PAM-CRASH, although well accepted in the automotive industry, are not yet established in the aircraft industry. The simulations have shown that very detailed geometrical models with suitable materials models and property data are required for good structural failure predictions. With large models the FE simulations must be run on fast workstations or main frame computers, and may require days of CPU time. Thus FE crash simulations are expensive in manpower and computers. At present it is not practical to carry out such a detailed simulation for a complete aircraft, such as the KRASH Lear Fan analysis reported above.

The main advantage of the FE simulation is that it is a complete predictive tool - being based only on a geometry model with appropriate materials constitutive laws and property data. Any testing required is on materials specimens, not at the substructure level. Where further work is still required is in the modelling of non-standard materials, particularly failure properties, and the implementation of these models into the FE codes. For composites under dynamic loads there are many possible failure modes such as crushing, fibre fracture, delamination, matrix shear, etc., and new materials models with associated test methods for measuring failure and damage parameters are currently active research areas.

For future vehicles - cars and aircraft - crash requirements will be of growing importance concerning operation licenses. Ultralight structural concepts will be required to account for low pollution and effective engine concepts with respect to environmental aspects. Therefore, composite crushing behaviour will be a matter of enlarged interest for the manufacturers. A lot of different composite element configurations were crash tested in the past and further tests must be carried out. Development of computer codes and approximation formulae must be continued. However, unrealistic emphasis on exaggerated exactness postulation should be avoided due to the typical scatter of response of crashing structures.

Acknowledgements

The author would like to thank H.D. Carden, Dr. K.E. Jackson and L.E. Jones of NASA Langley Research Center for providing the test pulse data for the complete Lear Fan aircraft and for the composite fuselage section. He also wants to thank his collegues Dr. A. Johnson, D. Kohlgrüber, M. Lützenburger for their essential contributions to numerical simulation work and H. Weissinger for performing the complete test work in the DLR drop test facility.

References

1. S.P. Desjardins et al, Aircraft Crash Survival Design Guide. USAAVSCOM TR 89-D-22A-E, Vol. I-IV, Dec. 1989.
2. Military Standard: Light Fixed Wing and Rotary Wing Aircraft Crash Resistance. MIL-STD-1290 A (AV), Sept. 1988.
3. D. Hull, A unified approach to progressive crushing of fibre-reinforced composite tubes. Composite Science and Technology, 40 (1991) 29-34.
4. G.L. Farley, Energy absorption capbility of composite tubes and beams. Ph. D. Thesis, Virginia Polytechnic Institute and State University, 1989.
5. P.H. Thornton & P.J. Edwards, Energy absorption in composite tubes. J. Composite Materials, 13 (1979), 247-262.
6. M. Maier, Experimentelle Untersuchung und numerische Simulation des Crashverhaltens von Faserverbundwerkstoffen. Dissertation, University of Kaiserslautern, 1990.
7. C.M. Kindervater, Energy absorbing qualities of fibre-reinforced composite tubes. Proc. AHS Nat.Specialist Meeting on Composite Structures, Philadelphia, March 1983.
8. J.D. Cronkhite & T.J. Berry, Investigation of the crash impact characteristics of helicopter composite structure. USA AVRADCOM-TR-82D-14, Feb. 1983.
9. H.G.S.J. Thuis & J.F.M. Wiggenraad, A tensor skin concept forcrashworthiness of helicopters in case of water impact, AHS 50th Annual Forum, Washington DC, May 1994.
10. A.F. Johnson & C.M. Kindervater & H.G.S.J. Thuis & J.F.M. Wiggenraad, Crash resistant composite subfloor structures for helicopters. AGARD FVP Symposium: Advances in Rotorcraft Technology, Ottawa, may 1996.
11. K.E. Jackson, S. Kellas, C.M. Kindervater and M. Lützenburger, Experimental and simulated crash responses of composite airframe structures. American Helicopter Society 50th Annual Forum, Washington, 1994.
12. M. Gamon, G. Wittlin and B. LaBarge, KRASH 85 User's Guide - Input/Output Format, DOT/FAA/CT-85/10, July 1985 (revised May 1986).
13. DRI/KRASH Version 9601 - Users Manual, Dynamic Response Inc., Sherman Oaks, California, USA, January 1996.
14. G. Wittlin, M.B. Rapaport, Naval Rotorcraft Water Impact Crash Simulation Using Program KRASH. American Helicopter Society 49th Annual Forum, St. Louis Missouri, May 19-21, 1993.
15. C.M. Kindervater, A. Gietl & R. Müller, Crash investigations with sub-components of a helicopter lower airframe section. AGARD-CP-443, 1988:
16. L.E. Jones and H.D. Carden, Overview of structural behaviour and occupant responses from a crash test of a composite airplane. Paper 951168, SAE Aviation Meeting, Wichita, Kansas, May 1995.
17. LS-DYNA3D, Livermore Software Technology Corp., Livermore, CA 94550.
18. PAM-CRASH, Engineering Systems International GmbH, D-65760 Eschborn.
19. IMT-2002, Crashworthiness for Commercial Aircraft, EU RTD Project, 1993-1995.

20. H.D. Carden and S. Kellas, Energy absorbing beam design for composite aircraft subfloors.Proc. 34th SDM Conf., La Jolla, CA, April 1993.
21. A.F. Johnson, Modelling the crash response of a composite aircraft section. ICCM-10, Whistler, Canada, 1995.
22. E. Haug and A. De Rouvray, Crash response of composite structures. Ch. 7 "Structural Crashworthiness and Failure", (ed) N. Jones and T. Wierzbicki, Elsevier, London, 1993.
23. C.M. Kindervater and H. Georgi, Composite strength and energy absorption. Ch. 6 "Structural Crashworthiness and Failure", (ed) N. Jones and T. Wierzbicki, Elsevier, London, 1993.
24. D.H. Laananen, Computer simulation of an aircraft seat and occupant in a crash environment - SOM-LA/SOM-TA User Manual. Report DOT/FAA/CT-90/4, May 1991.

CURRENT ISSUES REGARDING AIRCRAFT CRASH INJURY PROTECTION

H. M. LANKARANI
Associate Professor and Bombardier/Learjet Fellow
Mechanical Engineering Department
and National Institute for Aviation Research
Wichita State University
Wichita, KS 67260-0035
USA

Abstract

This paper addresses several issues related to the crash protection of aircraft occupants. The origin and nature of the current crashworthiness standards are first described. A number of test facilities, from sophisticated full-scale airframe drop tower facilities to the facilities for sled-testing of the seats, restraint systems, and anthropomorphic test devices or dummies (ATD's), and also the ATD component testing are then presented. A description of the analysis codes, specifically the occupant models, and the latest modifications on these models are provided. Some of the current aircraft crash safety issues are then discussed. One global issue in the performance and certification of aircraft is the problem of protection of occupants from the head impact onto the interior structures. Latest developments in that area are presented. Another current area of research is the development of energy-absorbing seats capable of complying with the newly proposed criteria for the commuter type aircraft. The small size of these aircraft results in stiffer structures and consequently higher impact loads on the occupants. Side-facing seats in business jets and in the lounge areas of some of the transport aircraft have also presented new challenge in certification, and hence the latest advancements in that area of research are also discussed.

1. Introduction

The design of aircraft seats and interior has traditionally received little attention to the crashworthiness performance but rather to the aesthetics, comfort, appearance, and weight. Aircraft crashes were thought of being unsurvivable. This was in part due to the lack of understanding and definition of the response of an aircraft subjected to a crash event and lack of definitive or consistent crash dynamic design standards [1]. This trend continued until the field observations of many crash scenes demonstrated that with improvements in the structural integrity and energy absorption of some, specially small, airframes, and also design of safer passenger seats, restraint systems and surroundings,

J. A. C. Ambrósio et al. (eds.),
Crashworthiness of Transportation Systems: Structural Impact and Occupant Protection, 579–612.

many of the aircraft crashes would have been survivable. Detailed studies of the accident investigation reports established the fact that aircraft passengers survivability could be greatly improved if crashworthiness was initially included in the design stage of the aircraft [2].

Chandler's work in his book, *Aircraft Crashworthiness* [3], and his paper [2], provides an extensive historical background on the development of crash injury protection in civil aviation, a brief summary of which is presented here. The earliest form of aircraft crash injury protection device was a restraint system used around 1910 by the U.S. Army airplanes [4]. The systems was adapted from a leather trunk strap by the local harness maker for the calvary, and it was to hold the aviator in place. However, it usually broke upon impact or caused internal injures. Because of this, most pilots had strong objections to strapping themselves into their seats, also fearing of being trapped in burning aircraft, even though fire played a small part in actual crash deaths. Even with the development of new designs of quick-release mechanisms for the seat belts, the pilots most often released the belts themselves just before landing so there would be no risk of entrapment. World War I brought a widespread and organized use of airplanes, and provided a systematic policy of accident prevention. In 1918, the *General Rules and Regulations Governing Flying on Individual Fields* became effective that included the following two crash protection requirements [5]. All machines must be equipped with safety belts for pilots and passengers (item 17). Always use the belts. In case of accident, do not release belt until after accident. It will probably save injury, especially if the machine turn over (item 18).

Between the wars, hundreds of airplanes left from the war were sold in factions of their cost, and with that there was an increase in fatal aircraft accidents. The U.S. Congress responded by imposing the Kelly Air Mail Act (1925) to foster air commerce and to promote a system of safe airways and airports. The *Air Commerce Regulations* (1926 and modified in 1928) specifically required a safety seat belt, capable of withstanding a load of 1000 pounds [6].

In 1942, DeHaven started a Crash Injury Research (CIR) program at Cornell University Medical College. From his investigations of crash accident data, he concluded that it was doubtful that shoulder harness, what is known as "passive protection" now, would greatly improve crash safety; however, a stronger seat belt and improved cabin installation provided a higher degree of crash protection [7]. In 1950, DeHaven and Criswold developed a new safety harness consisting of a combined diagonal and safety belt, and later added an inertia reel for taking up the slack. Based on the efforts of DeHaven, an Aviation Crash Injury Research (AvCIR) program was then started at Cornell College in 1952. The name charged to Aviation Safety Engineering and Research (AvSER) in 1963, emphasizing activities toward engineering aspects of crashworthiness design. The program developed a *Crash Survival Design Guide* (CSDG), first published in 1967, which provided design data and guidelines that could be used to incorporate crashworthiness features in airplanes. AvSER later conducted full-scale crash tests of large transport-type aircraft, based on contacts awarded by Federal Aviation Administration (FAA) [8,9]. The National Advisory Committee on Aeronautics (NACA) initiated a comprehensive study of airplane crash problems, and published the data from several crash tests of large transport and cargo airplanes [10]. These test data indicated that the crash environment could be represented by a fairly long, low-amplitude acceleration with a superimposed high acceleration secondary pulse. Martin Eiband's curve for vertical acceleration tolerance is perhaps the best known of the reports generated by the NACA program [11]. The major aircraft crashworthiness program by National Aeronautics and Space Administration (NASA)/NACA began in 1973. The objectives of the program were to determine the effects of impact speed, flight path angle, roll angle, and ground condition on the dynamic response of airplane structures, seats, restraints systems, and occupants during crashes in which the airplane structure retained sufficient volume and integrity to permit occupant survival. Several airframes were swung from a large gantry onto concrete on soil surface [12]. A sled test program, in cooperation with the FAA Civil Aeromedical Institute (CAMI), was also initiated, which still continues seat testing and development [13].

Much of the currents regulatory standards are based on the efforts of the FAA and NASA in the 1970's, which initiated a wide rage of research and development programs to address the crashworthiness characteristics of the transport category aircraft, small general aviation aircraft, and rotorcraft. These programs represented a concentrated effort to analyze the aircraft behavior and the occupant characteristics through interrelated studies of accident data, dynamic analyses of crash events, full-scale aircraft impact tests, and aircraft seat tests. The results of these studies formed the basis for the development of crashworthiness design standards for civil aircraft [1]. These requirements are defined in the *Federal Aviation Regulations* (FARs) Parts 23, 25, and 27 for general aviation aircraft, transport aircraft, and rotorcraft respectively [14-16]. In general, the FARs contain two distinct dynamic test conditions. Test-1 conditions, illustrated in Figure 1 and described in Table 1, require a seat inclination of 60 degrees in pitch and a mean velocity change of no less than 30 ft/sec (20 mph = 9 m/s), and is intended to evaluate the means provided to reduce the spinal loading and related injuries produced by the combined vertical/horizontal load environment typically generated by an aircraft crash event. Test-2 conditions, illustrate in Figure 2 and described in Table 2, require the inclination of the seat on the track by 10 degrees in yaw direction and a mean velocity change of no less than 42 ft/sec (29 mph = 13 m/s) in the longitudinal direction. To account for a reasonable floor warpage level that may occur during a crash, one of the seat tracks is misaligned ny 10 degrees in pitch and the other one by 10 degrees in roll. The Test-2 conditions are intended to provide an assessment of the seat structural performance and the occupant restraint system. The deceleration pulse in both tests are triangular shaped.

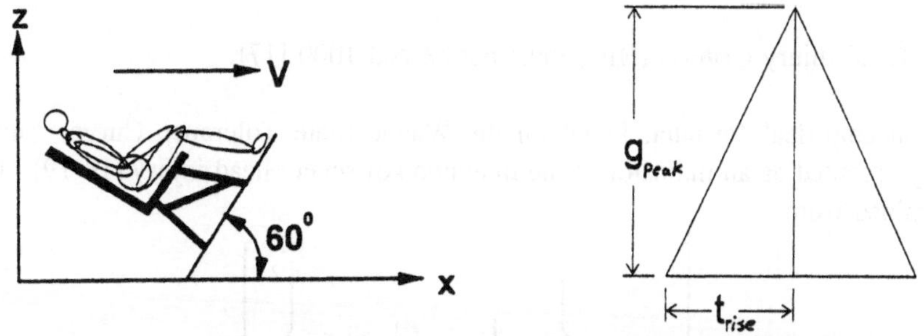

Figure 1. Federal Aviation Regulations Test-1 configuration.

TABLE 1. Test-1 dynamic test requirements (combined vertical/longitudinal test).

	Part 23	Part 25	Part 27,29
Velocity Change V (ft/sec)	31[1]/31[2]/31[3]	35	30
Peak Acceleration (G's)	19[1]/15[2]/32[3]	14	30
Rise time to peak (sec)	0.05[1]/0.06[2]/0.03[3]	0.08	0.031

[1] Pilot, [2] Passenger, [3] Commuter (proposed NPRM 93-71)

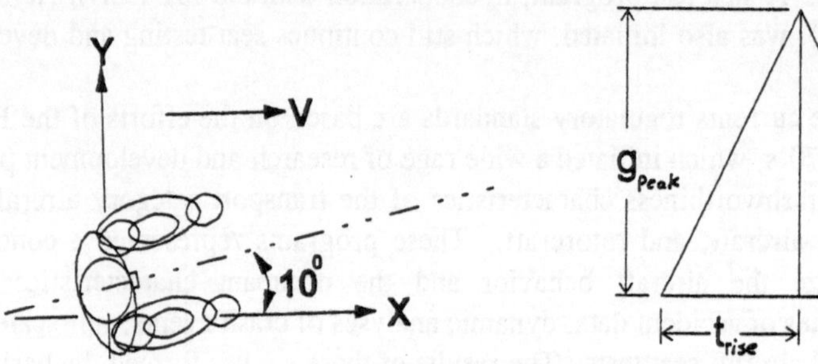

Figure 2. Federal Aviation Regulations Test-2 configuration.

TABLE 2. Test 2 dynamic test requirements (longitudinal with yaw components test).

	Part 23	Part 25	Part 27,29
Velocity Change V (ft/sec)	42	44	42
Peak Acceleration (G's)	26[1]/21[2]	16	18.4
Rise time to peak (sec)	0.05[1]/0.06[2]	0.09	0.071

[1] Pilot, [2] Passenger

The pass/fail criteria in the FARs include the following.

a. Maximum compressive load measured between the pelvis and the lumbar spine of the Hybrid II ATD (per FAR's) must not exceed 1500 pounds (6675 Newtons).

b. Loads in the individual straps must not exceed 1750 pounds (7785 Newtons) for pilot and 2000 pounds (8,900 Newtons) for passengers.

c. The Head Injury Criteria (HIC) must not exceed 1000 [17].

HIC is an empirical formula, based on the Wayne State Tolerance Curve, and is generally accepted as an indicator of the likelihood of severe head injury [18-19]. HIC is determined from

$$\mathrm{HIC} = \left[(t_2 - t_1) \left\{ \frac{1}{(t_2 - t_1)} \int_{t_1}^{t_2} a(t)\,dt \right\}^{2.5} \right]_{max} , \tag{1}$$

where "a(t)" is the resultant acceleration of the head center of gravity in G's, "t_1" is an arbitrary time in the pulse, and "t_2" for a given "t_1", is a time in the pulse which maximizes the HIC. *If this index is less than 1000, the situation is considered not to be life threatening.* The HIC is a method for defining an acceptable limit, i.e., the maximum value of the HIC should not exceed 1000. The HIC has been a part of the *Federal Motor Vehicle Safety standards* (FMVSS) [20] and the FARs [17].

The design of aircraft for improvement of the crashworthiness requires the knowledge and integration of several items. These items include the specifications and standards, human tolerance, injury criteria, energy absorption concepts in airframe design, seat legs, seat pan, seat cushion, restraint systems, surrounding structures, fire safety, economics considerations, and ergonomics considerations. This paper starts by describing the different crashworthiness testing methodologies, including the full-scale aircraft tests such as the NASA Langley's IDRF (Impact Dynamics Research Facility) gantry swing test set up [21] and the FAA drop tower facility [22]; sled test set ups such as the ones in CAMI [23], Navy's NADC sled [24], National Institute for Aviation Research (NIAR) [25], and Simula's drop tower facility [26]; and component tests such as the simplified HIC test setups. A review of the analysis codes for modelling both the occupants and structural crash responses are then presented. The codes include SOM-LA/TA (Seat Occupant Model/Light Aircraft-Transport Aircraft) [27,28], ATB (Articulated Total Body) [29], and MADYMO (Mathematical and Dynamic Model of Occupant) [30]. Latest modifications in these analyses tools for better prediction of the occupant protection and possible injuries to an occupant in an aircraft crash are briefly described. The modifications include: development of a nonlinear finite element model of the lumbar spine and inclusion in the SOM-LA/TA code for better prediction of the lumbar loads; development of a discrete-parameter model of the cervical spine for better prediction of the cervical spine injuries in compression and tension, forward and lateral bending, and axial torsion modes; and finally development of hydrodynamic and finite element models of the skull-brain for better prediction of brain injuries. Some of the current issues regarding aircraft crash safety will then be examined including: the simplified HIC component test setups for certification of head impact protection for seats facing bulkhead; compliance with the crash safety criteria for commuter type aircraft having smaller size and not enough under-floor space for energy-absorption through plastic deformation; and certification of side-facing seats.

2. Experimental Procedures

In the 1970's, the FAA and NASA embarked on a full-scale small airplane impact test program. Those impact tests were conducted at the NASA Langley Impact Dynamics Research Facilities (IDRF) [21]. This is basically a pendulum-type gantry test facility, as shown in Figure 3, that can subject a test aircraft to both vertical and longitudinal velocity changes or decelerations. Some of the objectives of that test program were to determine and quantify the dynamic response of airplane structures, seats, and occupants during a crash, and to determine the effect of flight parameters at impact on loads and on structural damage. The results of those tests provided the framework for a number of additional impact test programs for small general aviation airplanes, rotorcraft, and transport category airplanes. They also provided the basis for the seat dynamic performance standards for small general aviation airplanes.

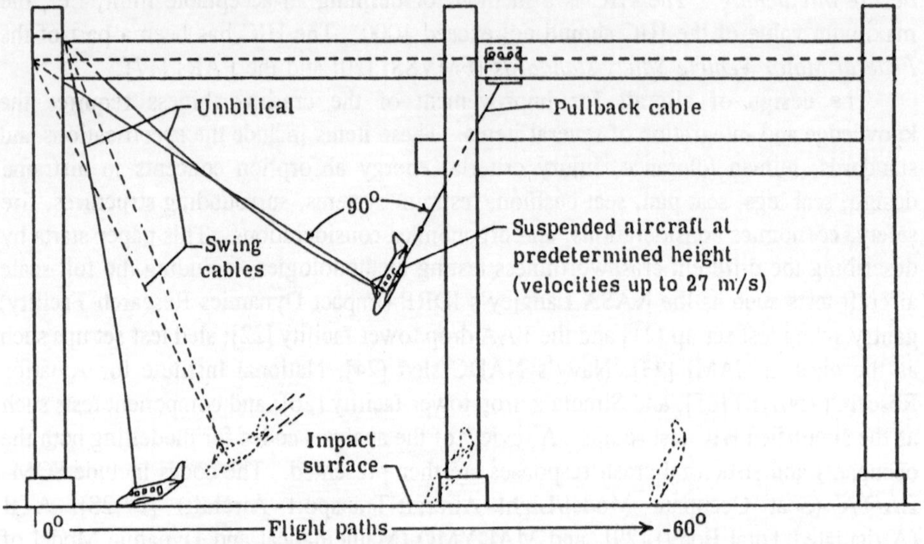

Figure 3. NASA's Impact Dynamic Research Facilities for full-scale aircraft crash testing.

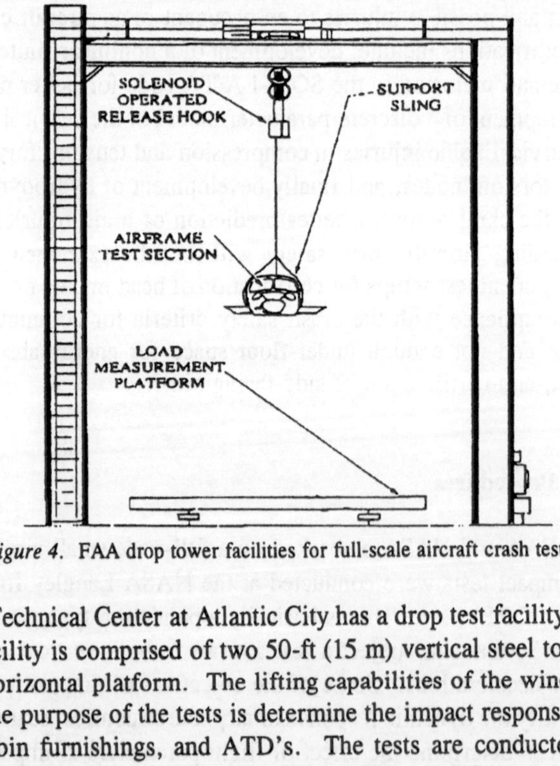

Figure 4. FAA drop tower facilities for full-scale aircraft crash testing.

The FAA Technical Center at Atlantic City has a drop test facility, shown in Figure 4 [22]. The facility is comprised of two 50-ft (15 m) vertical steel towers connected at the tops by a horizontal platform. The lifting capabilities of the winch is rated at 13.6 klb (60 kN). The purpose of the tests is determine the impact responses of the fuselage, cabin floor, cabin furnishings, and ATD's. The tests are conducted to simulate the vertical velocity component of a potentially survivable crash impact condition. Most of the recent tests on the drop tower facility has been concentrated on the proposed dynamic tests requirements for Part 23 commuter category aircraft. The airframe is dropped from a ceratin heights (around 11 ft or 3.4 m) onto a rigid surface at a level airplane attitude with a nominal impact velocity of about 27 ft/sec (8 m/s). The airframe acceleration, resultant structural deformations, effectiveness of energy absorbing seat designs in providing occupant impact injury protection are the basic objective of these tests.

There are a number sled facilities dedicated to certification and research on aircraft seats, restraint systems, and ATD's. CAMI has a horizontal impact sled, shown schematically in Figure 5 [23]. The sled is accelerated to the desired velocity on parallel circular rails using a cable and pulley system attached to a falling weight. As the sled coasts at a constant velocity into contact with wires in the brake, the wires are pulled through rollers fixed on each side of the track. The action of pulling brake wires through the rollers creates a deceleration force. The deceleration pulse shape is controlled by the number, spacing, and lengths of wires placed in the brake. Accelerometer data in the ATD cavities as well as high speed photometric procedures are used in the sled tests.

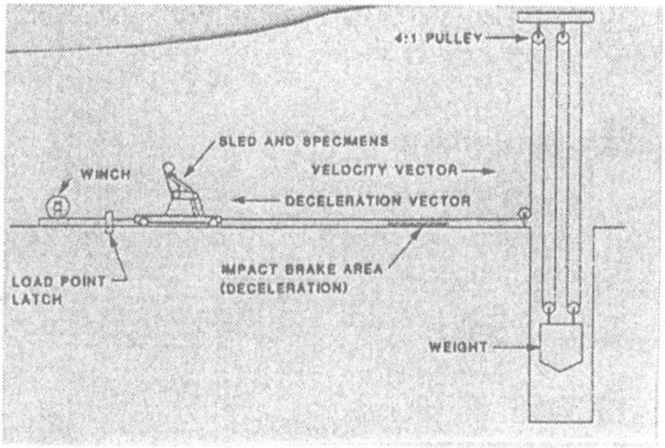

Figure 5. CAMI horizontal impact sled facilities.

Figure 6. The acceleration sled test facility at NADC.

Navy's impact sled test facility at NADC is an accelerator-type sled [24]. The system, shown in Figure 6, is pneumatically accelerated. The system contains the sled, rails, actuator, high pressure air supply, nitrogen supply, event sequencer, and control console. The actuator provides the acceleration pulse for the test. Friction and brakes decelerate the sled after the pulse. The pulse is controlled by varying a metering pin, set length, load length, set pressure, and load pressure. The pulse generated by the system is usually a smooth acceleration curve resembling a half-sine.

NIAR also has deceleration impact sled test facilities for dynamic testing and evaluation of the response characteristics of ATD's, seats, restraint systems, and occupant surrounding [25]. The horizontal impact test sled, shown in Figure 7, moves on a 78-ft (24 m) track and has a maximum speed of 80 ft/sec (55 mph = 24 m/s). It runs on plastic shoes and is pneumatically propelled. The sled can attain its maximum acceleration within fifty feet of its travel. Steel straps are placed at the end of the track to bring the motion of the sled to an end. Several ATD's including 50th percentile Hybrid II, 5th percentile female, 95th percentile male, and SID ATD's are available. In addition to the data from the accelerometers, the displacement, velocity and acceleration of the ATD are recorded and evaluated using a high speed video system and optical target tracking techniques.

Figure 7. Impact test sled at NIAR.

Simula Corporation has a drop tower facility for the purpose of crash response characteristics of seats, restraint systems, and ATD's [26]. The system, shown in Figure 8, is gravity driven, and is decelerated to a particular desired pulse shape via a pyramid of layers of composite material layered on a concrete surface. The breaking of the fibers provides the energy absorption needed to decelerate the dropped seat/ATD system.

Figure 8. Simula drop tower facility.

3. Occupant Modelling

3.1. OCCUPANT MODELS

To improve aircraft crash safety, conditions critical to occupants' survival during a crash must be known. A large number of possible aircraft crash environments exists, and the impact sled testing may not be possible nor feasible for some configurations. Cost and time are also other burdens of testing procedures. Furthermore, prior to testing, multiple simulations (analyses) must be conducted to better define the experimental testing program. Rigorous analytical techniques are necessary for design of crashworthy aircraft airframes, seats, occupant surroundings, and restraint systems. Validated analytical models also reduce the necessity of fabrication of design modifications. Analysis codes that model the dynamic response of a human or ATD during a crash are known as gross motion simulators. These models are comprised basically of kinematically connected body segments with joint stiffnesses and contact forces between penetrating segments or segments contact with the surrounding. Some of the existing body gross motion simulators include: ROS, CAL2D, HSRI, MVMA2D, SIMULA, PROMETHEUS, CAL3D, UCIN, SOM-LA/TA, ATB, MADYMO [31]. Currently, the codes mostly used for reconstruction of aircraft crash scenarios are SOM-LA/TA, ATB, and MADYMO, a brief description of each is provided next.

Program SOM-LA/SOM-TA (Seat Occupant Model-Light Aircraft/ Transport Aircraft) is a widely used code, developed under FAA sponsorship [27,28]. It has a dynamic model of occupant and a finite-element model of seat structure. The occupant models are of multi-body type. The two-dimensional model has 7-segment and 11 DOF, and it also has a force elements in torso and neck, as shown in Figure 9. The three-dimensional model has 11-segment and 29 DOF. The joint resistances are described in terms of displacement limiting moment, viscous damping, and muscular resistance in rotor cup, elbow, pelvic, d knee joints. Contact surfaces are described in terms of spheres, ellipsoids, etc., and resulting contact forces are generated at the seat back, head-rest, seat cushion, and floor. Data for 50th percentile human and Hybrid II models are available. The type of restraint systems available are: lap belt, lap belt tie-down and negative-G strap, and single or double shoulder harness. The seat is modeled as either rigid, energy-absorbing, or nonlinear finite-element model. An updated Lagrangian method with large plastic deformations, buckling of bending members and material nonlinearality is used in the finite element modelling of the seat. For numerical solution, joint coordinate formulation as well as a predictor/corrector (Adams-Bashforth/Adams-Moulton) type with adjustable step-size and filtering, are used.

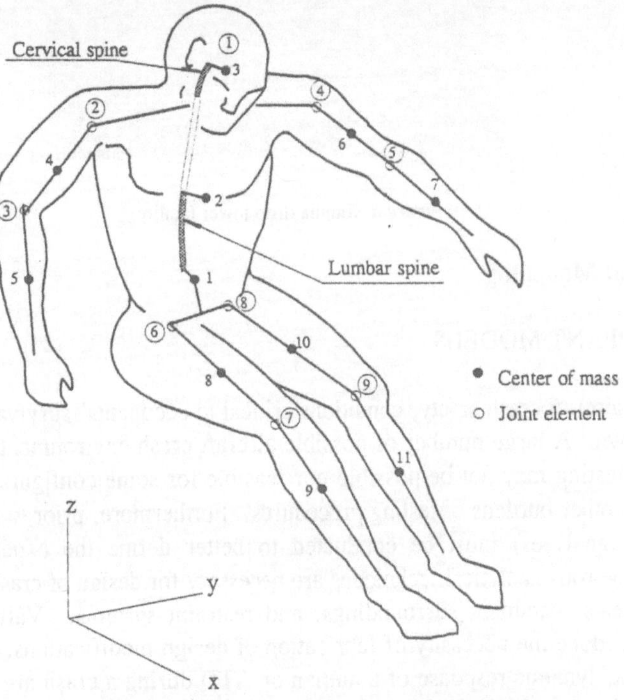

Figure 9. Occupant model of SOM-LA/TA code.

The ATB (Articulate Total Body) has been developed in Armstrong Aerospace Medical Research Laboratory (AAMRL) at Wright Patterson Airforce Base (WPAFB) [29]. It started as CVS (Crash Victim Simulator) described earlier, and was later modified by Calspan [32]. It was used in the prediction of aircraft seat ejection with windblast exposure. Most of the current use at WPAFB is for automobile rollover. ATB is a stand-alone occupant model. The occupant models are of multi-body type, and include an occupant database for children (2-19 years) , 50th percentile adult human male and female, 50th percentile sitting and standing, Hybrid III ATD, 50th percentile Hybrid II ATD, and user-defined dimensions. The user can specify a maximum of 30 segments. The hybrid II ATD model has 15 segments and 14 joints, shown in Figure 10. Injury measure evaluated are HIC, Gadd Severity Index (GSI), and lumbar load.

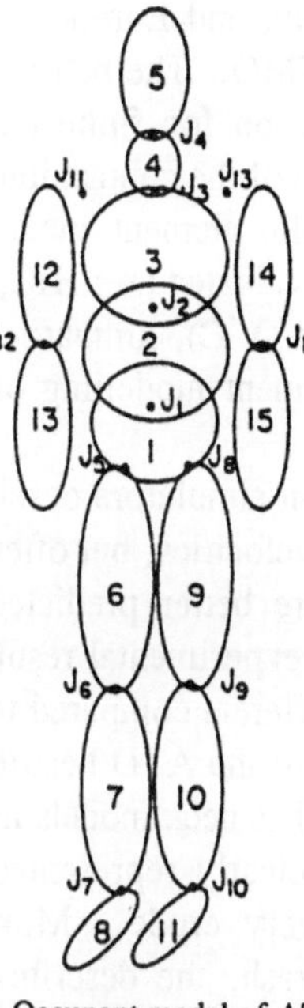

Figure 10. Occupant model of ATB.

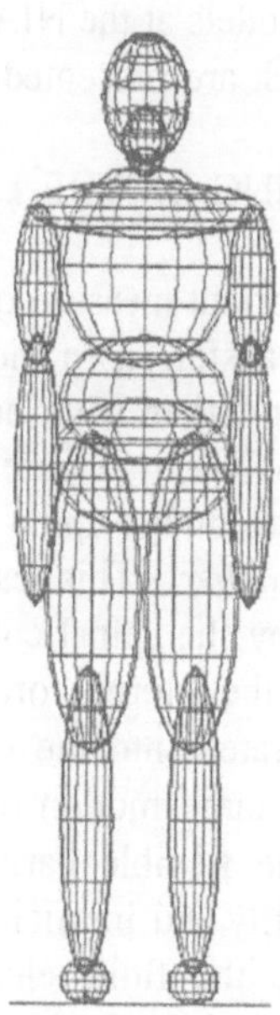

Figure 11. 50th percentile Hybrid III ATD model in MADYMO.

Program MADYMO is currently widely used in the automobile industry, and is developed by TNO Road-vehicle Research, in Netherlands [30]. It can simulate motion of multibody systems and assembly of finite elements. Both 2D and 3D models are available. Occupant models include: standard 50th percentile Hybrid II and Hybrid III ATD's, human data of the same size, 9 month old child dummy, 3 year old child dummy, 6 year old child dummy, and European side impact ATD. Figure 11 shows the Hybrid III model in MADYMO. The belt model accounts for belt slack, rupture, and sliding over torso. Provision for finite element modelling of the belt is also available. A detailed description of the contact interactions between bodies and restraint system is also available. Finite element modelling of the body segments can be conducted. The injury criteria evaluated are: HIC, GSI, 3 ms criteria, Thoracic Trauma Index (TTI(d)), Viscous Criteria (V^*C), lumbar load, femur load, and axial loads in all segments. Nonlinear finite element modelling of the surrounding structures are also available.

In general, the gross motion simulators described have a reasonable prediction of positions (e.g., head path) and velocities, but often do not exhibit good correlations for accelerations. Accelerations are better predicted by MADYMO than the ATB and SOM-LA/TA, compared to the experimental results. HIC prediction for bulkhead seat tests is most of the time very different compared to the experimental results, in part due to the lack of proper definition of the ATD head/bulkhead contact compliance and also the volatile nature of the HIC. The neck models in these codes are simplified and forces produced by the neck are not clearly represented. Belt attachment model for SOM-LA/TA and MADYMO are very crude. More importantly, when applied to the reconstruction of an aircraft crash, the described codes do not exhibit a reasonable prediction of the lumbar load under the FAR Test-1 conditions. Some modifications have been made on the occupant models at the NIAR for better prediction of the lumbar and cervical spine responses, which are presented next.

3.2. KINTO-STATIC MODELLING OF THE LUMBAR SPINE

As mentioned earlier, an important measure of injury to an occupant in an aircraft accident is the amount of load transferred to the occupant lumbar spine [17]. The occupant models described earlier either have no flexibility provided in the lumbar region an represent the chest-to-pelvis by a pure kinematic joint, or utilize a typical load-deflection formula of an intact lumbar spine vertebral column to predict the axial load and moments acting on the lumbar. This results in an incorrect representation of the dynamic responses produced by the spinal column during a impact tests of Test-1 configuration. In order to predict the internal forces and moments in the spine, a finite element model has been incorporated into the occupant model of SOM-LA/TA. In essence, the relation between the large motion of the occupant segments such as the limbs and the deformation of the flexible parts such as spine is established. To overcome the problems of instability and inefficiency, a kineto-static methodology has been developed, which combines the finite element methods with rigid multibody dynamics principles [33]. A description of this methodology is provided next.

In finite element analysis, the loads will generally be thought of as being applied, while the displacements are thought of as resulting. However, in analysis of multibody responses, it often happens that we have mixed boundary conditions. An example of a multibody system consisting of three bodies, two rigid bodies connected by a flexible body, is shown in Figure 12. Here, body "i" represents the pelvis, body "j" the thorax, and body "k" the lumbar spine.

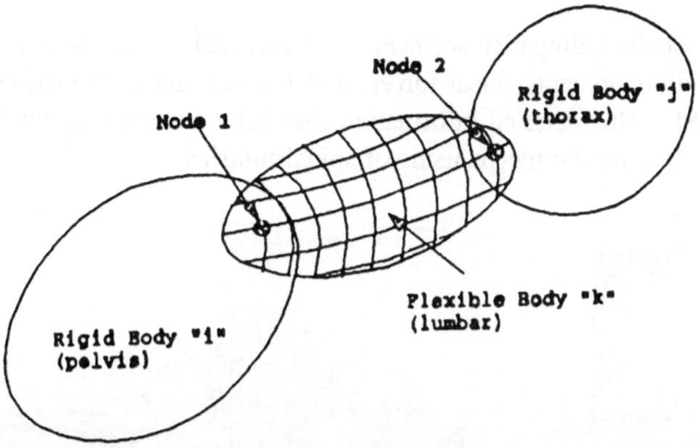

Figure 12. Flexible body connecting two rigid bodies.

While performing a dynamic analysis, the configuration of the rigid bodies and hence the displacements of nodes 1 and 2 of the flexible body are known. The problem is to find the deformed shape of the entire structure and also the forces and moments acting on the rigid bodies by the deformable body. The kinto-static approach is formulated by rearranging external forces, displacements, and the structural stiffness matrix at every time step. To solve the structural equilibrium equations for the unknown forces and displacements, let us partition the known (k) and unknown (u) variables as

$$\begin{bmatrix} \mathbf{F}_u \\ \mathbf{F}_k \end{bmatrix} = \begin{bmatrix} \mathbf{K}_{kk} & \mathbf{K}_{ku} \\ \mathbf{K}_{uk} & \mathbf{K}_{uu} \end{bmatrix} \begin{bmatrix} \delta\mathbf{q}_k \\ \delta\mathbf{q}_u \end{bmatrix} \tag{2}$$

where matrices $\mathbf{K}_{kk}$, $\mathbf{K}_{ku}$, $\mathbf{K}_{uk}$ and $\mathbf{K}_{uu}$ are sub-matrices of rearranged structural stiffness matrix $\mathbf{K}$, $\mathbf{F}_k$ is a vector of known forces acting on the spine, $\mathbf{F}_u$ a vector of unknown forces acting at the ends of lumbar spine, $\delta\mathbf{q}_u$ a vector of unknown incremental displacements of lumbar spine from the previous time step, and $\delta\mathbf{q}_k$ known incremental displacements of nodes 1 and 2 from the previous time step calculated from the rigid body dynamics. The unknown displacements and forces can be determined as

$$\delta\mathbf{q}_u = \mathbf{K}_{uu}^{-1}(\mathbf{F}_k - \mathbf{K}_{uk}\,\delta\mathbf{q}_k) \tag{3}$$

$$\mathbf{F}_u = \mathbf{K}_{kk}\delta\mathbf{q}_k + \mathbf{K}_{ku}\delta\mathbf{q}_u\ . \tag{4}$$

The forces $\mathbf{F}_u$ with some additional damping are included in the rigid body dynamics equations of motion. Using this methodology, a nonlinear finite element model of the lumbar spine was developed for both a 50th percentile human and a Hybrid II ATD. The lumbar spine model was then incorporated into the gross motion simulator SOM-LA/TA. Impact sled tests of an ATD under Test-1 condition (60 degree pitch) for a general aviation (Part 23.562) were conducted on the NIAR impact sled, as shown in Figure 13. Analytical results were then correlated with the experimental results, a sample of which are shown in Figure 14. As observed, the occupant model with a nonlinear finite element model of the lumbar spine has a better prediction of the lumbar loads in a crash scenario. With this extended occupant model, the gross motion of the occupant segments including displacement, velocities and accelerations as well as spinal axial loads, bending moments, shear forces, nodal forces, and deformation time histories can be evaluated. This detailed information also helps in assessing the level of spinal injury and determining the mechanisms of spinal injuries.

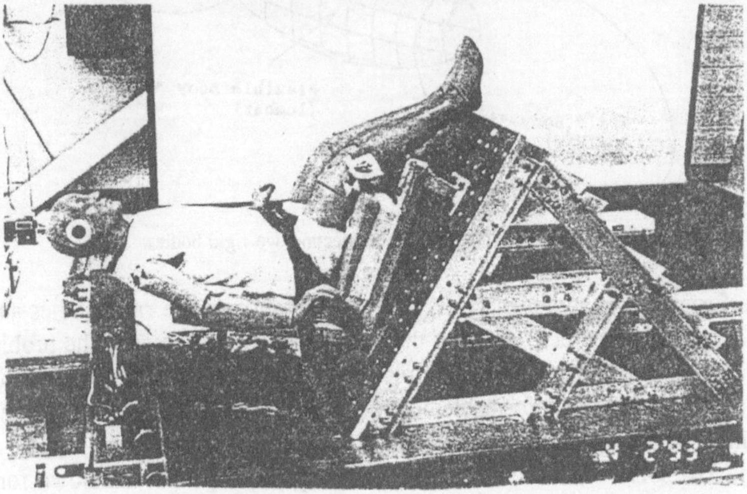

Figure 13. A general aviation iron crew seat under FAR Test-1 configuration at NIAR.

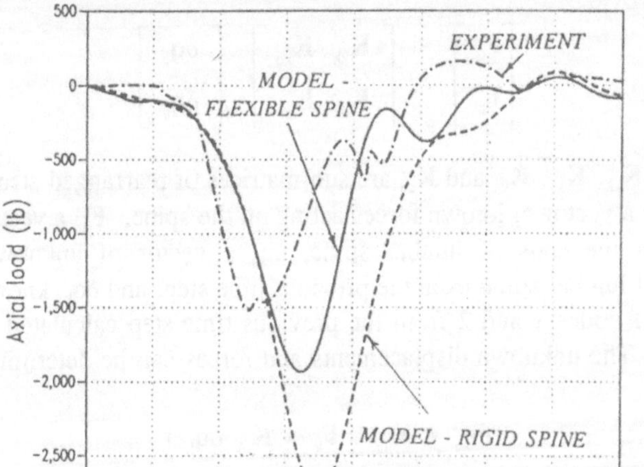

Figure 14. Comparison of lumbar loads on ATD from analyses and experiment.

3.3. MULTIBODY MODELLING OF THE CERVICAL SPINE

The human and ATD head-neck models have been constructed using an efficient discretized approach in order to better predict the neck motion and possible injuries in a crash. The human cervical spine model is shown in Figure 15. Both two- and three-dimensional models have been generated. The two-dimensional model has 9 degrees of freedom, and allows simulation of the forward extension and flexion. The three dimensional model has 36 degrees of freedom, and can be used for simulating the forward flexion and extension, lateral bending, axial rotation, and axial tension/compression. The model includes the skull, brain, and cervical spine elements C_1 through C_7, all connected by a number of joints in conjunction with nonlinear translational and rotational spring-dampers at the joints. The model shown in Figure 15, is based on the detailed cervical model of Merrill et al. [34] and Arabyan [35].

For validation, the model was used to reconstruct the longitudinal sled tests of volunteers conducted at the AAMRL of WPAFB. Tests were conducted using human volunteers at 6G and 15G, and the subjects were tightly belted using a very stiff symmetrical shoulder harness so that the head neck motion were pronouncedly related to the sled motion. Accelerometers were mounted on each subject's first thoracic vertebrae T_1 and head. A sample of the longitudinal (x) and lateral (y) components of acceleration measured at the base of the neck T_1 for the 15G test run is shown in Figure 16. This acceleration was used as an input to the head and cervical spine model. The head resultant translational and angular accelerations from the test and also from the head-neck model are shown in Figure 17(a) and (b). As observed, the cervical spine model has a good representation of the experimental tests conducted. The test outcome in the neck corridors of injuries are also illustrated in Figure 17(c), which indicate that the conducted tests were close to the borderline of injuring the neck of the subjects. In addition to the forward flexion/extension, a number of other tests including lateral bending tests were also conducted under a sled 5G test at AAMRL. The performance of the model in the lateral direction as compared to the experimental results is shown in Figure 18. The interested reader is referred to the reference [36] for more detailed description of the model and further results.

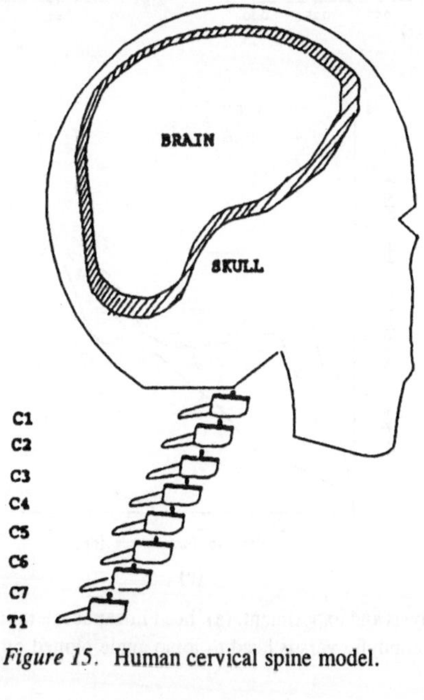

Figure 15. Human cervical spine model.

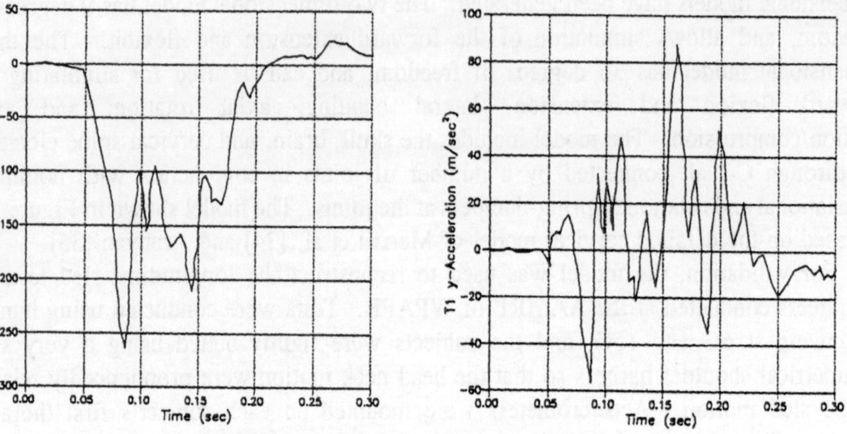

Figure 16. Input to cervical spine model (15G test).

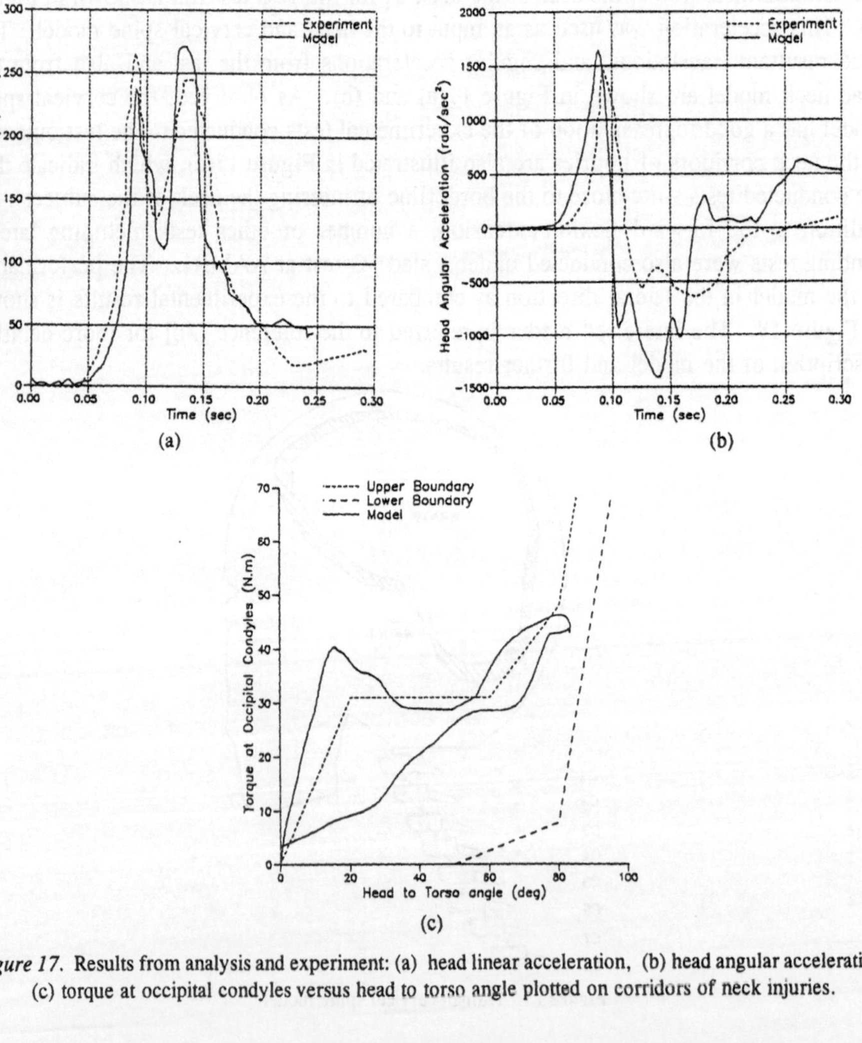

Figure 17. Results from analysis and experiment: (a) head linear acceleration, (b) head angular acceleration, (c) torque at occipital condyles versus head to torso angle plotted on corridors of neck injuries.

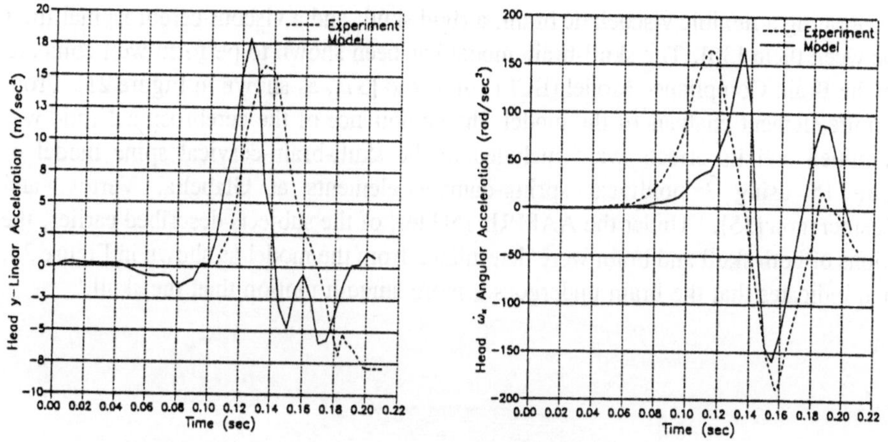

Figure 18. Sample lateral linear and angular accelerations from the lateral impact 5G tests results.

3.3. MODELLING SKULL-BRAIN INTERACTION

Another modification to the occupant model is the construction of a brain model, based on the principles of hydrodynamics (squeeze effects). The basic model, representing the skull-brain interaction of Figure 19, studies the effects of two spherical bodies with fluid in between them. For such a model, the pressure p and corresponding reaction force f developed inside are calculated using

$$p = \frac{3\mu_v V}{(c/R)^3 R\epsilon}\left[\frac{1}{1 - \epsilon\cos(\theta)^2} - 1\right] \tag{5}$$

$$f = \frac{6\pi\mu_v VR}{(c/R)^3}\left[\frac{1}{\epsilon^3}\ln(1 - \epsilon) + \frac{1}{1 - \epsilon} - \frac{1}{2\epsilon}\right] \tag{6}$$

where μ_v is the viscosity of the fluid; V the velocity of the inner spherical body; R the radius of the inner spherical body; ϵ the eccentricity ratio; c the radial clearance, and θ the angular coordinate measured from the maximum film thickness.

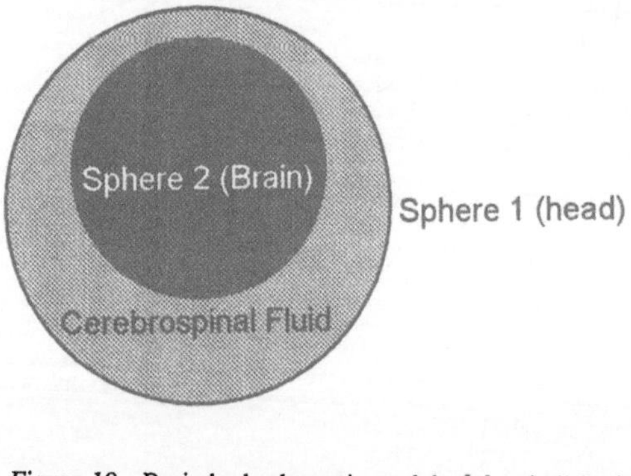

Figure 19. Basic hydrodynamic model of the skull-brain.

A more detailed model is the finite element model shown in Figure 20, incorporating a flexible viscoelatic brain, a rigid skull, and a viscous cerebrospinal fluid in between them [36]. The skull-brain model has been shown to perform well compare with the Brain Compliance Model (BCM) of Viano [37], as shown in Figure 21. From the finite element analysis of the model, the compliance of the cerebrospinal fluid was determined and its effects were included in the skull-brain-cervical spine model of Figure 15 using 3 nonlinear spring-damper elements at Glabella, Vertex, and Ophistocranon [35]. Under the AAMRL 15G test of the subjects described earlier, the motions of both skull and brain were determined from the model as shown in Figure 22, which indicates that the brain undergoes a more smooth motion than the skull.

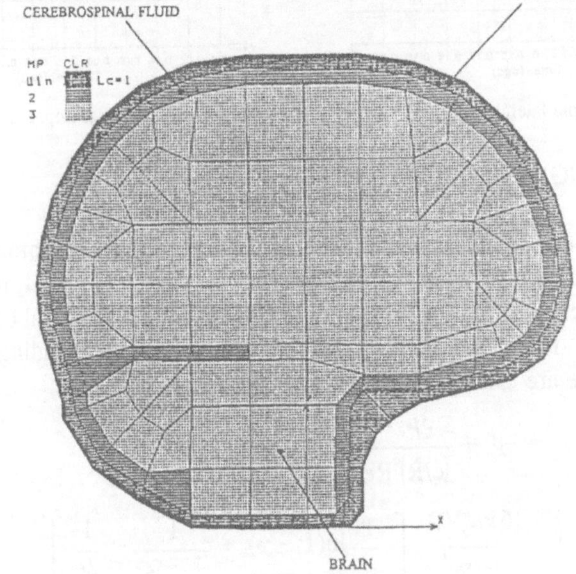

Figure 20. Skull-brain interaction using finite element method.

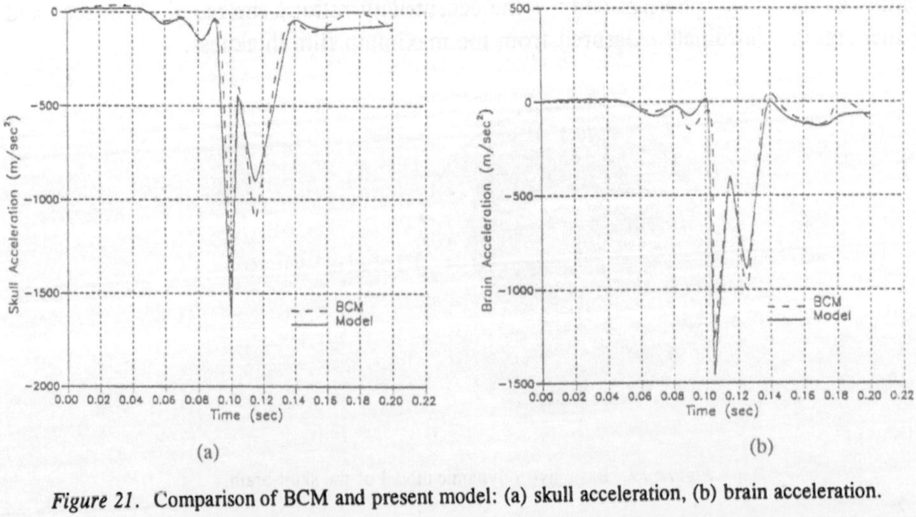

Figure 21. Comparison of BCM and present model: (a) skull acceleration, (b) brain acceleration.

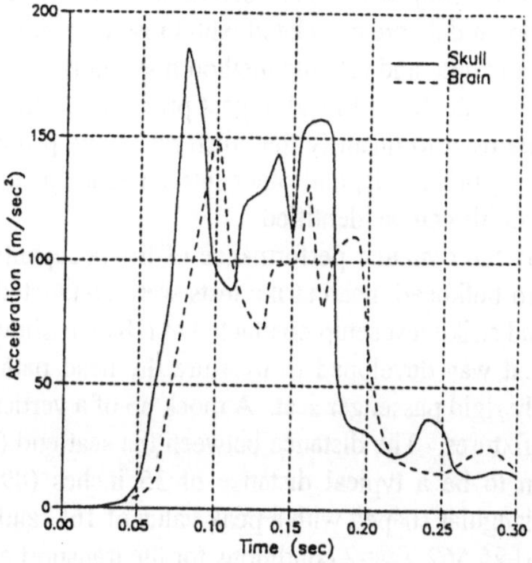

Figure 22. **Skull and brain accelerations for the 15G longitudinal test.**

4. Head Impact Protection for Seats Facing Bulkhead

In sections 4, 5, and 6, some of the latest areas of concern and research related to aircraft crashworthiness are described. The seats facing bulkheads, interior walls, class dividers, lavatory, or instrument panels of aircraft have presented problems demonstrating the compliance with the HIC requirement. Protection of aircraft occupants from head impact onto these surfaces as well as front seats (row-to-row) head impact with fixed back and breaking seat backs, and also the impact of crew with the glare shield is becoming a global and important safety issue. Due to the additional complications arising from the seat, restraint system, configuration of the occupant, and the articulation of the body segments on the HIC, the seat and airframe manufacturers still have many difficulties with the certification of seats having the HIC requirements of 1000. Full-scale ATD/seat/restraint sled-testing has proven to be costly, time-consuming, and to provide too much scatter on the HIC. The seat and airframe manufacturers, certification and research facilities, and the FAA are all interested in coming up with alternative testing procedures that may replace the full-scale sled testing procedures for the bulkhead seats [38]. The reasons for this interest are that the element testing is more controllable; is more repeatable; reduces the elapsed time for evaluation of designs, materials, and other significant parameters; reduces the cost of test hardware; and reduces the test facilities cost. The CAMI "bowling Ball" test has proven that it is possible to replace the full-scale tests with element tests and obtain reasonable comparison between the two [39]. However, the setup cannot produce a pure component of velocity of impact, changing the configuration of the head impact is rather difficult, the weight of the ball has a large effect on HIC, and the pass/fail criteria is not reliable. Other setups, such as the MGA's "linear impactor" [40] used at Boeing and the MTS "head restraint test system" [41] are currently being studied. Even though these systems occupy much less space, a pure component of the velocity with no acceleration or deceleration is difficult to produce. The most suitable setup for element or component testing is still the subject of much debate until all the different methods are tested and their corresponding data are correlated with those from full-scale sled-tests. It is prudent to come up with a simple and reliable method that can be used as an alternative to full-scale sled tests for the evaluation of head impact protection. Furthermore, component testing and analysis can be used to quantify the crush resistance properties of the padding materials on the back seats, bulkheads, and other interior structures, from which suitable properties of padding materials can be identified.

In order to study the dynamic performance of the occupant/seat/restraint system during a head impact onto bulkhead, head strike tests were conducted at both CAMI [42] and NIAR. A sample head strike test setup conducted at NIAR is shown in Figure 23. The protocol in the experiment was developed to measure the head path and velocity of the ATD restrained in a highly rigid passenger seat. A mock up of a vertical wall in front of the seat was included in the fixtures. The distance between the seat end (seat reference point) and the wall was chosen to be a typical distance of 35 inches (89 cm), and the input deceleration pulse was triangular shaped with a peak value of 16G and a rise time of 90 ms according to the FAR Part 25.562 Test-2 conditions for the transport aircraft category [15]. Several different pads and thicknesses were tests. A sample of the sled tests results is shown in Figure 24. The figure shows the sled deceleration pulse and also the resultant acceleration of the ATD head onto the panel with a 0.5-in Confor CF-42 padding material. The HIC for the test was measured to be 1425. The resultant and normal component of the velocity of the ATD head right before impact on the wall was measured from the high-speed photography of the head strike path and also from the integration of the accelerometer data. The resultant velocity of the ATD head onto the bulkhead at the time impact was measured to be 59 ft/sec (40 mph = 18 m/s), and the normal component of velocity was 51 ft/s (35 mph = 15 m/s).

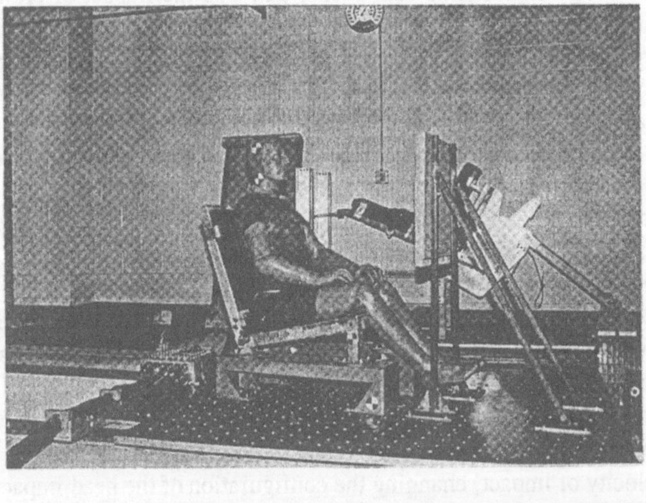

Figure 23. Sled test set up for occupant head impact with aircraft bulkhead.

A series of analytical simulations were also conducted using the gross motion simulators SOM-LA/TA [26,27] and its modification described carlier to duplicate the experiments. The simulations also showed that the normal velocity of the ATD head impact would be as high as 51 ft/sec (15 m/s) right before impacting the wall/bulkhead. To match the experimental results on the HIC, a nonlinear, viscoelastic-type contact force model was developed and used [43,44]. The simulations however showed that the HIC is very sensitive to the factors defining the bulkhead compliance, and although the simulations have a reasonable representation of the motion of the ATD body parts before impact, determination of the physical properties, force-deflection-rate properties, of the bulkhead limits their use. Hence, it is important to develop a simple and reliable method that may replace the full-scale sled tests for the evaluation of the head impact protection.

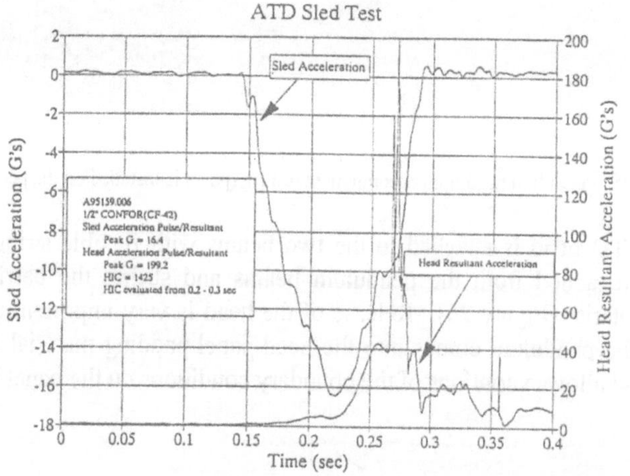

Figure 24. **Sled and the ATD head accelerations for the full-scale sled testing of the seat/ATD/restraint/bulkhead system.**

Component tests were then carried out duplicating the ATD head impact velocity on the bulkhead. The component test rig, designed and fabricated at NIAR, has a length of 22 ft (6.7 m), and is capable of attaining a maximum velocity of 53 ft/s (36 mph = 16 m/s). For a component test, the ATD head is raised to the required and is then released. The pendulum is brought to rest by a motion arrester pendulum, a locking mechanism, and a set of shock absorbers. Any desired value of impact velocity can be obtained by suitably adjusting the inclination angle Φ of the pendulum, as shown in Figures 25 and 26, and letting it fall freely under the action of gravity. Two types of tests can be carried out with this setup.

1. A panel with padding material is fixed loosely onto the pendulum beams (Figure 25). The panel strikes the ATD head, either mounted on a pole fixed to the ground or it strikes the entire seated ATD, and it then gets released. This method allows testing of both the ATD head and the ATD/seat/restraint system. However, it cannot be used for varying boundary conditions on the panel structure. It is a suitable method for obtaining localized responses of padding materials that are mounted on stiff walls.

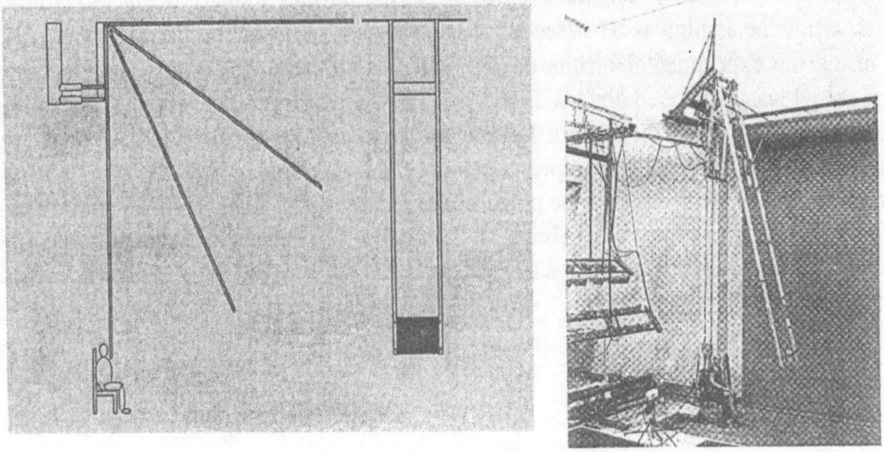

Figure 25. Pendulum head impact component test rig (panel is installed on the pendulum).

2. The ATD head is attached to the two beams with flexible strings such that the head gets detached from the pendulum beams and strikes the barrier (panel) with padding material (Figure 26). Release of the head is very important as otherwise the inertia of the pendulum comes into the head-panel-padding material impact system. This method allows variations of the boundary conditions on the panel structures.

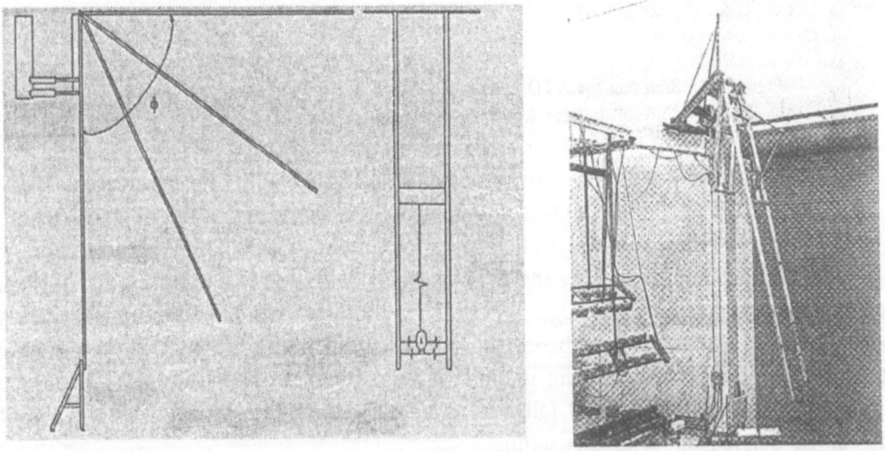

Figure 26. Pendulum head impact component test rig (ATD head is mounted on the pendulum).

Tests were conducted to duplicate the full-scale test conducted earlier. The same head-strike panel was used in the component test. The same pads used in the sled tests were also considered for the component tests. To achieve a normal head impact velocity of 51 ft/sec, the pendulum was released from an angle of 147 degrees from the vertical axis downward. To assess the repeatability of the procedures, component tests were repeated under the same conditions. Figure 27 shows the head acceleration of the ATD from two component tests with the 0.5-in Confor at 51 ft/s. As observed from the head acceleration profiles, the conducted tests are highly repeatable. The HIC recorded from the two tests were 1652 and 1527, providing a HIC repeatability of within 8 percent. Comparison of the component tests results with the full-scale tests results shows that the component tests predicts the HIC from the full-scale sled tests of 1425 with around 11 percent accuracy, but as yet does not match the acceleration profile well because of the different panel sizes tested under the sled and component tests.

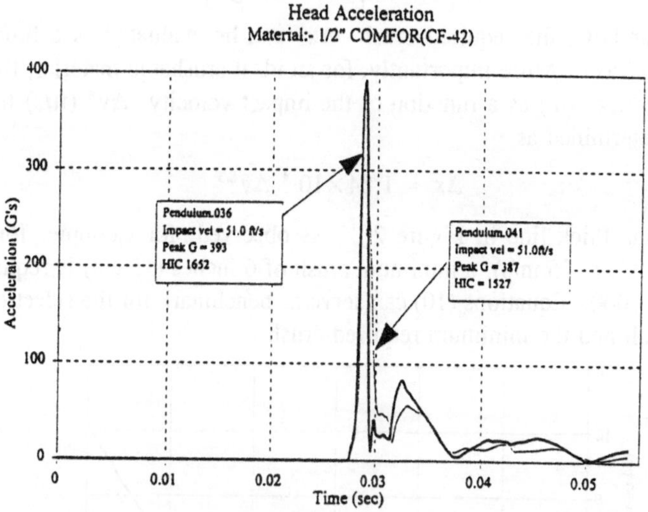

Figure 27. Results from two component tests of 0.5-in Confor and velocity of 51 ft/s.

The tests results have shown the potential for component testing if all the parameters of the tests are chosen properly. It must be kept in mind that these results have been obtained for one set of defined conditions and using the configuration of the tests shown in Figure 5 with the panel mounted on the pendulum. A number of further tests, both component and full-scale must be conducted for different categories of aircraft and function of the occupants (passenger or crew), different speeds (angles of drop, etc., a data base be developed for the tests results, and the results on the HIC should be statistically correlated with the sled tests and other component test setups.

It is also possible to determine the required crush remaining on the panel as a function of impact speed, if the panel is made of an ideal crushable material [45,46]. Such material is represented by a constant deceleration "a", where

$$a = \frac{\Delta v}{\Delta t \, g} \tag{7}$$

in which "Δv" is the velocity change or the impact velocity, "Δt" is the duration of the contact, and "g" is the gravitational acceleration. The velocity change "Δv" would be linear for an ideal crushable material, and the maximum crush "Δx" can be determined by the area under the linear v-t diagram as

$$\Delta x = \frac{1}{2} \Delta v \, \Delta t \tag{8}$$

Eliminating Δv from equations (2) and (3) and substituting for "a(t)" in (1) for HIC results in

$$\Delta x = \Delta v^{8/3} \left[\frac{1}{8\,(HIC)^2\, g^5} \right]^{1/3} \tag{9}$$

For a particular HIC, the required crush "Δx" can be evaluated as a function of the impact velocity "Δv". More importantly, for an ideal crushable material, the minimum required crush "Δx" (in) as a function of the impact velocity "Δv" (ft/s) for a HIC of 1000 can be determined as

$$\Delta x = 1.84 \times 10^{-4}\, \Delta v^{8/3} \tag{10}$$

as shown by the thick line in Figure 28. As observed, for example, for an impact velocity of 51 ft/sec (15 m/s), a minimum crush of 6 inches (15 cm) is required to keep the HIC under 1000. Equations (10) can serve as benchmark for the selection of proper pads on the wall and the minimum required crush.

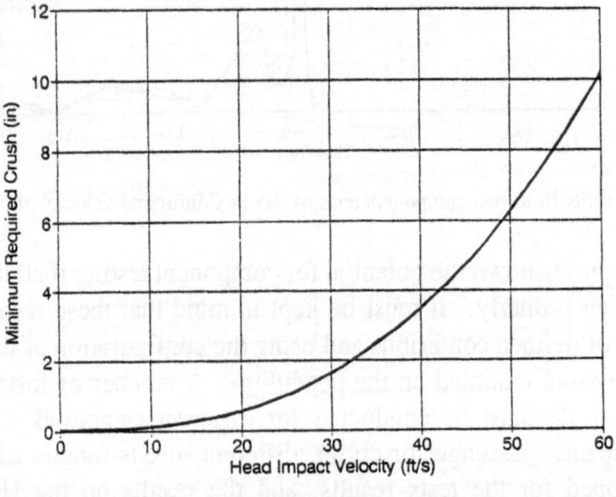

Figure 28. Minimum required crush resulting in a HIC of less than 1000 for an ideal fully crushable material.

For better definition of the true behavior of a panel-pad, representing the bulkhead, and its nonlinear behavior, nonlinear finite element analyses of the different panel/pads were also conducted to study the crush response of the panel as a result of a head impact. HIC response surfaces, generated as functions of yield strength, Young's modulus, and plastic deformation contour slope indicated that the HIC is most sensitive to the Young's modulus of the pad.

5. Proposed Rule for Certification of Commuter Aircraft

The FAA's analysis of commuter aircraft accidents and ongoing research has indicated that the crashworthiness capabilities of smaller aircraft may be questionable [47]. The small size of these aircraft resulting in a stiff structure and consequently higher impact loads experienced by the occupants is the subject of much of the current research for the enhancement of crashworthiness standards for these commuter type aircraft. In 1993, the FAA issued a Notice of Proposed Rule Making (NPRM) 93-71 to increase the deceleration pulse amplitude under Test 1 to 32G for the commuter type aircraft. To meet this condition, the seat design must exploit the energy absorption potential for its structural components. Energy absorbing components may include the seat legs, seat pan, and seat cushion. The intent is to design the seat so that it strikes well beyond the elastic limit to absorb the energy of the impact. To date, no seat has yet been able to pass the proposed criteria with an acceptable limit on the lumbar load (1500 pounds). At NIAR, alternative seat legs have been designed and analyzed to meet the 32G requirement. Two types of energy absorbing seat legs, S-shape and Y-shape seat legs, shown in Figures 29(a) and (b), were designed and analyzed.

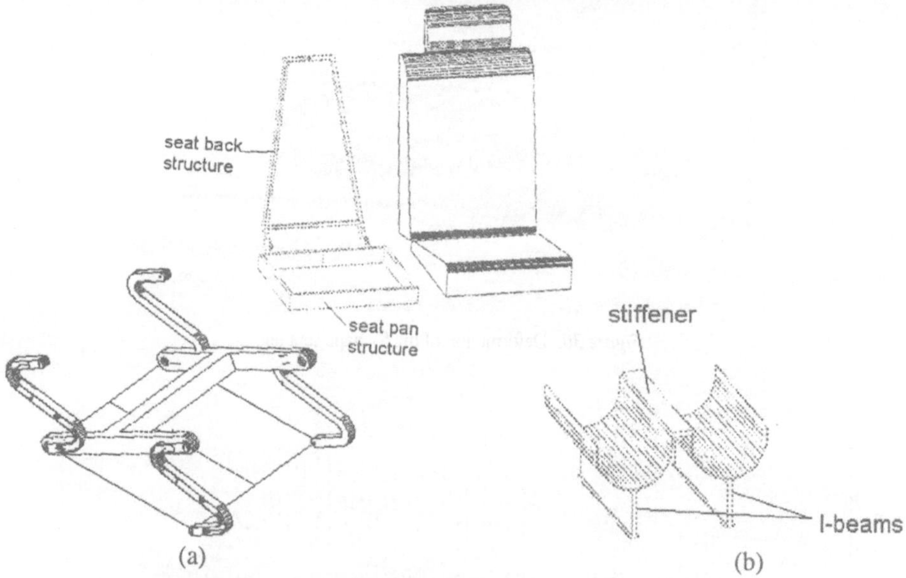

Figure 29. (a) S-shape and (b) Y-shape seat leg designs.

Nonlinear finite element models of the seat legs were constructed to study the effect of static loading on the seat leg designs, a sample of which is shown in Figure 30 for the S-shape seat leg. Both MADYMO and SOM-LA/TA codes were then utilized to model the occupant model (multibody model), restraint system (lap belt), seat back, seat pan, and seat legs (finite element model). The alternative energy absorbing seat leg designs resulted in the peak lumbar load of 1300 and 1370 pounds respectively for the S-shape and the Y-shape seat legs as compared to 1850 pounds obtained for a straight leg seat design. Analytical results on the lumbar loads for a 50th percentile Hybrid II ATD have shown that practical seats can be designed which could keep the lumbar load below 1500 pounds for the 32G commuter test requirement. For illustration, a graphical response of the Hybrid II ATD on the S-shape seat leg and under the proposed 32G condition is shown in Figure 31. A seat with the S-shape leg design was fabricated and tested on the impact sled of the NIAR [48], and was also used in a drop test of a commuter aircraft at the FAA Technical Center at Atlantic City. The sled tests results showed promising results on the capabilities of the seat legs in dissipating energy through plastic deformations. The results for the drop test however, were not as favorable as expected and the seat pitched forward, because the seat legs were not properly fixed to the aircraft floor and sidewall.

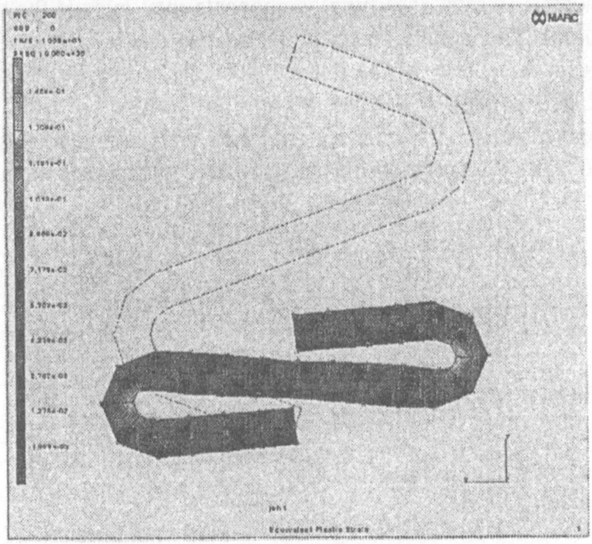

Figure 30. Deformation of the S-shape seat leg.

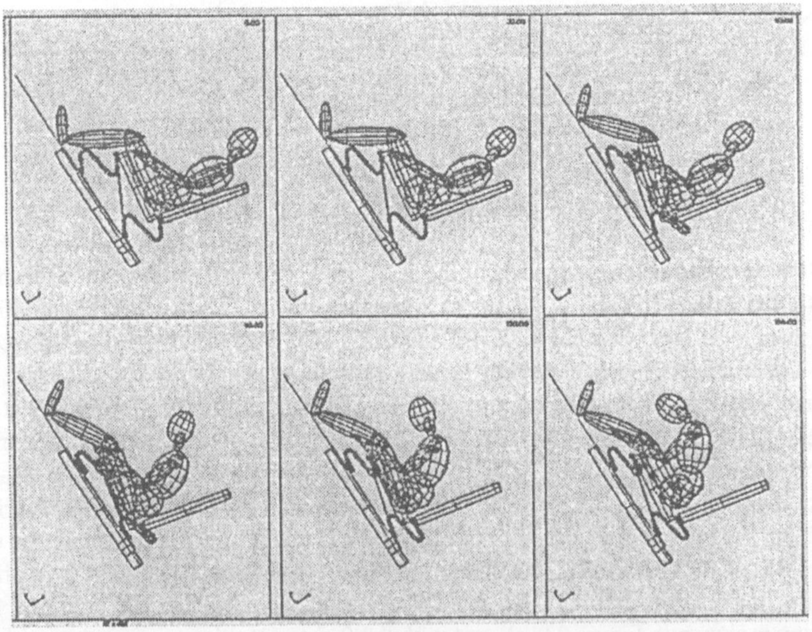

Figure 31. Response of a Hybrid II ATD on a S-shape seat under the 32G criteria for commuter aircraft.

In order to consider the response of a wide range of occupants seated on these new designs, a small 5th percentile female ATD and a large 95th percentile male ATD were also studied, shown in Figure 32. The lumbar loads from the crash analysis using MADYMO and under the proposed 32G rule for commuter type aircraft under Test-1 are shown in Figure 33 for the three seat leg designs and the three different-size occupants. The figure shows that the peak lumbar load decreases for the small subject and increased for the large subject, yet all fell below the 1500 ponds for the S-shape seat leg design, although no definitive injury criteria yet exists for the limit lumbar loads on different size ATD's. The analysis has shown that the energy-absorbing seat leg designs could reduce the lumbar load as much as 40 percent compared to the seats with straight legs.

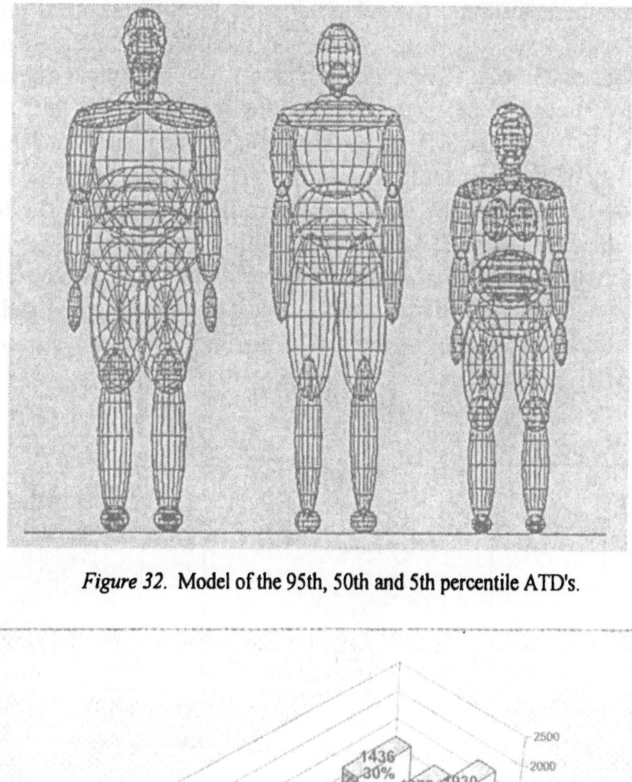

Figure 32. Model of the 95th, 50th and 5th percentile ATD's.

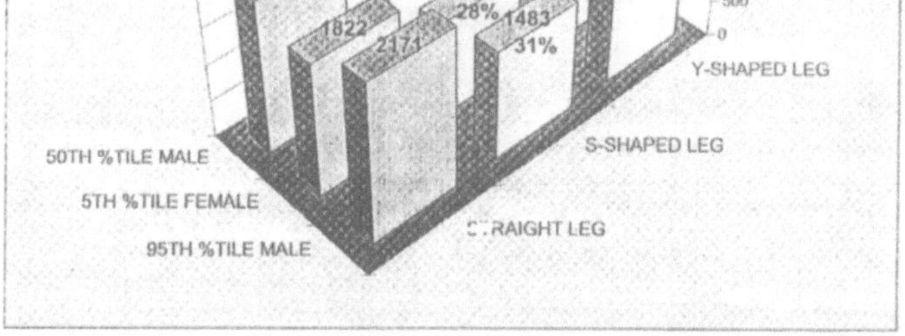

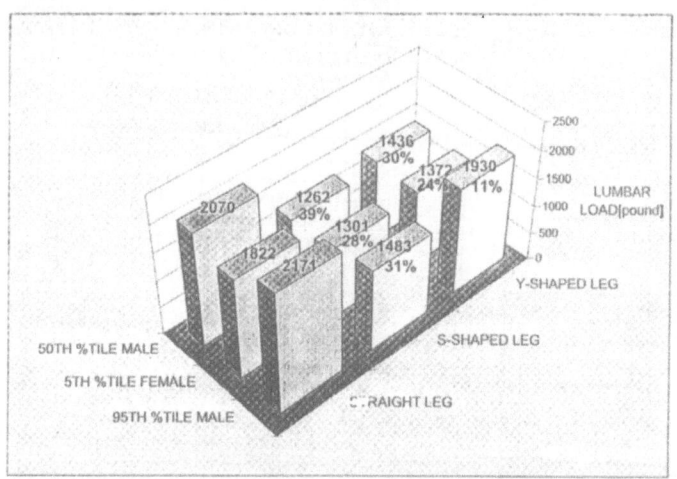

Figure 33. Lumbar load results from different seat leg designs and different ATD's.

6. Side-facing Seats

In recent years, there has been an increase in the number of private or business aircraft. The interiors of these aircraft are unique, most being custom made. Some of these aircraft are equipped with couch type (side-facing) seats. The seats are designed to accommodate a single or up to three passengers. Passengers seated on side-facing seats experience different dynamic response compared to those on forward- or aft-facing seats in an event of crash. In side-facing seats, the occupant dynamic response depends not only on the seat and geometry and type of the restraint system, but also on the position of possible divider panel (bulkhead) and also the position or number of occupants on the couch. Certification of the side-facing seat has become mandatory under the FAR 23.785, stating that these seats must provide the same level of occupant protection as a forward or aft-facing seats with a safety belt and shoulder harness, and provide the protection provisions of 23.562. Compliance with these new requirements has presented new challenges for the airframe, seat, and interior manufacturers [49]. Analysis, conducted at NIAR using MADYMO in order to gain better insight into the problem, is presented here. Both 50th percentile Hybrid II ATD (requirement of the FAR's) and US DOT SID (Side Impact Dummy) were considered in the analysis. The response of each ATD was studied for unbelted occupants and for several different geometries and types of restraints including lap belts, asymmetrical shoulder belt, symmetrical shoulder belt, and side airbag. Two types of bulkhead, extending beyond the seat and flush with the seat, were considered. Samples of the results obtained from analysis are shown in Figures 34 and 35.

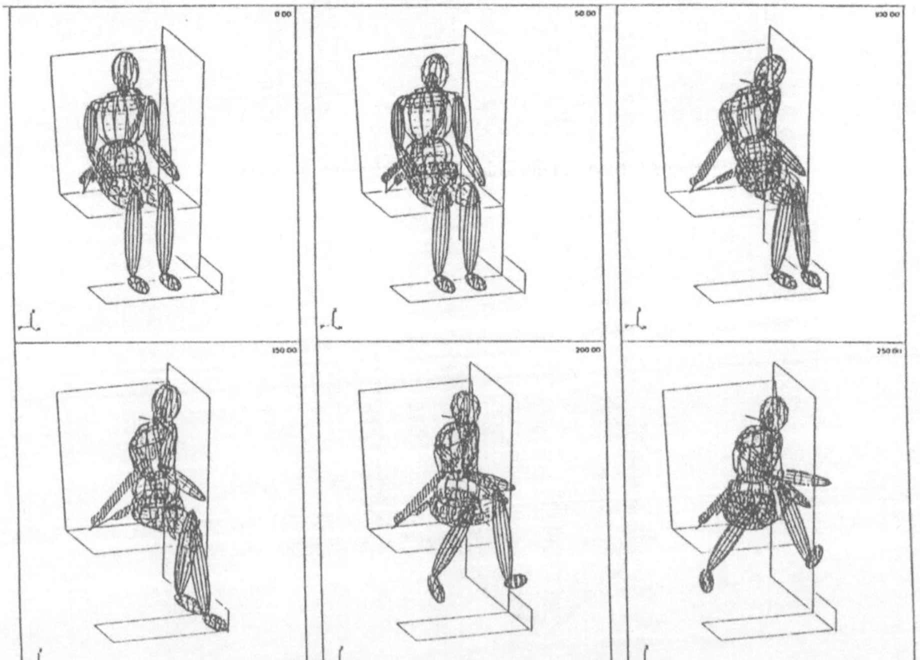

Figure 34. Kinematics of a Hybrid II with a lap and asymmetrical shoulder harness on a side-facing seat.

Injury parameters, including the HIC limit of 1000, lumbar spine load of 1500 pounds, Thoracic Trauma Index (TTI(d) of 85G limit to evaluate injuries to the rib cage or thorax, Viscous Criteria (V*C) with a limit of 1.5 m/s to evaluate injuries to the vital organs of the chest [50], neck moment limits of 504 lb-in in extension and 1680 lb-in in compression, axial load limit of approximately 250 lb in compressive and 650 lb in tensile loading [51], and finally shoulder harness load of 1750 pounds on one strap, were used. Figure 34 describes the kinematics of the Hybrid II ATD with a three-point restraint system (lap and diagonal shoulder belt) on a side facing seat under the FAR Part 23 Test-2 condition. This particular configuration was tested in a CAMI sled test [51], and the results from analysis correlated well with the CAMI sled test results. Figure 35 describes the kinematics of the SID restrained by a symmetrical shoulder harness and side airbag mounted on the panel.

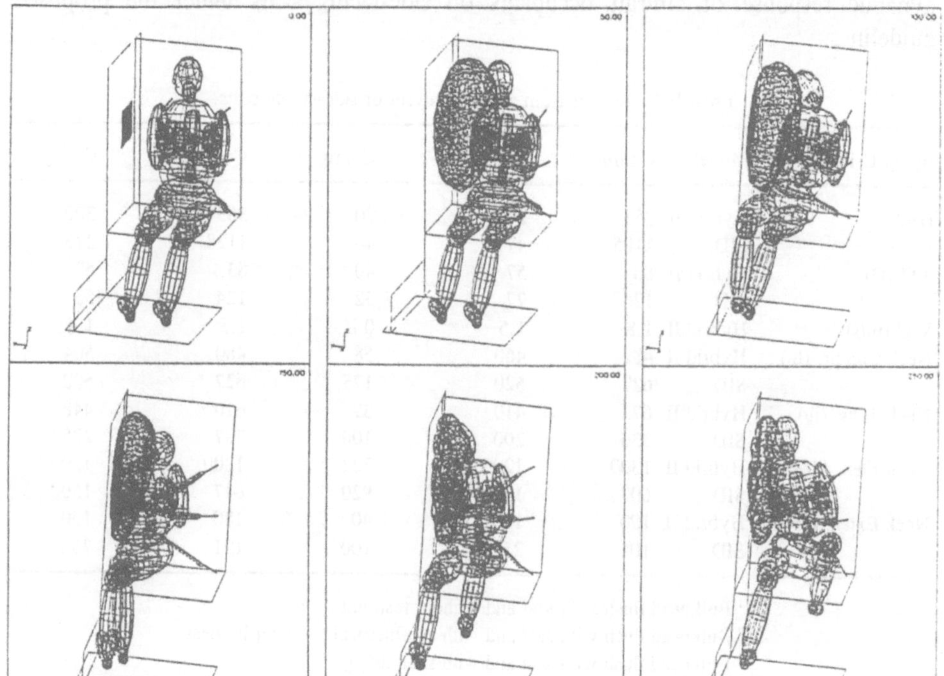

Figure 35. Kinematics of a SID with a lap and symmetrical shoulder harness and airbag on a side-facing seat.

A summary of all the analyses and their corresponding results are shown in Table 3. In situations where there was no dropped aisle and no restraint on the ATD, extreme flailing of the lower extremities occurred, increasing the likelihood of submarining. The HIC was much higher for SID than the one for Hybrid II which was within the tolerance limit. In the absence of any restraint, both SID and Hybrid II showed fatal injuries to the head. SID, having a much better biofidelity in side impact, showed fatal injuries to the chest, but the Viscous Criteria was within the limit. The asymmetrical shoulder belt reduced the G-load on the head, chest, and pelvis; however, it resulted in high neck loads and moments tending to strangle the occupant at times. A symmetrical shoulder harness reduced the G-loads on the occupant segments, but the neck loads were still close to the tolerances. The use of a side airbag caused all the injury parameters to fall below the allowable injury tolerances. Finally, the tests with extended bulkhead showed lower level of injury than with flush bulkhead. In conclusion, it can be stated that whenever the divider panel does not extend beyond the seat, provision should be made to retrain the legs, such as providing a dropped aisle. The lap belt might cause severe flailing of the occupant causing injuries to the occupant's chest and internal organs. The use of an asymmetrical shoulder harness must be avoided especially for the passengers seated in the middle of the couch. A lap belt with symmetrical shoulder harness, and if economically feasible, and airbag in the bulkhead, should provide the optimum level of safety for passengers seated on side-facing seats. Finally, impact sled tests must also be conducted under these recommendations to further evaluate the dynamic response of aircraft occupants on side-facing seats under the proposed guidelines.

TABLE 3. Results from all the analyses of side-facing seats.

Injury Criteria	Model	Config. 1[1]	Config. 2[2]	Config. 3[3]	Config. 4[4]	Config. 5[5]
HIC	Hybrid II	355	299	20	353	300
	SID	1105	471	48	1128	273
TTI (G)	Hybrid II	86	57	40	83	57
	SID	126	77	32	124	55
V*C (m/s)	Hybrid II	1.8	1.5	0.8	1.5	1.4
Neck Compr. (lb)	Hybrid II	482	460	58	480	504
	SID	627	520	175	627	500
Neck Tens. (lb)	Hybrid II	672	410	39	610	448
	SID	336	200	100	397	225
Neck Flex. (lb-in)	Hybrid II	1300	1205	322	1200	1200
	SID	607	1510	920	647	1500
Neck Exten. (lb-in)	Hybrid II	190	100	40	180	150
	SID	101	250	100	121	250

[1] Bulkhead flush with seat and without restraint
[2] Bulkhead flush with seat and with symmetrical shoulder harness
[3] Bulkhead flush with seat and with side airbag
[4] Bulkhead extended beyond the seat and without restraint
[5] Bulkhead extended beyond the seat and with symmetrical shoulder harness

7. Conclusion

The design of aircraft for improvement of the crashworthiness requires the knowledge and integration of several items. These items include the specifications and standards, human tolerance, injury criteria, energy absorption concepts in airframe design, seat legs, seat pan, seat cushion, restraint systems, surrounding structures, fire safety, economics, and ergonomics. This paper addressed some of these issues. A brief background on the origin and nature of the current crashworthiness standards was presented. Different testing methodologies were described including the full-scale aircraft tests such as the NASA Langley's IDRF (Impact Dynamics Research Facility) gantry-type test setup and the FAA's drop tower facility, sled test setups such as the Civil Aeromedical Institute (CAMI) and National Institute for Aviation Research (NIAR) facilities, and component tests such as the simplified HIC test setups. A review of the analysis codes for crashworthiness was performed for modelling both occupant and structural crash responses. Latest modifications in these codes for better prediction of occupant crash responses, specifically better prediction of the lumbar loads, cervical spine response, and skull-brain interaction, were presented.

Some of the current issues regarding aircraft crash safety were then examined. One particular global issue in performance and certification of aircraft is the problem of protection of occupants from head impact onto the interior structures. A description of a non-sled procedure for component or element testing of an anthropomorphic test dummy (ATD) head was presented in this study. The system is more controllable, more repeatable, less time-consuming, and less costly than the conventional sled testing procedures. This simple and reliable method may be used to better define and reduce the number of full-scale sled tests for the evaluation of head impact protection. The developed system is a pendulum-type head impact component testing device at NIAR. The outcome of this research would be to identify type and thickness of padding materials as well as boundary conditions of the panel structures that would be certifiable for the head impact protection. Other possible solutions to alleviate the problem may be requiring shoulder harness for passengers seated next to the bulkhead, eliminating a row of seating, turning the seat around (aft-facing), or installing airbags on the bulkhead specifically designed for the protection of the occupants under the FAR's.

The seats for commuter type aircraft have presented problems complying with the lumbar load limit of 1000 pounds under the proposed 32G rule. Alternative seat leg designs, instead of the straight seat legs, were analyzed under the proposed rule. In order to assess the response of a wide range of occupants, a 95th percentile male, a 50th percentile male, and a 5th percentile female ATD's were analyzed under the proposed rule. The results showed that the incorporation of energy-absorbing seat legs may lower the lumbar load as much as 40 percent. Further research in this area is necessary in analyzing the stability of these structures as they undergo large plastic deformations. Further sled and drop tests also need to be conducted on these alternative energy-absorbing seat leg designs to support the claims based on the analysis. In addition, other elements, including the seat pan and the seat cushion must be further studied to incorporate even more energy-absorption capability to the seats designed for commuter type aircraft.

A study was also presented to evaluate the dynamic response of occupants of side-facing seats, typical of some business jets. These seats were shown to impose the occupant to different dynamic behavior. Lap belt may impose the occupants to too much flailing and sever head injuries, and asymmetrical shoulder harness might tend to strangle the occupants. Symmetrical shoulder harness and airbag for seats next to bulkhead seem to perform best in protecting the occupants in a crash. Further studies, including analysis of couch-type seats with multiple occupants and sled testing of these side-facing seats must also be conducted to expand the knowledge of the occupant response on those side-facing seats.

Although, crash protection of occupants in the survivable configurations has seen much improvement in the last decade, a number of other issues need to be addressed. These issues include: the use of integrated or add-on child seats and restraint systems, energy-absorption characteristics of the overhead bins as related to certification of aircraft, future use and certification of large double-decker transport aircraft, bird strike problem, evacuation in crashes that may pose post-crash fire, quantification of non-life threatening injuries, lack of suitable test ATD for testing under FAR's, and the lack of injury data for a wide range of occupants sizes.

8. References

1. Olcott, J.W. (1995) The development of dynamic performance standards for federal aviation aircraft seats, Business and Commercial Aviation.
2. Chandler, R.F. (1993) Crash injury protection in civil aviation, in A.M. Nahum and J.W. Melvin (eds.), *Accidental Injury Biomechanics and Prevention*, Springer-Verlag, New York, 151-185.
3. Chandler, R.F. (1995) *Aircraft Crashworthiness*, Society of Automotive Engineers, PT-50.
4. Villard, H.S. (1968) *Contact!* Bonanza Books, New York.
5. Army. (1920) Aviation medicine in the A.E.F., Document No. 1004, Office of Director of Air Service, War Department, Washington, DC.
6. CAA. (1929) Airworthiness requirement of air commerce regulations, Aeronautics Bulletin No. 7-A, Department of Commerce, Aeronautics Branch, Washington, DC.
7. DeHaven, H. (1945) Relationship of injuries to structure in survivable aircraft accidents, NRC Report 440, National Research Council, Washington, DC.
8. Reed, W.H., Robertson, S.H., Weinberg, L.T.W. and Tyndall, L.H. (1965) Full scale dynamic crash test of a Douglas DC-7 aircraft, FAA-ADS-37, Federal Aviation Agency Aircraft Development Service, Washington, DC.
9. Reed, W.H., Robertson, S.H., Weinberg, L.T.W. and Tyndall, L.H. (1965) Full scale dynamic crash test of a Lockheed Constellation model 1649 aircraft, Federal Aviation Agency Aircraft Development Service, Washington, DC.
10. Preston, G.M., Preman, G.J. (1958) Accelerations in transport airplane crashes, NACA TN 4158, NACA Lewis Flight Propulsion Laboratory, Cleveland, OH.
11. Eiband, A.M. (1959) Human tolerance to rapidly applied accelerations: a summary of the literature, NASA Lewis Research Center, Cleveland, OH.
12. Thompson, R.G., Carden, H.D., and Hayduk, R.J. (1984) Survey of NASA research on crash dynamics. NASA Technical Paper 2298, Langley Research Center, Hampton, VA.
13. Chandler, R.F. (1983) Crash injury protection research at CAMI, SAE 830746, Society of Automotive Engineers, Warrendale, PA.
14. Title 14 U.S. code of Federal Regulations, (1988) Part 23, Amendment 23-39, Section 23.562, published in the Federal Register of August 14, 1988, effective date of September 14, 1988.
15. Title 14 U.S. code of Federal Regulations, (1988) Part 25, Amendment 25-64, Section 25.562, published in the Federal Register of May 17, 1988, effective date of June 16, 1988.
16. Title 14 U.S. code of Federal Regulations, (1989) Part 27, Amendment 27-25, Section 27.562, published in the Federal Register of November 13, 1989, effective date of December 13, 1989.
17. SAE. (1990) AS8049, Performance standards for seats in civil rotorcraft and transport airplanes, Aerospace Standards.
18. Gurdjian, E.S., Lissner, H.R., Latimer, F.R., Haddad, B.F., and Webster, J.E. (1953) Quantitative determination of acceleration and intercranial pressure in experimental head injury, *Neurology* 3, 417-423.
19. Gurdjian, E.S., Roberts, V.L., Thomas, L.M. (1964) Tolerance curves of acceleration and intercranial pressure and protective index in experimental head injury, *J. of Trauma* 6, 600.

20. Department of Transportation (1971) Occupant crash protection - Head Injury Criterion, NHTSA Docket Number 69-7, Notice 19, S6.2 of MVSS 208, March.
21. Vaughen, V.L., and Alfaro-Bou, E. (1976) Impact dynamic research facilities for full-scale aircraft crash testing, NASA Tech. Note, D-8179, National Aeronautics and Space Administration, Washington, DC.
22. Soltis, S., and McGwire, R. (1995) Small airplane vertical impact test program, SAE General, Corporate, and Regional Aviation Meeting and Exposition, Wichita, KS, SAE Paper No. 951162.
23. Chandler, R.F., and Gowdy, V. (1983) Loads measured in passenger seat tests. Memorandum Report AAC-119-81-8A, Protection and Survival Lab., FAA Civil Aeromedical Institute, Oklahoma City, OK.
24. Navy. (1996) NADC sled test facility, Philadelphia, PA.
25. NIAR. (1996) Sled test facilities, National Institute for Aviation Brochure, Wichita, KS.
26. Desjardins, S.P., and Shane, S.J. (1991) Structural testing of transport seats, Aircraft Interiors Conference, Wichita, KS.
27. Laananen, D.H., Bolukbasi, A.O., and Coltman, J.W. (1982) Computer simulation of an aircraft seat and occupant in a crash environment, Vol. - Technical Report, TR-82401, Simula, Inc., Tempe, AZ; DOT/FAA/CT-82/33-1, US Department of Transportation, Federal Aviation Administration Technical Center, Atlantic city Airport, NJ.
28. Laananen, D.H. (1991) Computer simulation of aircraft seat and occupant(s) in a crash environment, program SOM-LA/SOM-TA user's manual, DOT/FAA/CT-90/4, US Department of Transportation, Federal Aviation Administration Technical Center, Atlantic city Airport, NJ.
29. Obergefell, L.A., Gardner, T.R., Fleck, J.T. (1988) Articulated total body model enhancements, Vol. 2, User's Guide, Report No. AAMRL-TR-88-043, Armstrong Aerospace Medical Research Laboratory, Wright Patterson Airforce Base, Dayton, OH.
30. TNO. (1992) Madymo user's manual, Version 5.0, TNO, Delft, Netherlands.
31. Lankarani, H.M., Ma, D., and Menon, R. (1995) Impact dynamics of multibody systems and application to crash responses of aircraft occupant/structure, in M.S. Pereira and J.A.C. Ambrosio (eds.), *Computational Dynamics in Multibody Systems*, Kluwer Academic Publishers, Dordrecht, 239-265. Selected papers presented at the NATO Advanced Science Institute on Computer Aide Analysis of Rigid and Flexible Mechanical Systems (1993), Volume II, 527-552.
32. Bartz, J.A. (1971) A three-dimensional computer simulation of a motor vehicle crash victim, Calspan Report No. VJ-2978-V1.
33. Ma, D., and Lankarani, H.M. (1995) A nonlinear finite-element approach for kineto-static analysis of multibody systems, *International J. of Nonlinear Dynamics*, Kluwer Academic Publishers **8**, 237-250.
34. Merrill, T., Goldsmith, W., and Deng, Y.C. (1984) Three-dimensional lumped parameter head-neck model due to impact and impulsive loading, *J. of Biomechanics* **17**, No. 2, 81-95.
35. Arabyan, A., and Tsai, D. (1991) A multibody dynamic model of the human head and neck, Technical Report No. CAEL-91-1, University of Arizona, Tucson, AZ.
36. Menon, R. (1995) Multibody and finite-element approaches in the development of a model for the human head-neck-torso system, PhD Dissertation, Wichita State University, Wichita, KS.
37. Viano, D.C. (1988) Biomechanics of head injury - toward a theory linking head dynamic motion, brain tissue deformation, and neural trauma, SAE Technical Paper 8810708, Society of Automotive Engineers.
38. SAE Seat Ad Hoc Committee Meeting (1994) Development of a standard to evaluate head impact by element test, Civil Aeromedical Institute, Oklahoma city, OK, March.
39. Gowdy, V. (1995) Evaluation of the instrumented ball impact procedure to assess head impact protection in airplanes, SAE Paper 951166, Society of Automotive Engineers, SAE General, Corporate and Regional Aviation Meeting and Exposition, Wichita, KS.
40. MGA. (1996) Head linear impactor, MGA Corporation, Company Brochure.
41. MTS. (1996) Automated head restraint test system, MTS Systems Corporation, Company Brochures.
42. Gowdy, V., and DeWeese, R. (1991) Evaluation of head impact kinematics for passengers seated behind interior walls, Aircraft Interiors Conference, Wichita, KS.
43. Lankarani, H.M., Ma, D., and Menon, R. (1994) Multibody dynamics of aircraft occupants seated behind interior walls, *International J. of Nonlinear Dynamics* **6**, 237-246.
44. Lankarani, H.M., Ma, D., and Menon, R. (1992) Occupant dynamic responses for evaluation of compliance characteristics of aircraft bulkheads, ASME Advance in Design Automation, DE-Vol. 44-3, 391-398.

45. Lemmon, D.R., and Huston, R.L. (1994) Automobile hood/fender design optimization for improved pedestrian head impact protection," ASME Advances in Design Automation, DE-Vol. 2, 569-577.
46. Richards, M. (1992) 16G research on crashworthiness of transport airplane seats, Southern California Safety Institute Meeting, Pasadena, CA.
47. Gowdy, V. (1990) Development of crashworthy seats for commuter aircraft, Aircraft Interiors Conference, Wichita, KS.
48. Hooper, S.J., Holmes, M.W., and Margariti, A. (1995) The design and development of an energy-absorbing commuter seat, SAE Paper No. 951163, Society of Automotive Engineers, SAE General, Corporate, and Regional Meeting and Exposition, Wichita, KS.
49. Campbell, K.C. (1995) Dynamic testing of side-facing seats, SAE Paper No. 951165, Society of Automotive Engineers, SAE General, Corporate, and Regional Meeting and Exposition, Wichita, KS.
50. Viano, D.C., and Lau, I.V. (1988) A viscous tolerance criteria for soft tissue injury assessment, *J. of Biomechanics* 21, No. 5, 387-399.
51. Shams, T., Zhao, Y.M., and Rangarajan, N. (1995) Safety of side-facing seats in general aviation aircraft, SAE Paper No. 951164, Society of Automotive Engineers, SAE General, Corporate, and Regional Meeting and Exposition, Wichita, KS.

PART VIII
Conclusions and Future Trends

CRASHWORTHINESS OF TRANSPORTATION SYSTEMS: CURRENT ISSUES AND FUTURE TRENDS

J.A.C. AMBRÓSIO[1], W. ABRAMOWICZ[2], N. JONES[3], A. KING[4]
[1] *IDMEC Pólo IST,*
Avenida Rovisco Pais, 1096 Lisboa CODEX, Portugal
[2] *Department of Mechanical Engineering,*
University of Liverpool, Brownlow St., Liverpool L69 3GH, U.K.
[3] *Institute of Fundamental Technological Research,*
Polish Academy of Sciences, Swietokszyska 21, 00-049 Warsaw, Poland
[4] *Bioengineering Center,*
Wayne State University, Detroit, Michigan, U.S.A.

Abstract

This chapter synthesizes the final conclusions of the Advanced Study Institute as discussed in a panel session. Special emphasis is given to the actual state-of-art on crashworthiness design and analysis concerning its different aspects of injury biomechanics, occupant kinematics, structural impact, new materials, restraint systems, design methodologies, numerical analysis tools or experimental procedures. For each one of these aspects the necessities and trends for future developments are outlined.

1. Introduction

During a two week period, in the Summer of 1996, leading scientists, researchers and industrial experts discussed the state-of-art of various areas contributing to the crashworthiness of transportation systems. The programme of the meeting, based on lectures by leading experts and invited presentations delivered by specialists, lead to an important interaction between all participants on all themes contributing to crashworthiness topics. In a panel session, with the aid of the lecturers and participants in the Institute, a balance of the works was made, the state-of-art of crashworthiness was synthesized and trends for future developments were outlined.

The crashworthiness of transportation systems involves basically two major areas: the structural behavior of the vehicle and its components; the occupant kinematics and injury biomechanics. In both areas numerical models for simulation in terms of the finite element method, multibody dynamics or hybrid procedures play important roles.

J. A. C. Ambrósio et al. (eds.),
Crashworthiness of Transportation Systems: Structural Impact and Occupant Protection, 615–624.

Also, the experimental testing used to validate the models described, to achieve a full understanding of integrated prototype behaviors or simply to collect the data for the numerical methodologies are also essential for a solid comprehension of both the biomechanics and structural crashworthiness. Ultimately not only the experimental procedures can bring light into the shortcomings of the numerical techniques but also the numerical simulation methodologies give raise to better experimental testing.

In the last instance, the benefits of new systems developed to dissipate the vehicle kinetic energy in terms of controlled plastic deformation or to control occupant kinematics in terms of restraint systems are evaluated through a statistical analysis of data from real life accidents. Though only in some cases this data analysis can be correlated directly with the weaknesses of both numerical and experimental procedures it can be directly related with the technological developments in the areas that contribute to crashworthiness. The analysis of real accidents data gives directions for the developer to come out with better designs, for the legislator to put in place more efficient regulations and ultimately for the researcher to develop better analysis and design tools that enable the developer and legislator to anticipate the occurrence of problems in the safety systems. In order for the statistical data concerned with real life accidents to be credible it is necessary that the medical evidence, in what injuries are concerned, and forensic evidence, in case of casualties, relate the occupant injury with the vehicles and their protection systems.

2. Occupant Kinematics and Occupant Biomechanics

Currently, major developments in safety systems are directed to control occupant kinematics in order to avoid contact with unfriendly surfaces while maintaining the accelerations below the dangerous thresholds. This is the case of restraint systems such as airbags or seat belts. Through statistical data it is clear that most of the serious injuries and fatalities are related with brain, head, neck or spinal cord injuries.

Brain injury is a major problem because its injury mechanisms or its responses are not fully understood yet. New brain models are being developed and advanced computer models are becoming more popular, though there is no reliable characterization of the biological materials. In neck injuries, there has been quite good work done in some universities concerning the mechanisms of catastrophic spinal cord injury. However, its prevention still poses a serious problem. Take for instance the extreme case of a motorcyclist crashing at about 160 kph where no work can lead to the prevention of these injuries. More research groups are looking into the whiplash problem in order to obtain validated models for this kind of neck response, typical of lower speed crashes.

Little is known on the behavior of upper extremities of the body during crash and their injury mechanisms. Only recently, some work has been done, especially because of the problems posed with the airbag and side airbag. There is the need for a full range of research on the upper extremities. For the lower extremities of the human body only now more detailed work is being carried on. It is necessary to identify detailed mechanisms for these extremities and detailed models of the tissues and surrogates.

The use of mathematical models of human surrogates in automotive, train and aerospace crashworthiness research must be pursued. At this point there is no argument that vehicles should be safer for humans rather than for dummies. The human body models are still in their infancy, there are still many problems to be solved. There must be a coordinated effort to develop experimentally validated models. Models can easily be created, however, to make them reliable, usable and accepted by the community is a major task. The development of models for the entire human body is an effort that will necessarily involve different institutions. Cooperation in this area between different institutions and different expertise is essential to lead to unified developments. The example of parallel developments leading to different side impact regulations and dummies in the United States and in Europe, must be avoided in this case.

The actual measures of injury risk and potential for real-world injury do not take into account many of the new results obtained in recent years and need to be improved or replaced. Measures such as HIC or SI are sums of the head acceleration that do not take into account for instance the head angular acceleration. Nevertheless, they have been the more important measures that guided the development of the actual restraint systems. To understand injuries it is necessary to achieve a deeper knowledge of their mechanisms. Issues such as the definition of skeletal injuries, or the overall injury being controlled by a sum of small injuries or on the relative importance between shear forces and hydrostatic pressure with respect to injury, can only be resolved with an enlarged collaboration between the engineering and the medical communities.

3. Structural Impact

Most of the problems in structural impact are common in the field of crashworthiness of cars, buses, trains or airplanes. Many of the specialists in crashworthiness use, in a broad basis, what can be defined as quasi-static methods of analysis without fully understanding the range of validity of these methods. However, these techniques provide a good understanding of the physics of the problems, they are efficient for design, cheap to use and credible. These methods have a very high potential and can be improved in order to describe properly the transition between different modes and to overcome their shortcomings. As the study of crash cases in different scenarios are simulated or tested for higher and higher velocities these quasi-static methods will not remain valid. For instance, for high speed train crashes and airplane crashes, extra care must be taken in the analysis of the results obtained.

The sophistication of the finite element method has progressed well beyond the accuracy of the input data that can be provided to those codes. Some of the points where this is visible are:

Material properties, particularly for large strains

Structural failure criteria
Strain rate conditions, particularly for large strains
Residual stresses
Manufacturing imperfections
Loading characteristics
Scaling

There is a large amount of information that is really not defined clearly enough as the input for finite element codes and for some simulation codes based on other formulations. This class of very large codes, which require complete information of the problem, are very good for checking the final design but are not very useful in the initial design stages. Simplified methods, such as those based on multibody dynamics or macro elements, are still needed. Though they do give a good physical understanding of the problems during the preliminary design they are more difficult to master. It might be that finite element programs are relatively simple to operate, but they are notoriously difficult be used properly in the sense that they incorporate the interaction between many nonlinear phenomena, some of which that are not well understood.

Integrated environments for analysis and design are necessary such as those produced already by some of the larger manufacturers of automobiles and aircraft. Also here, the problem of cooperation between different communities is fundamental in order to integrate these tools to make them truly flexible.

4. Numerical Methodologies for Analysis and Design

From an industrial point of view, the crashworthiness of land, air and water transportation systems is achieved by a combination of testing and simulations using various numerical techniques. The later have evolved to be an integral part of the design process of structural components and occupant restraint systems where many iterations are necessary to meet the product objectives. Crashworthiness aims at maintaining the vehicle structural integrity while reducing occupant harm to the extend possible in survivable crash environments. Numerical simulations evolved from one-dimensional lumped parameter spring-mass systems with few degrees of freedom, developed in the late 1960's, to current three-dimensional models with over 500,000 degrees of freedom. Models of full-scale vehicles, occupants, dummies, restraint systems or structural components are being used on a broad basis under different crash scenarios. Integrated models using biomechanical models of the occupants with and without restraint systems are now appearing as standard practice. The finite element method, multibody dynamics, macro-elements and hybrid methodologies are the numerical tool in use today. The maximization of the vehicle performance in terms of crashworthiness requires the use of the most appropriate numerical procedure during the different design and analysis phases, according to the amount of information available.

The finite element method is today the most suitable numerical procedure for crashworthiness analysis of metallic structures where the geometry and material properties are known, due to the flexibility and possible detail of the models. However, few, as yet unsolved issues, limit the effectiveness and timeliness of the technology. Among which are: the model build time, especially for complex vehicle structures; modeling of nonmetallic soft materials such as rubbers, plastics, polyurethane foams or human tissues; strain rate effects in different materials. At the moment the use of strain rate effects appears to be arbitrary where it is used in component simulations and rarely in full vehicle simulations.

The multibody dynamics has emerged in the past decade as a numerical methodology well accepted for its flexibility and for the wide range of models that can be simulated. Not only the large rigid body motion of the system components is described in a manner consistent with the kinematic restrictions between these but also the deformation of the system components is described either by linear and nonlinear elements or by finite elements. The incorporation in these models of experimental results of component testing or of the finite element analysis of structural components, using the concept of plastic hinges, has made them a primary tool for design or for conceptual modeling of vehicles and biomechanical systems such as dummies. Some work is still necessary to improve the multibody methodologies for crashworthiness analysis and design in what the modeling of nonlinear flexible bodies is concerned. For this purpose: interaction with nonlinear finite element codes is necessary; a better description of the velocity discontinuities during impact is required; a general form to deal with the contact problem must be found; advanced kinematic joints and contact surfaces between bodies must be developed for models of the human body. There is a need for the standardization of data format for multibody codes as well as for finite element codes. It is conceivable in the near future that the data formats will coalesce into a unified standard making it much easier for someone to migrate from one type of code to another. Once this is available the use of different tools of increasing complexity in industry will be much easier.

The macro element approach to the analysis and design of crashworthy systems is specially appealing in the field of crushing mechanics were large distortions are expected. Elements such as superfolding or supertearing element has been successfully applied to the analysis and design of different transportation vehicles. This methodology is extremely simple to use and does not require complex input data. Therefore, this methodology is specially useful in the early design stages when the design concept undergoes frequent modifications and fast, but still accurate, predictive techniques are of primary importance to practicing engineers.

The macro element approach is based on the global space-time formulation and in contrast to the standard finite element method does not require massive computing. Instead, it involves a manipulation of incompatible elements with different space-time shape functions and variable number of degrees of freedom. Therefore, the object oriented programming techniques are widely used in the implementation of macro algorithms. This, in turn, results in the incompatibility of macro element and finite element codes. Further efforts are necessary in order to effectively implement macro elements as an integral and compatible part of finite element codes. This unification should establish an integrated environment for design and analysis of crashworthy structures that will provide for a smooth transition from simplified to complex computational models during the entire design process. Furthermore, some aspects of the theoretical formulation still require basic studies. In particular, the generalization of the method to the case of non-metallic materials and the development of standard methods for the analysis of crushing response under arbitrary loading constitute challenging subjects of fundamental research in upcoming years.

The use of numerical simulation in worldwide safety regulations is an issue that certainly will arise in the near future. The development of reliable simulation techniques should harmonize various safety standards and regulations and move certification techniques towards a balanced combination of experimental and numerical modeling. For example, advanced human body models are expected to be used in simulation replacing dummy models provided that the actual problems on their modeling are solved. However, a substantial effort is required to standardize the international requirements and to establish widely accepted numerical simulation techniques.

More robust and user friendly simulation packages capable of a smooth exchange of data between various CAD/CAM packages and computational tools of increasing complexity are urgently needed by industrial end users. In the current level of development, this can be done by developing consistent standard data exchange channels that effectively streamline input and output data of different numerical tools. A transparent integration of macro element, hybrid, multibody and finite element codes with advanced optimization techniques in the crashworthiness design and analysis is a goal for future developments.

5. Accident Data and Forensic Analysis

The final test grounds for any new system or structural changes in vehicles are the real life accidents, in what crashworthiness is concerned. Drastic reductions in the number of fatalities have been observed in recent years as a direct result of the combined use of airbags and seatbelts. However, there are still over 250,000 people killed and 2,500,000 injured every year as a result of vehicle accidents. These numbers may even be worse if the crash injury reporting systems limitations taken into account. In most countries these are based on police reports that tend to under report cases of minor injuries and deaths are only accounted for, when traffic related, if they occur at the site of the accident or within a certain time period afterwards. Many debilitating headaches and backpains which may develop over a period of time as a result of accidents are often not included in the statistics.

Statistically, brain injuries have dropped so much that it is no longer an issue in frontal impact. In side impact, there are situations where the person's head may protrude out of the car and hit an object or a pole and consequently side airbags start to be used.

Statistical data also shows that injuries to the lower extremities are becoming more important. Though great advances have been made in engineering safety over the years, the diminishing returns dictate a much greater effort to reach the same level of advances in the future. It is expected that advances towards intelligent restraint systems will be made. Issues such as the occupant size, sitting position, biomechanical variations or changing crash exposure are challenges for research and development of new restraint systems able to recognize these variations. However, methods of observing the occupants in vehicles will always have to deal with legal issues related with privacy.

As the basis for major trends for development is this statistical data has to be reliable. For instance it is easy for rib fractures due to a cardiac reactivation to be confused with injuries due to seat belt use. A close collaboration between medical doctors and engineers is very important not only to improve crash data reporting of police and hospital data but also to identify basic mechanisms of injury.

6. Industrial Needs

When a new vehicle design is started from scratch not much is known, besides parts of the envelope, i.e., the outside of the vehicle. From this starting point it is necessary to make designs that meet all the safety standards as well as the company internal biomechanics standards and others. It takes from three to four years for a new car to be designed and into production and about half of that time for a new train. As a result, models that may be extremely complete and accurate will not play an important role in the actual design of the vehicle if they are not fast enough to use. Though, there is the need for accurate models, decisions in the industry can be based on having designs that are less than 100% optimal. Consequently, there is a time and place for all technologies in the industry.

Concept modeling, that is generally related to the use of simplified numerical techniques, plays a dominant role in industry. Among these procedures, the use of multibody or lumped parameter models and models of beam elements falls into this category of concept modeling. These methods require a good understanding of the physics of the problem and can only be used by experts in the field of application. A certain controversy between finite element users and the users of so-called simpler techniques arises from the fact that less experienced users construct full scale finite element models even with information clearly insufficient to produce a reliable model. One of the reasons for this situation is that, with today's graphics, parts can be seen, their shapes and the way they look in the cars can be evaluated. That kind of modeling, though yielding very good results at times, is time consuming due to the complexity of the models that take a long time to build and to execute. The physical understanding of the phenomena is more difficult to grasp by complex finite element models. The trend in many companies is to train people in concept modeling before releasing them into finite element modeling so that they obtain a fair understanding of the crash phenomena.

Along similar lines, there is a lot of biomechanics research in the transportation industry. There is the need to make cars, trains or aircraft better and safer to avoid different types of injuries. For that purpose, models of dummies are being used instead of models of the human being. One reason is that industry cannot really afford to have biomechanical models of humans with an excess of 100,000 degrees of freedom. Another reason is that the actual knowledge of the constitutive relations of biological materials are not yet fully understood and, consequently, there are still reservations relative to the biofidelity of these models. What these biomechanics models can do in a first phase in the future is to provide much more data so that the dummies themselves can have a much higher biofidelity. On a second phase it is possible that these models can become smaller in terms of their degrees of freedom and still provide all the results that are available today with the dummies. Moreover, it is necessary that the models of real humans allow about twenty analyses a day before they can be integrated into the design process of a car for instance.

In the design of new products for improved crashworthiness behavior, the injury measures play an extremely important role. For instance, looking into what has happened with car design, especially with airbags, the industry has relied on the use of HIC. In spite of the HIC being a poor injury criteria, the number of head injuries has been reduced significantly. New brain models can provide a better understanding of the injury mechanisms and eventually they will lead to better injury criteria. Also a better understanding of the mechanisms of injury of the upper and lower extremities is necessary to diminish the number of injuries in vehicles. This is probably one of the most important areas of research and development because in comparison with the head injuries not as much as been done.

The aircraft industry is well ahead of all other industries in the use of composites. Such technology can be transferred more rapidly to the automotive industry particularly when the cost of these materials become much less expensive. At this time there is work going on for the next generation of cars, the so-called supercar project, where it is intended to use composites.

7. Aerospace Specific Issues

The requirements for crashworthiness design are very tough to be setup in the aircraft industry. It takes many years from first consideration to their enforcement. Probably due to competition, the car manufacturers have more strict requirements already and so they are more visible. In the aircraft industry, crashworthiness design has to rely more on the simulation of numerical models than in extensive testing due to the costs involved. For instance, regulations relative to aircraft seats are very strict because in minor crashes when the structural integrity of the plane is assured there is still considerable danger for the occupant when the seats collapse. This is a design case where testing for crashworthiness can be done easily. Other types of regulations are expected to be introduced in the coming years.

The aircraft industry will face in the near future the challenge of different aircraft sizes with very specific safety problems. For instance, main airliners with a capacity of about 800 people are expected. One of the possibilities today is to extend the passenger capacity by, for example, using the lower deck for seating. In the actual airplane design this region of the structure has an important role of energy dissipation during crash landings and consequently new constructive philosophies and eventually new materials need to be employed. Other important challenges are expected in the aircraft industry.

As far as simulation is concerned, the aircraft industry relies basically on hybrid models and finite elements. New approaches such as multibody dynamics and macro elements have a very high potential because in order to simulate a full-size airplane it is still not feasible to use a complete finite element model. So far the hybrid approach is the only workable numerical method for the simulation of the complete aircraft. The global and local coupling may be much more cost effective in the future or to develop the finite element tools so that they can be used more in the sense of a hybrid simulation tool.

The use of composite materials in the aerospace industry is a reality today and it is expected to increase steadily in the future. The description of the behavior of these materials and their constitutive models is still a challenge in the simulation tools. There is the need to generate complete databases where the knowledge available is collected. Some of these databases are available today, but they are not linked together and the information cannot be exchanged between them. Helicopters are well in front in the use of composite materials field. There is here already a good experience and there is a good potential for exchange and transfer of technology to other industries.

8. Conclusions

Noteworth advances in crashworthiness have been achieved in the past years due to improvements in the vehicle structural response, compliance of the vehicle interior and exterior and restraint systems to control the occupant kinematics. Statistical data shows that the concurrent use of different protection systems leads to the increase of their efficiency, for instance the simultaneous use of seat belts and airbags, with the consequent minimization of injury. Integrated methodologies of analysis based on finite elements, multibody dynamics, simplified elements and experimental testing are now possible for both structural impact and occupant kinematics. From the design phases of new systems until their complete analysis, different tools of increasing complexity can be used as more information on the final design becomes available.

It is foreseen in the future that the improvement in occupant protection of transportation systems will occur using human models in the numerical design and analysis tools on the one hand and by improvements in the biofidelity of dummies on the other hand. A more advanced characterization of live tissue, in terms of their constitutive equations, is therefore necessary in order to make them suitable for use in numerical simulation tools. Ultimately, decisions on new designs, concerned with their crashworthy behavior, are based on the predicted potential for occupant injury. More advanced injury criteria, that incorporate the actual knowledge on the injury mechanisms are necessary. New advanced models for the brain, head-neck, spinal chord and body extremities are being developed with that purpose in mind.

The use of new materials in transportation vehicles will increase in the future once these become an economic alternative to the materials used today. A better description of the behavior and constitutive equations of materials such as composites, foams or plastics is necessary in order for methodologies such as the finite element method to provide reliable results. An area that requires further investigation is the identification of dynamic material properties at large plastic strains for all methods of analysis which use them.

Acknowledgments

The authors acknowledge the contribution of Dr. Tom Khalil, Dr. David Viano, Prof. Chris Kindervater and Dr. Claude Tarriere towards the contents of this manuscript.

Index